# EUREKA MATH™

## A Story of Functions

## Algebra II, Module 3
## Exponential and Logarithmic Functions

JOSSEY-BASS™
A Wiley Brand

Cover design by Chris Clary



Published by Jossey-Bass
A Wiley Brand
One Montgomery Street, Suite 1000, San Francisco, CA 94104-4594—www.josseybass.com



ISBN: 978-1-118-81157-3

Printed in the United States of America
FIRST EDITION
*PB Printing* 10 9 8 7 6 5 4 3 2 1

When do you know you really understand something? One test is to see if you can explain it to someone else—well enough that *they* understand it. Eureka Math routinely requires students to "turn and talk" and explain the math they learned to their peers.

That is because the goal of Eureka Math (which you may know as the EngageNY math modules) is to produce students who are not merely literate, but fluent, in mathematics. By fluent, we mean not just knowing what process to use when solving a problem but understanding why that process works.

Here's an example. A student who is fluent in mathematics can do far more than just name, recite, and apply the Pythagorean theorem to problems. She can explain why $a^2 + b^2 = c^2$ is true. She not only knows the theorem can be used to find the length of a right triangle's hypotenuse, but can apply it more broadly—such as to find the distance between any two points in the coordinate plane, for example. She also can see the theorem as the glue joining seemingly disparate ideas including equations of circles, trigonometry, and vectors.

By contrast, the student who has merely memorized the Pythagorean theorem does not know why it works and can do little more than just solve right triangle problems by rote. The theorem is an abstraction—not a piece of knowledge, but just a process to use in the limited ways that she has been directed. For her, studying mathematics is a chore, a mere memorizing of disconnected processes.

Eureka Math provides much more. It offers students math knowledge that will serve them well beyond any test. This fundamental knowledge not only makes wise citizens and competent consumers, but it gives birth to budding physicists and engineers. Knowing math deeply opens vistas of opportunity.

A student becomes fluent in math—as they do in any other subject—by following a course of study that builds their knowledge of the subject, logically and thoroughly. In Eureka Math, concepts flow logically from PreKindergarten through high school. The "chapters" in the story of mathematics are "A Story of Units" for the elementary grades, followed by "A Story of Ratios" in middle school and "A Story of Functions" in high school.

This sequencing is joined with a mix of new and old methods of instruction that are proven to work. For example, we utilize an exercise called a "sprint" to develop students' fluency with standard algorithms (routines for adding, subtracting, multiplying, and dividing whole numbers and fractions). We employ many familiar models and tools such as the number line and tape diagrams (aka bar diagrams). A newer model highlighted in the curriculum is the number bond (illustrated below), which clearly shows how numbers are comprised of other numbers.

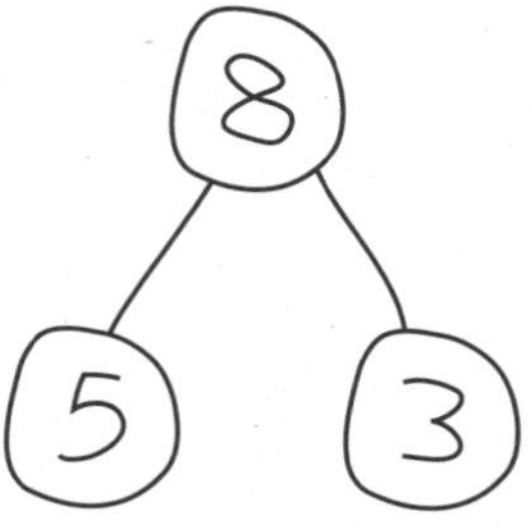

Eureka Math is designed to help accommodate different types of classrooms and serve as a resou for educators, who make decisions based on the needs of students. The "vignettes" of teacher-student interactions included in the curriculum are not scripts, but exemplars illustrating methods of instruct recommended by the teachers who have crafted our curricula.

Eureka Math has been adopted by districts from East Meadows, NY to Lafayette, LA to Chula Vis CA. At Eureka Math we are excited to have created the most transparent math curriculum in history—every lesson, all classwork, and every problem is available online.

Many of us have less than joyful memories of learning mathematics: lots of memorization, lots of rules to follow without understanding, and problems that didn't make any sense. What if a curriculum came along that gave children a chance to avoid that math anxiety and replaced it with authentic understanding, excitement, and curiosity? Like a NY educator attending one of our trainings said: "Wh didn't I learn mathematics this way when I was a kid? It is so much easier than the way I learned it!"

Eureka!

Lynne Mun
Washington
September 2

# Mathematics Curriculum

e of Contents[1]

# onential and Logarithmic Functions



---

son is ONE day, and ONE day is considered a 45-minute period.

# Algebra II • Module 3

# Exponential and Logarithmic Functions

## OVERVIEW

In this module, students synthesize and generalize what they have learned about a variety of function families. They extend the domain of exponential functions to the entire real line (**N-RN.A.1**) and then extend their work with these functions to include solving exponential equations with logarithms (**F-LE.A.4**). They explore (with appropriate tools) the effects of transformations on graphs of exponential and logarithmic functions. They notice that the transformations on a graph of a logarithmic function relate to the logarithmic properties (**F-BF.B.3**). Students identify appropriate types of functions to model a situation. They adjust parameters to improve the model, and they compare models by analyzing appropriateness of fit and making judgments about the domain over which a model is a good fit. The description of modeling as, "the process of choosing and using mathematics and statistics to analyze empirical situations, to understand them better, and to make decisions," is at the heart of this module. In particular, through repeated opportunities in working through the modeling cycle (see page 61 of the CCLS), students acquire the insight that the same mathematical or statistical structure can sometimes model seemingly different situations.

This module builds on the work in Algebra I, Modules 3 and 5, where students first modeled situations using exponential functions and considered which type of function would best model a given real world situation. The module also introduces students to the extension standards relating to inverse functions and composition of functions to further enhance student understanding of logarithms.

Topic E is a culminating project spread out over several lessons where students consider applying their knowledge to financial literacy. They plan a budget, consider borrowing money to buy a car and a home, study paying off a credit card balance, and finally, decide how they could accumulate one million dollars.

## Focus Standards

### Extend the properties of exponents to rational exponents.

**N-RN.A.1** Explain how the definition of the meaning of rational exponents follows from extending the properties of integer exponents to those values, allowing for a notation for radicals in terms of rational exponents. *For example, we define* $5^{\frac{1}{3}}$ *to be the cube root of* $5$ *because we want* $\left(5^{\frac{1}{3}}\right)^3 = 5^{\left(\frac{1}{3}\right)^3}$ *to hold, so* $\left(5^{\frac{1}{3}}\right)^3$ *must equal* $5$.

**N-RN.A.2**[2] Rewrite expressions involving radicals and rational exponents using the properties of exponents.

[2] Including expressions where either base or exponent may contain variables.

## Reason quantitatively and use units to solve problems.

**N-Q.A.2**[3] Define appropriate quantities for the purpose of descriptive modeling.★

## Write expressions in equivalent forms to solve problems.

**A-SSE.B.3**[4] Choose and produce an equivalent form of an expression to reveal and explain properties of the quantity represented by the expression.★

c. Use the properties of exponents to transform expressions for exponential functions. *For example the expression* $1.15^t$ *can be rewritten as* $\left(1.15^{\frac{1}{12}}\right)^{12t} \approx 1.012^{12t}$ *to reveal the approximate equivalent monthly interest rate if the annual rate is* $15\%$.

**A-SSE.B.4**[5] Derive the formula for the sum of a finite geometric series (when the common ratio is not 1), and use the formula to solve problems. *For example, calculate mortgage payments.*★

## Create equations that describe numbers or relationships.

**A-CED.A.1**[6] Create equations and inequalities in one variable and use them to solve problems. *Include equations arising from linear and quadratic functions, and simple rational and exponential functions.*★

## Represent and solve equations and inequalities graphically.

**A-REI.D.11**[7] Explain why the $x$-coordinates of the points where the graphs of the equations $y = f(x)$ and $y = g(x)$ intersect are the solutions of the equation $f(x) = g(x)$; find the solutions approximately, e.g., using technology to graph the functions, make tables of values, or find successive approximations. Include cases where $f(x)$ and/or $g(x)$ are linear, polynomial, rational, absolute value, exponential, and logarithmic functions.★

## Understand the concept of a function and use function notation.

**F-IF.A.3**[8] Recognize that sequences are functions, sometimes defined recursively, whose domain is a subset of the integers. *For example, the Fibonacci sequence is defined recursively by* $f(0) = f(1) = 1,\ f(n+1) = f(n) + f(n-1)$ *for* $n \geq 1$.

---

[3] This standard will be assessed in Algebra II by ensuring that some modeling tasks (involving Algebra II content or securely held content from previous grades and courses) require the student to create a quantity of interest in the situation being described (i.e., this is not provided in the task). For example, in a situation involving periodic phenomena, the student might autonomously decide that amplitude is a key variable in a situation, and then choose to work with peak amplitude.
[4] Tasks have a real-world context. As described in the standard, there is an interplay between the mathematical structure of the expression and the structure of the situation, such that choosing and producing an equivalent form of the expression reveals something about the situation. In Algebra II, tasks include exponential expressions with rational or real exponents.
[5] This standard includes using the summation notation symbol.
[6] Tasks have a real-world context. In Algebra II, tasks include exponential equations with rational or real exponents, rational functions, and absolute value functions.
[7] In Algebra II, tasks may involve any of the function types mentioned in the standard.
[8] This standard is Supporting Content in Algebra II. This standard should support the Major Content in F-BF.2 for coherence.

## Interpret functions that arise in applications in terms of the context.

**F-IF.B.4**[9] For a function that models a relationship between two quantities, interpret key features of graphs and tables in terms of the quantities, and sketch graphs showing key features given a verbal description of the relationship. *Key features include: intercepts; intervals where the function is increasing, decreasing, positive, or negative; relative maximums and minimums; symmetries; end behavior; and periodicity.*★

**F-IF.B.5** Relate the domain of a function to its graph and, where applicable, to the quantitative relationship it describes. *For example, if the function $h(n)$ gives the number of person-hours it takes to assemble $n$ engines in a factory, then the positive integers would be an appropriate domain for the function.*★

**F-IF.B.6**[9] Calculate and interpret the average rate of change of a function (presented symbolically or as a table) over a specified interval. Estimate the rate of change from a graph.★

## Analyze functions using different representations

**F-IF.C.7** Graph functions expressed symbolically and show key features of the graph, by hand in simple cases and using technology for more complicated cases.★

e. Graph exponential and logarithmic functions, showing intercepts and end behavior, and trigonometric functions, showing period, midline, and amplitude.

**F-IF.C.8**[10] Write a function defined by an expression in different but equivalent forms to reveal and explain different properties of the function.

b. Use the properties of exponents to interpret expressions for exponential functions. *For example, identify percent rate of change in functions such as $y = (1.02)^t$, $y = (0.97)^t$, $y = (1.01)^{12t}$, $y = (1.2)^{\frac{t}{10}}$, and classify them as representing exponential growth or decay.*

**F-IF.C.9**[11] Compare properties of two functions each represented in a different way (algebraically, graphically, numerically in tables, or by verbal descriptions). *For example, given a graph of one quadratic function and an algebraic expression for another, say which has the larger maximum.*

## Build a function that models a relationship between two quantities

**F-BF.A.1** Write a function that describes a relationship between two quantities.★

a. Determine an explicit expression, a recursive process, or steps for calculation from a context.[12]

---

[9] Tasks have a real-world context. In Algebra II, tasks may involve polynomial, exponential, logarithmic, and trigonometric functions.

[10] Tasks include knowing and applying $A = Pe^{rt}$ and $A = P\left(1+\frac{r}{n}\right)^{nt}$.

[11] In Algebra II, tasks may involve polynomial, exponential, logarithmic, and trigonometric functions

[12] Tasks have a real-world context. In Algebra II, tasks may involve linear functions, quadratic functions, and exponential functions.

b. Combine standard function types using arithmetic operations. *For example, build a function that models the temperature of a cooling body by adding a constant function to a decaying exponential, and relate these functions to the model.*[13]

**F-BF.A.2** Write arithmetic and geometric sequences both recursively and with an explicit formula, use them to model situations, and translate between the two forms.★

### Build new functions from existing functions.

**F-BF.B.3**[14] Identify the effect on the graph of replacing $f(x)$ by $f(x) + k$, $k\, f(x)$, $f(kx)$, and $f(x + k)$ for specific values of $k$ (both positive and negative); find the value of $k$ given the graphs. Experiment with cases and illustrate an explanation of the effects on the graph using technology. *Include recognizing even and odd functions from their graphs and algebraic expressions for them.*

**F-BF.B.4** Find inverse functions.

a. Solve an equation of the form $f(x) = c$ for a simple function $f$ that has an inverse and write an expression for the inverse. *For example,* $f(x) = 2\,x^3$ *or* $f(x) = (x + 1)/(x - 1)$ *for* $x \neq 1$.

### Construct and compare linear, quadratic, and exponential models and solve problems.

**F-LE.A.2**[15] Construct linear and exponential functions, including arithmetic and geometric sequences, given a graph, a description of a relationship, or two input-output pairs (include reading these from a table).★

**F-LE.A.4**[16] For exponential models, express as a logarithm the solution to $ab^{ct} = d$ where $a$, $c$, and $d$ are numbers and the base $b$ is 2, 10, or $e$; evaluate the logarithm using technology.★

### Interpret expressions for functions in terms of the situation they model.

**F-LE.B.5**[17] Interpret the parameters in a linear or exponential function in terms of a context.★

## Foundational Standards

### Use properties of rational and irrational numbers.

**N-RN.B.3** Explain why the sum or product of two rational numbers is rational; that the sum of a rational number and an irrational number is irrational; and that the product of a nonzero rational number and an irrational number is irrational.

---

[13] Combining functions also includes composition of functions.
[14] In Algebra II, tasks may involve polynomial, exponential, logarithmic, and trigonometric functions. Tasks may involve recognizing even and odd functions.
[15] In Algebra II, tasks will include solving multi-step problems by constructing linear and exponential functions.
[16] Students learn terminology that logarithm without a base specified is base 10 and that natural logarithm always refers to base $e$.
[17] Tasks have a real-world context. In Algebra II, tasks include exponential functions with domains not in the integers.

## Interpret the structure of expressions.

**A-SSE.A.2** Use the structure of an expression to identify ways to rewrite it. *For example, see $x^4 - y^4$ as $(x^2)^2 - (y^2)^2$, thus recognizing it as a difference of squares that can be factored as $(x^2 - y^2)(x^2 + y^2)$.*

## Create equations that describe numbers or relationships.

**A-CED.A.2** Create equations in two or more variables to represent relationships between quantities; graph equations on coordinate axes with labels and scales.★

**A-CED.A.4** Rearrange formulas to highlight a quantity of interest, using the same reasoning as in solving equations. *For example, rearrange Ohm's law $V = IR$ to highlight resistance $R$.*★

## Represent and solve equations and inequalities graphically.

**A-REI.D.10** Understand that the graph of an equation in two variables is the set of all its solutions plotted in the coordinate plane, often forming a curve (which could be a line).

## Understand the concept of a function and use function notation.

**F-IF.A.1** Understand that a function from one set (called the domain) to another set (called the range) assigns to each element of the domain exactly one element of the range. If $f$ is a function and $x$ is an element of its domain, then $f(x)$ denotes the output of $f$ corresponding to the input $x$. The graph of $f$ is the graph of the equation $y = f(x)$.

**F-IF.A.2** Use function notation, evaluate functions for inputs in their domains, and interpret statements that use function notation in terms of a context.

## Construct and compare linear, quadratic, and exponential models and solve problems.

**F-LE.A.1** Distinguish between situations that can be modeled with linear functions and with exponential functions.★

b. Recognize situations in which one quantity changes at a constant rate per unit interval relative to another.

c. Recognize situations in which a quantity grows or decays by a constant percent rate per unit interval relative to another.

**F-LE.A.3** Observe using graphs and tables that a quantity increasing exponentially eventually exceeds a quantity increasing linearly, quadratically, or (more generally) as a polynomial function.★

# Focus Standards for Mathematical Practice

**MP.1** **Make sense of problems and persevere in solving them.** Students make sense of rational and real number exponents and in doing so are able to apply exponential functions to solve problems involving exponential growth and decay for continuous domains such as time. They explore logarithms numerically and graphically to understand their meaning and how they can be used to solve exponential equations. Students have multiple opportunities to make connections between information presented graphically, numerically, and algebraically and search for similarities between these representations to further understand the underlying mathematical properties of exponents and logarithms. When presented with a wide variety of information related to financial planning, students make sense of the given information and use appropriate formulas to effectively plan for a long-term budget and savings plan.

**MP.2** **Reason abstractly and quantitatively.** Students consider appropriate units when exploring the properties of exponents for very large and very small numbers. They reason about quantities when solving a wide variety of problems that can be modeled using logarithms or exponential functions. Students relate the parameters in exponential expressions to the situations they model. They write and solve equations and then interpret their solutions within the context of a problem.

**MP.4** **Model with mathematics.** Students use exponential functions to model situations involving exponential growth and decay. They model the number of digits needed to assign identifiers using logarithms. They model exponential growth using a simulation with collected data. The application of exponential functions and logarithms as a means to solve an exponential equation is a focus of several lessons that deal with financial literacy and planning a budget. Here, students must make sense of several different quantities and their relationships as they plan and prioritize for their future financial solvency.

**MP.7** **Look for and make use of structure.** Students extend the laws of exponents for integer exponents to rational and real number exponents. They connect how these laws are related to the properties of logarithms and understand how to rearrange an exponential equation into logarithmic form. Students analyze the structure of exponential and logarithmic functions to understand how to sketch graphs and see how the properties relate to transformations of these types of functions. They analyze the structure of expressions to reveal properties such as recognizing when a function models exponential growth versus decay. Students use the structure of equations to understand how to identify an appropriate solution method.

**MP.8** **Look for and express regularity in repeated reasoning.** Students discover the properties of logarithms and the meaning of a logarithm by investigating numeric examples. They develop formulas that involve exponentials and logarithms by extending patterns and examining tables and graphs. Students generalize transformations of graphs of logarithmic functions by examining several different cases.

# Terminology

## New or Recently Introduced Terms

- **Arithmetic Series** (An *arithmetic series* is a series whose terms form an arithmetic sequence.)
- **Geometric Series** (A *geometric series* is a series whose terms form a geometric sequence.)
- **Invertible Function** (Let $f$ be a function whose domain is the set $X$, and whose image is the set $Y$. Then $f$ is *invertible* if there exists a function $g$ with domain $Y$ and image $X$ such that $f$ and $g$ satisfy the property:

  For all $x \in X$ and $y \in Y$, $f(x) = y$ if and only if $g(y) = x$.

  The function $g$ is called the *inverse* of $f$, and is denoted $f^{-1}$.

  The way to interpret the property is to look at all pairs $(x, y) \in X \times Y$: If the pair $(x, y)$ makes $f(x) = y$ a true equation, then $g(y) = x$ is a true equation. If it makes $f(x) = y$ a false equation, then $g(y) = x$ is false. If that happens for each pair in $X \times Y$, then $f$ and $g$ are invertible and are inverses of each other.)
- **Logarithm** (If three numbers, $L$, $b$, and $x$ are related by $x = b^L$, then $L$ is the *logarithm base $b$ of $x$*, and we write $L = \log_b(x)$. That is, the value of the expression $\log_b(x)$ is the power of $b$ needed to be equivalent to $x$.

  Valid values of $b$ as a base for a logarithm are $0 < b < 1$ and $b > 1$.)
- **Series** (Let $a_1, a_2, a_3, a_4, \ldots$ be a sequence of numbers. A sum of the form

  $$a_1 + a_2 + a_3 + \cdots + a_n$$

  for some positive integer $n$ is called a *series, or finite series,* and is denoted $S_n$. The $a_i$'s are called the *terms* of the series. The number that the series adds to is called the *sum* of the series.

  Sometimes $S_n$ is called the $n^{th}$ *partial sum*.)
- ***e*** (Euler's number, $e$, is an irrational number that is approximately equal to $e \approx 2.7182818284590$.)
- **Σ** (The Greek letter sigma, $\Sigma$, is used to represent the sum. There is no rigid way to use $\Sigma$ to represent a summation, but all notations generally follow the same rules. We will discuss the most common way it is used. Given the sequence $a_1, a_2, a_3, a_4, \ldots$, we can write the sum of the first $n$ terms of the sequence using the expression:

  $$\sum_{k=1}^{n} a_k \text{ .)}$$

## Familiar Terms and Symbols[18]

- Compound Interest
- Exponential Decay
- Exponential Expression
- Exponential Growth
- Scientific Notation

[18] These are terms and symbols students have seen previously.

## Suggested Tools and Representations

- Graphing calculator or Desmos online calculator simulation
- Wolfram Alpha Software
- GeoGebra or Geometer's Sketchpad Software
- Excel or other spreadsheet software, such as Calc (part of the OpenOffice suite)

## Assessment Summary

| Assessment Type | Administered | Format | Standards Addressed |
|---|---|---|---|
| Mid-Module Assessment Task | After Topic B | Constructed response with rubric | N-RN.A.1, N-RN.A.2, N-Q.A.2, A.CED.A.1, F-IF.B.6, F-BF.A.1a, F-LE.A.4 |
| End-of-Module Assessment Task | After Topic E | Constructed response with rubric | A-SSE.B.3c, A-SSE.B.4, A-CED.A.1, A-REI.D.11, F-IF.A.3, F-IF.B.4, F-IF.B.5, F-IF.B.6, F-IF.C.7e, F-IF.C.8b, F-IF.C.9, F-BF-A.1a, F-BF.A.1b, F-BF.A.2, F-BF.B.3, F-BF.B.4a, F-LE.A.2, F-LE.A.4, F-LE.B.5 |

# Mathematics Curriculum

## Topic A:

# Real Numbers

**N-RN.A.1, N-RN.A.2, N-Q.A.2, F-IF.B.6, F-BF.A.1a, F-LE.A.2**

| | | |
|---|---|---|
| **Focus Standards:** | N-RN.A.1 | Explain how the definition of the meaning of rational exponents follows from extending the properties of integer exponents to those values, allowing for a notation for radicals in terms of rational exponents. *For example, we define* $5^{\frac{1}{3}}$ *to be the cube root of* 5 *because we want* $\left(5^{\frac{1}{3}}\right)^3 = 5^{\left(\frac{1}{3}\right)^3}$ *to hold, so* $\left(5^{\frac{1}{3}}\right)^3$ *must equal* 5. |
| | N-RN.A.2 | Rewrite expressions involving radicals and rational exponents using the properties of exponents. |
| | N-Q.A.2 | Define appropriate quantities for the purpose of descriptive modeling.★ |
| | F-IF.6 | Calculate and interpret the average rate of change of a function (presented symbolically or as a table) over a specified interval. Estimate the rate of change from a graph.★ |
| | F-BF.A.1a | Write a function that describes a relationship between two quantities.★ a. Determine an explicit expression, a recursive process, or steps for calculation from a context |
| | F-LE.A.2 | Construct linear and exponential functions, including arithmetic and geometric sequences, given a graph, a description of a relationship, or two input-output pairs (include reading these from a table).★ |
| **Instructional Days:** | 6 | |
| **Lesson 1:** | Integer Exponents (E)[1] | |
| **Lesson 2:** | Base 10 and Scientific Notation (P) | |
| **Lesson 3:** | Rational Exponents—What are $2^{\frac{1}{2}}$ and $2^{\frac{1}{3}}$? (S) | |
| **Lesson 4:** | Properties of Exponents and Radicals (P) | |
| **Lesson 5:** | Irrational Exponents—What are $2^{\sqrt{2}}$ and $2^{\pi}$? (S) | |
| **Lesson 6:** | Euler's Number, *e* (P) | |

[1] Lesson Structure Key: **P**-Problem Set Lesson, **M**-Modeling Cycle Lesson, **E**-Exploration Lesson, **S**-Socratic Lesson

In Topic A, students prepare to generalize what they know about various function families by examining the behavior of exponential functions. One goal of the module is to show that the domain of the exponential function, $f(x) = b^x$, where $b$ is a positive number not equal to $1$, is all real numbers. In Lesson 1, students review and practice applying the laws of exponents to expressions in which the exponents are integers. Students first tackle a challenge problem on paper folding that is related to exponential growth and then apply and practice applying the laws of exponents to rewriting algebraic expressions. They experiment, create a table of values, observe patterns, and then generalize a formula to represent different measurements in the folded stack of paper as specified in **F-LE.A.2**. They also use the laws of exponents to work with very large and very small numbers.

Lesson 2 sets the stage for the introduction of base $10$ logarithms in Topic B of the module by reviewing how to express numbers using scientific notation, how to compute using scientific notation, and how to use the laws of exponents to simplify those computations in accordance with **N-RN.A.2**. Students should gain a sense of the change in magnitude when different powers of $10$ are compared. The activities in these lessons prepare students for working with quantities that increase in magnitude by powers of $10$ and by showing them the usefulness of exponent properties when performing arithmetic operations. Similar work will be done in later lessons relating to logarithms. Exercises on distances between planets in the solar system and on comparing magnitudes in other real-world contexts provide additional practice with arithmetic operations on numbers written using scientific notation.

Lesson 3 begins with students examining the graph of $y = 2^x$ and estimating the values as a means of extending their understanding of integer exponents to rational exponents. The examples are generalized to $2^{\frac{1}{n}}$ before generalizing further to $2^{\frac{m}{n}}$. As the domain of the identities involving exponents is expanded, it is important to maintain consistency with the properties already developed. Students work specifically to make sense that $2^{\frac{1}{2}} = \sqrt{2}$ and $2^{\frac{1}{3}} = \sqrt[3]{2}$ to develop the more general concept that $2^{\frac{1}{n}} = \sqrt[n]{2}$. The lesson demonstrates how people develop mathematics (a) to be consistent with what is already known and (b) to make additional progress. Additionally, students practice MP.7 as they extend the rules for integer exponents to rules for rational exponents (**N-RN.A.1**).

Lesson 4 continues the discussion of properties of exponents and radicals, and students continue to practice MP.7 as they extend their understanding of exponents to all rational numbers and for all positive real bases as specified in **N-RN.A.1**. Students rewrite expressions involving radicals and rational exponents using the properties of exponents (**N-RN.A.2**). The notation $x^{\frac{1}{n}}$ specifically indicates the principal root of $x$: the positive root when $n$ is even and the real-valued root when $n$ is odd. To avoid inconsistencies in our later work with logarithms, $x$ is required to be positive.

Lesson 5 revisits the work of Lesson 3 and extends student understanding of the domain of the exponential function $f(x) = b^x$, where $b$ is a positive real number, from the rational numbers to all real numbers through the process of considering what it means to raise a number to an irrational exponent (such as $2^{\sqrt{2}}$). In many ways, this lesson parallels the work students did in Lesson 3 to make a solid case for why the laws of exponents hold for all rational number exponents. The recursive procedure that students employ in this lesson aligns with **F-BF.A.1a**. This lesson is important both because it helps to portray mathematics as a coherent body of knowledge that makes sense and because we need to make sure that students understand that logarithms can be irrational numbers. Essentially, we need to guarantee that exponential and logarithmic functions are continuous functions. Students take away from these lessons an understanding that the domain of exponents in the laws of exponents does indeed extend to all real numbers rather than integers, as defined previously in Grade 8.

Lesson 6 is a modeling lesson in which students practice MP.4 when they find an exponential function to model the amount of water in a tank after $t$ seconds when the height of the water is constantly doubling or tripling, and apply **F-IF.B.6** as they explore the average rate of change of the height of the water over smaller and smaller intervals. If we denote the height of the water in the tank at time $t$ seconds by $H(t) = b^t$, then the average rate of change of the height of the water on an interval $[T, T + \varepsilon]$ is approximated by $\frac{H(T+\varepsilon)-H(t)}{\varepsilon} \approx c \cdot H(T)$. Students calculate that if the height of the water is doubling each second, then $c \approx 0.69$, and if the height of the water is tripling each second, then $c \approx 1.1$. Students discover Euler's number, $e$, by applying repeated reasoning (MP.8) and numerically approximating the base $b$ for which the constant $c$ is equal to 1. Euler's number will be used extensively in the future and occurs in many different applications.

# Lesson 1: Integer Exponents

## Student Outcomes

- Students review and practice applying the properties of exponents for integer exponents.
- Students model a real-world scenario involving exponential growth and decay.

## Lesson Notes

To fully understand exponential functions and their use in modeling real-world situations, students must be able to extend the properties of integer exponents to rational and real numbers. In previous grades, students established the properties of exponents for integer exponents and worked with radical expressions and irrational numbers such as $\sqrt{2}$.

In this module, we use the properties of exponents to show that for any positive real number $b$, the domain of the exponential function $f(x) = b^x$ is all real numbers. In Algebra I, students primarily worked with exponential functions where the domain was limited to a set of integers. In the latter part of this module, students are introduced to logarithms, which allow them to find real number solutions to exponential equations. Students come to understand how logarithms simplify computation, particularly with different measuring scales.

Much of the work in this module relies on the ability to reason quantitatively and abstractly (MP.2), to make use of structure (MP.7), and to model with mathematics (MP.4). Lesson 1 begins with a challenge problem where students are asked to fold a piece of paper in half 10 times and construct exponential functions based on their experience (**F-LE.A.2**). It is physically impossible to fold a sheet of notebook paper in half more than seven or eight times; the difficulty lies in the thickness of the paper compared to the resulting area when the paper is folded. To fold a piece of paper in half more than seven or eight times requires either a larger piece of paper, a very thin piece of paper, or a different folding scheme, such as accordion folding. In 2001, a high school student, Britney Gallivan, successfully folded a very large piece of paper in half 12 times and derived a mathematical formula to determine how large a piece of paper would be needed to successfully accomplish this task (http://pomonahistorical.org/12times.htm). Others have tried the problem as well, including the hosts of the television show Mythbusters (https://www.youtube.com/watch?v=kRAEBbotuIE) and students at St. Mark's High School in Massachusetts (http://www.newscientist.com/blogs/nstv/2012/01/paper-folding-limits-pushed.html). You may wish to share one of these resources with your class at the close of the lesson or after they have completed the Exploratory Challenge. After this challenge, which reintroduces students to exponential growth, students review the properties of exponents for integer exponents and apply them to the rewriting of algebraic expressions (**N-RN.A.2**). The lesson concludes with fluency practice where students apply properties of exponents to rewrite expressions in a specified form.

You may want to have the following materials on hand in case students want to explore this problem in more detail: access to the internet, chart paper, cash register tape, a roll of toilet paper, origami paper, tissue paper or facial tissues, and rulers.

## Classwork

Students begin this lesson by predicting whether they can fold a piece of paper in half 10 times, how tall the folded paper will be, and whether or not the area of paper showing on top is smaller or larger than a postage stamp. They will explore the validity of their predictions in the Exploratory Challenge that follows.

### Opening Exercise (3 minutes)

Give students a short amount of time to think about and write individual responses. Lead a short discussion with the entire class to poll students on their responses. Record solutions on the board or chart paper for later reference. At this point, most students will probably say that they *can* fold the paper in half 10 times. The sample responses shown below are NOT correct and represent possible initial student responses. Some students may be familiar with this challenge, having seen it discussed on a television program or on the Internet and, consequently, will say that you cannot fold a piece of notebook paper in half 10 times. Accept all responses and avoid excessive explaining or justifying of answers at this point.

*Scaffolding:*

- Demonstrate folding the paper once or twice during the discussion to illustrate what the questions are asking and what is meant by *how thick* the folded paper will be.
- Ask advanced learners to provide justification for their claim about the thickness of the folded paper.

**Opening Exercise**

**Can you fold a piece of notebook paper in half $10$ times?**

***Answers will vary. Although incorrect, many students may initially answer "Yes."***

**How thick will the folded paper be?**

***Answers will vary. The following is a typical student guess: It will be about $1$ cm.***

**Will the area of the paper on the top of the folded stack be larger or smaller than a postage stamp?**

***It will be smaller because I will be dividing the rectangle in half $10$ times, and since a piece of paper is about $8.5$ in. by $11$ in., it will be very small when divided in half that many times.***

### Discussion (2 minutes)

Students should brainstorm ideas for further exploring these questions to come up with a more precise answer to these questions. At this point, some students will likely be folding a piece of notebook paper. On chart paper, record ideas for additional information needed to answer the original questions more precisely.

- How can you be sure of your answers?
  - *We could actually fold a piece of paper and measure the height and area on top of the stack. We could determine the thickness of a sheet of notebook paper and then multiply it by the number of folds. We could find the area of the original paper and divide it by $2$ successively for each fold.*

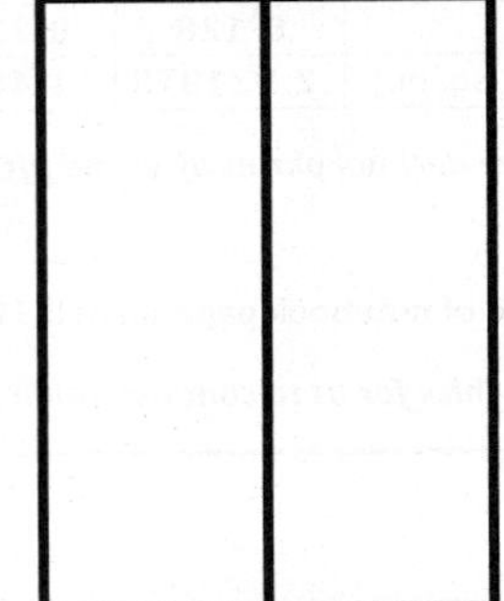

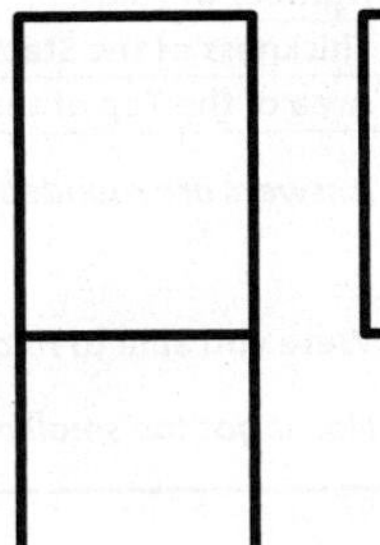

MP.1

- What additional information is needed to solve this problem?
  - *We need to know the thickness of the paper, the dimensions of the original piece of paper, and a consistent way to fold the paper.*
  - *We need to know the size of a postage stamp.*
- How will you organize your work?
  - *We can make a table.*

## Exploratory Challenge (20 minutes)

Students should work on this challenge problem in small groups. Groups can use the suggested scaffolding questions on the student pages, or you can simply give each group a piece of chart paper and access to appropriate tools such as a ruler, different types of paper, a computer for researching the thickness of a sheet of paper, etc., and start them on the task. Have them report their results on their chart paper. Student solutions will vary from group to group. Sample responses have been provided below for a standard 8.5 in. by 11 in. piece of paper that has been folded in half as shown below. The size of a small postage stamp is $\frac{7}{8}$ in. by 1 in.

**Exploratory Challenge**

a. **What are the dimensions of your paper?**

*The dimensions are* $8.5$ in. *by* $11$ in.

b. **How thick is one sheet of paper? Explain how you decided on your answer.**

*A ream of paper is* $500$ *sheets. It is about* $2$ in. *high. Dividing* $2$ *by* $500$ *would give a thickness of a piece of paper to be approximately* $0.004$ in.

c. **Describe how you folded the paper.**

*First, we folded the paper in half so that it was* $8.5$ in. *by* $5.5$ in.*; then, we rotated the paper and folded it again so that it was* $5.5$ in. *by* $4.25$ in.*; then, we rotated the paper and folded it again, and so on.*

d. **Record data in the following table based on the size and thickness of your paper.**

| Number of Folds | 0 | 1 | 2 | 3 | 4 |
|---|---|---|---|---|---|
| Thickness of the Stack (in.) | 0.004 | 0.008 | 0.016 | 0.032 | 0.064 |
| Area of the Top of the Stack (sq. in.) | 93.5 | 46.75 | 23.375 | 11.6875 | 5.84375 |

| Number of Folds | 5 | 6 | 7 | 8 | 9 | 10 |
|---|---|---|---|---|---|---|
| Thickness of the Stack (in.) | 0.128 | 0.256 | 0.512 | 1.024 | 2.048 | 4.096 |
| Area of the Top of the Stack (sq. in.) | 2.921875 | 1.461 | 0.730 | 0.365 | 0.183 | 0.091 |

*Answers are rounded to three decimal places after the fifth fold.*

e. **Were you able to fold a piece of notebook paper in half** $10$ **times? Why or why not?**

*No. It got too small and too thick for us to continue folding it.*

Debrief after part (e) by having groups present their solutions so far. At this point, students should realize that it is impossible to fold a sheet of notebook paper in half 10 times. If groups wish to try folding a larger piece of paper, such as a piece of chart paper, or a different thickness of paper, such as a facial tissue, or using a different folding technique, such as an accordion fold, then allow them to alter their exploration. You may wish to have tissue paper, facial tissues, a roll of toilet paper, chart paper, or cash register tape on hand for student groups to use in their experiments.

> *Scaffolding:*
> If students are struggling to develop the formulas in their groups, you can complete the rest of this challenge as a whole class. Students will have many opportunities to model using exponential functions later in this module. You can write the height and thickness as products of repeated twos to help students see the pattern. For example, after three folds, the height would be $T \cdot 2 \cdot 2 \cdot 2$, where $T$ is the thickness of the paper.

After students have made adjustments to their model and tested it, have them write formulas to predict the height and area after 10 folds and explain how these answers compare to their original predictions. Students worked with exponential functions and geometric sequences in Algebra I. Since this situation involves doubling or halving, most groups should be able to write a formula. When debriefing this next section with the entire class, help students to write a well-defined formula. Did they specify the meaning of their variables? Did they specify a domain if they used function notation? They may not have used the same variables shown in the solutions below and should be using specific values for the thickness and area of the paper based on their assumptions during modeling. Students will likely be surprised by these results.

- How thick would the stack be if you could fold it 10 times?
- Is the area of the top of the stack smaller or larger than a postage stamp?
- How do these answers compare to your prediction?

> **f. Create a formula that approximates the height of the stack after $n$ folds.**
>
> ***Our formula is $H(n) = T(2)^n$, where $T$ is the thickness of the paper, and $H(n)$ is the height after $n$ folds. In this case, $T = 0.004$ in.***
>
> **g. Create a formula that will give you the approximate area of the top after $n$ folds.**
>
> ***Our formula is $A(n) = A_0\left(\frac{1}{2}\right)^n$, where $A_0$ is the area of the original piece of paper, and $A(n)$ is the area of the top after $n$ folds. In this case, $A_0 = 93.5$ sq. in.***
>
> **h. Answer the original questions from the Opening Exercise. How do the actual answers compare to your original predictions?**
>
> ***It was impossible to fold the paper more than 7 times. Using our model, if we could fold the paper $10$ times, it would be just over $4$ in. thick and less than $\frac{1}{10}$ sq. in., which is much smaller than the area of a postage stamp. Our predictions were inaccurate because we did not consider how drastically the sizes change when successively doubling or halving measurements.***

Student groups should present their solutions again. If it did not come up earlier, ask students to consider how they might increase the likelihood that they could fold a piece of paper in half more than seven or eight times.

- What are some ways to increase the likelihood that you could successfully fold a piece of paper in half more than seven or eight times?
  - *You could use a thinner piece of paper. You could use a larger piece of paper. You could try different ways of folding the paper.*

Brittney Gallivan, the high school student who solved this problem in 2001, first folded a very thin sheet of gold foil in half over seven times and then successfully folded an extremely large piece of paper in half 12 times at a local shopping mall. In 2011, students at St. Mark's High School in Massachusetts folded miles of taped together toilet paper in half 13 times.

## Example 1 (5 minutes): Using the Properties of Exponents to Rewrite Expressions

In this example, you will show students how to represent their expressions using powers of 2. Model this using the folding of a 10 in. by 10 in. square sheet of gold foil. The thickness of gold foil is 0.28 millionths of a meter. The information in this problem is based on the first task Britney accomplished: folding a sheet of gold foil in half twelve times. Of course, her teacher then modified the task and required her to actually use paper. (http://www.abc.net.au/science/articles/2005/12/21/1523497.htm). The goal of this example is to remind students of the meaning of integer exponents.

> *Scaffolding:*
>
> Use a vocabulary notebook or an anchor chart posted on the wall to remind students of vocabulary words associated with exponents.
>
> 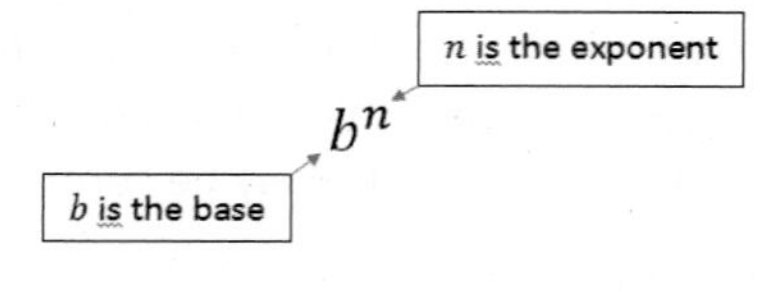
> 

Many students will likely see the area sequence as successive divisions by two. For students who are struggling to make sense of the meaning of a negative exponent, you can model rewriting the expressions in the last column as follows:

$$\frac{100}{2} = 100\left(\frac{1}{2}\right) = 100\left(\frac{1}{2}\right)^1 = 100 \cdot 2^{-1};$$

$$\frac{100}{4} = 100\left(\frac{1}{4}\right) = 100\left(\frac{1}{2}\right)^2 = 100 \cdot 2^{-2};$$

$$\frac{100}{8} = 100\left(\frac{1}{8}\right) = 100\left(\frac{1}{2}\right)^3 = 100 \cdot 2^{-3}.$$

**Example 1: Using the Properties of Exponents to Rewrite Expressions**

**The table below displays the thickness and area of a folded square sheet of gold foil. In 2001, Britney Gallivan, a California high school junior, successfully folded this size sheet of gold foil in half 12 times to earn extra credit in her mathematics class.**

**Rewrite each of the table entries as a multiple of a power of 2.**

| Number of Folds | Thickness of the Stack (Millionths of a Meter) | Thickness Using a Power of 2 | Area of the Top (Square Inches) | Area Using a Power of 2 |
|---|---|---|---|---|
| $0$ | $0.28$ | $0.28 \cdot 2^0$ | $100$ | $100 \cdot 2^0$ |
| $1$ | $0.56$ | $0.28 \cdot 2^1$ | $50$ | $100 \cdot 2^{-1}$ |
| $2$ | $1.12$ | $0.28 \cdot 2^2$ | $25$ | $100 \cdot 2^{-2}$ |
| $3$ | $2.24$ | $0.28 \cdot 2^3$ | $12.5$ | $100 \cdot 2^{-3}$ |
| $4$ | $4.48$ | $0.28 \cdot 2^4$ | $6.25$ | $100 \cdot 2^{-4}$ |
| $5$ | $8.96$ | $0.28 \cdot 2^5$ | $3.125$ | $100 \cdot 2^{-5}$ |
| $6$ | $17.92$ | $0.28 \cdot 2^6$ | $1.5625$ | $100 \cdot 2^{-6}$ |

As you model this example with students, take the opportunity to discuss the fact that exponentiation with positive integers can be thought of as repeated multiplication by the base, whereas exponentiation with negative integers can be thought of as repeated division by the base. For example,

$$4^{24} = \underbrace{4 \cdot 4 \cdot 4 \cdot \dots \cdot 4}_{24\text{ times}} \quad \text{and} \quad 4^{-24} = \frac{1}{\underbrace{4 \cdot 4 \cdot 4 \cdot \dots \cdot 4}_{24\text{ times}}}.$$

Alternatively, you can describe the meaning of a negative exponent when the exponent is an integer as repeated multiplication by the reciprocal of the base. For example,

$$4^{-24} = \underbrace{\left(\frac{1}{4}\right)\left(\frac{1}{4}\right)\left(\frac{1}{4}\right)\cdots\left(\frac{1}{4}\right)}_{24\text{ times}}.$$

Interpreting exponents as repeated multiplication or division only makes sense for integer exponents. However, the properties of exponents do apply for any real number exponent.

## Example 2 (5 minutes): Applying the Properties of Exponents to Rewrite Expressions

Transition into this example by explaining that many times when we work with algebraic and numeric expressions that contain exponents, it is advantageous to rewrite them in different forms (MP.7). One obvious advantage of exponents is that they shorten the length of an expression involving repeated multiplication. Imagine always having to write out ten 2s if you wanted to express the number 1024 as a power of 2. While the exponent notation gives us a way to express repeated multiplication succinctly, the properties of exponents also provide a way to make computations with exponents more efficient. Share the properties of exponents below. Have students record these in their math notebooks.

> *Scaffolding:*
>
> Students needing additional practice with exponents can use the following numeric examples that mirror the algebraic expressions in Example 1. Have students work them side-by-side to help see the structure of the expressions.
>
> - Write each expression in the form $kb^n$, where $k$ is a real number and $b$ and $n$ are integers.
>
> $$(5 \cdot 2^7)(-3 \cdot 2^2)$$
> $$\frac{3 \cdot 4^5}{(2 \cdot 4)^4}$$
> $$\frac{3}{(5^2)^{-3}}$$
> $$\frac{3^{-3} \cdot 3^4}{3^8}$$

The Properties of Exponents

For nonzero real numbers $x$ and $y$ and all integers $a$ and $b$, the following properties hold.

$$x^a \cdot x^b = x^{a+b}$$
$$(x^a)^b = x^{ab}$$
$$(xy)^a = x^a y^a$$
$$\frac{x^a}{x^b} = x^{a-b}$$

(Note: Most cases of the properties listed above hold when $x = 0$ or $y = 0$. The only cases that cause problems are when values of the variables result in the expression $0^0$ or division by 0.) Ask students to discuss with a partner different ways to rewrite the following expressions in the form of $kx^n$, where $k$ is a real number and $n$ is an integer. Have students share out their responses with the entire class. For each problem, model both approaches. As you model, make sure to have students verbalize the connections between the methods. Ask volunteers to explain why the rules hold.

**Example 2: Applying the Properties of Exponents to Rewrite Expressions**

**Rewrite each expression in the form of $kx^n$, where $k$ is a real number, $n$ is an integer, and $x$ is a nonzero real number.**

a. $5x^5 \cdot -3x^2$

***Method 1: Apply the definition of an exponent and properties of algebra.***

$$5x^5 \cdot -3x^2 = 5 \cdot -3 \cdot x^5 \cdot x^2 = -15 \cdot (x \cdot x \cdot x \cdot x \cdot x) \cdot (x \cdot x) = -15\, x^7$$

***Method 2: Apply the rules of exponents and the properties of algebra.***

$$5x^5 \cdot -3x^2 = 5 \cdot -3 \cdot x^5 \cdot x^2 = -15 \cdot x^{5+2} = -15\, x^7$$

b. $\dfrac{3x^5}{(2x)^4}$

***Method 1: Apply the definition of an exponent and properties of algebra.***

$$\frac{3x^5}{(2x)^4} = \frac{3 \cdot x \cdot x \cdot x \cdot x \cdot x}{2x \cdot 2x \cdot 2x \cdot 2x} = \frac{3 \cdot x \cdot x \cdot x \cdot x \cdot x}{2 \cdot 2 \cdot 2 \cdot 2 \cdot x \cdot x \cdot x \cdot x} = \frac{3x}{16} \cdot \frac{x}{x} \cdot \frac{x}{x} \cdot \frac{x}{x} \cdot \frac{x}{x} = \frac{3x}{16} \cdot 1 \cdot 1 \cdot 1 \cdot 1 = \frac{3}{16}x$$

***Method 2: Apply the rules of exponents and the properties of algebra.***

$$\frac{3x^5}{2^4x^4} = \frac{3}{16}x^{5-4} = \frac{3}{16}x$$

c. $\dfrac{3}{(x^2)^{-3}}$

***Method 1: Apply the definition of an exponent and properties of algebra.***

$$\frac{3}{(x^2)^{-3}} = \frac{3}{\left(\frac{1}{x^2}\right)\left(\frac{1}{x^2}\right)\left(\frac{1}{x^2}\right)} = \frac{3}{\frac{1}{x} \cdot \frac{1}{x} \cdot \frac{1}{x} \cdot \frac{1}{x} \cdot \frac{1}{x} \cdot \frac{1}{x}} = \frac{3}{\frac{1}{x^6}} = 3 \cdot x^6 = 3x^6$$

***Method 2: Apply the properties of exponents.***

$$\frac{3}{(x^2)^{-3}} = \frac{3}{x^{2 \cdot -3}} = \frac{3}{x^{-6}} = 3x^6$$

d. $\dfrac{x^{-3}x^4}{x^8}$

***Method 1: Apply the definition of an exponent and properties of algebra.***

$$\frac{x^{-3}x^4}{x^8} = \frac{\frac{1}{x} \cdot \frac{1}{x} \cdot \frac{1}{x} \cdot x \cdot x \cdot x \cdot x}{x \cdot x \cdot x \cdot x \cdot x \cdot x \cdot x \cdot x} = \frac{\frac{x}{x} \cdot \frac{x}{x} \cdot \frac{x}{x} \cdot x}{x \cdot x \cdot x \cdot x \cdot x \cdot x \cdot x \cdot x} = \frac{x}{x} \cdot \frac{1}{x \cdot x \cdot x \cdot x \cdot x \cdot x \cdot} = x^{-7}$$

***Method 2: Apply the properties of exponents.***

$$\frac{x^{-3}x^4}{x^8} = \frac{x^{-3+4}}{x^8} = \frac{x^1}{x^8} = x^{1-8} = x^{-7}$$

After seeing these examples, students should begin to understand why the properties of exponents are so useful for working with exponential expressions. Show both methods as many times as necessary in order to reinforce the properties with students so that they will ultimately rewrite most expressions like these by inspection, using the properties rather than expanding exponential expressions.

### Exercises 1–5 (5 minutes)

The point of these exercises is to force students to use the properties of exponents to rewrite expressions in the form $kx^n$. Typical high school textbooks ask students to write expressions with non-negative exponents. However, in advanced mathematics classes, students need to be able to fluently rewrite expressions in different formats. Students who continue on to study higher-level mathematics such as calculus will need to rewrite expressions in this format in order to quickly apply a common derivative rule. The last two exercises are not feasible to work out by expanding the exponent. Use them to assess whether or not students are able to apply the rules fluently even when larger numbers or variables are involved.

**Exercises 1–5**

**Rewrite each expression in the form of $kx^n$, where $k$ is a real number and $n$ is an integer. Assume $x \neq 0$.**

1. $2x^5 \cdot x^{10}$

   $2x^{15}$

2. $\dfrac{1}{3x^8}$

   $\dfrac{1}{3}x^{-8}$

3. $\dfrac{6x^{-5}}{x^{-3}}$

   $6x^{-5-(-3)} = 6x^{-2}$

4. $\left(\dfrac{3}{x^{-22}}\right)^{-3}$

   $(3x^{22})^{-3} = 3^{-3}x^{-66} = \dfrac{1}{27}x^{-66}$

5. $(x^2)^n \cdot x^3$

   $x^{2n} \cdot x^3 = x^{2n+3}$

### Closing (2 minutes)

Have students respond individually in writing or with a partner to the following questions.

- How can the properties of exponents help us to rewrite expressions?
  - *They make the process less tedious, especially when the exponents are very large or very small integers.*
- Why are the properties of exponents useful when working with large or small numbers?
  - *You can quickly rewrite expressions without having to rewrite each power of the base in expanded form.*

**Lesson Summary**

**The Properties of Exponents**
**For real numbers $x$ and $y$ with $x \neq 0$, $y \neq 0$, and all integers $a$ and $b$, the following properties hold.**

$$x^a \cdot x^b = x^{a+b}$$

$$(x^a)^b = x^{ab}$$

$$(xy)^a = x^a y^a$$

$$\frac{1}{x^a} = x^{-a}$$

$$\frac{x^a}{x^b} = x^{a-b}$$

$$\left(\frac{x}{y}\right)^a = \frac{x^a}{y^a}$$

$$x^0 = 1$$

## Exit Ticket (3 minutes)

Name ______________________________ Date ____________

# Lesson 1: Integer Exponents

## Exit Ticket

The following formulas for paper folding were discovered by Britney Gallivan in 2001 when she was a high school junior. The first formula determines the minimum width, $W$, of a square piece of paper of thickness $T$ needed to fold it in half $n$ times, alternating horizontal and vertical folds. The second formula determines the minimum length, $L$, of a long rectangular piece of paper of thickness $T$ needed to fold it in half $n$ times, always folding perpendicular to the long side.

$$W = \pi \cdot T \cdot 2^{\frac{3(n-1)}{2}} \qquad L = \frac{\pi T}{6}(2^n + 4)(2^n - 1)$$

1. Notebook paper is approximately $0.004$ in. thick. Using the formula for the width $W$, determine how wide a square piece of notebook paper would need to be to successfully fold it in half 13 times, alternating horizontal and vertical folds.

2. Toilet paper is approximately $0.002$ in. thick. Using the formula for the length $L$, how long would a continuous sheet of toilet paper have to be to fold it in half 12 times, folding perpendicular to the long edge each time?

3. Use the properties of exponents to rewrite each expression in the form $kx^n$. Then evaluate the expression for the given value of $x$.

   a. $2x^3 \cdot \frac{5}{4}x^{-1}$; $x = 2$

   b. $\frac{9}{(2x)^{-3}}$; $x = -\frac{1}{3}$

## Exit Ticket Sample Solutions

**The following formulas for paper folding were discovered by Britney Gallivan in 2001 when she was a high school junior. The first formula determines the minimum width, $W$, of a square piece of paper of thickness $T$ needed to fold it in half $n$ times, alternating horizontal and vertical folds. The second formula determines the minimum length, $L$, of a long rectangular piece of paper of thickness $T$ needed to fold it in half $n$ times, always folding perpendicular to the long side.**

$$W = \pi \cdot T \cdot 2^{\frac{3(n-1)}{2}} \qquad L = \frac{\pi T}{6}(2^n + 4)(2^n - 1)$$

1. **Notebook paper is approximately $0.004$ in. thick. Using the formula for the width $W$, determine how wide a square piece of notebook paper would need to be to successfully fold it in half $13$ times.**

   *The paper would need to be approximately* $3294.2$ in. *wide:* $W = \pi\, T\, 2^{\frac{3(13-1)}{2}} = \pi(0.004)2^{18} \approx 3294.199$.

2. **Toilet paper is approximately $0.002$ in. thick. Using the formula for the length $L$, how long would a continuous sheet of toilet paper have to be to fold it in half $12$ times?**

   *The paper would have to be approximately* $17{,}581.92$ in. *long, which is approximately* $0.277$ mi.*:*

   $$L = \left(\frac{\pi(0.002)}{6}\right)(2^{12} + 4)(2^{12} - 1) = \pi\left(\frac{1}{300}\right)(4100)(4095) = 55965\,\pi \approx 17{,}581.92.$$

3. **Use the properties of exponents to rewrite each expression in the form $kx^n$. Then evaluate the expression for the given value of $x$.**

   a. $2x^3 \cdot \frac{5}{4}x^{-1};\ x = 2$

   $$2\left(\frac{5}{4}\right)x^3x^{-1} = \frac{5}{2}x^2$$

   *When* $x = 2$, *the expression has the value* $\frac{5}{2}(2)^2 = 10$.

   b. $\frac{9}{(2x)^{-3}};\ x = -\frac{1}{3}$

   $$\frac{9}{2^{-3}x^{-3}} = 72x^3$$

   *When* $x = -\frac{1}{3}$, *the expression has the value* $72\left(-\frac{1}{3}\right)^3 = -\frac{8}{3}$.

## Problem Set Sample Solutions

1. **Suppose your class tried to fold an unrolled roll of toilet paper. It was originally $4$ in. wide and $30$ ft. long. Toilet paper is approximately $0.002$ in. thick.**

   a. **Complete each table and represent the area and thickness using powers of $2$.**

| Number of Folds $n$ | Thickness After $n$ Folds (in.) |
|---|---|
| $0$ | $0.002 = 0.002 \cdot 2^0$ |
| $1$ | $0.004 = 0.002 \cdot 2^1$ |
| $2$ | $0.008 = 0.002 \cdot 2^2$ |
| $3$ | $0.016 = 0.002 \cdot 2^3$ |
| $4$ | $0.032 = 0.002 \cdot 2^4$ |
| $5$ | $0.064 = 0.002 \cdot 2^5$ |
| $6$ | $0.128 = 0.002 \cdot 2^6$ |

| Number of Folds $n$ | Area on Top After $n$ Folds ($\text{in}^2$) |
|---|---|
| $0$ | $1440 = 1440 \cdot 2^0$ |
| $1$ | $720 = 1440 \cdot 2^{-1}$ |
| $2$ | $360 = 1440 \cdot 2^{-2}$ |
| $3$ | $180 = 1440 \cdot 2^{-3}$ |
| $4$ | $90 = 1440 \cdot 2^{-4}$ |
| $5$ | $45 = 1440 \cdot 2^{-5}$ |
| $6$ | $22.5 = 1440 \cdot 2^{-6}$ |

   b. **Create an algebraic function that describes the area in square inches after $n$ folds.**

   $A(n) = 1440 \cdot 2^{-n}$, *where $n$ is a positive integer.*

   c. **Create an algebraic function that describes the thickness in inches after $n$ folds.**

   $T(n) = 0.002 \cdot 2^{n}$, *where $n$ is a positive integer.*

2. **In the Exit Ticket, we saw the formulas below. The first formula determines the minimum width, $W$, of a square piece of paper of thickness $T$ needed to fold it in half $n$ times, alternating horizontal and vertical folds. The second formula determines the minimum length, $L$, of a long rectangular piece of paper of thickness $T$ needed to fold it in half $n$ times, always folding perpendicular to the long side.**

$$W = \pi \cdot T \cdot 2^{\frac{3(n-1)}{2}} \qquad L = \frac{\pi T}{6}(2^n + 4)(2^n - 1)$$

**Use the appropriate formula to verify why it is possible to fold a $10$ inch by $10$ inch sheet of gold foil in half $13$ times. Use $0.28$ millionths of a meter for the thickness of gold foil.**

*Given that the thickness of the gold foil is $0.28$ millionths of a meter, we have*

$$\frac{0.28}{1{,}000{,}000}\,\text{m} \cdot \frac{100\text{ cm}}{1\text{ m}} = 0.000028\text{ cm} \cdot \frac{1\text{ in}}{2.54\text{ cm}} = 0.00001102\text{ in.}$$

*Using the formula*

$$W = \pi\, T\, 2^{\frac{3(n-1)}{2}}$$

*with $n = 13$ and $T = 0.00001102$ in., we get*

$$W = \pi(0.00001102)2^{\frac{3(13-1)}{2}}\text{ in.} \approx 9.1\text{ in.}$$

3. Use the formula from the Exit Ticket to determine if you can fold an unrolled roll of toilet paper in half more than $10$ times. Assume that the thickness of a sheet of toilet paper is approximately $0.002$ in. and that one roll is $102$ ft. long.

*First convert feet to inches.* $102\text{ ft.} = 1224\text{ in.}$

*Then, substitute $0.002$ and $10$ into the formula for $T$ and $n$, respectively.*

$$L = \frac{\pi(0.002)}{6}(2^{10}+4)(2^{10}-1) = 1101.3$$

*The roll is just long enough to fold in half $10$ times.*

4. Apply the properties of exponents to rewrite expressions in the form $kx^n$, where $n$ is an integer and $x \neq 0$.

   a. $(2x^3)(3x^5)(6x)^2$

      $2 \cdot 3 \cdot 36x^{3+5+2} = 216x^{10}$

   b. $\dfrac{3x^4}{(-6x)^{-2}}$

      $3x^4 \cdot 36x^2 = 108x^6$

   c. $\dfrac{x^{-3}x^5}{3x^4}$

      $\dfrac{1}{3}x^{-3+5-4} = \dfrac{1}{3}x^{-2}$

   d. $5(x^3)^{-3}(2x)^{-4}$

      $\dfrac{5}{16}x^{-9+(-4)} = \dfrac{5}{16}x^{-13}$

   e. $\left(\dfrac{x^2}{4x^{-1}}\right)^{-3}$

      $\dfrac{x^{-6}}{4^{-3}x^3} = 64x^{-6-3} = 64x^{-9}$

5. Apply the properties of exponents to verify that each statement is an identity.

   a. $\dfrac{2^{n+1}}{3^n} = 2\left(\dfrac{2}{3}\right)^n$ for integer values of $n$.

      $\dfrac{2^{n+1}}{3^n} = \dfrac{2^n 2^1}{3^n} = \dfrac{2 \cdot 2^n}{3^n} = 2 \cdot \left(\dfrac{2}{3}\right)^n$

   b. $3^{n+1} - 3^n = 2 \cdot 3^n$ for integer values of $n$.

      $3^{n+1} - 3^n = 3^n \cdot 3^1 - 3^n = 3^n(3-1) = 3^n \cdot 2 = 2 \cdot 3^n$

   c. $\dfrac{1}{(3^n)^2} \cdot \dfrac{4^n}{3} = \dfrac{1}{3}\left(\dfrac{2}{3}\right)^{2n}$ for integer values of $n$.

      $\dfrac{1}{(3^n)^2} \cdot \dfrac{4^n}{3} = \dfrac{1}{3^{2n}} \cdot \dfrac{(2^2)^n}{3} = \dfrac{1 \cdot 2^{2n}}{3 \cdot 3^{2n}} = \dfrac{1}{3} \cdot \left(\dfrac{2}{3}\right)^{2n}$

6. Jonah was trying to rewrite expressions using the properties of exponents and properties of algebra for nonzero values of $x$. In each problem, he made a mistake. Explain where he made a mistake in each part and provide a correct solution.

**Jonah's Incorrect Work**

a. $(3x^2)^{-3} = -9x^{-6}$

b. $\dfrac{2}{3x^5} = 6x^{-5}$

c. $\dfrac{2x-x^3}{3x} = \dfrac{2}{3} - x^3$

*In part (a), he multiplied* $3$ *by the exponent* $-3$. *The correct solution is* $3^{-3}x^{-6} = \frac{1}{27}x^{-6}$.

*In part (b), he multiplied* $2$ *by* $3$ *when he rewrote* $x^5$. *The* $3$ *should remain in the denominator of the expression. The correct solution is* $\frac{2}{3}x^{-5}$.

*In part (c), he only divided the first term by* $3x$, *but he should have divided both terms by* $3x$. *The correct solution is* $\frac{2x}{3x} - \frac{x^3}{3x} = \frac{2}{3} - \frac{x^2}{3}$.

7. If $x = 5a^4$, and $a = 2b^3$, express $x$ in terms of $b$.

*By the substitution property, if* $x = 5a^4$, *and* $a = 2b^3$, *then* $x = 5(2b^3)^4$. *Rewriting the right side in an equivalent form gives* $x = 80b^{12}$.

8. If $a = 2b^3$, and $b = -\frac{1}{2}c^{-2}$, express $a$ in terms of $c$.

*By the substitution property, if* $a = 2b^3$, *and* $b = -\frac{1}{2}c^{-2}$, *then* $a = 2\left(-\frac{1}{2}c^{-2}\right)^3$. *Rewriting the right side in an equivalent form gives* $a = -\frac{1}{4}c^{-6}$.

9. If $x = 3y^4$, and $y = \frac{s}{2x^3}$, show that $s = 54y^{13}$.

*Rewrite the equation* $y = \frac{s}{2x^3}$ *to isolate the variable* $s$.

$$y = \frac{s}{2x^3}$$
$$2x^3y = s$$

*By the substitution property, if* $s = 2x^3y$, *and* $x = 3y^4$, *then* $s = 2(3y^4)^3 \cdot y$. *Rewriting the right side in an equivalent form gives* $s = 2 \cdot 27y^{12} \cdot y = 54y^{13}$.

10. Do the following without a calculator.

a. Express $8^3$ as a power of $2$.

$$8^3 = (2^3)^3 = 2^9$$

b. Divide $4^{15}$ by $2^{10}$.

$$\frac{4^{15}}{2^{10}} = \frac{2^{30}}{2^{10}} = 2^{20} \quad \textit{or} \quad \frac{4^{15}}{2^{10}} = \frac{4^{15}}{4^5} = 4^{10}$$

11. Use powers of 2 to help you perform each calculation.

a. $\frac{2^7 \cdot 2^5}{16}$

$\frac{2^7 \cdot 2^5}{16} = \frac{2^7 \cdot 2^5}{2^4} = 2^{7+5-4} = 2^8 = 256$

b. $\frac{512000}{320}$

$\frac{512000}{320} = \frac{512 \cdot 1000}{32 \cdot 10} = \frac{2^9}{2^5} \cdot 100 = 2^4 \cdot 100 = 1600$

12. Write the first five terms of each of the following recursively-defined sequences.

a. $a_{n+1} = 2a_n, a_1 = 3$

$\{3, 6, 12, 24, 48\}$

b. $a_{n+1} = (a_n)^2, a_1 = 3$

$\{3, 9, 81, 6561, 43046721\}$

c. $a_{n+1} = 2(a_n)^2, a_1 = x$, where $x$ is a real number. Write each term in the form $kx^n$.

$\{x, 2x^2, 8x^4, 128x^8, 32768x^{16}\}$

d. $a_{n+1} = 2(a_n)^{-1}, a_1 = y, (y \neq 0)$. Write each term in the form $kx^n$.

$\{y, 2y^{-1}, y, 2y^{-1}, y\}$

13. In Module 1, you established the identity

$(1 - r)(1 + r + r^2 + \cdots + r^{n-1}) = 1 - r^n$, where $r$ is a real number and $n$ is a positive integer.

Use this identity to find explicit formulas as specified below.

a. Rewrite the given identity to isolate the sum $1 + r + r^2 + \cdots r^{n-1}$ for $r \neq 1$.

$(1 + r + r^2 + \cdots + r^{n-1}) = \frac{1 - r^n}{1 - r}$

b. Find an explicit formula for $1 + 2 + 2^2 + 2^3 + \cdots + 2^{10}$.

$\frac{1 - 2^{11}}{1 - 2} = 2^{11} - 1$

c. Find an explicit formula for $1 + a + a^2 + a^3 + \cdots + a^{10}$ in terms of powers of $a$.

$\frac{1 - a^{11}}{1 - a}$

d. Jerry simplified the sum $1 + a + a^2 + a^3 + a^4 + a^5$ by writing $1 + a^{15}$. What did he do wrong?

*He assumed that when you add terms with the same base, you also add the exponents. You only add the exponents when you multiply like bases.*

e. **Find an explicit formula for $1 + 2a + (2a)^2 + (2a)^3 + \cdots + (2a)^{12}$ in terms of powers of $a$.**

$$\frac{1-(2a)^{13}}{1-2a}$$

f. **Find an explicit formula for $3 + 3(2a) + 3(2a)^2 + 3(2a)^3 + \ldots + 3(2a)^{12}$ in terms of powers of $a$. Hint: Use part (e).**

$$3 \cdot \frac{1-(2a)^{13}}{1-2a}$$

g. **Find an explicit formula for $P + P(1+r) + P(1+r)^2 + P(1+r)^3 + \cdots + P(1+r)^{n-1}$ in terms of powers of $(1+r)$.**

$$P \cdot \frac{1-(1+r)^n}{1-(1+r)} = P \cdot \frac{1-(1+r)^n}{-r}$$

Note to the teacher: Problem 3, part (g) will be important for the financial lessons that occur near the end of this module.

# Lesson 2: Base 10 and Scientific Notation

## Student Outcomes

- Students review place value and scientific notation.
- Students use scientific notation to compute with large numbers.

## Lesson Notes

This lesson reviews how to express numbers using scientific notation. Students first learn about scientific notation in Grade 8 where they express numbers using scientific notation (**8.EE.A.3**) and compute with and compare numbers expressed using scientific notation (**8.EE.A.4**). Refer to Grade 8, Module 1, Topic B to review the approach to introducing scientific notation and its use in calculations with very large and very small numbers. This lesson sets the stage for the introduction of base $10$ logarithms later in this module by focusing on the fact that every real number can be expressed as product of a number between $1$ and $10$ and a power of $10$. In the paper folding activity in the last lesson, students worked with some very small and very large numbers. This lesson opens with these numbers to connect these lessons. Students also compute with numbers using scientific notation and discuss how the properties of exponents can simplify these computations (**N-RN.A.2**). In both the lesson and the problem set, students define appropriate quantities for the purpose of descriptive modeling (**N-Q.A.2**). The lesson includes a demonstration that reinforces the point that using scientific notation is convenient when working with very large or very small numbers and helps students gain some sense of the change in magnitude when we compare different powers of $10$. This is an excellent time to watch the 9-minute classic film "Powers of 10" by Charles and Ray Eames, available at https://www.youtube.com/watch?v=0fKBhvDjuy0, which clearly illustrates the effect of adding another zero. The definition of scientific notation from Grade 8, Module 1, Lesson 9 is included after Example 1. You may want to allow students to use a calculator for this lesson.

## Classwork

### Opening (2 minutes)

In the last lesson, one of the examples gave the thickness of a sheet of gold foil as $0.28$ millionths of a meter. The Exit Ticket had students calculate the size of a square sheet of paper that could be folded in half thirteen times, and it was very large. We will use these numbers in the Opening Exercise. Before students begin the Opening Exercise, briefly remind them of these numbers that they saw in the previous lesson, and tell them that this lesson will provide them with a way to conveniently represent very small or very large numbers. If you did not have an opportunity to share one of the news stories in Lesson 1, you could do that at this time as well.

### Opening Exercise (5 minutes)

Students should work these exercises independently and then share their responses with a partner. Ask one or two students to explain their calculations to the rest of the class. Be sure to draw out the meaning of 0.28 millionths of a meter and how that can be expressed as a fraction. Check to make sure students know the place value for large numbers like billions. Students should be able to work the first exercise without a calculator, but on the second exercise, students should definitely use a calculator. If they do not have access to a calculator, give them the number squared and simply have them write the rounded value.

**Opening Exercise**

**In the last lesson, you worked with the thickness of a sheet of gold foil (a very small number) and some very large numbers that gave the size of a piece of paper that actually could be folded in half more than 13 times.**

a. **Convert $0.28$ millionths of a meter to centimeters and express your answer as a decimal number.**

$$\frac{0.28}{1,000,000}\text{ m}\cdot\frac{100\text{ cm}}{1\text{ m}}=\frac{28}{1,000,000}\text{ cm}=0.000028\text{ cm}$$

b. **The length of a piece of square notebook paper that could be folded in half 13 times was $3294.2$ in. Use this number to calculate the area of a square piece of paper that could be folded in half $14$ times. Round your answer to the nearest million.**

$$(2\cdot 3294.2)^2=43407014.56$$

*Rounded to the nearest million, the number is* $43,000,000$.

c. **Sort the following numbers into products and single numeric expressions. Then match the equivalent expressions without using a calculator.**

| $3.5\times 10^5$ | $-6$ | $-6\times 10^0$ | $0.6$ | $3.5\times 10^{-6}$ |
|---|---|---|---|---|
| $3,500,000$ | $350,000$ | $6\times 10^{-1}$ | $0.0000035$ | $3.5\times 10^6$ |

*Products:* $3.5\times 10^5$, $-6\times 10^0$, $3.5\times 10^{-6}$, $6\times 10^{-1}$, *and* $3.5\times 10^6$

*Single numeric expressions:* $-6$, $0.6$, $3500000$, $350000$, *and* $0.0000035$

$3.5\times 10^5$ *is equal to* $350,000$.

$-6\times 10^0$ *is equal to* $-6$.

$6\times 10^{-1}$ *is equal to* $0.6$.

$3.5\times 10^6$ *is equal to* $3,500,000$.

$3.5\times 10^{-6}$ *is equal to* $0.0000035$.

As you review these solutions, point out that very large and very small numbers require us to include many digits to indicate the place value as shown in Exercises 1 and 2. Also, based on Exercise 3, it appears that you can use integer powers of 10 to express a number as a product.

### Example 1 (7 minutes)

Write the following statement on the board and ask students to consider whether or not they believe it is true. Have them discuss their thoughts with a partner, and then ask for volunteers to explain their thinking.

Students should explain that every nonzero decimal number can be expressed as the product of a number between 1 and 10 and a power of 10.

MP.3

- Think of an example of a decimal number between 1 and 10. Write it down.
  - $2.5$
- Think of a power of 10. Write it down.
  - $100$ *or* $10^2$
- What does the word product mean?
  - *It means the result of multiplying two numbers together.*
- Compute the product of the two numbers you wrote down.
  - $2.5 \cdot 10^2 = 2500$

*Scaffolding:*

- Have advanced learners write the number 245 as the product of a number and a power of 10 three different ways.
- To challenge advanced learners, have them make a convincing argument regarding the truth of a statement such as the following:

  Every decimal number can be expressed as the product of another decimal number and a power of 10.

First, have students share their answers with a partner. Then, put a few of the examples on the board. Finally, demonstrate how to reverse the process to express the following numbers as the product of a number between 1 and 10 and a power of 10. At this point, you can explain to students that when we write numbers in this fashion, we say that they are written using scientific notation. This is an especially convenient way to represent extremely large or extremely small numbers that otherwise would have many zero digits as place holders. Make sure to emphasize that this notation is simply rewriting a numerical expression in a different form, which helps us to quickly determine the size of the number. Students may need to be reminded of place value for numbers less than 1 and writing equivalent fractions whose denominators are powers of 10. The solutions demonstrate how any number can be expressed using what we call scientific notation.

**Example 1**

**Write each number as a product of a decimal number between 1 and 10 and a power of 10.**

a. $\mathbf{234,000}$

$\mathbf{2.34 \cdot 100,000 = 2.34 \times 10^5}$

b. $\mathbf{0.0035}$

$$\mathbf{\frac{35}{10000} = \frac{3.5}{1000} = 3.5 \cdot \frac{1}{1000} = 3.5 \times 10^{-3}}$$

c. $\mathbf{532,100,000}$

$\mathbf{5.321 \cdot 10,000,000 = 5.321 \times 10^8}$

d. $\mathbf{0.0000000012}$

$$\mathbf{\frac{12}{10,000,000,000} = \frac{1.2}{1,000,000,000} = 1.2 \cdot \frac{1}{1,000,000,000} = 1.2 \times 10^{-9}}$$

e. $3.331$

$3.331 \cdot 1 = 3.331 \times 10^0$

Students may recall scientific notation from previous grades. Take time to review the definition of scientific notation provided below.

- Our knowledge of the integer powers of 10 enable us to understand the concept of scientific notation.
- Consider the estimated number of stars in the universe: $6 \times 10^{22}$. This is a 23-digit *whole number* with the **leading digit** (the leftmost digit) 6 followed by 22 zeros. When it is written in the form $6 \times 10^{22}$, it is said to be expressed in *scientific notation*.
- A positive, finite decimal[1] $s$ is said to be written in **scientific notation** if it is expressed as a product $d \times 10^n$, where $d$ is a finite decimal so that $1 \leq d < 10$, and $n$ is an integer. That is, $d$ is a finite decimal with only a single, nonzero digit to the left of the decimal point. The integer $n$ is called the **order of magnitude**[2] of the decimal $d \times 10^n$.

A positive, finite decimal $s$ is said to be written in scientific notation if it is expressed as a product $d \times 10^n$, where $d$ is a finite decimal number so that $1 \leq d < 10$, and $n$ is an integer.

The integer $n$ is called the *order of magnitude* of the decimal $d \times 10^n$.

## Exercises 1–6 (4 minutes)

Students should work these exercises independently as you monitor their progress. Encourage students to work quickly and begin generalizing a process for quickly writing numbers using scientific notation (such as counting the number of digits between the leading digit and the ones digit). After a few minutes, share the solutions with students so they can check their work.

**Exercises 1–6**

**For Exercises 1–6, write each decimal in scientific notation.**

1. $532,000,000$

   $5.32 \times 10^8$

2. $0.0000000000000000123$ (16 zeros after the decimal place)

   $1.23 \times 10^{-17}$

3. $8,900,000,000,000,000$ (14 zeros after the 9)

   $8.9 \times 10^{15}$

4. $0.00003382$

   $3.382 \times 10^{-5}$

[1] Recall that every whole number is a finite decimal.

[2] Sometimes the value of $10^n$ is known as the order of magnitude of the number $d \times 10^n$, but we will use $n$.

5. $34,000,000,000,000,000,000,000,000$ (24 zeros after the $4$)

   $3.4 \times 10^{25}$

6. $0.0000000000000000000004$ (21 zeros after the decimal place)

   $4 \times 10^{-22}$

To help students quickly write these problems using scientific notation, the number of zeros is written above for each problem. Be very careful that students are not using this number as the exponent on the base 10. Lead a discussion to clarify that difference for all students who make this careless mistake.

## Exercises 7–8 (5 minutes)

After students practice writing numbers in scientific notation, you can really drive the point home that scientific notation is useful when working with very large or very small numbers by showing this demonstration: http://joshworth.com/dev/pixelspace/pixelspace_solarsystem.html, which illustrates just how far the planets in our solar system are from each other. After the demonstration, write down the distances between Earth and the Sun, between Jupiter and the Sun, and between Pluto and the Sun on the board, and have students work with a partner to answer Exercise 7. Be sure to mention that these distances are averages; he distances between the planets and the Sun are constantly changing as the planets complete their orbits. The average distance from the Sun to Earth is 151,268,468 km. The average distance from the Sun to Jupiter is 780,179,470 km. The average distance between the Sun and Pluto is 5,908,039,124 km. In these exercises, students will round the distances to the nearest tenth to minimize all the writing and help them focus more readily on the magnitude of the numbers relative to one another.

**Exercises 7–8**

7. **Approximate the average distances between the Sun and Earth, Jupiter, and Pluto. Express your answers in scientific notation ($d \times 10^n$), where $d$ is rounded to the nearest tenth.**

   a. **Sun to Earth:**

      $1.5 \times 10^8$ km

   b. **Sun to Jupiter:**

      $7.8 \times 10^8$ km

   c. **Sun to Pluto:**

      $5.9 \times 10^9$ km

   d. **Earth to Jupiter:**

      $780,179,470$ km $- 151,268,468$ km $= 628,911,002$ km

      $6.3 \times 10^8$ km

   e. **Jupiter to Pluto:**

      $5,908,039,124 - 780,179,470 = 5,127,859,654$ km

      $5.1 \times 10^9$ km

8. **Order the numbers in Exercise 7 from smallest to largest. Explain how writing the numbers in scientific notation helps you to quickly compare and order them.**

*The numbers from smallest to largest are* $1.5\times 10^8$, $6.3\times 10^8$, $7.8\times 10^8$, $5.1\times 10^9$, *and* $5.9\times 10^9$. *The power of* $10$ *helps to quickly sort the numbers by their order of magnitude, and then it is easy to quickly compare the numbers with the same order of magnitude because they are only written as a number between one and ten.*

## Example 2 (10 minutes): Arithmetic Operations With Numbers Written Using Scientific Notation

Model the solutions to the following example problems. Be sure to emphasize that final answers should be expressed using the scientific notation convention of a number between 1 and 10 and a power of 10. On part (a), you may need to provide some additional scaffolding if students are struggling to rewrite the numbers using the same order of magnitude. Have students practice writing a number as a product of a power of 10 in three different ways. For example, $15{,}000 = 1.5\times 10^4 = 15\times 10^3 = 150\times 10^2$, etc. The lessons of Algebra I, Module 1, Topic B provide some suggestions and fluency exercises if students need additional practice on arithmetic operations with numbers in scientific notation. Be sure that students understand that the properties of exponents allow you to quickly perform the indicated operations.

*Scaffolding:*

- For Example 2, part (a), students will want to add the exponents as they do when multiplying numbers written using scientific notation. Take time to discuss the differences in the three expressions if you notice students making this type of mistake.

**Example 2: Arithmetic Operations with Numbers Written Using Scientific Notation**

a. $(2.4\times 10^{20}) + (4.5\times 10^{21})$

$(2.4\times 10^{20}) + (45\times 10^{20}) = 47.4\times 10^{20} = 4.74\times 10^{21}$

b. $(7\times 10^{-9})(5\times 10^{5})$

$(7\cdot 5)\times(10^{-9}\cdot 10^{5}) = 35\times 10^{-4} = 3.5\times 10^{-3}$

c. $\dfrac{1.2\times 10^{15}}{3\times 10^{7}}$

$\dfrac{1.2}{3}\times 10^{15-7} = 0.4\times 10^{8} = 4\times 10^{7}$

Debrief with the questions designed to help students see that the order of magnitude and the properties of exponents greatly simplify calculations.

- How do the properties of exponents help to simplify these calculations?
- How can you quickly estimate the size of your answer?

## Exercises 9–11 (5 minutes)

**Exercises 9–11**

9. **Perform the following calculations without rewriting the numbers in decimal form.**

   a. $(1.42 \times 10^{15}) - (2 \times 10^{13})$

   $142 \times 10^{13} - 2 \times 10^{13} = (142 - 2) \times 10^{13} = 140 \times 10^{13} = 1.4 \times 10^{15}$

   b. $(1.42 \times 10^{15})(2.4 \times 10^{13})$

   $(1.42 \cdot 2.4) \times (10^{15} \cdot 10^{13}) = 3.408 \times 10^{28}$

   c. $\dfrac{1.42 \times 10^{-5}}{2 \times 10^{13}}$

   $\dfrac{1.42 \times 10^{-5}}{2 \times 10^{13}} = 0.71 \times 10^{-5-13} = 0.71 \times 10^{-18} = 7.1 \times 10^{-19}$

10. **Estimate how many times farther Jupiter is from the Sun than Earth is from the Sun. Estimate how many times farther Pluto is from the Sun than Earth is from the Sun.**

    *Earth is approximately* $1.5 \times 10^{8}$ km *from the Sun, and Jupiter is approximately* $7.8 \times 10^{8}$ km *from the Sun. Therefore, Jupiter is about* $5$ *times as far from the Sun as Earth is from the sun. Pluto is approximately* $5.9 \times 10^{9}$ km *from the Sun. Therefore, Pluto is approximately* $40$ *times as far from the Sun as Earth is, since*

    $$\frac{59 \times 10^{8}}{1.5 \times 10^{8}} = 59 \div 1.5 \approx 40.$$

11. **Estimate the distance between Earth and Jupiter and between Jupiter and Pluto.**

    *The distance between Earth and Jupiter is approximately* $(7.8 - 1.5) \times 10^{8}$ km, *which is equal to* $6.3 \times 10^{8}$ km. *The distance between Jupiter and Pluto is approximately* $(59 - 7.8) \times 10^{8} = 51.2 \times 10^{8} = 5.12 \times 10^{9}$ km.

## Closing (3 minutes)

Have students discuss the following question with a partner and record the definition of scientific notation in their mathematics notebook. Debrief by asking a few students to share their responses with the entire class.

- List two advantages of writing numbers using scientific notation.
  - *You do not have to write as many zeros when working with very large or very small numbers, and you can quickly multiply and divide numbers using the properties of exponents.*

## Exit Ticket (4 minutes)

Name ______________________________ Date ______________

# Lesson 2: Base 10 and Scientific Notation

## Exit Ticket

1. A sheet of gold foil is $0.28$ millionths of a meter thick. Write the thickness of a gold foil sheet measured in centimeters using scientific notation.

2. Without performing the calculation, estimate which expression is larger. Explain how you know.

$$(4 \times 10^{10})(2 \times 10^{5}) \quad \text{and} \quad \frac{4 \times 10^{12}}{2 \times 10^{-4}}$$

## Exit Ticket Sample Solutions

1. **A sheet of gold foil is $0.28$ millionths of a meter thick. Write the thickness of a gold foil sheet measured in centimeters using scientific notation.**

   *The thickness is* $0.28 \times 10^{-6}$ m. *In scientific notation, the thickness of a gold foil sheet is* $2.8 \times 10^{-7}$ m, *which is* $2.8 \times 10^{-5}$ cm.

2. **Without performing the calculation, estimate which expression is larger. Explain how you know.**

$$(4 \times 10^{10})(2 \times 10^{5}) \quad \text{and} \quad \frac{4 \times 10^{12}}{2 \times 10^{-4}}$$

   *The order of magnitude on the first expression is* $15$*, and the order of magnitude on the second expression is* $16$*. The product and quotient of the number between* $1$ *and* $10$ *in each expression is a number between* $1$ *and* $10$*. Therefore, the second expression is larger than the first one.*

## Problem Set Sample Solutions

1. **Write the following numbers used in these statements in scientific notation. (Note: Some of these numbers have been rounded.)**

   a. **The density of helium is $0.0001785$ grams per cubic centimeter.**

      $1.785 \times 10^{-4}$

   b. **The boiling point of gold is $5200°$F.**

      $5.2 \times 10^{3}$

   c. **The speed of light is $186,000$ miles per second.**

      $1.86 \times 10^{5}$

   d. **One second is $0.000278$ hours.**

      $2.78 \times 10^{-4}$

   e. **The acceleration due to gravity on the Sun is $900$ ft/s$^2$.**

      $9 \times 10^{2}$

   f. **One cubic inch is $0.0000214$ cubic yards.**

      $2.14 \times 10^{-5}$

   g. **Earth's population in 2012 was $7,046,000,000$ people.**

      $7.046 \times 10^{9}$

   h. **Earth's distance from the Sun is $93,000,000$ miles.**

      $9.3 \times 10^{7}$

i. Earth's radius is $4,000$ miles.

$4 \times 10^3$

j. The diameter of a water molecule is $0.000000028$ cm.

$2.8 \times 10^{-8}$

2. Write the following numbers in decimal form. (Note: Some of these numbers have been rounded.)

a. A light year is $9.46 \times 10^{15}$ m.

$9,460,000,000,000,000$

b. Avogadro's number is $6.02 \times 10^{23}\ \text{mol}^{-1}$.

$602,000,000,000,000,000,000,000$

c. The universal gravitational constant is $.674 \times 10^{-11}\ \text{N}\left(\frac{\text{m}}{\text{kg}}\right)^2$.

$0.00000000006674$

d. Earth's age is $4.54 \times 10^9$ years.

$4,540,000,000$

e. Earth's mass is $5.97 \times 10^{24}$ kg.

$5,970,000,000,000,000,000,000,000$

f. A foot is $1.9 \times 10^{-4}$ miles.

$0.00019$

g. The population of China in 2014 was $1.354 \times 10^9$ people.

$1,354,000,000$

h. The density of oxygen is $1.429 \times 10^{-4}$ grams per liter.

$0.0001429$

i. The width of a pixel on a smartphone is $7.8 \times 10^{-2}$ mm.

$0.078$

j. The wavelength of light used in optic fibers is $1.55 \times 10^{-6}$ m.

$0.00000155$

3. State the necessary value of $n$ that will make each statement true.

a. $0.000027 = 2.7 \times 10^{n}$

$-5$

b. $-3.125 = -3.125 \times 10^{n}$

$0$

c. $7,540,000,000 = 7.54 \times 10^{n}$

$9$

d. $0.033 = 3.3 \times 10^{n}$

$-2$

e. $15 = 1.5 \times 10^{n}$

$1$

f. $26,000 \times 200 = 5.2 \times 10^{n}$

$6$

g. $3000 \times 0.0003 = 9 \times 10^{n}$

$-1$

h. $0.0004 \times 0.002 = 8 \times 10^{n}$

$-7$

i. $\dfrac{16000}{80} = 2 \times 10^{n}$

$2$

j. $\dfrac{500}{0.002} = 2.5 \times 10^{n}$

$5$

Perform the following calculations without rewriting the numbers in decimal form.

k. $(2.5 \times 10^{4}) + (3.7 \times 10^{3})$

$2.87 \times 10^{4}$

l. $(6.9 \times 10^{-3}) - (8.1 \times 10^{-3})$

$-1.2 \times 10^{-3}$

m. $(6 \times 10^{11})(2.5 \times 10^{-5})$

$1.5 \times 10^{7}$

n. $\dfrac{4.5 \times 10^{8}}{2 \times 10^{10}}$

$2.25 \times 10^{-2}$

4. **The wavelength of visible light ranges from $650$ nanometers to $850$ nanometers, where $1\text{ nm} = 1 \times 10^{-7}$ cm. Express the wavelength range of visible light in centimeters.**

*Convert $650$ nm to centimeters:* $(6.5 \times 10^{2})(1 \times 10^{-7}) = 6.5 \times 10^{-5}$

*The wavelength of visible light in centimeters is* $6.5 \times 10^{-5}$ cm *to* $8.5 \times 10^{-5}$ cm.

5. **In 1694, the Dutch scientist Antonie van Leeuwenhoek was one of the first scientists to see a red blood cell in a microscope. He approximated that a red blood cell was "$25,000$ times as small as a grain of sand." Assume a grain of sand is $\frac{1}{2}$ mm wide and a red blood cell is approximately $7$ micrometers wide. One micrometer is $1 \times 10^{-6}$ m. Support or refute Leeuwenhoek's claim. Use scientific notation in your calculations.**

*Convert millimeters to meters:* $(5 \times 10^{-1})(1 \times 10^{-3}) = 5 \times 10^{-4}$. *A medium size grain of sand measures* $5 \times 10^{-4}$ m *across. Similarly, a red blood cell is approximately* $7 \times 10^{-6}$ m *across. If you divide these numbers, you get*

$$\frac{5 \times 10^{-4}}{7 \times 10^{-6}} = 0.714 \times 10^{2} = 7.14 \times 10^{1}.$$

*So, a red blood cell is* $71.4$ *times as small as a grain of sand. Leeuwenhoek's claim was off by approximately a factor of* $350$.

6. **When the Mars Curiosity Rover entered the atmosphere of Mars on its descent in 2012, it was traveling roughly $13,200$ mph. On the surface of Mars, its speed averaged $0.00073$ mph. How many times faster was the speed when it entered the atmosphere than its typical speed on the planet's surface? Use scientific notation in your calculations.**

$$\frac{1.32 \times 10^{4}}{7.3 \times 10^{-4}} = 0.18 \times 10^{8} = 1.8 \times 10^{7}$$

*The speed when it entered the atmosphere is greater than its surface speed by an order of magnitude of* $7$.

7. **Earth's surface is approximately $70\%$ water. There is no water on the surface of Mars, and its diameter is roughly half of Earth's diameter. Assume both planets are spherical. The surface area of a sphere is given by the formula $SA = 4\pi r^{2}$, where $r$ is the radius of the sphere. Which has more land mass, Earth or Mars? Use scientific notation in your calculations.**

*The surface area of Earth is* $4\pi(4000)^{2} \approx 2 \times 10^{8}$ sq. mi., *and the surface area of Mars is* $4\pi(2000)^{2} \approx 5 \times 10^{7}$ sq. mi. *Thirty percent of Earth's surface area is approximately* $6 \times 10^{7}$ sq. mi. *Earth has more land mass by approximately* $20\%$.

8. **There are approximately $25$ trillion $(2.5 \times 10^{13})$ red blood cells in the human body at any one time. A red blood cell is approximately $7 \times 10^{-6}$ m wide. Imagine if you could line up all your red blood cells end to end. How long would the line of cells be? Use scientific notation in your calculations.**

$(2.5 \times 10^{13})(7 \times 10^{-6}) = 1.75 \times 10^{8}$ m. *That means the line of cells would be* $1.75 \times 10^{5}$ km *long. One mile is equivalent to* $1.6$ km, *so the line of blood cells measures* $\frac{1.75 \times 10^{5}}{1.6}$ km $\approx 109,375$ km, *which is almost halfway to the moon!*

9. Assume each person needs approximately $100$ sq. ft. of living space. Now imagine that we are going to build a giant apartment building that will be $1$ mile wide and $1$ mile long to house all the people in the United States, estimated to be $313.9$ million people in 2012. If each floor of the apartment building is $10$ ft. high; how tall will the apartment building be?

*We need* $(3.139 \times 10^{8})(100) = 3.139 \times 10^{10}\ \text{ft}^2$ *of living space.* $1\ \text{mi}^2 = 5280^2\ \text{ft}^2 = 27878400\ \text{ft}^2$.

*Next, divide the total number of square feet by the number of square feet per floor to get the number of needed floors.*

$$\frac{3.139 \times 10^{10}}{2.78784 \times 10^{7}} \approx 1.126 \times 10^{3}$$

*Multiplying the number of floors by* $10$ ft. *per floor gives a height of* $11,260$ ft.*, which is approximately* $2.13$ mi.

# Lesson 3: Rational Exponents—What are $2^{\frac{1}{2}}$ and $2^{\frac{1}{3}}$?

## Student Outcomes

- Students will calculate quantities that involve positive and negative rational exponents.

## Lesson Notes

Students extend their understanding of integer exponents to rational exponents by examining the graph of $f(x) = 2^x$ and estimating the values of $2^{\frac{1}{2}}$ and $2^{\frac{1}{3}}$. The lesson establishes the meaning of these numbers in terms of radical expressions and these form the basis of how we define expressions of the form $b^{\frac{1}{n}}$ before generalizing further to expressions of the form $b^{\frac{m}{n}}$, where $b$ is a positive real number and $m$ and $n$ are integers, with $n \neq 0$ (**N-RN.A.1**). The lesson and Problem Set provide fluency practice in applying the properties of exponents to expressions containing rational exponents and radicals (**N-RN.A.2**). In the following lesson, students will verify that the definition of an expression with rational exponents, $b^{\frac{m}{n}} = \sqrt[n]{b^m}$, is consistent with the remaining exponential properties. The lesson begins with students creating a graph of a simple exponential function, $f(x) = 2^x$, which they studied in Module 3 of Algebra I. In that module, students also learned about geometric sequences and their relationship to exponential functions, which is a concept that we revisit at the end of this module. In Algebra I, students worked with exponential functions with integer domains. This lesson, together with the subsequent Lessons 4 and 5, helps students understand why the domain of an exponential function $f(x) = b^x$, where $b$ is positive number and $b \neq 1$, is the set of real numbers. To do so we must establish what it means to raise $b$ to a rational power and, in Lesson 5, to any real number power.

## Classwork

### Opening (1 minutes)

In Algebra I, students worked with geometric sequences and simple exponential functions. Remind them that in Lesson 1, we created formulas based on repeatedly doubling and halving a number when we modeled folding a piece of paper. We reviewed how to use the properties of exponents for expressions that had integer exponents.

*Scaffolding:*

- Encourage students to create a table of values to help them construct the graph.
- For advanced learners, have them repeat these exercises with the function $f(x) = 3^x$ and ask them to estimate $3^{\frac{1}{2}}$ and $3^{\frac{1}{3}}$.

### Opening Exercise (5 minutes)

Have students graph $f(x) = 2^x$ for each integer $x$ from $x = -2$ to $x = 5$ without using a graphing utility or calculator on the axes provided. Discuss the pattern of points and ask students to connect the points in a way that produces a smooth curve. Students should work these two problems independently and check their solutions with a partner if time permits before you lead a whole class discussion to review the solutions.

**Opening Exercise**

MP.3

a. **What is the value of $2^{\frac{1}{2}}$? Justify your answer.**

*A possible student response follows: I think it will be around* $1.5$ *because* $2^0 = 1$ *and* $2^1 = 2$.

b. **Graph $f(x) = 2^x$ for each integer $x$ from $x = -2$ to $x = 5$. Connect the points on your graph with a smooth curve.**

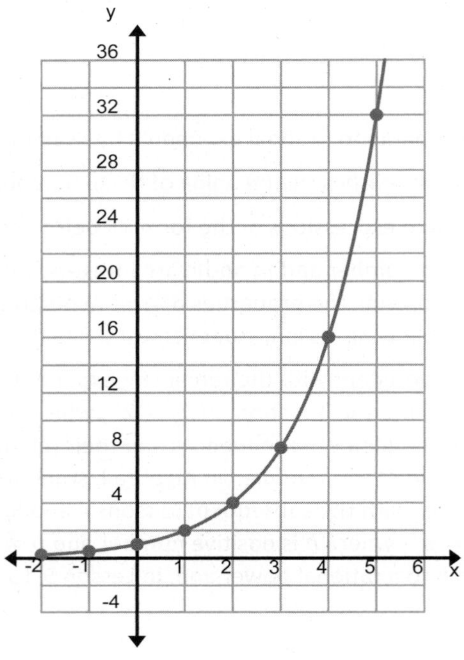


Ask for a few volunteers to explain their reasoning for their answers to Opening Exercise part (a). Then, debrief these two exercises by leading a short discussion.

- The directions in the Opening Exercise said to connect the points with a smooth curve. What does it imply about the domain of a function when we connect points that we have plotted?
  - *That the domain of the function includes the points between the points that we plotted.*
- How does your graph support or refute your answer to the first exercise? Do you need to modify your answer to the question: What is the value of $2^{\frac{1}{2}}$? Why or why not?
  - *If the domain is all real numbers, then the value of* $2^{\frac{1}{2}}$ *will be the* $y$*-coordinate of the point on the graph where* $x = \frac{1}{2}$. *From the graph it looks like the value is somewhere between* $1$ *and* $2$. *The scaling on this graph is not detailed enough for me to accurately refine my answer yet.*

Transition to the next set of exercises by telling the students that we can better estimate the value of $2^{\frac{1}{2}}$ by looking at a graph that is scaled more precisely. Have students proceed to the next questions. They should work in small groups or with a partner on these exercises.

The graph of $f(x) = 2^x$ shown below for the next exercises will appear in the student materials but you could also display the graph using technology. Have students estimate the value of $2^{\frac{1}{2}}$ and $2^{\frac{1}{3}}$ from the graph. Students should work by themselves or in pairs to work through the following questions without using a calculator.

**The graph at right shows a close-up view of $f(x) = 2^x$ for $-0.5 < x < 1.5$.**

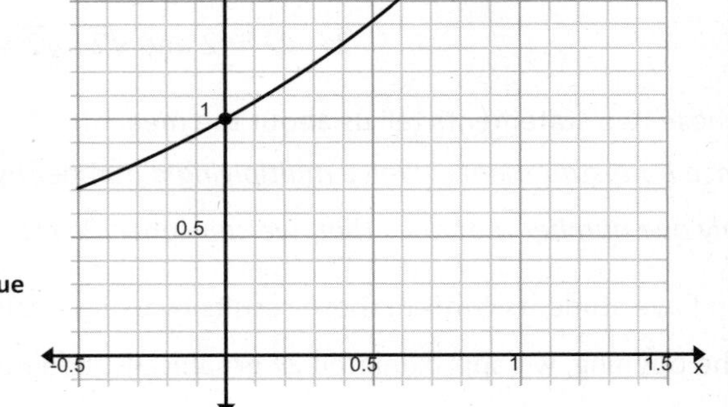


c. **Find two consecutive integers that are over and underestimates of the value of $2^{\frac{1}{2}}$.**

$2^0 < 2^{\frac{1}{2}} < 2^1$

$1 < 2^{\frac{1}{2}} < 2$

d. **Does it appear that $2^{\frac{1}{2}}$ is halfway between the integers you specified in Exercise 1?**

***No, it looks like $2^{\frac{1}{2}}$ is a little less than halfway between $1$ and $2$.***

e. **Use the graph of $f(x) = 2^x$ to estimate the value of $2^{\frac{1}{2}}$.**

$2^{\frac{1}{2}} \approx 1.4$

f. **Use the graph of $f(x) = 2^x$ to estimate the value of $2^{\frac{1}{3}}$.**

$2^{\frac{1}{3}} \approx 1.25$

## Discussion (9 minutes)

Before getting into the point of this lesson, which is to connect rational exponents to radical expressions, revisit the initial question with students.

- What is the value of $2^{\frac{1}{2}}$? Does anyone want to adjust their initial guess?
  - *Our initial guess was a little too big. It seems like $1.4$ might be a better answer.*
- How could we make a better guess?
  - *We could look at the graph with even smaller increments for the scale using technology.*

> *Scaffolding:*
> - If needed, you can demonstrate the argument using perfect squares. For example, use a base of $4$ instead of a base of $2$.
>
> Show that $\left(4^{\frac{1}{2}}\right)^2 = 4$ and $\left(\sqrt{4}\right)^2 = 4$.

If time permits, you can zoom in further on the graph of $f(x) = 2^x$ using a graphing calculator or other technology either by examining a graph or a table of values of $x$ closer and closer to $\frac{1}{2}$.

Next, we will make the connection that $2^{\frac{1}{2}} = \sqrt{2}$. Walk students through the following questions, providing guidance as needed. Students proved that there was only one positive number that squared to 2 in Geometry, Module 2. You may need to remind them of this with a bit more detail if they are struggling to follow this argument.

- Assume for the moment that whatever $2^{\frac{1}{2}}$ means that it satisfies $b^m \cdot b^n = b^{m+n}$. Working with this assumption what is the value of $2^{\frac{1}{2}} \cdot 2^{\frac{1}{2}}$?
  - *It would be* $2$ *because* $2^{\frac{1}{2}} \cdot 2^{\frac{1}{2}} = 2^{\frac{1}{2}+\frac{1}{2}} = 2^1 = 2$.
- What unique positive number squares to $2$? That is, what is the only positive number that when multiplied by itself is equal to $2$?
  - *By definition, we call the unique positive number that squares to* $2$ *the square root of* $2$ *and we write* $\sqrt{2}$.

MP.7

Write the following statements on the board and ask your students to compare them and think about what they must tell us about the meaning of $2^{\frac{1}{2}}$.

$$2^{\frac{1}{2}} \cdot 2^{\frac{1}{2}} = 2 \text{ and } \sqrt{2} \cdot \sqrt{2} = 2$$

- What do these two statements tell us about the meaning of $2^{\frac{1}{2}}$.
  - *Since both statements involve multiplying a number by itself and getting* $2$*, and we know that there is only one number that does that, we can conclude that* $2^{\frac{1}{2}} = \sqrt{2}$.

At this point, you can have students confirm these results by using a calculator to approximate both $2^{\frac{1}{2}}$ and $\sqrt{2}$ to several decimal places. In the opening, we approximated $2^{\frac{1}{2}}$ graphically, and now we have shown it to be an irrational number.

Next ask students to think about the meaning of $2^{\frac{1}{3}}$ using a similar line of reasoning.

- Assume that whatever $2^{\frac{1}{3}}$ means it will satisfy $b^m \cdot b^n = b^{m+n}$. What is the value of $\left(2^{\frac{1}{3}}\right)\left(2^{\frac{1}{3}}\right)\left(2^{\frac{1}{3}}\right)$?
  - *The value is* $2$ *because* $\left(2^{\frac{1}{3}}\right)\left(2^{\frac{1}{3}}\right)\left(2^{\frac{1}{3}}\right) = 2^{\frac{1}{3}+\frac{1}{3}+\frac{1}{3}} = 2^1 = 2$.

MP.7

- What is the value of $\sqrt[3]{2} \cdot \sqrt[3]{2} \cdot \sqrt[3]{2}$?
  - *The value is* $2$ *because* $\sqrt[3]{2} \cdot \sqrt[3]{2} \cdot \sqrt[3]{2} = \left(\sqrt[3]{2}\right)^3 = 2$.
- What appears to be the meaning of $2^{\frac{1}{3}}$?
  - *Since both the exponent expression and the radical expression involve multiplying a number by itself three times and the result is equal to* $2$*, we know that* $2^{\frac{1}{3}} = \sqrt[3]{2}$.

*Scaffolding:*

- If needed, you can demonstrate the argument using perfect cubes. For example, use a base of $8$ instead of a base of $2$.

  Show $\left(8^{\frac{1}{3}}\right)^3 = 8$ and $\left(\sqrt[3]{8}\right)^3 = 8$.

Students can also confirm using a calculator that the decimal approximations of $2^{\frac{1}{3}}$ and $\sqrt[3]{2}$ are the same. Next, we ask them to generalize their findings.

- Can we generalize this relationship? Does $2^{\frac{1}{4}} = \sqrt[4]{2}$? Does $2^{\frac{1}{10}} = \sqrt[10]{2}$? What is $2^{\frac{1}{n}}$, for any positive integer $n$? Why?
  - $2^{\frac{1}{n}} = \sqrt[n]{2}$ *because*

MP.8

$$\left(2^{\frac{1}{n}}\right)^n = \underbrace{\left(2^{\frac{1}{n}}\right)\left(2^{\frac{1}{n}}\right)\cdots\left(2^{\frac{1}{n}}\right)}_{n \text{ times}} = 2^{\overbrace{\frac{1}{n}+\frac{1}{n}+\cdots+\frac{1}{n}}^{n \text{ times}}} = 2^1 = 2.$$

Have students confirm these using a calculator as well checking to see if the decimal approximations of $2^{\frac{1}{n}}$ and $\sqrt[n]{2}$ are the same for different values of $n$ such as $4, 5, 6, 10, \ldots$. Be sure to share the generalization shown above on the board to help students understand why it makes sense to define $2^{\frac{1}{n}}$ to be $\sqrt[n]{2}$.

However, be careful to not stop here; there is a problem with our reasoning if we do not define $\sqrt[n]{2}$. In previous courses, only square roots and cube roots were defined.

We first need to define the $n$th root of a number; there may be more than one, as in the case where $2^2 = 4$ and $(-2)^2 = 4$. We say that both $-2$ and $2$ are square roots of $4$. However, we give priority to the positive-valued square root, and we say that $2$ is the *principal square root* of $4$. We often just refer to *the* square root of $4$ when we mean the *principal* square root of $4$. The definition of $n^{\text{th}}$ root presented below is consistent with allowing complex $n^{\text{th}}$ roots, which students will encounter in college if they pursue engineering or higher mathematics. If we allow complex $n^{\text{th}}$ roots, there are three cube roots of 2: $\sqrt[3]{2}$, $\sqrt[3]{2}\left(-\frac{1}{2}+\frac{\sqrt{3}}{2}i\right)$, and $\sqrt[3]{2}\left(-\frac{1}{2}-\frac{\sqrt{3}}{2}i\right)$, and we refer to the real number $\sqrt[3]{2}$ as the principal cube root of $2$. There is no need to discuss this with any but the most advanced students.

The $\sqrt[n]{2}$ is the positive real number $a$ such that $a^n = 2$. In general, if $a$ is positive, then the $n^{\text{th}}$ root of $a$ exists for any positive integer $n$, and if $a$ is negative, then the $n^{\text{th}}$ root of $a$ exists only for odd integers $n$. This even/odd condition is handled subtly in the definition below; the $n^{\text{th}}$ root exists only if there is already an exponential relationship $b = a^n$.

Present the following definitions to students and have them record them in their notes.

> **$n^{\text{TH}}$ ROOT OF A NUMBER:** Let $a$ and $b$ be numbers, and let $n \geq 2$ be a positive integer. If $b = a^n$, then $a$ is an *$n^{th}$ root of $b$*. If $n = 2$, then the root is a called a *square root*. If $n = 3$, then the root is called a *cube root*.
>
> **PRINCIPAL $n^{\text{TH}}$ ROOT OF A NUMBER:** Let $b$ be a real number that has at least one real $n^{\text{th}}$ root. The *principal $n^{th}$ root of $b$* is the real $n^{\text{th}}$ root that has the same sign as $b$ and is denoted by a radical symbol: $\sqrt[n]{b}$.
>
> Every positive number has a unique principal $n^{\text{th}}$ root. We often refer to the principal $n^{\text{th}}$ root of $b$ as just the *$n^{th}$ root of $b$*. The $n^{\text{th}}$ root of $0$ is $0$.

Students have already learned about square and cube roots in previous grades. In Module 1 and at the beginning of this lesson, students worked with radical expressions involving cube and square roots. Explain that the $n^{\text{th}}$ roots of a number satisfy the same properties of radicals learned previously. Have students record these properties in their notes.

> If $a \geq 0$, $b \geq 0$ ($b \neq 0$ when $b$ is a denominator) and $n$ is a positive integer, then
>
> $$\sqrt[n]{ab} = \sqrt[n]{a} \cdot \sqrt[n]{b} \quad \text{and} \quad \sqrt[n]{\frac{a}{b}} = \frac{\sqrt[n]{a}}{\sqrt[n]{b}}.$$

Background information regarding $n^{\text{th}}$ roots and their uniqueness is provided below. You may wish to share this with your advanced learners or the entire class if you extend this lesson to an additional day.

The existence of the principal $n^{\text{th}}$ root of a real number $b$ is a consequence of the fundamental theorem of algebra: Consider the polynomial function $f(x) = x^n - b$. When $n$ is odd, we know that $f$ has at least one real zero because the graph of $f$ must cross the $x$-axis. That zero is a positive number which (after showing that it is the *ONLY* real zero) is the $n^{\text{th}}$ root. The case for when $n$ is even follows a similar argument.

To show uniqueness of the $n^{\text{th}}$ root, suppose there are two $n^{\text{th}}$ roots of a number $b$, $x$, and $y$ such that $x > 0$, $y > 0$, $x^n = b$, and $y^n = b$. Then $x^n - y^n = b - b = 0$. The expression $x^n - y^n$ factors (see Lesson 7 in Module 1).

$$\begin{aligned} 0 &= x^n - y^n \\ &= (x-y)(x^{n-1} + x^{n-2}y + x^{n-3}y^2 + \cdots + xy^{n-2} + y^{n-1}) \end{aligned}$$

Since both $x$ and $y$ are positive, the second factor is never zero. Thus, for $x^n - y^n = 0$, we must have $x - y = 0$ and it follows that $x = y$. Thus, there is only one $n^{\text{th}}$ root of $b$.

A proof of the first radical property is shown below for teacher background information. You may wish to share this proof with your advanced learners or the entire class if you extend this lesson to an additional day.

Prove that $\sqrt[n]{ab} = \sqrt[n]{a} \cdot \sqrt[n]{b}$.

Let $x \geq 0$ be the number such that $x^n = a$ and let $y \geq 0$ be the number such that $y^n = b$, so that $x = \sqrt[n]{a}$ and $y = \sqrt[n]{b}$. Then, by a property of exponents, $(xy)^n = x^n y^n = ab$. Thus, $xy$ must be the $n^{\text{th}}$ root of $ab$. Writing this using our notation gives

$$\sqrt[n]{ab} = xy = \sqrt[n]{a} \cdot \sqrt[n]{b}.$$

After students have recorded this information in their notes, you can proceed with Example 1 and Exercise 1.

### Example 1 (3 minutes)

This example will familiarize students with the wording in the definition presented above.

**Example 1**

a. **What is the $4^{\text{th}}$ root of $16$?**

   $x^4 = 16$ *when* $x = 2$ *because* $2^4 = 16$. *Thus,* $\sqrt[4]{16} = 2$.

b. **What is the cube root of $125$?**

   $x^3 = 125$ *when* $x = 5$ *because* $5^3 = 125$. *Thus,* $\sqrt[3]{125} = 5$.

c. **What is the $5^{\text{th}}$ root of $100,000$?**

   $x^5 = 100,000$ *when* $x = 10$ *because* $10^5 = 100,000$. *Thus,* $\sqrt[5]{100,000} = 10$.

## Exercise 1 (2 minutes)

In these brief exercises, students will work with definition of $n^{\text{th}}$ roots and the multiplication property presented above. Have students check their work with a partner and briefly discuss as a whole class any questions that arise.

**Exercise 1**

1. **Evaluate each expression.**

   a. $\sqrt[4]{81}$

   $3$

   b. $\sqrt[5]{32}$

   $2$

   c. $\sqrt[3]{9} \cdot \sqrt[3]{3}$

   $\sqrt[3]{27} = 3$

   d. $\sqrt[4]{25} \cdot \sqrt[4]{100} \cdot \sqrt[4]{4}$

   $\sqrt[4]{10,000} = 10$

*Scaffolding:*

If needed, continue to support students that struggle with abstraction by including additional numeric examples.

## Discussion (8 minutes)

Return to the question posted in the title of this lesson: What are $2^{\frac{1}{2}}$ and $2^{\frac{1}{3}}$? Now we know the answer, $2^{\frac{1}{2}} = \sqrt{2}$ and $2^{\frac{1}{3}} = \sqrt[3]{2}$. So far, we have given meaning to $2^{\frac{1}{n}}$ by equating $2^{\frac{1}{n}} = \sqrt[n]{2}$. Ask students if they believe these results extend to any base $b > 0$.

- We just did some exercises where the $b$-value was a number different from 2. In our earlier work, was there something special about using 2 as the base of our exponential expression? Would these results generalize to expressions of the form $3^{\frac{1}{n}}$? $7^{\frac{1}{n}}$? $10^{\frac{1}{n}}$? $b^{\frac{1}{n}}$, for any positive real number $b$?
  - *There is nothing inherently special about the base 2 in the above discussion. These results should generalize to expressions of the form $b^{\frac{1}{n}}$ for any positive real number base b because we have defined an $n^{th}$ root for any positive base $b$.*

Now that we know what the $n^{\text{th}}$ root of a number $b$, $\sqrt[n]{b}$, means, the work we did earlier with base 2 suggests that $b^{\frac{1}{n}}$ should also be defined as the $n^{\text{th}}$ root of $b$. Discuss this definition with your class. If the class is unclear on the definition, do some numerical examples: $(-32)^{\frac{1}{5}} = -2$ because $(-2)^5 = -32$, but $(-16)^{\frac{1}{4}}$ does not exist because there is no principal $4^{\text{th}}$ root of a negative number.

For a real number $b$ and a positive integer $n$, define $b^{\frac{1}{n}}$ to be the principal $n^{\text{th}}$ root of $b$ when it exists. That is, $b^{\frac{1}{n}} = \sqrt[n]{b}$.

Note that this definition holds for any real number $b$ if $n$ is an odd integer and for positive real numbers $b$ if $n$ is an even integer. You may choose to emphasize this with your class. Thus, when $b$ is negative and $n$ is an odd integer, the expression $b^{\frac{1}{n}}$ will be negative. If $n$ is an even integer, then we must restrict $b$ to positive real numbers only, and $b^{\frac{1}{n}}$ will be positive.

In the next lesson, we see that with this definition $b^{\frac{1}{n}}$ satisfies all the usual properties of exponents so it makes sense to define it in this way.

At this point, you can also revisit our original question with the students one more time.

- What is the value of $2^{\frac{1}{2}}$? What does it mean? What is the value of $b^{\frac{1}{2}}$ for any positive number $b$? How are radicals related to rational exponents?
  - *We now know that $2^{\frac{1}{2}}$ is equal to $\sqrt{2}$. In general $b^{\frac{1}{2}} = \sqrt{b}$ for any positive real number $b$. A number that contains a radical can be expressed using rational exponents in place of the radical.*

As you continue the discussion, we extend the definition of exponential expressions with exponents of the form $\frac{1}{n}$ to any positive rational exponent. We begin by considering an example: What do we mean by $2^{\frac{3}{4}}$? Give students a few minutes to respond individually in writing to this question on their student pages and then have them discuss their reasoning with a partner. Make sure to correct any blatant misconceptions and to clarify incomplete thinking as you lead the discussion that follows.

MP.3

**Discussion**

**If $2^{\frac{1}{2}} = \sqrt{2}$ and $2^{\frac{1}{3}} = \sqrt[3]{2}$, what does $2^{\frac{3}{4}}$ equal? Explain your reasoning.**

***Student solutions and explanations will vary. One possible solution would be $2^{\frac{3}{4}} = \left(2^{\frac{1}{4}}\right)^3$, so it must mean that $2^{\frac{3}{4}} = \left(\sqrt[4]{2}\right)^3$. Since the properties of exponents and the meaning of an exponent made sense with integers and now for rational numbers in the form $\frac{1}{n}$, it would make sense that they would work all rational numbers too.***

Now that we have a definition for exponential expressions of the form $b^{\frac{1}{n}}$, use the discussion below to define $b^{\frac{m}{n}}$, where $m$ and $n$ are both integers, $n \neq 0$, and $b$ is a positive real number. Make sure students understand that our interpretation of $b^{\frac{m}{n}}$ must be consistent with the exponent properties (which hold for integer exponents) and the definition of $b^{\frac{1}{n}}$.

- How can we rewrite the exponent of $2^{\frac{3}{4}}$ using integers and rational numbers in the form $\frac{1}{n}$?
  - *We can write $2^{\frac{3}{4}} = \left(2^{\frac{1}{4}}\right)^3$ or we can write $2^{\frac{3}{4}} = (2^3)^{\frac{1}{4}}$.*
- Now apply our definition of $b^{\frac{1}{n}}$.
  - $2^{\frac{3}{4}} = \left(2^{\frac{1}{4}}\right)^3 = \left(\sqrt[4]{2}\right)^3$ *or* $2^{\frac{3}{4}} = (2^3)^{\frac{1}{4}} = \sqrt[4]{2^3} = \sqrt[4]{8}$

- Does this make sense? If $2^{\frac{3}{4}} = \sqrt[4]{8}$, then if we raise $2^{\frac{3}{4}}$ to the fourth power, we should get $8$. Does this happen?
  - $\left(2^{\frac{3}{4}}\right)^4 = \left(2^{\frac{3}{4}}\right)\left(2^{\frac{3}{4}}\right)\left(2^{\frac{3}{4}}\right)\left(2^{\frac{3}{4}}\right) = 2^{\left(4\cdot\frac{3}{4}\right)} = 2^3 = 8.$
  - *So, $8$ is the product of four equal factors, which we denote by $2^{\frac{3}{4}}$. Thus, $2^{\frac{3}{4}} = \sqrt[4]{8}$.*

Take a few minutes to allow students to think about generalizing their work above to $2^{\frac{m}{n}}$ and then to $b^{\frac{m}{n}}$. Have them write a response to the following questions and share it with a partner before proceeding as a whole class.

- Can we generalize this result? How would you define $2^{\frac{m}{n}}$, for positive integers $m$ and $n$?
  - *Conjecture: $2^{\frac{m}{n}} = \sqrt[n]{2^m}$, or equivalently, $2^{\frac{m}{n}} = \left(\sqrt[n]{2}\right)^m$.*
- Can we generalize this result to any positive real number base $b$? What is $3^{\frac{m}{n}}$? $7^{\frac{m}{n}}$? $10^{\frac{m}{n}}$? $b^{\frac{m}{n}}$?
  - *There is nothing inherently special about the base $2$ in the above Discussion. These results should generalize to expressions of the form $b^{\frac{m}{n}}$ for any positive real number base $b$.*
  - *Then we are ready to define $b^{\frac{m}{n}} = \sqrt[n]{b^m}$ for positive integers $m$ and $n$ and positive real numbers $b$.*

This result is summarized in the box below.

For any positive integers $m$ and $n$, and any real number $b$ for which $b^{\frac{1}{n}}$ exists, we define

$$b^{\frac{m}{n}} = \sqrt[n]{b^m}, \text{ which is equivalent to } b^{\frac{m}{n}} = \left(\sqrt[n]{b}\right)^m$$

Note that this property holds for any real number $b$ if $n$ is an odd integer. You may choose to emphasize this with your class. When $b$ is negative and $n$ is an odd integer, the expression $b^{\frac{m}{n}}$ will be negative. If $n$ is an even integer then we must restrict $b$ to positive real numbers only.

## Exercises 2–8 (4 minutes)

In these exercises, students use the definitions above to rewrite and evaluate expressions. Have students check their work with a partner and briefly discuss as a whole class any questions that arise.

**Exercises 2–12**

**Rewrite each exponential expression as an $n^{\text{th}}$ root.**

**2.** $3^{\frac{1}{2}}$

$3^{\frac{1}{2}} = \sqrt{3}$

**3.** $11^{\frac{1}{5}}$

$11^{\frac{1}{5}} = \sqrt[5]{11}$

4. $\left(\frac{1}{4}\right)^{\frac{1}{5}}$

$$\left(\frac{1}{4}\right)^{\frac{1}{5}} = \sqrt[5]{\frac{1}{4}}$$

5. $6^{\frac{1}{10}}$

$$6^{\frac{1}{10}} = \sqrt[10]{6}$$

**Rewrite the following exponential expressions as equivalent radical expressions. If the number is rational, write it without radicals or exponents.**

6. $2^{\frac{3}{2}}$

$$2^{\frac{3}{2}} = \sqrt{2^3} = 2\sqrt{2}$$

7. $4^{\frac{5}{2}}$

$$4^{\frac{5}{2}} = \sqrt{4^5} = \left(\sqrt{4}\right)^5 = 2^5 = 32$$

8. $\left(\frac{1}{8}\right)^{\frac{5}{3}}$

$$\left(\frac{1}{8}\right)^{\frac{5}{3}} = \sqrt[3]{\left(\frac{1}{8}\right)^5} = \left(\sqrt[3]{\frac{1}{8}}\right)^5 = \left(\frac{1}{2}\right)^5 = \frac{1}{32}$$

## Exercise 9 (3 minutes)

In this exercise, we ask students to consider a negative rational exponent. Have students work directly with a partner and ask them to use thinking similar to the preceding discussion. Correct and extend student thinking as you review the solution.

9. **Show why the following statement is true:**

$$2^{-\frac{1}{2}} = \frac{1}{2^{\frac{1}{2}}}$$

***Student solutions and explanations will vary. One possible solution would be***

$$2^{-\frac{1}{2}} = \left(2^{\frac{1}{2}}\right)^{-1} = \left(\sqrt{2}\right)^{-1} = \frac{1}{\sqrt{2}} = \frac{1}{2^{\frac{1}{2}}}$$

***Since $\sqrt{2}$ is a real number the properties of exponents hold and we have defined $b^{\frac{1}{n}}$. We can show these two expressions are the same.***

Share the following property with your class and show how the work they did in the previous exercises supports this conclusion. Should you choose to you can verify these properties using an argument similar to the ones presented earlier for the meaning of $b^{\frac{m}{n}}$.

For any positive integers $m$ and $n$, and any real number $b$ for which $b^{\frac{1}{n}}$ exists, we define

$$b^{-\frac{m}{n}} = \frac{1}{\sqrt[n]{b^m}}$$

or, equivalently,

$$b^{-\frac{m}{n}} = \frac{1}{\left(\sqrt[n]{b}\right)^m}.$$

## Exercises 10–12 (3 minutes)

**Rewrite the following exponential expressions as equivalent radical expressions. If the number is rational, write it without radicals or exponents.**

**10.** $4^{-\frac{3}{2}}$

$$4^{-\frac{3}{2}} = \frac{1}{\sqrt{4^3}} = \frac{1}{\left(\sqrt{4}\right)^3} = \frac{1}{8}$$

**11.** $27^{-\frac{2}{3}}$

$$27^{-\frac{2}{3}} = \frac{1}{27^{\frac{2}{3}}} = \frac{1}{\left(\sqrt[3]{27}\right)^2} = \frac{1}{3^2} = \frac{1}{9}$$

**12.** $\left(\frac{1}{4}\right)^{-\frac{1}{2}}$

*We have* $\left(\frac{1}{4}\right)^{-\frac{1}{2}} = \left(\sqrt{\frac{1}{4}}\right)^{-1} = \left(\frac{1}{2}\right)^{-1} = 2$. *Alternatively,* $\left(\frac{1}{4}\right)^{-\frac{1}{2}} = \left(\left(\frac{1}{4}\right)^{-1}\right)^{\frac{1}{2}} = (4)^{\frac{1}{2}} = \sqrt{4} = 2$.

## Closing (3 minutes)

Have students summarize the key points of the lesson in writing. Circulate around the classroom to informally assess understanding and provide assistance. Their work should reflect the summary provided below.

**Lesson Summary**

$n^{\text{TH}}$ ROOT OF A NUMBER: Let $a$ and $b$ be numbers, and let $n \geq 2$ be a positive integer. If $b = a^n$, then $a$ is an $n^{th}$ *root of* $b$. If $n = 2$, then the root is a called a square root. If $n = 3$, then the root is called a cube root.

PRINCIPAL $n^{\text{TH}}$ ROOT OF A NUMBER: Let $b$ be a real number that has at least one real $n^{\text{th}}$ root. The *principal* $n^{th}$ *root of* $b$ is the real $n^{\text{th}}$ root that has the same sign as $b$, and is denoted by a radical symbol: $\sqrt[n]{b}$.

Every positive number has a unique principal $n^{\text{th}}$ root. We often refer to the principal $n^{\text{th}}$ root of $b$ as just *the* $n^{th}$ *root of* $b$. The $n^{\text{th}}$ root of $0$ is $0$.

For any positive integers $m$ and $n$, and any real number $b$ for which the principal $n^{\text{th}}$ root of $b$ exists, we have

$$b^{\frac{1}{n}} = \sqrt[n]{b}$$

$$b^{\frac{m}{n}} = \sqrt[n]{b^m} = \left(\sqrt[n]{b}\right)^m$$

$$b^{-\frac{m}{n}} = \frac{1}{\sqrt[n]{b^m}}.$$

## Exit Ticket (4 minutes)

Name ______________________________ Date ______________

# Lesson 3: Rational Exponents—What Are $2^{\frac{1}{2}}$ and $2^{\frac{1}{3}}$?

## Exit Ticket

1. Write the following exponential expressions as equivalent radical expressions.

   a. $2^{\frac{1}{2}}$

   b. $2^{\frac{3}{4}}$

   c. $3^{-\frac{2}{3}}$

2. Rewrite the following radical expressions as equivalent exponential expressions.

   a. $\sqrt{5}$

   b. $2\sqrt[4]{3}$

   c. $\frac{1}{\sqrt[3]{16}}$

3. Provide a written explanation for each question below.

   a. Is it true that $\left(4^{\frac{1}{2}}\right)^3 = (4^3)^{\frac{1}{2}}$? Explain how you know.

   b. Is it true that $\left(1000^{\frac{1}{3}}\right)^3 = (1000^3)^{\frac{1}{3}}$? Explain how you know.

   c. Suppose that $m$ and $n$ are positive integers and $b$ is a real number so that the principal $n^{\text{th}}$ root of $b$ exists. In general does $\left(b^{\frac{1}{n}}\right)^m = (b^m)^{\frac{1}{n}}$? Provide at least one example to support your claim.

## Exit Ticket Sample Solutions

1. **Rewrite the following exponential expressions as equivalent radical expressions.**

   a. $2^{\frac{1}{2}}$

   $2^{\frac{1}{2}} = \sqrt{2}$

   b. $2^{\frac{3}{4}}$

   $2^{\frac{3}{4}} = \sqrt[4]{2^3} = \sqrt[4]{8}$

   c. $3^{-\frac{2}{3}}$

   $3^{-\frac{2}{3}} = \frac{1}{\sqrt[3]{3^2}} = \frac{1}{\sqrt[3]{9}}$

2. **Rewrite the following radical expressions as equivalent exponential expressions.**

   a. $\sqrt{5}$

   $\sqrt{5} = 5^{\frac{1}{2}}$

   b. $2\sqrt[4]{3}$

   $2\sqrt[4]{3} = \sqrt[4]{2^4 \cdot 3} = \sqrt[4]{48} = 48^{\frac{1}{4}}$

   c. $\frac{1}{\sqrt[3]{16}}$

   $\frac{1}{\sqrt[3]{16}} = (2^4)^{-\frac{1}{3}} = 2^{-\frac{4}{3}}$

   $\frac{1}{\sqrt[3]{16}} = (16)^{-\frac{1}{3}}$

3. **Provide a written explanation for each question below.**

MP.3 & MP.7

   a. **Is it true that $\left(4^{\frac{1}{2}}\right)^3 = (4^3)^{\frac{1}{2}}$? Explain how you know.**

   $\left(4^{\frac{1}{2}}\right)^3 = \left(\sqrt{4}\right)^3 = 2^3 = 8$

   $(4^3)^{\frac{1}{2}} = 64^{\frac{1}{2}} = \sqrt{64} = 8$

   ***So the first statement is true.***

   b. **Is it true that $\left(1000^{\frac{1}{3}}\right)^3 = (1000^3)^{\frac{1}{3}}$? Explain how you know.**

   ***Similarly the left and right sides of the second statement are equal to one another.***

   $\left(1000^{\frac{1}{3}}\right)^3 = \left(\sqrt[3]{1000}\right)^3 = 10^3 = 1000$

   $(1000^3)^{\frac{1}{3}} = (1000000000)^{\frac{1}{3}} = 1000$

c. **Suppose that $m$ and $n$ are positive integers and $b$ is a real number so that the principal $n^{\text{th}}$ root of $b$ exists. In general does $\left(b^{\frac{1}{n}}\right)^m = (b^m)^{\frac{1}{n}}$? Provide at least one example to support your claim.**

*Thus, it appears the statement is true and it also holds for when the exponents are integers. Based on our other work in this lesson, this property should extend to rational exponents as well.*

*Here is another example,*

$\left(8^{\frac{1}{3}}\right)^2 = (2)^2 = 4$

*and*

$(8^2)^{\frac{1}{3}} = (64)^{\frac{1}{3}} = 4.$

*Thus,*

$\left(8^{\frac{1}{3}}\right)^2 = (8^2)^{\frac{1}{3}}.$

## Problem Set Sample Solutions

1. **Select the expression from (A), (B), and (C) that correctly completes the statement.**

| | | (A) | (B) | (C) |
|---|---|---|---|---|
| a. | $x^{\frac{1}{3}}$ is equivalent to ________. | $\frac{1}{3}x$ | $\sqrt[3]{x}$ | $\frac{3}{x}$ |
| b. | $x^{\frac{2}{3}}$ is equivalent to ________. | $\frac{2}{3}x$ | $\sqrt[3]{x^2}$ | $\left(\sqrt{x}\right)^3$ |
| c. | $x^{-\frac{1}{4}}$ is equivalent to ________. | $-\frac{1}{4}x$ | $\frac{4}{x}$ | $\frac{1}{\sqrt[4]{x}}$ |
| d. | $\left(\frac{4}{x}\right)^{\frac{1}{2}}$ is equivalent to ________. | $\frac{2}{x}$ | $\frac{4}{x^2}$ | $\frac{2}{\sqrt{x}}$ |

a. *(B)*

b. *(B)*

c. *(C)*

d. *(C)*

2. **Identify which of the expressions (A), (B), and (C) are equivalent to the given expression.**

| | | (A) | (B) | (C) |
|---|---|---|---|---|
| a. | $16^{\frac{1}{2}}$ | $\left(\frac{1}{16}\right)^{-\frac{1}{2}}$ | $8^{\frac{2}{3}}$ | $64^{\frac{3}{2}}$ |
| b. | $\left(\frac{2}{3}\right)^{-1}$ | $-\frac{3}{2}$ | $\left(\frac{9}{4}\right)^{\frac{1}{2}}$ | $\frac{27^{\frac{1}{3}}}{6}$ |

a. *(A) and (B)*

b. *(B) only*

3. **Rewrite in radical form. If the number is rational, write it without using radicals.**

a. $6^{\frac{3}{2}}$

$\sqrt{216}$

b. $\left(\frac{1}{2}\right)^{\frac{1}{4}}$

$\sqrt[4]{\frac{1}{2}}$

c. $3(8)^{\frac{1}{3}}$

$3\sqrt[3]{8} = 6$

d. $\left(\frac{64}{125}\right)^{-\frac{2}{3}}$

$\left(\sqrt[3]{\frac{125}{16}}\right)^2 = \frac{25}{16}l$

e. $81^{-\frac{1}{4}}$

$\frac{1}{\sqrt[4]{81}} = \frac{1}{3}$

4. **Rewrite the following expressions in exponent form.**

a. $\sqrt{5}$

$5^{\frac{1}{2}}$

b. $\sqrt[3]{5^2}$

$5^{\frac{2}{3}}$

c. $\sqrt{5^3}$

$5^{\frac{3}{2}}$

d. $\left(\sqrt[3]{5}\right)^2$

$5^{\frac{2}{3}}$

5. **Use the graph of $f(x) = 2^x$ shown to the right to estimate the following powers of 2.**

a. $2^{\frac{1}{4}}$ $\approx 1.2$

b. $2^{\frac{2}{3}}$ $\approx 1.6$

c. $2^{\frac{3}{4}}$ $\approx 1.7$

d. $2^{0.2}$ $\approx 1.15$

e. $2^{1.2}$ $\approx 2.3$

f. $2^{-\frac{1}{5}}$ $\approx 0.85$

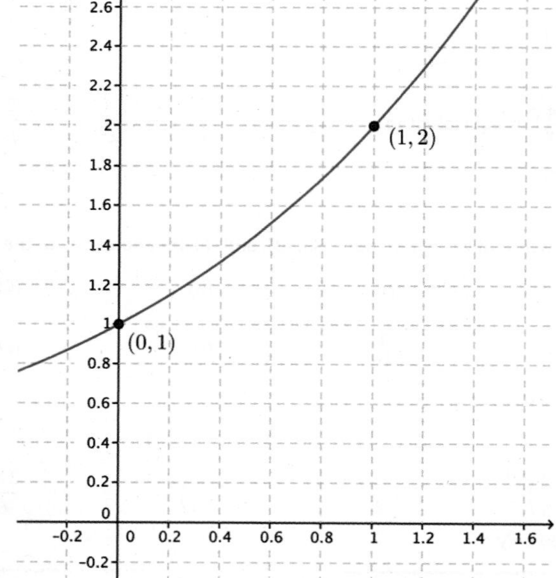

6. Rewrite each expression in the form $kx^n$, where $k$ is a real number, $x$ is a positive real number, and $n$ is rational.

a. $\sqrt[4]{16x^3}$

$2x^{\frac{3}{4}}$

b. $\frac{5}{\sqrt{x}}$

$5x^{-\frac{1}{2}}$

c. $\sqrt[3]{1/x^4}$

$x^{-\frac{4}{3}}$

d. $\frac{4}{\sqrt[3]{8x^3}}$

$2x^{-1}$

e. $\frac{27}{\sqrt{9x^4}}$

$9x^{-2}$

f. $\left(\frac{125}{x^2}\right)^{-\frac{1}{3}}$

$\frac{1}{5}x^{\frac{2}{3}}$

7. Find a value of $x$ for which $2x^{\frac{1}{2}} = 32$.

$256$

8. Find a value of $x$ for which $x^{\frac{4}{3}} = 81$.

$27$

9. If $x^{\frac{3}{2}} = 64$, find the value of $4x^{-\frac{3}{4}}$.

$x = 16$, *so* $4(16)^{-\frac{3}{4}} = 4(2)^{-3} = \frac{4}{8} = \frac{1}{2}$.

10. If $b = \frac{1}{9}$, evaluate the following expressions.

a. $b^{-\frac{1}{2}}$

$\left(\frac{1}{9}\right)^{-\frac{1}{2}} = 9^{\frac{1}{2}} = 3$

b. $b^{\frac{5}{2}}$

$\left(\frac{1}{9}\right)^{\frac{5}{2}} = \left(\frac{1}{3}\right)^5 = \frac{1}{243}$

c. $\sqrt[3]{3b^{-1}}$

$\sqrt[3]{3\left(\frac{1}{9}\right)^{-1}} = \sqrt[3]{27} = 3$

11. Show that each expression is equivalent to $2x$. Assume $x$ is a positive real number.

    a. $\sqrt[4]{16x^4}$

    $\sqrt[4]{16} \cdot \sqrt[4]{x^4} = 2x$

    b. $\dfrac{\left(\sqrt[3]{8x^3}\right)^2}{\sqrt{4x^2}}$

    $\dfrac{(2x)^2}{(4x^2)^{\frac{1}{2}}} = \dfrac{4x^2}{2x} = 2x$

    c. $\dfrac{6x^3}{\sqrt[3]{27x^6}}$

    $\dfrac{6x^3}{3x^{\frac{6}{3}}} = \dfrac{6x^3}{3x^2} = 2x$

MP.3

12. Yoshiko said that $16^{\frac{1}{4}} = 4$ because $4$ is one-fourth of $16$. Use properties of exponents to explain why she is or is not correct.

    *Yoshiko's reasoning is not correct. By our exponent properties,* $\left(16^{\frac{1}{4}}\right)^4 = 16^{\left(\frac{1}{4}\right)\cdot 4} = 16^1 = 16$, *but* $4^4 = 256$. *Since* $\left(16^{\frac{1}{4}}\right)^4 \neq 4^4$, *we know that* $16^{\frac{1}{4}} \neq 4$.

13. Jefferson said that $8^{\frac{4}{3}} = 16$ because $8^{\frac{1}{3}} = 2$ and $2^4 = 16$. Use properties of exponents to explain why he is or is not correct.

    *Jefferson's reasoning is correct. We know that* $8^{\frac{4}{3}} = \left(8^{\frac{1}{3}}\right)^4$, *so* $8^{\frac{4}{3}} = 2^4$, *and thus* $8^{\frac{4}{3}} = 16$.

14. Rita said that $8^{\frac{2}{3}} = 128$ because $8^{\frac{2}{3}} = 8^2 \cdot 8^{\frac{1}{3}}$, so $8^{\frac{2}{3}} = 64 \cdot 2$, and then $8^{\frac{2}{3}} = 128$. Use properties of exponents to explain why she is or is not correct.

    *Rita's reasoning is not correct because she did not apply the properties of exponents correctly. She should also realize that raising* $8$ *to a positive power less than* $1$ *will produce a number less than* $8$. *The correct calculation is below.*

$$\begin{aligned} 8^{\frac{2}{3}} &= \left(8^{\frac{1}{3}}\right)^2 \\ &= 2^2 \\ &= 4 \end{aligned}$$

15. **Suppose for some positive real number $a$ that $\left(a^{\frac{1}{4}} \cdot a^{\frac{1}{2}} \cdot a^{\frac{1}{4}}\right)^2 = 3$.**

   a. **What is the value of $a$?**

$$\left(a^{\frac{1}{4}} \cdot a^{\frac{1}{2}} \cdot a^{\frac{1}{4}}\right)^2 = 3$$
$$\left(a^{\frac{1}{4}+\frac{1}{2}+\frac{1}{4}}\right)^2 = 3$$
$$(a^1)^2 = 3$$
$$a^2 = 3$$
$$a = \sqrt{3}$$

   b. **Which exponent properties did you use to find your answer to part (a)?**

   ***We used the properties $b^n \cdot b^m = b^{m+n}$ and $(b^m)^n = b^{mn}$.***

16. **In the lesson, you made the following argument:**

$$\begin{aligned}\left(2^{\frac{1}{3}}\right)^3 &= 2^{\frac{1}{3}} \cdot 2^{\frac{1}{3}} \cdot 2^{\frac{1}{3}} \\ &= 2^{\frac{1}{3}+\frac{1}{3}+\frac{1}{3}} \\ &= 2^1 \\ &= 2.\end{aligned}$$

**Since $\sqrt[3]{2}$ is a number so that $\left(\sqrt[3]{2}\right)^3 = 2$ and $2^{\frac{1}{3}}$ is a number so that $\left(2^{\frac{1}{3}}\right)^3 = 2$, you concluded that $2^{\frac{1}{3}} = \sqrt[3]{2}$. Which exponent property was used to make this argument?**

***We used the property $b^n \cdot b^m = b^{m+n}$. (Students may also mention the uniqueness of $n^{\text{th}}$ roots.)***

# Lesson 4: Properties of Exponents and Radicals

## Student Outcomes

- Students rewrite expressions involving radicals and rational exponents using the properties of exponents.

## Lesson Notes

In Lesson 1, students reviewed the properties of exponents for integer exponents before establishing the meaning of the $n^{\text{th}}$ root of a positive real number and how it can be expressed as a rational exponent in Lesson 3. In Lesson 4, students extend properties of exponents that applied to expressions with integer exponents to expressions with rational exponents. In each case, the notation $b^{\frac{1}{n}}$ specifically indicates the principal root (e.g., $2^{\frac{1}{2}}$ is $\sqrt{2}$, as opposed to $-\sqrt{2}$).

This lesson extends students' thinking using the properties of radicals and the definitions from Lesson 3 so that they can see why it makes sense that the properties of exponents hold for any rational exponents (**N-RN.A.1**). Examples and exercises work to establish fluency with the properties of exponents when the exponents are rational numbers and emphasize rewriting expressions and evaluating expressions using the properties of exponents and radicals (**N-RN.A.2**).

## Classwork

### Opening (2 minutes)

Students revisit the properties of square roots and cube roots studied in Module 1 to remind them that we extended those to any $n^{\text{th}}$ root in Lesson 3. So, they are now ready to verify that the properties of exponents hold for rational exponents.

Draw students' attention to a chart posted prominently on the wall or to their notebooks where the properties of exponents and radicals are displayed, including those developed in Lesson 3.

Remind students of the description of exponential expressions of the form $b^{\frac{m}{n}}$, which they will be making use of throughout the lesson:

Let $b$ be any positive real number, and $m$, $n$ be any integers with $n > 0$; then $b^{\frac{m}{n}} = \sqrt[n]{b^m}$ and $b^{\frac{m}{n}} = \left(\sqrt[n]{b}\right)^m$.

*Scaffolding:*

- Throughout the lesson, remind students of past properties of integer exponents and radicals either through an anchor chart posted on the wall or by writing relevant properties as they come up. Included is a short list of previous properties used in this module.
- For all real numbers $a, b > 0$, and all integers $m, n$:

$$b^m \cdot b^n = b^{m+n}$$
$$(b^m)^n = b^{mn}$$
$$(ab)^m = a^m \cdot b^m$$
$$b^{-m} = \frac{1}{b^m}$$
$$\sqrt[n]{b} = b^{\frac{1}{n}}$$
$$\sqrt[n]{a} \cdot \sqrt[n]{b} = \sqrt[n]{ab}$$
$$\sqrt[n]{b^n} = \left(\sqrt[n]{b}\right)^n = b$$
$$\sqrt[n]{b^m} = \left(\sqrt[n]{b}\right)^m = b^{\frac{m}{n}}.$$

## Opening Exercise (5 minutes)

These exercises briefly review content from Module 1 and the last lesson.

**Opening Exercise**

**Write each exponent as a radical, and then use the definition and properties of radicals to write that expression as an integer.**

a. $7^{\frac{1}{2}} \cdot 7^{\frac{1}{2}}$

$\sqrt{7} \cdot \sqrt{7} = \sqrt{49} = 7$

b. $3^{\frac{1}{3}} \cdot 3^{\frac{1}{3}} \cdot 3^{\frac{1}{3}}$

$\sqrt[3]{3} \cdot \sqrt[3]{3} \cdot \sqrt[3]{3} = \sqrt[3]{9} \cdot \sqrt[3]{3} = \sqrt[3]{27} = 3$

c. $12^{\frac{1}{2}} \cdot 3^{\frac{1}{2}}$

$\sqrt{12} \cdot \sqrt{3} = \sqrt{12 \cdot 3} = \sqrt{36} = 6$

d. $\left(64^{\frac{1}{3}}\right)^{\frac{1}{2}}$

$\sqrt{\sqrt[3]{64}} = \sqrt{4} = 2$

To transition from the Opening Exercise to Example 1, ask students to write the first two problems above in exponent form. Then, ask them to discuss with a partner whether or not it would be true in general that $b^{\frac{m}{n}} \cdot b^{\frac{p}{q}} = b^{\frac{m}{n}+\frac{p}{q}}$ for positive real numbers $b$ where $m$, $n$, $p$, and $q$ are integers with $n \neq 0$ and $q \neq 0$.

MP.3 & MP.8

- How could you write the $\sqrt{7} \cdot \sqrt{7} = 7$ with rational exponents? How about $\sqrt[3]{3} \cdot \sqrt[3]{3} \cdot \sqrt[3]{3} = 3$?
  - $7^{\frac{1}{2}} \cdot 7^{\frac{1}{2}} = 7^1$ *and* $3^{\frac{1}{3}} \cdot 3^{\frac{1}{3}} \cdot 3^{\frac{1}{3}} = 3^1$
- Based on these examples, is the exponent property $b^m \cdot b^n = b^{m+n}$ valid when $m$ and $n$ are rational numbers? Explain how you know.
  - *Since the exponents on the left side of each statement add up to the exponents on the right side, it appears to be true. However, the right side exponent was always* 1. *If we work with* $\sqrt[3]{8} \cdot \sqrt[3]{8} = \sqrt[3]{64}$ *and write it in exponent form, we would get (after noting* $\sqrt[3]{64} = \sqrt[3]{8^2}$*) that* $8^{\frac{1}{3}} \cdot 8^{\frac{1}{3}} = 8^{\frac{2}{3}}$. *So, it appears to be true in general. Note that examples alone do not prove that a mathematical statement is always true.*

In the rest of this lesson, we will make sense of these observations in general to extend the properties of exponents to rational numbers by applying the definition of the $n$ root *of* $b$ and the properties of radicals introduced in Lesson 3.

## Examples 1–3 (10 minutes)

In the previous lesson, we assumed that the exponent property $b^m b^n = b^{m+n}$ for positive real numbers $b$ and integers $m$ and $n$ would also hold for rational exponents when the exponents were of the form $\frac{1}{n}$, where $n$ was a positive integer. This example will help students see that the property below makes sense for any rational exponent:

$$b^{\frac{m}{n}} \cdot b^{\frac{p}{q}} = b^{\frac{m}{n}+\frac{p}{q}}, \text{ where } m, n, p, \text{ and } q \text{ are integers with } n > 0 \text{ and } q > 0.$$

Perhaps model Example 1 below and then have students work with a partner on Example 2. Make sure students include a justification for each step in the problem. When you get to Example 3, be sure to use the following discussion questions to guide students.

MP.7 & MP.8

- How can we write these expressions using radicals?
  - *In Lesson 3, we learned that $b^{\frac{1}{n}} = \sqrt[n]{b}$ and $b^{\frac{m}{n}} = \sqrt[n]{b^m}$ for positive real numbers $b$ and positive integers $m$ and $n$.*
- Which properties help us to write the expression as a single radical?
  - *The property of radicals that states $\sqrt[n]{a} \cdot \sqrt[n]{b} = \sqrt[n]{ab}$ for positive real numbers $a$ and $b$ and positive integer $n$, and the property of exponents that states $b^m \cdot b^n = b^{m+n}$ for positive real numbers $b$ and integers $m$ and $n$.*
- How do we rewrite this expression in exponent form?
  - *In Lesson 3, we related radicals and rational exponents by $b^{\frac{m}{n}} = \sqrt[n]{b^m}$.*
- What makes Example 3 different from Examples 1 and 2?
  - *The exponents have different denominators, so when we write the expression in radical form, the roots are not the same, and we cannot apply the property that $\sqrt[n]{a} \cdot \sqrt[n]{b} = \sqrt[n]{ab}$.*
- Can you think of a way to rewrite the problem so it looks like the first two problems?
  - *We can write the exponents as equivalent fractions with the same denominator.*

**Examples 1–3**

**Write each expression in the form $b^{\frac{m}{n}}$ for positive real numbers $b$ and integers $m$ and $n$ with $n > 0$ by applying the properties of radicals and the definition of $n^{\text{th}}$ root.**

**1.** $b^{\frac{1}{4}} \cdot b^{\frac{1}{4}}$

***By the definition of nth root,***

$b^{\frac{1}{4}} \cdot b^{\frac{1}{4}} = \sqrt[4]{b} \cdot \sqrt[4]{b}.$

$= \sqrt[4]{b \cdot b}$ ***By the properties of radicals and properties of exponents***

$= \sqrt[4]{b^2}$

$= b^{\frac{2}{4}}$ ***By the definition of $b^{\frac{m}{n}}$***

***The rational number $\frac{2}{4}$ is equal to $\frac{1}{2}$. Thus,***

$b^{\frac{1}{4}} \cdot b^{\frac{1}{4}} = b^{\frac{1}{2}}.$

*Scaffolding:*

- Throughout the lesson, you can create parallel problems to demonstrate that these problems work with numerical values as well.
- For example, in part (a) substitute 4 for $b$.
- In part (b), substitute a perfect cube such as 8 or 27 for $b$.

**2.** $b^{\frac{1}{3}} \cdot b^{\frac{4}{3}}$

$b^{\frac{1}{3}} \cdot b^{\frac{4}{3}} = \sqrt[3]{b} \cdot \sqrt[3]{b^4}$ *By the definition of* $b^{\frac{1}{n}}$ *and* $b^{\left(\frac{m}{n}\right)}$

$= \sqrt[3]{b \cdot b^4}$ *By the properties of radicals and properties of exponents*

$= \sqrt[3]{b^5}$

$= b^{\frac{5}{3}}$ *By the definition of* $b^{\frac{m}{n}}$

*Thus,*

$b^{\frac{1}{3}} \cdot b^{\frac{4}{3}} = b^{\frac{5}{3}}.$

**3.** $b^{\frac{1}{5}} \cdot b^{\frac{3}{4}}$

*Write the exponents as equivalent fractions with the same denominator.*

$$b^{\frac{1}{5}} \cdot b^{\frac{3}{4}} = b^{\frac{4}{20}} \cdot b^{\frac{15}{20}}$$

*Rewrite in radical form.*

$$= \sqrt[20]{b^4} \cdot \sqrt[20]{b^{15}}$$

*Rewrite as a single radical expression.*

$$= \sqrt[20]{b^4 \cdot b^{15}}$$
$$= \sqrt[20]{b^{19}}$$

*Rewrite in exponent form using the definition.*

$$= b^{\frac{19}{20}}$$

*Thus,*

$$b^{\frac{1}{5}} \cdot b^{\frac{3}{4}} = b^{\frac{19}{20}}.$$

- Now add the exponents in each example. What is $\frac{1}{4} + \frac{1}{4}$? $\frac{1}{3} + \frac{4}{3}$? $\frac{1}{5} + \frac{3}{4}$?
  - $\frac{1}{4} + \frac{1}{4} = \frac{1}{2}$, $\frac{1}{3} + \frac{4}{3} = \frac{5}{3}$, *and* $\frac{1}{5} + \frac{3}{4} = \frac{19}{20}$.
- What do you notice about these sums and the value of the exponent when we rewrote each expression?
  - *The sum of the exponents was equal to the exponent of the answer.*

Based on these examples, particularly the last one, it seems reasonable to extend the properties of exponents to hold when the exponents are any rational number. Thus, we can state the following property.

- For any integers $m$, $n$, $p$ and $q$, with $n > 0$ and $q > 0$, and any real numbers $b$ so that $b^{\frac{1}{n}}$ and $b^{\frac{1}{q}}$ are defined,
$$b^{\frac{m}{n}} \cdot b^{\frac{p}{q}} = b^{\frac{m}{n}+\frac{p}{q}}.$$

Have students copy this property into their notes along with the ones listed below. You can also write these properties on a piece of chart paper and display them in your classroom. These properties are listed in the lesson summary.

In a similar fashion, the other properties of exponents can be extended to hold for any rational exponents as well.

- For any integers $m$, $n$, $p$, and $q$, with $n > 0$ and $q > 0$, and any real numbers $a$ and $b$ so that $a^{\frac{1}{n}}$, $b^{\frac{1}{n}}$, and $b^{\frac{1}{q}}$ are defined,

$$b^{\frac{m}{n}} = \sqrt[n]{b^m}$$

$$\left(b^{\frac{1}{n}}\right)^n = b$$

$$(b^n)^{\frac{1}{n}} = b$$

$$(ab)^{\frac{m}{n}} = a^{\frac{m}{n}} \cdot b^{\frac{m}{n}}$$

$$\left(b^{\frac{m}{n}}\right)^{\frac{p}{q}} = b^{\frac{mp}{nq}}$$

$$b^{-\frac{m}{n}} = \frac{1}{b^{\frac{m}{n}}}.$$

At this point, you might have your class look at the opening exercise again and ask them which property could be used to simplify each problem.

For advanced learners, a derivation of the property we explored in Example 1 is provided below.

Rewrite $b^{\frac{m}{n}}$ and $b^{\frac{p}{q}}$ as equivalent exponential expressions in which the exponents have the same denominator, and apply the definition of the $b^{\frac{m}{n}}$ as the $n^{\text{th}}$ root.

By the definition of $b^{\frac{m}{n}}$ and then using properties of algebra, we can rewrite the exponent to be $\frac{m}{n} + \frac{p}{q}$.

$$\begin{aligned} b^{\frac{m}{n}} \cdot b^{\frac{p}{q}} &= b^{\frac{mq}{nq}} b^{\frac{np}{nq}} \\ &= \sqrt[nq]{b^{mq}} \cdot \sqrt[nq]{b^{np}} \\ &= \sqrt[nq]{b^{mq} \cdot b^{np}} \\ &= \sqrt[nq]{b^{mq+np}} \\ &= b^{\frac{mq+np}{nq}} \\ &= b^{\frac{mq}{nq}+\frac{np}{nq}} \\ &= b^{\frac{m}{n}+\frac{p}{q}} \end{aligned}$$

### Exercises 1–4 (6 minutes)

Have students work with a partner or in small groups to complete these exercises. Students are rewriting expressions with rational exponents using the properties presented above. As students work, emphasize that we do not need to write these expressions using radicals since we have just established that we believe that the properties of exponents hold for rational numbers. In the last two exercises, students will have to use their knowledge of radicals to rewrite the answers without exponents.

*Scaffolding:*

- When students get to these exercises, you may need to remind them that it is often easier to rewrite $b^{\frac{m}{n}}$ as $\left(\sqrt[n]{b}\right)^m$ when you are trying to evaluate radical expressions.
- For students who struggle with arithmetic, you can provide a scientific calculator, but be sure to encourage them to show the steps using the properties.

**Exercises 1–4**

**Write each expression in the form $b^{\frac{m}{n}}$. If a numeric expression is a rational number, then write your answer without exponents.**

1. $b^{\frac{2}{3}} \cdot b^{\frac{1}{2}}$

   $b^{\frac{2}{3}+\frac{1}{2}} = b^{\frac{7}{6}}$

2. $\left(b^{-\frac{1}{5}}\right)^{\frac{2}{3}}$

   $b^{-\frac{1}{5}\cdot\frac{2}{3}} = b^{-\frac{2}{15}}$

3. $64^{\frac{1}{3}} \cdot 64^{\frac{3}{2}}$

   $$\begin{aligned} 64^{\frac{1}{3}+\frac{3}{2}} &= 64^{\frac{11}{6}} \\ &= \left(\sqrt[6]{64}\right)^{11} \\ &= 2^{11} \\ &= 2048 \end{aligned}$$

4. $\left(\frac{9^3}{4^2}\right)^{\frac{3}{2}}$

   $$\begin{aligned} \left(\frac{9^3}{4^2}\right)^{\frac{3}{2}} &= \frac{9^{\frac{9}{2}}}{4^3} \\ &= \frac{\left(\sqrt[2]{9}\right)^9}{64} \\ &= \frac{3^9}{64} \\ &= \frac{19683}{64} \end{aligned}$$

## Example 4 (5 minutes)

- We can rewrite radical expressions using properties of exponents. There are other methods for rewriting radical expressions, but this example models using the properties of exponents. Often, textbooks and exams give directions to simplify an expression, which is vague unless we specify what it means. We want students to develop fluency in applying the properties, so the directions here say to rewrite in a specific fashion.

**Example 4**

**Rewrite the radical expression $\sqrt{48x^5y^4z^2}$ so that no perfect square factors remain inside the radical.**

$$\begin{aligned}\sqrt{48\cdot x^5\cdot y^4\cdot z^2} &= (4^2\cdot 3\cdot x^5\cdot y^4\cdot z^2)^{\frac{1}{2}}\\ &= 4^{\frac{2}{2}}\cdot 3^{\frac{1}{2}}\cdot x^{\frac{5}{2}}\cdot y^{\frac{4}{2}}\cdot z^{\frac{2}{2}}\\ &= 4\cdot 3^{\frac{1}{2}}\cdot x^{2+\frac{1}{2}}\cdot y^2\cdot z\\ &= 4x^2y^2z\cdot(3x)^{\frac{1}{2}}\\ &= 4x^2y^2z\sqrt{3x}\end{aligned}$$

- Although this process may seem drawn out, once it has been practiced, most of the steps can be internalized and expressions are quickly rewritten using this technique.

## Exercise 5 (5 minutes)

Students should work individually or in pairs on this exercise.

**Exercise 5**

5. **If $x = 50$, $y = 12$, and $z = 3$, the following expressions are difficult to evaluate without using properties of radicals or exponents (or a calculator). Use the definition of rational exponents and properties of exponents to rewrite each expression in a form where it can be easily evaluated, and then use that exponential expression to find the value.**

   a. $\sqrt{8x^3y^2}$

   $$\begin{aligned}\sqrt{8x^3y^2} &= 2^{\frac{3}{2}}x^{\frac{3}{2}}y^{\frac{2}{2}}\\ &= 2xy\cdot(2x)^{\frac{1}{2}}\end{aligned}$$

   *Evaluating, we get* $2(50)(12)(2\cdot 50)^{\frac{1}{2}} = 100\cdot 12\cdot 10 = 12{,}000$.

   b. $\sqrt[3]{54y^7z^2}$

   $$\begin{aligned}\sqrt[3]{54y^7z^2} &= 27^{\frac{1}{3}}\cdot 2^{\frac{1}{3}}\cdot y^{\frac{7}{3}}\cdot z^{\frac{2}{3}}\\ &= 3y^2\cdot(2yz^2)^{\frac{1}{3}}\end{aligned}$$

   *Evaluating, we get* $3(12)^2(2\cdot 12\cdot 3^2)^{\frac{1}{3}} = 3(144)(216)^{\frac{1}{3}} = 3\cdot 144\cdot 6 = 2592$.

### Exercise 6 (5 minutes)

This exercise will remind students that rational numbers can be represented in decimal form and will give them a chance to work on their numeracy skills. Students should work on this exercise with a partner or in their groups to encourage dialogue and debate. Have a few students demonstrate their results to the entire class. There is more than one possible approach, so when you debrief, try to share different approaches that show varied reasoning. Conclude with one or two strong arguments. Students can confirm their reasoning using a calculator.

**Exercise 6**

6. **Order these numbers from smallest to largest. Explain your reasoning.**

$16^{2.5}$ $9^{3.1}$ $32^{1.2}$

*$16^{2.5}$ is between $256$ and $4,096$. We can rewrite $16^{2.5} = (2^4)^{2.5}$, which is $2^{10}$.*

*$32^{1.2}$ is between $32$ and $1,024$. We can rewrite $32^{1.2} = (2^5)^{1.2}$, which is $2^6$.*

*$9^{3.6}$ is larger than $9^3 = 729$.*

*Thus, $32^{1.2}$ is clearly the smallest number, but we need to determine if $9^{3.6}$ is greater than or less than $1,024$. To do this, we know that $9^{3.1} = 9^{3+0.6} = 9^3 \cdot 9^{0.6}$. This means that $9^{3.6} > 9^3 \cdot 9^{0.5}$, and $9^3 \cdot 9^{0.5} = 729 \cdot 3$, which is greater than $1,024$.*

*Thus, the numbers in order from smallest to largest are $32^{1.2}$, $16^{2.5}$, and $9^{3.6}$.*

### Closing (2 minutes)

Have students summarize the definition and properties of rational exponents and any important ideas from the lesson by creating a list of what they have learned so far about the properties of exponents and radicals. Circulate around the classroom to informally assess understanding. Reinforce the properties of exponents listed below.

**Lesson Summary**

**The properties of exponents developed in Grade 8 for integer exponents extend to rational exponents.**

**That is, for any integers $m$, $n$, $p$, and $q$, with $n > 0$ and $q > 0$, and any real numbers $a$ and $b$ so that $a^{\frac{1}{n}}$, $b^{\frac{1}{n}}$, and $b^{\frac{1}{q}}$ are defined, we have the following properties of exponents:**

$$b^{\frac{m}{n}} \cdot b^{\frac{p}{q}} = b^{\frac{m}{n}+\frac{p}{q}}$$
$$b^{\frac{m}{n}} = \sqrt[n]{b^m}$$
$$\left(b^{\frac{1}{n}}\right)^n = b$$
$$(b^n)^{\frac{1}{n}} = b$$
$$(ab)^{\frac{m}{n}} = a^{\frac{m}{n}} \cdot b^{\frac{m}{n}}$$
$$\left(b^{\frac{m}{n}}\right)^{\frac{p}{q}} = b^{\frac{mp}{nq}}$$
$$b^{-\frac{m}{n}} = \frac{1}{b^{\frac{m}{n}}}.$$

### Exit Ticket (5 minutes)

Name ______________________________ Date ______________

# Lesson 4: Properties of Exponents and Radicals

## Exit Ticket

1. Find the exact value of $9^{\frac{11}{10}} \cdot 9^{\frac{2}{5}}$ without using a calculator.

2. Justify that $\sqrt[3]{8} \cdot \sqrt[3]{8} = \sqrt{16}$ using the properties of exponents in at least two different ways.

## Exit Ticket Sample Solutions

1. Find the exact value of $9^{\frac{11}{10}} \cdot 9^{\frac{2}{5}}$ without using a calculator.

$$\begin{aligned} 9^{\frac{11}{10}} \cdot 9^{\frac{2}{5}} &= 9^{\frac{11}{10}+\frac{2}{5}} \\ &= 9^{\frac{15}{10}} \\ &= 9^{\frac{3}{2}} \\ &= \left(\sqrt[2]{9}\right)^3 \\ &= 27 \end{aligned}$$

2. Justify that $\sqrt[3]{8} \cdot \sqrt[3]{8} = \sqrt{16}$ using the properties of exponents in at least two different ways.

$$\begin{aligned} 8^{\frac{1}{3}} \cdot 8^{\frac{1}{3}} &= 8^{\frac{2}{3}} \\ &= (2^3)^{\frac{2}{3}} \\ &= 2^2 \\ &= 2^{\frac{4}{2}} \\ &= (2^4)^{\frac{1}{2}} \\ &= \sqrt{16} \end{aligned}$$

$$\begin{aligned} 16^{\frac{1}{2}} &= (4 \cdot 4)^{\frac{1}{2}} \\ &= 4^{\frac{1}{2}} \cdot 4^{\frac{1}{2}} \\ &= 2 \cdot 2 \\ &= 8^{\frac{1}{3}} \cdot 8^{\frac{1}{3}} \\ &= \sqrt[3]{8} \cdot \sqrt[3]{8} \end{aligned}$$

## Problem Set Sample Solutions

1. Evaluate each expression if $a = 27$ and $b = 64$.

a. $\sqrt[3]{a}\sqrt{b}$

$\sqrt[3]{27} \cdot \sqrt{64} = 3 \cdot 8 = 24$

b. $\left(3\sqrt[3]{a}\sqrt{b}\right)^2$

$(3 \cdot 3 \cdot 8)^2 = 5184$

c. $\left(\sqrt[3]{a} + 2\sqrt{b}\right)^2$

$(3 + 2 \cdot 8)^2 = 361$

d. $a^{-\frac{2}{3}} + b^{\frac{3}{2}}$

$\dfrac{1}{\left(\sqrt[3]{27}\right)^2} + \left(\sqrt{64}\right)^3 = \dfrac{1}{9} + 512 = 512\dfrac{1}{9}$

e. $\left(a^{-\frac{2}{3}} \cdot b^{\frac{3}{2}}\right)^{-1}$

$\left(\dfrac{1}{9} \cdot 512\right)^{-1} = \dfrac{9}{512}$

f. $\left(a^{-\frac{2}{3}} - \frac{1}{8}b^{\frac{3}{2}}\right)^{-1}$

$\left(\dfrac{1}{9} - \dfrac{1}{8} \cdot 512\right)^{-1} = \left(-\dfrac{575}{9}\right)^{-1} = -\dfrac{9}{575}$

2. Rewrite each expression so that each term is in the form $kx^n$, where $k$ is a real number, $x$ is a positive real number, and $n$ is a rational number.

a. $x^{-\frac{2}{3}} \cdot x^{\frac{1}{3}}$

*$x^{-\frac{1}{3}}$*

b. $2x^{\frac{1}{2}} \cdot 4x^{-\frac{5}{2}}$

*$8x^{-2}$*

c. $\dfrac{10x^{\frac{1}{3}}}{2x^2}$

*$5x^{-\frac{5}{3}}$*

d. $\left(3x^{\frac{1}{4}}\right)^{-2}$

*$\frac{1}{9}x^{-\frac{1}{2}}$*

e. $x^{\frac{1}{2}}\left(2x^2 - \frac{4}{x}\right)$

*$2x^{\frac{5}{2}} - 4x^{-\frac{1}{2}}$*

f. $\sqrt[3]{\dfrac{27}{x^6}}$

*$3x^{-2}$*

g. $\sqrt[3]{x} \cdot \sqrt[3]{-8x^2} \cdot \sqrt[3]{27x^4}$

*$-6x^{\frac{7}{3}}$*

h. $\dfrac{2x^4 - x^2 - 3x}{\sqrt{x}}$

*$2x^{\frac{7}{2}} - x^{\frac{3}{2}} - 3x^{\frac{1}{2}}$*

i. $\dfrac{\sqrt{x} - 2x^{-3}}{4x^2}$

*$\frac{1}{4}x^{-\frac{3}{2}} - \frac{1}{2}x^{-5}$*

3. Show that $\left(\sqrt{x} + \sqrt{y}\right)^2$ is not equal to $x^1 + y^1$ when $x = 9$ and $y = 16$.

*When $x = 9$ and $y = 16$, the two expressions are $\left(\sqrt{9} + \sqrt{16}\right)^2$ and $9 + 16$. The first expression simplifies to $49$, and the second simplifies to $25$. The two expressions are not equal.*

4. Show that $\left(x^{\frac{1}{2}} + y^{\frac{1}{2}}\right)^{-1}$ is not equal to $\dfrac{1}{x^{\frac{1}{2}}} + \dfrac{1}{y^{\frac{1}{2}}}$ when $x = 9$ and $y = 16$.

*When $x = 9$ and $y = 16$, the two expressions are $\left(\sqrt{9} + \sqrt{16}\right)^{-1}$ and $\frac{1}{\sqrt{9}} + \frac{1}{\sqrt{16}}$. The first expression is $\frac{1}{7}$, and the second one is $\frac{1}{3} + \frac{1}{4} = \frac{7}{12}$. The two expressions are not equal.*

5. From these numbers, select (a) one that is negative, (b) one that is irrational, (c) one that is not a real number, and (d) one that is a perfect square:

$$3^{\frac{1}{2}} \cdot 9^{\frac{1}{2}}, 27^{\frac{1}{3}} \cdot 144^{\frac{1}{2}}, 64^{\frac{1}{3}} - 64^{\frac{2}{3}}, \text{ and } \left(4^{-\frac{1}{2}} - 4^{\frac{1}{2}}\right)^{\frac{1}{2}}.$$

*The first number, $3^{\frac{1}{2}} \cdot 9^{\frac{1}{2}}$, is irrational, the second number, $27^{\frac{1}{3}} \cdot 144^{\frac{1}{2}}$, is a perfect square, the third number, $64^{\frac{1}{3}} - 64^{\frac{2}{3}}$, is negative, and the last number, $\left(4^{-\frac{1}{2}} - 4^{\frac{1}{2}}\right)^{\frac{1}{2}}$, is not a real number.*

6. Show that the expression $2^n \cdot 4^{n+1} \cdot \left(\frac{1}{8}\right)^n$ is equal to $4$.

   $2^n \cdot 2^{2n+2} \cdot 2^{-3n} = 2^2 = 4$

7. Express each answer as a power of $10$.

   a. Multiply $10^n$ by $10$.

      $10^n \cdot 10 = 10^{n+1}$

   b. Multiply $\sqrt{10}$ by $10^n$.

      $10^{\frac{1}{2}} \cdot 10^n = 10^{\frac{1}{2}+n}$

   c. Square $10^n$.

      $(10^n)^2 = 10^{2n}$

   d. Divide $100 \cdot 10^n$ by $10^{2n}$.

      $\frac{100 \cdot 10^n}{10^{2n}} = 10^{2+n-2n} = 10^{2-n}$

   e. Show that $10^n = 11 \cdot 10^n - 10^{n+1}$.

      $$\begin{aligned} 11 \cdot 10^n - 10^{n+1} &= 11 \cdot 10^n - 10 \cdot 10^n \\ &= 10^n(11 - 10) \\ &= 10^n \cdot 1 \\ &= 10^n \end{aligned}$$

8. Rewrite each of the following radical expressions as an equivalent exponential expression in which each variable occurs no more than once.

   a. $\sqrt{8x^2y}$

      $$\begin{aligned} \sqrt{8x^2y} &= 2^{\frac{2}{2}}x^{\frac{2}{2}}(2y)^{\frac{1}{2}} \\ &= 2x \cdot (2y)^{\frac{1}{2}} \\ &= 2^{\frac{3}{2}}\, x\, y^{\frac{1}{2}} \end{aligned}$$

   b. $\sqrt[5]{96x^3y^{15}z^6}$

      $$\begin{aligned} \sqrt[5]{96x^3y^{15}z^6} &= (32 \cdot 3 \cdot x^3 \cdot y^{15} \cdot z^6)^{\frac{1}{5}} \\ &= 32^{\frac{1}{5}} \cdot 3^{\frac{1}{5}} \cdot x^{\frac{3}{5}} \cdot y^{\frac{15}{5}} \cdot z^{\frac{6}{5}} \\ &= 2 \cdot 3^{\frac{1}{5}} \cdot x^{\frac{3}{5}} \cdot y^3 \cdot z^{\frac{6}{5}} \end{aligned}$$

9. Use properties of exponents to find two integers that are upper and lower estimates of the value of $4^{1.6}$.

   $$4^{1.5} < 4^{1.6} < 4^2$$

   $4^{1.5} = 2^3 = 8$ *and* $4^2 = 16$, *so* $8 < 4^{1.6} < 16$

10. Use properties of exponents to find two integers that are upper and lower estimates of the value of $8^{2.3}$.

$$8^2 < 8^{2.3} < 8^{2+\frac{1}{3}}$$

$8^2 = 64$ *and* $8^{\frac{1}{3}} = 2$, *so* $8^{2+\frac{1}{3}} = 8^2 \cdot 8^{\frac{1}{3}} = 128$. *Thus,* $64 < 8^{2.3} < 128$.

11. Kepler's third law of planetary motion relates the average distance, $a$, of a planet from the Sun to the time, $t$, it takes the planet to complete one full orbit around the Sun according to the equation $t^2 = a^3$. When the time, $t$, is measured in Earth years, the distance, $a$, is measured in astronomical units (AU). (One AU is equal to the average distance from Earth to the Sun.)

a. Find an equation for $t$ in terms of $a$ and an equation for $a$ in terms of $t$.

$$t^2 = a^3$$
$$t = a^{\frac{3}{2}}$$
$$a = t^{\frac{2}{3}}$$

b. Venus takes about $0.616$ Earth years to orbit the Sun. What is its average distance from the Sun?

$a = (0.616)^{\frac{2}{3}} \approx 0.724 \text{ AU}$

c. Mercury is an average distance of $0.387$ AU from the Sun. About how long is its orbit in Earth years?

$t = (0.387 \text{ AU})^{\frac{3}{2}} \approx 0.241 \text{ year}$

# Lesson 5: Irrational Exponents—What Are $2^{\sqrt{2}}$ and $2^{\pi}$?

## Student Outcomes

- Students approximate the value of quantities that involve positive irrational exponents.
- Students extend the domain of the function $f(x) = b^x$ for positive real numbers $b$ to all real numbers.

## Lesson Notes

Our goal today is to define 2 to an irrational power. We already have a definition for 2 to a rational power $\frac{p}{q}$: $2^{\frac{p}{q}} = \sqrt[q]{2^p}$, but irrational numbers cannot be written as "an integer divided by an integer." By defining 2 to an irrational power, we will be able to state definitively that for any positive real number $b > 0$, the domain of the function $f(x) = b^x$ is all real numbers. This is an important result and one that is necessary for us to proceed to the study of logarithms. The lesson provides a new way to reinforce standard **8.NS.A.2** when students determine a recursive process for calculation from a context (**F-BF.A.1a**) when they use rational approximations of irrational numbers to approximate first $\sqrt{2}$ and then $2^{\sqrt{2}}$. Extending rational exponents to real exponents is an application of **N-RN.A.1**. The foundational work done in this lesson with exponential expressions will be extended to logarithms in later lessons so that logarithmic functions in base 2, 10, and $e$ are well-defined and can be used to solve exponential equations. Understanding the domain of exponential functions will also allow students to correctly graph exponential and logarithmic functions in Topic C. The work done in Lesson 5 will also help demystify irrational numbers, which will ease the introduction to Euler's number, $e$, in Lesson 6.

## Classwork

### Opening (5 minutes)

Use the Opening to recall the definitions of rational and irrational numbers and solicit examples and characteristics from the class. Randomly select students to explain what they know about rational and irrational numbers. Then, make a list including examples and characteristics of both. Alternatively, have students give you rational and irrational numbers, make a class list, and then have students generalize characteristics of rational and irrational numbers in their notebooks. Rational and irrational numbers along with some characteristics and examples are described below.

RATIONAL NUMBER: A *rational number* is a number that can be represented as $\frac{p}{q}$ where $p$ and $q$ are integers with $q \neq 0$.

IRRATIONAL NUMBER: An *irrational number* is a real number that cannot be represented as $\frac{p}{q}$ for any integers $p$ and $q$ with $q \neq 0$.

- What are some characteristics of rational numbers?
  - *A rational number can be represented as a finite or repeating decimal; that is, a rational number can be written as a fraction.*
- What are some characteristics of irrational numbers?
  - *An irrational number cannot be represented as a finite or repeating decimal, so it must be represented symbolically or as an infinite, nonrepeating decimal.*

- What are some examples of irrational numbers?
  - $\sqrt{2}, \pi, \sqrt[3]{17}$
- We usually assume that the rules we develop for rational numbers hold true for irrational numbers, but what could something like $2^{\sqrt{2}}$ or $2^{\pi}$ mean?
  - *Solicit ideas from the class. Students may consider numbers like this to be between rational exponents or "filling the gaps" from rational exponents.*
- Let's find out more about exponents raised to irrational powers and how we can get a handle on their values.

### Exercise 1 (8 minutes)

Have students work on the following exercises independently or in pairs. Students will need to use calculators. After students finish, debrief them with the questions that follow the exercises.

**Exercise 1**

a. **Write the following finite decimals as fractions (you do not need to reduce to lowest terms).**

$$\mathbf{1, \quad 1.4, \quad 1.41, \quad 1.414, \quad 1.4142, \quad 1.41421}$$

$$1.4 = \frac{14}{10}$$
$$1.41 = \frac{141}{100}$$
$$1.414 = \frac{1414}{1000}$$
$$1.4142 = \frac{14142}{10000}$$
$$1.41421 = \frac{141421}{100000}$$

b. **Write $2^{1.4}$, $2^{1.41}$, $2^{1.414}$, and $2^{1.4142}$ in radical form ($\sqrt[n]{2^m}$).**

$$2^{1.4} = 2^{14/10} = \sqrt[10]{2^{14}}$$
$$2^{1.41} = 2^{141/100} = \sqrt[100]{2^{141}}$$
$$2^{1.414} = 2^{1414/1000} = \sqrt[1000]{2^{1414}}$$
$$2^{1.4142} = 2^{14142/10000} = \sqrt[10000]{2^{14142}}$$
$$2^{1.41421} = 2^{141421/100000} = \sqrt[100000]{2^{141421}}$$

c. **Compute a decimal approximation to 5 decimal places of the radicals you found in part (b) using your calculator. For each approximation, underline the digits that are also in the previous approximation, starting with $2.00000$ done for you below. What do you notice?**

$$2^1 = 2 = 2.00000$$

$2^{1.4} = \sqrt[10]{2^{14}} \approx \underline{2}.63901$

$2^{1.41} = \sqrt[100]{2^{141}} \approx \underline{2.6}5737$

$2^{1.414} = \sqrt[1000]{2^{1414}} \approx \underline{2.66}474$

$2^{1.4142} = \sqrt[10000]{2^{14142}} \approx \underline{2.665}11$

$2^{1.41421} = \sqrt[100000]{2^{141421}} \approx \underline{2.6651}4$

***More and more of the digits are the same.***

Note to teacher: Students cannot find $2^{1414}$ on most calculators due to the number being $426$ digits long. They will need to calculate $\left(\sqrt[1000]{2}\right)^{1414}$ instead. At this point it may be a good time to switch to using the decimal approximation within the exponent, reminding students that the calculator is evaluating the decimal by using the radical form, that is, $b^{\frac{m}{n}} = \sqrt[n]{b^m}$. Ideally, a student will suggest using the decimal exponent first. If roots are used, make sure that the root is taken before the exponent for large exponents. Examples and possible solutions throughout the lesson assume that roots are used so the true meaning of rational exponents is emphasized.

- Why are more of the digits the same? How are the exponents in each power changing?
  - *A new digit is included in the exponent each time:* $1.4$, $1.4\underline{1}$, $1.41\underline{4}$, $1.414\underline{2}$, $1.4142\underline{1}$.
- .8 If we kept including more digits, what do you conjecture will happen to the decimal approximations?
  - *A greater and greater number of digits in each approximation would remain the same.*
- Let's see!

## Example 1 (6 minutes)

Students should already be aware that rational exponents are defined using roots and exponents.

Write a decimal approximation for $2^{1.4142135}$.

$2^{1.4142135}$ is the $10{,}000{,}000$th root of $2^{14142135}$

Remember to take the root first. We get

$$2^{1.4142135} \approx \underline{2.66514}.$$

- Can anyone tell the class what the exponents $1.4, 1.41, 1.414, \ldots$ approximate?

Hopefully, one student will say $\sqrt{2}$, but if not, ask them to find $\sqrt{2}$ on their calculator and ask again.

- Yes, $\sqrt{2} \approx 1.414213562$.

  The goal of this lesson is to find a meaning for $2^{\sqrt{2}}$. We now know enough to discuss both the problem and solution to defining $2$ to an irrational power such as $\sqrt{2}$.

- First, the problem: Each time we took a better finite decimal approximation of the irrational number $2^{\sqrt{2}}$, we needed to take a greater $n^{\text{th}}$ root. However, an irrational number has an infinite number of digits in its decimal expansion. We cannot take an $\infty^{\text{th}}$ root! In particular, while we have always assumed $2^{\sqrt{2}}$ and $2^{\pi}$ existed (because when we show the graph of $f(x) = 2^x$, we drew a solid curve—not one with "holes" at $x = \sqrt{2}, \pi$, etc.), we do not as of yet have a way to define what $2^{\sqrt{2}}$ and $2^{\pi}$ really are.
- Fortunately, our beginning exercise suggests a solution using a limit process (much the way we defined the area of a circle in Geometry Module 3 in terms of limits).
- Let $a_k$ stand for the term of the sequence of finite decimal approximations of $\sqrt{2}$ with $k$ digits after the decimal point:

$$\{1, 1.4, 1.41, 1.414, 1.4142, 1.41421, 1.414213, 1.4142135, \ldots\},$$

  and label these as $a_0 = 1,\ a_1 = 1.4,\ a_2 = 1.41,\ a_3 = 1.414$. Then define $2^{\sqrt{2}}$ to be the limit of the values of $2^{a_k}$. Thus,

$$2^{a_k} \to 2^{\sqrt{2}} \text{ as } k \to \infty.$$

The important point to make to students is that each $2^{a_k}$ can be computed since each $a_k$ is a rational number and therefore has a well-defined value in terms of $n^{\text{th}}$ roots.

This is how calculators and computers are programmed to compute approximations of $2^{\sqrt{2}}$. Try it: The calculator says that $2^{\sqrt{2}} \approx 2.66514414$.

## Exercise 2 (5 minutes)

Students should attempt the following exercise independently or in pairs. After the exercise, use the Discussion to debrief and informally assess understanding.

**Exercise 2**

a. **Write six terms of a sequence that a calculator can use to approximate $2^{\pi}$. (Hint: $\pi = 3.14159\ldots$)**

   $\{2^3, 2^{3.1}, 2^{3.14}, 2^{3.141}, 2^{3.1415}, 2^{3.14159}, \ldots\}$

b. **Compute $2^{3.14} = \sqrt[100]{2^{314}}$ and $2^{\pi}$ on your calculator. In which digit do they start to differ?**

   $2^{3.14} = \sqrt[100]{2^{314}} \approx 8.81524$
   $2^{\pi} \approx 8.82497$
   ***They start to differ in the hundredths place.***

c. **How could you improve the accuracy of your estimate of $2^{\pi}$?**

   ***Include more digits of the decimal approximation of $\pi$ in the exponent.***

*Scaffolding:*

- Have advanced students give the most accurate estimate they can for part (b). Most calculators can provide an additional three to four decimal places of $\pi$. For reference, $\pi \approx 3.14159265358979323846$.
- Another option for advanced students is to discuss the sequence of upper bounds of $\pi$ $\{4, 3.2, 3.15, 3.142, 3.1416, \ldots\}$ and whether this will provide an accurate estimate of $2^{\pi}$.

**Discussion (10 minutes)**

- Why does the sequence $2^3, 2^{3.1}, 2^{3.14}, 2^{3.141}, 2^{3.1415}, \ldots$ get closer and closer to $2^\pi$?

Allow students to make some conjectures, but be sure to go through the reasoning below.

- We can trap $2^\pi$ in smaller and smaller intervals, each one contained in the previous interval.

Write the following incomplete inequalities on the board and ask students to help you complete them before continuing. Mention that in this process, we are *squeezing* $\pi$ between two rational numbers that are each getting closer and closer to the value of $\pi$.

$$3 < \pi < 4$$
$$3.1 < \pi < ? \qquad 3.2$$
$$3.14 < \pi < ? \qquad 3.15$$
$$3.141 < \pi < ? \qquad 3.142$$
$$3.1415 < \pi < ? \qquad 3.1416$$
$$\vdots$$

- Since $3 < \pi < 4$, and the function $f(x) = 2^x$ increases, we know that $2^3 < \pi < 2^4$. Likewise, we can use the smaller intervals that contain $\pi$ to find smaller intervals that contain $2^\pi$. In this way, we can squeeze $2^\pi$ between rational powers of 2.

Now, have students use calculators to estimate the endpoints of each interval created by the upper and lower estimates of the values of $2^\pi$ and write the numerical approximations of each interval on the board so students can see the endpoints of the intervals getting closer together, squeezing the value of $2^\pi$ between them. Record values to four decimal places.

| | *Decimal Form* |
|---|---|
| $2^3 < 2^\pi < 2^4$ | $8.0000 < 2^\pi < 16.0000$ |
| $2^{3.1} < 2^\pi < 2^{3.2}$ | $8.5742 < 2^\pi < 9.1896$ |
| $2^{3.14} < 2^\pi < 2^{3.15}$ | $8.8152 < 2^\pi < 8.8766$ |
| $2^{3.141} < 2^\pi < 2^{3.142}$ | $8.8214 < 2^\pi < 8.8275$ |
| $2^{3.1415} < 2^\pi < 2^{3.1416}$ | $8.8244 < 2^\pi < 8.8250$ |
| $\vdots$ | $\vdots$ |

- What is the approximate value of $2^\pi$? How many digits of this number do we know?
  - *Because our upper and lower estimates agree to two decimal places, our best approximation is* $2^\pi \approx 8.82$.
- How could we get a more accurate estimate of $2^\pi$?
  - *Use more and more digits of $\pi$ as exponents.*

- As the exponents get closer to the value of $\pi$, what happens to the size of the interval?
  - *The intervals get smaller; the endpoints of the interval get closer together.*
- What does every interval share in common?
  - *Every interval contains* $2^{\pi}$.
- The only number that is guaranteed to be contained in every interval is $2^{\pi}$. (Emphasize this fact.)
- There was nothing special about our choice of 2 in this discussion, or $\sqrt{2}$ or $\pi$. In fact, with a little more work, we could define $\pi^{\sqrt{2}}$ using the same ideas.

## Closing (6 minutes)

Ask students to respond to the following questions either in writing or with a partner. Use this as an opportunity to informally assess understanding. The summative point of the lesson is that the domain of an exponential function $f(x) = b^x$ is *all real numbers*, so emphasize the final question below.

- For any positive real number $b > 0$ and any rational number $r$, how do we define $b^r$?
  - *If $r$ is rational, then $r = \frac{p}{q}$ for some integers $p$ and $q$. Then $b^r = \sqrt[q]{b^p}$. For example, $5^{\frac{2}{3}} = \sqrt[3]{5^2}$.*
- For any positive real number $b > 0$ and any irrational number $r$, how do we define $b^r$?
  - *If $r$ is irrational, $b^r$ is the limit of the values $a^{b_n}$ where $b_n$ is the finite decimal approximation of $b$ to $n$ decimal places.*
- If $b$ is any positive real number, then consider the function $f(x) = b^x$. How is $f(x)$ defined if $x$ is a rational number?
  - *If $x$ is a rational number, then there are integers $p$ and $q$ so that $x = \frac{p}{q}$. Then $f(x) = b^{\frac{p}{q}} = \sqrt[q]{b^p}$.*
- How is $f(x)$ defined if $x$ is an irrational number?
  - *If $x$ is an irrational number, we find a sequence of rational numbers $\{a_0, a_1, a_2, \ldots\}$ that gets closer and closer to $x$. Then the sequence $\{b^{a_0}, b^{a_1}, b^{a_2}, \ldots\}$ approaches $f(x)$.*
- What is the domain of the exponential function $f(x) = b^x$?
  - *The domain of the function $f(x) = b^x$ is all real numbers.*

## Exit Ticket (5 minutes)

Name ______________________________ Date ______________

# Lesson 5: Irrational Exponents—What Are $2^{\sqrt{2}}$ and $2^{\pi}$?

## Exit Ticket

Use the process outlined in the lesson to approximate the number $2^{\sqrt{3}}$. Use the approximation $\sqrt{3} \approx 1.7320508$.

a. Find a sequence of five intervals that contain $\sqrt{3}$ whose endpoints get successively closer to $\sqrt{3}$.

b. Find a sequence of five intervals that contain $2^{\sqrt{3}}$ whose endpoints get successively closer to $2^{\sqrt{3}}$. Write your intervals in the form $2^r < 2^{\sqrt{3}} < 2^s$ for rational numbers $r$ and $s$.

c. Use your calculator to find approximations to four decimal places of the endpoints of the intervals in part (b).

d. Based on your work in part (c) what is your best estimate of the value of $2^{\sqrt{3}}$?

## Exit Ticket Sample Solutions

Use the process outlined in the lesson to approximate the number $2^{\sqrt{3}}$. Use the approximation $\sqrt{3} \approx 1.7320508$.

a. Find a sequence of five intervals that contain $\sqrt{3}$ whose endpoints get successively closer to $\sqrt{3}$.

$$1 < \sqrt{3} < 2$$
$$1.7 < \sqrt{3} < 1.8$$
$$1.73 < \sqrt{3} < 1.74$$
$$1.732 < \sqrt{3} < 1.733$$
$$1.7320 < \sqrt{3} < 1.7321$$

b. Find a sequence of five intervals that contain $2^{\sqrt{3}}$ whose endpoints get successively closer to $2^{\sqrt{3}}$. Write your intervals in the form $2^r < 2^{\sqrt{3}} < 2^s$ for rational numbers $r$ and $s$.

$$2^1 < 2^{\sqrt{3}} < 2^2$$
$$2^{1.7} < 2^{\sqrt{3}} < 2^{1.8}$$
$$2^{1.73} < 2^{\sqrt{3}} < 2^{1.74}$$
$$2^{1.732} < 2^{\sqrt{3}} < 2^{1.733}$$
$$2^{1.7320} < 2^{\sqrt{3}} < 2^{1.7321}$$

c. Use your calculator to find approximations to four decimal places of the endpoints of the intervals in part (b).

$$2.0000 < 2^{\sqrt{3}} < 4.0000$$
$$3.2490 < 2^{\sqrt{3}} < 3.4822$$
$$3.3173 < 2^{\sqrt{3}} < 3.3404$$
$$3.3219 < 2^{\sqrt{3}} < 3.3242$$
$$3.3219 < 2^{\sqrt{3}} < 3.3221$$

d. Based on your work in part (c) what is your best estimate of the value of $2^{\sqrt{3}}$?

$$2^{\sqrt{3}} \approx 3.322$$

## Problem Set Sample Solutions

1. Is it possible for a number to be both rational and irrational?

   *No. Either the number can be written as $\frac{p}{q}$ for integers $p$ and $q$ or it cannot. If it can, the number is rational. If it cannot, the number is irrational.*

2. Use properties of exponents to rewrite the following expressions as a number or an exponential expression with only one exponent.

   a. $\left(2^{\sqrt{3}}\right)^{\sqrt{3}}$ $\quad = 8$

b. $\left(\sqrt{2}^{\sqrt{2}}\right)^{\sqrt{2}} = 2$

c. $\left(3^{1+\sqrt{5}}\right)^{1-\sqrt{5}} = \frac{1}{81}$

d. $3^{\frac{1+\sqrt{5}}{2}} \cdot 3^{\frac{1-\sqrt{5}}{2}} = 3$

e. $3^{\frac{1+\sqrt{5}}{2}} \div 3^{\frac{1-\sqrt{5}}{2}} = 3^{\sqrt{5}}$

f. $3^{2\cos^2(x)} \cdot 3^{2\sin^2(x)} = 9$

3.

a. **Between what two integer powers of 2 does $2^{\sqrt{5}}$ lie?**

$2^2 < 2^{\sqrt{5}} < 2^3$

b. **Between what two integer powers of 3 does $3^{\sqrt{10}}$ lie?**

$3^3 < 3^{\sqrt{10}} < 3^4$

c. **Between what two integer powers of 5 does $5^{\sqrt{3}}$ lie?**

$5^1 < 5^{\sqrt{3}} < 5^2$

4. **Use the process outlined in the lesson to approximate the number $2^{\sqrt{5}}$. Use the approximation $\sqrt{5} \approx 2.23606798$.**

a. **Find a sequence of five intervals that contain $\sqrt{5}$ whose endpoints get successively closer to $\sqrt{5}$.**

$$2 < \sqrt{5} < 3$$
$$2.2 < \sqrt{5} < 2.3$$
$$2.23 < \sqrt{5} < 2.24$$
$$2.236 < \sqrt{5} < 2.237$$
$$2.2360 < \sqrt{5} < 2.2361$$

b. **Find a sequence of five intervals that contain $2^{\sqrt{5}}$ whose endpoints get successively closer to $2^{\sqrt{5}}$. Write your intervals in the form $2^r < 2^{\sqrt{5}} < 2^s$ for rational numbers $r$ and $s$.**

$$2^2 < 2^{\sqrt{5}} < 2^3$$
$$2^{2.2} < 2^{\sqrt{5}} < 2^{2.3}$$
$$2^{2.23} < 2^{\sqrt{5}} < 2^{2.24}$$
$$2^{2.236} < 2^{\sqrt{5}} < 2^{2.237}$$
$$2^{2.2360} < 2^{\sqrt{5}} < 2^{2.2361}$$

c. **Use your calculator to find approximations to four decimal places of the endpoints of the intervals in part (b).**

$$4.0000 < 2^{\sqrt{5}} < 8.0000$$
$$4.5948 < 2^{\sqrt{5}} < 4.9246$$
$$4.6913 < 2^{\sqrt{5}} < 4.7240$$
$$4.7109 < 2^{\sqrt{5}} < 4.7142$$
$$4.7109 < 2^{\sqrt{5}} < 4.7112$$

d. **Based on your work in part (c), what is your best estimate of the value of $2^{\sqrt{5}}$?**

$$2^{\sqrt{5}} \approx 4.711$$

e. **Can we tell if $2^{\sqrt{5}}$ is rational or irrational? Why or why not?**

***No. We do not have enough information to determine whether $2^{\sqrt{5}}$ has a repeated pattern in its decimal representation or not.***

5. **Use the process outlined in the lesson to approximate the number $3^{\sqrt{10}}$. Use the approximation $\sqrt{10} \approx 3.1622777$.**

a. **Find a sequence of five intervals that contain $3^{\sqrt{10}}$ whose endpoints get successively closer to $3^{\sqrt{10}}$. Write your intervals in the form $3^r < 3^{\sqrt{10}} < 3^s$ for rational numbers $r$ and $s$.**

$$3^3 < 3^{\sqrt{10}} < 3^4$$
$$3^{3.1} < 3^{\sqrt{10}} < 3^{3.2}$$
$$3^{3.16} < 3^{\sqrt{10}} < 3^{3.17}$$
$$3^{3.162} < 3^{\sqrt{10}} < 3^{3.163}$$
$$3^{3.1622} < 3^{\sqrt{10}} < 3^{3.1623}$$

b. **Use your calculator to find approximations to four decimal places of the endpoints of the intervals in part (a).**

$$9.0000 < 3^{\sqrt{10}} < 81.0000$$
$$30.1353 < 3^{\sqrt{10}} < 33.6347$$
$$32.1887 < 3^{\sqrt{10}} < 32.5443$$
$$32.2595 < 3^{\sqrt{10}} < 32.2949$$
$$32.2666 < 3^{\sqrt{10}} < 32.2701$$

c. **Based on your work in part (b), what is your best estimate of the value of $3^{\sqrt{10}}$?**

$$3^{\sqrt{10}} \approx 32.27$$

6. **Use the process outlined in the lesson to approximate the number $5^{\sqrt{7}}$. Use the approximation $\sqrt{7} \approx 2.64575131$.**

a. **Find a sequence of seven intervals that contain $5^{\sqrt{7}}$ whose endpoints get successively closer to $5^{\sqrt{7}}$. Write your intervals in the form $5^r < 5^{\sqrt{7}} < 5^s$ for rational numbers $r$ and $s$.**

$$5^2 < 5^{\sqrt{7}} < 5^3$$
$$5^{2.6} < 5^{\sqrt{7}} < 5^{2.7}$$
$$5^{2.64} < 5^{\sqrt{7}} < 5^{2.65}$$
$$5^{2.645} < 5^{\sqrt{7}} < 5^{2.646}$$
$$5^{2.6457} < 5^{\sqrt{7}} < 5^{2.6458}$$
$$5^{2.64575} < 5^{\sqrt{7}} < 5^{2.64576}$$
$$5^{2.645751} < 5^{\sqrt{7}} < 5^{2.645752}$$

b. **Use your calculator to find approximations to four decimal places of the endpoints of the intervals in part (a).**

$$25.0000 < 5^{\sqrt{7}} < 125.0000$$
$$65.6632 < 5^{\sqrt{7}} < 77.1292$$
$$70.0295 < 5^{\sqrt{7}} < 71.1657$$
$$70.5953 < 5^{\sqrt{7}} < 70.7090$$
$$70.6749 < 5^{\sqrt{7}} < 70.6862$$
$$70.6805 < 5^{\sqrt{7}} < 70.6817$$
$$70.6807 < 5^{\sqrt{7}} < 70.6808$$

c. **Based on your work in part (b), what is your best estimate of the value of $5^{\sqrt{7}}$?**

$$5^{\sqrt{7}} \approx 70.681$$

7. **Can the value of an irrational number raised to an irrational power ever be rational?**

*Yes. For instance, in part (b) above, $\sqrt{2}$ is irrational and the number $\sqrt{2}^{\sqrt{2}}$ is either irrational or rational. If it is rational, then this is an example of an irrational number raised to an irrational power that is rational. If it is not, then $\sqrt{2}^{\sqrt{2}}$ is irrational and part (b) is an example of an irrational number raised to an irrational power that is rational.*

# Lesson 6: Euler's Number, $e$

## Student Outcomes

- Students write an exponential function that represents the amount of water in a tank after $t$ seconds if the height of the water doubles every $10$ seconds.
- Students discover Euler's number $e$ by numerically approaching the constant for which the height of water in a tank equals the rate of change of the height of the water in the tank.
- Students calculate the average rate of change of a function.

## Lesson Notes

Leonhard Euler (pronounced "Oiler"), 1707–1783, was a prolific Swiss mathematician and physicist who made many important contributions to mathematics such as much of our modern terminology and notation, including function notation and popularizing the use of $\pi$ to represent the circumference of a circle divided by its diameter. Euler also discovered many properties of the irrational number $e$, which is now known as Euler's number. Euler's number naturally occurs in various applications, and a comparison can be made to $\pi$, which also occurs naturally. During the lesson, students determine an explicit expression for the height of water in a water tank from its context (**F-BF.A.1a**) and calculate the average rate of change over smaller and smaller intervals to create a sequence that converges to $e$ (**F-IF.B.6**). It is important to stress that the water tank exploration is a way to *define* $e$. Yes it is remarkable, but when students discover it, the teacher's reaction should not be "Ta da! It's magic!" Instead, the teacher should stress that students have defined this special constant (similar to how $\pi$ is defined as the ratio of any circle's circumference to its diameter) that will be used extensively in the near future and occurs in many different applications.

## Classwork

### Exercises 1–3 (8 minutes)

In these exercises, students find exponential equations that model the increasing height of water in a cylindrical tank as it doubles over a fixed time interval. These preliminary exercises will lead to the discovery of Euler's number, $e$, at the end of the lesson. As a demonstration, show students the $47$-second video in which the height of water in a tank doubles repeatedly until it fills the tank completely; note how long it takes for the height to appear to change at all. Although this situation is contrived, it provides a good visual representation of the power of exponential growth. This is a good time to discuss constraints and how quantities cannot realistically increase exponentially without bound due to physical constraints to the growth. In this case, the water tank has a finite volume and there is only a finite amount of water on the planet. For population growth, the main constraint is availability of such resources as food and land.

After watching the video, students may work individually or in pairs. Point out to the students that the growth shown in the video happened much more quickly than it will in the problems below, but the underlying concept is the same. Students should be prepared to share their solutions with the class.

Exercises 1–3

1. Assume that there is initially $1$ cm of water in the tank and the height of the water doubles every $10$ seconds. Write an equation that could be used to calculate the height $H(t)$ of the water in the tank at any time $t$.

   *The height of the water at time t seconds can be modeled by* $H(t) = 2^{t/10}$.

2. How would the equation in Exercise 1 change if …

   a. The initial depth of water in the tank was $2$ cm?

      $H(t) = 2 \cdot 2^{t/10}$

   b. The initial depth of water in the tank was $\frac{1}{2}$ cm?

      $H(t) = \frac{1}{2} \cdot 2^{t/10}$

   c. The initial depth of water in the tank was $10$ cm?

      $H(t) = 10 \cdot 2^{t/10}$

   d. The initial depth of water in the tank was $A$ cm, for some positive real number $A$?

      $H(t) = A \cdot 2^{t/10}$

3. How would the equation in Exercise 2, part (d) change if …

   a. The height tripled every ten seconds?

      $H(t) = A \cdot 3^{t/10}$

   b. The height doubled every five seconds?

      $H(t) = A \cdot 2^{t/5}$

   c. The height quadrupled every second?

      $H(t) = A \cdot 4^{t}$

   d. The height halved every ten seconds?

      $H(t) = A \cdot (0.5^{t/10})$

*Scaffolding:*

Struggling students can create a table of water depths to visualize the accumulation of water. Since the doubling happens every 10 seconds, have them deduce the exponent by asking, "How many times would doubling occur in 30 seconds? One minute?"

| Time (s) | Depth (cm) |
|---|---|
| 0 | 1 |
| 10 | 2 |
| 20 | 4 |
| 30 | 8 |

## Discussion (2 minutes)

Students have worked informally with the average rate of change of a function before in Modules 3 and 4 of Algebra I. For the next examples, we will need the following definition. Go through this definition and post it on the board or in another prominent place before beginning the next example. Students will continue to work with the average rate of change of a function in the problem set.

AVERAGE RATE OF CHANGE: Given a function $f$ whose domain contains the interval of real numbers $[a, b]$ and whose range is a subset of the real numbers, the *average rate of change on the interval* $[a, b]$ is defined by the number:

$$\frac{f(b) - b(a)}{b - a}$$

## Example 1 (4 minutes)

Use this example to model the process of finding the average rate of change of the height of the water that is increasing according to one of the exponential functions in our hypothetical scenario. The students will repeat this calculation in the exercises that follow. The Student Materials contain the images below of the three water tanks, but not the accompanying formulas.

**Example 1**

1. **Consider two identical water tanks, each of which begins with a height of water $1$ cm and fills with water at a different rate. Which equations can be used to calculate the height of water in each tank at time $t$? Use $H_1$ for tank 1 and $H_2$ for tank 2.**

| The height of the water in TANK 1 doubles every second. | The height of the water in TANK 2 triples every second. |
|---|---|
| $H_1(t) = 2^t$ | $H_2(t) = 3^t$ |

a. **If both tanks start filling at the same time, which one fills first?**

*Tank 2 will fill first because the level is rising more quickly.*

b. **We want to know the average rate of change of the height of the water in these tanks over an interval that starts at a fixed time $T$ as they are filling up. What is the formula for the average rate of change of a function $f$ on an interval $[a, b]$?**

$$\frac{f(b) - f(a)}{b - a}$$

c. **What is the formula for the average rate of change of the function $H_1$ on an interval $[a, b]$?**

$$\frac{H_1(b) - H_1(a)}{b - a}$$

d. **Let's calculate the average rate of change of the function $H_1$ on the interval $[T, T + 0.1]$, which is an interval one-tenth of a second long starting at an unknown time $T$.**

$$\begin{aligned}\frac{H_1(T+0.1) - H_1(T)}{T + 0.1 - T} &= \frac{(2^{T+0.1}) - (2^T)}{0.1} \\ &= \frac{2^T \cdot 2^{0.1} - 2^T}{0.1} \\ &= \frac{2^T(2^{0.1} - 1)}{0.1} \\ &\approx 2^T(0.717735) \\ &\approx 0.717735\, H_1(T)\end{aligned}$$

- So, the average rate of change of the height function is a multiple of the value of the function. This means that the speed at which the height is changing at time $T$ depends on the depth of water at that time. On average, over the interval $[T, T+0.1]$, the water in tank 1 rises at a rate of approximately $0.717735\ H_1(T)$ centimeters per second.
- Let's say that at time $T$ there is a height of 5 cm of water in the tank. Then after one-tenth of a second, the height of the water would increase by $\frac{1}{10}(0.717735(5)) \approx 0.3589$ cm. But, if there is a height of 20 cm of water in the tank, after one-tenth of a second the height of the water would increase by $\frac{1}{10}(0.717735(20)) \approx 1.4355$ cm.

## Exercises 4–5 (10 minutes)

Students will need to use calculators to compute the numerical constants in the exercises below.

**Exercises 4–8**

4. **For the second tank, calculate the average change in the height, $H_2$, from time $T$ seconds to $T+0.1$ seconds. Express the answer as a number times the value of the original function at time $T$. Explain the meaning of these findings.**

$$\begin{aligned}\frac{H_2(T+0.1)-H_2(T)}{0.1} &= \frac{3^{T+0.1}-3^T}{0.1}\\ &= \frac{3^T\cdot 3^{0.1}-3^T}{0.1}\\ &= \frac{3^T(3^{0.1}-1)}{0.1}\\ &\approx \frac{3^T(0.116123)}{0.1}\\ &\approx 1.16123\cdot 3^T\\ &\approx 1.16123\ \cdot H_2(T)\end{aligned}$$

*On average, over the time interval $[T, T+0.1]$, the water in tank 2 rises at a rate of approximately $1.16123H_2(T)$ centimeters per second.*

5. **For each tank, calculate the average change in height from time $T$ seconds to $T+0.001$ seconds. Express the answer as a number times the value of the original function at time $T$. Explain the meaning of these findings.**

*Tank 1:*

$$\begin{aligned}\frac{H_1(T+0.001)-H_1(T)}{0.001} &= \frac{2^{T+0.001}-2^T}{0.001}\\ &= \frac{2^T\cdot 2^{0.001}-2^T}{0.001}\\ &= \frac{2^T(2^{0.001}-1)}{0.001}\\ &\approx \frac{2^T(0.000693)}{0.001}\\ &\approx 0.69339\cdot 2^T\\ &\approx 0.69339\cdot H_1(T)\end{aligned}$$

*On average, over the time interval $[T, T+0.001]$, the water in tank 1 rises at a rate of approximately $0.693387H_1(T)$ centimeters per second.*

***Tank 2:***

$$\begin{aligned}\frac{H_2(T+0.001)-H_2(T)}{0.001} &= \frac{3^{T+0.001}-3^T}{0.001}\\ &= \frac{3^T\cdot 3^{0.001}-3^T}{0.001}\\ &= \frac{3^T(3^{0.001}-1)}{0.001}\\ &\approx \frac{3^T(0.00110)}{0.001}\\ &\approx 1.09922\cdot 3^T\\ &\approx 1.09922\cdot H_2(T)\end{aligned}$$

***Over the time interval*** $[T, T+0.001]$***, the water in tank 2 rises at an average rate of approximately*** $1.09922H_2(T)$ ***centimeters per second.***

## Exercises 6–8 (12 minutes)

The following exercises will lead to discovery of the constant $e$ that occurs naturally in many situations we can model mathematically. Looking at the results of the previous three exercises, if the height of the water doubles, then the expression for the average rate of change contains a factor less than one. If the height of the water triples, then the expression for the average rate of change contains a factor greater than one. Under what conditions will the expression for the average rate of change contain a factor of exactly one? Answering this question leads us to $e$.

**6. In Exercise 5, the average rate of change of the height of the water in tank 1 on the interval $[T, T+0.01]$ can be described by the expression $c_1\cdot 2^T$, and the average rate of change of the height of the water in tank 2 on the interval $[T, T+0.01]$ can be described by the expression $c_2\cdot 3^T$. What are approximate values of $c_1$ and $c_2$?**

$c_1 \approx 0.69339$ ***and*** $c_2 \approx 1.09922$

**7. As an experiment, let's look for a value of $b$ so that if the height of the water can be described by $H(t) = b^t$, then the expression for the average of change on the interval $[T, T+0.01]$ is $1\cdot H(T)$.**

**a. Write out the expression for the average rate of change of $H(t) = b^t$ on the interval $[T, T+0.01]$.**

$$\frac{H_b(T+0.001)-H_b(T)}{0.001}$$

**b. Set your expression in part (a) equal to $1\cdot H(T)$ and reduce to an expression involving a single $b$.**

$$\begin{aligned}\frac{H_b(T+0.001)-H_b(T)}{0.001} &= 1\cdot H_b(T)\\ \frac{b^{T+0.001}-b^T}{0.001} &= b^t\\ \frac{b^T(b^{0.001}-1)}{0.001} &= b^T\\ b^{0.001}-1 &= 0.001\\ b^{0.001} &= 1.001\end{aligned}$$

c. **Now we want to find the value of $b$ that satisfies the equation you found in part (b), but we do not have a way to explicitly solve this equation. Look back at Exercise 6; which two consecutive integers have $b$ between them?**

***We are looking for the base of the exponent that produces a rate of change on a small interval near t that is $1 \cdot H(t)$. When that base is $2$, the value of the rate is roughly $0.69H(t)$. When the base is $3$, the value of the rate is roughly $1.1H$. Since $0.69 < 1 < 1.1$, the base we are looking for is somewhere between $2$ and $3$.***

d. **Use your calculator and a guess-and-check method to find an approximate value of $b$ to 2 decimal places.**

***Students may choose to use a table such as the following. Make sure that students are maintaining enough decimal places of $b^{0.001}$ to determine which value is closest to $0.001$.***

| $b$ | $b^{0.001}$ | $b$ | $b^{0.001}$ |
|---|---|---|---|
| $2.0$ | $1.00069$ | $2.70$ | $1.000994$ |
| $2.1$ | $1.00074$ | $2.71$ | $1.000997$ |
| $2.2$ | $1.00079$ | $2.72$ | $\underline{1.001001}$ |
| $2.3$ | $1.00083$ | $2.73$ | $1.001005$ |
| $2.4$ | $1.00088$ | $2.74$ | |
| $2.5$ | $1.00092$ | $2.75$ | |
| $2.6$ | $1.00096$ | $2.76$ | |
| $2.7$ | $\underline{1.00099}$ | $2.77$ | |
| $2.8$ | $1.00103$ | $2.78$ | |
| $2.9$ | $1.00107$ | $2.79$ | |
| $3.0$ | $1.00110$ | $2.80$ | |

*Then $b \approx 2.72$.*

8. **Verify that for the value of $b$ found in Exercise 7, $\frac{H_b(T+0.001)-H_b(T)}{0.001} = H_b(T)$, where $H_b(T) = b^T$.**

$$\begin{aligned}\frac{H_b(T+0.001)-H_b(T)}{0.001} &= \frac{2.72^{T+0.001}-2.72^{0.001}}{0.001}\\ &= \frac{2.72^T(2.72^{0.001}-1)}{0.001}\\ &\approx \frac{2.72^T(0.001000)}{0.001}\\ &\approx 1.00 \cdot 2.72^T\\ &\approx 1.00 \cdot H_b(T)\end{aligned}$$

***When the height of the water increases by a factor of $2.72$ units per second, the height at any time is equal to the rate of change of height at that time.***

## Discussion (2 minutes)

If there is time, perform the calculation of $b$ several more times, over smaller and smaller time intervals and finding more and more digits of $b$. If not, then just present students with the fact below.

- What happens to the value of $b$?
  - *If we were to keep finding the average rate of change of the function $H_b$ on smaller and smaller time intervals and solving the equation $H_b(t) = A \cdot b^t$, we would find that the height of the water increases by a factor that gets closer and closer to the number* $2.7182818284$ ....
- The number that this process leads to is called Euler's number, and is denoted by $e$. Like $\pi$, $e$ is an irrational number so it cannot be accurately represented by a decimal expansion. The approximation of $e$ to 13 decimal places is $e \approx 2.7182818284590$.

- Like $\pi$, $e$ is important enough to merit inclusion on scientific calculators. Depending on the calculator, $e$ may appear alone, as the base of an exponential expression $e^x$, or both. Find the $e$ button on your calculator and experiment with its use. Make sure you can use your calculator to provide an approximation of $e$ and use the button to calculate $e^2$ and $2e$.

## Closing (4 minutes)

Summarize the lesson with the students and ensure the first two points below are addressed. Have students highlight what they think is important about the lesson in writing or with a partner. Use this as an opportunity to informally assess learning.

- We just discovered the number $e$, which is important in the world of mathematics. It naturally occurred in our water tank exploration. It also occurs naturally in many other applications, such as finance and population growth.
- Just as we can create and use an exponential function $f(x) = 2^x$ or $f(x) = 10^x$, we can also create and use an exponential function $f(x) = e^x$. The interesting thing about the exponential function base $e$ is that the rate of change of this function at a value $a$ is the same as the value of this function at $a$.
- Euler's number will surface in a variety of different places in your future exposure to mathematics and you will see how it is one of the numbers on which much of the mathematics we practice is based.

**Lesson Summary**

- **Euler's number, $e$, is an irrational number that is approximately equal to $e \approx 2.7182818284590$.**
- **AVERAGE RATE OF CHANGE: Given a function $f$ whose domain contains the interval of real numbers $[a, b]$ and whose range is a subset of the real numbers, the *average rate of change on the interval* $[a, b]$ is defined by the number**

$$\frac{f(b) - b(a)}{b - a}.$$

## Exit Ticket (3 minutes)

Name ______________________________ Date ______________

# Lesson 6: Euler's Number, $e$

## Exit Ticket

1. Suppose that water is entering a cylindrical water tank so that the initial height of the water is 3 cm and the height of the water doubles every 30 seconds. Write an equation of the height of the water at time $t$ seconds.

2. Explain how the number $e$ arose in our exploration of the average rate of change of the height of the water in the water tank.

## Exit Ticket Sample Solutions

1. **Suppose that water is entering a cylindrical water tank so that the initial height of the water is $3$ cm and the height of the water doubles every $30$ seconds. Write an equation of the height of the water at time $t$ seconds.**

   $H(t) = 3\left(2^{\frac{t}{30}}\right)$

2. **Explain how the number $e$ arose in our exploration of the average rate of change of the height of the water in the water tank.**

   *We first noticed that if the water level in the tank was doubling every second, then the average rate of change of the height of the water was roughly $0.69$ times the height of the water at that time. And if the water level in the tank was tripling every second, then the average rate of change of the height of the water was roughly $1.1$ times the height of the water at that time. When we went looking for a base $b$ so that the average rate of change of the height of the water was $1.0$ times the height of the water at that time, we found that the base was roughly $e$. Calculating the average rate of change over shorter intervals gave a better approximation of $e$.*

## Problem Set Sample Solutions

Problems 1–5 address other occurrences of $e$ and some fluency practice with the number $e$, and the remaining problems focus on the average rate of change of a function. The last two problems are extension problems that introduce some ideas of calculus with the familiar formulas for the area and circumference of a circle and the volume and surface area of a sphere.

1. **The product $4 \cdot 3 \cdot 2 \cdot 1$ is called $4$ *factorial* and is denoted by $4!$. Then $10! = 10 \cdot 9 \cdot 8 \cdot 7 \cdot 6 \cdot 5 \cdot 4 \cdot 3 \cdot 2 \cdot 1$, and for any positive integer $n$, $n! = n(n-1)(n-2)\cdots 3 \cdot 2 \cdot 1$.**

   a. **Complete the following table of factorial values:**

| $n$ | $1$ | $2$ | $3$ | $4$ | $5$ | $6$ | $7$ | $8$ |
|---|---|---|---|---|---|---|---|---|
| $n!$ | $1$ | $2$ | $6$ | $24$ | $120$ | $720$ | $5040$ | $40320$ |

   b. **Evaluate the sum $1 + \frac{1}{1!}$.**

   $2$

   c. **Evaluate the sum $1 + \frac{1}{1!} + \frac{1}{2!}$.**

   $2.5$

   d. **Use a calculator to approximate the sum $1 + \frac{1}{1!} + \frac{1}{2!} + \frac{1}{3!}$ to 7 decimal places. Do not round the fractions before evaluating the sum.**

   $\frac{8}{3} \approx 2.6666667$

   e. **Use a calculator to approximate the sum $1 + \frac{1}{1!} + \frac{1}{2!} + \frac{1}{3!} + \frac{1}{4!}$ to 7 decimal places. Do not round the fractions before evaluating the sum.**

   $\frac{65}{24} \approx 2.7083333$

f. Use a calculator to approximate sums of the form $1+\frac{1}{1!}+\frac{1}{2!}+\cdots+\frac{1}{k!}$ to 7 decimal places for $k=5,6,7,8,9,10$. Do not round the fractions before evaluating the sums with a calculator.

*If* $k=5$*, the sum is* $\frac{163}{60}\approx 2.1766667$.

*If* $k=6$*, the sum is* $\frac{1957}{720}\approx 2.7180556$.

*If* $k=7$*, the sum is* $\frac{685}{252}\approx 2.7182540$.

*If* $k=8$*, the sum is* $\frac{109601}{40320}\approx 2.7182788$.

*If* $k=9$*, the sum is* $\frac{98461}{36288}\approx 2.7182815$.

*If* $k=10$*, the sum is* $\frac{9864101}{3628800}\approx 2.7182818$.

g. Make a conjecture about the sums $1+\frac{1}{1!}+\frac{1}{2!}+\cdots+\frac{1}{k!}$ for positive integers $k$ as $k$ increases in size.

*It seems that as k gets larger, the sums* $1+\frac{1}{1!}+\frac{1}{2!}+\cdots+\frac{1}{k!}$ *get closer to e.*

h. Would calculating terms of this sequence ever yield an exact value of $e$? Why or why not?

*No. The number e is irrational so it cannot be written as a quotient of integers. Any finite sum* $1+\frac{1}{1!}+\frac{1}{2!}+\cdots+\frac{1}{k!}$ *can be expressed as a single rational number with denominator* $k!$*, so the sums are all rational numbers. However, the more terms that are calculated, the closer to e the sum becomes, so these sums provide better and better rational number approximations of e.*

2. Consider the sequence given by the function $a_n=\left(1+\frac{1}{n}\right)^n$, where $n\geq 1$ is an integer.

a. Use your calculator to approximate the first 5 terms of this sequence to 7 decimal places.

$$a_1=\left(1+\frac{1}{1}\right)^1=2$$

$$a_2=\left(1+\frac{1}{2}\right)^2=2.25$$

$$a_3=\left(1+\frac{1}{3}\right)^3\approx 2.3703704$$

$$a_4=\left(1+\frac{1}{4}\right)^4\approx 2.4414063$$

$$a_5=\left(1+\frac{1}{5}\right)^5=2.4883200$$

b. Does it appear that this sequence settles near a particular value?

*No, the numbers get bigger, but we cannot tell if it keeps getting bigger or settles on or near a particular value.*

c. **Use a calculator to approximate the following terms of this sequence to 7 decimal places.**

i. $a_{100}$ $= 2.7081383$

ii. $a_{1000}$ $= 2.7169239$

iii. $a_{10,000}$ $= 2.7181459$

iv. $a_{100,000}$ $= 2.7182682$

v. $a_{1,000,000}$ $= 2.7182805$

vi. $a_{10,000,000}$ $= 2.7182816$

vii. $a_{100,000,000}$ $= 2.7182818$

d. **Does it appear that this sequence settles near a particular value?**

*Yes, it appears that as $n$ gets really large (at least $100,000,000$), the terms $a_n$ of the sequence settle near the value of $e$.*

e. **Compare the results of this exercise with the results of Problem 1. What do you observe?**

*It took about $10$ terms of the sum in Problem 1 to see that the sum settled at the value $e$, but it takes $100,000,000$ terms of the sequence in this problem to see that the sum settles at the value $e$.*

3. **If $x = 5a^4$ and $a = 2e^3$, express $x$ in terms of $e$ and approximate to the nearest whole number.**

*If $x = 5a^4$ and $a = 2e^3$, then $x = 5(2e^3)^4$. Rewriting the right side in an equivalent form gives $x = 80e^{12} \approx 13020383$.*

4. **If $a = 2b^3$ and $b = -\frac{1}{2}e^{-2}$, express $a$ in terms of $e$ and approximate to four decimal places.**

*If $a = 2b^3$ and $b = -\frac{1}{2}e^{-2}$, then $a = 2\left(-\frac{1}{2}e^{-2}\right)^3$. Rewriting the right side in an equivalent form gives $a = -\frac{1}{4}e^{-6} \approx -0.0006$.*

5. **If $x = 3e^4$ and $e = \frac{s}{2x^3}$, show that $s = 54e^{13}$ and approximate $s$ to the nearest whole number.**

*Rewrite the equation $e = \frac{s}{2x^3}$ to isolate the variable $s$.*

$$e = \frac{s}{2x^3}$$

$$2x^3e = s$$

*By the substitution property, if $s = 2x^3e$ and $x = 3e^4$, then $s = 2(3e^4)^3 \cdot e$. Rewriting the right side in an equivalent form gives $s = 2 \cdot 27e^{12} \cdot e = 54e^{13} \approx 23890323$.*

6. **The following graph shows the number of barrels of oil produced by the Glenn Pool well in Oklahoma from 1910 to1916.**

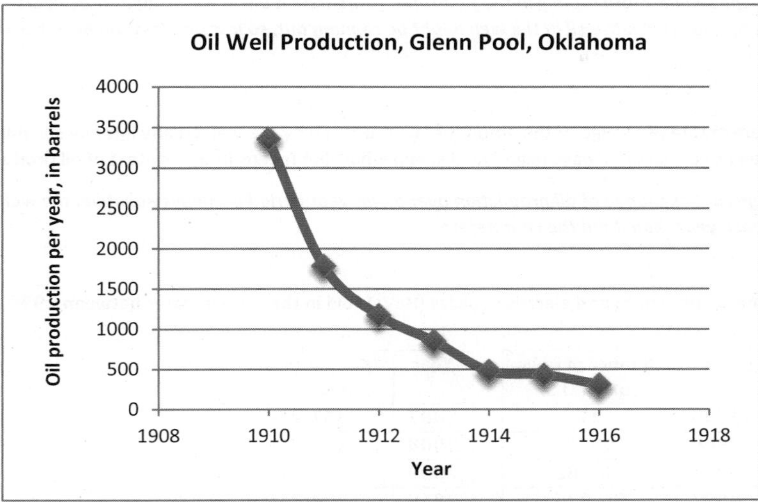


**Source: Cutler, Willard W., Jr. Estimation of Underground Oil Reserves by Oil-Well Production Curves, U.S. Department of the Interior, 1924.**

a. **Estimate the average rate of change of the amount of oil produced by the well on the interval $[1910, 1916]$ and explain what that number represents.**

*Student responses will vary based on how they read the points on the graph. Over the interval* $[1910, 1916]$, *the average rate of change is roughly*

$$\frac{300 - 3200}{1916 - 1910} = -\frac{2900}{6} \approx -483.33.$$

*This says that the production of the well decreased by an average of about* $483$ *barrels of oil each year between* $1910$ *and* $1916$.

b. **Estimate the average rate of change of the amount of oil produced by the well on the interval $[1910, 1913]$ and explain what that number represents.**

*Student responses will vary based on how they read the points on the graph. Over the interval* $[1910, 1913]$, *the average rate of change is roughly*

$$\frac{800 - 3200}{1913 - 1910} = -\frac{2400}{3} = -800.$$

*This says that the production of the well decreased by an average of about* $800$ *barrels of oil per year between* $1910$ *and* $1913$.

c. **Estimate the average rate of change of the amount of oil produced by the well on the interval $[1913, 1916]$ and explain what that number represents.**

*Student responses will vary based on how they read the points on the graph. Over the interval* $[1913, 1916]$, *the average rate of change is roughly*

$$\frac{300 - 800}{1916 - 1913} = -\frac{500}{3} \approx -166.67.$$

*This says that the production of the well decreased by an average of about* $166.67$ *barrels of oil per year between* $1913$ *and* $1916$.

d. **Compare your results for the rates of change in oil production in the first half and the second half of the time period in question in parts (b) and (c). What do those numbers say about the production of oil from the well?**

*The production dropped much more rapidly in the first three years than it did in the second three years. Looking at the graph, it looks like the oil in the well might be running out, so less and less can be extracted each year.*

e. **Notice that the average rate of change of the amount of oil produced by the well on any interval starting and ending in two consecutive years is always negative. Explain what that means in the context of oil production.**

*Because the average rate of change of oil production over a one-year period is always negative, the well is producing less oil each year than it did the year before.*

7. **The following table lists the number of hybrid electric vehicles (HEVs) sold in the United States between $1999$ and $2013$.**

| Year | Number of HEVs sold in U.S. | Year | Number of HEVs sold in U.S. |
|---|---|---|---|
| 1999 | 17 | 2007 | 352,274 |
| 2000 | 9350 | 2008 | 312,386 |
| 2001 | 20,282 | 2009 | 290,271 |
| 2002 | 36,035 | 2010 | 274,210 |
| 2003 | 47,600 | 2011 | 268,752 |
| 2004 | 84,199 | 2012 | 434,498 |
| 2005 | 209,711 | 2013 | 495,685 |
| 2006 | 252,636 | | |

**Source: U.S. Department of Energy, Alternative Fuels and Advanced Vehicle Data Center, 2013**

a. **During which one-year interval is the average rate of change of the number of HEVs sold the largest? Explain how you know.**

*The average rate of change of the number of HEVs sold is largest during* $[2011, 2012]$ *because the number of HEVs sold increases by the largest amount between those two years.*

b. **Calculate the average rate of change of the number of HEVs sold on the interval $[2003, 2004]$ and explain what that number represents.**

*On the interval* $[2003, 2004]$*, the average rate of change in sales of HEVs is* $\frac{84{,}199-47{,}600}{2004-2003} = 36{,}599$*. This means that during this one-year period, HEVs sere selling at a rate of* $36{,}599$ *vehicles per year.*

c. **Calculate the average rate of change of the number of HEVs sold on the interval $[2003, 2008]$ and explain what that number represents.**

*On the interval* $[2003, 2008]$*, the average rate of change in sales of HEV is* $\frac{312{,}386-47{,}600}{2008-2003} = 52{,}957.2$*. This means that during this five-year period, HEVs were selling at an average rate of* $52{,}957$ *vehicles per year.*

d. **What does it mean if the average rate of change of the number of HEVs sold is negative?**

*If the average rate of change of the vehicles sold is negative, then the sales are declining. This means that fewer cards were sold than in the previous year.*

**Extension:**

8. **The formula for the area of a circle of radius $r$ can be expressed as a function $A(r) = \pi r^2$.**

   a. **Find the average rate of change of the area of a circle on the interval $[4, 5]$.**

$$\frac{A(5) - A(4)}{5 - 4} = \frac{25\pi - 16\pi}{1} = 9\pi$$

   b. **Find the average rate of change of the area of a circle on the interval $[4, 4.1]$.**

$$\frac{A(4.1) - A(4)}{4.1 - 4} = \frac{16.81\pi - 16\pi}{0.1} = 8.1\pi$$

   c. **Find the average rate of change of the area of a circle on the interval $[4, 4.01]$.**

$$\frac{A(4.01) - A(4)}{4.01 - 4} = \frac{16.0801\pi - 16\pi}{0.01} = 8.01\pi$$

   d. **Find the average rate of change of the area of a circle on the interval $[4, 4.001]$.**

$$\frac{A(4.001) - A(4)}{4.001 - 4} = \frac{16.008001\pi - 16\pi}{0.001} = 8.001\pi$$

   e. **What is happening to the average rate of change of the area of the circle as the interval gets smaller and smaller?**

   ***The average rate of change of the area of the circle appears to be getting close to $8\pi$.***

   f. **Find the average rate of change of the area of a circle on the interval $[4, 4 + h]$ for some small positive number $h$.**

$$\begin{aligned}\frac{A(4+h) - A(4)}{(4+h) - 4} &= \frac{(4+h)^2\pi - 16\pi}{h} \\ &= \frac{(16 + 8h + h^2)\pi - 16\pi}{h} \\ &= \frac{1}{h}(8h + h^2)\pi \\ &= (8 + h)\pi\end{aligned}$$

   g. **What happens to the average rate of change of the area of the circle on the interval $[4, 4 + h]$ as $h \to 0$? Does this agree with your answer to part (d)? Should it agree with your answer to part (e)?**

   ***As $h \to 0$, $8 + h \to 8$, so as $h$ gets smaller, the average rate of change approaches $8$. This agrees with my response to part (e), and it should because as $h \to 0$, the interval $[4, 4 + h]$ gets smaller.***

   h. **Find the average rate of change of the area of a circle on the interval $[r_0, r_0 + h]$ for some positive number $r_0$ and some small positive number $h$.**

$$\begin{aligned}\frac{A(r_0+h) - A(r_0)}{(r_0+h) - r_0} &= \frac{(r_0+h)^2\pi - r_0^2\pi}{h} \\ &= \frac{(r_0^2 + 2r_0h + h^2)\pi - r_0^2\pi}{h} \\ &= \frac{1}{h}(2r_0h + h^2)\pi \\ &= (2r_0 + h)\pi\end{aligned}$$

i. **What happens to the average rate of change of the area of the circle on the interval $[r_0, r_0 + h]$ as $h \to 0$? Do you recognize the resulting formula?**

*As $h \to 0$, the expression for the average rate of change becomes $2\pi r_0$, which is the circumference of the circle with radius $r_0$.*

9. **The formula for the volume of a sphere of radius $r$ can be expressed as a function $V(r) = \frac{4}{3}\pi r^3$. As you work through these questions, you will see the pattern develop more clearly if you leave your answers in the form of a coefficient times $\pi$. Approximate the coefficient to five decimal places.**

a. **Find the average rate of change of the volume of a sphere on the interval $[2, 3]$.**

$$\frac{V(3) - V(2)}{3 - 2} = \frac{\frac{4}{3} \cdot 27\pi - \frac{4}{3} \cdot 8\pi}{1} = \frac{4}{3} \cdot 19\pi \approx 25.33333\pi$$

b. **Find the average rate of change of the volume of a sphere on the interval $[2, 2.1]$.**

$$\frac{V(2.1) - V(2)}{2.1 - 2} = \frac{\frac{4}{3}\pi(2.1^3 - 8)}{0.1} = 16.81333\pi$$

c. **Find the average rate of change of the volume of a sphere on the interval $[2, 2.01]$.**

$$\frac{V(2.01) - V(2)}{2.01 - 2} = \frac{\frac{4}{3}\pi(2.01^3 - 8)}{0.01} = 16.08010\pi$$

d. **Find the average rate of change of the volume of a sphere on the interval $[2, 2.001]$.**

$$\frac{V(2.001) - V(2)}{2.001 - 2} = \frac{\frac{4}{3}\pi(2.001^3 - 8)}{0.001} = 16.00800\pi$$

e. **What is happening to the average rate of change of the volume of a sphere as the interval gets smaller and smaller?**

*The average rate of change of the volume of the sphere appears to be getting close to $16\pi$.*

f. **Find the average rate of change of the volume of a sphere on the interval $[2, 2 + h]$ for some small positive number $h$.**

$$\begin{aligned}\frac{V(2+h) - V(2)}{(2+h) - 2} &= \frac{\frac{4}{3}\pi((2+h)^3 - 8)}{h} \\ &= \frac{4}{3}\pi \cdot \frac{1}{h}(8 + 12h + 6h^2 + h^3 - 8) \\ &= \frac{4\pi}{3h}(12h + 6h^2 + h^3) \\ &= \frac{4\pi}{3}(12 + 6h + h^2)\end{aligned}$$

g. **What happens to the average rate of change of the volume of a sphere on the interval $[2, 2 + h]$ as $h \to 0$? Does this agree with your answer to part (e)? Should it agree with your answer to part (e)?**

*As $h \to 0$, the polynomial $12 + 6h + h^2 \to 12$. Then the average rate of change approaches $\frac{4\pi}{3} \cdot 12 = 16$. This agrees with my response to part (e), and it should because as $h \to 0$, the interval $[2, 2 + h]$ gets smaller.*

h. **Find the average rate of change of the volume of a sphere on the interval $[r_0, r_0 + h]$ for some positive number $r_0$ and some small positive number $h$.**

$$\frac{V(r_0 + h) - V(r_0)}{(r_0 + h) - r_0} = \frac{\frac{4}{3}\pi\left((r_0 + h)^3 - r_0^3\right)}{h}$$
$$= \frac{4}{3}\pi \cdot \frac{1}{h}\left(r_0^3 + 3r_0^2h + 3r_0h^2 + h^3 - r_0^3\right)$$
$$= \frac{4\pi}{3h}\left(3r_0^2h + 3r_0h^2 + h^3\right)$$
$$= \frac{4\pi}{3}\left(3r_0^2 + 3r_0h + h^2\right)$$

i. **What happens to the average rate of change of the volume of a sphere on the interval $[r_0, r_0 + h]$ as $h \to 0$? Do you recognize the resulting formula?**

*As $h \to 0$, the expression for the average rate of change becomes $4\pi r_0^2$, which is the surface area of the sphere with radius $r_0$.*

# Topic B:

# Logarithms

**N-Q.A.2, A-CED.A.1, F-BF.A.1a, F-LE.A.4**

| | | |
|---|---|---|
| **Focus Standards:** | N-Q.A.2 | Define appropriate quantities for the purpose of descriptive modeling.★ |
| | A-CED.A.1 | Create equations and inequalities in one variable and use them to solve problems. Include equations arising from linear and quadratic functions, and simple rational and exponential functions.★ |
| | F-BF.A.1a | Write a function that describes a relationship between two quantities.★<br>a. Determine an explicit expression, a recursive process, or steps for calculation from a context. |
| | F-LE.A.4 | For exponential models, express as a logarithm the solution to $ab^{ct} = d$ where $a$, $c$, and $d$ are numbers and the base $b$ is $2$, $10$, or $e$; evaluate the logarithm using technology.★ |
| **Instructional Days:** | 9 | |
| **Lesson 7:** | Bacteria and Exponential Growth (S)[1] | |
| **Lesson 8:** | The "WhatPower" Function (P) | |
| **Lesson 9:** | Logarithms—How Many Digits Do You Need? (E) | |
| **Lesson 10:** | Building Logarithmic Tables (P) | |
| **Lesson 11:** | The Most Important Property of Logarithms (P) | |
| **Lesson 12:** | Properties of Logarithms (P) | |
| **Lesson 13:** | Changing the Base (P) | |
| **Lesson 14:** | Solving Logarithmic Equations (P) | |
| **Lesson 15:** | Why Were Logarithms Developed? (P) | |

[1] Lesson Structure Key: **P**-Problem Set Lesson, **M**-Modeling Cycle Lesson, **E**-Exploration Lesson, **S**-Socratic Lesson

The lessons covered in Topic A familiarize students with the laws and properties of real-valued exponents. In Topic B, students extend their work with exponential functions to include solving exponential equations numerically and to develop an understanding of the relationship between logarithms and exponentials. To model the growth of bacteria and populations, Lesson 7 introduces students to simple exponential equations whose solutions do not follow from equating exponential terms of similar bases. To solve those equations, an algorithmic numerical approach is employed (**F-BF.A.1a**). Students work to develop progressively better approximations for the solutions to equations whose solutions are irrational numbers. In doing this, students increase their understanding of the real number system and truly begin to understand what it means for a number to be irrational. Students learn that some simple exponential equations can be solved exactly without much difficulty but that we lack mathematical tools to solve other equations whose solutions must be approximated numerically.

Lesson 8 begins with the logarithmic function disguised as the more intuitive "WhatPower" function, whose behavior is studied as a means of introducing how the function works and what it does to expressions. Students find the power needed to raise a base $b$ in order to produce a given number. The lesson ends with students defining the term *logarithm base b*. Lesson 8 is just a first introduction to logarithms in preparation for solving exponential equations per **F-LE.A.4**; students neither use tables nor look at graphs in this lesson. Instead, they simply develop the ideas and notation of logarithmic expressions, leaving many ideas to be explored later in the module.

Just as population growth is a natural example to use to give context to exponential growth, Lesson 9 gives context to logarithmic calculation through the example of assigning unique identification numbers to a group of people. In this lesson, students consider the meaning of the logarithm in the context of calculating the number of digits needed to create student ID numbers, phone numbers, and social security numbers, in accordance with **N-Q.A.2**. This gives students a real-world context for the abstract idea of a logarithm; in particular, students observe that a base 10 logarithm provides a way to keep track of the number of digits used in a number in our base 10 system.

Lessons 10–15 develop both the theory of logarithms and procedures for solving various forms of exponential and logarithmic equations. In Lessons 10 and 11, students discover the logarithmic properties by completing carefully structured logarithmic tables and answering sets of directed questions. Throughout these two lessons, students look for the structure in the table and use that structure to extract logarithmic properties (MP.7). Using the structure of the logarithmic expression together with the logarithmic properties to rewrite an expression aligns with the foundational standard **A-SSE.A.2**. While the logarithmic properties are not themselves explicitly listed in the standards, standard **F-LE.A.4** cannot be adequately met without an understanding of how to apply logarithms to solve exponential equations, and the seemingly odd behavior of graphs of logarithmic functions (**F-IF.C.7e**) cannot be adequately explained without an understanding of the properties of logarithms. In particular, in Lesson 11, students discover the "most important property of logarithms": for positive real numbers $x$ and $y$, $\log(xy) = \log(x) + \log(y)$. Students also discover the pattern $\log_b\left(\frac{1}{x}\right) = -\log_b(x)$ that leads to conjectures about additional properties of logarithms.

Lesson 12 continues the consideration of properties of the logarithm function, while remaining focused solely on base 10 logarithms. Its centerpiece is the demonstration of basic properties of logarithms such as the power, product, and quotient properties, which allows students to practice MP.3 and **A-SSE.A.2**, providing justification in terms of the definition of logarithm and the properties already developed. In this lesson, students begin to learn how to solve exponential equations, beginning with base 10 exponential equations that can be solved by taking the common logarithm of both sides of the equation.

Lesson 13 again focuses on the structure of expressions (**A-SSE.A.2**), as students change logarithms from one base to another. It begins by showing students how they can make that change and then develops properties of logarithms for the general base $b$. The students are introduced to the use of a calculator instead of a table in finding logarithms, and then *natural logarithms* are defined: $\ln(x) = \log_e(x)$. One goal of the lesson, in addition to introducing the base $e$ for logarithms, is to explain why, for finding logarithms to any base, the calculator has only LOG and LN keys. In this lesson, students learn to solve exponential equations with any base by the application of an appropriate logarithm. Lessons 12 and 13 both address **F-LE.A.4**, solving equations of the form $ab^{ct} = d$, as do later lessons in the module.

Lesson 14 includes the first introduction to solving logarithmic equations. In this lesson, students apply the definition of the logarithm to rewrite logarithmic equations in exponential form, so the equations must first be rewritten in the form $\log_b(X) = c$, for an algebraic expression $X$ and some constant $c$. Solving equations in this way requires that students think deeply about the definition of the logarithm and how logarithms interact with exponential expressions. Although solving logarithmic equations is not listed explicitly in the standards, this skill is implicit in standard **A-REI.D.11**, which has students solve equations of the form $f(x) = g(x)$ where $f$ and $g$ can be logarithmic functions. Additionally, logarithmic equations provide a greater context in which to study both the properties of logarithms and the definition, both of which are needed to solve the equations listed in **F-LE.A.4**.

Topic B concludes with Lesson 15, in which students learn a bit of the history of how and why logarithms first appeared. The materials for this lesson contain a base 10 logarithm table that should be copied and distributed to the class. Although modern technology has made logarithm tables functionally obsolete, there is still value in understanding the historical development of logarithms. Logarithms were critical to the development of astronomy and navigation in the days before computing machines, and this lesson presents a rationale for the pre-technological advantage afforded to scholars by the use of logarithms. In this lesson, the case is finally made that logarithm functions are one-to-one (without explicitly using that terminology): if $\log_b(X) = \log_b(Y)$, then $X = Y$. In alignment with **A-SSE.A.2**, this fact not only validates the use of tables to look up anti-logarithms, but also allows exponential equations to be solved with logarithms on both sides of the equation.

# Lesson 7: Bacteria and Exponential Growth

## Student Outcomes

- Students solve simple exponential equations numerically.

## Lesson Notes

The lessons in Topic A familiarized students with the laws and properties of real-valued exponents. Topic B introduces the logarithm and develops logarithmic properties through exploration of logarithmic tables, primarily in base $10$. This lesson introduces simple exponential equations whose solutions do not follow from equating exponential terms of equal bases. Because we have no sophisticated tools for solving exponential equations until we introduce logarithms in later lessons, we use numerical methods to approximate solutions to exponential equations, a process which asks students to determine a recursive process from a context to solve $2^x = 10$ (**F-BF.A.1a**, **F-BF.B.4a**, **A-CED.A.1**). Students will have many opportunities to solve such equations algebraically throughout the module, using both the technique of equating exponents of exponential expressions with the same base and logarithms. The goals of this lesson are to help students understand (1) why logarithms are useful by introducing a situation (i.e., solving $2^x = 10$) offering students no option other than numerical methods to solve it, (2) that it is often possible to solve equations numerically by trapping the solution through better and better approximation, and (3) that the better and better approximations are converging on a (possibly) irrational number.

Exponential equations are used frequently to model bacteria and population growth, and both of those scenarios occur in this lesson.

## Classwork

### Opening Exercise (6 minutes)

In this exercise, students work in groups to solve simple exponential equations that can be solved by rewriting the expressions on each side of the equation as a power of the same base and equating exponents. It is also possible for students to use a table of values to solve these problems numerically; either method is valid and both should be discussed at the end of the exercise. Asking students to solve equations of this type demands that they think deeply about the meaning of exponential expressions. Because students have not solved exponential equations previously, the exercises are scaffolded to begin very simply and progress in difficulty; the early exercises may be merely solved by inspection. When students are finished, ask for volunteers to share their solutions on the board and discuss different solution methods.

**Opening Exercise**

**Work with your partner or group to solve each of the following equations for $x$.**

a. $2^x = 2$

$x = 1$

b. $2^x = 2^3$

$x = 3$

*Scaffolding:*

Encourage struggling students to make a table of values of the powers of 2 to use as a reference for these exercises.

c. $2^x = 16$

$2^x = 2^4$
$x = 4$

d. $2^x - 64 = 0$

$2^x = 64$
$2^x = 2^6$
$x = 6$

e. $2^x - 1 = 0$

$2^x = 1$
$2^x = 2^0$
$x = 0$

f. $2^{3x} = 64$

$2^{3x} = 2^6$
$3x = 6$
$x = 2$

g. $2^{x+1} = 32$

$2^{x+1} = 2^5$
$x + 1 = 5$
$x = 4$

*Scaffolding:*

Give early finishers a more challenging equation where both bases need to be changed such as $4^{2x} = 8^{x+3}$.

## Discussion (3 minutes)

This discussion should emphasize that the equations in the opening exercise have straightforward solutions because both sides can be expressed in terms of the common base 2 with an exponent.

MP.7

- How did the structure of the expressions in these equations allow you to solve them easily?
  - *Both sides of the equations could be written as exponential expressions with base* 2.
- Suppose the opening exercise had asked us to solve the equation $2^x = 10$ instead of the equation $2^x = 8$. Why is it far more difficult to solve the equation $2^x = 10$?
  - *We do not know how to express* 10 *as a power of* 2. *In the Opening Exercise, it is straightforward that* 8 *can be expressed as* $2^3$.
- Can we find two integers that are over and under estimates of the solution to $2^x = 10$? That is, can we find $a$ and $b$ so that $a < x < b$?
  - *Yes; the unknown* $x$ *is between* 3 *and* 4 *because* $2^3 < 10 < 2^4$.
- In the next example, we will use a calculator (or other technology) to find a more accurate estimate of the solutions to $2^x = 10$.

## Example (12 minutes)

The purpose of this exercise is to numerically pinpoint the solution $d$ to the equation $2^t = 10$ by squeezing the solution between numbers that get closer and closer together. We start with $3 < d < 4$, then find that $2^{3.3} < 10$ and $10 < 2^{3.4}$, so we must have $3.3 < d < 3.4$. Continuing with this logic, we squeeze $3.32 < d < 3.33$, and then $3.321 < d < 3.322$. The point of this exercise is that we can continue squeezing $d$ between numbers with more and more digits, meaning that we have an approximation of $d$ to greater and greater accuracy.

In the Student Materials, the tables for the Discussion below are presented next to each other, but they are spread out here so you can see how they fit into the discussion.

**Example**

**The *Escherichia coli* bacteria (commonly known as *E. coli*), reproduces once every 30 minutes, meaning that a colony of *E. coli* can double every half hour. *Mycobacterium tuberculosis* has a generation time in the range of 12 to 16 hours. Researchers have found evidence that suggests certain bacteria populations living deep below the surface of the earth may grow at extremely slow rates, reproducing once every several thousand years. With this variation in bacterial growth rates, it is reasonable that we assume a 24-hour reproduction time for a hypothetical bacteria colony in the next example.**

**Suppose we have a bacteria colony that starts with 1 bacterium, and the population of bacteria doubles every day.**

**What function $P$ can we use to model the bacteria population on day $t$?**

***$P(t) = 2^t$, for real numbers $t \geq 0$.***

Have the students volunteer values of $P(t)$ to help you complete the following table.

| $t$ | $P(t)$ |
|---|---|
| 1 | 2 |
| 2 | 4 |
| 3 | 8 |
| 4 | 16 |
| 5 | 32 |

**How many days will it take for the bacteria population to reach 8?**

***It will take 3 days, because $P(3) = 2^3 = 8$.***

**How many days will it take for the bacteria population to reach 16?**

***It will take 4 days, because $P(4) = 2^4 = 16$.***

**Roughly how long will it take for the population to reach 10?**

***Between 3 and 4 days; the number $d$ so that $2^d = 10$.***

**We already know from our previous discussion that if $2^d = 10$, then $3 < d < 4$, and the table confirms that. At this point, we have an underestimate of 3 and an overestimate of 4 for $d$. How can we find better under and over estimates for $d$?**

***(Note to teacher: Once students respond, have them volunteer values to complete the table.)***

***Calculate the values of $2^{3.1}$, $2^{3.2}$, $2^{3.3}$, etc., until we find two consecutive values that have 10 between them.***

| $t$ | $P(t)$ |
|---|---|
| 3.1 | 8.574 |
| 3.2 | 9.190 |
| 3.3 | 9.849 |
| 3.4 | 10.556 |

MP.8

**From our table, we now know another set of under and over estimates for the number $d$ that we seek. What are they?**

*We know $d$ is between $3.3$ and $3.4$. That is, $3.3 < d < 3.4$.*

**Continue this process of "squeezing" the number $d$ between two numbers until you are confident you know the value of $d$ to two decimal places.**

| $t$ | $P(t)$ |
|---|---|
| $3.31$ | $9.918$ |
| $3.32$ | $9.987$ |
| $3.33$ | $10.056$ |

| $t$ | $P(t)$ |
|---|---|
| $3.321$ | $9.994$ |
| $3.322$ | $10.001$ |
| | |

*Since $3.321 < d < 3.322$, and both numbers round to $3.32$, we can say that $d \approx 3.32$. We see that the population reaches $10$ after $3.32$ days, i.e., $2^{3.32} \approx 10$.*

**What if we had wanted to find $d$ to $5$ decimal places?**

*Keep squeezing $d$ between under and over estimates until they agree to the first $5$ decimal places. (Note to teacher: To $5$ decimal places, $3.321928 < d < 3.321929$, so $d \approx 3.32193$.)*

**To the nearest minute, when does the population of bacteria become $10$?**

*It takes $3.322$ days, which is roughly $3$ days, $7$ hours and $43$ minutes.*

## Discussion (2 minutes)

- Could we repeat the same process to find the time required for the bacteria population to reach $20$ (or $100$ or $500$)?
  - *Yes, we could start by determining between which two integers the solution to the equation $2^t = 20$ must lie and then continue the same process to find the solution.*
- Could we achieve the same level of accuracy as we did in the example? Could we make our solution more accurate?
  - *Yes, we can continue to repeat the process and eventually "trap" the solution to as many decimal places as we would like.*

Lead students to the idea that for any positive number $x$, we can repeat the process above to approximate the exponent $L$ so that $2^L = x$ to as many decimal places of accuracy as we would like. Likewise, we can approximate an exponent so that we can write the number $x$ as a power of $10$, or a power of $3$, or a power of $e$, or a power of any positive number other than $1$.

Note that there is a little bit of a theoretical hole here that will be filled in later when the logarithm function is introduced. In this lesson, we are only finding a rational approximation to the value of the exponent, which is the logarithm and is generally an irrational number. That is, in this example we are not truly writing $10$ as a power of $2$, but we are only finding a close approximation. If students question this subtle point, let them know that later in the module we will have definitive ways to write any positive number exactly as a power of the base.

## Exercise (8 minutes)

Divide students into groups of 2 or 3, and assign each group a different equation to solve from the list below. Students should repeat the process of the Example to solve these equations by squeezing the solution between more and more precise under and over estimates. Record the solutions in a way that students can see the entire list either written on poster board, written on the whiteboard, or projected through the document camera.

**Exercise**

**Use the method from the Example to approximate the solution to the equations below to two decimal places.**

a. $2^x = 1000$ $x \approx 9.97$

b. $3^x = 1000$ $x \approx 6.29$

c. $4^x = 1000$ $x \approx 4.98$

d. $5^x = 1000$ $x \approx 4.29$

e. $6^x = 1000$ $x \approx 3.85$

f. $7^x = 1000$ $x \approx 3.55$

g. $8^x = 1000$ $x \approx 3.32$

h. $9^x = 1000$ $x \approx 3.14$

i. $11^x = 1000$ $x \approx 2.88$

j. $12^x = 1000$ $x \approx 2.78$

k. $13^x = 1000$ $x \approx 2.69$

l. $14^x = 1000$ $x \approx 2.62$

m. $15^x = 1000$ $x \approx 2.55$

n. $16^x = 1000$ $x \approx 2.49$

## Discussion (2 minutes)

MP.3

- Do you observe a pattern in the solutions to the equations in Exercise 1?
  - *Yes, the larger the base, the smaller the solution.*
- Why would that be?
  - *The larger the base, the smaller the exponent needs to be in order to reach* 1000.

## Closing (4 minutes)

Have students respond to the following questions individually in writing or orally with a partner.

- Explain when a simple exponential equation, such as those we have seen today, can be solved exactly using our current methods.
  - *If both sides of the equation can be written as exponential expressions with the same base, then the equation can be solved exactly.*
- When a simple exponential equation cannot be solved by hand, what can we do?
  - *Give crude under and over estimates for the solution using integers.*
  - *Use a calculator to find increasingly accurate over and under estimates to the solution until we are satisfied.*

## Exit Ticket (8 minutes)

Name ______________________________ Date ______________

# Lesson 7: Bacteria and Exponential Growth

## Exit Ticket

Loggerhead turtles reproduce every 2–4 years, laying approximately 120 eggs in a clutch. Using this information, we can derive an approximate equation to model the turtle population. As is often the case in biological studies, we will count only the female turtles. If we start with a population of one female turtle in a protected area, and assume that all turtles survive, we can roughly approximate the population of female turtles by $T(t) = 5^t$. Use the methods of the Example to find the number of years, $Y$, it will take for this model to predict that there will be $300$ female turtles.

## Exit Ticket Sample Solutions

**Loggerhead turtles reproduce every 2–4 years, laying approximately $120$ eggs in a clutch. Using this information, we can derive an approximate equation to model the turtle population. As is often the case in biological studies, we will count only the female turtles. If we start with a population of one female turtle in a protected area, and assume that all turtles survive, we can roughly approximate the population of female turtles by $T(t) = 5^t$. Use the methods of the Example to find the number of years, $Y$, it will take for this model to predict that there will be $300$ female turtles.**

*Since $5^3 = 125$ and $5^4 = 625$, we know that $3 < Y < 4$.*

*Since $5^{3.5} \approx 279.5084$ and $5^{3.6} \approx 328.3160$, we know that $3.5 < Y < 3.6$.*

*Since $5^{3.54} \approx 298.0944$ and $5^{3.55} \approx 302.9308$, we know that $3.54 < Y < 3.55$.*

*Since $5^{3.543} \approx 299.5372$ and $5^{3.544} \approx 300.0196$, we know that $3.543 < Y < 3.544$.*

*Thus, to two decimal places, we have $Y \approx 3.54$. So, it will take roughly $3\frac{1}{2}$ years for the population to grow to $300$ female turtles.*

## Problem Set Sample Solutions

The Problem Set gives students an opportunity to practice using the numerical methods for approximating solutions to exponential equations that they have established in this lesson.

1. **Solve each of the following equations for $x$ using the same technique as was used in the Opening Exercises.**

   a. $2^x = 32$
   
   $x = 5$

   b. $2^{x-3} = 2^{2x+5}$
   
   $x = -8$

   c. $2^{x^2-3x} = 2^{-2}$
   
   $x = 1$ *or* $x = 2$

   d. $2^x - 2^{4x-3} = 0$
   
   $x = 1$

   e. $2^{3x} \cdot 2^5 = 2^7$
   
   $x = \frac{2}{3}$

   f. $2^{x^2-16} = 1$
   
   $x = 4$ *or* $x = -4$

   g. $3^{2x} = 27$
   
   $x = \frac{3}{2}$

   h. $3^{\frac{2}{x}} = 81$
   
   $x = \frac{1}{2}$

   i. $\frac{3^{x^2}}{3^{5x}} = 3^6$
   
   $x = 6$ *or* $x = -1$

2. **Solve the equation $\frac{2^{2x}}{2^{x+5}} = 1$ algebraically using two different initial steps as directed below.**

   a. **Write each side as a power of $2$.**

   $2^{2x-(x+5)} = 2^0$
   
   $x - 5 = 0$
   
   $x = 5$

   b. **Multiply both sides by $2^{x+5}$.**

   $2^{2x} = 2^{x+5}$
   
   $2x = x + 5$
   
   $x = 5$

3. **Find consecutive integers that are under and over estimates of the solutions to the following exponential equations.**

a. $2^x = 20$

$2^4 = 16$ *and* $2^5 = 25$, *so* $4 < x < 5$.

b. $2^x = 100$

$2^6 = 64$ *and* $2^7 = 128$, *so* $6 < x < 7$.

c. $3^x = 50$

$3^3 = 27$ *and* $3^4 = 81$, *so* $3 < x < 4$.

d. $10^x = 432,901$

$10^5 = 100,000$ *and* $10^6 = 1,000,000$, *so* $5 < x < 6$.

e. $2^{x-2} = 750$

$2^9 = 512$ *and* $2^{10} = 1,024$, *so* $9 < x - 2 < 10$; *thus,* $11 < x < 12$.

f. $2^x = 1.35$

$2^0 = 1$ *and* $2^1 = 2$, *so* $0 < x < 1$.

4. **Complete the following table to approximate the solution to $10^x = 34,198$ to two decimal places.**

| $t$ | $P(t)$ |
|---|---|
| 1 | 10 |
| 2 | 100 |
| 3 | 1,000 |
| 4 | 10,000 |
| 5 | 100,000 |
| | |

| $t$ | $P(t)$ |
|---|---|
| 4.1 | 12,589.254 |
| 4.2 | 15,848.932 |
| 4.3 | 19,952.623 |
| 4.4 | 25,118.864 |
| 4.5 | 31,622.777 |
| 4.6 | 39,810.717 |

| $t$ | $P(t)$ |
|---|---|
| 4.51 | 32,359.366 |
| 4.52 | 33,113.112 |
| 4.53 | 33,884.416 |
| 4.54 | 34,673.685 |
| | |
| | |

| $t$ | $P(t)$ |
|---|---|
| 4.531 | 33,962.527 |
| 4.532 | 34,040.819 |
| 4.533 | 34,119.291 |
| 4.534 | 34,197.944 |
| 4.535 | 34,276.779 |
| | |

$10^x = 34,198$

$10^{4.53} \approx 34,198$

5. Complete the following table to approximate the solution to $2^x = 18$ to two decimal places.

| $t$ | $P(t)$ | $t$ | $P(t)$ | $t$ | $P(t)$ | $t$ | $P(t)$ |
|---|---|---|---|---|---|---|---|
| 1 | 2 | 4.1 | 17.1484 | 4.11 | 17.2677 | 4.161 | 17.8890 |
| 2 | 4 | 4.2 | 18.3792 | 4.12 | 17.3878 | 4.162 | 17.9014 |
| 3 | 8 | | | 4.13 | 17.5087 | 4.163 | 17.9138 |
| 4 | 16 | | | 4.14 | 17.6305 | 4.164 | 17.9262 |
| 5 | 32 | | | 4.15 | 17.7531 | 4.165 | 17.9387 |
| | | | | 4.16 | 17.8766 | 4.166 | 17.9511 |
| | | | | 4.17 | 18.0009 | 4.167 | 17.9635 |
| | | | | | | 4.168 | 17.9760 |
| | | | | | | 4.169 | 17.9884 |
| | | | | | | 4.170 | 18.0009 |

$2^x = 18$

$2^{4.17} \approx 18$

6. Approximate the solution to $5^x = 5555$ to four decimal places.

*Since* $5^5 = 3125$ *and* $5^6 = 15,625$, *we know that* $5 < x < 6$.

*Since* $5^{5.3} \approx 5064.5519$ *and* $5^{5.4} \approx 5948.9186$, *we know that* $5.3 < x < 5.4$.

*Since* $5^{5.35} \approx 5488.9531$ *and* $5^{5.36} \approx 5578.0092$, *we know that* $5.35 < x < 5.36$.

*Since* $5^{5.357} \approx 5551.1417$ *and* $5^{5.358} \approx 5560.0831$, *we know that* $5.357 < x < 5.358$.

*Since* $5^{5.3574} \approx 5554.7165$ *and* $5^{5.3575} \approx 5555.6106$, *we know that* $5.3574 < x < 5.3575$.

*Since* $5^{5.35743} \approx 5554.9847$ *and* $5^{5.35744} \approx 5555.0741$, *we know that* $5.35743 < x < 5.35744$.

*Thus, the approximate solution to this equation to two decimal places is* $5.3574$.

7. A dangerous bacterial compound forms in a closed environment but is immediately detected. An initial detection reading suggests the concentration of bacteria in the closed environment is one percent of the fatal exposure level. This bacteria is known to double in growth (double in concentration in a closed environment) every hour and can be modeled by the function $P(t) = 100 \cdot 2^t$, where $t$ is measured in hours.

   a. In the function $P(t) = 100 \cdot 2^t$, what does the $100$ mean? What does the $2$ mean?

      *The* $100$ *represents the initial population of bacteria, which is* $1\%$ *of the fatal level. This means that the fatal level occurs when* $P(t) = 10,000$. *The base* $2$ *represents the growth rate of the bacteria; it doubles every hour.*

   b. Doctors and toxicology professionals estimate that exposure to two-thirds of the bacteria's fatal concentration level will begin to cause sickness. Without consulting a calculator or other technology, offer a rough time limit for the inhabitants of the infected environment to evacuate in order to avoid sickness in the doctors' estimation. Note that immediate evacuation is not always practical, so offer extra evacuation time if it is affordable.

      *The bacteria level is dangerous when* $P(t) = 100 \cdot 2^t = \frac{2}{3}(10,000) \approx 6666.67$.

      *Since* $2^6 = 64$, $P(6) \approx 6400$, *so inhabitants of the infected area should evacuate within* $6$ *hours to avoid sickness.*

c. **A more conservative approach is to evacuate the infected environment before bacteria concentration levels reach one-third of fatal levels. Without consulting a calculator or other technology, offer a rough time limit for evacuation in this circumstance.**

*Under these guidelines, The bacteria level is dangerous when* $P(t) = 100 \cdot 2^t = \frac{1}{3}(10,000) \approx 3333.33$.
*Since* $2^5 = 32$, $P(5) \approx 3200$, *so the conservative approach is to recommend evacuation within* $5$ *hours.*

d. **Use the method of the Example to approximate when the infected environment will reach fatal levels (**$100\%$**) of bacteria concentration, to the nearest minute.**

*We need to approximate the solution to* $100 \cdot 2^t = 10,000$, *which is equivalent to solving* $2^t = 100$.

| $t$ | $2^t$ |
|---|---|
| 1 | 2 |
| 2 | 4 |
| 3 | 8 |
| 4 | 16 |
| 5 | 32 |
| 6 | 64 |
| 7 | 128 |

| $t$ | $2^t$ |
|---|---|
| 6.1 | 68.5935 |
| 6.2 | 73.5167 |
| 6.3 | 78.7932 |
| 6.4 | 84.4485 |
| 6.5 | 90.5097 |
| 6.6 | 97.0059 |
| 6.7 | 103.9683 |

| $t$ | $2^t$ |
|---|---|
| 6.61 | 97.6806 |
| 6.62 | 98.3600 |
| 6.63 | 99.0442 |
| 6.64 | 99.7331 |
| 6.65 | 100.4268 |
| | |
| | |

| $t$ | $2^t$ |
|---|---|
| 6.641 | 99.8022 |
| 6.642 | 99.8714 |
| 6.643 | 99.9407 |
| 6.644 | 100.0010 |
| | |
| | |
| | |

| $t$ | $2^t$ |
|---|---|
| 6.6436 | 99.9822 |
| 6.6437 | 99.9892 |
| 6.6438 | 99.9961 |
| 6.6439 | 100.0030 |
| | |
| | |
| | |

*Inhabitants need to evacuate within* $6.644$ *hours, which is approximately* $6$ *hours and* $39$ *minutes.*

# Lesson 8: The "WhatPower" Function

## Student Outcomes

- Students calculate a simple logarithm using the definition.

## Lesson Notes

The term *logarithm* is foreign and can be intimidating, so we begin the lesson with a more intuitive function, the "WhatPower" function, which is a simple renaming of the logarithm function. Do not explain this function to students directly, but let them figure out what the function does. The first two exercises have already been solved to provide a hint of how the "WhatPower" function works.

This lesson is just the first introduction to logarithms, and the work done here will prepare students to solve exponential equations of the form $ab^{ct} = d$ (**F-LE.A.4**) and use logarithms to model relationships between two quantities (**F-BF.B.4a**) in later lessons. In the next lessons, students will create logarithm tables to discover some of the basic properties of logarithms before continuing on to look at the graphs of logarithmic functions, and then to finally modeling logarithmic data. In this lesson, we develop the ideas and notation of logarithmic expressions, leaving many ideas to be explored later in the module.

## Classwork

### Opening Exercise (12 minutes)

Allow students to work in pairs or small groups to complete these exercises. Do not explain this function to students directly, but allow them to struggle to figure out what this new "WhatPower" function means and how to evaluate these expressions. When there is about two minutes left, instruct groups that have not finished parts (a)–(p) to skip to part (q) so that all groups have time to think about and state the definition of this function. You may choose to collect the groups' definitions on paper and share some or all of them with the class using the document camera. We will work on refining this definition through the lessons; in particular, we are interested in the allowable values of the base $b$.

**Opening Exercise**

**Evaluate each expression. The first two have been completed for you.**

a. $\text{WhatPower}_2(8) = 3$

   ***3, because*** $2^3 = 8$

b. $\text{WhatPower}_3(9) = 2$

   ***2, because*** $3^2 = 9$

c. $\text{WhatPower}_6(36) =$ ______

   ***2, because*** $6^2 = 36$

d. $\text{WhatPower}_2(32) =$ ______

**5, *because* $2^5 = 32$**

e. $\text{WhatPower}_{10}(1000) =$ ______

**3, *because* $10^3 = 1000$**

f. $\text{WhatPower}_{10}(1,000,000) =$ ______

**6, *because* $10^6 = 1,000,000$**

g. $\text{WhatPower}_{100}(1,000,000) =$ ______

**3, *because* $100^3 = 1,000,000$**

h. $\text{WhatPower}_4(64) =$ ______

**3, *because* $4^3 = 64$**

i. $\text{WhatPower}_2(64) =$ ______

**6, *because* $2^6 = 64$**

j. $\text{WhatPower}_9(3) =$ ______

**$\frac{1}{2}$, *because* $9^{\frac{1}{2}} = 3$**

k. $\text{WhatPower}_5(\sqrt{5}) =$ ______

**$\frac{1}{2}$, *because* $5^{\frac{1}{2}} = \sqrt{5}$**

l. $\text{WhatPower}_{\frac{1}{2}}\left(\frac{1}{8}\right) =$ ______

**3, *because* $\left(\frac{1}{2}\right)^3 = \frac{1}{8}$**

m. $\text{WhatPower}_{42}(1) =$ ______

**0, *because* $42^0 = 1$**

n. $\text{WhatPower}_{100}(0.01) =$ ______

**$-1$, *because* $100^{-1} = 0.01$**

o. $\text{WhatPower}_2\left(\frac{1}{4}\right) =$ ______

**$-2$, *because* $2^{-2} = \frac{1}{4}$**

p. $\text{WhatPower}_{\frac{1}{4}}(2) =$ ______

$-\frac{1}{2}$, *because* $\left(\frac{1}{4}\right)^{-\frac{1}{2}} = 4^{\frac{1}{2}} = 2$

MP.8

q. **With your group members, write a definition for the function $\text{WhatPower}_b$, where $b$ is a number.**

*The value of* $\text{WhatPower}_b$ *is the number you need to raise b to in order to get* $x$*. That is, if* $b^L = x$*, then* $L = \text{WhatPower}_b(x)$.

## Discussion (3 minutes)

Discuss the definitions the students created, but do not settle on an official definition just yet. To reinforce the idea of how this function works, ask students a series of "WhatPower" questions, writing the expressions on the board or the document camera, and reading $\text{WhatPower}_b(x)$ as "What power of $b$ is $x$?" Be sure that students are visually seeing the odd structure of this notation and hearing the question "What power of $b$ is $x$?" to reinforce the meaning of this function that depends on both the parameter $b$ and the variable $x$.

- $\text{WhatPower}_2(16)$
  - $4$
- $\text{WhatPower}_2(4)$
  - $2$
- $\text{WhatPower}_2(\sqrt{2})$
  - $\frac{1}{2}$
- $\text{WhatPower}_2(1)$
  - $0$
- $\text{WhatPower}_2\left(\frac{1}{8}\right)$
  - $-3$

## Exercises 1–9 (8 minutes)

The point of this set of exercises is for students to determine which real numbers $b$ make sense as a base for the $\text{WhatPower}_b$ function. Have students complete this exercise in pairs or small groups, and allow time for students to debate.

**Exercises 1–9**

**Evaluate the following expressions and justify your answers.**

1. $\text{WhatPower}_7(49)$

   $\text{WhatPower}_7(49) = 2$ *because* $7^2 = 49$.

2. $\text{WhatPower}_0(7)$

   $\text{WhatPower}_0(7)$ *does not make sense because there is no power of* $0$ *that will produce* $7$.

3. $\mathbf{WhatPower_5(1)}$

   $\mathbf{WhatPower_5(1) = 0}$ *because* $\mathbf{5^0 = 1}$.

4. $\mathbf{WhatPower_1(5)}$

   $\mathbf{WhatPower_1(5)}$ *does not exist because for any exponent* $L$, $1^L = 1$, *so there is no power of* $1$ *that will produce* $5$.

5. $\mathbf{WhatPower_{-2}(16)}$

   $\mathbf{WhatPower_{-2}(16) = 4}$ *because* $(-2)^4 = 16$.

6. $\mathbf{WhatPower_{-2}(32)}$

   $\mathbf{WhatPower_{-2}(32)}$ *does not make sense because there is no power of* $-2$ *that will produce* $32$.

7. $\mathbf{WhatPower_{\frac{1}{3}}(9)}$

   $\mathbf{WhatPower_{\frac{1}{3}}(9) = -2}$ *because* $\left(\frac{1}{3}\right)^{-2} = 9$.

8. $\mathbf{WhatPower_{-\frac{1}{3}}(27)}$

   $\mathbf{WhatPower_{-\frac{1}{3}}(27)}$ *does not make sense because there is no power of* $-\frac{1}{3}$ *that will produce* $27$.

9. **Describe the allowable values of $b$ in the expression $\mathrm{WhatPower}_b(x)$. When can we define a function $f(x) = \mathrm{WhatPower}_b(x)$? Explain how you know.**

   *If $b = 0$ or $b = 1$, then the expression $\mathrm{WhatPower}_b(x)$ does not make sense. If $b < 0$, then the expression $\mathrm{WhatPower}_b(x)$ makes sense for some values of $x$ but not for others, so we cannot define a function $f(x) = \mathrm{WhatPower}_b(x)$ if $b < 0$. Thus, we can define the function $f(x) = \mathrm{WhatPower}_b(x)$ if $b > 0$ and $b \neq 1$.*

## Discussion (5 Minutes)

Ask student groups to share their responses to Exercise 9, in which they determined which values of $b$ are allowable in the $\mathrm{WhatPower}_b$ function. By the end of this Discussion, be sure that all groups understand that we need to restrict $b$ so that either $0 < b < 1$ or $b > 1$. Then, continue on to rename the WhatPower function to its true name, the logarithm base $b$.

- What we are calling the "WhatPower" function is known by the mathematical term logarithm, built from the Greek word *logos* (pronounced lo-gohs), meaning ratio, and *arithmos* (pronounced uh-rith-mohs), meaning number. The number $b$ is the base of the logarithm, and we denote the logarithm base $b$ of $x$ (which means the power to which we raise $b$ to get $x$) by $\log_b(x)$. That is, whenever you see $\log_b(x)$, think of $\mathrm{WhatPower}_b(x)$.

MP.8 Discuss the definition shown in the Frayer diagram below. Ask students to articulate the definition in their own words to a partner and then share some responses. Have students work with a partner to fill in the remaining parts of the diagram and then share responses as a class. Provide some sample examples and non-examples as needed to illustrate some of the characteristics of logarithms.

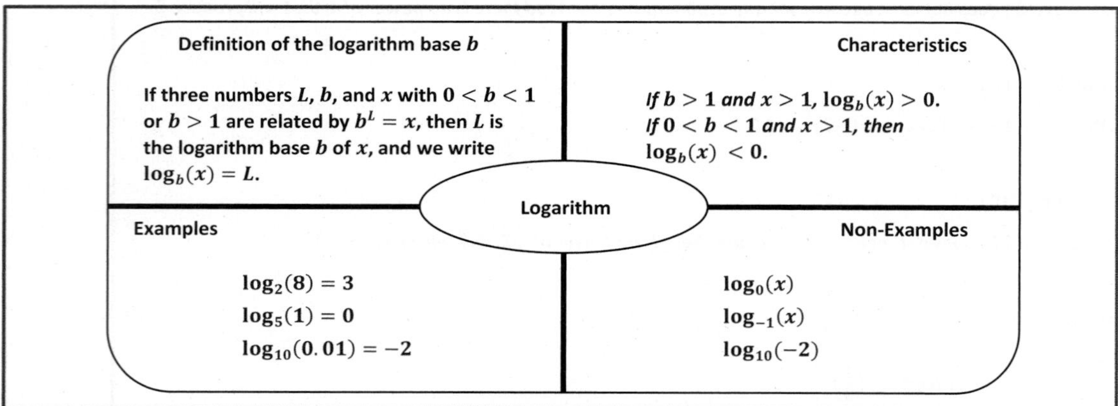


- What are some examples of logarithms?
  - $\log_2(4) = 2$
  - $\log_3(27) = 3$
  - $\log_{10}(0.10) = -1$
- What are some non-examples?
  - $\log_0(4)$
  - $\log_1(4)$
- Why can't $b = 0$? Why can't $b = 1$?
  - *If there is a number* $L$ *so that* $\log_0(4) = L$, *then* $0^L = 4$. *But, there is no number* $L$ *such that* $0^L$ *is* $4$, *so this does not make sense. Similar reasoning can be applied to* $\log_1(4)$.
- Is $\log_5(25)$ a valid example?
  - *Yes.* $\log_5(25) = 2$ *because* $5^2 = 25$.
- Is $\log_5(-25)$ a valid example?
  - *No. There is no number* $L$ *such that* $5^L = -25$. *It is impossible to raise a positive base to an exponent and get a negative value.*
- Is $\log_5(0)$ a valid example?
  - *No. There is no number* $L$ *such that* $5^L = 0$. *It is impossible to raise a positive base to an exponent and get an answer of* $0$.
- So what are some characteristics of logarithms?
  - *The base* $b$ *must be a positive number not equal to* $1$. *The input must also be a positive number. The output may be any real number (positive, negative, or* $0$*).*

### Examples (4 Minutes)

Lead the class through the computation of the following logarithms. These have all been computed in the Opening Exercise using the WhatPower terminology.

**Examples**

1. $\log_2(8) = 3$

   ***3, because $2^3 = 8$.***

2. $\log_3(9) = 2$

   ***2, because $3^2 = 9$.***

3. $\log_6(36) =$ ______

   ***2, because $6^2 = 36$.***

4. $\log_2(32) =$ ______

   ***5, because $2^5 = 32$.***

5. $\log_{10}(1000) =$ ______

   ***3, because $10^3 = 1000$.***

6. $\log_{42}(1) =$ ______

   ***0, because $42^0 = 1$.***

7. $\log_{100}(0.01) =$ ______

   ***−1, because $100^{-1} = 0.01$.***

8. $\log_2\left(\frac{1}{4}\right) =$ ______

   ***−2, because $2^{-2} = \frac{1}{4}$.***

### Exercise 10 (6 minutes)

Have students complete this exercise alone or in pairs.

*Scaffolding:*

- If students are struggling with notation, give them examples where they convert between logarithmic and exponential form.
- Use this chart as a visual support.

| Logarithmic form | Exponential form |
|---|---|
| $\log_8(64) = 2$ | |
| | $8^{-2} = \frac{1}{64}$ |
| $\log_{64}(4) = \frac{1}{3}$ | |

**Exercise 10**

**10. Compute the value of each logarithm. Verify your answers using an exponential statement.**

a. $\log_2(32)$

   ***$\log_2(32) = 5$, because $2^5 = 32$.***

b. $\log_3(81)$

   ***$\log_3(81) = 4$, because $3^4 = 81$.***

c. $\log_9(81)$

   ***$\log_9(81) = 2$, because $9^2 = 81$.***

d. $\log_5(625)$

   ***$\log_5(625) = 4$, because $5^4 = 625$.***

e. $\log_{10}(1,000,000,000)$

   ***$\log_{10}(1,000,000,000) = 9$, because $10^9 = 1,000,000,00$.***

f. $\log_{1000}(1,000,000,000)$

   ***$\log_{1000}(1,000,000,000) = 3$, because $1000^3 = 1,000,000,000$.***

g. $\log_{13}(13)$

   ***$\log_{13}(13) = 1$, because $13^1 = 13$.***

h. $\log_{13}(1)$

   ***$\log_{13}(1) = 0$, because $13^0 = 1$.***

i. $\log_9(27)$

$\log_9(27) = \frac{3}{2}$, *because* $9^{\frac{3}{2}} = 3^3 = 27$.

j. $\log_7(\sqrt{7})$

$\log_7(\sqrt{7}) = \frac{1}{2}$, *because* $7^{\frac{1}{2}} = \sqrt{7}$.

k. $\log_{\sqrt{7}}(7)$

$\log_{\sqrt{7}}(7) = 2$, *because* $\left(\sqrt{7}\right)^2 = 7$.

l. $\log_{\sqrt{7}}\left(\frac{1}{49}\right)$

$\log_{\sqrt{7}}\left(\frac{1}{49}\right) = -4$, *because* $\left(\sqrt{7}\right)^{-4} = \frac{1}{\left(\sqrt{7}\right)^4} = \frac{1}{49}$.

m. $\log_x(x^2)$

$\log_x(x^2) = 2$, *because* $(x)^2 = x^2$.

## Closing (2 minutes)

Ask students to summarize the important parts of the lesson, either in writing, to a partner, or as a class. Use this as an opportunity to informally assess understanding of the lesson. The following are some important summary elements.

**Lesson Summary**

- If three numbers, $L$, $b$, and $x$ are related by $x = b^L$, then $L$ is the *logarithm base* $b$ of $x$, and we write $\log_b(x)$. That is, the value of the expression $L = \log_b(x)$ is the power of $b$ needed to obtain $x$.
- Valid values of $b$ as a base for a logarithm are $0 < b < 1$ and $b > 1$.

## Exit Ticket (5 minutes)

Name ______________________________ Date ______________

# Lesson 8: The "WhatPower" Function

## Exit Ticket

1. Explain why we need to specify $0 < b < 1$ and $b > 1$ as valid values for the base $b$ in the expression $\log_b(x)$.

2. Calculate the following logarithms.

   a. $\log_5(25)$

   b. $\log_{10}\left(\frac{1}{100}\right)$

   c. $\log_9(3)$

## Exit Ticket Sample Solutions

1. Explain why we need to specify $0 < b < 1$ and $b > 1$ as valid values for the base $b$ in the expression $\log_b(x)$.

   *If $b = 0$, then $\log_0(x) = L$ means that $0^L = x$, which cannot be true if $x \neq 0$.*

   *If $b = 1$, then $\log_1(x) = L$ means that $1^L = x$, which cannot be true if $x \neq 1$.*

   *If $b < 0$, then $\log_b(x) = L$ makes sense for some but not all values of $x > 0$; for example, if $b = -2$ and $x = 32$, there is no power of $-2$ that would produce $32$, so $\log_{-2}(32)$ does not makes sense.*

   *Thus, if $b \leq 0$ or $b = 1$, then for many values of $x$, the expression $\log_b(x)$ does not make sense.*

2. Calculate the following logarithms.

   a. $\log_5(25)$

   $\log_5(25) = 2$

   b. $\log_{10}\left(\frac{1}{100}\right)$

   $\log_{10}\left(\frac{1}{100}\right) = -2$

   c. $\log_9(3)$

   $\log_9(3) = \frac{1}{2}$

## Problem Set Sample Solutions

In this introduction to logarithms, the students are only asked to find simple logarithms base $b$, in which the logarithm is an integer or simple fraction, and the expression can be calculated by inspection.

1. Rewrite each of the following in the form $\text{WhatPower}_b(x) = L$.

   a. $3^5 = 243$

   $\text{WhatPower}_3(243) = 5$

   b. $6^{-3} = \frac{1}{216}$

   $\text{WhatPower}_6\left(\frac{1}{216}\right) = -3$

   c. $9^0 = 1$

   $\text{WhatPower}_9(1) = 0$

2. Rewrite each of the following in the form $\log_b(x) = L$.

   a. $16^{\frac{1}{4}} = 2$

   $\log_{16}(2) = \frac{1}{4}$

   b. $10^3 = 1{,}000$

   $\log_{10}(1{,}000) = 3$

   c. $b^k = r$

   $\log_b(r) = k$

3. Rewrite each of the following in the form $b^L = x$.

   a. $\log_5(625) = 4$

   $5^4 = 625$

   b. $\log_{10}(0.1) = -1$

   $10^{-1} = 0.1$

   c. $\log_{27} 9 = \frac{2}{3}$

   $27^{\frac{2}{3}} = 9$

4. **Consider the logarithms base 2. For each logarithmic expression below, either calculate the value of the expression, or explain why the expression does not make sense.**

   a. $\log_2(1024)$

      $10$

   b. $\log_2(128)$

      $7$

   c. $\log_2(\sqrt{8})$

      $\frac{3}{2}$

   d. $\log_2\left(\frac{1}{16}\right)$

      $-4$

   e. $\log_2(0)$

      ***This does not make sense. There is no value of $L$ so that $2^L = 0$.***

   f. $\log_2\left(-\frac{1}{32}\right)$

      ***This does not make sense. There is no value of $L$ so that $2^L$ is negative.***

5. **Consider the logarithms base 3. For each logarithmic expression below, either calculate the value of the expression, or explain why the expression does not make sense.**

   a. $\log_3(243)$

      $5$

   b. $\log_3(27)$

      $3$

   c. $\log_3(1)$

      $0$

   d. $\log_3\left(\frac{1}{3}\right)$

      $-1$

   e. $\log_3(0)$

      ***This does not make sense. There is no value of $L$ so that $3^L = 0$.***

   f. $\log_3\left(-\frac{1}{3}\right)$

      ***This does not make sense. There is no value of $L$ so that $3^L < 0$.***

6. Consider the logarithms base 5. For each logarithmic expression below, either calculate the value of the expression, or explain why the expression does not make sense.

   a. $\log_5(3125)$

      $5$

   b. $\log_5(25)$

      $2$

   c. $\log_5(1)$

      $0$

   d. $\log_5\left(\frac{1}{25}\right)$

      $-2$

   e. $\log_5(0)$

      *This does not make sense. There is no value of $L$ so that $5^L = 0$.*

   f. $\log_5\left(-\frac{1}{25}\right)$

      *This does not make sense. There is no value of $L$ so that $5^L$ is negative.*

7. Is there any positive number $b$ so that the expression $\log_b(0)$ makes sense? Explain how you know.

   *No, there is no value of $L$ so that $b^L = 0$. I know $b$ has to be a positive number. A positive number raised to an exponent never equals $0$.*

8. Is there any positive number $b$ so that the expression $\log_b(-1)$ makes sense? Explain how you know.

   *No, since $b$ is positive, there is no value of $L$ so that $b^L$ is negative. A positive number raised to an exponent never has a negative value.*

9. Verify each of the following by evaluating the logarithms.

   a. $\log_2(8) + \log_2(4) = \log_2(32)$

      $3 + 2 = 5$

   b. $\log_3(9) + \log_3(9) = \log_3(81)$

      $2 + 2 = 4$

   c. $\log_4(4) + \log_4(16) = \log_4(64)$

      $1 + 2 = 3$

   d. $\log_{10}(10^3) + \log_{10}(10^4) = \log_{10}(10^7)$

      $3 + 4 = 7$

10. Looking at the results from Problem 9, do you notice a trend or pattern? Can you make a general statement about the value of $\log_b(x) + \log_b(y)$?

    *The sum of two logarithms of the same base is found by multiplying the input values, $\log_b(x) + \log_b(y) = \log_b(xy)$ (Note to teacher: Do not evaluate this answer harshly. This is just a preview of a property that students will learn later in the module.)*

11. To evaluate $\log_2(3)$, Autumn reasoned that since $\log_2(2) = 1$ and $\log_2(4) = 2$, $\log_2(3)$ must be the average of $1$ and $2$ and, therefore, $\log_2(3) = 1.5$. Use the definition of logarithm to show that $\log_2(3)$ cannot be $1.5$. Why is her thinking not valid?

    *According to the definition of logarithm, $\log_2(3) = 1.5$ only if $2^{1.5} = 3$. According to the calculator, $2^{1.5} \approx 2.828$, so $\log_2(3)$ cannot be $1.5$. Autumn was assuming that the outputs would follow a linear pattern, but since the outputs are exponents, the relationship is not linear.*

12. Find the value of each of the following.

    a. If $x = \log_2(8)$ and $y = 2^x$, find the value of $y$.

       $y = 8$

    b. If $\log_2(x) = 6$, find the value of $x$.

       $x = 64$

    c. If $r = 2^6$ and $s = \log_2(r)$, find the value of $s$.

       $s = 6$

# Lesson 9: Logarithms—How Many Digits Do You Need?

## Student Outcomes

- Students use logarithms to determine how many characters are needed to generate unique identification numbers in different scenarios.
- Students understand that logarithms are useful when relating the number of digits in a number to the magnitude of the number and that base 10 logarithms are useful when measuring quantities that have a wide range of values such as the magnitude of earthquakes, volume of sound, and pH levels in chemistry.

## Lesson Notes

In this lesson, students learn that logarithms are useful in a wide variety of situations but have extensive application when we want to generate a list of unique identifiers for a population of a given size (**N-Q.A.2**). This application of logarithms is used in computer programming, when determining how many digits are needed in a phone number to have enough unique numbers for a population, and more generally when assigning a scale to any quantity that has a wide range of values.

In this lesson, students make sense of a simple scenario and then see how it can be applied to other real world situations (MP.1 and MP.2). They observe and extend patterns to formulate a model (MP.7 and MP.4). They reason about and make sense of situations in context and use logarithms to draw conclusions regarding different real world scenarios (MP.3 and MP.4).

## Classwork

### Opening Exercise (2 minutes)

Remind students that the WhatPower expressions are called logarithms, and announce that we will be using logarithms to help us make sense of and solve some real world problems.

Students briefly convert two WhatPower expressions into a logarithmic expression and evaluate the result.

**Opening Exercise**

a. **Evaluate $\text{WhatPower}_2(8)$. State your answer as a logarithm and evaluate it.**

$\log_2(8) = 3$

b. **Evaluate $\text{WhatPower}_5(625)$. State your answer as a logarithm and evaluate it.**

$\log_5(625) = 4$

If students struggle with these exercises, you may want to plan for some additional practice on problems like those found in Lesson 7.

## Exploratory Challenge (15 minutes)

Divide students up into small groups, and give them about ten minutes to work through the questions that follow. As you circulate around the room, encourage students to be systematic when assigning IDs to the club members. If a group is stuck, you can ask questions to help move the group along.

- Remember, we only want to use A's and B's, but two-character IDs only provide enough for four people. Can you provide an example of a three-character ID using only the letters A and B?
  - *A three-character ID might be ABA or AAA.*
- How many different IDs would three characters make? How could you best organize your results?
  - *Three characters would make 8 IDs because I could take the four I already have and add an A or B onto the end. The best way to organize it is to take the existing two-character IDs and add an A and then take those existing two-character IDs again and add a B.*

**Exploratory Challenge**

**Autumn is starting a new club with eight members including herself. She wants everyone to have a secret identification code made up of only A's and B's. For example, using two characters, her ID code could be AB, which also happens to be her initials.**

a. **Using A's and B's, can Autumn assign each club member a unique two character ID using only A's and B's? Justify your answer. Here's what Autumn has so far.**

| Club Member Name | Secret ID |
|---|---|
| Autumn | AA |
| Kris | |
| Tia | |
| Jimmy | |

| Club Member Name | Secret ID |
|---|---|
| Robert | |
| Jillian | |
| Benjamin | |
| Scott | |

***No, she cannot assign a unique 2-character ID to each member. The only codes available are AA, BA, AB, and BB, and there are eight people.***

b. **Using A's and B's, how many characters would be needed to assign each club member a unique ID code? Justify your answer by showing the IDs you would assign to each club member by completing the table above (adjust Autumn's ID if needed).**

***You would need three characters in each code. A completed table is shown below. Students could assign one of the unique codes to any club member so this is not the only possible solution.***

| Club Member Name | Secret ID |
|---|---|
| Autumn | *AAA* |
| Kris | *BAA* |
| Tia | *ABA* |
| Jimmy | *BBA* |

| Club Member Name | Secret ID |
|---|---|
| Robert | *AAB* |
| Jillian | *BAB* |
| Benjamin | *ABB* |
| Scott | *BBB* |

MP.7 & MP.4

When the club grew to $16$ members, Autumn started noticing a pattern.

Using A's and B's:

i. Two people could be given a secret ID with $1$ letter: A and B.

ii. Four people could be given a secret ID with $2$ letters: AA, BA, AB, BB.

iii. Eight people could be given a secret ID with $3$ letters: AAA, BAA, ABA, BBA, AAB, BAB, ABB, BBB.

c. Complete the following statement and list the secret IDs for the $16$ people.

$16$ people could be given a secret ID with _________ letters using A's and B's.

*$16$ people could be given a secret ID with $4$ characters. Notice the original members have their original three-character code with an A added to the end. Then, the newer members have the original three-character codes with a B added to the end.*

| Club Member Name | Secret ID |
|---|---|
| Autumn | *AAAA* |
| Kris | *BAAA* |
| Tia | *ABAA* |
| Jimmy | *BBAA* |
| Robert | *AABA* |
| Jillian | *BABA* |
| Benjamin | *ABBA* |
| Scott | *BBBA* |

| Club Member Name | Secret ID |
|---|---|
| Gwen | *AAAB* |
| Jerrod | *BAAB* |
| Mykel | *ABAB* |
| Janette | *BBAB* |
| Nellie | *AABB* |
| Serena | *BABB* |
| Ricky | *ABBB* |
| Mia | *BBBB* |

d. Describe the pattern in words. What type of function could be used to model this pattern?

*The number of people in the club is a power of $2$. The number of characters needed to generate a unique ID using only two characters is the exponent of the power of $2$.*

$$\log_2(2) = 1$$
$$\log_2(4) = 2$$
$$\log_2(8) = 3$$
$$\log_2(16) = 4$$

*A logarithm function could be used to model this pattern. For $16$ people, you will need a four-character ID.*

To debrief this Exploratory Challenge, have different groups explain how they arrived at their solutions. If a group does not demonstrate an efficient way to organize its answers when the club membership increases, be sure to show it to the class. For example, the $16$ group member IDs were generated by adding an A onto the end of the original $8$ IDs and then adding a B onto the end of the original $8$ IDs, as shown in the solutions above.

## Exercises 1–2 (3 minutes)

Give students a few minutes to answer these questions individually or in groups. Check to see if students are using logarithms when they explain their solutions. If they are not, be sure to review the answers with the entire class using logarithm notation.

**Exercises 1–2**

**In the previous problems, the letters A and B were like the digits in a number. A four-digit ID for Autumn's club could be any four-letter arrangement of A's and B's because in her ID system, the only digits are the letters A and B.**

1. **When Autumn's club grows to include more than 16 people, she will need five digits to assign a unique ID to each club member. What is the maximum number of people that could be in the club before she needs to switch to a six-digit ID? Explain your reasoning.**

   *Since $\log_2(32) = 5$ and $\log_2(64) = 6$, she will need to switch to a six-digit ID when the club gets more than $32$ members.*

2. **If Autumn has $256$ members in her club, how many digits would she need to assign each club member a unique ID using only A's and B's? Show how you got your answers.**

   *She will need $8$ digits because $\log_2(256) = 8$.*

## Discussion (10 minutes)

Computers store keyboard characters, such as 1, 5, X, x, Q, @, /, and &, using an identification system much like Autumn's system called ASCII, which stands for American Standard Code for Information Interchange. We pronounce the acronym ASCII as "as-key." Each character in a font list on the computer is given an ID that a computer can recognize. A computer is essentially a lot of electrical switches, which can be in one of two states, ON or OFF, just like Autumn's A's and B's.

There are usually $256$ characters in a font list, so using the solution to Exercise 2 above, a computer needs eight positions or digits to encode each character in a font list.

- For example, the standard ASCII code for uppercase P is 01010000. If A is zero and B is 1, how would uppercase P be encoded using Autumn's code?
  - *Using A's and B's in Autumn's code, this would be ABABAAAA.*
- How would the computer read this code?
  - *The computer reads the code as "on, off, on, off, on, on, on, on."*

If time permits, a quick Internet search for the term *ASCII* will return web pages where you can view the standard code for different keyboard symbols. Each character in the ASCII code is called a *bit*, and the entire 8-character code is called a *byte*. Each byte is made up of eight bits, and each byte describes a unique character in the font list such as a P, p, %, _, 4, etc.

- When computer saves a basic text document and reports that it is 3,242 bytes, what do you think that means?
  - *It means that there would be $3{,}242$ letters, symbols, spaces, punctuation marks, etc. in the document.*

We have seen how to create unique IDs using two letters for both Autumn's secret club and to encode text characters in a way that is readable by a computer using ASCII code. Next, we will examine why a logarithm really is the right operation to describe the number of characters or digits needed to create unique identifiers for people by exploring some real-world situations where people are assigned a number.

*Scaffolding:*

Use a word wall or a chart to provide a visual reference for English language learners using the academic terms related to computers.

**ASCII** (American Standard Code for Information Exchange, acronym pronounced "as-key")

**Bit** (one of eight positions in a byte)

**Byte** (a unique eight-character identifier for each ASCII symbol in font list)

*Scaffolding:*

Ask advanced learners to quickly estimate how many digits would be needed to generate ID numbers for the number of students enrolled in your school or in your school district. For example, a school district with more than $10{,}000$ students would need at least a five-digit ID.

## Example 1 (5 minutes)

This is a simplified example that switches students to using base 10 logarithms because we are going to be assigning IDs using the digits 0–9. Give students a few minutes to think about this answer and have them discuss their ideas with a partner. Most students will likely say let the first person be 0, the next person be 1, and so on up to 999. Then, tie the solution to logarithms. Since there are 10 symbols (digits), we can use $\log_{10}(1000) = 3$ to find the answer, which just counts the number of digits needed to count to 999.

- How can a logarithm help you determine the solution quickly?
  - *The logarithm counts the number of digits needed because each time we add another digit to our numbers, we are increasing by a factor of* 10. *For example,* $1 = 10^0$, $10 = 10^1$, $100 = 10^2$, *etc.*

> **Example 1**
>
> **A thousand people are given unique identifiers made up of the digits $0, 1, 2, \dots, 9$. How many digits would be needed for each ID number?**
>
> ***You would just need three digits:*** $\{000, 001, 002, \dots, 099, 100, 101, \dots 998, 999\}$.

Using logarithms, we need to determine the value of $\log(1000)$, which is 3. This will quickly tell us the number of digits needed to uniquely identify any range of numbers. You can follow up by asking students to extend their thinking.

- When would you need to switch from four to five digits to assign unique numbers to a population?
  - *You could assign up to* $10^4$ *people a four-digit ID, which would be* 10,000 *people. Once you exceeded that number, you would need five digits to assign each person a unique number.*

## Exercises 3–4 (5 minutes)

Students should return to their small groups to work these exercises. Have different groups present their solutions to the whole class after a few minutes. Discuss different approaches, and make sure that students see the power of using a logarithm to help them quickly solve or justify a solution to the problem.

MP.1 & MP.3

> **Exercises 3–4**
>
> 3. **There are approximately 317 million people in the United States. Compute and use $\log(100,000,000)$ and $\log(1,000,000,000)$ to explain why Social Security numbers are 9 digits long.**
>
>    ***We know that*** $\log(100,000,000) = 8$, ***which is the number of digits needed to assign an ID to 100 million people. We know that*** $\log(1,000,000,000) = 9$, ***which is the number of digits needed to assign an ID to 1 billion people.***
>
>    ***The United States government will not need to increase the number of digits in a Social Security number until the United States population reaches one billion.***
>
> 4. **There are many more telephones than the number of people in the United States because of people having home phones, cell phones, business phones, fax numbers, etc. Assuming we need at most 10 billion phone numbers in the United States, how many digits would be needed so that each phone number is unique? Is this reasonable? Explain.**
>
>    ***Since*** $\log(10,000,000,000) = 10$, ***you would need a ten-digit phone number in order to have ten billion unique numbers. Phone numbers in the United States are 10 digits long. If you divide 10 billion by 317 million (the number of people in the United States), that would allow for approximately 31 phone numbers per person. That is plenty of numbers for individuals to have more than one number, leaving many additional numbers for businesses and the government.***

## Closing (2 minutes)

Ask students to respond to the following statements in writing or with a partner. Share a few answers to close the lesson before students begin the Exit Ticket. Preview other situations where logarithms are useful, such as the Richter scale for measuring the magnitude of an earthquake.

- To increase the value of $\log_2(x)$ by $1$, you would multiply $x$ by $2$. To increase the value of $\log_{10}(x)$ by $1$, you would multiply $x$ by $10$. How does this idea apply to the situations in today's lesson?
  - *We saw that each time the population of Autumn's club doubled, we needed to increase the total number of digits needed for the ID numbers by $1$. We saw that since the population of the United States was between $100$ million and $1$ billion, we only needed $9$ digits ($\log(1{,}000{,}000{,}000)$) to generate a Social Security number.*
- Situations like the ones in today's lesson can be modeled with logarithms. Can you think of a situation besides the ones we discussed today where it would make sense to use logarithms?
  - *Any time a measurement can take on a wide range of values, such as the magnitude of an earthquake or the volume (decibel) level of sounds, a logarithm could be used to model the situation.*

## Exit Ticket (3 minutes)

Name ______________________________________________ Date ____________________

# Lesson 9: Logarithms—How Many Digits Do You Need?

## Exit Ticket

A brand new school district needs to generate ID numbers for its student body. The district anticipates a total enrollment of $75{,}000$ students within the next ten years. Will a five-digit ID number comprising the symbols $0, 1, \dots, 9$ be enough? Explain your reasoning.

## Exit Ticket Sample Solutions

**A brand new school district needs to generate ID numbers for its student body. The district anticipates a total enrollment of $75,000$ students within the next ten years. Will a five-digit ID number comprised of the symbols $0, 1, \dots, 9$ be enough? Explain your reasoning.**

*$\log(10,000) = 4$ and $\log(100,000) = 5$, so $5$ digits should be enough. However, students who enter school at the kindergarten level in the tenth year of this numbering scheme would need to keep their IDs for $13$ years. Dividing $75,000$ by $13$ shows there would be roughly $6,000$ students per grade. Adding that many students per year would take the number of needed IDs at any one time over $100,000$ in just a few more years. The district should probably use a six-digit ID number.*

## Problem Set Sample Solutions

1. **The student body president needs to assign each officially sanctioned club on campus a unique ID number for purposes of tracking expenses and activities. She decides to use the letters A, B, and C to create a unique three-character code for each club.**

   a. **How many clubs can be assigned a unique ID according to this proposal?**

   *Since $\log_3(27) = 3$, the president could have $27$ clubs according to this proposal.*

   b. **There are actually over $500$ clubs on campus. Assuming the student body president still wants to use the letters A, B, and C, how many characters would be needed to generate a unique ID for each club?**

   *We need to estimate $\log_3(500)$. Since $3^5 = 243$ and $3^6 = 729$, she could use a six-character combination of letters and have enough unique IDs for up to $729$ clubs.*

2. **Can you use the numbers $1, 2, 3$, and $4$ in a combination of four digits to assign a unique ID to each of $500$ people? Explain your reasoning.**

$$\log_4(4) = 1$$
$$\log_4(16) = 2$$
$$\log_4(64) = 3$$
$$\log_4(256) = 4$$
$$\log_4(1024) = 5$$

   *No, you would need to use a five-digit ID using combinations of 1s, 2s, 3s and 4s such as $11111$ or $12341$, or you could use the numbers $1$ to $5$ in four characters such as $1231, 1232, 1233, 1234, 1235$, etc. because $\log_5(625) = 4$.*

3. **Automobile license plates typically have a combination of letters ($26$) and numbers ($10$). Over time, the state of New York has used different criteria to assign vehicle license plate numbers.**

   a. **From $1973$ to $1986$, the state used a 3-letter and 4-number code where the three letters indicated the county where the vehicle was registered. Essex County had $13$ different 3-letter codes in use. How many cars could be registered to this county?**

   *Since $\log(10,000) = 4$, the 4-digit code could be used to register up $10,000$ vehicles. Multiply that by $13$ different county codes, and up to $130,000$ vehicles could be registered to Essex County.*

MP.1

b. **Since $2001$, the state has used a $3$-letter and $4$-number code but no longer assigns letters by county. Is this coding scheme enough to register $10$ million vehicles?**

*Since $\log_{26}(x) = 3$ when $x = 26^3 = 17,576$, there are $17,576$ three-letter codes. Since $\log(10000) = 4$, there are $10,000$ four-digit codes. Multiply $17576 \cdot 10,000 = 175,760,000$, and we see that this scheme generates over $100$ million license plate numbers.*

4. **The Richter scale uses common (base $10$) logarithms to assign a magnitude to an earthquake that is based on the amount of force released at the earthquake's source as measured by seismographs in various locations.**

a. **Explain the difference between an earthquake that is assigned a magnitude of $5$ versus one assigned a magnitude of $7$.**

*The difference between $5$ and $7$ is $2$, so a magnitude of $7$ would be $10^2$, or $100$, times greater force.*

b. **A magnitude $2$ earthquake can usually be felt by multiple people who are located near the earthquake's origin. The largest recorded earthquake was magnitude $9.5$ in Chile in $1960$. How many times greater force than a magnitude $2$ earthquake was the largest recorded earthquake?**

*The difference between $2$ and $9.5$ is $7.5$, so it would be about $10^{7.5}$ times the force. This is approximately $31$ million times greater force.*

c. **What would be the magnitude of an earthquake whose force was $1000$ times greater than a magnitude $4.3$ quake?**

*Since $10^3 = 1000$, the magnitude would be the sum of $4.3$ and $3$, which is $7.3$.*

MP.2 & MP.3

5. **Sound pressure level is measured in decibels (dB) according to the formula $L = 10\log\left(\frac{I}{I_0}\right)$, where $I$ is the intensity of the sound and $I_0$ is a reference intensity that corresponds to a barely perceptible sound.**

a. **Explain why this formula would assign $0$ decibels to a barely perceptible sound.**

*If we let $I = I_0$, then*

$$L = 10\log\left(\frac{I}{I_0}\right)$$
$$L = 10\log(1)$$
$$L = 10 \cdot 0$$
$$L = 0.$$

*Therefore, the reference intensity is always $0$ dB.*

b. **Decibel levels above $120$ dB can be painful to humans. What would be the intensity that corresponds to this level?**

$$120 = 10\log\left(\frac{I}{I_0}\right)$$
$$1.2 = \log\left(\frac{I}{I_0}\right)$$
$$\frac{I}{I_0} = 10^{1.2}$$
$$I \approx 15.8 I_0$$

*From this equation, we can see that the intensity is about $16$ times greater than barely perceptible sound.*

# Lesson 10: Building Logarithmic Tables

## Student Outcomes

- Students construct a table of logarithms base 10 and observe patterns that indicate properties of logarithms.

## Lesson Notes

In the previous lesson, students were introduced to the concept of the logarithm by finding the power to which we need to raise a base $b$ in order to produce a given number, which we originally called the $\text{WhatPower}_b$ function. In this lesson and the next, students will build their own base 10 logarithm tables using their calculators. By taking the time to construct the table themselves (as opposed to being handed a pre-prepared table), students will have a better opportunity to observe patterns in the table and practice MP.7. These observed patterns will lead to formal statements of the properties of logarithms in upcoming lessons. Using logarithmic properties to rewrite logarithmic expressions satisfies the foundational standard **A-SSE.A.2**.

## Materials Needed

Students will need access to a calculator or other technological tool able to compute exponents and logarithms base 10.

## Classwork

### Opening Exercise (3 minutes)

In this quick Opening Exercise, we ask students to recall the $\text{WhatPower}_b$ function from the previous lesson and reinforce that the logarithm base $b$ is the formal name of the $\text{WhatPower}_b$ function. We will consider only base 10 logarithms in this lesson as we construct our table, so this Opening Exercise is constrained to base 10 calculations.

At the end of this exercise, announce to the students that the notation $\log(x)$ without the little $b$ in the subscript means $\log_{10}(x)$. This is called the *common logarithm*.

> *Scaffolding:*
>
> Prompt struggling students to restate the logarithmic equation $\log_{10}(10^3) = x$ as the exponential equation $10^x = 10^3$.

**Opening Exercise**

**Find the value of the following expressions without using a calculator.**

| | | | |
|---|---|---|---|
| $\text{WhatPower}_{10}(1000)$ | $= 3$ | $\log_{10}(1000)$ | $= 3$ |
| $\text{WhatPower}_{10}(100)$ | $= 2$ | $\log_{10}(100)$ | $= 2$ |
| $\text{WhatPower}_{10}(10)$ | $= 1$ | $\log_{10}(10)$ | $= 1$ |
| $\text{WhatPower}_{10}(1)$ | $= 0$ | $\log_{10}(1)$ | $= 0$ |
| $\text{WhatPower}_{10}\left(\frac{1}{10}\right)$ | $= -1$ | $\log_{10}\left(\frac{1}{10}\right)$ | $= -1$ |
| $\text{WhatPower}_{10}\left(\frac{1}{100}\right)$ | $= -2$ | $\log_{10}\left(\frac{1}{100}\right)$ | $= -2$ |

**Formulate a rule based on your results above: If $k$ is an integer, then $\log_{10}(10^k) =$ ______.**

$\log_{10}(10^k) = k$

## Example 1 (6 minutes)

In this example, we get our first glimpse of the property $\log_b(xy) = \log_b(x) + \log_b(y)$. Be careful not to give this formula away; by the end of the next lesson, students should have discovered it for themselves.

- Suppose that you are an astronomer, and you measure the distance to a star as $100{,}000{,}000{,}000{,}000$ miles. A second star is collinear with the first star and the Earth and is $1{,}000{,}000$ times farther away from Earth than the first star is. How many miles is the second star from Earth? Note: The figure is not to scale.

**Example 1**

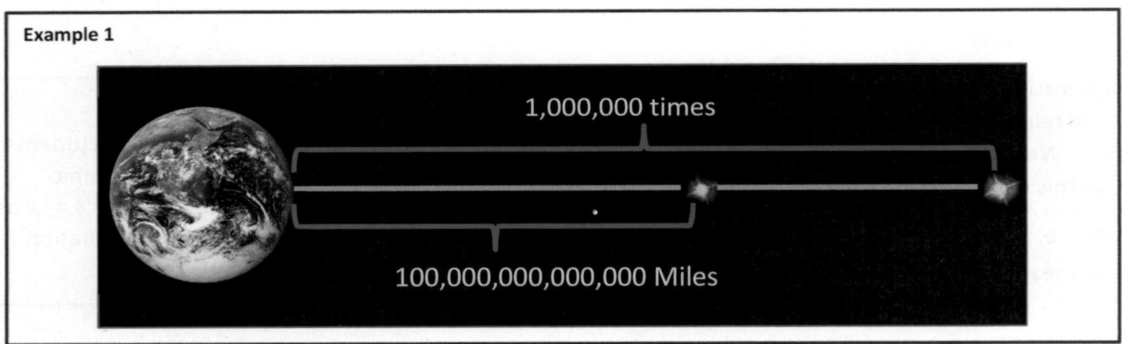


  - $(100{,}000{,}000{,}000{,}000)(1{,}000{,}000) = 100{,}000{,}000{,}000{,}000{,}000{,}000$, *so the second star is* $100$ *quintillion miles away from Earth.*
- How did you arrive at that figure?
  - *I counted the zeros; there are* $14$ *zeros in* $100{,}000{,}000{,}000{,}000$ *and* $6$ *zeros in* $1{,}000{,}000$, *so there must be* $20$ *zeros in the product.*
- Can we restate that in terms of exponents?
  - $(10^{14})(10^6) = 10^{20}$
- How are the exponents related?
  - $14 + 6 = 20$

- What are $\log(10^{14})$, $\log(10^{6})$, and $\log(10^{20})$?
  - $14$, $6$, *and* $20$
- In this case, can we state an equivalent expression for $\log(10^{14} \cdot 10^{6})$?
  - $\log(10^{14} \cdot 10^{6}) = \log(10^{14}) + \log(10^{6})$
- Why is this equation true?
  - $\log(10^{14} \cdot 10^{6}) = \log(10^{20}) = 20 = 14 + 6 = \log(10^{14}) + \log(10^{6})$
- Generalize to find an equivalent expression for $\log(10^{m} \cdot 10^{n})$ for integers $m$ and $n$. Why is this equation true?
  - $\log(10^{m} \cdot 10^{n}) = \log(10^{m+n}) = m + n = \log(10^{m}) + \log(10^{n})$
  - *This equation is true because when we multiply powers of* $10$ *together, the resulting product is a power of* $10$ *whose exponent is the sum of the exponents of the factors.*
- Keep this result in mind as we progress through the lesson.

## Exercises 1–6 (8 minutes)

Historically, logarithms were calculated using tables because there were no calculators or computers to do the work. Every scientist and mathematician kept a book of logarithmic tables on hand to use for calculation. It is very easy to find the value of a base 10 logarithm for a number that is a power of 10, but what about for the other numbers? In this exercise, students will find an approximate value of $\log(30)$ using exponentiation, the same way we approximated $\log_2(10)$ in Lesson 6. After this exercise, we will rely on the logarithm button on the calculator to compute logarithms base 10 for the remainder of this lesson. Emphasize to the students that logarithms are generally irrational numbers, so that the results produced by the calculator are only decimal approximations. As such, we should be careful to use the approximation symbol, $\approx$, when writing out a decimal expansion of a logarithm.

**Exercises**

1. **Find two consecutive powers of $10$ so that $30$ is between them. That is, find an integer exponent $k$ so that $10^{k} < 30 < 10^{k+1}$.**

   ***Since $10 < 30 < 100$, we have $k = 1$.***

2. **From your result in Exercise 1, $\log(30)$ is between which two integers?**

   ***Since $30$ is some power of $10$ between $1$ and $2$, $1 < \log(30) < 2$.***

3. **Find a number $k$ to one decimal place so that $10^{k} < 30 < 10^{k+0.1}$, and use that to find under and over estimates for $\log(30)$.**

   ***Since $10^{1.4} \approx 25.1188$ and $10^{1.5} \approx 31.6228$, we have $10^{1.4} < 30 < 10^{1.5}$. Then $1.4 < \log(30) < 1.5$, and $k \approx 1.4$.***

4. **Find a number $k$ to two decimal places so that $10^{k} < 30 < 10^{k+0.01}$, and use that to find under and over estimates for $\log(30)$.**

   ***Since $10^{1.47} \approx 29.5121$, and $10^{1.48} \approx 30.1995$, we have $10^{1.47} < 30 < 10^{1.48}$ so that $1.47 < \log(30) < 1.48$. So, $k \approx 1.47$.***

5. **Repeat this process to approximate the value of $\log(30)$ to 4 decimal places.**

*Since* $10^{1.477} \approx 29.9916$, *and* $10^{1.478} \approx 30.0608$, *we have* $10^{1.477} < 30 < 10^{1.478}$ *so that* $1.477 < \log(30) < 1.478$.

*Since* $10^{1.4771} \approx 29.9985$, *and* $10^{1.4772} \approx 30.0054$, *we have* $10^{1.4771} < 30 < 10^{1.4772}$ *so that* $1.4771 < \log(30) < 1.4772$.

*Since* $10^{1.47712} \approx 29.99991$, *and* $10^{1.47713} \approx 30.0006$, *we have* $10^{1.47712} < 30 < 10^{1.47713}$ *so that* $1.47712 < \log(30) < 1.47713$.

*Thus, to four decimal places,* $\log(30) \approx 1.4771$.

6. **Verify your result on your calculator, using the LOG button.**

*The calculator gives* $\log(30) \approx 1.477121255$.

## Discussion (1 minute)

In the next exercises, students will use their calculators to create a table of logarithms that they will analyze to look for patterns that will lead to the discovery of the logarithmic properties. The process of identifying and generalizing the observed patterns provides students with an opportunity to practice MP.7.

- Historically, since there were no calculators or computers, logarithms were calculated using a complicated algorithm involving multiple square roots. Thankfully, we have calculators and computers to do this work for us now.
- We will use our calculators to create a table of values of base 10 logarithms. Once the table is made, see what patterns you can observe.

## Exercises 7–10 (6 minutes)

Put students in pairs or small groups, but have students work individually to complete the table in Exercise 7. Before working on Exercises 8–10 in groups, have students check their tables against each other. You may need to remind students that $\log(x)$ means $\log_{10}(x)$.

7. **Use your calculator to complete the following table. Round the logarithms to 4 decimal places.**

| $x$ | $\log(x)$ |
|---|---|
| 1 | 0 |
| 2 | 0.3010 |
| 3 | 0.4771 |
| 4 | 0.6021 |
| 5 | 0.6990 |
| 6 | 0.7782 |
| 7 | 0.8451 |
| 8 | 0.9031 |
| 9 | 0.9542 |

| $x$ | $\log(x)$ |
|---|---|
| 10 | 1 |
| 20 | 1.3010 |
| 30 | 1.4771 |
| 40 | 1.6021 |
| 50 | 1.6990 |
| 60 | 1.7782 |
| 70 | 1.8451 |
| 80 | 1.9031 |
| 90 | 1.9542 |

| $x$ | $\log(x)$ |
|---|---|
| 100 | 2 |
| 200 | 2.3010 |
| 300 | 2.4771 |
| 400 | 2.6021 |
| 500 | 2.6990 |
| 600 | 2.7782 |
| 700 | 2.8451 |
| 800 | 2.9031 |
| 900 | 2.9542 |

8. **What pattern(s) can you see in the table from Exercise 7 as $x$ is multiplied by $10$? Write the pattern(s) using logarithmic notation.**

*I found the patterns* $\log(10x) = 1 + \log(x)$ *and* $\log(100x) = 2 + \log(x)$. *I also noticed that* $\log(100x) = 1 + \log(10x)$.

9. **What pattern would you expect to find for $\log(1000x)$? Make a conjecture and test it to see whether or not it appears to be valid.**

*I would guess that the values of* $\log(1000x)$ *will all start with 3. That is,* $\log(1000x) = 3 + \log(x)$. *This appears to be the case since* $\log(2000) \approx 3.3010$, $\log(500) \approx 3.6990$, *and* $\log(800) \approx 3.9031$.

10. **Use your results from Exercises 8 and 9 to make a conjecture about the value of $\log(10^k \cdot x)$ for any positive integer $k$.**

*It appears that* $\log(10^k \cdot x) = k + \log(x)$, *for any positive integer* $k$.

## Discussion (3 minutes)

Ask groups to share the patterns they observed in Exercise 8 and the conjectures they made in Exercises 9 and 10. Ensure that all students have the correct conjectures recorded in their notebooks or journals before continuing to the next set of exercises, which extend the result from Exercise 10 to all integers $k$ (and not just positive values of $k$).

## Exercises 11–14 (8 minutes)

In this set of exercises, students will discover a rule for calculating logarithms of the form $\log(10^k \cdot x)$, where $k$ is any integer. Have students again work individually to complete the table in Exercise 11 and to check their tables against each other before they proceed to discuss and answer Exercises 12–14 in groups.

*Scaffolding:*

If students are having difficulty seeing the pattern in the table for Exercise 12, nudge them to add together $\log(x)$ and $\log\left(\frac{x}{10}\right)$ for some values of $x$ in the table.

11. **Use your calculator to complete the following table. Round the logarithms to 4 decimal places.**

| $x$ | $\log(x)$ |
|---|---|
| $1$ | $0$ |
| $2$ | $0.3010$ |
| $3$ | $0.4771$ |
| $4$ | $0.6021$ |
| $5$ | $0.6990$ |
| $6$ | $0.7782$ |
| $7$ | $0.8451$ |
| $8$ | $0.9031$ |
| $9$ | $0.9542$ |

| $x$ | $\log(x)$ |
|---|---|
| $0.1$ | $-1$ |
| $0.2$ | $-0.6990$ |
| $0.3$ | $-0.5229$ |
| $0.4$ | $-0.3979$ |
| $0.5$ | $-0.3010$ |
| $0.6$ | $-0.2218$ |
| $0.7$ | $-0.1549$ |
| $0.8$ | $-0.0969$ |
| $0.9$ | $-0.0458$ |

| $x$ | $\log(x)$ |
|---|---|
| $0.01$ | $-2$ |
| $0.02$ | $-1.6990$ |
| $0.03$ | $-1.5229$ |
| $0.04$ | $-1.3979$ |
| $0.05$ | $-1.3010$ |
| $0.06$ | $-1.2218$ |
| $0.07$ | $-1.1549$ |
| $0.08$ | $-1.0969$ |
| $0.09$ | $-1.0458$ |

12. **What pattern(s) can you see in the table from Exercise 11? Write them using logarithmic notation.**

*I found the patterns* $\log(x) - \log\left(\frac{x}{10}\right) = 1$, *which can be written as* $\log\left(\frac{x}{10}\right) = -1 + \log(x)$, *and* $\log\left(\frac{x}{100}\right) = -2 + \log(x)$.

MP.7

13. **What pattern would you expect to find for $\log\left(\frac{x}{1000}\right)$? Make a conjecture, and test it to see whether or not it appears to be valid.**

*I would guess that the values of* $\log\left(\frac{x}{1000}\right)$ *will all start with* $-2$*, and that* $\log\left(\frac{x}{1000}\right) = -3 + \log(x)$*. This appears to be the case since* $\log(0.002) \approx -2.6990$*, and* $-2.6990 = -3 + 0.3010$*;* $\log(0.005) \approx -2.3010$*, and* $-2.3010 = -3 + 0.6990$*;* $\log(0.008) \approx -2.0969$*, and* $-2.0969 = -3 + 0.9031$*.*

14. **Combine your results from Exercises 10 and 12 to make a conjecture about the value of the logarithm for a multiple of a power of 10; that is, find a formula for $\log(10^k \cdot x)$ for any integer $k$.**

*It appears that* $\log(10^k \cdot x) = k + \log(x)$*, for any integer* $k$*.*

## Discussion (2 minutes)

Ask groups to share the patterns they observed in Exercise 12 and the conjectures they made in Exercises 13 and 14 with the class. Ensure that all students have the correct conjectures recorded in their notebooks or journals before continuing to the next example.

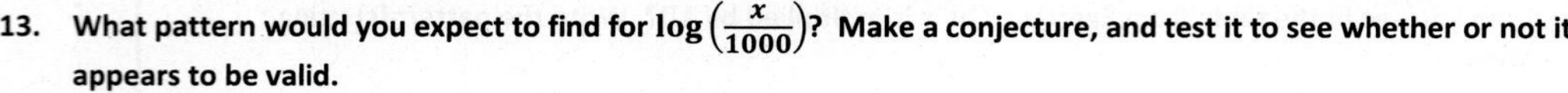

## Examples 2–3 (2 minutes)

Lead the class through these calculations. You may decide to let them work on Example 3 either alone or in groups after you have led them through Example 2.

Example 2

Use the logarithm tables and the rules we discovered to calculate $\log(40000)$ to 4 decimal places.

$$\begin{aligned}\log(40000) &= \log(10^4 \cdot 4)\\ &= 4 + \log(4)\\ &\approx 4.6021\end{aligned}$$

Example 3

Use the logarithm tables and the rules we discovered to calculate $\log(0.000004)$ to 4 decimal places.

$$\begin{aligned}\log(0.000004) &= \log(10^{-6} \cdot 4)\\ &= -6 + \log(4)\\ &\approx -5.3979\end{aligned}$$

## Closing (2 minutes)

Ask students to summarize the important parts of the lesson, either in writing, to a partner, or as a class. Use this as an opportunity to informally assess understanding of the lesson. The following are some important summary elements:

Lesson Summary

- The notation $\log(x)$ is used to represent $\log_{10}(x)$.
- For integers $k$, $\log(10^k) = k$.
- For integers $m$ and $n$, $\log(10^m \cdot 10^n) = \log(10^m) + \log(10^n)$.
- For integers $k$ and positive real numbers $x$, $\log(10^k \cdot x) = k + \log(x)$.

**Exit Ticket (4 minutes)**

Name ______________________________ Date ______________

# Lesson 10: Building Logarithmic Tables

## Exit Ticket

1. Use the log table below to approximate the following logarithms to four decimal places. Do not use a calculator.

| $x$ | $\log(x)$ |
|---|---|
| 1 | 0.0000 |
| 2 | 0.3010 |
| 3 | 0.4771 |
| 4 | 0.6021 |
| 5 | 0.6990 |

| $x$ | $\log(x)$ |
|---|---|
| 6 | 0.7782 |
| 7 | 0.8451 |
| 8 | 0.9031 |
| 9 | 0.9542 |
| 10 | 1.0000 |

a. $\log(500)$

b. $\log(0.0005)$

2. Suppose that $A$ is a number with $\log(A) = 1.352$.

a. What is the value of $\log(1000A)$?

b. Which of the following is true? Explain how you know.

i. $A < 0$

ii. $0 < A < 10$

iii. $10 < A < 100$

iv. $100 < A < 1000$

v. $A > 1000$

## Exit Ticket Sample Solutions

1. **Use the log table below to approximate the following logarithms to four decimal places. Do not use a calculator.**

| $x$ | $\log(x)$ |
|---|---|
| 1 | 0.0000 |
| 2 | 0.3010 |
| 3 | 0.4771 |
| 4 | 0.6021 |
| 5 | 0.6990 |

| $x$ | $\log(x)$ |
|---|---|
| 6 | 0.7782 |
| 7 | 0.8451 |
| 8 | 0.9031 |
| 9 | 0.9542 |
| 10 | 1.0000 |

a. $\log(500)$

$$\begin{aligned}\log(500) &= \log(10^2 \cdot 5)\\ &= 2 + \log(5)\\ &\approx 2.6990\end{aligned}$$

b. $\log(0.0005)$

$$\begin{aligned}\log(0.0005) &= \log(10^{-4} \cdot 5)\\ &= -4 + \log(5)\\ &\approx -3.3010\end{aligned}$$

2. **Suppose that $A$ is a number with $\log(A) = 1.352$.**

a. **What is the value of $\log(1000A)$?**

$\log(1000A) = \log(10^3A) = 3 + \log(A) = 4.352$

b. **Which of the following statements is true? Explain how you know.**

i. $A < 0$

ii. $0 < A < 10$

iii. $10 < A < 100$

iv. $100 < A < 1000$

v. $A > 1000$

*Because $\log(A) = 1.352 = 1 + 0.352$, $A$ is greater than $10$ and less than $100$. Thus, (iii) is true. In fact, from the table above, we can narrow the value of $A$ down to between $20$ and $30$ because $\log(20) \approx 1.3010$, and $\log(30) \approx 1.4771$.*

## Problem Set Sample Solutions

These problems should be solved without a calculator.

1. **Complete the following table of logarithms without using a calculator; then, answer the questions that follow.**

| $x$ | $\log(x)$ |
|---|---|
| $1,000,000$ | $6$ |
| $100,000$ | $5$ |
| $10,000$ | $4$ |
| $1000$ | $3$ |
| $100$ | $2$ |
| $10$ | $1$ |

| $x$ | $\log(x)$ |
|---|---|
| $0.1$ | $-1$ |
| $0.01$ | $-2$ |
| $0.001$ | $-3$ |
| $0.0001$ | $-4$ |
| $0.00001$ | $-5$ |
| $0.000001$ | $-6$ |

a. **What is $\log(1)$? How does that follow from the definition of a base $10$ logarithm?**

*Since $10^0 = 1$, we know that $\log(1) = 0$.*

b. **What is $\log(10^k)$ for an integer $k$? How does that follow from the definition of a base $10$ logarithm?**

*By the definition of the logarithm, we know that $\log(10^k) = k$.*

c. **What happens to the value of $\log(x)$ as $x$ gets really large?**

*For any $x > 1$, there exists $k > 0$ so that $10^k \leq x < 10^{k+1}$. As $x$ gets really large, $k$ gets large. Since $k \leq \log(x) < k + 1$, as $k$ gets large, $\log(x)$ gets large.*

d. **For $x > 0$, what happens to the value of $\log(x)$ as $x$ gets really close to zero?**

*For any $0 < x < 1$, there exists $k > 0$ so that $10^{-k} \leq x < 10^{-k+1}$. Then $-k \leq x < -k + 1$. As $x$ gets closer to zero, $k$ gets larger. Thus, $\log(x)$ is negative, and $|\log(x)|$ gets large as the positive number $x$ gets close to zero.*

2. **Use the table of logarithms below to estimate the values of the logarithms in parts (a)–(h).**

| $x$ | $\log(x)$ |
|---|---|
| $2$ | $0.3010$ |
| $3$ | $0.4771$ |
| $5$ | $0.6990$ |
| $7$ | $0.8451$ |
| $11$ | $1.0414$ |
| $13$ | $1.1139$ |

a. $\log(70,000)$

$4.8451$

b. $\log(0.0011)$

$-2.9586$

c. $\log(20)$

$1.3010$

d. $\log(0.00005)$

$-4.3010$

e. $\log(130,000)$

$5.1139$

f. $\log(3000)$

$3.4771$

g. $\log(0.07)$

$-1.1549$

h. $\log(11,000,000)$

$7.0414$

3. **If $\log(n) = 0.6$, find the value of $\log(10n)$.**

$\log(10n) = 1.6$

4. **If $m$ is a positive integer and $\log(m) \approx 3.8$, how many digits are there in $m$? Explain how you know.**

*Since $3 < \log(m) < 4$, we know $1,000 < m < 10,000$; therefore, $m$ has $4$ digits.*

5. **If $m$ is a positive integer and $\log(m) \approx 9.6$, how many digits are there in $m$? Explain how you know.**

*Since $9 < \log(m) < 10$, we know $10^9 < m < 10^{10}$; therefore, $m$ has $10$ digits.*

6. **Vivian says $\log(452,000) = 5 + \log(4.52)$, while her sister Lillian says that $\log(452,000) = 6 + \log(0.452)$. Which sister is correct? Explain how you know.**

*Both sisters are correct. Since $452,000 = 4.52 \cdot 10^5$, we can write $\log(452,000) = 5 + \log(4.52)$. However, we could also write $452,000 = 0.452 \cdot 10^6$, so $\log(452,000) = 6 + \log(0.452)$. Both calculations give $\log(452,000) \approx 5.65514$.*

7. **Write the logarithm base $10$ of each number in the form $k + \log(x)$, where $k$ is the exponent from the scientific notation, and $x$ is a positive real number.**

a. $2.4902 \times 10^4$

$4 + \log(2.4902)$

b. $2.58 \times 10^{13}$

$13 + \log(2.58)$

c. $9.109 \times 10^{-31}$

$-31 + \log(9.109)$

8. **For each of the following statements, write the number in scientific notation, and then write the logarithm base 10 of that number in the form $k + \log(x)$, where $k$ is the exponent from the scientific notation, and $x$ is a positive real number.**

a. **The speed of sound is $1116$ ft/s..**

$1116 = 1.116 \times 10^3$, *so* $\log(1116) = 3 + \log(1.116)$.

b. **The distance from Earth to the Sun is 93 million miles.**

$93,000,000 = 9.3 \times 10^7$, *so* $\log(93,000,000) = 7 + \log(9.3)$.

c. **The speed of light is $29,980,000,000\ \mathrm{cm/s}$.**

$29,980,000,000 = 2.998 \times 10^{10}$, *so* $\log(29,980,000,000) = 10 + \log(2.998)$.

d. **The weight of the earth is $5,972,000,000,000,000,000,000,000\ \mathrm{kg}$.**

$5,972,000,000,000,000,000,000,000 = 5.972 \times 10^{24}$, *so*
$\log(5,972,000,000,000,000,000,000,000) = 24 + \log(5.972)$.

e. **The diameter of the nucleus of a hydrogen atom is $0.00000000000000175\ \mathrm{m}$.**

$0.00000000000000175 = 1.75 \times 10^{-15}$, *so* $\log(0.00000000000000175) = -15 + \log(1.75)$.

f. **For each part (a)–(e), you have written each logarithm in the form $k + \log(x)$, for integers $k$ and positive real numbers $x$. Use a calculator to find the values of the expressions $\log(x)$. Why are all of these values between 0 and 1?**

$\log(1.116) \approx 0.047664$

$\log(9.3) \approx 0.968483$

$\log(2.998) \approx 0.476832$

$\log(5.972) \approx 0.77612$

$\log(1.75) \approx 0.243038$

*These values are all between* $0$ *and* $1$ *because* $x$ *is between* $1$ *and* $10$. *We can rewrite* $1 < x < 10$ *as* $10^0 < x < 10^1$. *If we write* $x = 10^L$ *for some exponent* $L$, *then* $10^0 < 10^L < 10^1$, *so* $0 < L < 1$. *This exponent* $L$ *is the base* $10$ *logarithm of* $x$.

# Lesson 11: The Most Important Property of Logarithms

## Student Outcomes

- Students construct a table of logarithms base 10 and observe patterns that indicate properties of logarithms.

## Lesson Notes

In the previous lesson, students discovered that for logarithms base $10$, $\log(10^k \cdot x) = k + \log(x)$. In this lesson, we extend this result to develop the most important property of logarithms: $\log(xy) = \log(x) + \log(y)$. Additionally, students will discover the reciprocal property of logarithms: $\log\left(\frac{1}{x}\right) = -\log(x)$. Students will continue to hone their skills at observing and generalizing patterns in this lesson as they create tables of logarithms and observe patterns, practicing MP.8. In the next lesson, these logarithmic properties will be formalized and generalized for any base, but for this lesson we will solely focus on logarithms base $10$. Understanding deeply the properties of logarithms will help prepare students to rewrite expressions based on their structure (**A-SSE.A.2**), solve exponential equations (**F-LE.A.4**), and interpret transformations of graphs of logarithmic functions (**F-BF.B.3**).

## Materials Needed

Students will need access to a calculator or other technological tool able to compute exponents and logarithms base 10.

## Classwork

### Opening (1 minute)

In the previous lesson, students discovered the logarithmic property $\log(10^k \cdot x) = k + \log(x)$, which is a special case of the additive property $\log(xy) = \log(x) + \log(y)$ that they will discover today. The Opening Exercise reminds students of how we can use this property to compute logarithms of numbers not in our table. By the end of today's lesson, students will be able to calculate any logarithm base $10$ using just a table of values of $\log(x)$ for prime integers $x$. The only times in this lesson that calculators should be used is to create the tables in Exercises 1 and 6.

## Opening Exercise (4 minutes)

Students should complete this exercise without the use of a calculator.

**Opening Exercise**

**Use the logarithm table below to calculate the specified logarithms.**

| $x$ | $\log(x)$ |
|---|---|
| $1$ | $0$ |
| $2$ | $0.3010$ |
| $3$ | $0.4771$ |
| $4$ | $0.6021$ |
| $5$ | $0.6990$ |
| $6$ | $0.7782$ |
| $7$ | $0.8451$ |
| $8$ | $0.9031$ |
| $9$ | $0.9542$ |

a. $\log(80)$

$\log(80) = \log(10^1 \cdot 8) = 1 + \log(8) \approx 1.9031$

b. $\log(7000)$

$\log(7000) = \log(10^3 \cdot 7) = 3 + \log(7) \approx 3.8451$

c. $\log(0.00006)$

$\log(0.00006) = \log(10^{-5} \cdot 6) = -5 + \log(6) \approx -4.2218$

d. $\log(3.0 \times 10^{27})$

$\log(3.0 \times 10^{27}) = \log(10^{27} \cdot 3) = 27 + \log(3) \approx 27.4771$

e. $\log(9.0 \times 10^k)$ **for an integer** $k$

$\log(9.0 \times 10^k) = \log(10^k \cdot 9) = k + \log(9) \approx k + 0.9542$

*Scaffolding:*

- For struggling students, model the decomposition of 80 with the whole class before starting or with a small group.

$$\begin{aligned}\log(80) &= \log(10 \cdot 8)\\ &= \log(10) + \log(8)\\ &= 1 + \log(8)\end{aligned}$$

- Ask advanced students to write an expression for $\log(10^k \cdot x)$ independently.

## Discussion (3 minutes)

Use this discussion to review the formulas discovered in the previous lesson. Tell the class that they will be using more logarithmic tables to discover some other interesting properties of logarithms.

In the next set of exercises, we want students to discover the additive property of logarithms, which is not as readily apparent as the patterns observed yesterday. Plant the seed of the idea by restating the previous property in the additive format.

- What was the formula we developed in the last class?
  - $\log(10^k \cdot x) = k + \log(x)$

- What is the value of $\log(10^k)$?
  - $\log(10^k) = k$
- So, what is another way we can write the formula $\log(10^k \cdot x) = k + \log(x)$?
  - $\log(10^k \cdot x) = \log(10^k) + \log(x)$
- Keep this statement of the formula in mind as you progress through the next set of exercises.

### Exercises 1–5 (6 minutes)

Students may be confused by the fact that the formulas do not appear to be exact–for example, the table shows that $\log(4) = 0.6021$, and $2\log(2) = 0.6020$. If this question arises, remind students that since we have made approximations to irrational numbers, there is some error in rounding off the decimal expansions to four decimal places. Students should question this in part (f) of Exercise 2.

**Exercises 1–5**

**1. Use your calculator to complete the following table. Round the logarithms to four decimal places.**

| $x$ | $\log(x)$ |
|---|---|
| 1 | 0 |
| 2 | $0.3010$ |
| 3 | $0.4771$ |
| 4 | $0.6021$ |
| 5 | $0.6990$ |
| 6 | $0.7782$ |
| 7 | $0.8451$ |
| 8 | $0.9031$ |
| 9 | $0.9542$ |

| $x$ | $\log(x)$ |
|---|---|
| 10 | $1.0000$ |
| 12 | $1.0792$ |
| 16 | $1.2041$ |
| 18 | $1.2553$ |
| 20 | $1.3010$ |
| 25 | $1.3979$ |
| 30 | $1.4771$ |
| 36 | $1.5563$ |
| 100 | $2.0000$ |

**2. Calculate the following values. Do they appear anywhere else in the table?**

a. $\log(2) + \log(4)$

***We see that*** $\log(2) + \log(4) \approx 0.9031$***, which is approximately*** $\log(8)$.

b. $\log(2) + \log(6)$

***We see that*** $\log(2) + \log(6) \approx 1.0792$***, which is approximately*** $\log(12)$.

c. $\log(3) + \log(4)$

***We see that*** $\log(3) + \log(4) \approx 1.0792$***, which is approximately*** $\log(12)$.

d. $\log(6) + \log(6)$

***We see that*** $\log(6) + \log(6) \approx 1.5663$***, which is approximately*** $\log(36)$.

e. $\log(2) + \log(18)$

***We see that*** $\log(2) + \log(18) \approx 1.5663$***, which is approximately*** $\log(36)$.

f. $\log(3) + \log(12)$

*We see that* $\log(3) + \log(12) \approx 1.5664$*, which is approximately* $\log(36)$.

MP.8

**3. What pattern(s) can you see in Exercise 2 and the table from Exercise 1? Write them using logarithmic notation.**

*I found the pattern* $\log(xy) = \log(x) + \log(y)$.

**4. What pattern would you expect to find for $\log(x^2)$? Make a conjecture, and test it to see whether or not it appears to be valid.**

*I would guess that* $\log(x^2) = \log(x) + \log(x) = 2\log(x)$*. This is verified by the fact that* $\log(4) \approx 0.6021 \approx 2\log(2)$, $\log(9) \approx 0.9542 \approx 2\log(3)$, $\log(16) \approx 1.2041 \approx 2\log(4)$*, and* $\log(25) \approx 1.3980 \approx 2\log(5)$.

**5. Make a conjecture for a logarithm of the form $\log(xyz)$, where $x$, $y$, and $z$ are positive real numbers. Provide evidence that your conjecture is valid.**

*It appears that* $\log(xyz) = \log(x) + \log(y) + \log(z)$*. This is due to applying the property from Exercise 3 twice.*

$$\begin{aligned}\log(xyz) &= \log(xy \cdot z)\\ &= \log(xy) + \log(z)\\ &= \log(x) + \log(y) + \log(z)\end{aligned}$$

*OR*

*It appears that* $\log(xyz) = \log(x) + \log(y) + \log(z)$*. We can see that*

$\log(18) \approx 1.2553 \approx 0.3010 + 0.3010 + 0.4771 \approx \log(2) + \log(2) + \log(3)$,
$\log(20) \approx 1.3010 \approx 0.3010 + 0.3010 + 0.6990 \approx \log(2) + \log(2) + \log(5)$*, and*
$\log(36) \approx 1.5563 \approx 0.3010 + 0.4771 + 0.7782 \approx \log(2) + \log(3) + \log(6)$.

## Discussion (2 minutes)

Ask groups to share the patterns and conjectures they formed in Exercises 3–5 with the class; emphasize that the pattern discovered in Exercise 3 is the *most important property* of logarithms base 10. Ensure that all students have the correct statements recorded in their notebooks or journals before continuing to the next example.

## Example 1 (5 minutes)

Lead the class through these four logarithmic calculations, relying only on the values in the table from Exercise 1. Notice that since we do not have a value for $\log(11)$ in the table, we do not have enough information to calculate $\log(121)$. Allow students to figure this out for themselves.

**Example 1**

**Use the logarithm table from Exercise 1 to approximate the following logarithms:**

a. $\log(14)$

$\log(14) = \log(2) + \log(7) \approx 0.3010 + 0.8451$*, so* $\log(14) \approx 1.1461$.

b. $\log(35)$

$\log(35) = \log(5) + \log(7) \approx 0.6990 + 0.8451$*, so* $\log(35) \approx 1.5441$.

c. $\log(72)$

$\log(72) = \log(8) + \log(9) \approx 0.9031 + 0.9542$, *so* $\log(72) \approx 1.8573$.

d. $\log(121)$

$\log(121) = \log(11) + \log(11)$, ***but we do not have a value for*** $\log(11)$ ***in the table, so we cannot evaluate*** $\log(121)$.

## Discussion (3 minutes)

- Suppose we are building a logarithm table, and we have already approximated the values of $\log(2)$ and $\log(3)$. What other values of $\log(x)$ for $4 \leq x \leq 20$ can we approximate by applying the additive property developed in Exercise 5?
  - *Since the table contains* $\log(2)$ *and* $\log(3)$, *we can figure out approximations of* $\log(4)$, $\log(6)$, $\log(8)$, $\log(9)$, $\log(12)$, *and* $\log(18)$ *since the only factors of* $4$, $6$, $8$, $9$, $12$, $16$, *and* $18$ *are* $2$ *and* $3$.
- In order to develop the entire logarithm table for all integers between 1 and 20, what is the smallest set of logarithmic values that we need to know?
  - *We need to know the values of the logarithms for the prime numbers:* $2, 3, 5, 7, 11, 13, 17$, *and* $19$.
- Why does the additive property make sense based on what we know about exponents?
  - *We know that when you multiply two powers of the same base together, the exponents are added. For example,* $10^4 \cdot 10^5 = 10^{4+5} = 10^9$, *so* $\log(10^4 \cdot 10^5) = \log(10^9) = 9$.

## Exercises 6–8 (7 minutes)

Have students again work individually to complete the table in Exercise 6 and to check their tables against each other before they proceed to discuss and answer Exercise 7 in groups. Ensure that there is enough time for a volunteer to present justification for the conjecture in Exercise 8.

*Scaffolding:*

Ask students who struggle with decimal to fraction conversion to convert the decimal values in the second table to fractions before looking for the pattern, or present the table with values as fractions.

**Exercises 6–8**

6. **Use your calculator to complete the following table. Round the logarithms to four decimal places.**

| $x$ | $\log(x)$ |
|---|---|
| $2$ | $0.3013$ |
| $4$ | $0.6021$ |
| $5$ | $0.6990$ |
| $8$ | $0.9031$ |
| $10$ | $1.0000$ |
| $16$ | $1.2041$ |
| $20$ | $1.3010$ |
| $50$ | $1.6990$ |
| $100$ | $2.0000$ |

| $x$ | $\log(x)$ |
|---|---|
| $0.5$ | $-0.3013$ |
| $0.25$ | $-0.6021$ |
| $0.2$ | $-0.6990$ |
| $0.125$ | $-0.9031$ |
| $0.1$ | $-1.0000$ |
| $0.0625$ | $-1.2041$ |
| $0.05$ | $-1.3010$ |
| $0.02$ | $-1.6990$ |
| $0.01$ | $-2.0000$ |

MP.8

7. **What pattern(s) can you see in the table from Exercise 6? Write a conjecture using logarithmic notation.**

*For any real number* $x > 0$, $\log\left(\frac{1}{x}\right) = -\log(x)$.

8. **Use the definition of logarithm to justify the conjecture you found in Exercise 7.**

*If* $\log\left(\frac{1}{x}\right) = a$ *for some number* $a$*, then* $10^a = \frac{1}{x}$*. So,* $10^{-a} = x$*, and thus* $\log(x) = -a$*. We then have* $\log\left(\frac{1}{x}\right) = -\log(x)$.

## Discussion (3 minutes)

Ask groups to share the conjecture they formed in Exercise 7 with the class. Ensure that all students have the correct conjecture recorded in their notebooks or journals before continuing to the next example.

## Example 2 (5 minutes)

Lead the class through these calculations. You may decide to let them work either alone or in groups on parts (b)-(d) after you have led them through part (a).

**Example 2**

**Use the logarithm tables and the rules we discovered to estimate the following logarithms to four decimal places.**

a. $\log(2100)$

$$\begin{aligned}\log(2100) &= \log(10^2 \cdot 21)\\ &= 2 + \log(21)\\ &= 2 + \log(3) + \log(7)\\ &\approx 2 + 0.4771 + 0.8451\\ &\approx 3.3222\end{aligned}$$

b. $\log(0.00049)$

$$\begin{aligned}\log(0.00049) &= \log(10^{-5} \cdot 49)\\ &= -5 + \log(49)\\ &= -5 + \log(7) + \log(7)\\ &\approx -5 + 0.8451 + 0.8451\\ &\approx -3.3098\end{aligned}$$

c. $\log(42{,}000{,}000)$

$$\begin{aligned}\log(42{,}000{,}000) &= \log(10^6 \cdot 42)\\ &= 6 + \log(42)\\ &= 6 + \log(6) + \log(7)\\ &\approx 6 + 0.7782 + 0.8451\\ &\approx 7.6233\end{aligned}$$

d. $\log\left(\frac{1}{640}\right)$

$$\begin{aligned}\log\left(\frac{1}{640}\right) &= -\log(640)\\ &= -(\log(10\cdot 64))\\ &= -(1+\log(64))\\ &= -(1+\log(8)+\log(8))\\ &\approx -(1+0.9031+0.9031)\\ &\approx -2.8062\end{aligned}$$

## Closing (2 minutes)

Ask students to summarize the important parts of the lesson, either in writing, to a partner, or as a class. Use this as an opportunity to informally assess understanding of the lesson. The following are some important summary elements:

**Lesson Summary**

- **The notation $\log(x)$ is used to represent $\log_{10}(x)$.**
    - **The most important property of logarithms base 10 is that for positive real numbers $x$ and $y$,**
      $$\log(xy) = \log(x) + \log(y).$$
- **For positive real numbers $x$,**
  $$\log\left(\frac{1}{x}\right) = -\log(x).$$

## Exit Ticket (4 minutes)

Name ______________________________ Date ______________

# Lesson 11: The Most Important Property of Logarithms

## Exit Ticket

1. Use the table below to approximate the following logarithms to four decimal places. Do not use a calculator.

   a. $\log(9)$

| $x$ | $\log(x)$ |
|---|---|
| 2 | 0.3010 |
| 3 | 0.4771 |
| 5 | 0.6990 |
| 7 | 0.8451 |

   b. $\log\left(\frac{1}{15}\right)$

   c. $\log(45{,}000)$

2. Suppose that $k$ is an integer, $a$ is a positive real number, and you know the value of $\log(a)$. Explain how to find the value of $\log(10^k \cdot a^2)$.

## Exit Ticket Sample Solutions

1. **Use the table below to approximate the following logarithms to four decimal places. Do not use a calculator.**

| $x$ | $\log(x)$ |
|---|---|
| $2$ | $0.3010$ |
| $3$ | $0.4771$ |
| $5$ | $0.6990$ |
| $7$ | $0.8451$ |

   a. $\log(9)$

$$\begin{aligned}\log(9) &= \log(3) + \log(3)\\ &\approx 0.4771 + 0.4771\\ &\approx 0.9542\end{aligned}$$

   b. $\log\left(\frac{1}{15}\right)$

$$\begin{aligned}\log\left(\frac{1}{15}\right) &= -\log(15)\\ &= -(\log(3) + \log(5))\\ &\approx -(0.4771 + 0.6990)\\ &\approx -1,1761\end{aligned}$$

   c. $\log(45,000)$

$$\begin{aligned}\log(45,000) &= \log(10^3 \cdot 45)\\ &= 3 + \log(45)\\ &= 3 + \log(5) + \log(9)\\ &\approx 3 + 0.6990 + 0.9542\\ &\approx 4.6532\end{aligned}$$

2. **Suppose that $k$ is an integer, $a$ is a positive real number, and you know the value of $\log(a)$. Explain how to find the value of $\log(10^k \cdot a^2)$.**

   ***Applying the rule for the logarithm of a number multiplied by a power of $10$, and then the rule for the logarithm of a product, we have***

$$\begin{aligned}\log(10^k \cdot a^2) &= k + \log(a^2)\\ &= k + \log(a) + \log(a)\\ &= k + 2\log(a).\end{aligned}$$

## Problem Set Sample Solutions

All of the exercises in this problem set should be completed without the use of a calculator.

1. **Use the table of logarithms at right to estimate the value of the logarithms in parts (a)–(h).**

| $x$ | $\log(x)$ |
|---|---|
| $2$ | $0.30$ |
| $3$ | $0.48$ |
| $5$ | $0.70$ |
| $7$ | $0.85$ |
| $11$ | $1.04$ |
| $13$ | $1.11$ |

   a. $\log(25)$

   $1.40$

   b. $\log(27)$

   $1.44$

   c. $\log(33)$

   $1.52$

   d. $\log(55)$

   $1.74$

e. $\log(63)$

$1.81$

f. $\log(75)$

$1.88$

g. $\log(81)$

$1.92$

h. $\log(99)$

$2.00$

2. Use the table of logarithms at right to estimate the value of the logarithms in parts (a)–(f).

a. $\log(350)$

$2.55$

b. $\log(0.0014)$

$-2.85$

c. $\log(0.077)$

$-1.11$

d. $\log(49,000)$

$4.70$

e. $\log(1.69)$

$0.22$

f. $\log(6.5)$

$0.81$

3. Use the table of logarithms at right to estimate the value of the logarithms in parts (a)–(f).

a. $\log\left(\frac{1}{30}\right)$

$-1.48$

b. $\log\left(\frac{1}{35}\right)$

$-1.55$

c. $\log\left(\frac{1}{40}\right)$

$-1.60$

d. $\log\left(\frac{1}{42}\right)$

$-1.63$

e. $\log\left(\frac{1}{50}\right)$

$-1.70$

f. $\log\left(\frac{1}{64}\right)$

$-1.80$

4. Reduce each expression to a single logarithm of the form $\log(x)$.

a. $\log(5) + \log(7)$

$\log(35)$

b. $\log(3) + \log(9)$

$\log(27)$

c. $\log(15) - \log(5)$

$\log(3)$

d. $\log(8) + \log\left(\frac{1}{4}\right)$

$\log(2)$

5. Use properties of logarithms to write the following expressions involving logarithms of only prime numbers.

a. $\log(2500)$

$2 + 2\log(5)$

b. $\log(0.00063)$

$-5 + 2\log(3) + \log(7)$

c. $\log(1250)$

$1 + 3\log(5)$

d. $\log(26,000,000)$

$6 + \log(2) + \log(13)$

6. Use properties of logarithms to show that $\log(26) = \log(2) - \log\left(\frac{1}{13}\right)$.

$$\begin{aligned}\log(2) - \log\left(\frac{1}{13}\right) &= \log(2) - \log(13^{-1}) \\ &= \log(2) + \log(13) \\ &= \log(26)\end{aligned}$$

7. Use properties of logarithms to show that $\log(3) + \log(4) + \log(5) - \log(6) = 1$.

*There are multiple ways to solve this problem.*

$$\begin{aligned}\log(3) + \log(4) + \log(5) - \log(6) &= \log(3) + \log(4) + \log(5) + \log\left(\frac{1}{6}\right) \\ &= \log\left(3 \cdot 4 \cdot 5 \cdot \frac{1}{6}\right) \\ &= \log(10) \\ &= 1\end{aligned}$$

*OR*

$$\begin{aligned}\log(3) + \log(4) + \log(5) &= \log(60) \\ &= \log(10 \cdot 6) \\ &= \log(10) + \log(6) \\ &= 1 + \log(6)\end{aligned}$$

$$\log(3) + \log(4) + \log(5) - \log(6) = 1$$

8. Use properties of logarithms to show that $-\log(3) = \log\left(\frac{1}{2} - \frac{1}{3}\right) + \log(2)$.

$$\begin{aligned}\log\left(\frac{1}{2} - \frac{1}{3}\right) + \log(2) &= \log\left(\frac{1}{6}\right) + \log(2) \\ &= -\log(6) + \log(2) \\ &= -\big(\log(2) + \log(3)\big) + \log(2) \\ &= -\log(3)\end{aligned}$$

9. Use properties of logarithms to show that $\log\left(\frac{1}{3} - \frac{1}{4}\right) + \left(\log\left(\frac{1}{3}\right) - \log\left(\frac{1}{4}\right)\right) = -2\log(3)$.

$$\begin{aligned}\log\left(\frac{1}{3} - \frac{1}{4}\right) + \left(\log\left(\frac{1}{3}\right) - \log\left(\frac{1}{4}\right)\right) &= \log\left(\frac{1}{12}\right) + \log\left(\frac{1}{3}\right) - \log\left(\frac{1}{4}\right) \\ &= -\log(12) - \log(3) + \log(4) \\ &= -\big(\log(3) + \log(4)\big) - \log(3) + \log(4) \\ &= -2\log(3)\end{aligned}$$

# Lesson 12: Properties of Logarithms

## Student Outcomes

- Students justify properties of logarithms using the definition and properties already developed.

## Lesson Notes

In this lesson, students work exclusively with logarithms base $10$; generalization of these results to a generic base $b$ will occur in the next lesson. The opening of this lesson, which echoes homework from Lesson 11, is meant to launch a consideration of some properties of the common logarithm function. The centerpiece of the lesson is the demonstration of six basic properties of logarithms theoretically using the properties of exponents instead of numerical approximation as has been done in prior lessons. In the Problem Set, students will apply these properties to calculating logarithms, rewriting logarithmic expressions, and solving exponential equations base 10 (**A-SSE.A.2**, **F-LE.A.4**).

## Classwork

### Opening Exercise (5 minutes)

Students should work in groups of two or three on this exercise. These exercises serve to remind students of the "most important property" of logarithms and prepare them for justifying the properties later in the lesson. Verify that students are breaking up the logarithm, using the property, and evaluating the logarithm at known values (e.g., $0.1$, $10$, $100$).

*Scaffolding:*

- Ask students who are having trouble with any part of this exercise, "How is the number in parentheses related to $2$?" Follow that with the question, "So how might you find its logarithm given that you know $\log(2)$?"
- Ask students to factor each number into powers of $10$ and factors of $2$ before splitting the factors using $\log(xy) = \log(x) + \log(y)$. Students still struggling can be given additional products to break down before finding the approximations of their logarithms.

$$4 = 2 \cdot 2$$
$$40 = 10^1 \cdot 2 \cdot 2$$
$$0.4 = 10^{-1} \cdot 2 \cdot 2$$
$$400 = 10^2 \cdot 2 \cdot 2$$
$$0.04 = 10^{-2} \cdot 2 \cdot 2$$

- Advanced students may be challenged with a more general version of part (c):

$$\log(2^k).$$

This can be explored by having students find $\log(2^5)$, $\log(2^6)$, etc.

**Opening Exercise**

**Use the approximation $\log(2) \approx 0.3010$ to approximate the values of each of the following logarithmic expressions.**

a. $\log(20)$

$$\begin{aligned}\log(20) &= \log(10 \cdot 2) \\ &= \log(10) + \log(2) \\ &\approx 1 + 0.301 \\ &\approx 1.301\end{aligned}$$

b. $\log(0.2)$

$$\begin{aligned}\log(0.2) &= \log(0.1 \cdot 2) \\ &= \log(0.1) + \log(2) \\ &\approx -1 + 0.3010 \\ &\approx -0.6990\end{aligned}$$

c. $\log(2^4)$

$$\begin{aligned}\log(2^4) &= \log(2\cdot 2\cdot 2\cdot 2)\\ &= \log(2\cdot 2) + \log(2\cdot 2)\\ &= \log(2)+\log(2)+\log(2)+\log(2)\\ &\approx 4\cdot(0.3010)\\ &\approx 1.2040\end{aligned}$$

## Discussion (4 minutes)

Discuss the properties of logarithms used in the Opening Exercises.

- In all three parts of the Opening Exercise, we used the important property $\log(xy) = \log(x) + \log(y)$.
- What are some other properties we used?
  - *We also used* $\log(10) = 1$ *and* $\log(0.1) = -1$.
- It can be helpful to look further at properties of expressions involving logarithms.

## Example (6 minutes)

Recall that, by definition, $L = \log(x)$ means $10^L = x$. Consider some possible values of $x$ and $L$, noting that $x$ cannot be a negative number. What is $L$ …

- When $x = 1$?
  - $L = 0$
- When $x = 0$?
  - *The logarithm* $L$ *is not defined. There is no exponent of* $10$ *that yields a value of* $0$.
- When $x = 10^9$?
  - $L = 9$
- When $x = 10^n$?
  - $L = n$
- When $x = \sqrt[3]{10}$?
  - $L = \frac{1}{3}$

## Exercises 1–6 (15 minutes)

Students should work in groups of two or three on each exercise. The first three should be straightforward in view of the definition of base 10 logarithms. Exercise 4 may look somewhat odd, but it, too, follows directly from the definition. Exercises 5 and 6 are more difficult, which is why the hints are supplied. When all properties have been established, groups might be asked to show their explanations to the rest of the class as time permits.

Exercises

**For Exercises 1–6, explain why each statement below is a property of base 10 logarithms.**

1. **Property 1: $\log(1) = 0$.**

   *Because $L = \log(x)$ means $10^L = x$, then when $x = 1$, $L = 0$.*

2. **Property 2: $\log(10) = 1$.**

   *Because $L = \log(x)$ means $10^L = x$, then when $x = 10$, $L = 1$.*

3. **Property 3: For all real numbers $r$, $\log(10^r) = r$.**

   *Because $L = \log(x)$ means $10^L = x$, then when $x = 10^r$, $L = r$.*

4. **Property 4: For any $x > 0$, $10^{\log(x)} = x$.**

   *Because $L = \log(x)$ means $10^L = x$, then $x = 10^{\log(x)}$.*

MP.3

5. **Property 5: For any positive real numbers $x$ and $y$, $\log(x \cdot y) = \log(x) + \log(y)$.**
   **Hint: Use an exponent rule as well as property 4.**

   *By the rule $a^b \cdot a^c = a^{b+c}$, $10^{\log(x)} \cdot 10^{\log(y)} = 10^{\log(x)+\log(y)}$.*

   *By property 4, $10^{\log(x)} \cdot 10^{\log(y)} = x \cdot y$.*

   *Therefore, $x \cdot y = 10^{\log(x)+\log(y)}$. Again, by property 4, $x \cdot y = 10^{\log(x \cdot y)}$.*

   *Then, $10^{\log(x \cdot y)} = 10^{\log(x)+\log(y)}$; so, the exponents must be equal, and $\log(x \cdot y) = \log(x) + \log(y)$.*

6. **Property 6: For any positive real number $x$ and any real number $r$, $\log(x^r) = r \cdot \log(x)$.**
   **Hint: Again, use an exponent rule as well as property 4.**

   *By the rule $(a^b)^c = a^{bc}$, $10^{k\log(x)} = \left(10^{\log(x)}\right)^k$.*

   *By property 4, $\left(10^{\log(x)}\right)^r = x^r$.*

   *Therefore, $x^r = 10^{r\log(x)}$. Again, by property 4, $x^r = 10^{\log(x^r)}$.*

   *Then, $10^{\log(x^r)} = 10^{r\log(x)}$; so, the exponents must be equal, and $\log(x^r) = r \cdot \log(x)$.*

*Scaffolding:*

Establishing the logarithmic properties relies on the exponential laws. Make sure that students have access to the exponential laws either through a poster displayed in the classroom or through notes in their notebooks.

*Scaffolding:*

Students in groups that struggle with Exercises 3–6 should be encouraged to check the property with numerical values for $k$, $x$, $m$, and $n$. The check may suggest an explanation.

## Exercises 7–10 (8 minutes)

These exercises bridge the gap between the abstract properties of logarithms and computational problems like those in the Problem Set. Allow students to work alone, in pairs, or in small groups as you see fit. Circulate to ensure that students are applying the properties correctly. Calculators are not needed for these exercises and should not be used. In Exercises 9 and 10, students need to know that the logarithm is well-defined; that is, for positive real numbers $X$ and $Y$, if $X = Y$, then $\log(X) = \log(Y)$. This is why we can "take the log of both sides" of an equation in order to bring down an exponent and solve the equation. In these last two exercises, students need to choose an appropriate base for the logarithm to apply to solve the equation. Any logarithm will work to solve the equations if applied properly, so students may find equivalent answers that appear to be different from those listed here.

7. Apply properties of logarithms to rewrite the following expressions as a single logarithm or number.

a. $\frac{1}{2}\log(25) + \log(4)$

$\log(5) + \log(4) = \log(20)$

b. $\frac{1}{3}\log(8) + \log(16)$

$\log(2) + \log(2^4) = \log(32)$

c. $3\log(5) + \log(0.8)$

$\log(125) + \log(0.8) = \log(100) = 2$

8. Apply properties of logarithms to rewrite each expression as a sum of terms involving numbers, $\log(x)$, and $\log(y)$.

a. $\log(3x^2y^5)$

$\log(3) + 2\log(x) + 5\log(y)$

b. $\log\left(\sqrt{x^7y^3}\right)$

$\frac{7}{2}\log(x) + \frac{3}{2}\log(y)$

9. In mathematical terminology, logarithms are well defined because if $X = Y$, then $\log(X) = \log(Y)$ for $X, Y > 0$. This means that if you want to solve an equation involving exponents, you can apply a logarithm to both sides of the equation, just as you can take the square root of both sides when solving a quadratic equation. You do need to be careful not to take the logarithm of a negative number or zero.

Use the property stated above to solve the following equations.

a. $10^{10x} = 100$

$$\log(10^{10x}) = \log(100)$$
$$10x = 2$$
$$x = \frac{1}{5}$$

b. $10^{x-1} = \frac{1}{10^{x+1}}$

$$\log(10^{x-1}) = -\log(10^{x+1})$$
$$x - 1 = -(x + 1)$$
$$2x = 0$$
$$x = 0$$

c. $100^{2x} = 10^{3x-1}$

$$\log(100^{2x}) = \log(10^{3x-1})$$
$$2x\log(100) = (3x - 1)$$
$$4x = 3x - 1$$
$$x = -1$$

10. **Solve the following equations.**

a. $10^x = 2^7$

$$\log(10^x) = \log(2^7)$$
$$x = 7\log(2)$$

b. $10^{x^2+1} = 15$

$$\log\left(10^{x^2+1}\right) = \log(15)$$
$$x^2 + 1 = \log(15)$$
$$x = \pm\sqrt{\log(15) - 1}$$

c. $4^x = 5^3$

$$\log(4^x) = \log(5^3)$$
$$x\log(4) = 3\log(5)$$
$$x = \frac{3\log(5)}{\log(4)}$$

## Closing (2 minutes)

Point out that for each property 1–6, we have established that the property holds, so we can use these properties in our future work with logarithms. The Lesson Summary might be posted in the classroom for at least the rest of the module. The Exit Ticket asks the students to show that properties 7 and 8 hold.

**Lesson Summary**

**We have established the following properties for base 10 logarithms, where $x$ and $y$ are positive real numbers and $r$ is any real number:**

1. $\log(1) = 0$
2. $\log(10) = 1$
3. $\log(10^r) = r$
4. $10^{\log(x)} = x$
5. $\log(x \cdot y) = \log(x) + \log(y)$
6. $\log(x^r) = r \cdot \log(x)$

**Additional properties not yet established are the following:**

1. $\log\left(\frac{1}{x}\right) = -\log(x)$
2. $\log\left(\frac{x}{y}\right) = \log(x) - \log(y)$

**Also, logarithms are well defined, meaning that for $X, Y > 0$, if $X = Y$, then $\log(X) = \log(Y)$.**

## Exit Ticket (5 minutes)

Name ______________________________ Date ______________

# Lesson 12: Properties of Logarithms

## Exit Ticket

1. State as many of the six properties of logarithms as you can.

2. Use the properties of logarithms to show that $\log\left(\frac{1}{x}\right) = -\log(x)$ for all $x > 0$.

3. Use the properties of logarithms to show that $\log\left(\frac{x}{y}\right) = \log(x) - \log(y)$ for $x > 0$ and $y > 0$.

## Exit Ticket Sample Solutions

1. **State as many of the six properties of logarithms as you can.**

   $\log(1) = 0$

   $\log(10) = 1$

   $\log(10^r) = r$

   $10^{\log(x)} = x$

   $\log(x \cdot y) = \log(x) + \log(y)$

   $\log(x^r) = r \cdot \log(x)$

2. **Use the properties of logarithms to show that $\log\left(\frac{1}{x}\right) = -\log(x)$ for $x > 0$.**

   *By property 6, $\log(x^k) = k \cdot \log(x)$.*

   *Let $k = -1$, then for $x > 0$, $\log(x^{-1}) = (-1) \cdot \log(x)$, which is equivalent to $\log\left(\frac{1}{x}\right) = -\log(x)$.*

   *Thus, for any $x > 0$, $\log\left(\frac{1}{x}\right) = -\log(x)$.*

3. **Use the properties of logarithms to show that $\log\left(\frac{x}{y}\right) = \log(x) - \log(y)$ for $x > 0$ and $y > 0$.**

   *By property 5, $\log(x \cdot y) = \log(x) + \log(y)$.*

   *By Problem 2 above, for $y > 0$, $\log(y^{-1}) = (-1) \cdot \log(y)$.*

   *Therefore,*

$$\begin{aligned}\log\left(\frac{x}{y}\right) &= \log(x) + \log\left(\frac{1}{y}\right)\\ &= \log(x) + (-1)\log(y)\\ &= \log(x) - \log(y).\end{aligned}$$

   *Thus, for any $x > 0$ and $y > 0$, $\log\left(\frac{x}{y}\right) = \log(x) - \log(y)$.*

## Problem Set Sample Solutions

Problems 1–7 give students an opportunity to practice using the properties they have established in this lesson, and in the remaining problems, students apply base 10 logarithms to solve simple exponential equations.

1. **Use the approximate logarithm values below to estimate each of the following logarithms. Indicate which properties you used.**

   $\log(2) = 0.3010$ $\quad$ $\log(3) = 0.4771$

   $\log(5) = 0.6990$ $\quad$ $\log(7) = 0.8451$

   a. $\log(6)$

      *Using property 5,*

      $\log(6) = \log(3) + \log(2) \approx 0.7781.$

b. $\log(15)$

*Using property 5,*

$\log(15) = \log(3) + \log(5) \approx 1.1761.$

c. $\log(12)$

*Using properties 5 and 6,*

$\log(12) = \log(3) + \log(2^2) = \log(3) + 2\log(2) \approx 1.0791.$

d. $\log(10^7)$

*Using property 3,*

$\log(10^7) = 7.$

e. $\log\left(\frac{1}{5}\right)$

*Using property 7,*

$\log\left(\frac{1}{5}\right) = -\log(5) \approx -0.6990.$

f. $\log\left(\frac{3}{7}\right)$

*Using property 8,*

$\log\left(\frac{3}{7}\right) = \log(3) - \log(7) \approx -0.368.$

g. $\log(\sqrt[4]{2})$

*Using property 6,*

$\log(\sqrt[4]{2}) = \log\left(2^{\frac{1}{4}}\right) = \frac{1}{4}\log(2) \approx 0.0753.$

2. Let $\log(X) = r$, $\log(Y) = s$, and $\log(Z) = t$. Express each of the following in terms of $r$, $s$, and $t$.

a. $\log\left(\frac{X}{Y}\right)$

$r - s$

b. $\log(YZ)$

$s + t$

c. $\log(X^r)$

$r^2$

d. $\log(\sqrt[3]{Z})$

$\frac{t}{3}$

e. $\log\left(\sqrt[4]{\frac{Y}{Z}}\right)$

$\frac{s-t}{4}$

f. $\log(XY^2Z^3)$

$r + 2s + 3t$

3. Use the properties of logarithms to rewrite each expression in an equivalent form containing a single logarithm.

a. $\log\left(\frac{13}{5}\right) + \log\left(\frac{5}{4}\right)$

$\log\left(\frac{13}{4}\right)$

b. $\log\left(\frac{5}{6}\right) - \log\left(\frac{2}{3}\right)$

$\log\left(\frac{5}{4}\right)$

c. $\frac{1}{2}\log(16) + \log(3) + \log\left(\frac{1}{4}\right)$

$\log(3)$

4. Use the properties of logarithms to rewrite each expression in an equivalent form containing a single logarithm.

a. $\log(\sqrt{x}) + \frac{1}{2}\log\left(\frac{1}{x}\right) + 2\log(x)$

$\log(x^2)$

b. $\log(\sqrt[5]{x}) + \log(\sqrt[5]{x^4})$

$\log(x)$

c. $\log(x) + 2\log(y) - \frac{1}{2}\log(z)$

$\log\left(\frac{xy^2}{\sqrt{z}}\right)$

d. $\frac{1}{3}(\log(x) - 3\log(y) + \log(z))$

$\log\left(\sqrt[3]{\frac{xz}{y^3}}\right)$

e. $2(\log(x) - \log(3y)) + 3(\log(z) - 2\log(x))$

$\log\left(\left(\frac{x}{3y}\right)^2\right) + \log\left(\left(\frac{z}{x^2}\right)^3\right) = \log\left(\frac{z^3}{9y^2x^4}\right)$

5. Use properties of logarithms to rewrite the following expressions in an equivalent form containing only $\log(x)$, $\log(y)$, $\log(z)$, and numbers.

a. $\log\left(\frac{3x^2y^4}{\sqrt{z}}\right)$

$\log(3) + 2\log(x) + 4\log(y) - \frac{1}{2}\log(z)$

b. $\log\left(\frac{42\sqrt[3]{xy^7}}{x^2z}\right)$

$\log(42) - \frac{5}{3}\log(x) + \frac{7}{3}\log(y) - \log(z)$

c. $\log\left(\frac{100x^2}{y^3}\right)$

$2 + 2\log(x) - 3\log(y)$

d. $\log\left(\sqrt{\frac{x^3y^2}{10z}}\right)$

$\frac{1}{2}(3\log(x) + 2\log(y) - 1 - \log(z))$

e. $\log\left(\frac{1}{10x^2z}\right)$

$-1 - 2\log(x) - \log(z)$

6. Express $\log\left(\frac{1}{x} - \frac{1}{x+1}\right) + \left(\log\left(\frac{1}{x}\right) - \log\left(\frac{1}{x+1}\right)\right)$ as a single logarithm for positive numbers $x$.

$$\begin{aligned}\log\left(\frac{1}{x} - \frac{1}{x+1}\right) + \left(\log\left(\frac{1}{x}\right) - \log\left(\frac{1}{x+1}\right)\right) &= \log\left(\frac{1}{x(x+1)}\right) + \log\left(\frac{1}{x}\right) - \log\left(\frac{1}{x+1}\right)\\ &= -\log(x(x+1)) - \log(x) + \log(x+1)\\ &= -\log(x) - \log(x+1) - \log(x) + \log(x+1)\\ &= -2\log(x)\end{aligned}$$

7. Show that $\log(x + \sqrt{x^2-1}) + \log(x - \sqrt{x^2-1}) = 0$ for $x \geq 1$.

$$\begin{aligned}\log\left(x + \sqrt{x^2-1}\right) + \log\left(x - \sqrt{x^2-1}\right) &= \log\left(\left(x + \sqrt{x^2-1}\right)\left(x - \sqrt{x^2-1}\right)\right)\\ &= \log\left(x^2 - \left(\sqrt{x^2-1}\right)^2\right)\\ &= \log(x^2 - x^2 + 1)\\ &= \log(1)\\ &= 0\end{aligned}$$

8. If $xy = 10^{3.67}$, find the value of $\log(x) + \log(y)$.

$$\begin{aligned}xy &= 10^{3.67}\\ 3.67 &= \log(xy)\\ \log(xy) &= 3.67\\ \log(x) + \log(y) &= 3.67\end{aligned}$$

9. Solve the following exponential equations by taking the logarithm base 10 of both sides. Leave your answers stated in terms of logarithmic expressions.

a. $10^{x^2} = 320$

$$\begin{aligned}\log\left(10^{x^2}\right) &= \log(320)\\ x^2 &= \log(320)\\ x &= \pm\sqrt{\log(320)}\end{aligned}$$

b. $10^{\frac{x}{8}} = 300$

$$\log\left(10^{\frac{x}{8}}\right) = \log(300)$$
$$\frac{x}{8} = \log(10^2 \cdot 3)$$
$$\frac{x}{8} = 2 + \log(3)$$
$$x = 16 + 8\log(3)$$

c. $10^{3x} = 400$

$$\log(10^{3x}) = \log(400)$$
$$3x \cdot \log(10) = \log(10^2 \cdot 4)$$
$$3x \cdot 1 = 2 + \log(4)$$
$$x = \frac{1}{3}\left(2 + \log(4)\right)$$

d. $5^{2x} = 200$

$$\log(5^{2x}) = \log(200)$$
$$2x \cdot \log(5) = \log(100) + \log(2)$$
$$2x = \frac{2 + \log(2)}{\log(5)}$$
$$x = \frac{2 + \log(2)}{2\log(5)}$$

e. $3^x = 7^{-3x+2}$

$$\log(3^x) = \log\left(7^{-3x+2}\right)$$
$$x\log(3) = (-3x + 2)\log(7)$$
$$x\log(3) + 3x\log(7) = 2\log(7)$$
$$x\left(\log(3) + 3\log(7)\right) = 2\log(7)$$
$$x = \frac{2\log(7)}{\log(3) + 3\log(7)} = \frac{\log(49)}{\log(3) + \log(343)} = \frac{\log(49)}{\log(1029)}$$

***(Any of the three equivalent forms given above are acceptable answers.)***

10. Solve the following exponential equations.

a. $10^x = 3$

$x = \log(3)$

b. $10^y = 30$

$y = \log(30)$

c. $10^z = 300$

$z = \log(300)$

d. Use the properties of logarithms to justify why $x$, $y$, and $z$ form an arithmetic sequence whose constant difference is $1$.

*Since* $y = \log(30)$, $y = \log(10 \cdot 3) = 1 + \log(10) = 1 + x$.
*Similarly,* $z = 2 + \log(3) = 2 + x$.

*Thus, the sequence* $x, y, z$ *is the sequence* $\log(3)$, $1 + \log(3)$, $2 + \log(3)$, *and these numbers form an arithmetic sequence whose first term is* $\log(3)$ *with a constant difference of* $1$.

11. Without using a calculator, explain why the solution to each equation must be a real number between $1$ and $2$.

a. $11^x = 12$

$12$ *is greater than* $11^1$ *and less than* $11^2$, *so the solution is between* $1$ *and* $2$.

b. $21^x = 30$

$30$ *is greater than* $21^1$ *and less than* $21^2$, *so the solution is between* $1$ *and* $2$.

c. $100^x = 2000$

$100^2 = 10000$, *and* $2000$ *is less than that, so the solution is between* $1$ *and* $2$.

d. $\left(\frac{1}{11}\right)^x = 0.01$

$\frac{1}{100}$ *is between* $\frac{1}{11}$ *and* $\frac{1}{121}$, *so the solution is between* $1$ *and* $2$.

e. $\left(\frac{2}{3}\right)^x = \frac{1}{2}$

$\left(\frac{2}{3}\right)^2 = \frac{4}{9}$, *and* $\frac{1}{2}$ *is between* $\frac{2}{3}$ *and* $\frac{4}{9}$, *so the solution is between* $1$ *and* $2$.

f. $99^x = 9000$

$99^2 = 9801$. *Since* $9000$ *is less than* $9801$ *and greater than* $99$, *the solution is between* $1$ *and* $2$.

12. Express the exact solution to each equation as a base $10$ logarithm. Use a calculator to approximate the solution to the nearest $1000^{\text{th}}$.

a. $11^x = 12$

$$\log(11^x) = \log(12)$$
$$x\log(11) = \log(12)$$
$$x = \frac{\log(12)}{\log(11)}$$
$$x \approx 1.036$$

b. $21^x = 30$

$$x = \frac{\log(30)}{\log(21)}$$
$$x \approx 1.117$$

c. $100^x = 2000$

$$x = \frac{\log(2000)}{\log(100)}$$
$$x \approx 1.651$$

d. $\left(\frac{1}{11}\right)^x = 0.01$

$$x = -\frac{2}{\log\left(\frac{1}{11}\right)}$$
$$x \approx 1.921$$

e. $\left(\frac{2}{3}\right)^x = \frac{1}{2}$

$$x = \frac{\log\left(\frac{1}{2}\right)}{\log\left(\frac{2}{3}\right)}$$
$$x \approx 1.710$$

f. $99^x = 9000$

$$x = \frac{\log(9000)}{\log(99)}$$
$$x \approx 1.981$$

13. **Show that the value of $x$ that satisfies the equation $10^x = 3 \cdot 10^n$ is $\log(3) + n$.**

*Substituting $x = \log(3) + n$ into $10^x$ and using properties of exponents and logarithms gives*

$$\begin{aligned} 10^x &= 10^{\log(3)+n} \\ &= 10^{\log(3)}10^n \\ &= 3 \cdot 10^n. \end{aligned}$$

*Thus, $x = \log(3) + n$ is a solution to the equations $10^x = 3 \cdot 10^n$.*

14. **Solve each equation. If there is no solution, explain why.**

a. $3 \cdot 5^x = 21$

$$\begin{aligned} 5^x &= 7 \\ \log(5^x) &= \log(7) \\ x\log(5) &= \log(7) \\ x &= \frac{\log(7)}{\log(5)} \end{aligned}$$

b. $10^{x-3} = 25$

$$\begin{aligned} \log(10^{x-3}) &= \log(25) \\ x &= 3 + \log(25) \end{aligned}$$

c. $10^x + 10^{x+1} = 11$

$$10^x(1+10) = 11$$
$$10^x = 1$$
$$x = 0$$

d. $8 - 2^x = 10$

$$-2^x = 2$$
$$2^x = -2$$

*There is no solution because $2^x$ is always positive for all real $x$*

15. Solve the following equation for $n$: $A = P(1+r)^n$.

$$A = P(1+r)^n$$
$$\log(A) = \log[(P(1+r)^n]$$
$$\log(A) = \log(P) + \log[(1+r)^n]$$
$$\log(A) - \log(P) = n\log(1+r)$$
$$n = \frac{\log(A) - \log(P)}{\log(1+r)}$$
$$n = \frac{\log\left(\frac{A}{P}\right)}{\log(1+r)}$$

The remaining questions establish a property for the logarithm of a sum. Although this is an application of the logarithm of a product, the formula does have some applications in information theory and can help with the calculations necessary to use tables of logarithms, which will be explored further in Lesson 15.

16. In this exercise, we will establish a formula for the logarithm of a sum. Let $L = \log(x+y)$, where $x, y > 0$.

a. Show $\log(x) + \log\left(1 + \frac{y}{x}\right) = L$. State as a property of logarithms after showing this is a true statement.

$$\log(x) + \log\left(1+\frac{y}{x}\right) = \log\left(x\left(1+\frac{y}{x}\right)\right)$$
$$= \log\left(x + \frac{xy}{x}\right)$$
$$= \log(x+y)$$
$$= L$$

*Therefore, for $x, y > 0$, $\log(x+y) = \log(x) + \log\left(1+\frac{y}{x}\right)$.*

b. Use part (a) and the fact that $\log(100) = 2$ to rewrite $\log(365)$ as a sum.

$$\log(365) = \log(100 + 265)$$
$$= \log(100) + \log\left(1 + \frac{265}{100}\right)$$
$$= \log(100) + \log(3.65)$$
$$= 2 + \log(3.65)$$

c. Rewrite $365$ in scientific notation, and use properties of logarithms to express $\log(365)$ as a sum of an integer and a logarithm of a number between $0$ and $10$.

$$\begin{aligned} 365 &= 3.65 \times 10^2 \\ \log(365) &= \log(3.65 \times 10^2) \\ &= \log(3.65) + \log(10^2) \\ &= 2 + \log(3.65) \end{aligned}$$

d. What do you notice about your answers to (b) and (c)?

*Separating $365$ into $100 + 265$ and using the formula for the logarithm of a sum is the same as writing $365$ in scientific notation and using formula for the logarithm of a product.*

e. Find two integers that are upper and lower estimates of $\log(365)$.

*Since $1 < 3.65 < 10$, we know that $0 < \log(3.65) < 1$. This tells us that $2 < 2 + \log(3.65) < 3$, so $2 < \log(365) < 3$.*

# Lesson 13: Changing the Base

## Student Outcomes

- Students understand how to change logarithms from one base to another.
- Students calculate logarithms with any base using a calculator that computes only logarithms base 10 and base $e$.
- Students justify properties of logarithms with any base.

## Lesson Notes

The lesson begins by showing how to change logarithms from one base to another and develops properties of logarithms for the general base $b$. We introduce the use of a calculator instead of a table to approximate logarithms, and then we define $\ln(x) = \log_e(x)$. One goal of the lesson is to explain why the calculator only has a LOG and an LN key. Students solve exponential equations by applying the appropriate logarithm (**F-LE.A.4**).

## Materials

Students will need access either to graphing calculators or computer software facilities such as the Wolfram|Alpha engine for finding logarithms with base 10 and base $e$.

## Classwork

### Example 1 (5 minutes)

The purpose of this example is to show how to find $\log_2(x)$ using $\log(x)$.

- We have been working primarily with base 10 logarithms, but in Lesson 7 we defined logarithms for any base $b$. For example, the number 2 might be the base. When logarithms have bases other than 10, it often helps to be able to rewrite the logarithm in terms of base-10 logarithms. Let $L = \log_2(x)$, and show that $L = \frac{\log(x)}{\log(2)}$.
  - *Let $L = \log_2(x)$.*
    *Then $2^L = x$.*
    *Taking the logarithm of each side, we get*
    $$\log(2^L) = \log(x)$$
    $$L \cdot \log(2) = \log(x)$$
    $$L = \frac{\log(x)}{\log(2)}.$$
    *Therefore, $\log_2(x) = \frac{\log(x)}{\log(2)}$.*

> *Scaffolding:*
>
> - Students who struggle with the first step of this example might need to be reminded of the definition of *logarithm* from Lesson 7: $L = \log_b(x)$ means $b^L = x$. Therefore, $L = \log_2(x)$ means $2^L = x$.
> - Advanced learners may want to immediately start with the second part of the example, converting $\log_b(x)$ into $\frac{\log(x)}{\log(b)}$. Alternatively, students may explore the scale change by $\log(b)$ by finding different values of $\log(b)$ and discussing the effects on the graph of $y = \log(x)$.

Remember that $\log(2)$ is a number, so this shows that $\log_2(x)$ is a rescaling of $\log(x)$.

- The example shows how we can convert $\log_2(x)$ to an expression involving $\log(x)$. More generally, suppose we are given a logarithm with base $b$. What is $\log_b(x)$ in terms of $\log(x)$?
  - *Let $L = \log_b(x)$.*
    *Then $b^L = x$.*
    *Taking the logarithm of each side, we get*
    $$\log(b^L) = \log(x)$$
    $$L \cdot \log(b) = \log(x)$$
    $$L = \frac{\log(x)}{\log(b)}.$$
    *Therefore, $\log_b(x) = \frac{\log(x)}{\log(b)}$.*
- This equation not only allows us to change from $\log_b(x)$ to $\log(x)$, but to change the base in the other direction as well: $\log(x) = \log_b(x) \cdot \log(b)$.

## Exercise 1 (3 minutes)

The first exercise deals with the general formula for changing the base of a logarithm. It follows the same pattern as Example 1. Take time for students to share their results from Exercise 1 in a class discussion before moving on to Exercise 2 so that all students understand how the base of a logarithm is changed. Ask students to work in pairs on this exercise.

*Scaffolding:*
If students have difficulty with Exercise 1, they should review the argument in Example 1, noting that it deals with base 10, whereas this exercise generalizes that base to $a$.

**Exercises**

1. **Assume that $x$, $a$, and $b$ are all positive real numbers, so that $a \neq 1$ and $b \neq 1$. What is $\log_b(x)$ in terms of $\log_a(x)$? The resulting equation allows us to change the base of a logarithm from $a$ to $b$.**

   ***Let $L = \log_b(x)$. Then $b^L = x$. Taking the logarithm base $a$ of each side, we get***
   $$\log_a(b^L) = \log_a(x)$$
   $$L \cdot \log_a(b) = \log_a(x)$$
   $$L = \frac{\log_a(x)}{\log_a(b)}.$$
   ***Therefore, $\log_b(x) = \frac{\log_a(x)}{\log_a(b)}$.***

## Discussion (2 minutes)

Ask a student to present the solution to Exercise 1 to the class to ensure that all students understand how to change the base of a logarithm and how the formula comes from the definition of the logarithm as an exponential equation. Be sure that students record the formula in their notebooks.

Change of Base Formula for Logarithms

If $x$, $a$, and $b$ are all positive real numbers with $a \neq 1$ and $b \neq 1$, then
$$\log_b(x) = \frac{\log_a(x)}{\log_a(b)}.$$

## Exercise 2 (2 minutes)

In the second exercise, students practice changing bases. They will need a calculator with the ability to calculate logarithms base 10. Later in the lesson, students will need to calculate natural logarithms as well. Students should work in pairs on this exercise, with one student using the calculator and the other keeping track of the computation. Students should share their results for Exercise 2 in a class discussion before moving on to Exercise 3.

**2. Approximate each of the following logarithms to four decimal places. Use the LOG key on your calculator rather than logarithm tables, first changing the base of the logarithm to 10 if necessary.**

a. $\log(3^2)$

$\log(3^2) = \log(9) \approx 0.9542$

*Therefore,* $\log(3^2) \approx 0.9542$.

*OR*

$\log(3^2) = 2\log(3) \approx 2 \cdot 0.4771 \approx 0.9542$

*Therefore,* $\log(3^2) \approx 0.9542$.

b. $\log_3(3^2)$

$\log_3(3^2) = \dfrac{2\log(3)}{\log(3)} = 2$

*Therefore,* $\log_3(3^2) = 2.0000$.

c. $\log_2(3^2)$

$\log_2(3^2) = \log_2(9) = \dfrac{\log(9)}{\log(2)} \approx 3.1699$

*Therefore,* $\log_2(3^2) \approx 3.1699$.

> *Scaffolding:*
> Students who are not familiar with the LOG key on the calculator can check how it works by finding the following:
> $\log(1) = 0$;
> $\log(10) = 1$;
> $\log(10^3) = 3$.

## Exercise 3 (8 minutes)

**3. In Lesson 12, we justified a number of properties of base 10 logarithms. Working in pairs, justify the following properties of base-$b$ logarithms.**

a. $\log_b(1) = 0$

*Because* $L = \log_b(x)$ *means* $b^L = x$, *then when* $x = 1$, $L = 0$.

b. $\log_b(b) = 1$

*Because* $L = \log_b(x)$ *means* $b^L = x$, *then when* $x = b$, $L = 1$.

c. $\log_b(b^r) = r$

*Because* $L = \log_b(x)$ *means* $b^L = x$, *then when* $x = b^r$, $L = r$.

d. $b^{\log_b(x)} = x$

*Because* $L = \log_b(x)$ *means* $b^L = x$, *then* $x = b^{\log_b(x)}$.

> *Scaffolding:*
> By working in pairs, students should be able to reconstruct the arguments they used in Lesson 10. If they have trouble, they should be encouraged to use the definition and properties already justified.

e. $\log_b(x \cdot y) = \log_b(x) + \log_b(y)$

*By the rule* $a^q \cdot a^r = a^{q+r}$, $b^{\log_b(x)} \cdot b^{\log_b(y)} = b^{\log_b(x)+\log_b(y)}$.

*By property 4,* $b^{\log_b(x)} \cdot b^{\log_b(y)} = x \cdot y$.

*Therefore,* $x \cdot y = b^{\log_b(x)+\log_b(y)}$. *By property 4 again,* $x \cdot y = b^{\log_b(x \cdot y)}$.

*So, the exponents must be equal, and* $\log_b(x \cdot y) = \log_b(x) + \log_b(y)$.

f. $\log_b(x^r) = r \cdot \log_b(x)$

*By the rule* $(a^q)^r = a^{qr}$, $b^{r\log_b(x)} = \left(b^{\log_b(x)}\right)^r$.

*By property 4,* $\left(b^{\log_b(x)}\right)^r = x^r$.

*Therefore,* $x^r = b^{kr\log_b(x)}$. *By property 4 again,* $x^r = b^{\log_b(x^r)}$.

*So, the exponents must be equal, and* $\log_b(x^r) = r \cdot \log_b(x)$.

g. $\log_b\left(\frac{1}{x}\right) = -\log_b(x)$

*By property 6,* $\log_b(x^k) = k \cdot \log_b(x)$.

*Let* $k = -1$*, then for* $x \neq 0$, $\log_b(x^{-1}) = (-1) \cdot \log_b(x)$.

*Thus,* $\log_b\left(\frac{1}{x}\right) = -\log_b(x)$.

h. $\log_b\left(\frac{x}{y}\right) = \log_b(x) - \log_b(y)$

*By property 5,* $\log_b(x \cdot y) = \log_b(x) + \log_b(y)$.

*By property 7, for* $y \neq 0$, $\log_b(y^{-1}) = (-1) \cdot \log_b(y)$.

*Therefore,* $\log_b\left(\frac{x}{y}\right) = \log_b(x) - \log_b(y)$.

## Discussion (2 minutes)

The topic is the definition of the natural logarithm. Students often misinterpret the symbol "ln" as the word "in". Emphasize that the notation is an *L* followed by an *N*, which comes from the French for natural logarithm: *le logarithme naturel*.

- Recall Euler's number $e$ from Lesson 5, which is an irrational number approximated by $e \approx 2.71828 \ldots$. This number plays an important role in many parts of mathematics, and it is frequently used as the base of logarithms. When $e$ is taken as the base, the logarithm of a number $x$ is abbreviated as $\ln(x)$. Because $e$ is so often used to model growth and change in the natural world, it is not surprising that $\ln(x)$ is called the natural logarithm of $x$.
- Specifically, we say $\ln(x) = \log_e(x)$. What is the value of $\ln(1)$?
  - $\ln(1) = 0$
- What is the value of $\ln(e)$? The value of $\ln(e^2)$? Of $\ln(e^3)$?
  - $\ln(e) = 1$, $\ln(e^2) = 2$, *and* $\ln(e^3) = 3$.

- Because scientists primarily use logarithms base 10 and base $e$, calculators only have two logarithm buttons; LOG for calculating $\log(x)$ and LN for calculating $\ln(x)$. With the change of base formula, you can use either the common logarithm (base 10) or the natural logarithm (base $e$) to calculate the value of a logarithm with any allowable base $b$, so technically we only need one of those two buttons. However, each base has important uses, so most calculators are able to calculate logarithms in either base.

### Exercise 4 (3 minutes)

Exercise 4 introduces the natural logarithm. Students will need a calculator with an LN key. They should work in pairs on these exercises, with one student using the calculator and the other keeping track of the computation. They should share their results for Exercise 4 in a class discussion before moving on.

*Scaffolding:*

Students who are not familiar with the LN key on the calculator can check how it works by finding the following:

$\ln(1) = 0;$
$\ln(e) = 1;$
$\ln(e^3) = 3.$

**4. Find each of the following to four decimal places. Use the LN key on your calculator rather than a table.**

**a.** $\ln(3^2)$

$\ln(3^2) = \ln(9) \approx 2.1972$

**b.** $\ln(2^4)$

$\ln(2^4) = \ln(16) \approx 2.7726$

### Example 2 (4 minutes)

This example introduces more complicated expressions involving logarithms and showcases the power of logarithms in rearranging logarithmic expressions. Students have done exercises like this in their homework in prior lessons for base 10 logarithms, so this example and the following exercises demonstrate how the same procedures apply to logarithms with any base.

- Write as an expression containing only one logarithm: $\ln(k^2) + \ln\left(\frac{1}{k^2}\right) - \ln(\sqrt{k})$.
  - $\ln(k^2) + \ln\left(\frac{1}{k^2}\right) - \ln(\sqrt{k}) = 2\ln(k) - 2\ln(k) - \frac{1}{2}\cdot\ln(k) = -\frac{1}{2}\ln(k)$
- Therefore, $\ln(k^2) + \ln\left(\frac{1}{k^2}\right) - \ln(\sqrt{k}) = -\frac{1}{2}\ln(k)$.

### Exercises 5–6 (7 minutes)

Exercise 5 follows Example 3 by introducing somewhat more complicated expressions to be simplified that involve natural logarithms. In Exercise 5, students condense a sum of logarithmic expressions to an expression containing only one logarithm, while in Exercise 6, students take a single complicated logarithm and break it up into simpler parts. Students should work in pairs on these exercises, sharing their results in a class discussion before the closing.

5. **Write as a single logarithm.**

a. $\ln(4) - 3\ln\left(\frac{1}{3}\right) + \ln(2)$

$$\begin{aligned}\ln(4) - 3\ln\left(\frac{1}{3}\right) + \ln(2) &= \ln(4) + \ln(3^3) + \ln(2)\\ &= \ln(4 \cdot 3^3 \cdot 2)\\ &= \ln(216)\\ &= \ln(6^3)\\ &= 3\ln(6)\end{aligned}$$

***Any of the last three expressions is an acceptable final answer.***

b. $\ln(5) + \frac{3}{5}\ln(32) - \ln(4)$

$$\begin{aligned}\ln(5) + \frac{3}{5}\ln(32) - \ln(4) &= \ln(5) + \ln(8) - \ln(4)\\ &= \ln(5 \cdot 8) - \ln(4)\\ &= \ln\left(\frac{40}{4}\right)\\ &= \ln(10)\end{aligned}$$

***Therefore,*** $\ln(5) + \frac{3}{5}\ln(32) - \ln(4) = \ln(10)$.

6. **Write each expression as a sum or difference of constants and logarithms of simpler terms.**

a. $\ln\left(\frac{\sqrt{5x^3}}{e^2}\right)$

$$\begin{aligned}\ln\left(\frac{\sqrt{5x^3}}{e^2}\right) &= \ln(\sqrt{5}) + \ln\left(\sqrt{x^3}\right) - \ln(e^2)\\ &= \frac{1}{2}\ln(5) + \frac{3}{2}\ln(x) - 2\end{aligned}$$

b. $\ln\left(\frac{(x+y)^2}{x^2+y^2}\right)$

$$\begin{aligned}\ln\left(\frac{(x+y)^2}{x^2+y^2}\right) &= \ln(x+y)^2 - \ln(x^2+y^2)\\ &= 2\ln(x+y) - \ln(x^2+y^2)\end{aligned}$$

***The point of this simplification is that neither of these terms can be simplified further.***

## Closing (4 minutes)

Have students summarize the lesson by discussing the following questions and coming to a consensus before students record the answers in their notebook.

- What is the definition of the logarithm base $b$?
  - *If there exist numbers b, L, and x so that* $b^L = x$, *then* $L = \log_b(x)$.
- What does $\ln(x)$ represent?
  - *The notation* $\ln(x)$ *represents the logarithm of x base e; that is,* $\ln(x) = \log_e(x)$.

- How can we use a calculator to approximate a logarithm to a base other than $10$ or $e$?
  - *Use the change of base formula to convert a logarithm with base $b$ to one with base $10$ or base $e$; then, use the appropriate calculator button.*

**Lesson Summary**

**We have established a formula for changing the base of logarithms from $b$ to $a$:**

$$\log_b(x) = \frac{\log_a(x)}{\log_a(b)}.$$

**In particular, the formula allows us to change logarithms base $b$ to common or natural logarithms, which are the only two kinds of logarithms that calculators compute:**

$$\log_b(x) = \frac{\log(x)}{\log(b)} = \frac{\ln(x)}{\ln(b)}.$$

**We have also established the following properties for base $b$ logarithms. If $x$, $y$, $a$ and $b$ are all positive real numbers with $a \neq 1$ and $b \neq 1$ and $r$ is any real number, then:**

$$\log_b(1) = 0$$

$$\log_b(b) = 1$$

$$\log_b(b^r) = r$$

$$b^{\log_b(x)} = x$$

$$\log_b(x \cdot y) = \log_b(x) + \log_b(y)$$

$$\log_b(x^r) = r \cdot \log_b(x)$$

$$\log_b\left(\frac{1}{x}\right) = -\log_b(x)$$

$$\log_b\left(\frac{x}{y}\right) = \log_b(x) - \log_b(y)$$

## Exit Ticket (5 minutes)

Name ________________________________ Date ______________

# Lesson 13: Changing the Base

## Exit Ticket

1. Are there any properties that hold for base $10$ logarithms that would not be valid for the logarithm base $e$? Why? Are there any properties that hold for base $10$ logarithms that would not be valid for some positive base $b$, such that $b \neq 1$?

2. Write each logarithm as an equivalent expression involving only logarithms base $10$.
   a. $\log_3(25)$

   b. $\log_{100}(x^2)$

3. Rewrite each expression as an equivalent expression containing only one logarithm.
   a. $3\ln(p+q) - 2\ln(q) - 7\ln(p)$

   b. $\ln(xy) - \ln\left(\frac{x}{y}\right)$

## Exit Ticket Sample Solutions

1. **Are there any properties that hold for base $10$ logarithms that would not be valid for the logarithm base $e$? Why? Are there any properties that hold for base $10$ logarithms that would not be valid for some positive base $b$, such that $b \neq 1$?**

   ***No. Any property that is true for a base $10$ logarithm will be true for a base e logarithm. The only difference between a common logarithm and a natural logarithm is a scale change, because $\log(x) = \frac{\ln(x)}{\ln(10)}$, and $\ln(x) = \frac{\log(x)}{\log(e)}$.***

   ***Since $\log_b(x) = \frac{\log(x)}{\log(b)}$, we would only encounter a problem if $\log(b) = 0$, but this only happens when $b = 1$, and $1$ is not a valid base for logarithms.***

2. **Write each logarithm as an equivalent expression involving only logarithms base $10$.**

   a. $\log_3(25)$

   $$\log_3(25) = \frac{\log(25)}{\log(3)}$$

   b. $\log_{100}(x^2)$

   $$\begin{aligned}\log_{100}(x^2) &= \frac{\log(x^2)}{\log(100)}\\ &= \frac{2\log(x)}{2}\\ &= \log(x)\end{aligned}$$

3. **Rewrite each expression as an equivalent expression containing only one logarithm.**

   a. $3\ln(p+q) - 2\ln(q) - 7\ln(p)$

   $$\begin{aligned}3\ln(p+q) - 2\ln(q) - 7\ln(p) &= \ln((p+q)^3) - \left(\ln(q^2) + \ln(p^7)\right)\\ &= \ln((p+q)^3) - \ln(q^2p^7)\\ &= \ln\left(\frac{(p+q)^3}{q^2p^7}\right)\end{aligned}$$

   b. $\ln(xy) - \ln\left(\frac{x}{y}\right)$

   $$\begin{aligned}\ln(xy) - \ln\left(\frac{x}{y}\right) &= \ln(x) + \ln(y) - \ln(x) + \ln(y)\\ &= 2\ln(y)\\ &= \ln(y^2)\end{aligned}$$

   ***Therefore, $\ln(xy) - \ln\left(\frac{x}{y}\right)$ is equivalent to both $2\ln(y)$ and $\ln(y^2)$.***

## Problem Set Sample Solutions

1. Evaluate each of the following logarithmic expressions, approximating to four decimal places if necessary. Use the LN or LOG key on your calculator rather than a table.

   a. $\log_8(16)$

   $$\begin{aligned}\log_8(16) &= \frac{\log(16)}{\log(8)}\\ &= \frac{\log(2^4)}{\log(2^3)}\\ &= \frac{4\cdot\log(2)}{3\cdot\log(2)}\\ &= \frac{4}{3}\end{aligned}$$

   *Therefore,* $\log_8(16) = \frac{4}{3}$.

   b. $\log_7(11)$

   $$\begin{aligned}\log_7(11) &= \frac{\log(11)}{\log(7)}\\ &\approx 1.2323\end{aligned}$$

   *Therefore,* $\log_7(11) \approx 1.2323$.

   c. $\log_3(2) + \log_2(3)$

   $$\begin{aligned}\log_3(2) + \log_2(3) &= \frac{\log(2)}{\log(3)} + \frac{\log(3)}{\log(2)}\\ &\approx 2.2159\end{aligned}$$

   *Therefore,* $\log_3(2) + \log_2(3) \approx 2.2159$.

2. Use logarithmic properties and the fact that $\ln(2) \approx 0.69$ and $\ln(3) \approx 1.10$ to approximate the value of each of the following logarithmic expressions. Do not use a calculator.

   a. $\ln(e^4)$

   $$\begin{aligned}\ln(e^4) &= 4\ln(e)\\ &= 4\end{aligned}$$

   *Therefore,* $\ln(e^4) = 4$.

   b. $\ln(6)$

   $$\begin{aligned}\ln(6) &= \ln(2) + \ln(3)\\ &\approx 0.69 + 1.10\\ &\approx 1.79\end{aligned}$$

   *Therefore,* $\ln(6) \approx 1.79$.

c. $\ln(108)$

$$\begin{aligned}\ln(108) &= \ln(4\cdot 27)\\ &= \ln(4)+\ln(27)\\ &\approx 2\ln(2)+3\ln(3)\\ &\approx 1.38+3.30\\ &\approx 4.68\end{aligned}$$

*Therefore,* $\ln(108)\approx 4.68$.

d. $\ln\left(\frac{8}{3}\right)$

$$\begin{aligned}\ln\left(\frac{8}{3}\right) &= \ln(8)-\ln(3)\\ &= \ln(2^3)-\ln(3)\\ &\approx 3(0.69)-1.10\\ &\approx 0.97\end{aligned}$$

*Therefore,* $\ln\left(\frac{8}{3}\right)\approx 0.97$.

3. **Compare the values of $\log_{\frac{1}{9}}(10)$ and $\log_9\left(\frac{1}{10}\right)$ without using a calculator.**

*Using the change of base formula,*

$$\begin{aligned}\log_{\frac{1}{9}}(10) &= \frac{\log_9(10)}{\log_9\left(\frac{1}{9}\right)}\\ &= \frac{\log_9(10)}{-1}\\ &= -\log_9(10)\\ &= \log_9\left(\frac{1}{10}\right).\end{aligned}$$

*Thus,* $\log_{\frac{1}{9}}(10)=\log_9\left(\frac{1}{10}\right)$.

4. **Show that for any positive numbers $a$ and $b$ with $a\neq 1$ and $b\neq 1$, $\log_a(b)\cdot\log_b(a)=1$.**

*Using the change of base formula,*

$$\log_a(\mathrm{b})=\frac{\log_b(\mathrm{b})}{\log_b(\mathrm{a})}=\frac{1}{\log_b(\mathrm{a})}.$$

*Thus,*

$$\log_a(b)\cdot\log_b(a)=\frac{1}{\log_b(a)}\cdot\log_b(a)=1.$$

5. **Express $x$ in terms of $a$, $e$, and $y$ if $\ln(x)-\ln(y)=2a$.**

$$\begin{aligned}\ln(x)-\ln(y) &= 2a\\ \ln\left(\frac{x}{y}\right) &= 2a\\ \frac{x}{y} &= e^{2a}\\ x &= y\,e^{2a}\end{aligned}$$

6. **Rewrite each expression in an equivalent form that only contains one base 10 logarithm.**

   a. $\log_2(800)$

   $$\frac{\log(800)}{\log(2)} = \frac{\log(2^3)+2}{\log(2)} = 3 + \frac{2}{\log(2)}$$

   b. $\log_x\left(\frac{1}{10}\right)$, **for positive real values of** $x \neq 1$

   $$\frac{\log\left(\frac{1}{10}\right)}{\log(x)} = -\frac{1}{\log(x)}$$

   c. $\log_5(12,500)$

   $$\frac{\log(5^3 \cdot 10^2)}{\log(5)} = \frac{3\log(5)+2}{\log(5)} = 3 + \frac{2}{\log(5)}$$

   d. $\log_3(0.81)$

   $$\frac{\log\left(\frac{81}{100}\right)}{\log(3)} = \frac{4\log(3)-2}{\log(3)} = 4 - \frac{2}{\log(3)}$$

7. **Write each number in terms of natural logarithms, and then use the properties of logarithms to show that it is a rational number.**

   a. $\log_9(\sqrt{27})$

   $$\frac{\ln(\sqrt{27})}{\ln(9)} = \frac{\ln\left(3^{\frac{3}{2}}\right)}{\ln(3^2)} = \frac{\frac{3}{2}\ln(3)}{2\ln(3)} = \frac{3}{4}$$

   b. $\log_8(32)$

   $$\frac{\ln(32)}{\ln(8)} = \frac{\ln(2^5)}{\ln(2^3)} = \frac{5}{3}$$

   c. $\log_4\left(\frac{1}{8}\right)$

   $$\frac{\ln\left(\frac{1}{8}\right)}{\ln(4)} = \frac{\ln(2^{-3})}{\ln(2^2)} = -\frac{3}{2}$$

8. **Write each expression as an equivalent expression with a single logarithm. Assume $x$, $y$, and $z$ are positive real numbers.**

   a. $\ln(x) + 2\ln(y) - 3\ln(z)$

   $$\ln\left(\frac{xy^2}{z^3}\right)$$

b. $\frac{1}{2}(\ln(x+y) - \ln(z))$

$\ln\left(\sqrt{\frac{x+y}{z}}\right)$

c. $(x+y) + \ln(z)$

$(x+y)\ln(e) + \ln(z) = \ln(e^{x+y}) + \ln(z) = \ln(e^{x+y} \cdot z)$

9. **Rewrite each expression as sums and differences in terms of $\ln(x)$, $\ln(y)$, and $\ln(z)$.**

a. $\ln(xyz^3)$

$\ln(x) + \ln(y) + 3\ln(z)$

b. $\ln\left(\frac{e^3}{xyz}\right)$

$3 - \ln(x) - \ln(y) - \ln(z)$

c. $\ln\left(\sqrt{\frac{x}{y}}\right)$

$\frac{1}{2}(\ln(x) - \ln(y))$

10. **Solve the following equations in terms of base 5 logarithms. Then, use the change of base properties and a calculator to estimate the solution to the nearest $1000^{\text{th}}$. If the equation has no solution, explain why.**

a. $5^{2x} = 20$

$2x = \log_5(20)$

$x = \frac{1}{2}\log_5(20)$

$x = \frac{\log(20)}{2\log(5)}$

$x \approx 0.931$

b. $75 = 10 \cdot 5^{x-1}$

$7.5 = 5^{x-1}$

$x = \log_5(7.5) + 1$

$x \approx 2.252$

c. $5^{2+x} - 5^x = 10$

$5^x(5^2 - 1) = 10$

$5^x = \frac{10}{24}$

$x = \log_5\left(\frac{10}{24}\right)$

$x \approx -0.544$

d. $5^{x^2} = 0.25$

$x^2 = \log_5(0.25)$

$x^2 = \dfrac{\log(0.25)}{\log(5)}$

*This equation has no real solution because* $\dfrac{\log(0.25)}{\log(5)}$ *is negative.*

11. In Lesson 6, you discovered that $\log(x \cdot 10^k) = k + \log(x)$ by looking at a table of logarithms. Use the properties of logarithms to justify this property for an arbitrary base $b > 0$ with $b \neq 1$. That is, show that $\log_b(x \cdot b^k) = k + \log_b(x)$.

$$\begin{aligned}\log_b(x \cdot b^k) &= \log_b(x) + \log_b(b^k)\\ &= k + log_b(x)\end{aligned}$$

MP.3

12. Larissa argued that since $\log_2(2) = 1$ and $\log_2(4) = 2$, then it must be true that $\log_2(3) = 1.5$. Is she correct? Explain how you know.

*Larissa is not correct. According to the calculator and the change of base formula,* $\log_2(3) = \frac{\log(3)}{\log(2)} \approx 1.585$. *If* $\log_2(x) = 1.5$ *then* $2^{1.5} = x$, *so* $x = \sqrt{8} = 2\sqrt{2}$. *Since* $3 \neq 2\sqrt{2}$, *Larissa's calculation is not correct. Larissa is assuming that the logarithm function behaves like a linear function, which it does not.*

13. Extension: Suppose that there is some positive number $b$ so that

$$\log_b(2) = 0.36$$
$$\log_b(3) = 0.57$$
$$\log_b(5) = 0.84.$$

a. Use the given values of $\log_b(2)$, $\log_b(3)$, and $\log_b(5)$ to evaluate the following logarithms.

i. $\log_b(6)$

$$\begin{aligned}\log_b(6) &= \log_b(2 \cdot 3)\\ &= \log_b(2) + \log_b(3)\\ &= 0.36 + 0.57\\ &= 0.93\end{aligned}$$

ii. $\log_b(8)$

$$\begin{aligned}\log_b(8) &= \log_b(2^3)\\ &= 3 \cdot \log_b(2)\\ &= 3 \cdot 0.36\\ &= 1.08\end{aligned}$$

iii. $\log_b(10)$

$$\begin{aligned}\log_b(10) &= \log_b(2 \cdot 5)\\ &= \log_b(2) + \log_b(5)\\ &= 0.36 + 0.84\\ &= 1.20\end{aligned}$$

iv. $\log_b(600)$

$$\begin{aligned}\log_b(600) &= \log_b(6 \cdot 100)\\ &= \log_b(6) + \log_b(100)\\ &= 0.93 + 2\log_b(10)\\ &= 0.93 + 2(1.20)\\ &= 0.93 + 2.40\\ &= 3.33\end{aligned}$$

b. **Use the change of base formula to convert $\log_b(10)$ to base $10$, and solve for $b$. Give your answer to four decimal places.**

*From part (iii) above, $\log_b(10) = 1.20$. Then,*

$$\begin{aligned}1.20 &= \log_b(10)\\ 1.20 &= \frac{\log_{10}(10)}{\log_{10}(b)}\\ 1.20 &= \frac{1}{\log_{10}(b)}\\ \frac{1}{1.20} &= \log_{10}(b)\\ b &= 10^{\frac{1}{1.20}}\\ b &\approx 6.8129.\end{aligned}$$

14. **Solve the following exponential equations.**

a. $2^{3x} = 16$

$$\begin{aligned}\log_2(2^{3x}) &= \log_2(16)\\ 3x &= 4\\ x &= \frac{4}{3}\end{aligned}$$

b. $2^{x+3} = 4^{3x}$

$$\begin{aligned}\log_2(2^{x+3}) &= \log_2(4^{3x})\\ x+3 &= 3x \cdot \log_2(4)\\ x+3 &= 3x \cdot 2\\ 5x &= 3\\ x &= \frac{3}{5}\end{aligned}$$

c. $3^{4x-2} = 27^{x+2}$

$$\begin{aligned}\log_3(3^{4x-2}) &= \log_3(27^{x+2})\\ (4x-2)\log_3(3) &= (x+2)\log_3(27)\\ 4x-2 &= 3(x+2)\\ 4x-2 &= 3x+6\\ x &= 8\end{aligned}$$

d. $4^{2x} = \left(\frac{1}{4}\right)^{3x}$

$$\log_4(4^{2x}) = \log_4\left(\left(\frac{1}{4}\right)^{3x}\right)$$
$$2x\log_4(4) = 3x\log_4\left(\frac{1}{4}\right)$$
$$2x = 3x(-1)$$
$$5x = 0$$
$$x = 0$$

e. $5^{0.2x+3} = 625$

$$\log_5(5^{0.2x+3}) = \log_5(625)$$
$$(0.2x+3)\log_5(5) = \log_5(5^4)$$
$$0.2x + 3 = 4$$
$$0.2x = 1$$
$$x = 5$$

15. Solve each exponential equation.

a. $3^{2x} = 81$

$x = 2$

b. $6^{3x} = 36^{x+1}$

$x = 2$

c. $625 = 5^{3x}$

$x = \frac{4}{3}$

d. $25^{4-x} = 5^{3x}$

$x = \frac{8}{5}$

e. $32^{x-1} = \frac{1}{2}$

$x = \frac{4}{5}$

f. $\frac{4^{2x}}{2^{x-3}} = 1$

$x = -1$

g. $\frac{1}{8^{2x-4}} = 1$

$x = 2$

h. $2^x = 81$

$x = \frac{\ln(81)}{\ln(2)}$

i. $8 = 3^x$

$x = \frac{\ln(8)}{\ln(3)}$

j. $6^{x+2} = 12$

$x = -2 + \frac{\log(12)}{\log(6)}$

k. $10^{x+4} = 27$

$x = -4 + \log(27)$

l. $2^{x+1} = 3^{1-x}$

$x = \frac{\log(3) - \log(2)}{\log(2) + \log(3)}$

m. $3^{2x-3} = 2^{x+4}$

$x = \frac{4\log(2) + 3\log(3)}{3\log(3) - \log(2)}$

n. $e^{2x} = 5$

$x = \frac{\ln(5)}{2}$

o. $e^{x-1} = 6$

$x = -1 + 3\ln(2)$

16. In Problem 9(e) of Lesson 12, you solved the equation $3^x = 7^{-3x+2}$ using the logarithm base 10.

a. Solve $3^x = 7^{-3x+2}$ using the logarithm base 3.

$$\begin{aligned} \log_3(3^x) &= \log_3(7^{-3x+2}) \\ x &= (-3x+2)\log_3(7) \\ x &= -3x\log_3(7) + 2\log_3(7) \\ x + 3x\log_3(7) &= 2\log_3(7) \\ x(1 + 3\log_3(7)) &= 2\log_3(7) \\ x &= \frac{2\log_3(7)}{1 + 3\log_3(7)} \end{aligned}$$

b. Apply the change of base formula to show that your answer to part (a) agrees with your answer to Problem 9(e) of Lesson 12.

*Changing from base 3 to base 10, we see that*

$$\log_3(7) = \frac{\log(7)}{\log(3)}.$$

*Then,*

$$\begin{aligned} \frac{2\log_3(7)}{1 + 3\log_3(7)} &= \frac{2\left(\frac{\log(7)}{\log(3)}\right)}{1 + 3\left(\frac{\log(7)}{\log(3)}\right)} \\ &= \frac{2\log(7)}{\log(3) + 3\log(7)}, \end{aligned}$$

*which was the answer from Problem 9(e) of Lesson 12.*

c. Solve $3^x = 7^{-3x+2}$ using the logarithm base 7.

$$\begin{aligned} \log_7(3^x) &= \log_7(7^{-3x+2}) \\ x\log_7(3) &= -3x + 2 \\ 3x + x\log_7(3) &= 2 \\ x(3 + \log_7(3)) &= 2 \\ x &= \frac{2}{3 + \log_7(3)} \end{aligned}$$

d. Apply the change of base formula to show that your answer to part (c) also agrees with your answer to Problem 9(e) of Lesson 12.

*Changing from base 7 to base 10, we see that*

$$\log_7(3) = \frac{\log(3)}{\log(7)}.$$

*Then,*

$$\begin{aligned} \frac{2}{3 + \log_7(3)} &= \frac{2}{3 + \frac{\log(3)}{\log(7)}} \\ &= \frac{2\log(7)}{3\log(7) + \log(3)}, \end{aligned}$$

*which was the answer from Problem 9(e) of Lesson 12.*

17. **Pearl solved the equation $2^x = 10$ as follows:**

$$\log(2^x) = \log(10)$$
$$x\log(2) = 1$$
$$x = \frac{1}{\log(2)}.$$

**Jess solved the equation $2^x = 10$ as follows:**

$$\log_2(2^x) = \log_2(10)$$
$$x\log_2(2) = \log_2(10)$$
$$x = \log_2(10).$$

MP.3

**Is Pearl correct? Is Jess correct? Explain how you know.**

***Both Pearl and Jess are correct. If we take Jess's solution and apply the change of base formula, we have***

$$\begin{aligned} x &= \log_2(10) \\ &= \frac{\log(10)}{\log(2)} \\ &= \frac{1}{\log(2)}. \end{aligned}$$

***Thus, the two solutions are equivalent, and both students are correct.***

# Lesson 14: Solving Logarithmic Equations

## Student Outcomes

- Students solve simple logarithmic equations using the definition of logarithm and logarithmic properties.

## Lesson Notes

In this lesson, students will solve simple logarithmic equations by first putting them into the form $\log_b(Y) = L$, where $Y$ is an expression, and $L$ is a number for $b = 2, 10$, and $e$, and then using the definition of logarithm to rewrite the equation in the form $b^L = Y$. Students will be able to evaluate logarithms without technology by selecting an appropriate base; solutions are provided with this in mind. In Lesson 15, students will learn the technique of solving exponential equations using logarithms of any base without relying on the definition. Students will need to use the properties of logarithms developed in prior lessons to rewrite the equations in an appropriate form before solving (**A-SSE.A.2**, **F-LE.A.4**). The lesson starts with a few fluency exercises to reinforce the logarithmic properties before moving on to solving equations.

## Classwork

### Opening Exercise (3 minutes)

The following exercises provide practice with the definition of the logarithm and prepare students for the method of solving logarithmic equations that follows. Encourage students to work alone on these exercises, but allow students to work in pairs if necessary.

*Scaffolding:*

- Remind students of the main properties that they will be using by writing the following on the board:
  $\log_b(x) = L$ means $b^L = x$;
  $\log_b(xy) = \log_b(x) + \log_b(y)$;
  $\log_b\left(\frac{x}{y}\right) = \log_b(x) - \log_b(y)$;
  $\log_b(x^r) = r \cdot \log_b(x)$;
  $\log_b\left(\frac{1}{x}\right) = -\log(x)$.
- Consistently using a visual display of these properties throughout the module will be helpful.

**Opening Exercise**

**Convert the following logarithmic equations to exponential form:**

a. $\log(10{,}000) = 4$ $\qquad 10^4 = 10{,}000$

b. $\log(\sqrt{10}) = \frac{1}{2}$ $\qquad 10^{\frac{1}{2}} = \sqrt{10}$

c. $\log_2(256) = 8$ $\qquad 2^8 = 256$

d. $\log_4(256) = 4$ $\qquad 4^4 = 256$

e. $\ln(1) = 0$ $\qquad e^0 = 1$

f. $\log(x+2) = 3$ $\qquad x + 2 = 10^3$

## Examples 1–3 (6 minutes)

Students should be ready to take the next step from converting logarithmic equations to exponential form to solving the resulting equation. Use your own judgment on whether or not students will need to see a teacher-led example or can attempt to solve these equations in pairs. Anticipate that students will neglect to check for extraneous solutions in these examples, and after the examples, lead the discussion to the existence of an extraneous solution in Example 3.

**Examples 1–3**

**Write each of the following equations as an equivalent exponential equation, and solve for $x$.**

1. $\log(3x+7)=0$

$$\log(3x+7)=0$$
$$10^0=3x+7$$
$$1=3x+7$$
$$x=-2$$

2. $\log_2(x+5)=4$

$$\log_2(x+5)=4$$
$$2^4=x+5$$
$$16=x+5$$
$$x=11$$

3. $\log(x+2)+\log(x+5)=1$

$$\log(x+2)+\log(x+5)=1$$
$$\log((x+2)(x+5))=1$$
$$(x+2)(x+5)=10^1$$
$$x^2+7x+10=10$$
$$x^2+7x=0$$
$$x(x+7)=0$$
$$x=0 \text{ or } x=-7$$

***However, if $x=-7$, then $(x+2)=-5$, and $(x+5)=-2$, so both logarithms in the equation are undefined. Thus, $-7$ is an extraneous solution, and only $0$ is a valid solution to the equation.***

## Discussion (4 minutes)

Ask students to volunteer their solutions to the equations in the Opening Exercise. This line of questioning is designed to allow students to decide that there is an extraneous solution to Example 3. If the class has already discovered this fact, you may opt to accelerate or skip this discussion.

- What is the solution to the equation in Example 1?
  - $-2$
- What is the result if you evaluate $\log(3x+7)$ at $x=-2$? Did you find a solution?
  - $\log(3(-2)+7)=\log(1)=0$, *so* $-2$ *is a solution to* $\log(3x+7)=0$.

- What is the solution to the equation in Example 2?
  - $11$
- What is the result if you evaluate $\log_2(x+5)$ at $x = 11$? Did you find a solution?
  - $\log_2(11+5) = \log_2(16) = 4$, *so* $11$ *is a solution to* $\log_2(x+5) = 4$.
- What is the solution to the equation in Example 3?
  - *There were two solutions:* $0$ *and* $-7$.
- What is the result if you evaluate $\log(x+2) + \log(x+5)$ at $x = 0$? Did you find a solution?
  - $\log(2) + \log(5) = \log(2 \cdot 5) = \log(10) = 1$, *so* $0$ *is a solution to* $\log(x+2) + \log(x+5) = 1$.
- What is the result if you evaluate $\log(x+2) + \log(x+5)$ at $x = -7$? Did you find a solution?
  - $\log(-7+2)$ *and* $\log(-7+5)$ *are not defined because* $-7+2$ *and* $-7+5$ *are negative. Thus,* $-7$ *is not a solution to the original equation.*
- What is the term we use for an apparent solution to an equation that fails to solve the original equation?
  - *It is called an extraneous solution.*
- Remember to look for extraneous solutions, and exclude them when you find them.

### Exercise 1 (4 minutes)

Allow students to work in pairs or small groups to think about the exponential equation below. This equation can be solved rather simply by an application of the logarithmic property $\log_b(x^r) = r\log_b(x)$. However, if students do not see to apply this logarithmic property, it can become algebraically difficult.

**Exercise 1**

1. **Drew said that the equation $\log_2[(x+1)^4] = 8$ cannot be solved because he expanded $(x+1)^4 = x^4 + 4x^3 + 6x^2 + 4x + 1$ and realized that he cannot solve the equation $x^4 + 4x^3 + 6x^2 + 4x + 1 = 2^8$. Is he correct? Explain how you know.**

   ***If we apply the logarithmic properties, this equation is solvable.***

$$\begin{aligned} \log_2[(x+1)^4] &= 8 \\ 4\log_2(x+1) &= 8 \\ \log_2(x+1) &= 2 \\ x+1 &= 2^2 \\ x &= 3 \end{aligned}$$

   ***Check:** If $x = 3$, then $\log_2[(3+1)^4] = 4\log_2(4) = 4 \cdot 2 = 8$, so $3$ is a solution to the original equation.*

### Exercises 2–4 (6 minutes)

Students should work on these three exercises independently or in pairs to help develop fluency with these types of problems. Circulate around the room and remind students to check for extraneous solutions as necessary.

**Solve the equations in Exercises 2–4 for $x$.**

2. $\ln((4x)^5) = 15$

$$5 \cdot \ln(4x) = 15$$
$$\ln(4x) = 3$$
$$e^3 = 4x$$
$$x = \frac{e^3}{4}$$

*Check: Since* $4\left(\frac{e^3}{4}\right) > 0$, *we know that* $\ln\left(\left(4 \cdot \frac{e^3}{5}\right)^5\right)$ *is defined. Thus,* $\frac{e^3}{4}$ *is the solution to the equation.*

3. $\log((2x+5)^2) = 4$

$$2 \cdot \log(2x+5) = 4$$
$$\log(2x+5) = 2$$
$$10^2 = 2x+5$$
$$100 = 2x+5$$
$$95 = 2x$$
$$x = \frac{95}{2}$$

*Check: Since* $2\left(\frac{95}{2}\right) + 5 \neq 0$, *we know that* $\log\left(\left(2 \cdot \frac{95}{2} + 5\right)^2\right)$ *is defined.*

*Thus,* $\frac{95}{2}$ *is the solution to the equation.*

4. $\log_2((5x+7)^{19}) = 57$

$$19 \cdot \log_2(5x+7) = 57$$
$$\log_2(5x+7) = 3$$
$$2^3 = 5x+7$$
$$8 = 5x+7$$
$$1 = 5x$$
$$x = \frac{1}{5}$$

*Check: Since* $5\left(\frac{1}{5}\right) + 7 > 0$, *we know that* $\log_2\left(5 \cdot \frac{1}{5} + 7\right)$ *is defined.*

*Thus,* $\frac{1}{5}$ *is the solution to this equation.*

## Example 4 (4 minutes)

In Examples 2 and 3, students encounter more difficult logarithmic equations, and in Example 3, they encounter extraneous solutions. After each example, debrief the students to informally assess their understanding and provide guidance to align their understanding with the concepts. Some sample questions are included with likely student responses. Remember to have students check for extraneous solutions in all cases.

$$\log(x+10) - \log(x-1) = 2$$
$$\log\left(\frac{x+10}{x-1}\right) = 2$$
$$\frac{x+10}{x-1} = 10^2$$
$$x + 10 = 100(x-1)$$
$$99x = 110$$
$$x = \frac{10}{9}$$

- Is $\frac{10}{9}$ a valid solution? Explain how you know.
  - *Yes;* $\log\left(\frac{10}{9}+10\right)$ *and* $\log\left(\frac{10}{9}-1\right)$ *are both defined, so* $\frac{10}{9}$ *is a valid solution.*
- Why could we not rewrite the original equation in exponential form using the definition of the logarithm immediately?
  - *The equation needs to be in the form* $\log_b(Y) = L$ *before using the definition of a logarithm to rewrite it in exponential form, so we had to use the logarithmic properties to combine terms first.*

**Example 5 (3 minutes)**

Make sure students verify the solutions in Example 5 because there is an extraneous solution.

$$\log_2(x+1) + \log_2(x-1) = 3$$
$$\log_2\big((x+1)(x-1)\big) = 3$$
$$\log_2(x^2-1) = 3$$
$$2^3 = x^2 - 1$$
$$0 = x^2 - 9$$
$$0 = (x-3)(x+3)$$

Thus, $x = 3$ or $x = -3$. We need to check these solutions to see if they are valid.

- Is 3 a valid solution?
  - $\log_2(3+1) + \log_2(3-1) = \log_2(4) + \log_2(2) = 2 + 1 = 3$*, so 3 is a valid solution.*
- Is $-3$ a valid solution?
  - *Because* $-3+1 = -2$, $\log_2(-3+1) = \log_2(-2)$ *is undefined, so* $-3$ *not a valid solution. The value* $-3$ *is an extraneous solution, and this equation has only one solution:* 3.
- What should we look for when examining a solution to see if it is extraneous in logarithmic equations?
  - *We cannot take the logarithm of a negative number or* 0*, so any solution that would result in the input to a logarithm being negative or* 0 *cannot be included in the solution set for the equation.*

## Exercises 5–9 (8 minutes)

Have students work on these exercises individually to develop fluency with solving logarithmic equations. Circulate throughout the classroom to informally assess understanding and provide assistance when needed.

**Exercises 5–9**

**Solve the logarithmic equations in Exercises 5–9, and identify any extraneous solutions.**

5. $\log(x^2 + 7x + 12) - \log(x + 4) = 0$

$$\log\left(\frac{x^2 + 7x + 12}{x + 4}\right) = 0$$
$$\frac{x^2 + 7x + 12}{x + 4} = 10^0$$
$$\frac{x^2 + 7x + 12}{x + 4} = 1$$
$$x^2 + 7x + 12 = x + 4$$
$$0 = x^2 + 6x + 8$$
$$0 = (x + 4)(x + 2)$$
$$x = -4 \text{ or } x = -2$$

*Check: If $x = -4$, then $\log(x + 4) = \log(0)$, which is undefined. Thus, $-4$ is an extraneous solution. Therefore, the only solution is $-2$.*

6. $\log_2(3x) + \log_2(4) = 4$

$$\log_2(3x) + 2 = 4$$
$$\log_2(3x) = 2$$
$$2^2 = 3x$$
$$4 = 3x$$
$$x = \frac{4}{3}$$

*Check: Since $\frac{4}{3} > 0$, $\log_2\left(3 \cdot \frac{4}{3}\right)$ is defined.*

*Therefore, $\frac{4}{3}$ is a valid solution.*

7. $2\ln(x + 2) - \ln(-x) = 0$

$$\ln((x + 2)^2) - \ln(-x) = 0$$
$$\ln\left(\frac{(x + 2)^2}{-x}\right) = 0$$
$$1 = \frac{(x + 2)^2}{-x}$$
$$-x = x^2 + 4x + 4$$
$$0 = x^2 + 5x + 4$$
$$0 = (x + 4)(x + 1)$$
$$x = -4 \text{ or } x = -1$$

*Check: Thus, we get $x = -4$ or $x = -1$ as solutions to the quadratic equation. However, if $x = -4$, then $\ln(x + 2) = \ln(-2)$, so $-4$ is an extraneous solution. Therefore, the only solution is $-1$.*

8. $\log(x) = 2 - \log(x)$

$$\begin{aligned} \log(x) + \log(x) &= 2 \\ 2 \cdot \log(x) &= 2 \\ \log(x) &= 1 \\ x &= 10 \end{aligned}$$

*Check: Since $10 > 0$, $\log(10)$ is defined.*

*Therefore, $10$ is a valid solution to this equation.*

9. $\ln(x+2) = \ln(12) - \ln(x+3)$

$$\begin{aligned} \ln(x+2) + \ln(x+3) &= \ln(12) \\ \ln((x+2)(x+3)) &= \ln(12) \\ (x+2)(x+3) &= 12 \\ x^2 + 5x + 6 &= 12 \\ x^2 + 5x - 6 &= 0 \\ (x-1)(x+6) &= 0 \\ x = 1 \text{ or } x &= -6 \end{aligned}$$

*Check: If $= -6$, then the expressions $\ln(x+2)$ and $\ln(x+3)$ are undefined.*

*Therefore, the only valid solution to the original equation is $1$.*

## Closing (3 minutes)

Have students summarize the process they use to solve logarithmic equations in writing. Circulate around the classroom to informally assess student understanding.

- *If an equation can be rewritten in the form $\log_b(Y) = L$ for an expression $Y$ and a number $L$, then apply the definition of the logarithm to rewrite as $b^L = Y$. Solve the resulting exponential equation and check for extraneous solutions.*
- *If an equation can be rewritten in the form $\log_b(Y) = \log_b(Z)$ for expressions $Y$ and $Z$, then the fact that the logarithmic functions are one-to-one gives $Y = Z$. Solve this resulting equation, and check for extraneous solutions.*

## Exit Ticket (4 minutes)

Name ______________________________ Date ____________

# Lesson 14: Solving Logarithmic Equations

## Exit Ticket

Find all solutions to the following equations. Remember to check for extraneous solutions.

1. $5\log_2(3x+7)=0$

2. $\log(x-1)+\log(x-4)=1$

## Exit Ticket Sample Solutions

**Find all solutions to the following equations. Remember to check for extraneous solutions.**

1. $\log_2(3x+7) = 4$

$$\log_2(3x+7) = 4$$
$$3x+7 = 2^4$$
$$3x = 16 - 7$$
$$x = 3$$

*Since $3(3)+7 > 0$, we know $3$ is a valid solution to the equation.*

2. $\log(x-1) + \log(x-4) = 1$

$$\log\big((x-1)(x-4)\big) = 1$$
$$\log(x^2 - 5x + 4) = 1$$
$$x^2 - 5x + 4 = 10$$
$$x^2 - 5x - 6 = 0$$
$$(x-6)(x+1) = 0$$
$$x = 6 \text{ or } x = -1$$

*Check: Since the left side is not defined for $x = -1$, this is an extraneous solution.*

*Therefore, the only valid solution is $6$.*

## Problem Set Sample Solutions

1. **Solve the following logarithmic equations.**

a. $\log(x) = \frac{5}{2}$

$$\log(x) = \frac{5}{2}$$
$$x = 10^{\frac{5}{2}}$$
$$x = 100\sqrt{10}$$

*Check: Since $100\sqrt{10} > 0$, we know $\log\left(100\sqrt{10}\right)$ is defined.*

*Therefore, the solution to this equation is $100\sqrt{10}$.*

b. $5\log(x+4) = 10$

$$\log(x+4) = 2$$
$$x+4 = 10^2$$
$$x+4 = 100$$
$$x = 96$$

*Check: Since $96+4 > 0$, we know $\log(96+4)$ is defined.*

*Therefore, the solution to this equation is $96$.*

c. $\log_2(1-x) = 4$

$$1 - x = 2^4$$
$$x = -15$$

*Check: Since* $1-(-15) > 0$, *we know* $\log_2(1-(-15))$ *is defined.*

*Therefore, the solution to this equation is* $-15$.

d. $\log_2(49x^2) = 4$

$$\log_2[(7x)^2] = 4$$
$$2 \cdot \log_2(7x) = 4$$
$$\log_2(7x) = 2$$
$$7x = 2^2$$
$$x = \frac{4}{7}$$

*Check: Since* $49\left(\frac{4}{7}\right)^2 > 0$, *we know* $\log_2\left(49\left(\frac{4}{7}\right)^2\right)$ *is defined.*

*Therefore, the solution to this equation is* $\frac{4}{7}$.

e. $\log_2(9x^2 + 30x + 25) = 8$

$$\log_2[(3x+5)^2] = 8$$
$$2 \cdot \log_2(3x+5) = 8$$
$$\log_2(3x+5) = 4$$
$$3x + 5 = 2^4$$
$$3x + 5 = 16$$
$$3x = 11$$
$$x = \frac{11}{3}$$

*Check: Since* $9\left(\frac{11}{3}\right)^2 + 30\left(\frac{11}{3}\right) + 25 = 256$, *and* $256 > 0$, $\log_2\left(9\left(\frac{11}{3}\right)^2 + 30\left(\frac{11}{3}\right) + 25\right)$ *is defined.*

*Therefore, the solution to this equation is* $\frac{11}{3}$.

2. Solve the following logarithmic equations.

a. $\ln(x^6) = 36$

$$6 \cdot \ln(x) = 36$$
$$\ln(x) = 6$$
$$x = e^6$$

*Check: Since* $e^6 > 0$, *we know* $\ln((e^6)^6)$ *is defined.*

*Therefore, the only solution to this equation is* $e^6$.

b. $\log[(2x^2 + 45x - 25)^5] = 10$

$$5 \cdot \log(2x^2 + 45x - 25) = 10$$
$$\log(2x^2 + 45x - 25) = 2$$
$$2x^2 + 45x - 25 = 10^2$$
$$2x^2 + 45x - 125 = 0$$
$$2x^2 + 50x - 5x - 125 = 0$$
$$2x(x + 25) - 5(x + 25) = 0$$
$$(2x - 5)(x + 25) = 0$$

*Check: Since $2x^2 + 45x - 25 > 0$ for $x = -25$, and $x = \frac{5}{2}$, we know the left side is defined at these values.*

*Therefore, the two solutions to this equation are $-25$ and $\frac{5}{2}$.*

c. $\log[(x^2 + 2x - 3)^4] = 0$

$$4\log(x^2 + 2x - 3) = 0$$
$$\log(x^2 + 2x - 3) = 0$$
$$x^2 + 2x - 3 = 10^0$$
$$x^2 + 2x - 3 = 1$$
$$x^2 + 2x - 4 = 0$$

$$x = \frac{-2 \pm \sqrt{4 + 16}}{2}$$
$$= -1 \pm \sqrt{5}$$

*Check: Since $x^2 + 2x - 3 = 1$ when $x = -1 + \sqrt{5}$ or $x = -1 - \sqrt{5}$, we know the logarithm is defined for these values of $x$.*

*Therefore, the two solutions to the equation are $-1 + \sqrt{5}$ and $-1 - \sqrt{5}$.*

3. Solve the following logarithmic equations.

a. $\log(x) + \log(x - 1) = \log(3x + 12)$

$$\log(x) + \log(x - 1) = \log(3x + 12)$$
$$\log\big(x(x - 1)\big) = \log(3x + 12)$$
$$x(x - 1) = 3x + 12$$
$$x^2 - 4x - 12 = 0$$
$$(x + 2)(x - 6) = 0$$

*Check: Since $\log(-2)$ is undefined, $-2$ is an extraneous solution.*

*Therefore, the only solution to this equation is 6.*

b. $\ln(32x^2) - 3\ln(2) = 3$

$$\ln(32x^2) - \ln(2^3) = 3$$
$$\ln\left(\frac{32x^2}{8}\right) = 3$$
$$4x^2 = e^3$$
$$x^2 = \frac{e^3}{4}$$
$$x = \frac{\sqrt{e^3}}{2} \text{ or } x = -\frac{\sqrt{e^3}}{2}$$

***Check:*** *Since the value of $x$ in the logarithmic expression is squared, $\ln(32x^2)$ is defined for any non-zero value of $x$.*

*Therefore, both $\frac{\sqrt{e^3}}{2}$ and $-\frac{\sqrt{e^3}}{2}$ are valid solutions to this equation.*

c. $\log(x) + \log(-x) = 0$

$$\log(x(-x)) = 0$$
$$\log(-x^2) = 0$$
$$-x^2 = 10^0$$
$$x^2 = -1$$

*Since there is no real number $x$ so that $x^2 = -1$, there is no solution to this equation.*

d. $\log(x+3) + \log(x+5) = 2$

$$\log((x+3)(x+5)) = 2$$
$$(x+3)(x+5) = 10^2$$
$$x^2 + 8x + 15 - 100 = 0$$
$$x^2 + 8x - 85 = 0$$

$$x = \frac{-8 \pm \sqrt{64 + 340}}{2}$$
$$= -4 \pm \sqrt{101}$$

***Check:*** *The left side of the equation is not defined for $x = -4 - \sqrt{101}$, but it is for $x = -4 + \sqrt{101}$.*

*Therefore, the only solution to this equation is $x = -4 + \sqrt{101}$.*

e. $\log(10x + 5) - 3 = \log(x - 5)$

$$\log(10x + 5) - \log(x - 5) = 3$$
$$\log\left(\frac{10x + 5}{x - 5}\right) = 3$$
$$\frac{10x + 5}{x - 5} = 10^3$$
$$\frac{10x + 5}{x - 5} = 1000$$
$$10x + 5 = 1000x - 5000$$
$$5005 = 990x$$
$$x = \frac{91}{18}$$

*Check: Both sides of the equation are defined for* $x = \frac{91}{18}$.

*Therefore, the solution to this equation is* $\frac{91}{18}$.

f. $\log_2(x) + \log_2(2x) + \log_2(3x) + \log_2(36) = 6$

$$\log_2(x \cdot 2x \cdot 3x \cdot 36) = 6$$
$$\log_2(6^3x^3) = 6$$
$$\log_2[(6x)^3] = 6$$
$$3 \cdot \log_2(6x) = 6$$
$$\log_2(6x) = 2$$
$$6x = 2^2$$
$$x = \frac{2}{3}$$

*Check: Since* $\frac{2}{3} > 0$, *all logarithmic expressions in this equation are defined for* $x = \frac{2}{3}$.

*Therefore, the solution to this equation is* $\frac{2}{3}$.

4. **Solve the following equations.**

a. $\log_2(x) = 4$

$16$

b. $\log_6(x) = 1$

$6$

c. $\log_3(x) = -4$

$\frac{1}{81}$

d. $\log_{\sqrt{2}}(x) = 4$

$4$

e. $\log_{\sqrt{5}}(x) = 3$

$5\sqrt{5}$

f. $\log_3(x^2) = 4$

$9, -9$

g. $\log_2(y^{-3}) = 12$

$\frac{1}{16}$

h. $\log_3(8x + 9) = 4$

$9$

i. $2 = \log_4(3x - 2)$

$6$

j. $\log_5(3 - 2x) = 0$

$1$

k. $\ln(2x) = 3$

$\frac{e^3}{2}$

l. $\log_3(x^2 - 3x + 5) = 2$

$4, -1$

m. $\log((x^2 + 4)^5) = 10$

$4\sqrt{6}, -4\sqrt{6}$

n. $\log(x) + \log(x + 21) = 2$

$4$

o. $\log_4(x - 2) + \log_4(2x) = 2$

$4$

p. $\log(x) - \log(x + 3) = -1$

$\frac{1}{3}$

q. $\log_4(x + 3) - \log_4(x - 5) = 2$

$\frac{83}{15}$

r. $\log(x) + 1 = \log(x + 9)$

$1$

s. $\log_3(x^2 - 9) - \log_3(x + 3) = 1$

$6$

t. $1 - \log_8(x - 3) = \log_8(2x)$

$4$

u. $\log_2(x^2 - 16) - \log_2(x - 4) = 1$

*No solution*

v. $\log\left(\sqrt{(x + 3)^3}\right) = \frac{3}{2}$

$7$

w. $\ln(4x^2 - 1) = 0$

$\frac{1}{\sqrt{2}}, -\frac{1}{\sqrt{2}}$

x. $\ln(x + 1) - \ln(2) = 1$

$2e - 1$

# Lesson 15: Why Were Logarithms Developed?

## Student Outcomes

- Students use logarithm tables to calculate products and quotients of multi-digit numbers without technology.
- Students understand that logarithms were developed to speed up arithmetic calculations by reducing multiplication and division to the simpler operations of addition and subtraction.
- Students solve logarithmic equations of the form $\log(X) = \log(Y)$ by equating $X = Y$.

## Lesson Notes

This final lesson in Topic B includes two procedures that seem to be different but are closely related mathematically. First, students work with logarithm tables to see how applying logarithms simplified calculations in the days before computing machines and electronic technology. They also learn a bit of the history of how and why logarithms first appeared—a history often obscured when logarithmic functions are introduced as inverses of exponential functions. The last two pages of this document contain a base 10 table of logarithms that can be copied and distributed; such tables are also available on the Internet.

Then, students learn to solve the final type of logarithmic equation, $\log(X) = \log(Y)$, where $X$ and $Y$ are either real numbers or expressions that take on positive real values (**A.SSE.A.2**, **F-LE.A.4**). Using either technique requires that we know that the logarithm is a one-to-one function; that is, if $\log(X) = \log(Y)$, then $X = Y$. Students do not yet have the vocabulary to be told this directly, but we do state it as fact in this lesson, and they will further explore the idea of one-to-one functions in Precalculus. As with Lessons 10 and 12, this lesson involves only base 10 logarithms, but the problem set does require that students do some work with logarithms base $e$ and base 2. Remind students to check for extraneous solutions when solving logarithmic equations.

## Classwork

### Discussion (4 minutes): How to Read a Table of Logarithms

- For this lesson, we will pretend that we live in the time when logarithms were discovered, before there were calculators or computing machines. In this time, scientists, merchants, and sailors needed to make calculations for both astronomical observation and navigation. Logarithms made these calculations much easier, faster, and more accurate than calculation by hand. In fact, noted mathematician Pierre-Simon LaPlace (France, circa 1800) said that "[logarithms are an] admirable artifice which, by reducing to a few days the labour of many months, doubles the life of the astronomer, and spares him the errors and disgust inseparable from long calculations."
- A typical table of common logarithms, like the table at the end of this document, has many rows of numbers arranged in ten columns. The numbers in the table are decimals. In our table, they are given to four decimal places, and there are $90$ rows of them (some tables of logarithms have $900$ rows). Down the left-hand side of the table are the numbers from $1.0$ to $9.9$. Across the top of the table are the numbers from $0$ to $9$. To read the table, you locate the number whose logarithm you want using the numbers down the left of the table followed by the numbers across the top.

MP.5 & MP.6

- What does the number in the third row and second column represent (the entry for $1.21$)?
  - *The logarithm of* $1.21$*, which is* $0.0828$*.*
- The logarithm of numbers larger than $9.9$ and smaller than $1.0$ can also be found using this table. Suppose you want to find $\log(365)$. Is there any way we can rewrite this number to show a number between $1.0$ and $9.9$?
  - *Rewrite* $365$ *in scientific notation:* $3.65 \times 10^2$.
- Can we simplify $\log(3.65 \times 10^2)$?
  - *We can apply the formula for the logarithm of a product. Then, we have* $\log(10^2) + \log(3.65) = 2 + \log(3.65)$.
- Now, all that is left is to find the value of $\log(3.65)$ using the table. What is the value of $\log(365)$?
  - *The table entry is* $0.5623$*. That means* $\log(365) \approx 2 + 0.5623$*, so* $\log(365) \approx 2.5623$.
- How would you find $\log(0.365)$?
  - *In scientific notation,* $0.365 = 3.65 \times 10^{-1}$*. So, once again you would find the row for* $3.6$ *and the column for* $5$*, and you would again find the number* $0.5623$*. But this time, you would have* $\log(0.365) \approx -1 + 0.5623$*, so* $\log(0.365) \approx 0.4377$.

## Example 1 (7 minutes)

Students will multiply multi-digit numbers without technology, and then use a table of logarithms to find the same product using logarithms.

- Find the product $3.42 \times 2.47$ without using a calculator.
  - *Using paper and pencil, and without any rounding, students should get* $8.4474$*. The point is to show how much time the multiplication of multi-digit numbers can take.*
- How could we use logarithms to find this product?
  - *If we take the logarithm of the product, we can rewrite the product as a sum of logarithms.*
- Rewrite the logarithm of the product as the sum of logarithms.
  - $\log(3.42 \times 2.47) = \log(3.42) + \log(2.47)$
- Use the table of logarithms to look up the values of $\log(3.42)$ and $\log(2.47)$.
  - *According to the table,* $\log(3.42) \approx 0.5340$, and $\log(2.47) \approx 0.3927$.
- Approximate the logarithm $\log(3.42 \times 2.47)$.
  - *The approximate sum is*

$$\begin{aligned}\log(3.42 \times 2.47) &\approx 0.5340 + 0.3927\\ &\approx 0.9267.\end{aligned}$$

- What if there is more than one number that has a logarithm of $0.9267$? Suppose that there are two numbers $X$ and $Y$ that satisfy $\log(X) = 0.9267$ and $\log(Y) = 0.9267$. Then, $10^{0.9267} = X$, and $10^{0.9267} = Y$, so that $X = Y$. This means that there is only one number that has the logarithm $0.9267$. So, what is that number?

> *Scaffolding:*
> - Students may need to be reminded that if the logarithm is greater than 1, a power of 10 greater than 1 is involved, and only the decimal part of the number will be found in the table.
> - Struggling students should attempt a simpler product such as $1.20 \times 6.00$ to illustrate the process.
> - Advanced students may use larger or more precise numbers as a challenge. To multiply a product such as $34.293 \times 107.9821$, students will have to employ scientific notation and the property for the logarithm of a product.

- Can we find the exact number that has logarithm $0.9267$ using the table?
  - *The table says that* $\log(8.44) \approx 0.9263$, *and* $\log(8.45) \approx 0.9269$.
- Which is closer?
  - $\log(8.45) \approx 0.9267$
- Since $\log(3.42 \times 2.47) \approx \log(8.45)$, what can we conclude is an approximate value for $3.42 \times 2.47$?
  - *Since* $\log(3.42 \times 2.47) \approx \log(8.45)$, *we know that* $3.42 \times 2.47 \approx 8.45$.
- Does this agree with the product you found when you did the calculation by hand?
  - *Yes, by hand we found that the product is* $8.4474$, *which is approximately* $8.45$.

## Discussion (2 minutes)

In the above example, we showed that there was only one number that had logarithm $0.9267$. This result generalizes to any number and any base of the logarithm: If $\log_b(X) = \log_b(Y)$, then $X = Y$. We need to know this property both to use a logarithm table to look up values that produce a certain logarithmic value and to solve logarithmic equations later in the lesson.

> If $X$ and $Y$ are positive real numbers, or expressions that take on the value of positive real numbers, and $\log_b(X) = \log_b Y$, then $X = Y$.

## Example 2 (4 minutes)

This example is a continuation of the first example, with the addition of scientific notation to further explain the power of logarithms. Because much of the reasoning was explained in Example 1, this should take much less time to work through.

- Now, what if we needed to calculate $(3.42 \times 10^{14}) \times (5.76 \times 10^{12})$?
- Take the logarithm of this product, and find its approximate value using the logarithm table.
  - $$\begin{aligned}\log\left((3.42 \times 10^{14}) \times (5.76 \times 10^{12})\right) &= \log(3.42) + \log(10^{14}) + \log(5.76) + \log(10^{12}) \\ &= \log(3.42) + \log(5.76) + 14 + 12 \\ &\approx 0.5340 + 0.7604 + 26 \\ &\approx 27.2944\end{aligned}$$
- Look up $0.2944$ in the logarithm table.
  - *Since* $\log(1.97) \approx 0.2945$, *we can say that* $0.2944 \approx \log(1.97)$.
- How does that tell us which number has a logarithm approximately equal to $27.2944$?
  - $\log(1.97 \times 10^{27}) = 27 + \log(1.97)$, *so* $\log(1.97 \times 10^{27}) \approx 27.2944$.
- Finally, what is an approximate value of the product $(3.42 \times 10^{14}) \times (5.76 \times 10^{12})$?
  - $(3.42 \times 10^{14}) \times (5.76 \times 10^{12}) \approx 1.97 \times 10^{27}$.

### Example 3 (6 minutes)

- According to one estimate, the mass of the earth is roughly $5.28 \times 10^{24}$ kg, and the mass of the moon is about $7.35 \times 10^{22}$ kg. Without using a calculator but using the table of logarithms, find how many times greater the mass of earth is than the mass of the moon.
  - *Let $R$ be the ratio of the two masses. Then $R = \frac{5.28 \times 10^{24}}{7.35 \times 10^{22}} = \frac{5.28}{7.35} \cdot 10^2$*

    *Taking the logarithm of each side,*

$$\begin{aligned}\log(R) &= \log\left(\frac{5.28}{7.35} \cdot 10^2\right) \\ &= 2 + \log\left(\frac{5.28}{7.35}\right) \\ &= 2 + \log(5.28) - \log(7.35) \\ &\approx 2 + 0.7226 - 0.8663 \\ &\approx 1.8563.\end{aligned}$$

- Find $0.8563$ in the table entries to estimate $R$.
  - *In the table, $0.8563$ is closest to $\log(7.18)$.*
  - *So, $\log(71.8) \approx 1.8563$, and therefore, the mass of earth is approximately $71.8$ times that of the moon.*
- Logarithms turn out to be very useful in dealing with especially large or especially small numbers. Scientific notation was probably developed as an attempt to do arithmetic using logarithms. How does it help to have those numbers expressed in scientific notation if we are going to use a logarithm table to perform multiplication or division?
  - *Again, answers will differ, but students should at least recognize that scientific notation is helpful in working with very large or very small numbers. Using scientific notation, we can express each number as the product of a number between $1$ and $10$, and a power of $10$. Taking the logarithm of the number allows us to use properties of logarithms base $10$ to handle more easily any number $n$ where $0 < n < 1$ or $n \geq 10$. The logarithm of the number between $1$ and $10$ can be read from the table, and the exponent of the power of $10$ can then be added to it.*
- Whenever we have a number of the form $k \times 10^n$ where $n$ is an integer and $k$ is a number between 1.0 and 9.9, the logarithm of this number will always be $n + \log(k)$ and can be evaluated using a table of logarithms like the one included in this lesson.

### Discussion (4 minutes)

Logarithms were devised by the Scottish mathematician John Napier (1550–1617) with the help of the English mathematician Henry Briggs (1561–1630) to simplify arithmetic computations with multi-digit numbers by turning multiplication and division into addition and subtraction. The basic idea is that while a sequence of powers like $2^0, 2^1, 2^2, 2^3, 2^4, 2^5, \ldots$ is increasing multiplicatively, the sequence of its exponents is increasing additively. If numbers can be represented as the powers of a base, they can be multiplied by adding their exponents and divided by subtracting their exponents. Napier and Briggs published the first tables of what came to be called base 10 or common logarithms.

- It was Briggs's idea to base the logarithms on the number 10. Why do you think he made that choice?
  - *The number* 10 *is the base of our number system. So, taking* 10 *as the base of common logarithms makes hand calculations with logarithms easier. It is really the same argument that makes scientific notation helpful: Powers of* 10 *are easy to use in calculations.*

### Exercises 1–2 (12 minutes)

Now that we know that if two logarithmic expressions with the same base are equal, then the arguments inside of the logarithms are equal, and we can solve a wider variety of logarithmic equations without invoking the definition each time. Due to the many logarithmic properties that the students now know, there are multiple approaches to solving these equations. Discuss different approaches with the students and their responses to Exercise 2.

**Exercises 1–2**

1. **Solve the following equations. Remember to check for extraneous solutions because logarithms are only defined for positive real numbers.**

   a. $\log(x^2) = \log(49)$

   $$x^2 = 49$$
   $$x = 7 \text{ or } x = -7$$

   *Check: Both solutions are valid since* $7^2$ *and* $(-7)^2$ *are both positive numbers.*

   *The two solutions are* 7 *and* $-7$.

   b. $\log(x+1) + \log(x+2) = \log(7x-17)$

   $$\log((x+1)(x-2)) = \log(7x-17)$$
   $$(x+1)(x-2) = 7x-17$$
   $$x^2 - x - 2 = 7x - 17$$
   $$x^2 - 8x + 15 = 0$$
   $$(x-5)(x-3) = 0$$
   $$x = 3 \text{ or } x = 5$$

   *Check: Since* $x+1$, $x-2$, *and* $7x-17$ *are all positive for either* $x = 3$ *or* $x = 5$, *both solutions are valid.*

   *Thus, the solutions to this equation are* 3 *and* 5.

   c. $\log(x^2+1) = \log(x(x-2))$

   $$x^2 + 1 = x(x-2)$$
   $$= x^2 - 2x$$
   $$1 = -2x$$
   $$x = -\frac{1}{2}$$

   *Check: Both* $\left(-\frac{1}{2}\right)^2 + 1 > 0$ *and* $-\frac{1}{2}\left(-\frac{1}{2} - 2\right) > 0$, *so the solution* $-\frac{1}{2}$ *is valid.*

   *Thus,* $-\frac{1}{2}$ *is the only valid solution to this equation.*

*Scaffolding:*

If the class seems to be struggling with the process to solve logarithmic equations, then either encourage them to create a graphic organizer that summarizes the types of problems and approaches that they should use in each case, or hang one on the board for reference. A sample graphic organizer is included.

| Rewrite problem in the form… | |
|---|---|
| $\log_b(Y) = L$ | $\log_b(Y) = \log_b(Z)$ |
| Then… | |
| $b^L = Y$ | $Y = Z$ |

d. $\log(x+4) + \log(x-1) = \log(3x)$

$$\log\big((x+4)(x-1)\big) = \log(3x)$$
$$(x+4)(x-1) = 3x$$
$$x^2 + 3x - 4 = 3x$$
$$x^2 - 4 = 0$$
$$x = 2 \text{ or } x = -2$$

***Check:** Since $\log(3x)$ is undefined when $x = -2$, there is an extraneous solution of $x = -2$.*

*The only valid solution to this equation is 2.*

e. $\log(x^2 - x) - \log(x-2) = \log(x-3)$

$$\log(x-2) + \log(x-3) = \log(x^2 - x)$$
$$\log\big((x-2)(x-3)\big) = \log(x^2 - x)$$
$$(x-2)(x-3) = x^2 - x$$
$$x^2 - 5x + 6 = x^2 - x$$
$$4x = 6$$
$$x = \frac{3}{2}$$

***Check:** When $x = \frac{3}{2}$, we have $x - 2 < 0$, so $\log(x-2)$, $\log(x-3)$, and $\log(x^2 - x)$ are all undefined. So, the solution $x = \frac{3}{2}$ is extraneous.*

*There are no valid solutions to this equation.*

f. $\log(x) + \log(x-1) + \log(x+1) = 3\log(x)$

$$\log\big(x(x-1)(x+1)\big) = \log(x^3)$$
$$\log(x^3 - x) = \log(x^3)$$
$$x^3 - x = x^3$$
$$x = 0$$

*Since $\log(0)$ is undefined, $x = 0$ is an extraneous solution.*

*There are no valid solutions to this equation.*

g. $\log(x-4) = -\log(x-2)$

*Two possible approaches to solving this equation are shown.*

$$\log(x-4) = \log\left(\frac{1}{x-2}\right)$$
$$x - 4 = \frac{1}{x-2}$$
$$(x-4)(x-2) = 1$$
$$x^2 - 6x + 8 = 1$$
$$x^2 - 6x + 7 = 0$$
$$x = 3 \pm \sqrt{2}$$

$$\log(x-4) + \log(x-2) = 0$$
$$\log\big((x-4)(x-2)\big) = \log(1)$$
$$(x-4)(x-2) = 1$$
$$x^2 - 6x + 8 = 1$$
$$x^2 - 6x + 7 = 0$$
$$x = 3 \pm \sqrt{2}$$

***Check:** If $x = 3 - \sqrt{2}$, then $x < 2$, so $\log(x-2)$ is undefined. Thus, $3 - \sqrt{2}$ is an extraneous solution.*

*The only valid solution to this equation is $3 + \sqrt{2}$.*

> 2. **How do you know if you need to use the definition of logarithm to solve an equation involving logarithms as we did in Lesson 15 or if you can use the methods of this lesson?**
>
> ***If the equation involves only logarithmic expressions, then it can be reorganized to be of the form $\log(X) = \log(Y)$ and then solved by equating $X = Y$. If there are constants involved, then the equation can be solved by the definition.***

## Closing (2 minutes)

Ask students the following questions and after coming to a consensus, have students record the answers in their notebooks.

- How do we use a table of logarithms to compute a product of two numbers $x$ and $y$?
  - *We look up approximations to $\log(x)$ and $\log(y)$ in the table, add those logarithms, and then look up the sum in the table to extract the approximate product.*
- Does this process provide an exact answer? Explain how you know.
  - *It is only an approximation because the table only allows us to look up $x$ to two decimal places and $\log(x)$ to four decimal places.*
- How do we solve an equation in which every term contains a logarithm?
  - *We rearrange the terms to get an equation of the form $\log(X) = \log(Y)$, then equate $X = Y$, and solve from there.*
- How does that differ from solving an equation that contains constant terms?
  - *If an equation has constant terms, then we rearrange the equation to the form $\log(X) = c$, apply the definition of the logarithm, and solve from there.*

> **Lesson Summary**
>
> **A table of base $10$ logarithms can be used to simplify multiplication of multi-digit numbers:**
>
> 1. **To compute $A \times B$ for positive real numbers $A$ and $B$, look up the values $\log(A)$ and $\log(B)$ in the logarithm table.**
> 2. **Add $\log(A)$ and $\log(B)$. The sum can be written as $k + d$, where $k$ is an integer and $0 \leq d < 1$ is the decimal part.**
> 3. **Look back at the table and find the entry closest to the decimal part, $d$.**
> 4. **The product of that entry and $10^k$ is an approximation to $A \times B$.**
>
> **A similar process simplifies division of multi-digit numbers:**
>
> 1. **To compute $A \div B$ for positive real numbers $A$ and $B$, look up the values $\log(A)$ and $\log(B)$ in the logarithm table.**
> 2. **Calculate $\log(A) - \log(B)$. The difference can be written as $k + d$, where $k$ is an integer and $0 \leq d < 1$ is the decimal part.**
> 3. **Look back at the table to find the entry closest to the decimal part, $d$.**
> 4. **The product of that entry and $10^k$ is an approximation to $A \div B$.**
>
> **For any positive values $X$ and $Y$, if $\log_b(X) = \log_b(Y)$, we can conclude that $X = Y$. This property is the essence of how a logarithm table works, and it allows us to solve equations with logarithmic expressions on both sides of the equation.**

## Exit Ticket (4 minutes)

Name ______________________________ Date ______________

# Lesson 15: Why Were Logarithms Developed?

## Exit Ticket

The surface area of Jupiter is $6.14 \times 10^{10}\ \mathrm{km}^2$, and the surface area of Earth is $5.10 \times 10^{8}\ \mathrm{km}^2$. Without using a calculator but using the table of logarithms, find how many times greater the surface area of Jupiter is than the surface area of Earth.

## Exit Ticket Sample Solutions

**The surface area of Jupiter is $6.14 \times 10^{10}$ $\text{km}^2$, and the surface area of Earth is $5.10 \times 10^{8}$ $\text{km}^2$. Without using a calculator but using the table of logarithms, find how many times greater the surface area of Jupiter is than the surface area of Earth.**

*Let $R$ be the ratio of the two surface areas. Then, $R = \frac{6.14 \times 10^{10}}{5.10 \times 10^{8}} = \frac{6.14}{5.10} \cdot 10^2$.*

*Taking the logarithm of each side,*

$$\begin{aligned}\log(R) &= \log\left(\frac{6.14}{5.10} \cdot 10^2\right)\\ &= 2 + \log\left(\frac{6.14}{5.10}\right)\\ &= 2 + \log(6.14) - \log(5.10)\\ &\approx 2 + 0.7882 - 0.7076\\ &\approx 2.0806.\end{aligned}$$

*Find $0.0806$ in the table entries to estimate $R$.*

*Look up $0.0806$, which is closest to $\log(1.20)$. Note that $2 + 0.0806 \approx \log(100) + \log(1.20)$, so $\log(120) \approx 2.0806$. Therefore, the surface area of Jupiter is approximately $120$ times that of Earth.*

## Problem Set Sample Solutions

These problems give students additional practice using base 10 logarithms to perform arithmetic calculations and solve equations.

1. **Use the table of logarithms to approximate solutions to the following logarithmic equations.**

   a. $\log(x) = 0.5044$

   *In the table, $0.5044$ is closest to $\log(3.19)$, so $\log(x) \approx \log(3.19)$.*
   *Therefore, $x \approx 3.19$.*

   b. $\log(x) = -0.5044$ **(Hint: Begin by writing $-0.5044$ as $[(-0.5044) + 1] - 1$.)**

   $$\begin{aligned}\log(x) &= [(-0.5044) + 1] - 1\\ &= 0.4956 - 1\end{aligned}$$

   *In the table, $0.4956$ is closest to $\log(3.13)$, so*

   $$\begin{aligned}\log(x) &\approx \log(3.13) - 1\\ &\approx \log(3.13) - \log(10)\\ &\approx \log\left(\frac{3.13}{10}\right)\\ &\approx \log(0.313).\end{aligned}$$

   *Therefore, $x \approx 0.313$.*

   *Alternatively, $-0.5044$ is the opposite of $0.5044$, so $x$ is the reciprocal of the answer in part (a). Thus, $x \approx 3.19^{-1} \approx 0.313$.*

c. $\log(x) = 35.5044$

$$\begin{aligned}\log(x) &= 35 + 0.5044\\ &= \log(10^{35}) + 0.5044\\ &\approx \log(10^{35}) + \log(3.19)\\ &\approx \log(3.19 \times 10^{35})\end{aligned}$$

*Therefore,* $x \approx 3.19 \times 10^{35}$.

d. $\log(x) = 4.9201$

$$\begin{aligned}\log(x) &= 4 + 0.9201\\ &= \log(10^{4}) + 0.9201\\ &\approx \log(10^{4}) + \log(8.32)\\ &\approx \log(8.32 \times 10^{4})\end{aligned}$$

*Therefore,* $x \approx 83,200$.

2. Use logarithms and the logarithm table to evaluate each expression.

a. $\sqrt{2.33}$

$$\log\left((2.33)^{\frac{1}{2}}\right) = \frac{1}{2}\log(2.33) = \frac{1}{2}(0.3674) = 0.1837$$

*Thus,* $\log(\sqrt{2.33}) = 0.1837$*, and locating* $0.1837$ *in the logarithm table gives a value approximately* $1.53$. *Therefore,* $\sqrt{2.33} \approx 1.53$.

b. $13,500 \cdot 3,600$

$$\begin{aligned}\log(1.35 \cdot 10^4 \cdot 3.6 \cdot 10^3) &= \log(1.35) + \log(3.6) + 4 + 3\\ &\approx 0.1303 + 0.5563 + 7\\ &\approx 7.6866\end{aligned}$$

*Thus,* $\log(13,500 \cdot 3,600) \approx 7.6866$. *Locating* $0.6866$ *in the logarithm table gives a value of* $4.86$. *Therefore, the product is approximately* $4.86 \times 10^7$.

c. $\frac{7.2\times 10^9}{1.3\times 10^5}$

$$\log(7.2) + \log(10^9) - \log(1.3) - \log(10^5) \approx 0.8573 + 9 - 0.1139 - 5 \approx 4.7434$$

*Locating* $0.7434$ *in the logarithm table gives* $5.54$. *So, the quotient is approximately* $5.54 \times 10^4$.

3. Solve for $x$: $\log(3) + 2\log(x) = \log(27)$.

$$\begin{aligned}\log(3) + 2\log(x) &= \log(27)\\ \log(3) + \log(x^2) &= \log(27)\\ \log(3x^2) &= \log(27)\\ 3x^2 &= 27\\ x^2 &= 9\\ x &= 3\end{aligned}$$

4. Solve for $x$: $\log(3x) + \log(x+4) = \log(15)$.

$$\log(3x^2 + 12x) = \log(15)$$
$$3x^2 + 12x = 15$$
$$3x^2 + 12x - 15 = 0$$
$$3x^2 + 15x - 3x - 15 = 0$$
$$3x(x+5) - 3(x+5) = 0$$
$$(3x-3)(x+5) = 0$$

*Thus, $1$ and $-5$ solve the quadratic equation, but $-5$ is an extraneous solution. Hence, $1$ is the only solution.*

5. Solve for $x$.

a. $\log(x) = \log(y+z) + \log(y-z)$

$$\log(x) = \log(y+z) + \log(y-z)$$
$$\log(x) = \log\big((y+z)(y-z)\big)$$
$$\log(x) = \log(y^2 - z^2)$$
$$x = y^2 - z^2$$

b. $\log(x) = \big(\log(y) + \log(z)\big) + \big(\log(y) - \log(z)\big)$

$$\log(x) = \big(\log(y) + \log(z)\big) + \big(\log(y) - \log(z)\big)$$
$$\log(x) = 2\log(y)$$
$$\log(x) = \log(y^2)$$
$$x = y^2$$

6. If $x$ and $y$ are positive real numbers, and $\log(y) = 1 + \log(x)$, express $y$ in terms of $x$.

*Since $\log(10x) = 1 + \log(x)$, we see that $\log(y) = \log(10x)$. Then $y = 10x$.*

7. If $x$, $y$, and $z$ are positive real numbers, and $\log(x) - \log(y) = \log(y) - \log(z)$, express $y$ in terms of $x$ and $z$.

$$\log(x) - \log(y) = \log(y) - \log(z)$$
$$\log(x) + \log(z) = 2\log(y)$$
$$\log(xz) = \log(y^2)$$
$$xz = y^2$$
$$y = \sqrt{xz}$$

8. If $x$ and $y$ are positive real numbers, and $\log(x) = y\big(\log(y+1) - \log(y)\big)$, express $x$ in terms of $y$.

$$\log(x) = y\big(\log(y+1) - \log(y)\big)$$
$$\log(x) = y\left(\log\left(\frac{y+1}{y}\right)\right)$$
$$\log(x) = \log\left(\left(\frac{y+1}{y}\right)^y\right)$$
$$x = \left(\frac{y+1}{y}\right)^y$$

9. **If $x$ and $y$ are positive real numbers, and $\log(y) = 3 + 2\log(x)$, express $y$ in terms of $x$.**

*Since* $\log(1000x^2) = 3 + \log(x^2) = 3 + 2\log(x)$, *we see that* $\log(y) = \log(1000x^2)$. *Thus,* $y = 1000x^2$.

10. **If $x$, $y$, and $z$ are positive real numbers, and $\log(z) = \log(y) + 2\log(x) - 1$, express $z$ in terms of $x$ and $y$.**

*Since* $\log\left(\frac{x^2y}{10}\right) = \log(y) + 2\log(x) - 1$, *we see that* $\log(z) = \log\left(\frac{x^2y}{10}\right)$. *Thus,* $z = \frac{x^2y}{10}$.

11. **Solve the following equations.**

a. $\ln(10) - \ln(7-x) = \ln(x)$

$$\begin{aligned} \ln\left(\frac{10}{7-x}\right) &= \ln(x) \\ \frac{10}{7-x} &= x \\ 10 &= x(7-x) \\ x^2 - 7x + 10 &= 0 \\ (x-5)(x-2) &= 0 \\ x = 2 \text{ or } x &= 5 \end{aligned}$$

*Check: If* $x = 2$ *or* $x = 5$, *then the expressions* $x$ *and* $7 - x$ *are positive.*
*Thus, both* $2$ *and* $5$ *are valid solutions to this equation.*

b. $\ln(x+2) + \ln(x-2) = \ln(9x-24)$

$$\begin{aligned} \ln((x+2)(x-2)) &= \ln(9x-24) \\ x^2 - 4 &= 9x - 24 \\ x^2 - 9x + 20 &= 0 \\ (x-4)(x-5) &= 0 \\ x = 4 \text{ or } x &= 5 \end{aligned}$$

*Check: If* $x = 4$ *or* $x = 5$, *then the expressions* $x + 2$, $x - 2$, *and* $9x - 24$ *are all positive.*
*Thus, both* $4$ *and* $5$ *are valid solutions to this equation.*

c. $\ln(x+2) + \ln(x-2) = \ln(-2x-1)$

$$\begin{aligned} \ln((x+2)(x-2)) &= \ln(-2x-1) \\ \ln(x^2-4) &= \ln(-2x-1) \\ x^2 - 4 &= -2x - 1 \\ x^2 + 2x - 3 &= 0 \\ (x+3)(x-1) &= 0 \\ x = -3 \text{ or } x &= 1 \end{aligned}$$

*So,* $x = -3$ *or* $x = 1$, *but* $x = -3$ *makes the input to both logarithms on the left-hand side negative, and* $x = 1$ *makes the input to the second and third logarithms negative. Thus, there are no solutions to the original equation.*

12. **Suppose the formula $P = P_0(1+r)^t$ gives the population of a city $P$ growing at an annual percent rate $r$, where $P_0$ is the population $t$ years ago.**

   a. **Find the time $t$ it takes this population to double.**

   *Let $P = 2P_0$; then,*

$$2P_0 = P_0(1+r)^t$$
$$2 = (1+r)^t$$
$$\log(2) = \log(1+r)^t$$
$$\log(2) = t\log(1+r)$$
$$t = \frac{\log(2)}{\log(1+r)}.$$

   b. **Use the structure of the expression to explain why populations with lower growth rates take a longer time to double.**

   *If $r$ is a decimal between $0$ and $1$, then the denominator will be a number between $0$ and $\log(2)$. Thus, the value of $t$ will be large for small values of $r$ and getting closer to $1$ as $r$ increases.*

   c. **Use the structure of the expression to explain why the only way to double the population in one year is if there is a $100$ percent growth rate.**

   *For the population to double, we need to have $t = 1$. This happens if $\log(2) = \log(1+r)$, and then we have $2 = 1 + r$ and $r = 1$.*

13. **If $x > 0$, $a + b > 0$, $a > b$, and $\log(x) = \log(a+b) + \log(a-b)$, find $x$ in terms of $a$ and $b$.**

   *Applying properties of logarithms, we have*

$$\begin{aligned}\log(x) &= \log(a+b) + \log(a-b)\\ &= \log((a+b)(a-b))\\ &= \log(a^2 - b^2).\end{aligned}$$

   *So, $x = a^2 - b^2$.*

14. **Jenn claims that because $\log(1) + \log(2) + \log(3) = \log(6)$, then $\log(2) + \log(3) + \log(4) = \log(9)$.**

   a. **Is she correct? Explain how you know.**

   *Jenn is not correct. Even though $\log(1) + \log(2) + \log(3) = \log(1 \cdot 2 \cdot 3) = \log(6)$, the logarithm properties give $\log(2) + \log(3) + \log(4) = \log(2 \cdot 3 \cdot 4) = \log(24)$. Since $9 < 10$, we know that $\log(9) < 1$, and since $24 > 10$, we know that $\log(24) > 1$, so clearly $\log(9) \neq \log(24)$.*

b. If $\log(a)+\log(b)+\log(c)=\log(a+b+c)$, express $c$ in terms of $a$ and $b$. Explain how this result relates to your answer to part (a).

MP.3

$$\begin{aligned}\log(a)+\log(b)+\log(c)&=\log(a+b+c)\\ \log(abc)&=\log(a+b+c)\\ abc&=a+b+c\\ abc-c&=a+b\\ c(ab-1)&=a+b\\ c&=\frac{a+b}{ab-1}\end{aligned}$$

*If* $\log(2)+\log(3)+\log(4)$ *were equal to* $\log(9)$, *then we would have* $4=\frac{2+3}{2\cdot3-1}$. *However,* $\frac{2+3}{2\cdot3-1}=\frac{5}{5}=1\neq4$, *so we know that* $\log(2)+\log(3)+\log(4)\neq\log(9)$.

c. Find other values of $a$, $b$, and $c$ so that $\log(a)+\log(b)+\log(c)=\log(a+b+c)$.

*Many answers are possible; in fact, any positive values of* $a$ *and* $b$ *where* $ab\neq1$ *will produce* $c$ *so that* $\log(a)+\log(b)+\log(c)=\log(a+b+c)$. *One such answer is* $a=3$, $b=7$, *and* $c=\frac{1}{2}$.

15. In Problem 7 of the Lesson 12 Problem Set, you showed that for $x\geq1$, $\log\left(x+\sqrt{x^2-1}\right)+\log\left(x-\sqrt{x^2-1}\right)=0$. It follows that $\log\left(x+\sqrt{x^2-1}\right)=-\log\left(x-\sqrt{x^2-1}\right)$. What does this tell us about the relationship between the expressions $x+\sqrt{x^2-1}$ and $x-\sqrt{x^2-1}$?

*Since we know* $\log\left(x+\sqrt{x^2-1}\right)=-\log\left(x-\sqrt{x^2-1}\right)$, *and* $-\log\left(x-\sqrt{x^2-1}\right)=\log\left(\frac{1}{x-\sqrt{x^2-1}}\right)$, *we know that* $\log\left(x+\sqrt{x^2-1}\right)=\log\left(\frac{1}{x-\sqrt{x^2-1}}\right)$. *Then,* $x+\sqrt{x^2-1}=\frac{1}{x-\sqrt{x^2-1}}$. *We can verify that these expressions are reciprocals by multiplying them together:*

$$\begin{aligned}\left(x+\sqrt{x^2-1}\right)\left(x-\sqrt{x^2-1}\right)&=x^2+x\sqrt{x^2-1}-x\sqrt{x^2-1}-\left(\sqrt{x^2-1}\right)^2\\ &=x^2-(x^2-1)\\ &=1.\end{aligned}$$

16. Use the change of base formula to solve the following equations.

a. $\log(x)=\log_{100}(x^2-2x+6)$

$$\begin{aligned}\log(x)&=\frac{\log(x^2-2x+6)}{\log(100)}\\ \log(x)&=\frac{1}{2}\log(x^2-2x+6)\\ 2\log(x)&=\log(x^2-2x+6)\\ \log(x^2)&=\log(x^2-2x+6)\\ x^2&=x^2-2x+6\\ 2x&=6\\ x&=3\end{aligned}$$

*Since both sides of the equation are defined for* $x=3$, *the only solution to this equation is* $3$.

b. $\log(x-2) = \log_{100}(14-x)$

$$\log(x-2) = \frac{\log(14-x)}{\log(100)}$$
$$\log(x-2) = \frac{1}{2}\log(14-x)$$
$$2\log(x-2) = \log(14-x)$$
$$\log((x-2)^2) = \log(14-x)$$
$$(x-2)^2 = 14-x$$
$$x^2-4x+4 = 14-x$$
$$x^2-3x-10 = 0$$
$$(x-5)(x+2) = 0$$

*Thus, either $x = 5$ or $x = -2$. Since the left side of the equation is undefined when $x = -2$, but both sides are defined for $x = 5$, the only solution to the equation is $5$.*

c. $\log_2(x+1) = \log_4(x^2+3x+4)$

$$\log_2(x+1) = \log_4(x^2+3x+4)$$
$$\log_2(x+1) = \frac{\log_2(x^2+3x+4)}{\log_2(4)}$$
$$2\log_2(x+1) = \log_2(x^2+3x+4)$$
$$\log_2((x+1)^2) = \log_2(x^2+3x+4)$$
$$(x+1)^2 = x^2+3x+4$$
$$x^2+2x+1 = x^2+3x+4$$
$$2x+1 = 3x+4$$
$$x = -3$$

*Since the left side of the equation is undefined for $x = -3$, there is no solution to this equation.*

d. $\log_2(x-1) = \log_8(x^3-2x^2-2x+5)$

$$\log_2(x-1) = \frac{\log_2(x^3-2x^2-2x+5)}{\log_2(8)}$$
$$3\log_2(x-1) = \log_2(x^3-2x^2-2x+5)$$
$$\log_2((x-1)^3) = \log_2(x^3-2x^2-2x+5)$$
$$(x-1)^3 = x^3-2x^2-2x+5$$
$$x^3-3x^2+3x-1 = x^3-2x^2-2x+5$$
$$x^2-5x+6 = 0$$
$$(x-3)(x-2) = 0$$

*Since both sides of the equation are defined for $x = 3$ and $x = 2$, $2$ and $3$ are both valid solutions to this equation.*

17. Solve the following equation: $\log(9x) = \frac{2\ln(3)+\ln(x)}{\ln(10)}$.

***Rewrite the left-hand side using the change-of-base formula:***

$$\begin{aligned}\log(9x) &= \frac{\ln(9x)}{\ln(10)}\\ &= \frac{\ln(9)+\ln(x)}{\ln(10)}\\ &= \frac{\ln(3^2)+\ln(x)}{\ln(10)}.\end{aligned}$$

***Thus, the equation is true for all $x > 0$.***

# Common Logarithm Table

| N | 0 | 1 | 2 | 3 | 4 | 5 | 6 | 7 | 8 | 9 |
|---|---|---|---|---|---|---|---|---|---|---|
| 1.0 | 0.0000 | 0.0043 | 0.0086 | 0.0128 | 0.0170 | 0.0212 | 0.0253 | 0.0294 | 0.0334 | 0.0374 |
| 1.1 | 0.0414 | 0.0453 | 0.0492 | 0.0531 | 0.0569 | 0.0607 | 0.0645 | 0.0682 | 0.0719 | 0.0755 |
| 1.2 | 0.0792 | 0.0828 | 0.0864 | 0.0899 | 0.0934 | 0.0969 | 0.1004 | 0.1038 | 0.1072 | 0.1106 |
| 1.3 | 0.1139 | 0.1173 | 0.1206 | 0.1239 | 0.1271 | 0.1303 | 0.1335 | 0.1367 | 0.1399 | 0.1430 |
| 1.4 | 0.1461 | 0.1492 | 0.1523 | 0.1553 | 0.1584 | 0.1614 | 0.1644 | 0.1673 | 0.1703 | 0.1732 |
| 1.5 | 0.1761 | 0.1790 | 0.1818 | 0.1847 | 0.1875 | 0.1903 | 0.1931 | 0.1959 | 0.1987 | 0.2014 |
| 1.6 | 0.2041 | 0.2068 | 0.2095 | 0.2122 | 0.2148 | 0.2175 | 0.2201 | 0.2227 | 0.2253 | 0.2279 |
| 1.7 | 0.2304 | 0.2330 | 0.2355 | 0.2380 | 0.2405 | 0.2430 | 0.2455 | 0.2480 | 0.2504 | 0.2529 |
| 1.8 | 0.2553 | 0.2577 | 0.2601 | 0.2625 | 0.2648 | 0.2672 | 0.2695 | 0.2718 | 0.2742 | 0.2765 |
| 1.9 | 0.2788 | 0.2810 | 0.2833 | 0.2856 | 0.2878 | 0.2900 | 0.2923 | 0.2945 | 0.2967 | 0.2989 |
| 2.0 | 0.3010 | 0.3032 | 0.3054 | 0.3075 | 0.3096 | 0.3118 | 0.3139 | 0.3160 | 0.3181 | 0.3201 |
| 2.1 | 0.3222 | 0.3243 | 0.3263 | 0.3284 | 0.3304 | 0.3324 | 0.3345 | 0.3365 | 0.3385 | 0.3404 |
| 2.2 | 0.3424 | 0.3444 | 0.3464 | 0.3483 | 0.3502 | 0.3522 | 0.3541 | 0.3560 | 0.3579 | 0.3598 |
| 2.3 | 0.3617 | 0.3636 | 0.3655 | 0.3674 | 0.3692 | 0.3711 | 0.3729 | 0.3747 | 0.3766 | 0.3784 |
| 2.4 | 0.3802 | 0.3820 | 0.3838 | 0.3856 | 0.3874 | 0.3892 | 0.3909 | 0.3927 | 0.3945 | 0.3962 |
| 2.5 | 0.3979 | 0.3997 | 0.4014 | 0.4031 | 0.4048 | 0.4065 | 0.4082 | 0.4099 | 0.4116 | 0.4133 |
| 2.6 | 0.4150 | 0.4166 | 0.4183 | 0.4200 | 0.4216 | 0.4232 | 0.4249 | 0.4265 | 0.4281 | 0.4298 |
| 2.7 | 0.4314 | 0.4330 | 0.4346 | 0.4362 | 0.4378 | 0.4393 | 0.4409 | 0.4425 | 0.4440 | 0.4456 |
| 2.8 | 0.4472 | 0.4487 | 0.4502 | 0.4518 | 0.4533 | 0.4548 | 0.4564 | 0.4579 | 0.4594 | 0.4609 |
| 2.9 | 0.4624 | 0.4639 | 0.4654 | 0.4669 | 0.4683 | 0.4698 | 0.4713 | 0.4728 | 0.4742 | 0.4757 |
| 3.0 | 0.4771 | 0.4786 | 0.4800 | 0.4814 | 0.4829 | 0.4843 | 0.4857 | 0.4871 | 0.4886 | 0.4900 |
| 3.1 | 0.4914 | 0.4928 | 0.4942 | 0.4955 | 0.4969 | 0.4983 | 0.4997 | 0.5011 | 0.5024 | 0.5038 |
| 3.2 | 0.5051 | 0.5065 | 0.5079 | 0.5092 | 0.5105 | 0.5119 | 0.5132 | 0.5145 | 0.5159 | 0.5172 |
| 3.3 | 0.5185 | 0.5198 | 0.5211 | 0.5224 | 0.5237 | 0.5250 | 0.5263 | 0.5276 | 0.5289 | 0.5302 |
| 3.4 | 0.5315 | 0.5328 | 0.5340 | 0.5353 | 0.5366 | 0.5378 | 0.5391 | 0.5403 | 0.5416 | 0.5428 |
| 3.5 | 0.5441 | 0.5453 | 0.5465 | 0.5478 | 0.5490 | 0.5502 | 0.5514 | 0.5527 | 0.5539 | 0.5551 |
| 3.6 | 0.5563 | 0.5575 | 0.5587 | 0.5599 | 0.5611 | 0.5623 | 0.5635 | 0.5647 | 0.5658 | 0.5670 |
| 3.7 | 0.5682 | 0.5694 | 0.5705 | 0.5717 | 0.5729 | 0.5740 | 0.5752 | 0.5763 | 0.5775 | 0.5786 |
| 3.8 | 0.5798 | 0.5809 | 0.5821 | 0.5832 | 0.5843 | 0.5855 | 0.5866 | 0.5877 | 0.5888 | 0.5899 |
| 3.9 | 0.5911 | 0.5922 | 0.5933 | 0.5944 | 0.5955 | 0.5966 | 0.5977 | 0.5988 | 0.5999 | 0.6010 |
| 4.0 | 0.6021 | 0.6031 | 0.6042 | 0.6053 | 0.6064 | 0.6075 | 0.6085 | 0.6096 | 0.6107 | 0.6117 |
| 4.1 | 0.6128 | 0.6138 | 0.6149 | 0.6160 | 0.6170 | 0.6180 | 0.6191 | 0.6201 | 0.6212 | 0.6222 |
| 4.2 | 0.6232 | 0.6243 | 0.6253 | 0.6263 | 0.6274 | 0.6284 | 0.6294 | 0.6304 | 0.6314 | 0.6325 |
| 4.3 | 0.6335 | 0.6345 | 0.6355 | 0.6365 | 0.6375 | 0.6385 | 0.6395 | 0.6405 | 0.6415 | 0.6425 |
| 4.4 | 0.6435 | 0.6444 | 0.6454 | 0.6464 | 0.6474 | 0.6484 | 0.6493 | 0.6503 | 0.6513 | 0.6522 |
| 4.5 | 0.6532 | 0.6542 | 0.6551 | 0.6561 | 0.6571 | 0.6580 | 0.6590 | 0.6599 | 0.6609 | 0.6618 |
| 4.6 | 0.6628 | 0.6637 | 0.6646 | 0.6656 | 0.6665 | 0.6675 | 0.6684 | 0.6693 | 0.6702 | 0.6712 |
| 4.7 | 0.6721 | 0.6730 | 0.6739 | 0.6749 | 0.6758 | 0.6767 | 0.6776 | 0.6785 | 0.6794 | 0.6803 |
| 4.8 | 0.6812 | 0.6821 | 0.6830 | 0.6839 | 0.6848 | 0.6857 | 0.6866 | 0.6875 | 0.6884 | 0.6893 |
| 4.9 | 0.6902 | 0.6911 | 0.6920 | 0.6928 | 0.6937 | 0.6946 | 0.6955 | 0.6964 | 0.6972 | 0.6981 |
| 5.0 | 0.6990 | 0.6998 | 0.7007 | 0.7016 | 0.7024 | 0.7033 | 0.7042 | 0.7050 | 0.7059 | 0.7067 |
| 5.1 | 0.7076 | 0.7084 | 0.7093 | 0.7101 | 0.7110 | 0.7118 | 0.7126 | 0.7135 | 0.7143 | 0.7152 |
| 5.2 | 0.7160 | 0.7168 | 0.7177 | 0.7185 | 0.7193 | 0.7202 | 0.7210 | 0.7218 | 0.7226 | 0.7235 |
| 5.3 | 0.7243 | 0.7251 | 0.7259 | 0.7267 | 0.7275 | 0.7284 | 0.7292 | 0.7300 | 0.7308 | 0.7316 |
| 5.4 | 0.7324 | 0.7332 | 0.7340 | 0.7348 | 0.7356 | 0.7364 | 0.7372 | 0.7380 | 0.7388 | 0.7396 |

| N | 0 | 1 | 2 | 3 | 4 | 5 | 6 | 7 | 8 | 9 |
|---|---|---|---|---|---|---|---|---|---|---|
| 5.5 | 0.7404 | 0.7412 | 0.7419 | 0.7427 | 0.7435 | 0.7443 | 0.7451 | 0.7459 | 0.7466 | 0.7474 |
| 5.6 | 0.7482 | 0.7490 | 0.7497 | 0.7505 | 0.7513 | 0.7520 | 0.7528 | 0.7536 | 0.7543 | 0.7551 |
| 5.7 | 0.7559 | 0.7566 | 0.7574 | 0.7582 | 0.7589 | 0.7597 | 0.7604 | 0.7612 | 0.7619 | 0.7627 |
| 5.8 | 0.7634 | 0.7642 | 0.7649 | 0.7657 | 0.7664 | 0.7672 | 0.7679 | 0.7686 | 0.7694 | 0.7701 |
| 5.9 | 0.7709 | 0.7716 | 0.7723 | 0.7731 | 0.7738 | 0.7745 | 0.7752 | 0.7760 | 0.7767 | 0.7774 |
| 6.0 | 0.7782 | 0.7789 | 0.7796 | 0.7803 | 0.7810 | 0.7818 | 0.7825 | 0.7832 | 0.7839 | 0.7846 |
| 6.1 | 0.7853 | 0.7860 | 0.7868 | 0.7875 | 0.7882 | 0.7889 | 0.7896 | 0.7903 | 0.7910 | 0.7917 |
| 6.2 | 0.7924 | 0.7931 | 0.7938 | 0.7945 | 0.7952 | 0.7959 | 0.7966 | 0.7973 | 0.7980 | 0.7987 |
| 6.3 | 0.7993 | 0.8000 | 0.8007 | 0.8014 | 0.8021 | 0.8028 | 0.8035 | 0.8041 | 0.8048 | 0.8055 |
| 6.4 | 0.8062 | 0.8069 | 0.8075 | 0.8082 | 0.8089 | 0.8096 | 0.8102 | 0.8109 | 0.8116 | 0.8122 |
| 6.5 | 0.8129 | 0.8136 | 0.8142 | 0.8149 | 0.8156 | 0.8162 | 0.8169 | 0.8176 | 0.8182 | 0.8189 |
| 6.6 | 0.8195 | 0.8202 | 0.8209 | 0.8215 | 0.8222 | 0.8228 | 0.8235 | 0.8241 | 0.8248 | 0.8254 |
| 6.7 | 0.8261 | 0.8267 | 0.8274 | 0.8280 | 0.8287 | 0.8293 | 0.8299 | 0.8306 | 0.8312 | 0.8319 |
| 6.8 | 0.8325 | 0.8331 | 0.8338 | 0.8344 | 0.8351 | 0.8357 | 0.8363 | 0.8370 | 0.8376 | 0.8382 |
| 6.9 | 0.8388 | 0.8395 | 0.8401 | 0.8407 | 0.8414 | 0.8420 | 0.8426 | 0.8432 | 0.8439 | 0.8445 |
| 7.0 | 0.8451 | 0.8457 | 0.8463 | 0.8470 | 0.8476 | 0.8482 | 0.8488 | 0.8494 | 0.8500 | 0.8506 |
| 7.1 | 0.8513 | 0.8519 | 0.8525 | 0.8531 | 0.8537 | 0.8543 | 0.8549 | 0.8555 | 0.8561 | 0.8567 |
| 7.2 | 0.8573 | 0.8579 | 0.8585 | 0.8591 | 0.8597 | 0.8603 | 0.8609 | 0.8615 | 0.8621 | 0.8627 |
| 7.3 | 0.8633 | 0.8639 | 0.8645 | 0.8651 | 0.8657 | 0.8663 | 0.8669 | 0.8675 | 0.8681 | 0.8686 |
| 7.4 | 0.8692 | 0.8698 | 0.8704 | 0.8710 | 0.8716 | 0.8722 | 0.8727 | 0.8733 | 0.8739 | 0.8745 |
| 7.5 | 0.8751 | 0.8756 | 0.8762 | 0.8768 | 0.8774 | 0.8779 | 0.8785 | 0.8791 | 0.8797 | 0.8802 |
| 7.6 | 0.8808 | 0.8814 | 0.8820 | 0.8825 | 0.8831 | 0.8837 | 0.8842 | 0.8848 | 0.8854 | 0.8859 |
| 7.7 | 0.8865 | 0.8871 | 0.8876 | 0.8882 | 0.8887 | 0.8893 | 0.8899 | 0.8904 | 0.8910 | 0.8915 |
| 7.8 | 0.8921 | 0.8927 | 0.8932 | 0.8938 | 0.8943 | 0.8949 | 0.8954 | 0.8960 | 0.8965 | 0.8971 |
| 7.9 | 0.8976 | 0.8982 | 0.8987 | 0.8993 | 0.8998 | 0.9004 | 0.9009 | 0.9015 | 0.9020 | 0.9025 |
| 8.0 | 0.9031 | 0.9036 | 0.9042 | 0.9047 | 0.9053 | 0.9058 | 0.9063 | 0.9069 | 0.9074 | 0.9079 |
| 8.1 | 0.9085 | 0.9090 | 0.9096 | 0.9101 | 0.9106 | 0.9112 | 0.9117 | 0.9122 | 0.9128 | 0.9133 |
| 8.2 | 0.9138 | 0.9143 | 0.9149 | 0.9154 | 0.9159 | 0.9165 | 0.9170 | 0.9175 | 0.9180 | 0.9186 |
| 8.3 | 0.9191 | 0.9196 | 0.9201 | 0.9206 | 0.9212 | 0.9217 | 0.9222 | 0.9227 | 0.9232 | 0.9238 |
| 8.4 | 0.9243 | 0.9248 | 0.9253 | 0.9258 | 0.9263 | 0.9269 | 0.9274 | 0.9279 | 0.9284 | 0.9289 |
| 8.5 | 0.9294 | 0.9299 | 0.9304 | 0.9309 | 0.9315 | 0.9320 | 0.9325 | 0.9330 | 0.9335 | 0.9340 |
| 8.6 | 0.9345 | 0.9350 | 0.9355 | 0.9360 | 0.9365 | 0.9370 | 0.9375 | 0.9380 | 0.9385 | 0.9390 |
| 8.7 | 0.9395 | 0.9400 | 0.9405 | 0.9410 | 0.9415 | 0.9420 | 0.9425 | 0.9430 | 0.9435 | 0.9440 |
| 8.8 | 0.9445 | 0.9450 | 0.9455 | 0.9460 | 0.9465 | 0.9469 | 0.9474 | 0.9479 | 0.9484 | 0.9489 |
| 8.9 | 0.9494 | 0.9499 | 0.9504 | 0.9509 | 0.9513 | 0.9518 | 0.9523 | 0.9528 | 0.9533 | 0.9538 |
| 9.0 | 0.9542 | 0.9547 | 0.9552 | 0.9557 | 0.9562 | 0.9566 | 0.9571 | 0.9576 | 0.9581 | 0.9586 |
| 9.1 | 0.9590 | 0.9595 | 0.9600 | 0.9605 | 0.9609 | 0.9614 | 0.9619 | 0.9624 | 0.9628 | 0.9633 |
| 9.2 | 0.9638 | 0.9643 | 0.9647 | 0.9652 | 0.9657 | 0.9661 | 0.9666 | 0.9671 | 0.9675 | 0.9680 |
| 9.3 | 0.9685 | 0.9689 | 0.9694 | 0.9699 | 0.9703 | 0.9708 | 0.9713 | 0.9717 | 0.9722 | 0.9727 |
| 9.4 | 0.9731 | 0.9736 | 0.9741 | 0.9745 | 0.9750 | 0.9754 | 0.9759 | 0.9763 | 0.9768 | 0.9773 |
| 9.5 | 0.9777 | 0.9782 | 0.9786 | 0.9791 | 0.9795 | 0.9800 | 0.9805 | 0.9809 | 0.9814 | 0.9818 |
| 9.6 | 0.9823 | 0.9827 | 0.9832 | 0.9836 | 0.9841 | 0.9845 | 0.9850 | 0.9854 | 0.9859 | 0.9863 |
| 9.7 | 0.9868 | 0.9872 | 0.9877 | 0.9881 | 0.9886 | 0.9890 | 0.9894 | 0.9899 | 0.9903 | 0.9908 |
| 9.8 | 0.9912 | 0.9917 | 0.9921 | 0.9926 | 0.9930 | 0.9934 | 0.9939 | 0.9943 | 0.9948 | 0.9952 |
| 9.9 | 0.9956 | 0.9961 | 0.9965 | 0.9969 | 0.9974 | 0.9978 | 0.9983 | 0.9987 | 0.9991 | 0.9996 |

Name ______________________________ Date ______________

1. Use properties of exponents to explain why it makes sense to define $16^{\frac{1}{4}}$ as $\sqrt[4]{16}$.

2. Use properties of exponents to rewrite each expression as either an integer or as a quotient of integers $\frac{p}{q}$ to show the expression is a rational number.

   a. $\sqrt[4]{2}\,\sqrt[4]{8}$

   b. $\frac{\sqrt[3]{54}}{\sqrt[3]{2}}$

   c. $16^{\frac{3}{2}} \cdot \left(\frac{1}{27}\right)^{\frac{2}{3}}$

3. Use properties of exponents to rewrite each expression with only positive, rational exponents. Then find the numerical value of each expression when $x = 9$, $y = 8$, and $z = 16$. In each case, the expression evaluates to a rational number.

   a. $\sqrt{\dfrac{xy^2}{(x^3z)^{\frac{1}{2}}}}$

   b. $\sqrt[11]{y^2z^4}$

   c. $x^{-\frac{3}{2}}y^{\frac{4}{3}}z^{-\frac{3}{4}}$

4. We can use finite approximations of the decimal expansion of $\pi = 3.141519\ldots$ to find an approximate value of the number $3^{\pi}$.

   a. Fill in the missing exponents in the following sequence of inequalities that represents the recursive process of finding the value of $3^{\pi}$.

$$3^3 < 3^{\pi} < 3^4$$

$$3^{3.1} < 3^{\pi} < 3^{3.2}$$

$$3^{3.14} < 3^{\pi} < 3^{3.15}$$

$$3^{(\quad)} < 3^{\pi} < 3^{(\quad)}$$

$$3^{(\quad\;)} < 3^{\pi} < 3^{(\quad\;)}$$

   b. Explain how this recursive process leads to better and better approximations of the number $3^{\pi}$.

5. A scientist is studying the growth of a population of bacteria. At the beginning of her study, she has 800 bacteria. She notices that the population is quadrupling every hour.

   a. What quantities, including units, need to be identified to further investigate the growth of this bacteria population?

b. The scientist recorded the following information in her notebook, but she forgot to label each row. Label each row to show what quantities, including appropriate units, are represented by the numbers in the table, and then complete the table.

| | 0 | 1 | 2 | 3 | 4 |
|---|---|---|---|---|---|
| | 8 | 32 | 128 | | |

c. Write an explicit formula for the number of bacteria present after $t$ hours.

d. Another scientist studying the same population notices that the population is doubling every half an hour. Complete the table, and write an explicit formula for the number of bacteria present after $x$ half hours.

| Time, $t$ (hours) | $0$ | $\frac{1}{2}$ | $1$ | $\frac{3}{2}$ | $2$ | $\frac{5}{2}$ | $3$ |
|---|---|---|---|---|---|---|---|
| Time, $x$ (half-hours) | 0 | 1 | 2 | 3 | 4 | 5 | 6 |
| Bacteria (hundreds) | 8 | 16 | 32 | | | | |

e. Find the time, in hours, when there will be 5,120,000 bacteria. Express your answer as a logarithmic expression.

f. A scientist calculated the average rate of change for the bacteria in the first three hours to be 168. Which units should the scientist use when reporting this number?

6. Solve each equation. Express your answer as a logarithm, and then approximate the solution to the nearest thousandth.

   a. $3(10)^{-x} = \frac{1}{9}$

   b. $362\left(10^{\frac{t}{12}}\right) = 500$

   c. $(2)^{3x} = 9$

   d. $300e^{0.4t} = 900$

7. Because atoms and molecules are very small, they are counted in units of *moles*, where $1 \text{ mole} = 6.022 \times 10^{23}$. Concentration of molecules in a liquid is measured in units of moles per liter. The measure of the acidity of a liquid is called the pH of the liquid and is given by the formula

$$\text{pH} = -\log(H),$$

where $H$ is the concentration of hydrogen ions in units of moles per liter.

a. Water has a pH value of 7.0. How many hydrogen ions are in one liter of water?

b. If a liquid has a pH value larger than 7.0, does one liter of that liquid contain more or less hydrogen ions than one liter of water? Explain.

c. Suppose that liquid A is more acidic than liquid B, and their pH values differ by 1.2. What is the ratio of the concentration of hydrogen ions in liquid A to the concentration of hydrogen ions in liquid B?

8. A social media site is experiencing rapid growth. The table below shows values of $V$, the total number of unique visitors in each month, $t$, for a 6-month period of time. The graph shows the average minutes per visit to the site, $M$, in each month, $t$, for the same 6-month period of time.

| $t$, Month | 1 | 2 | 3 | 4 | 5 | 6 |
|---|---|---|---|---|---|---|
| $V(t)$, Number of Unique Visitors | 418,000 | 608,000 | 1,031,000 | 1,270,000 | 2,023,000 | 3,295,000 |

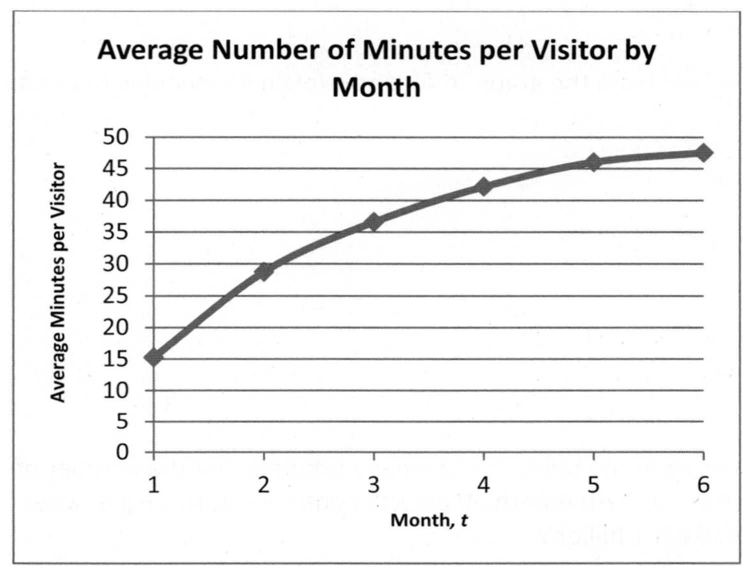


a. Between which two months did the site experience the most growth in total unique visitors? What is the average rate of change over this time interval?

b. Compute the value of $\frac{V(6)-V(1)}{6-1}$, and explain its meaning in this situation.

c. Between which two months did the average length of a visit change by the least amount? Estimate the average rate of change over this time interval.

d. Estimate the value of $\frac{M(3)-M(2)}{3-2}$ from the graph of $M$, and explain its meaning in this situation.

e. Based on the patterns they see in the table, the company predicts that the number of unique visitors will double each month after the sixth month. If growth continues at this pace, when will the number of unique visitors reach 1 billion?

| A Progression Toward Mastery | | | | | |
|---|---|---|---|---|---|
| Assessment Task Item | | **STEP 1** Missing or incorrect answer and little evidence of reasoning or application of mathematics to solve the problem. | **STEP 2** Missing or incorrect answer but evidence of some reasoning or application of mathematics to solve the problem. | **STEP 3** A correct answer with some evidence of reasoning or application of mathematics to solve the problem, <u>or</u> an incorrect answer with substantial evidence of solid reasoning or application of mathematics to solve the problem. | **STEP 4** A correct answer supported by substantial evidence of solid reasoning or application of mathematics to solve the problem. |
| **1** | **a**<br><br>**N-RN.A.1** | Student solution is incorrect or missing. There is little evidence of correct student thinking and little or no written explanation. | Student explanation shows some evidence of reasoning but contains a major omission or limited explanation to support the reasoning. | Student explanation is mathematically correct but communication is limited or may contain minor notation errors. | Student explanation is mathematically correct and clearly stated. Student explanation relies on properties and definitions. Student uses precise and accurate vocabulary and notation. |
| **2** | **a–c**<br><br>**N-RN.A.2** | Student work is inaccurate or missing. Student fails to express any of the expressions correctly as a rational number. | Student correctly rewrites only one of the expressions as a rational number. Student solutions show evidence of major errors or omissions in the work shown. | Student correctly rewrites each expression as a rational number, but the work shown is missing steps or contains minor notational errors.<br><u>OR</u><br>Student correctly rewrites two of three parts as rational numbers, and work shown is thorough and precise. | Student correctly rewrites each expression as a rational number and shows sufficient work to support the solution by applying properties and definitions. |
| **3** | **a–c**<br><br>**N-RN.A.2** | Student work is inaccurate or missing, and there is little or no evidence of understanding of how to apply the properties and definitions to rational exponents. | Student rewrites the expressions correctly but only evaluates one expression correctly.<br><u>OR</u><br>Student rewrites only one expression correctly but evaluates | Student solutions are largely correct with no more than one or two minor errors. Student solutions show sufficient work to demonstrate evidence of understanding the | Student rewrites each part correctly according to the problem specifications and then evaluates each expression correctly for the given values of the variables. |

| | | | | | |
|---|---|---|---|---|---|
| | | | the expressions without significant errors for the given values of the variable.<br>Student work indicates difficulty using proper notation. | properties and definitions, and student uses proper notation.<br>OR<br>Student correctly rewrites and evaluates the expressions for all three parts but shows little or no work to support reasoning. | Student solutions show sufficient work to demonstrate evidence of understanding the properties and definitions, and student uses proper notation. |
| **4** | **a–b**<br><br>**F-BF.A.1a**<br>**N-RN.A.1** | Student makes little or no attempt to describe a recursive process to estimate $3^{\pi}$. There is little or no written communication to support reasoning. | Student attempts to explain a recursive process with clear written communication, but solution contains one or more major errors or omissions in thinking. | Student solution is largely correct, but work shown and written communication could be more detailed and precise. | Student solution is correct, and work shown clearly outlines a recursive process for estimating $3^{\pi}$.<br>Student explains thinking in writing clearly and concisely, using proper notation and vocabulary to support work. |
| **5** | **a–b**<br><br>**N-Q.A.2** | Student work is inaccurate, missing, or off-task. | Student identifies one quantity of interest, but the solution fails to address proper units or includes more than one minor error. | Student identifies two quantities of interest, including units, in part (a), and labels the table in part (b) with no more than one minor error or omissions of units. | Student identifies two quantities of interest (number of bacteria and time), including units. Student correctly labels the table, including proper units. |
| | **c–d**<br><br>**F-LE.A.2**<br>**F-BF.A.1a** | Student does not write correct exponential formulas in explicit forms for either function. The work shown has several mathematical errors. | Student writes one of two exponential formulas correctly in explicit form. Solutions contain more than one minor error or show evidence of a major mathematical misconception. | Student writes two explicit exponential formulas that are largely correct with no more than one minor error. Student solutions may not clearly indicate how parameters were determined. | Student correctly writes two explicit exponential formulas to represent the amount of bacteria in each case. Student solution clearly shows evidence of how the parameters of each formula were determined. |
| | **e**<br><br>**F-LE.A.4**<br>**A-CED.A.1** | Student makes little or no attempt to write or solve an equation to answer this problem. | Student writes an incorrect equation, solves it correctly, and expresses the final answer as a logarithm.<br>OR<br>Student equation is correct, but the steps to solve the equation using logarithms show major mathematical errors in student thinking. | Student writes an equation of the form $f(t) = 512{,}000$ and solves it using logarithms. Final solution is not expressed as a logarithm, or the solution contains a minor error.<br>OR<br>Student writes an equation of the form | Student writes an equation of the form $f(t) = 512{,}000$, correctly solves it using logarithms, and expresses the final solution as a logarithm. |

| | | | | | |
|---|---|---|---|---|---|
| | | | | $f(t) = 512{,}000$ and solves it correctly using a graphical or numerical approach. | |
| | f<br><br>N-Q.A.2 | Student makes little or no attempt to answer the question. Work shown is off-task. | Student provides units that agree with the model used in part (c) but explanation is missing. | Student identifies units that do not agree with the model used in part (c). Explanation is otherwise correct. | Student clearly identifies the units and explains the answer. Units agree with the model student created in part (c). |
| 6 | a–d<br><br>F-LE.A.4 | Student fails to correctly solve any equations using logarithms, showing incomplete and incorrect work. | Student correctly solves two equations but shows little or no work and fails to provide the exact or estimated answer.<br>OR<br>Student correctly solves one or two equations using logarithms. Work shown indicates major mathematical errors. | Student correctly solves four equations using logarithms but may not express solutions as both logarithms and estimated to three decimal places.<br>OR<br>Student correctly solves three of four equations. Work is largely correct with only one or two minor errors. | Student correctly solves four equations using logarithms. Work shown uses proper notation, and solutions are expressed exactly as logarithms estimated to three decimal places. |
| 7 | a<br><br>A-CED.A.1 | Student makes little or no attempt to write and solve an equation to answer the question. Work shown contains several significant errors. | Student writes an equation equivalent to $-7 = \log(H)$ but makes significant mathematical errors in solving it. Student does not relate the solution back to the question that was asked. | Student writes and solves an equation equivalent to the equation $-7 = \log(H)$, but the solution contains minor errors, either in solving the equation or answering the question. | Student writes and solves an equation equivalent to the equation $7 = -\log(H)$ and uses the solution to answer the question correctly. |
| | b<br><br>N-Q.A.2 | Student solution is inaccurate and incomplete, showing student had significant difficulty thinking about the quantities in this problem. | Student is not able to correctly interpret the meaning of the $\mathrm{pH}$ values when compared to the $\mathrm{pH}$ of water. Student explanation gives evidence of partially correct thinking about the quantities in this problem. | Student uses the $\mathrm{pH}$ of water to correctly interpret the meaning of $\mathrm{pH}$ values more or less than $7.0$ in the context of this situation. Student explanation makes sense for the most part but could include more detail or may contain minor vocabulary or notation errors. | Student uses the $\mathrm{pH}$ of water to correctly interpret the meaning of $\mathrm{pH}$ values more or less than $7.0$ in the context of this situation. Student explanation clearly conveys correct thinking and uses proper vocabulary and notation. |

| | | | | | |
|---|---|---|---|---|---|
| | c<br><br>A-CED.A.1<br>N-Q.A.2 | Student makes little or no attempt to write and solve an equation. Student makes little or no attempt to relate work to the context of the situation. | Student does not write a correct equation but attempts to solve the problem. Student work and explanations show some evidence of correct thinking about the situation but contains some major mathematical errors. | Student correctly writes an equation with a solution that can be used to answer the question. Student attempts to solve the equation. Student work and explanations may contain minor mathematical errors. | Student correctly writes an equation with a solution that can be used to answer the question. Student solves the equation correctly. Student work and explanations clearly show correct thinking, with the solution related back to the situation. |
| 8 | a–d<br><br>F-IF.B.6 | Student is unsuccessful in calculating and interpreting average rate of change from either tables or graphs. Little or no work is shown, and the work that is present contains several mathematical errors. | Student calculates rate of change from tables and graphs correctly but struggles with a correct interpretation.<br>OR<br>Student solutions show major errors in calculating rate of change, but interpretations are largely accurate based on the values they calculated.<br>Student explanations may be limited. | Student calculates and interprets average rate of change from a table or graph in each portion of the problem.<br>Student solutions may contain a few minor errors that do not detract from the overall understanding of the problem. | Student successfully calculates and interprets average rate of change from a table or graph in each portion of the problem.<br>Student explanations and work shown are clear and concise, including proper notation and symbols. |
| | e<br><br>F-LE.A.4 | Student work is off-task, missing, or incorrect. Little or no attempt is made to write either a function or an equation that could be used to solve the problem. | Student fails to create an exponential function that accurately reflects the description in the problem. Student attempts to solve an equation of the form $f(x) = 1{,}000{,}000{,}000$, but the solution, while partially correct, contains some significant mathematical errors. | Student creates an exponential function that accurately reflects the description in the problem. Student writes and solves an equation of the form $f(x) = 1{,}000{,}000{,}000$. Student solution contains minor errors, and student may fail to explicitly interpret the solution in the context of the problem. | Student creates an exponential function that accurately reflects the description in the problem.<br>Student writes and correctly solves an equation of the form $f(x) = 1{,}000{,}000{,}000$, and interprets the solution in the context of the problem. |

Name ______________________________ Date ______________

1. Use properties of exponents to explain why it makes sense to define $16^{\frac{1}{4}}$ as $\sqrt[4]{16}$.

*We know that*

$$\left(16^{\frac{1}{4}}\right)^4 = 16^{\frac{1}{4}} \cdot 16^{\frac{1}{4}} \cdot 16^{\frac{1}{4}} \cdot 16^{\frac{1}{4}} = 16^{\frac{1}{4}+\frac{1}{4}+\frac{1}{4}+\frac{1}{4}} = 16^1 = 16,$$

*and* $2^4 = 16$.

*Since there is only one positive real number whose fourth power is 16, and that is* $\sqrt[4]{16} = 2$, *it makes sense to define* $16^{\frac{1}{4}} = \sqrt[4]{16}$.

2. Use properties of exponents to rewrite each expression as either an integer or as a quotient of integers $\frac{p}{q}$ to show the expression is a rational number.

a. $\sqrt[4]{2}\,\sqrt[4]{8}$

$$\sqrt[4]{2}\,\sqrt[4]{8} = 2^{\frac{1}{4}} \cdot 8^{\frac{1}{4}} = 2^{\frac{1}{4}} \cdot (2^3)^{\frac{1}{4}} = 2^{\frac{1}{4}} \cdot 2^{\frac{3}{4}} = 2^{\frac{1}{4}+\frac{3}{4}} = 2^1$$

*Thus,* $\sqrt[4]{2}\,\sqrt[4]{8} = 2$ *is a rational number.*

b. $\frac{\sqrt[3]{54}}{\sqrt[3]{2}}$

$$\frac{\sqrt[3]{54}}{\sqrt[3]{2}} = \frac{\left(54^{\frac{1}{3}}\right)}{2^{\frac{1}{3}}} = \frac{(2 \cdot 27)^{\frac{1}{3}}}{2^{\frac{1}{3}}} = \frac{2^{\frac{1}{3}} \cdot 27^{\frac{1}{3}}}{2^{\frac{1}{3}}} = 27^{\frac{1}{3}} = 3$$

*Thus,* $\frac{\sqrt[3]{54}}{\sqrt[3]{2}} = 3$ *is a rational number.*

c. $16^{\frac{3}{2}} \cdot \left(\frac{1}{27}\right)^{\frac{2}{3}}$

$$16^{\frac{3}{2}} \cdot \left(\frac{1}{27}\right)^{\frac{2}{3}} = \left(16^{\frac{1}{2}}\right)^3 \cdot \left(\frac{1}{\left(27^{\frac{1}{3}}\right)^2}\right) = 4^3 \cdot \frac{1}{3^2} = \frac{64}{9}$$

*Thus,* $16^{\frac{3}{2}} \cdot \left(\frac{1}{27}\right)^{\frac{2}{3}} = \frac{64}{9}$ *is a rational number.*

3. Use properties of exponents to rewrite each expression with only positive, rational exponents. Then find the numerical value of each expression when $x = 9$, $y = 8$, and $z = 16$. In each case, the expression evaluates to a rational number.

a. $\sqrt{\frac{xy^2}{x^3z^{\frac{1}{2}}}}$

$$\sqrt{\frac{xy^2}{x^3z^{\frac{1}{2}}}} = \left(\frac{x \cdot y^2}{x^3z^{\frac{1}{2}}}\right)^{\frac{1}{2}} = \left(\frac{y^2}{x^2\ z^{\frac{1}{2}}}\right)^{\frac{1}{2}} = \frac{y}{x\ z^{\frac{1}{4}}}$$

When $x = 9$, $y = 8$, and $z = 16$, we have

$$\sqrt{\frac{xy^2}{x^3z^{\frac{1}{2}}}} = \frac{8}{9\ 16^{\frac{1}{4}}} = \frac{8}{9 \cdot 2} = \frac{4}{9}.$$

b. $\sqrt[11]{y^2z^4}$

$$\sqrt[11]{y^2z^4} = (y^2z^4)^{\frac{1}{11}} = y^{\frac{2}{11}}z^{\frac{4}{11}}$$

When $y = 8$ and $z = 16$, we have

$$\sqrt[11]{y^2z^4} = 8^{\frac{2}{11}}16^{\frac{4}{11}} = (2^3)^{\frac{2}{11}} \cdot (2^4)^{\frac{4}{11}} = 2^{\frac{6}{11}} \cdot 2^{\frac{16}{11}} = 2^{\frac{22}{11}} = 2^2 = 4.$$

c. $x^{-\frac{3}{2}}y^{\frac{4}{3}}z^{-\frac{3}{4}}$

$$x^{-\frac{3}{2}}y^{\frac{4}{3}}z^{-\frac{3}{4}} = \frac{y^{\frac{4}{3}}}{x^{\frac{3}{2}} \cdot z^{\frac{3}{4}}}$$

When $x = 9$, $y = 8$, and $z = 16$, we have

$$x^{-\frac{3}{2}}y^{\frac{4}{3}}z^{-\frac{3}{4}} = \frac{8^{\frac{4}{3}}}{9^{\frac{3}{2}} \cdot 16^{\frac{3}{4}}} = \frac{(2^3)^{\frac{4}{3}}}{(3^2)^{\frac{3}{2}} \cdot (2^4)^{\frac{3}{4}}} = \frac{2^4}{3^3 \cdot 2^3} = \frac{2}{3^3} = \frac{2}{27}.$$

4. We can use finite approximations of the decimal expansion of $\pi = 3.141519\ldots$ to find an approximate value of the number $3^{\pi}$.

   a. Fill in the missing exponents in the following sequence of inequalities.

$$3^3 < 3^{\pi} < 3^4$$
$$3^{3.1} < 3^{\pi} < 3^{3.2}$$
$$3^{3.14} < 3^{\pi} < 3^{3.15}$$
$$3^{(3.141)} < 3^{\pi} < 3^{(3.142)}$$
$$3^{(3.1415)} < 3^{\pi} < 3^{(3.1416)}$$

   b. Explain how this recursive process leads to better and better approximations of the number $3^{\pi}$.

*We can get better and better approximations of $\pi$ by squeezing it between rational over and under estimates that use more and more digits of its decimal expansion.*

$$3 < \pi < 4$$
$$3.1 < \pi < 3.2$$
$$3.14 < \pi < 3.15$$
$$\vdots$$

*Then, we can estimate $3^{\pi}$ by squeezing it between expressions with rational exponents. Because we know how to calculate a number such as $3^{3.14} = 3^3 \cdot 3^{0.14} = 27\left(\sqrt[100]{3^{14}}\right)$, we have a method for calculating the over and under estimates of $3^{\pi}$.*

*Because $3 < \pi < 4$, we have $3^3 < 3^{\pi} < 3^4$. Thus, $27 < 3^{\pi} < 81$.*

*Because $3.1 < \pi < 3.2$, we have $3^{3.1} < 3^{\pi} < 3^{3.2}$. Thus, $30.1353 < 3^{\pi} < 33.6347$.*

*Because $3.14 < \pi < 3.15$, we have $3^{3.14} < 3^{\pi} < 3^{3.15}$. Thus, $31.4891 < 3^{\pi} < 31.8370$.*

*Continuing this process, we can approximate $3^{\pi}$ as closely as we want by starting with more and more digits of $\pi$ in the exponent.*

5. A scientist is studying the growth of a population of bacteria. At the beginning of her study, she has 800 bacteria. She notices that the population is quadrupling every hour.

   a. What quantities, including units, need to be identified to further investigate the growth of this bacteria population?

   *We need to have the initial population, $P_0$, in either units of single bacteria or hundreds of bacteria, the time, $t$, in hours, and the current population, $P(t)$ at time $t$, in the same units as the initial population $P_0$.*

   b. The scientist recorded the following information in her notebook, but she forgot to label each row. Label each row to show what quantities, including appropriate units, are represented by the numbers in the table, and then complete the table.

| *time, t (hours)* | 0 | 1 | 2 | 3 | 4 |
|---|---|---|---|---|---|
| *Population, P(t) (hundreds)* | 8 | 32 | 128 | *512* | *2048* |

   c. Write an explicit formula for the number of bacteria present after $t$ hours.

   *After t hours, there are $P(t) = 8(4^t)$ hundred bacteria present.*
   *(It is also acceptable to model this population by $P(t) = 800(4^t)$ single bacteria.)*

   d. Another scientist studying the same population notices that the population is doubling every half an hour. Complete the table, and write an explicit formula for the number of bacteria present after $x$ half hours.

| Time, $t$ (hours) | 0 | $\frac{1}{2}$ | 1 | $\frac{3}{2}$ | 2 | $\frac{5}{2}$ | 3 |
|---|---|---|---|---|---|---|---|
| Time, $x$ (half-hours) | 0 | 1 | 2 | 3 | 4 | 5 | 6 |
| Bacteria (hundreds) | 8 | 16 | 32 | 64 | 128 | 256 | 512 |

   *After x half-hours, there are $Q(x) = 8(2^x)$ hundred bacteria present.*
   *(It is also acceptable to model this population by $Q(x) = 800(2^x)$ single bacteria.)*

e. Find the time, in hours, when there will be 5,120,000 bacteria. Express your answer as a logarithmic expression.

*Students may choose to use the base-2 formula but will need to adjust the value of x, which counts half-hours, to t, which counts full hours, to correctly answer the question. Also, note that 5,120,000 is 51,200 hundred bacteria, so if students modeled the population using single bacteria instead of hundreds, they should solve $800(4^t) = 5{,}120{,}000$. Students may also solve this equation using the base-4 logarithm or base-2 logarithm, giving an equivalent solution that looks a little different..*

$$8(4^t) = 51200$$
$$4^t = 6400$$
$$t\log(4) = \log(6400)$$
$$t = \frac{\log(6400)}{\log(4)}$$

f. A scientist calculated the average rate of change for the bacteria in the first three hours to be 168. Which units should the scientist use when reporting this number? Explain how you know.

*The average rate of change over the first three hours is given by the formula $\frac{P(3) - P(0)}{3 - 0}$, which is a quotient of the number of bacteria in hundreds per hour. Thus, the unit should be reported as hundreds of bacteria per hour. Note that if students had modeled the population using single bacteria instead of hundreds, they should answer that the units are in bacteria per hour.*

6. Solve each equation. Express your answer as a logarithm, and then approximate the solution to the nearest thousandth.

a. $3(10)^{-x} = \frac{1}{9}$

$$3(10^{-x}) = \frac{1}{9}$$
$$\frac{1}{10^x} = \frac{1}{27}$$
$$10^x = 27$$
$$x = \log(27)$$
$$x \approx 1.431$$

b. $362\left(10^{\frac{t}{12}}\right) = 500$

$$362\left(10^{\frac{t}{12}}\right) = 500$$
$$10^{\frac{t}{12}} = \frac{500}{362}$$
$$\frac{t}{12} = \log\left(\frac{500}{362}\right)$$
$$t \approx 1.683$$

c. $(2)^{3x} = 9$

$$2^{3x} = 9$$
$$\log_2(2^{3x}) = \log_2(9)$$
$$3x = \log_2(9)$$
$$x = \frac{1}{3}\log_2(9) = \frac{1}{3}\frac{\log(9)}{\log(2)}$$
$$x \approx 1.057$$

d. $300e^{0.4t} = 900$

$$300e^{0.4t} = 900$$
$$e^{0.4t} = 3$$
$$\ln(e^{0.4t}) = \ln(3)$$
$$0.4t = \ln(3)$$
$$t = 2.5\ln(3)$$
$$t \approx 2.747$$

7. Because atoms and molecules are very small, they are counted in units of *moles*, where $1 \text{ mole} = 6.022 \times 10^{23}$ molecules. Concentration of molecules in a liquid is measured in units of moles per liter. The measure of the acidity of a liquid is called the pH of the liquid and is given by the formula

$$\text{pH} = -\log(H),$$

where $H$ is the concentration of hydrogen ions in units of moles per liter.

a. Water has a pH value of 7.0. How many hydrogen ions are in one liter of water?

$$7 = -\log(H)$$
$$-7 = \log(H)$$
$$H = 10^{-7}$$

*Thus, there are approximately $10^{-7}$ moles of hydrogen in one liter of water. If we multiply this by $6.022 \times 10^{23}$ ions per mole, we find that there are $6.022 \times 10^{16}$ hydrogen ions.*

b. If a liquid has a pH value larger than 7.0, does one liter of that liquid contain more or fewer hydrogen ions than one liter of water? Explain.

*A liquid with a pH value larger than 7.0 will contain fewer hydrogen ions than one liter of water because $H = \frac{1}{10^{pH}}$ and when the pH is larger than 7.0, the value of H will become smaller because the quantities H and $10^{pH}$ are inversely proportional to one another.*

c. Suppose that liquid A is more acidic than liquid B, and their pH values differ by 1.2. What is the ratio of the concentration of hydrogen ions in liquid A to the concentration of hydrogen ions in liquid B?

*Let $H_A$ be the concentration of hydrogen ions in liquid A, and let $H_B$ be the concentration of hydrogen ions in liquid B. Then, the difference of the pH values is $-\log(H_B) - (-\log(H_A)) = 1.2$. Solve this equation for $\frac{H_A}{H_B}$, the requested ratio.*

$$-\log(H_B) + \log(H_A) = 1.2$$
$$\log\left(\frac{H_A}{H_B}\right) = 1.2$$
$$\frac{H_A}{H_B} = 10^{1.2} \approx 15.85$$

*Liquid A contains approximately 16 times as many hydrogen ions as liquid B.*

8. A social media site is experiencing rapid growth. The table below shows values of $V$, and the total number of unique visitors in each month, $t$, for a 6-month period of time. The graph shows the average minutes per visit to the site, $M$, in each month, $t$, for the same 6-month period of time.

| $t$, Month | 1 | 2 | 3 | 4 | 5 | 6 |
|---|---|---|---|---|---|---|
| $V(t)$, Number of Unique Visitors | 418,000 | 608,000 | 1,031,000 | 1,270,000 | 2,023,000 | 3,295,000 |

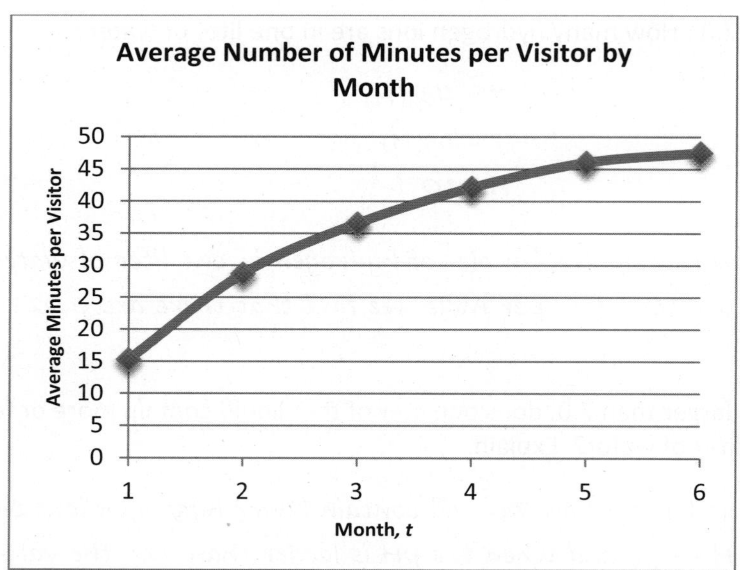


a. Between which two months did the site experience the most growth in total unique visitors? What is the average rate of change over this time interval?

$V(2) - V(1) = 190{,}000$   $V(3) - V(2) = 423{,}000$   $V(4) - V(3) = 239{,}000$
$V(5) - V(4) = 753{,}000$   $V(6) - V(5) = 1{,}272{,}000$

*The largest growth in the number of visitors occurs between months 5 and 6.*

b. Compute the value of $\frac{V(6)-V(1)}{6-1}$, and explain its meaning in this situation.

$$\frac{V(6) - V(1)}{6 - 1} = \frac{3{,}295{,}000 - 418{,}000}{5} = 575{,}400$$

*This means that the average monthly growth of visitors to the site between months 1 and 6 is 575,400 visitors per month.*

c. Between which two months did the average length of a visit change by the least amount? Estimate the average rate of change over this time interval.

*The two neighboring points that have the closest y-values are in months 5 and 6. Estimating values $M(6) \approx 47.5$ and $M(5) \approx 46$ from the graph, we see that the average rate of change over this interval is $\frac{M(6)-M(5)}{6-5} = \frac{47.5-46}{1} = 1.5$ minutes per visitor per month. (Students may read different values from the graph.)*

d. Estimate the value of $\frac{M(3)-M(2)}{3-2}$ from the graph of $M$, and explain its meaning in this situation.

*Estimating values $M(3) \approx 37$ and $M(2) \approx 28$ from the graph, we have $\frac{M(3)-M(2)}{3-2} = \frac{37-28}{1} = 9$, meaning that on average, each visit to the website increased by 9 minutes between months 2 and 3.*

e. Based on the patterns they see in the table, the company predicts that the number of unique visitors will double each month after the sixth month. If growth continues at this pace, when will the number of unique visitors reach 1 billion?

$$3{,}295{,}000(2)^t = 1{,}000{,}000{,}000$$
$$2^t = 303.49$$
$$\log(2^t) = \log(303.49)$$
$$t\log(2) = \log(303.49)$$
$$t = \frac{\log(303.49)}{\log(2)}$$
$$t \approx 8.25$$

*The number of unique visitors will reach 1 billion after 8.25 additional months have passed.*

# Topic C:

# Exponential and Logarithmic Functions and their Graphs

**F-IF.B.4, F-IF.B.5, F-IF.C.7e, F-BF.A.1a, F-BF.B.3, F-BF.B.4a, F-LE.A.2, F-LE.A.4**

| | | |
|---|---|---|
| **Focus Standards:** | F-IF.B.4 | For a function that models a relationship between two quantities, interpret key features of graphs and tables in terms of the quantities, and sketch graphs showing key features given a verbal description of the relationship. *Key features include: intercepts; intervals where the function is increasing, decreasing, positive, or negative; relative maximums and minimums; symmetries; end behavior; and periodicity.*★ |
| | F-IF.B.5 | Relate the domain of a function to its graph and, where applicable, to the quantitative relationship is describes. *For example, if the function $h(n)$ gives the number of person-hours it takes to assemble $n$ engines in a factory, then the positive integers would be an appropriate domain for the function.*★ |
| | F-IF.C.7e | Graph functions expressed symbolically and show key features of the graph, by hand in simple cases and using technology for more complicated cases.★<br>e. Graph exponential and logarithmic functions, showing intercepts and end behavior, and trigonometric functions, showing period, midline, and amplitude. |
| | F-BF.A.1a | Write a function that describes a relationship between two quantities.★<br>a. Determine an explicit expression, a recursive process, or steps for calculation from a context. |
| | F-BF.B.3 | Identify the effect on the graph of replacing $f(x)$ by $f(x) + k, k\,f(x), f(kx)$, and $f(x + k)$ for specific values of $k$ (both positive and negative); find the value of $k$ given the graphs. Experiment with cases and illustrate an explanation of the effects on the graph using technology. *Include recognizing even and odd functions from their graphs and algebraic expressions for them.* |

| | | |
|---|---|---|
| | F-BF.B.4a | Find inverse functions.<br>a. Solve an equation of the form $f(x) = c$ for a simple function $f$ that has an inverse and write an expression for the inverse. *For example,* $f(x) = 2x^3$ *or* $f(x) = (x+1)/(x-1)$ *for* $x \neq 1$. |
| | F-LE.A.2 | Construct linear and exponential functions, including arithmetic and geometric sequences, given a graph, a description of a relationship, or two input-output pairs (include reading these from a table). |
| | F-LE.A.4 | For exponential models, express as a logarithm the solution to $ab^{ct} = d$ where $a$, $c$, and $d$ are numbers, and the base $b$ is 2, 10, or $e$; evaluate the logarithm using technology. |
| **Instructional Days:** | 7 | |
| **Lesson 16:** | Rational and Irrational Numbers (S)[1] | |
| **Lesson 17:** | Graphing the Logarithm Function (P) | |
| **Lesson 18:** | Graphs of Exponential Functions and Logarithmic Functions (P) | |
| **Lesson 19:** | The Inverse Relationship between Logarithmic and Exponential Functions (P) | |
| **Lesson 20:** | Transformations of the Graphs of Logarithmic and Exponential Functions (E) | |
| **Lesson 21:** | The Graph of the Natural Logarithm Function (E) | |
| **Lesson 22:** | Choosing a Model (P) | |

The lessons covered in Topic A and Topic B build upon students' prior knowledge of the properties of exponents, exponential expression, and solving equations by extending the properties of exponents to all real number exponents and positive real number bases before introducing logarithms. This topic reintroduces exponential functions, introduces logarithmic functions, explains their inverse relationship, and explores the features of their graphs and how they can be used to model data.

Lesson 16 ties back to work in Topic A by helping students to further extend their understanding of the properties of real numbers, both rational and irrational (**N-RN.B.3**). This Algebra I standard is revisited in Algebra II so that students know and understand that the exponential functions are defined for all real numbers, and, thus, the graphs of the exponential functions can be represented by a smooth curve. Another consequence is that the logarithm functions are also defined for all positive real numbers. Lessons 17 and 18 introduce the graphs of logarithmic functions and exponential functions. Students compare the properties of graphs of logarithm functions for different bases and identify common features, which align with standards **F-IF.B.4**, **F-IF.B.5**, and **F-IF.C.7**. Students understand that if the range of this function is all real numbers, then some logarithms must be irrational. Students notice that $f(x) = b^x$ and $g(x) = \log_b(x)$ appear to be related via a reflection across the graph of the equation $y = x$.

Lesson 19 addresses standards **F-BF.B.4a** and **F-LE.A.4** while continuing the ideas introduced graphically in Lesson 18 to help students make the connection that the logarithmic function and the exponential function are inverses of each other. Inverses are introduced first by discussing operations and functions that can "undo" each other; then, students look at the graphs of pairs of these functions. The lesson ties the ideas back to reflections in the plane from geometry and illuminates why the graphs of inverse functions are

[1] Lesson Structure Key: **P**-Problem Set Lesson, **M**-Modeling Cycle Lesson, **E**-Exploration Lesson, **S**-Socratic Lesson

reflections of each other across the line given by $y = x$, developing these ideas intuitively without formalizing what it means for two functions to be inverses. Inverse functions will be addressed in greater detail in Precalculus.

During all of these lessons, connections are made to the properties of logarithms and exponents. The relationship between graphs of these functions, applying transformations to sketch graphs of functions, and the properties associated with these functions are linked in Lessons 20 and 21, showcasing standards **F-IF.C.7e** and **F-BF.B.3**. Students use properties and their knowledge of transformations to explain why two different functions such as $f(x) = \log(10x)$ and $g(x) = 1 + \log(x)$ have the same graph. Lesson 21 revisits the natural logarithm function, and students see how the change of base property of logarithms proves that we can write a logarithm function of any base $b$ as a vertical scaling of the natural logarithm function (or any other base logarithm function we choose).

Finally, in Lesson 22, students must synthesize knowledge across both Algebra I and Algebra II to decide whether a linear, quadratic, sinusoidal, or exponential function will best model a real-world scenario by analyzing the way in which we expect the quantity in question to change. For example, students need to determine whether or not to model daylight hours in Oslo, Norway, with a linear or a sinusoidal function because the data appears to be linear, but that does not make sense in context. They model the outbreak of a flu epidemic with an exponential function and a falling body with a quadratic function. In this lesson, the majority of the scenarios that require modeling are described verbally, and students determine an explicit expression for many of the functions in accordance with **F-BF.A.1a**, **F-LE.A.1**, and **F-LE.A.2**.

# Lesson 16: Rational and Irrational Numbers

## Student Outcomes

- Students interpret addition and multiplication of two irrational numbers in the context of logarithms and find better-and-better decimal approximations of the sum and product, respectively.
- Students work with and interpret logarithms with irrational values in preparation for graphing logarithmic functions.

## Lesson Notes

This foundational lesson revisits the fundamental differences between rational and irrational numbers. We begin by reviewing how to locate an irrational number on the number line by squeezing its infinite decimal expansion between two rational numbers, which students may recall from Grade 8, Module 7, Lesson 7. In preparation for graphing logarithm functions, the main focus of this lesson is to understand the process of locating values of logarithms on the number line (**F-IF.C.7e**). We then go a step further to understand this process in the context of adding two irrational logarithmic expressions (**N-RN.B.3**). Although students have addressed **N-RN.B.3** in Algebra I and have worked with irrational numbers in previous lessons in this module, such as Lesson 5, students need to fully understand how to sum two irrational logarithmic expressions in preparation for graphing logarithmic functions in the next lesson, in alignment with **F-IF.C.7e**. Students have been exposed to two approaches to adding rational numbers: a geometric approach by placing the numbers on the number line, as reviewed in Module 1, Lesson 24, and a numerical approach by applying an addition algorithm. In the Opening Exercise, students are asked to recall both methods for adding rational numbers. Both approaches fail with irrational numbers, and we locate the sum of two irrational numbers (or a rational and an irrational number) by squeezing its infinite decimal expansion between two rational numbers. Emphasize to students that since they have been performing addition for many, many years, we are more interested in the process of addition than in the result.

This lesson emphasizes mathematical practice standard MP.3 as students develop and then justify conjectures about sums and products of irrational logarithmic expressions.

## Classwork

### Opening Exercise (4 minutes)

Have students work in pairs or small groups on the sums below. Remind them that we are interested in the process of addition as much as obtaining the correct result. Ask groups to volunteer their responses at the end of the allotted time.

> *Scaffolding:*
>
> - Ask struggling students to represent a simpler sum, such as $\frac{1}{5}+\frac{1}{4}$.
> - Ask advanced students to explain how to represent the sum of two generic rational numbers, such as $\frac{a}{b}+\frac{c}{d}$, on the number line.

**Opening Exercise**

a. **Explain how to use a number line to add the fractions $\frac{7}{5}+\frac{9}{4}$.**

*First, we locate the point $\frac{7}{5}$ on the number line by dividing each unit into $5$ intervals of length $\frac{1}{5}$.*

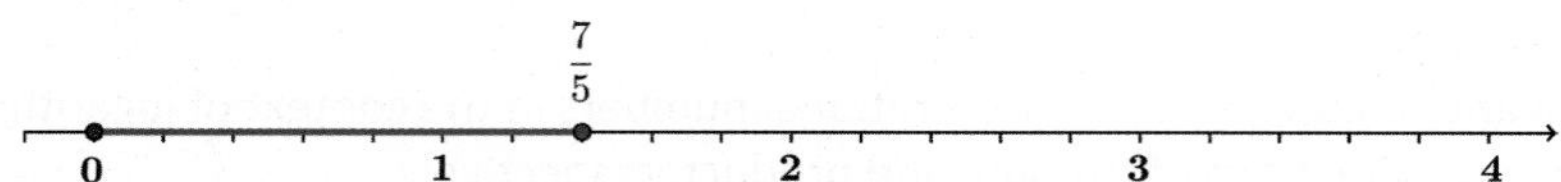


*Then, we locate the point $\frac{9}{4}$ on the number line by dividing each unit into $4$ intervals of length $\frac{1}{4}$.*

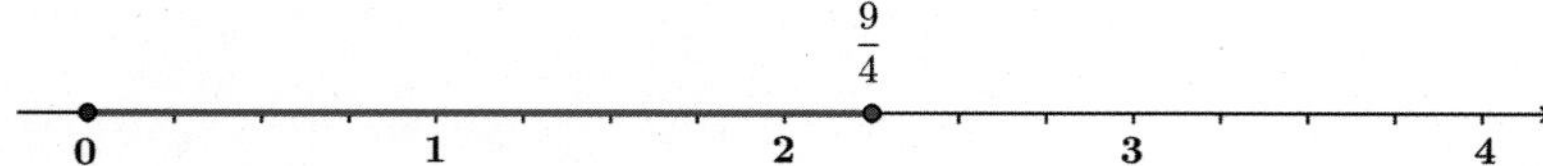


*To find the sum, we place the green segment of length $\frac{9}{4}$ end-to-end with the blue segment of length $\frac{7}{5}$, and the right end point of the green segment is the sum. Since the tick marks at units of $\frac{1}{4}$ and $\frac{1}{5}$ do not align, we make new tick marks that are $\frac{1}{20}$ apart. Then $\frac{7}{5}=\frac{28}{20}$ and $\frac{9}{4}=\frac{45}{20}$, so the sum is located at point $\frac{73}{20}$.*

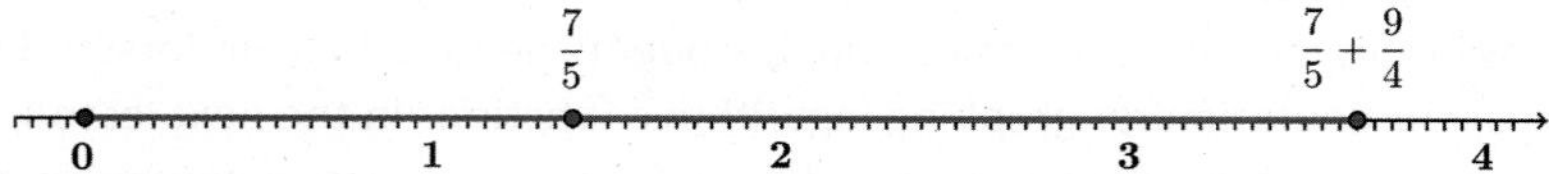


b. **Convert $\frac{7}{5}$ and $\frac{9}{4}$ to decimals, and explain the process for adding them together.**

*We know that $\frac{7}{5}=1.4$ and $\frac{9}{4}=2.25$. To add these numbers, we add a zero placeholder to $1.4$ to get $1.40$ so that each number has the same number of decimal places. Then, we line them up at the decimal place and add from right to left, carrying over a power of $10$ if needed (we do not need to carry for this sum).*

*Step 1: Working from right to left, we first add $0$ hundredths $+\ 5$ hundredths $=5$ hundredths.*

$$\begin{array}{r} 1.40 \\ +2.25 \\ \hline 5 \end{array}$$

*Step 2: Then we add $4$ tenths $+2$ tenths $=6$ tenths.*

$$\begin{array}{r} 1.40 \\ +2.25 \\ \hline 65 \end{array}$$

*Step 3: And, finally, we add $1$ one $+\ 2$ ones $=3$ ones.*

$$\begin{array}{r} 1.40 \\ +2.25 \\ \hline 3.65 \end{array}$$

## Discussion (8 minutes)

- How would we add two numbers such as $\pi+\frac{7}{5}$?
  - *Student answers will vary, but the point is that our algorithm, to add rational numbers in decimal form, does not apply to sums that involve irrational numbers.*
- Many of the numbers we have been working with are irrational numbers, such as $\sqrt{2}$, $e$, $\log(2)$, and $\pi$, meaning that we cannot write them as fractions. One of the key distinctions between rational and irrational numbers is that rational numbers can be expressed as a decimal that either terminates (such as $\frac{1}{8}=0.125$) or repeats infinitely (such as $\frac{1}{9}=0.11111\ldots$). Irrational numbers cannot be exactly represented by a decimal

expansion because the digits to the right of the decimal point never end and never repeat predictably. The best we can do to represent irrational numbers with a decimal expansion is to find an approximation.

- For example, consider the number $\pi$. What is the value of $\pi$?
  - $3.14$ *(Students may answer this question with varying degrees of accuracy.)*
- Is that the exact value of $\pi$?
  - *No. We cannot write a decimal expansion for the exact value of $\pi$.*
- What is the *last digit* of $\pi$?
  - *Since the decimal expansion for $\pi$ never terminates, there is no last digit of $\pi$.*

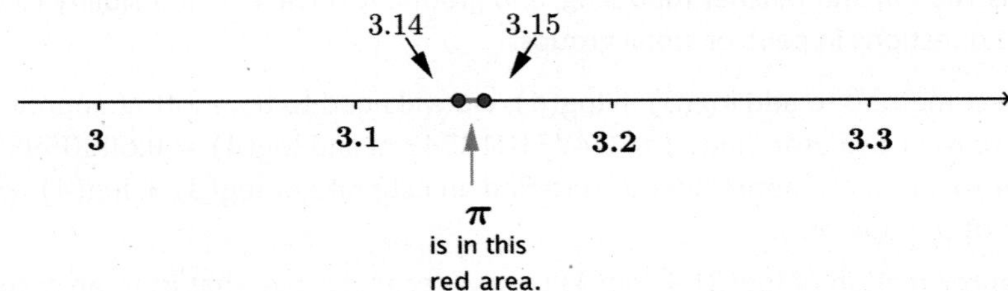


- Where is $\pi$ on the number line?
  - *The number $\pi$ is between $3.14$ and $3.15$.*
- What kind of numbers are $3.14$ and $3.15$?
  - *Rational numbers.*
- Can we get better under and over estimates than that?
  - *Yes, since $\pi \approx 3.14159$, we know that $3.1415 < \pi < 3.1416$.*

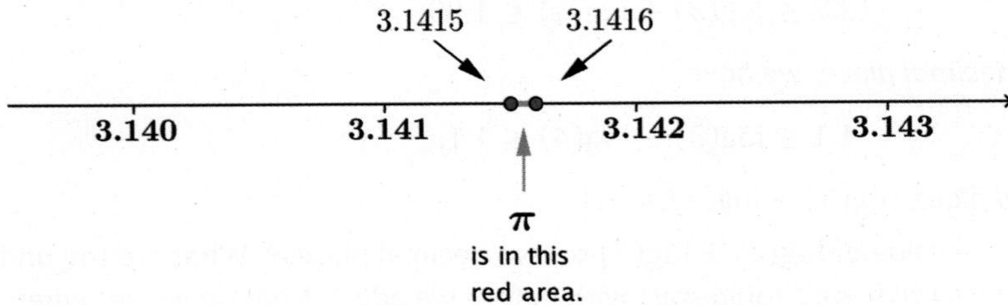


- What kind of numbers are $3.1415$ and $3.1416$?
  - *Rational numbers.*
- Can we find rational numbers with 10 digits that are good upper and lower bounds for $\pi$? How?
  - *Yes. Just take the expansion for $\pi$ on the calculator and round up and down.*
- So, how can we add two irrational numbers? Let's turn our attention back to logarithms.
- The numbers $\log(3)$ and $\log(4)$ are examples of irrational numbers. The calculator says that $\log(3) = 0.4771212547$ and $\log(4) = 0.6020599913$. What is wrong with those two statements?
  - *Since $\log(3)$ and $\log(4)$ are irrational numbers, their decimal expansions do not terminate so the calculator gives approximations and not exact values.*

- So, we really should write $\log(3) = 0.4771212547\ldots$ and $\log(4) = 0.6020599913\ldots$. This means there are more digits that we cannot see. What happens when we try to add these decimal expansions together numerically?
  - *There is no last digit for us to use to start our addition algorithm, which moves from right to left. We cannot even start adding these together using our usual method.*
- So, our standard algorithm for adding numbers fails. How can we add $\log(3) + \log(4)$?

## Example 1 (8 minutes)

Begin these examples with direct instruction and teacher modeling, and gradually release responsibility to the students when they are ready to tackle these questions in pairs or small groups.

- Since we do not have a direct method to add $\log(3) + \log(4)$, we will need to try another approach. Remember that according to the calculator, $\log(3) = 0.4771212547\ldots$ and $\log(4) = 0.6020599913\ldots$. While we could use the calculator to add these approximations to find an estimate of $\log(3) + \log(4)$, we are interested in making sense of the operation.
- What if we just need an approximation of $\log(3) + \log(4)$ to one decimal place, that is, to an accuracy of $10^{-1}$?

  If we do not need more accuracy than that, we can use $\log(3) \approx 0.477$ and $\log(4) \approx 0.602$. Then,

$$0.47 \leq \log(3) \leq 0.48;$$
$$0.60 \leq \log(4) \leq 0.61.$$

MP.3

- Based on these inequalities, what statement can we make about the sum $\log(3) + \log(4)$? Explain why you believe your statement is correct.
  - *Adding terms together, we have*

$$1.07 \leq \log(3) + \log(4) \leq 1.09.$$

*Rounding to one decimal place, we have*

$$1.1 \leq \log(3) + \log(4) \leq 1.1.$$

*So, to one decimal place,* $\log(3) + \log(4) \approx 1.1$.

- What if we wanted to find the value of $\log(3) + \log(4)$ to two decimal places? What are the under and over estimates for $\log(3)$ and $\log(4)$ that we should start with before we add? What do we get when we add them together?
  - *We should start with*

$$0.477 \leq \log(3) \leq 0.478;$$
$$0.602 \leq \log(4) \leq 0.603.$$

*Then we have*

$$1.079 \leq \log(3) + \log(4) \leq 1.081.$$

*To two decimal places,* $\log(3) + \log(4) \approx 1.08$.

- Now, find the value of $\log(3) + \log(4)$ to five decimal places, that is, to an accuracy of $10^{-5}$.
    - *We should start with*

$$0.477121 \leq \log(3) \leq 0.477122;$$
$$0.602059 \leq \log(4) \leq 0.602060.$$

    *Then we have*

$$1.079180 \leq \log(3) + \log(4) \leq 1.079182.$$

    *To five decimal places,* $\log(3) + \log(4) \approx 1.07918$.

- Now, find the value of $\log(3) + \log(4)$ to eight decimal places, that is, to an accuracy of $10^{-8}$.
    - *We should start with*

$$0.477121254 \leq \log(3) \leq 0.477121255$$
$$0.602059991 \leq \log(4) \leq 0.602059992.$$

    *Then we have*

$$1.079181245 \leq \log(3) + \log(4) \leq 1.079181247.$$

    *To eight decimal places,* $\log(3) + \log(4) \approx 1.07918125$.

- Notice that we are *squeezing* the actual value of $\log(3) + \log(4)$ between two rational numbers. Since we know how to plot a rational number on the number line (in theory, anyway), we can get really close to the location of the irrational number $\log(3) + \log(4)$ by squeezing it between two rational numbers.

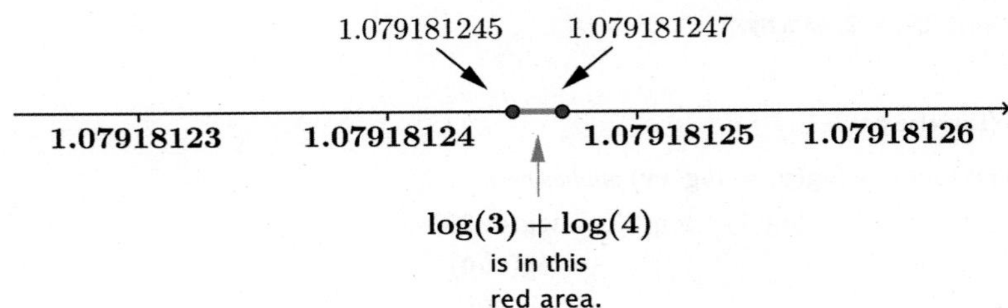


- Could we keep going? If we knew enough digits of the decimal expansions of $\log(3)$ and $\log(4)$, could we find an approximation of $\log(3) + \log(4)$ to 20 decimal places? Or 50 decimal places? Or 1,000 decimal places?
    - *Yes. There is no reason that we cannot continue this process, provided we know enough digits of the original irrational numbers* $\log(3)$ *and* $\log(4)$.
- Summarize what you have learned from this example in your notebook. (Allow students a minute or so to record the main points of this example.)

### Exercises 1–5 (8 minutes)

Have students complete these exercises in pairs or small groups. Expect students to react with surprise to the results: In the previous example, the sum $\log(3) + \log(4)$ was irrational, so the decimal expansion never terminated. In this set of exercises, the sum $\log(4) + \log(25)$ is rational with the exact value 2, and at each step, their estimate of the sum will be exact.

Exercises 1–5

1. According to the calculator, $\log(4) = 0.6020599913\ldots$ and $\log(25) = 1.3979400087\ldots$. Find an approximation of $\log(4) + \log(25)$ to one decimal place, that is, to an accuracy of $10^{-1}$.

$$0.60 < \log(4) < 0.61$$
$$1.39 < \log(25) < 1.40$$
$$1.99 < \log(4) + \log(25) < 2.01$$
$$\log(4) + \log(25) \approx 2.0$$

2. Find the value of $\log(4) + \log(25)$ to an accuracy of $10^{-2}$.

$$0.602 < \log(4) < 0.603$$
$$1.397 < \log(25) < 1.398$$
$$1.999 < \log(4) + \log(25) < 2.001$$
$$\log(4) + \log(25) \approx 2.00$$

3. Find the value of $\log(4) + \log(25)$ to an accuracy of $10^{-8}$.

$$0.602059991 \leq \log(4) \leq 0.602059992$$
$$1.397940008 \leq \log(25) \leq 1.397940009$$
$$1.999999999 \leq \log(4) + \log(25) \leq 2.000000001$$
$$\log(4) + \log(25) \approx 2.00000000$$

MP.3

4. Make a conjecture: Is $\log(4) + \log(25)$ a rational or an irrational number?

*It appears that* $\log(4) + \log(25) = 2$*, exactly.*

5. Why is your conjecture in Exercise 4 true?

*The logarithm rule that says* $\log(x) + \log(y) = \log(xy)$ *applies here.*

$$\begin{aligned}\log(4) + \log(25) &= \log(4 \cdot 25)\\ &= \log(100)\\ &= \log(10^2)\\ &= 2\end{aligned}$$

## Discussion (3 minutes)

- We have seen how we can squeeze the sum of two irrational numbers between rational numbers and get an approximation of the sum to whatever accuracy we want. What about multiplication?

MP.3

- Make a conjecture: Without actually calculating it, what is the value of $\log(4) \cdot \log(25)$?

Allow students time to discuss this question with a partner and write down a response before allowing students to put forth their ideas to the class. This would be an ideal time to use personal white boards, if you have them, for students to record and display their conjectures. See that students have a written record of this conjecture that will be disproven in the next set of exercises.

  - *Conjectures will vary; some might be* $\log(4) \cdot \log(25) = \log(29)$*, or* $\log(4) \cdot \log(25) = \log(100)$*.*

## Exercises 6–8 (6 minutes)

**Exercises 6–8**

**Remember that the calculator gives the following values: $\log(4) = 0.6020599913$ ... and $\log(25) = 1.3979400087$ ....**

6. **Find the value of $\log(4) \cdot \log(25)$ to three decimal places.**

$$0.6020 \leq \log(4) \leq 0.6021$$
$$1.3979 \leq \log(25) \leq 1.3980$$

$$0.8415358 \leq \log(4) \cdot \log(25) \leq 0.8417358$$

$$\log(4) \cdot \log(25) \approx 0.842$$

7. **Find the value of $\log(4) \cdot \log(25)$ to five decimal places.**

$$0.602059 \leq \log(4) \leq 0.602060$$
$$1.397940 \leq \log(25) \leq 1.397941$$

$$0.8416423585 \leq \log(4) \cdot \log(25) \leq 0.8416443585$$

$$\log(4) \cdot \log(25) \approx 0.84164$$

8. **Does your conjecture from the above discussion appear to be true?**

*No. The work from Exercise 6 shows that* $\log(4) \cdot \log(25) \neq \log(29)$, *and* $\log(4) \cdot \log(25) \neq \log(100)$.

## Closing (3 minutes)

Ask students to respond to the following prompts independently, and then have them share their responses with a partner. After students have a chance to write and discuss, go through the key points in the Lesson Summary below.

- List five rational numbers.
  - *Student responses will vary; possible responses include* $1, 10, \frac{3}{5}, 17$, *and* $-\frac{42}{13}$.
- List five irrational numbers.
  - *Student responses will vary; possible responses include* $\sqrt{3}, \pi, e, 1 + \sqrt{2}$, *and* $\pi^3$.
- Is 0 a rational or irrational number? Explain how you know.
  - *Since* $0$ *is an integer, we can write* $0 = \frac{0}{1}$, *which is a quotient of integers. Thus,* $0$ *is a rational number.*
- If a number is given as a decimal, how can you tell if it is a rational or an irrational number?
  - *If the decimal representation terminates or repeats at some point, then the number is rational and can be expressed as the quotient of two integers. Otherwise, the number is irrational.*

**Lesson Summary**

- **Irrational numbers occur naturally and frequently.**
- **The $n^{\text{th}}$ roots of most integers and rational numbers are irrational.**
- **Logarithms of most positive integers or positive rational numbers are irrational.**
- **We can locate an irrational number on the number line by trapping it between lower and upper approximations. The infinite process of squeezing the irrational number in smaller and smaller intervals locates exactly where the irrational number is on the number line.**
- **We can perform arithmetic operations such as addition and multiplication with irrational numbers using lower and upper approximations and squeezing the result of the operation in smaller and smaller intervals between two rational approximations to the result.**

## Exit Ticket (5 minutes)

Name ______________________________________________ Date ____________________

# Lesson 16: Rational and Irrational Numbers

## Exit Ticket

The decimal expansion of $e$ and $\sqrt{5}$ are given below.

$$e \approx 2.71828182 \ldots$$

$$\sqrt{5} \approx 2.23606797 \ldots$$

a. Find an approximation of $\sqrt{5} + e$ to three decimal places. Do not use a calculator.

b. Explain how you can locate $\sqrt{5} + e$ on the number line. How is this different from locating $2.6 + 2.7$ on the number line?

## Exit Ticket Sample Solutions

**The decimal expansion of $e$ and $\sqrt{5}$ are given below.**

$$e \approx 2.71828182\ldots$$
$$\sqrt{5} \approx 2.23606797\ldots$$

a. **Find an approximation of $\sqrt{5}+e$ to three decimal places. Do not use a calculator.**

$$2.2360 \leq \sqrt{5} \leq 2.2361$$
$$2.7182 \leq e \leq 2.7183$$
$$4.9542 \leq \sqrt{5}+e \leq 4.9544$$

*Thus, to three decimal places, $\sqrt{5}+e \approx 4.954$.*

b. **Explain how you can locate $\sqrt{5}+e$ on the number line. How is this different from locating $2.6+2.7$ on the number line?**

*We cannot locate $\sqrt{5}+e$ precisely on the number line because the sum is irrational, but we can get as close to it as we want by squeezing it between two rational numbers, $r_1$ and $r_2$, that differ only in the last decimal place, $r_1 \leq \sqrt{5}+e \leq r_2$. Since we can locate rational numbers on the number line, we can get arbitrarily close to the true location of $\sqrt{5}+e$ by starting with more and more accurate decimal representations of $\sqrt{5}$ and $e$. This differs from pinpointing the location of sums of rational numbers because we can precisely locate the sum $2.6+2.7=5.3$ by dividing the interval $[5,6]$ into $10$ parts of equal length $0.1$. Then, the point $5.3$ is located exactly at the point between the third and fourth parts.*

## Problem Set Sample Solutions

1. **Given that $\sqrt{5} \approx 2.2360679775$ and $\pi \approx 3.1415926535$, find the sum $\sqrt{5}+\pi$ to an accuracy of $10^{-8}$, without using a calculator.**

*From the estimations we are given, we know that*

$$2.236067977 < \sqrt{5} < 2.236067978$$
$$3.141592653 < \pi < 3.141592654.$$

*Adding these together gives*

$$5.377660630 < \sqrt{5}+\pi < 5.377660632.$$

*Then, to an accuracy of $10^{-8}$, we have*

$$\sqrt{5}+\pi \approx 5.37766063.$$

2. **Put the following numbers in order from least to greatest.**

$$\sqrt{2}, \pi, 0, e, \frac{22}{7}, \frac{\pi^2}{3}, 3.14, \sqrt{10}$$

$$0, \sqrt{2}, e, 3.14, \pi, \frac{22}{7}, \sqrt{10}, \frac{\pi^2}{3}$$

3. **Find a rational number between the specified two numbers.**

   a. $\frac{4}{13}$ and $\frac{5}{13}$

   *Many answers are possible. Since* $\frac{4}{13} = \frac{8}{26}$ *and* $\frac{5}{13} = \frac{10}{26}$, *we know that* $\frac{4}{13} < \frac{9}{26} < \frac{5}{13}$.

   b. $\frac{3}{8}$ and $\frac{5}{9}$

   *Many answers are possible. Since* $\frac{3}{8} = \frac{27}{72}$ *and* $\frac{5}{9} = \frac{40}{72}$, *we know that* $\frac{30}{72} = \frac{5}{12}$ *is between* $\frac{3}{8}$ *and* $\frac{5}{9}$.

   c. $1.7299999$ and $1.73$

   *Many answers are possible.* $1.7299999 < 1.72999995 < 1.73$.

   d. $\frac{\sqrt{2}}{7}$ and $\frac{\sqrt{2}}{9}$

   *Many answers are possible. Since* $\frac{\sqrt{2}}{9} \approx 0.157135$ *and* $\frac{\sqrt{2}}{7} \approx 0.202031$, *we know* $\frac{\sqrt{2}}{9} < 0.2 < \frac{\sqrt{2}}{7}$.

   e. $\pi$ and $\sqrt{10}$

   *Many answers are possible. Since* $\pi \approx 3.14159$ *and* $\sqrt{10} \approx 3.16228$, *we know* $\pi < 3.15 < \sqrt{10}$.

4. **Knowing that $\sqrt{2}$ is irrational, find an irrational number between $\frac{1}{2}$ and $\frac{5}{9}$.**

   *One such number is* $r\sqrt{2}$, *so* $\frac{1}{2} < r\sqrt{2} < \frac{5}{9}$. *Then we have* $\frac{1}{2\sqrt{2}} < r < \frac{5}{9\sqrt{2}}$. *Since* $\frac{1}{2\sqrt{2}} \approx 0.3536$ *and* $\frac{5}{9\sqrt{2}} \approx 0.3929$, *we can let* $r = 0.36$. *Then,* $0.36\sqrt{2}$ *is an irrational number between* $\frac{1}{2}$ *and* $\frac{5}{9}$.

5. **Give an example of an irrational number between $e$ and $\pi$.**

   *Many answers are possible, such as* $\frac{\pi+e}{2}$, $\sqrt{\pi e}$, *or* $\frac{10}{11}\pi$.

6. **Given that $\sqrt{2}$ is irrational, which of the following numbers are irrational?**

$$\frac{\sqrt{2}}{2}, 2+\sqrt{2}, \frac{\sqrt{2}}{2\sqrt{2}}, \frac{2}{\sqrt{2}}, \left(\sqrt{2}\right)^2$$

   *The numbers* $\frac{\sqrt{2}}{2}$, $2+\sqrt{2}$, *and* $\frac{2}{\sqrt{2}} = \sqrt{2}$ *are irrational. (Note that* $\frac{\sqrt{2}}{2\sqrt{2}} = \frac{1}{2}$ *and* $\left(\sqrt{2}\right)^2 = 2$ *are rational numbers.)*

7. **Given that $\pi$ is irrational, which of the following numbers are irrational?**

$$\frac{\pi}{2}, \frac{\pi}{2\pi}, \sqrt{\pi}, \pi^2$$

   *The numbers* $\frac{\pi}{2}$, $\sqrt{\pi}$, *and* $\pi^2$ *are irrational.*

8. **Which of the following numbers are irrational?**

$$1, 0, \sqrt{5}, \sqrt[3]{64}, e, \pi, \frac{\sqrt{2}}{2}, \frac{\sqrt{8}}{\sqrt{2}}, \cos\left(\frac{\pi}{3}\right), \sin\left(\frac{\pi}{3}\right)$$

*The numbers $\sqrt{5}$, $e$, $\pi$, $\frac{\sqrt{2}}{2}$, $\sin\left(\frac{\pi}{3}\right)$ are irrational.*

9. **Find two irrational numbers $x$ and $y$ so that their average is rational.**

*If $x = 1 + \sqrt{2}$ and $y = 3 - \sqrt{2}$, then $\frac{x+y}{2} = \frac{1}{2}\left((1+\sqrt{2}) + (3-\sqrt{2})\right) = 2$; so, the average of $x$ and $y$ is rational.*

10. **Suppose that $\frac{2}{3}x$ is an irrational number. Explain how you know that $x$ must be an irrational number. (Hint: What would happen if there were integers $a$ and $b$ so that $x = \frac{a}{b}$?)**

*If $x$ is rational, then there are integers $a$ and $b$ so that $x = \frac{a}{b}$. Then $\frac{2}{3}x = \frac{2a}{3b}$ is rational. Since we know that $\frac{2}{3}x$ is irrational, we cannot have $x = \frac{a}{b}$. Thus, $x$ must be an irrational number.*

11. **If $r$ and $s$ are rational numbers, prove that $r + s$ and $r - s$ are also rational numbers.**

*If $r$ and $s$ are rational numbers, then there exist integers $a$, $b$, $c$, $d$ with $b \neq 0$ and $d \neq 0$ so that $r = \frac{a}{b}$ and $s = \frac{c}{d}$. Then,*

$$\begin{aligned} r + s &= \frac{a}{b} + \frac{c}{d} \\ &= \frac{ad}{bd} + \frac{bc}{bd} \\ &= \frac{ad + bc}{bd} \end{aligned}$$

$$\begin{aligned} r - s &= \frac{a}{b} - \frac{c}{d} \\ &= \frac{ad - bc}{bd}. \end{aligned}$$

*Since $ad + bc$, $ad - bc$, and $bd$ are integers, $r + s$ and $r - s$ are rational numbers.*

12. **If $r$ is a rational number and $x$ is an irrational number, determine whether the following numbers are always rational, sometimes rational, or never rational. Explain how you know.**

   a. $r + x$

   *If $r + x = y$ and $y$ is rational, then $r - y = -x$ would be rational by Problem 10. Since $x$ is irrational, we know $-x$ is irrational, so $y$ cannot be rational. Thus, the sum $r + x$ is never rational.*

   b. $r - x$

   *If $r - x = y$ and $y$ is rational, then $r - y = x$ would be rational by Problem 10. Since $x$ is irrational, $y$ cannot be rational. Thus, the difference $r - x$ is never rational.*

   c. $rx$

   *If $rx = y$, $r \neq 0$, and $y$ is rational, then there are integers $a$, $b$, $c$, $d$ with $a \neq 0$, $b \neq 0$, and $d \neq 0$ so that $r = \frac{a}{b}$ and $y = \frac{c}{d}$. Then $x = \frac{y}{r} = \frac{cb}{ad}$, so $x$ is rational. Since $x$ was not rational, the only way that $rx$ can be rational is if $r = 0$. Thus, $rx$ is sometimes rational (in only one case).*

d. $x^r$

*If $x = \sqrt[r]{k}$ for some rational number $k$, then $x^r = k$ is rational. For example, $\left(\sqrt{5}\right)^2 = 5$ is rational. But, $\pi^r$ is never rational for any exponent $r$, so $x^r$ is sometimes rational.*

**13. If $x$ and $y$ are irrational numbers, determine whether the following numbers are always rational, sometimes rational, or never rational. Explain how you know.**

a. $x + y$

*This is sometimes rational. For example, $\pi + \sqrt{2}$ is irrational, but $\left(1 + \sqrt{2}\right) + \left(1 - \sqrt{2}\right) = 2$ is rational.*

b. $x - y$

*This is sometimes rational. For example, $\pi - \sqrt{2}$ is irrational, but $\left(1 + \sqrt{2}\right) + \left(2 + \sqrt{2}\right) = 3$ is rational.*

c. $xy$

*This is sometimes rational. For example, $\pi\sqrt{2}$ is irrational, but $\sqrt{2} \cdot \sqrt{8} = 4$ is rational.*

d. $\frac{x}{y}$

*This is sometimes rational. For example, $\frac{\pi}{\sqrt{2}}$ is irrational, but $\frac{\sqrt{8}}{\sqrt{2}} = 2$ is rational.*

# Lesson 17: Graphing the Logarithm Function

## Student Outcomes

- Students graph the functions $f(x) = \log(x)$, $g(x) = \log_2(x)$, and $h(x) = \ln(x)$ by hand and identify key features of the graphs of logarithmic functions.

## Lesson Notes

In this lesson, students work in pairs or small groups to generate graphs of $f(x) = \log(x)$, $g(x) = \log_2(x)$, or $h(x) = \log_5(x)$. Students compare the graphs of these three functions to derive the key features of graphs of general logarithmic functions for bases $b > 1$. Tables of function values are provided so that calculators are not needed in this lesson; all graphs should be drawn by hand. Students will relate the domain of the logarithmic functions to the graph in accordance with **F-IF.B.5**. After the graphs are generated and conclusions drawn about their properties, students use properties of logarithms to find additional points on the graphs. Continue to rely on the definition of the logarithm, which was stated in Lesson 8, and properties of logarithms developed in Lessons 12 and 13:

> **LOGARITHM:** If three numbers, $L$, $b$, and $x$ are related by $x = b^L$, then $L$ is the *logarithm base $b$ of $x$*, and we write $\log_b(x)$. That is, the value of the expression $L = \log_b(x)$ is the power of $b$ needed to obtain $x$. Valid values of $b$ as a base for a logarithm are $0 < b < 1$ and $b > 1$.

## Classwork

### Opening (1 minute)

Divide the students into pairs or small groups; ideally, the number of groups formed will be a multiple of three. Assign the function $f(x) = \log(x)$ to one-third of the groups, and refer to these groups as the 10-team. Assign the function $g(x) = \log_2(x)$ to the second third of the groups, and refer to these groups as the 2-team. Assign the function $h(x) = \log_5(x)$ to the remaining third of the groups, and refer to these groups as the 5-team.

*Scaffolding:*

- Struggling students will benefit from watching the teacher model the process of plotting points.
- Consider assigning struggling students to the 2-team because the function values are integers.
- Alternatively, assign advanced students to the 2-team and ask them to generate the graph of $y = \log_2(x)$ without the given table.

### Opening Exercise (8 minutes)

While student groups are creating the graphs and responding to the prompts that follow, circulate and observe student work. Select three groups to present their graphs and results at the end of the exercise.

**Opening Exercise**

**Graph the points in the table for your assigned function $f(x) = \log(x)$, $g(x) = \log_2(x)$, or $h(x) = \log_5(x)$ for $0 < x \leq 16$. Then, sketch a smooth curve through those points and answer the questions that follow.**

| 10-team $f(x) = \log(x)$ | |
|---|---|
| $x$ | $f(x)$ |
| 0.0625 | −1.20 |
| 0.125 | −0.90 |
| 0.25 | −0.60 |
| 0.5 | −0.30 |
| 1 | 0 |
| 2 | 0.30 |
| 4 | 0.60 |
| 8 | 0.90 |
| 16 | 1.20 |

| 2-team $g(x) = \log_2(x)$ | |
|---|---|
| $x$ | $g(x)$ |
| 0.0625 | −4 |
| 0.125 | −3 |
| 0.25 | −2 |
| 0.5 | −1 |
| 1 | 0 |
| 2 | 1 |
| 4 | 2 |
| 8 | 3 |
| 16 | 4 |

| 5-team $h(x) = \log_5(x)$ | |
|---|---|
| $x$ | $h(x)$ |
| 0.0625 | −1.72 |
| 0.125 | −1.29 |
| 0.25 | −0.86 |
| 0.5 | −0.43 |
| 1 | 0 |
| 2 | 0.43 |
| 4 | 0.86 |
| 8 | 1.29 |
| 16 | 1.72 |

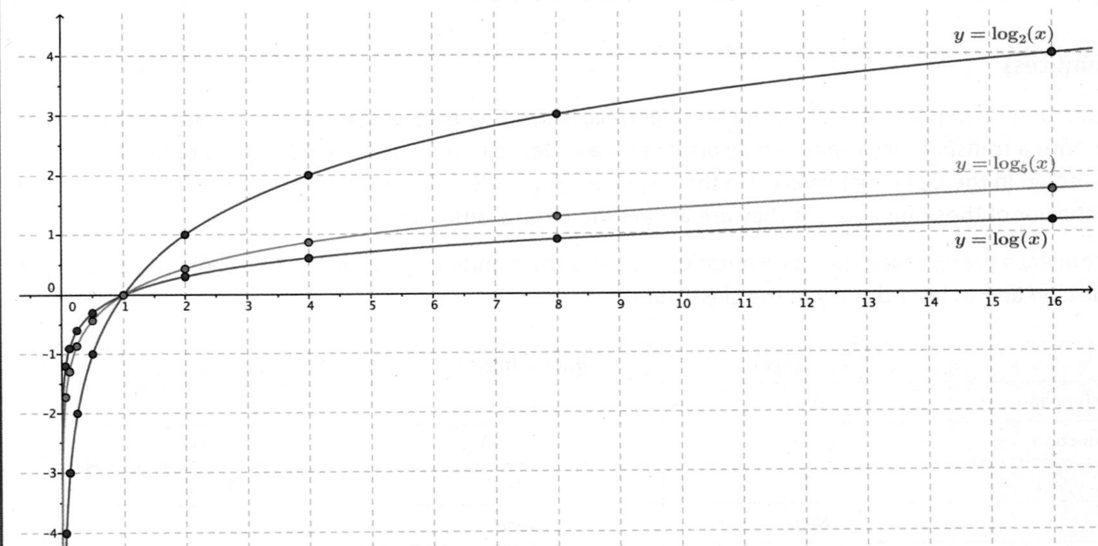


a. **What does the graph indicate about the domain of your function?**

*The domain of each of these functions is the positive real numbers, which can be stated as* $(0, \infty)$.

b. **Describe the $x$-intercepts of the graph.**

*There is one $x$-intercept at* $1$.

c. **Describe the $y$-intercepts of the graph.**

*There are no $y$-intercepts of this graph.*

d. **Find the coordinates of the point on the graph with $y$-value 1.**

*For the 10-team, this is* $(10, 1)$. *For the 2-team, this is* $(2, 1)$. *For the 5-team, this is* $(5, 1)$.

e. **Describe the behavior of the function as $x \to 0$.**

*As $x \to 0$, the function values approach negative infinity; that is, $f(x) \to -\infty$. The same is true for the functions $g$ and $h$.*

f. **Describe the end behavior of the function as $x \to \infty$.**

*As $x \to \infty$, the function values slowly increase. That is, $f(x) \to \infty$. The same is true for the functions $g$ and $h$.*

g. **Describe the range of your function.**

*The range of each of these functions is all real numbers, $(-\infty, \infty)$.*

h. **Does this function have any relative maxima or minima? Explain how you know.**

*Since the function values continue to increase, and there are no peaks or valleys in the graph, the function has no relative maxima or minima.*

## Presentations (5 minutes)

Select three groups of students to present each of the three graphs, projecting each graph through a document camera or copying the graph onto a transparency sheet and displaying on an overhead projector. Ask students to point out the key features they identified in the Opening Exercise on the displayed graph. If students do not mention it, emphasize that the long-term behavior of these functions is they are always increasing, although very slowly.

As representatives from each group make their presentations, record their findings on a chart. This chart can be used to help summarize the lesson and to later display in the classroom.

| | $f(x) = \log(x)$ | $g(x) = \log_2(x)$ | $h(x) = \log_5(x)$ |
|---|---|---|---|
| Domain of the function | $(0, \infty)$ | $(0, \infty)$ | $(0, \infty)$ |
| Range of the function | $(-\infty, \infty)$ | $(-\infty, \infty)$ | $(-\infty, \infty)$ |
| $x$-intercept | $1$ | $1$ | $1$ |
| $y$-intercept | *None* | *None* | *None* |
| Point with $y$-value 1 | $(10, 1)$ | $(2, 1)$ | $(5, 1)$ |
| Behavior as $x \to 0$ | $f(x) \to -\infty$ | $g(x) \to -\infty$ | $h(x) \to -\infty$ |
| End behavior as $x \to \infty$ | $f(x) \to \infty$ | $g(x) \to \infty$ | $h(x) \to \infty$ |

## Discussion (5 minutes)

Debrief the Opening Exercise by asking students to generalize the key features of the graphs $y = \log_b(x)$. If possible, display the graph of all three functions $f(x) = \log(x)$, $g(x) = \log_2(x)$, and $h(x) = \log_5(x)$ together on the same axes during this discussion.

We saw in Lesson 5 that the expression $2^x$ is defined for all real numbers $x$; therefore, the range of the function $g(x) = \log_2(x)$ is all real numbers. Likewise, the expressions $10^x$ and $5^x$ are defined for all real numbers $x$, so the range of the functions $f$ and $h$ are all real numbers. Notice that since the range is all real numbers in each case, there must be logarithms that are irrational. We saw examples of such logarithms in Lesson 16.

- What are the domain and range of the logarithm functions?
  - *The domain is the positive real numbers, and the range is all real numbers.*
- What do the three graphs of $f(x) = \log(x)$, $g(x) = \log_2(x)$, and $h(x) = \log_5(x)$ have in common?
  - *The graphs all cross the $x$-axis at $(1, 0)$.*
  - *None of the graphs intersect the $y$-axis.*
  - *They have the same end behavior as $x \to \infty$, and they have the same behavior as $x \to 0$.*
  - *The functions all increase quickly for $0 < x < 1$, then increase more and more slowly.*
- What do you expect the graph of $y = \log_3(x)$ will look like?
  - *It will look just like the other graphs, except that it will lie between the graphs of $y = \log_2(x)$ and $y = \log_5(x)$ because $2 < 3 < 5$.*
- What do you expect the graph of $y = \log_b(x)$ will look like for any number $b > 1$?
  - *It will have the same key features of the other graphs of logarithmic functions. As the value of b increases, the graph will flatten as $x \to \infty$.*

## Exercise 1 (8 minutes)

Keep students in the same groups for this exercise. Students will plot points and sketch the graph of $y = \log_{\frac{1}{b}}(x)$ for $b = 10$, $b = 2$, or $b = 5$, depending on whether they are on the 10-team, the 2-team, or the 5-team. Then, students will observe the relationship between their two graphs, justify the relationship using properties of logarithms, and generalize the observed relationship to graphs of $y = \log_b(x)$ and $y = \log_{\frac{1}{b}}(x)$ for $b > 0$, $b \neq 1$.

**Exercises**

**1. Graph the points in the table for your assigned function $r(x) = \log_{\frac{1}{10}}(x)$, $s(x) = \log_{\frac{1}{2}}(x)$, or $t(x) = \log_{\frac{1}{5}}(x)$ for $0 < x \leq 16$. Then, sketch a smooth curve through those points, and answer the questions that follow.**

| 10-team $r(x) = \log_{\frac{1}{10}}(x)$ | |
|---|---|
| $x$ | $r(x)$ |
| $0.0625$ | $1.20$ |
| $0.125$ | $0.90$ |
| $0.25$ | $0.60$ |
| $0.5$ | $0.30$ |
| $1$ | $0$ |
| $2$ | $-0.30$ |
| $4$ | $-0.60$ |
| $8$ | $-0.90$ |
| $16$ | $-1.20$ |

| 2-team $s(x) = \log_{\frac{1}{2}}(x)$ | |
|---|---|
| $x$ | $s(x)$ |
| $0.0625$ | $4$ |
| $0.125$ | $3$ |
| $0.25$ | $2$ |
| $0.5$ | $1$ |
| $1$ | $0$ |
| $2$ | $-1$ |
| $4$ | $-2$ |
| $8$ | $-3$ |
| $16$ | $-4$ |

| e-team $t(x) = \log_{\frac{1}{5}}(x)$ | |
|---|---|
| $x$ | $t(x)$ |
| $0.0625$ | $1.72$ |
| $0.125$ | $1.29$ |
| $0.25$ | $0.86$ |
| $0.5$ | $0.43$ |
| $1$ | $0$ |
| $2$ | $-0.43$ |
| $4$ | $-0.86$ |
| $8$ | $-1.29$ |
| $16$ | $-1.72$ |

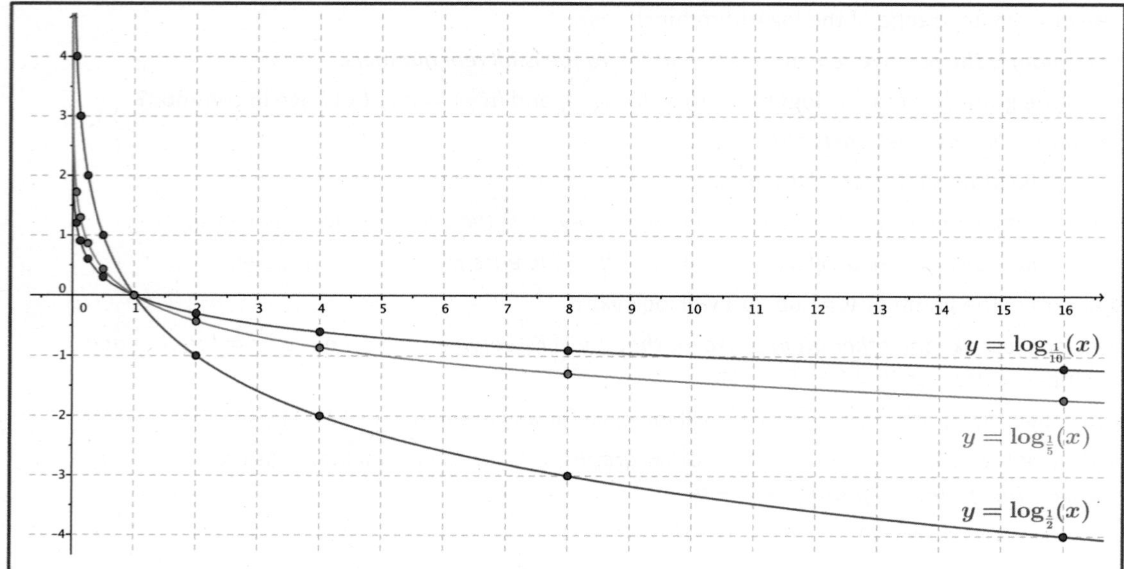


a. **What is the relationship between your graph in the Opening Exercise and your graph from this exercise?**

***The second graph is the reflection of the graph in the Opening Exercise across the $x$-axis.***

b. **Why does this happen? Use the change of base formula to justify what you have observed in part (a).**

***Using the change of base formula, we have*** $\log_{\frac{1}{2}}(x) = \frac{\log_2(x)}{\log_2\left(\frac{1}{2}\right)}$. ***Since*** $\log_2\left(\frac{1}{2}\right) = -1$, ***we have*** $\log_{\frac{1}{2}}(x) = \frac{\log_2(x)}{-1}$, ***so*** $\log_{\frac{1}{2}}(x) = -\log_2(x)$. ***Thus, the graph of*** $y = \log_{\frac{1}{2}}(x)$ ***is the reflection of the graph of*** $y = \log_2(x)$ ***across the $x$-axis.***

*Scaffolding:*

- Students struggling with the comparison of graphs may find it easier to draw the graphs on transparent plastic sheets and compare them that way.

## Discussion (4 minutes)

Ask students from each team to share their graphs results from part (a) of Exercise 1 with the class. During their presentations, complete the chart below.

| | $r(x) = \log_{\frac{1}{10}}(x)$ | $s(x) = \log_{\frac{1}{2}}(x)$ | $t(x) = \log_{\frac{1}{5}}(x)$ |
|---|---|---|---|
| Domain of the function | $(0, \infty)$ | $(0, \infty)$ | $(0, \infty)$ |
| Range of the function | $(-\infty, \infty)$ | $(-\infty, \infty)$ | $(-\infty, \infty)$ |
| $x$-intercept | $1$ | $1$ | $1$ |
| $y$-intercept | *None* | *None* | *None* |
| Point with $y$-value $-1$ | $(10, -1)$ | $(2, -1)$ | $(5, -1)$ |
| Behavior as $x \to 0$ | $r(x) \to \infty$ | $s(x) \to \infty$ | $t(x) \to \infty$ |
| End behavior as $x \to \infty$ | $r(x) \to \infty$ | $s(x) \to -\infty$ | $t(x) \to -\infty$ |

Then proceed to hold the following discussion.

- From what we have seen of these three sets of graphs of functions, can we state the relationship between the graphs of $y = \log_b(x)$ and $y = \log_{\frac{1}{b}}(x)$, for $b \neq 1$?
    - *If $b \neq 1$, then the graphs of $y = \log_b(x)$ and $y = \log_{\frac{1}{b}}(x)$ are reflections of each other across the $x$-axis.*
- Describe the key features of the graph of $y = \log_b(x)$ for $0 < b < 1$.
    - *The graph crosses the $x$-axis at $(1, 0)$.*
    - *The graph does not intersect the $y$-axis.*
    - *The graph passes through the point $(b, -1)$.*
    - *As $x \to 0$, the function values increase quickly; that is, $f(x) \to \infty$.*
    - *As $x \to \infty$, the function values continue to decrease; that is, $f(x) \to -\infty$.*
    - *There are no relative maxima or relative minima.*

## Exercises 2–3 (6 minutes)

Keep students in the same groups for this set of exercises. Students will plot points and sketch the graph of $y = \log_b(bx)$ for $b = 10$, $b = 2$, or $b = 5$, depending on whether they are on the 10-team, the 2-team, or the 5-team. Then, students will observe the relationship between their two graphs, justify the relationship using properties of logarithms, and generalize the observed relationship to graphs of $y = \log_b(x)$ and $y = \log_b(x)$ for $b > 0, b \neq 1$. If there is time at the end of these exercises, consider using GeoGebra or other dynamic geometry software to demonstrate the property illustrated in Exercise 3 below by graphing $y = \log_2(x)$, $y = \log_2(2x)$, and $y = 1 + \log_2(x)$ on the same axes.

Consider having students graph these functions on the same axes as used in the Opening Exercise.

**2. In general, what is the relationship between the graph of a function $y = f(x)$ and the graph of $y = f(kx)$ for a constant $k$?**

***The graph of $y = f(kx)$ is a horizontal scaling of the graph of $y = f(x)$.***

**3. Graph the points in the table for your assigned function $u(x) = \log(10x)$, $v(x) = \log_2(2x)$, or $w(x) = \log_5(5x)$ for $0 < x \leq 16$. Then sketch a smooth curve through those points, and answer the questions that follow.**

| 10-team $u(x) = \log(10x)$ | |
|---|---|
| $x$ | $u(x)$ |
| $0.0625$ | $-0.20$ |
| $0.125$ | $0.10$ |
| $0.25$ | $0.40$ |
| $0.5$ | $0.70$ |
| $1$ | $1$ |
| $2$ | $1.30$ |
| $4$ | $1.60$ |
| $8$ | $1.90$ |
| $16$ | $2.20$ |

| 2-team $v(x) = \log_2(2x)$ | |
|---|---|
| $x$ | $v(x)$ |
| $0.0625$ | $-3$ |
| $0.125$ | $-2$ |
| $0.25$ | $-1$ |
| $0.5$ | $0$ |
| $1$ | $1$ |
| $2$ | $2$ |
| $4$ | $3$ |
| $8$ | $4$ |
| $16$ | $5$ |

| 5-team $w(x) = \log_5(5x)$ | |
|---|---|
| $x$ | $w(x)$ |
| $0.0625$ | $-0.72$ |
| $0.125$ | $-0.29$ |
| $0.25$ | $0.14$ |
| $0.5$ | $0.57$ |
| $1$ | $1$ |
| $2$ | $1.43$ |
| $4$ | $1.86$ |
| $8$ | $2.29$ |
| $16$ | $2.72$ |

MP.7

a. **Describe a transformation that takes the graph of your team's function in this exercise to the graph of your team's function in the Opening Exercise.**

***The graph produced in this exercise is a vertical translation of the graph from the Opening Exercise by one unit upward.***

b. **Do your answers to Exercise 2 and part (a) agree? If not, use properties of logarithms to justify your observations in part (a).**

***The answers to Exercise 2 and part (a) do not appear to agree. However, because*** $\log_b(bx) = \log_b(b) + \log_b(x) = 1 + \log_b(x)$***, the graph of*** $y = \log_b(bx)$ ***and the graph of*** $y = 1 + \log_b(x)$ ***coincide.***

## Closing (3 minutes)

Ask students to respond to these questions in writing or orally to a partner.

- In which quadrants is the graph of the function $f(x) = \log_b(x)$ located?
  - *The first and fourth quadrants.*
- When $b > 1$, for what values of $x$ are the values of the function $f(x) = \log_b(x)$ negative?
  - *When* $b > 1$, $f(x) = \log_b(x)$ *is negative for* $0 < x < 1$.
- When $0 < b < 1$, for what values of $x$ are the values of the function $f(x) = \log_b(x)$ negative?
  - *When* $0 < b < 1$, $f(x) = \log_b(x)$ *is negative for* $x > 1$.
- What are the key features of the graph of a logarithmic function $f(x) = \log_b(x)$ when $b > 1$?
  - *The domain of the function is all positive real numbers, and the range is all real numbers. The* $x$*-intercept is 1, the graph passes through* $(b, 1)$ *and there is no* $y$*-intercept. As* $x \to 0$, $f(x) \to -\infty$ *quickly, and as* $x \to \infty$, $f(x) \to \infty$ *slowly.*
- What are the key features of the graph of a logarithmic function $f(x) = \log_b(x)$ when $0 < b < 1$?
  - *The domain of the function is the positive real numbers, and the range is all real numbers. The* $x$*-intercept is 1, the graph passes through* $(b, -1)$*, and there is no* $y$*-intercept. As* $x \to 0$, $f(x) \to \infty$ *quickly, and as* $x \to \infty$, $f(x) \to -\infty$ *slowly.*

**Lesson Summary**

**The function** $f(x) = \log_b(x)$ **is defined for irrational and rational numbers. Its domain is all positive real numbers. Its range is all real numbers.**

**The function** $f(x) = \log_b(x)$ **goes to negative infinity as** $x$ **goes to zero. It goes to positive infinity as** $x$ **goes to positive infinity.**

**The larger the base** $b$**, the more slowly the function** $f(x) = \log_b(x)$ **increases.**

**By the change of base formula,** $\log_{\frac{1}{b}}(x) = -\log_b(x)$.

## Exit Ticket (5 minutes)

Name ______________________________ Date ______________

# Lesson 17: Graphing the Logarithm Function

## Exit Ticket

Graph the function $f(x) = \log_3(x)$ without using a calculator, and identify its key features.

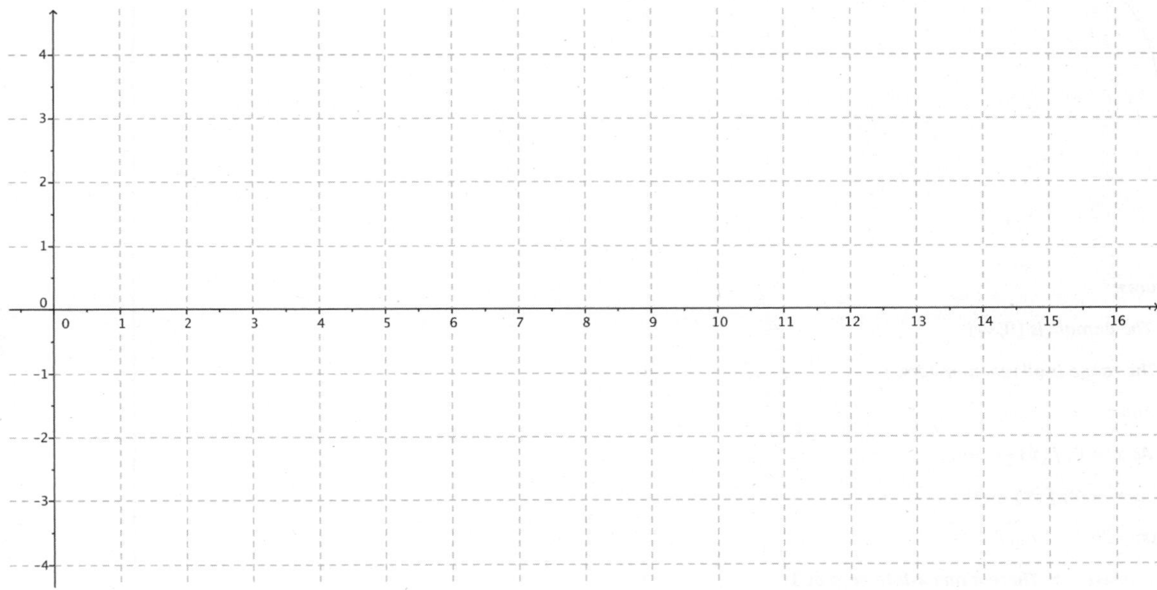

## Exit Ticket Sample Solutions

**Graph the function $f(x) = \log_3(x)$ without using a calculator, and identify its key features.**

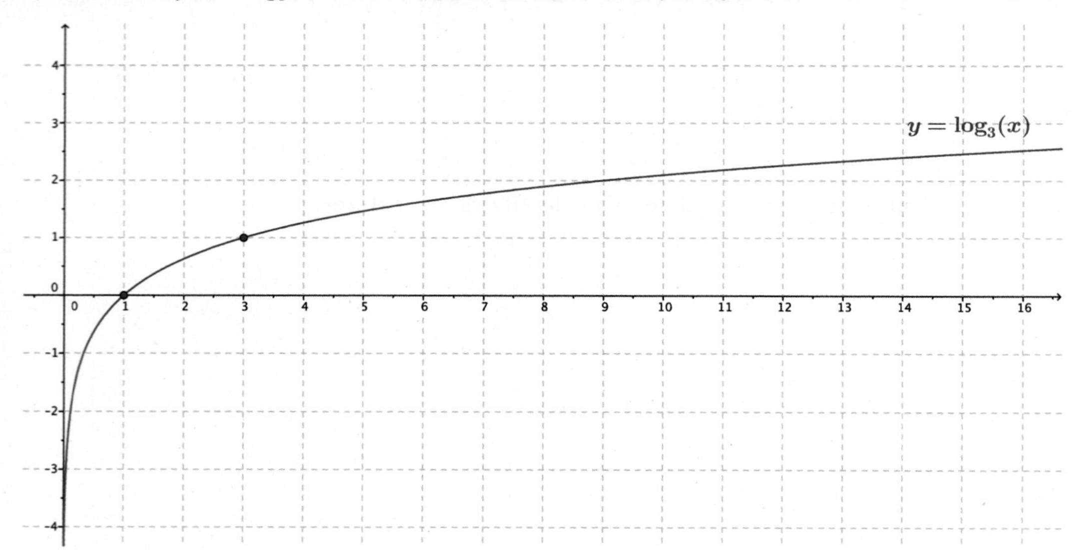


***Key features:***

*The domain is* $(0, \infty)$.

*The range is all real numbers.*

***End behavior:***

*As* $x \to 0$, $f(x) \to -\infty$.

*As* $x \to \infty$, $f(x) \to \infty$.

***Intercepts:***

$x$*-intercept: There is one* $x$*-intercept at* $1$.

$y$*-intercept: The graph does not cross the* $y$*-axis.*

*The graph passes through* $(3, 1)$.

## Problem Set Sample Solutions

For the Problem Set, students will need graph paper. They should not use calculators or other graphing technology. In Problems 2 and 3, students compare different representations of logarithmic functions. Problems 4–6 continue the reasoning from the lesson in which students observed the logarithmic properties through the transformations of logarithmic graphs.

Fluency problems 9–10 are a continuation of work done in Algebra I and are in this lesson to recall concepts that are required in Lesson 19. Similar review problems occur in the next lesson.

1. **The function $Q(x) = \log_b(x)$ has function values in the table at right.**

   a. **Use the values in the table to sketch the graph of $y = Q(x)$.**

| $x$ | $Q(x)$ |
|---|---|
| $0.1$ | $1.66$ |
| $0.3$ | $0.87$ |
| $0.5$ | $0.50$ |
| $1.00$ | $0.00$ |
| $2.00$ | $-0.50$ |
| $4.00$ | $-1.00$ |
| $6.00$ | $-1.29$ |
| $10.00$ | $-1.66$ |
| $12.00$ | $-1.79$ |

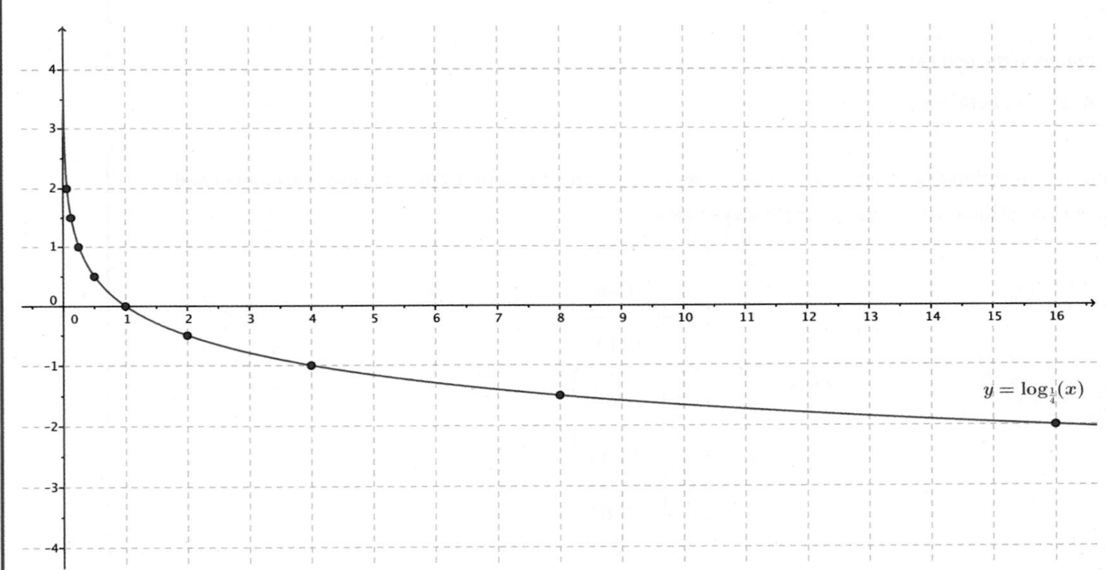


   b. **What is the value of $b$ in $Q(x) = \log_b(x)$? Explain how you know.**

   ***Because the point $(4, -1)$ is on the graph of $y = Q(x)$, we know $\log_b(4) = -1$, so $b^{-1} = 4$. It follows that $b = \frac{1}{4}$.***

   c. **Identify the key features in the graph of $y = Q(x)$.**

   ***Because $0 < b < 1$, the function values approach $\infty$ as $x \to 0$, and the function values approach $-\infty$ as $x \to \infty$. There is no $y$-intercept, and the $x$-intercept is $1$. The domain of the function is $(0, \infty)$, the range is $(-\infty, \infty)$, and the graph passes through $(b, 1)$.***

2. Consider the logarithmic functions $f(x) = \log_b(x)$, $g(x) = \log_5(x)$, where $b$ is a positive real number, and $b \neq 1$. The graph of $f$ is given at right.

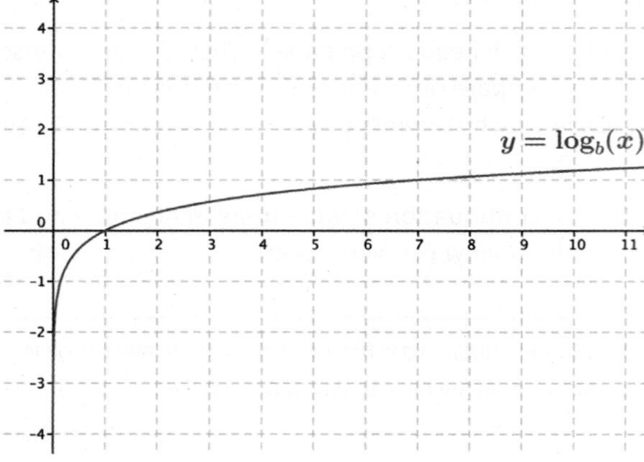


   a. Is $b > 5$, or is $b < 5$? Explain how you know.

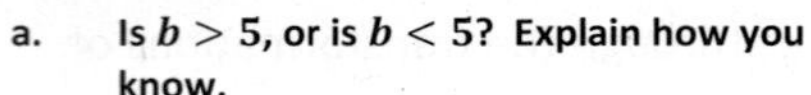

   *Since $f(7) = 1$, and $g(7) \approx 1.21$, the graph of $f$ lies below the graph of $g$ for $x \geq 1$. This means that $b$ is larger than 5, so we have $b > 5$. (Note: The actual value of $b$ is 7.)*

   b. Compare the domain and range of functions $f$ and $g$.

   *Functions $f$ and $g$ have the same domain, $(0, \infty)$, and the same range, $(-\infty, \infty)$.*

   c. Compare the $x$-intercepts and $y$-intercepts of $f$ and $g$.

   *Both $f$ and $g$ have an $x$-intercept at 1 and no $y$-intercepts.*

   d. Compare the end behavior of $f$ and $g$.

   *As $x \to \infty$, both $f(x) \to \infty$ and $g(x) \to \infty$.*

3. Consider the logarithmic functions $f(x) = \log_b(x)$ and $g(x) = \log_{\frac{1}{2}}(x)$, where $b$ is a positive real number and $b \neq 1$. A table of approximate values of $f$ is given below.

| $x$ | $f(x)$ |
|---|---|
| $\frac{1}{4}$ | $0.86$ |
| $\frac{1}{2}$ | $0.43$ |
| $1$ | $0$ |
| $2$ | $-0.43$ |
| $4$ | $-0.86$ |

   a. Is $b > \frac{1}{2}$, or is $b < \frac{1}{2}$? Explain how you know.

   *Since $g(2) = -1$, and $f(2) \approx -0.43$, the graph of $f$ lies above the graph of $g$ for $x \geq 1$. This means that $b$ is closer to 0 than $\frac{1}{2}$ is, so we have $b < \frac{1}{2}$. (Note: The actual value of $b$ is $\frac{1}{5}$.)*

   b. Compare the domain and range of functions $f$ and $g$.

   *Functions $f$ and $g$ have the same domain, $(0, \infty)$, and the same range, $(-\infty, \infty)$.*

c. **Compare the $x$-intercepts and $y$-intercepts of $f$ and $g$.**

*Both $f$ and $g$ have an $x$-intercept at $1$ and no $y$-intercepts.*

d. **Compare the end behavior of $f$ and $g$.**

*As $x \to \infty$, both $f(x) \to -\infty$ and $g(x) \to -\infty$.*

4. **On the same set of axes, sketch the functions $f(x) = \log_2(x)$ and $g(x) = \log_2(x^3)$.**

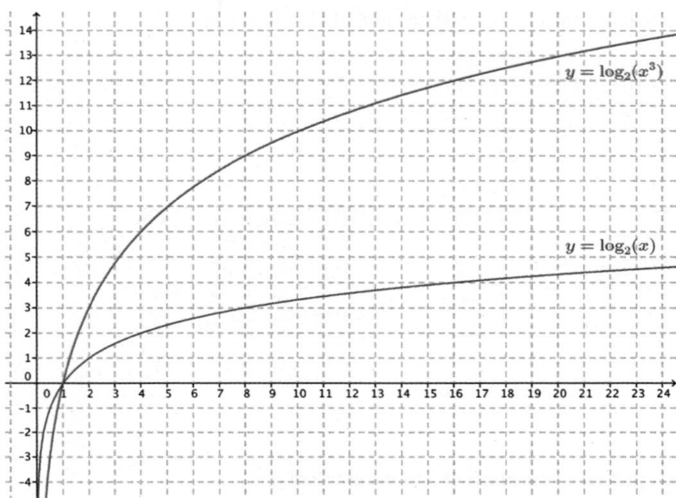


a. **Describe a transformation that takes the graph of $f$ to the graph of $g$.**

*The graph of $g$ is a vertical scaling of the graph of $f$ by a factor of $3$.*

b. **Use properties of logarithms to justify your observations in part (a).**

*Using properties of logarithms, we know that $g(x) = \log_2(x^3) = 3\log_2(x) = 3\,f(x)$. Thus, the graph of $f$ is a vertical scaling of the graph of $g$ by a factor of $3$.*

5. **On the same set of axes, sketch the functions $f(x) = \log_2(x)$ and $g(x) = \log_2\left(\frac{x}{4}\right)$.**

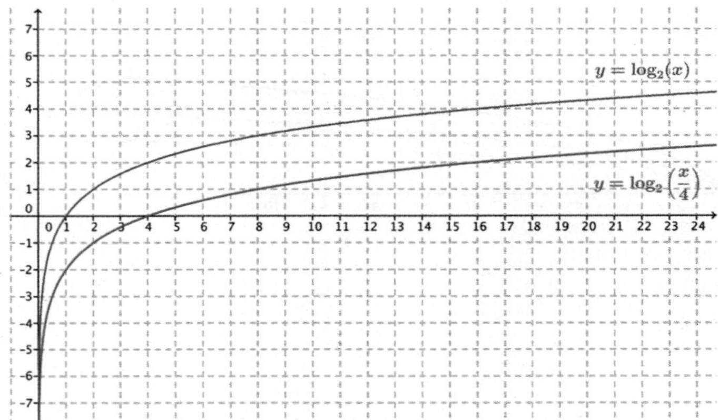

a. **Describe a transformation that takes the graph of $f$ to the graph of $g$.**

*The graph of $g$ is the graph of $f$ translated down by $2$ units.*

b. **Use properties of logarithms to justify your observations in part (a).**

*Using properties of logarithms, $g(x) = \log_2\left(\frac{x}{4}\right)\left(\frac{x}{4}\right) = \log_2(x) - \log_2(4) = f(x) - 2$. Thus, the graph of $g$ is a translation of the graph of $f$ down $2$ units.*

6. **On the same set of axes, sketch the functions $f(x) = \log_{\frac{1}{2}}(x)$ and $g(x) = \log_2\left(\frac{1}{x}\right)$.**

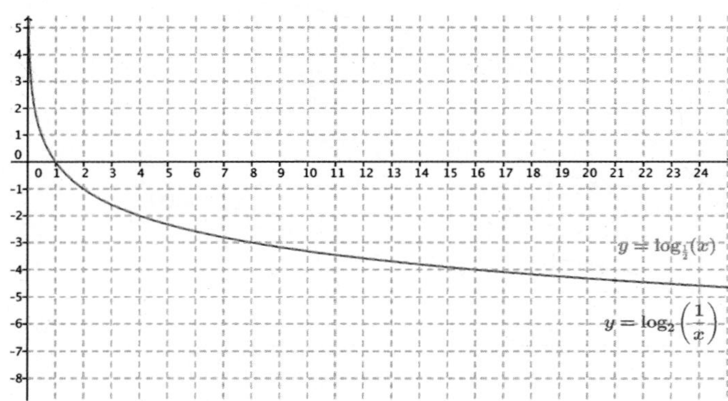


a. **Describe a transformation that takes the graph of $f$ to the graph of $g$.**

*These two graphs coincide, so the identity transformation takes the graph of $f$ to the graph of $g$.*

b. **Use properties of logarithms to justify your observations in part (a).**

*If $\log_{\frac{1}{2}}(x) = y$, then $\left(\frac{1}{2}\right)^y = x$, so $\frac{1}{x} = 2^y$. Then, $y = \log_2$; so, $\log_2\left(\frac{1}{x}\right) = \log_{\frac{1}{2}}(x)$; thus, $g(x) = f(x)$ for all $x > 0$.*

7. **The figure below shows graphs of the functions $f(x) = \log_3(x)$, $g(x) = \log_5(x)$, and $h(x) = \log_{11}(x)$.**

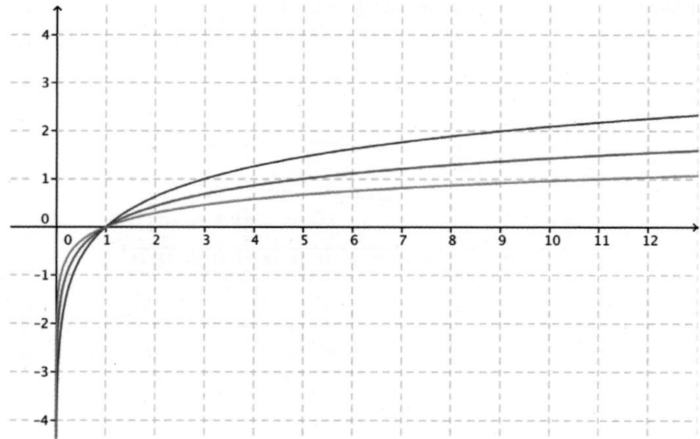

a. **Identify which graph corresponds to which function. Explain how you know.**

*The top graph (in blue) is the graph of $f(x) = \log_3(x)$, the middle graph (in green) is the graph of $g(x) = \log_5(x)$, and the lower graph (in red) is the graph of $h(x) = \log_{11}(x)$. We know this because the blue graph passes through the point $(3, 1)$, the green graph passes through the point $(5, 1)$, and the red graph passes through the point $(11, 1)$. We also know that the higher the value of the base $b$, the flatter the graph, so the graph of the function with the largest base, $11$, must be the red graph on the bottom, and the graph of the function with the smallest base, $3$, must be the blue graph on the top.*

b. **Sketch the graph of $k(x) = \log_7(x)$ on the same axes.**

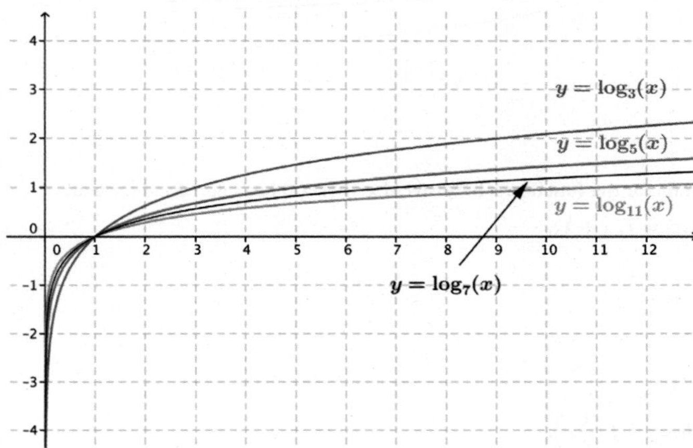


8. **The figure below shows graphs of the functions $f(x) = \log_{\frac{1}{3}}(x)$, $g(x) = \log_{\frac{1}{5}}(x)$, and $h(x) = \log_{\frac{1}{11}}(x)$.**

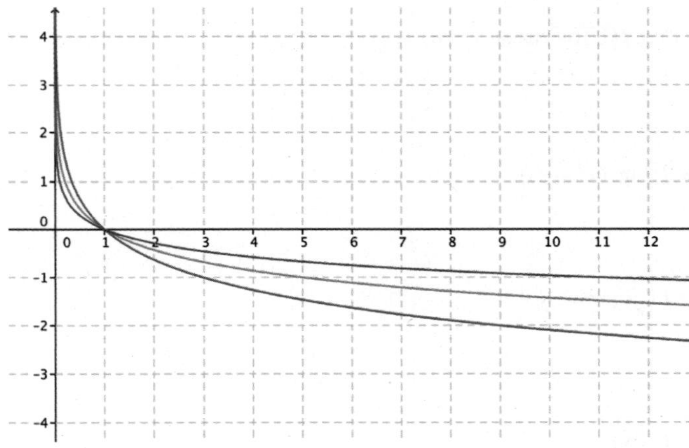

a. **Identify which graph corresponds to which function. Explain how you know.**

*The top graph (in blue) is the graph of $h(x) = \log_{\frac{1}{11}}(x)$, the middle graph (in red) is the graph of $g(x) = \log_{\frac{1}{5}}(x)$, and the lower graph is the graph of $f(x) = \log_{\frac{1}{3}}(x)$. We know this because the blue graph passes through the point $(11, -1)$, the red graph passes through the point $(5, -1)$, and the green graph passes through the point $(3, -1)$.*

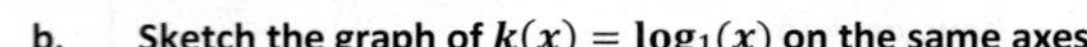

b. Sketch the graph of $k(x) = \log_{\frac{1}{7}}(x)$ on the same axes.

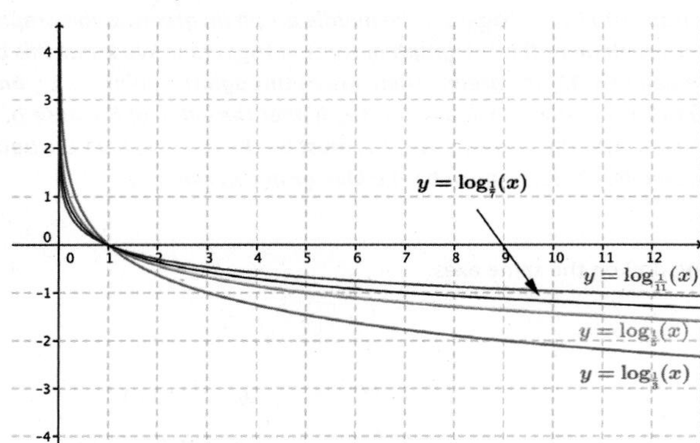


9. For each function $f$, find a formula for the function $h$ in terms of $x$. Part (a) has been done for you.

a. If $f(x) = x^2 + x$, find $h(x) = f(x+1)$.

$$\begin{aligned} h(x) &= f(x+1) \\ &= (x+1)^2 + (x+1) \\ &= x^2 + 3x + 2 \end{aligned}$$

b. If $f(x) = \sqrt{x^2 + \frac{1}{4}}$, find $h(x) = f\left(\frac{1}{2}x\right)$.

$$h(x) = \frac{1}{2}\sqrt{x^2 + 1}$$

c. If $f(x) = \log(x)$, find $h(x) = f(\sqrt[3]{10x})$ when $x > 0$.

$$h(x) = \frac{1}{3} + \frac{1}{3}log(x)$$

d. If $f(x) = 3^x$, find $h(x) = f(\log_3(x^2 + 3))$.

$$h(x) = x^2 + 3$$

e. If $f(x) = x^3$, find $h(x) = f\left(\frac{1}{x^3}\right)$ when $x \neq 0$.

$$h(x) = \frac{1}{x^6}$$

f. If $f(x) = x^3$, find $h(x) = f(\sqrt[3]{x})$.

$$h(x) = x$$

g. If $f(x) = \sin(x)$, find $h(x) = f(x + \frac{\pi}{2})$.

$$h(x) = \sin\left(x + \frac{\pi}{2}\right)$$

h. If $f(x) = x^2 + 2x + 2$, find $h(x) = f(\cos(x))$.

$$h(x) = (\cos(x))^2 + 2\cos(x) + 2$$

10. **For each of the functions $f$ and $g$ below, write an expression for (i) $f(g(x))$, (ii) $g(f(x))$, and (iii) $f(f(x))$ in terms of $x$. Part (a) has been done for you.**

a. $f(x) = x^2,\ g(x) = x + 1$

$$f(g(x)) = f(x+1) = (x+1)^2$$

$$g(f(x)) = g(x^2) = x^2 + 1$$

$$f(f(x)) = f(x^2) = (x^2)^2 = x^4$$

b. $f(x) = \frac{1}{4}x - 8,\ g(x) = 4x + 1$

*i.* $x - \frac{31}{4}$

*ii.* $x - 31$

*iii.* $\frac{1}{16}x - 10$

c. $f(x) = \sqrt[3]{x+1},\ g(x) = x^3 1$

*i.* $x$

*ii.* $x$

*iii.* $\sqrt[3]{\sqrt[3]{x+1} + 1}$

d. $f(x) = x^3,\ g(x) = \frac{1}{x}$

*i.* $\frac{1}{x^3}$

*ii.* $\frac{1}{x^3}$

*iii.* $x^9$

e. $f(x) = |x|,\ g(x) = x^2$

*i.* $|x^2|$ *or* $x^2$

*ii.* $(|x|)^2$ *or* $x^2$

*iii.* $|x|$

**Extension:**

11. **Consider the functions $f(x) = \log_2(x)$ and $(x) = \sqrt{x-1}$.**

 a. **Use a calculator or other graphing utility to produce graphs of $f(x) = \log_2(x)$ and $g(x) = \sqrt{x-1}$ for $x \leq 17$.**

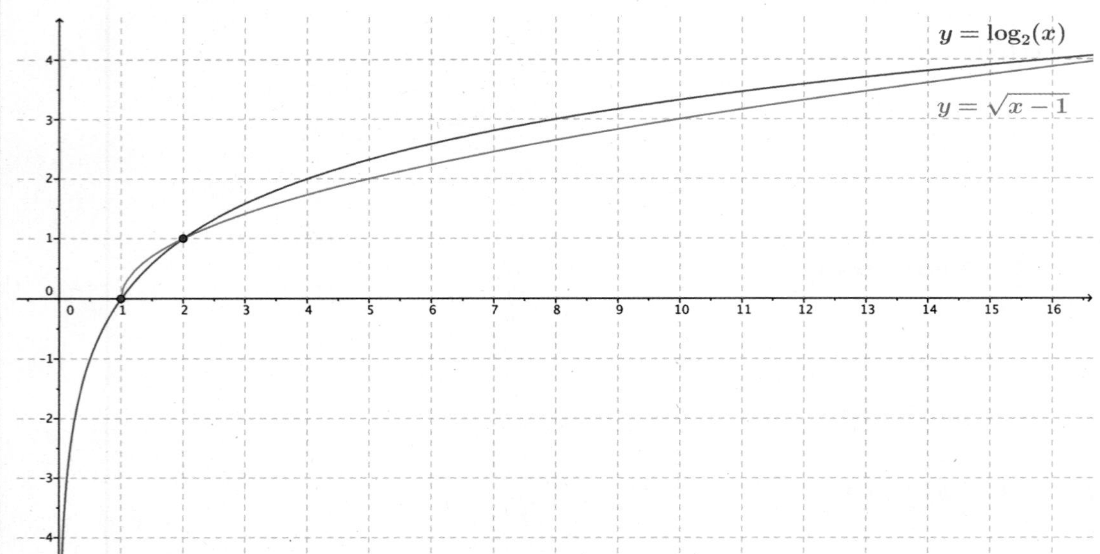


 b. **Compare the graph of the function $f(x) = \log_2(x)$ with the graph of the function $g(x) = \sqrt{x-1}$. Describe the similarities and differences between the graphs.**

 ***They are not the same, but they have a similar shape when $x \geq 1$. Both graphs pass through the points $(1, 0)$ and $(2, 1)$. Both functions appear to approach infinity slowly as $x \to \infty$.***

 ***The graph of $f(x) = \log_2(x)$ lies below the graph of $g(x) = \sqrt{x-1}$ on the interval $(1, 2)$, and the graph of $f$ appears to lie above the graph of $g$ on the interval $(2, \infty)$. The logarithm function $f$ is defined for $x > 0$, and the radical function $g$ is defined for $x \geq 1$. Both functions appear to slowly approach infinity as $x \to \infty$.***

 c. **Is it always the case that $\log_2(x) > \sqrt{x-1}$ for $x > 2$?**

 ***No, for $2 < x \leq 19$, $\log_2(x) > \sqrt{x-1}$. Between $19$ and $20$, the graphs cross again, and we have $\sqrt{x-1} > \log_2(x)$ for $x \geq 20$.***

12. Consider the functions $f(x) = \log_2(x)$ and $(x) = \sqrt[3]{x-1}$.

a. Use a calculator or other graphing utility to produce graphs of $f(x) = \log_2(x)$ and $h(x) = \sqrt[3]{x-1}$ for $x \leq 28$.

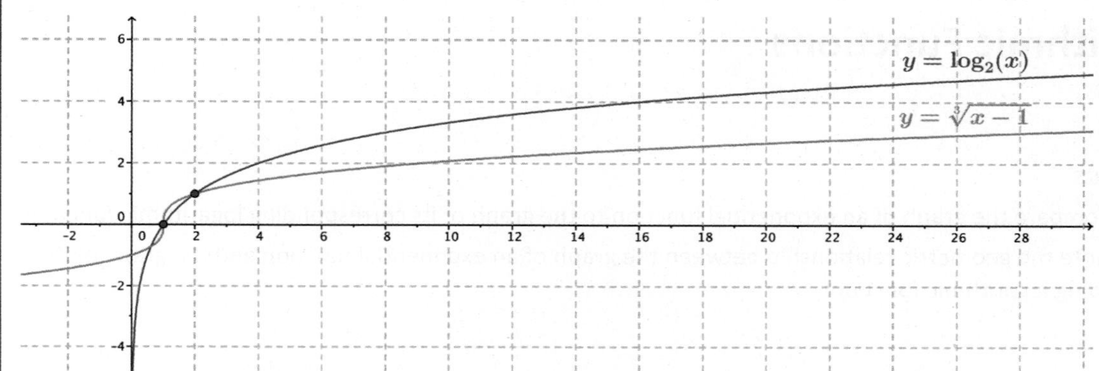


b. Compare the graph of the function $f(x) = \log_2(x)$ with the graph of the function $h(x) = \sqrt[3]{x-1}$. Describe the similarities and differences between the graphs.

*They are not the same, but they have a similar shape when $x \geq 1$. Both graphs pass through the points $(1, 0)$ and $(2, 1)$. Both functions appear to approach infinity slowly as $x \to \infty$.*

*The graph of $f(x) = \log_2(x)$ lies below the graph of $h(x) = \sqrt[3]{x-1}$ on the interval $(1, 2)$, and the graph of $f$ appears to lie above the graph of $h$ on the interval $(2, \infty)$. The logarithm function $f$ is defined for $x > 0$, and the radical function $h$ is defined for all real numbers $x$. Both functions appear to approach infinity slowly as $x \to \infty$.*

c. Is it always the case that $\log_2(x) > \sqrt[3]{x-1}$ for $x > 2$?

*No, if we extend the viewing window on the calculator, we see that the graphs cross again between 983 and 984. Thus, $\log_2(x) > \sqrt[3]{x-1}$ for $2 < x \leq 983$, and $\log_2(x) < \sqrt[3]{x-1}$ for $x \geq 984$.*

# Lesson 18: Graphs of Exponential Functions and Logarithmic Functions

## Student Outcomes

- Students compare the graph of an exponential function to the graph of its corresponding logarithmic function.
- Students note the geometric relationship between the graph of an exponential function and the graph of its corresponding logarithmic function.

## Lesson Notes

In the previous lesson, students practiced graphing transformed logarithmic functions and observed the effects of the logarithmic properties in the graphs. In this lesson, students graph the logarithmic functions along with their corresponding exponential functions. Be careful to ensure that the scale is the same on both axes so that the geometric relationship between the graph of the exponential function and the graph of the logarithmic function is apparent. Part of the focus of the lesson is for students to begin seeing that these functions are the inverses of each other—but without the teacher actually saying it yet. Encourage students to draw the graphs carefully so that they can see that the two graphs are reflections of each other about the diagonal. The asymptotic nature of the two functions may be discussed. (**F-IF.B.4**, **F-IF.C.7e**) The teacher is encouraged to consider using graphing software such as GeoGebra.

## Classwork

### Opening Exercise (5 minutes)

Allow students to work in pairs or small groups on the following exercise, in which they graph a few points on the curve $y = 2^x$, reflect these points over the diagonal line with the equation $y = x$, and analyze the result.

*Scaffolding:*

- Model the process of reflecting a set of points, such as $\triangle ABC$ with vertices $A(-3, 2)$, $B(-3, 7)$, and $C(2, 7)$, over the diagonal line $y = x$ before asking students to do the same.
- After the graph of $y = 2^x$ and its reflection are shown, ask advanced students, "If the first graph represents the points that satisfy $y = 2^x$, then what equation do the points on the reflected graph satisfy?"

**Opening Exercise**

**Complete the following table of values of the function $f(x) = 2^x$. We want to sketch the graph of $y = f(x)$ and then reflect that graph across the diagonal line with equation $y = x$.**

| $x$ | $y = 2^x$ | Point $(x, y)$ on the graph of $y = 2^x$ |
|---|---|---|
| $-3$ | $\frac{1}{8}$ | $\left(-3, \frac{1}{8}\right)$ |
| $-2$ | $\frac{1}{4}$ | $\left(-2, \frac{1}{4}\right)$ |
| $-1$ | $\frac{1}{2}$ | $\left(-1, \frac{1}{2}\right)$ |
| $0$ | $1$ | $(0, 1)$ |
| $1$ | $2$ | $(1, 2)$ |
| $2$ | $4$ | $(2, 4)$ |
| $3$ | $8$ | $(3, 8)$ |

On the set of axes below, plot the points from the table and sketch the graph of $y = 2^x$. Next, sketch the diagonal line with equation $y = x$, and then reflect the graph of $y = 2^x$ across the line.

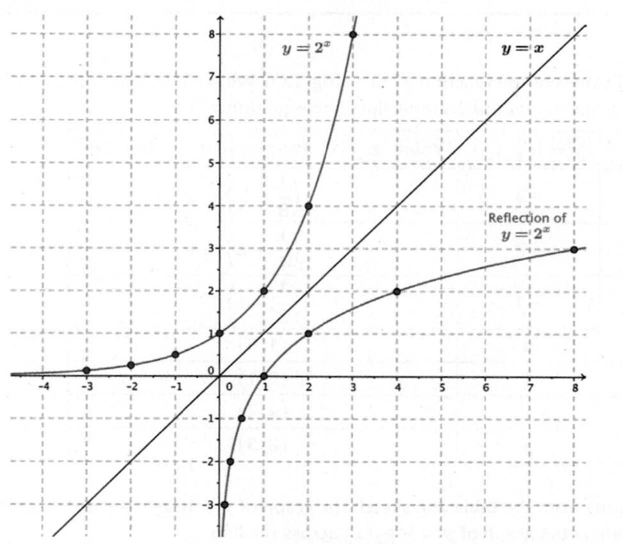


## Discussion (4 minutes)

Use the following discussion to reinforce the process by which a point is reflected across the diagonal line given by $y = x$ and the reasoning for why reflecting points on an exponential curve produces points on the corresponding logarithmic curve.

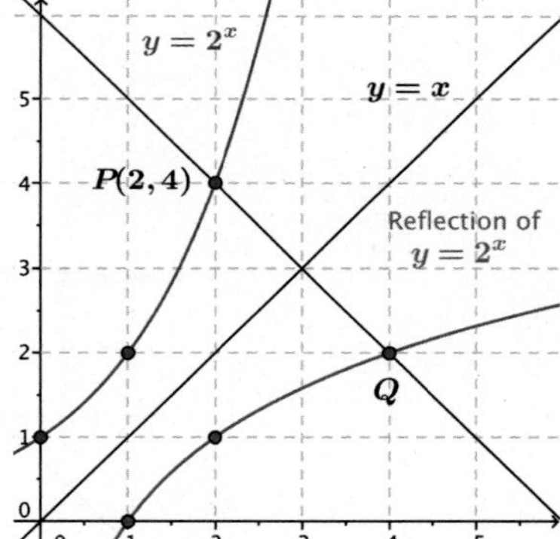


- How do we find the reflection of the point $P(2, 4)$ across the line given by $y = x$?
  - *Point $P(2, 4)$ is reflected to point $Q$ on the line through $(2, 4)$ that is perpendicular to the line given by $y = x$ so that points $P$ and $Q$ are equidistant from the diagonal line.*
- What is the slope of the line through $P$ and $Q$? Explain how you know. (*Draw the figure at right.*)
  - *The slope is $-1$ because this line is perpendicular to the diagonal line that has slope $1$.*
- We know that $P$ and $Q$ are the same distance from the diagonal line. What are the coordinates of the point $Q$?
  - *Point $Q$ has coordinates $(4, 2)$.*
- What are the coordinates of the reflection of the point $(1, 2)$ across the line given by $y = x$?
  - *The reflection of the point $(1, 2)$ is the point $(2, 1)$.*

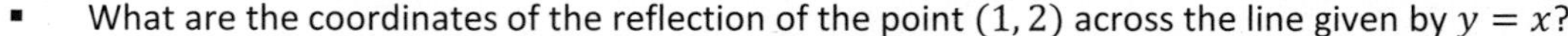

- What are the coordinates of the reflection of the point $(a, b)$ across the line given by $y = x$?
- When we reflect about the line with equation $y = x$, we actually switch the axes themselves by folding the plane along this line. Therefore, the reflection of the point $(a, b)$ is the point $(b, a)$.

## Exercise 1 (7 minutes)

**Exercises**

1. **Complete the following table of values of the function $g(x) = \log_2(x)$. We want to sketch the graph of $y = g(x)$ and then reflect that graph across the diagonal line with equation $y = x$.**

| $x$ | $y = \log_2(x)$ | Point $(x, y)$ on the graph of $y = \log_2(x)$ |
|---|---|---|
| $-\frac{1}{8}$ | $-3$ | $\left(\frac{1}{8}, -3\right)$ |
| $-\frac{1}{4}$ | $-2$ | $\left(\frac{1}{4}, -2\right)$ |
| $-\frac{1}{2}$ | $-1$ | $\left(\frac{1}{2}, -1\right)$ |
| $1$ | $0$ | $(1, 0)$ |
| $2$ | $1$ | $(2, 1)$ |
| $4$ | $2$ | $(4, 2)$ |
| $8$ | $3$ | $(8, 3)$ |

**On the set of axes below, plot the points from the table and sketch the graph of $y = \log_2(x)$. Next, sketch the diagonal line with equation $y = x$, and then reflect the graph of $y = \log_2(x)$ across the line.**

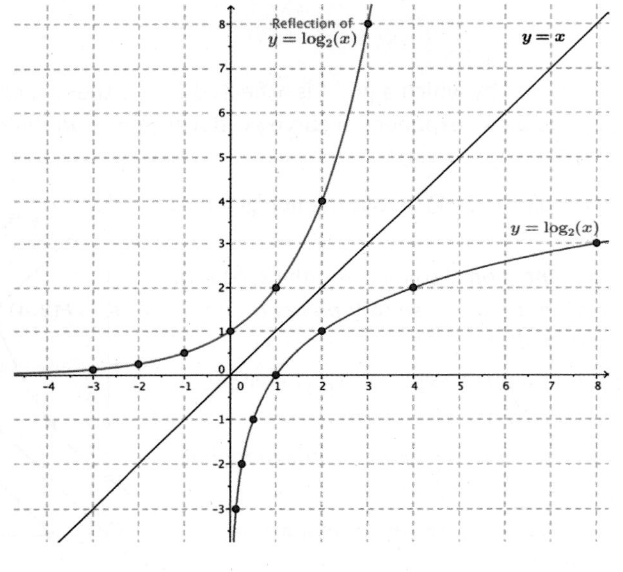


## Discussion (5 minutes)

This discussion makes clear that the reflection of the graph of an exponential function is the graph of a corresponding logarithmic function, and vice-versa.

- How do we find the reflection of the point $P(2, 4)$ across the line given by $y = x$?

MP.7

- What similarities do you notice about this exercise and the Opening Exercise?
  - *The points $(0, 1)$, $(2, 1)$, and $(4, 2)$ on the logarithmic graph are the reflections of the points we plotted on this first graph of $f(x) = 2^x$ across the diagonal line.*
  - *The point $(2, 4)$ on the graph of the exponential function is the reflection across the diagonal line of the*

*point $(4, 2)$ on the graph of the logarithm, and the point $(4, 2)$ on the graph of the logarithm function is the reflection across the diagonal line of the point $(2, 4)$ on the graph of the exponential function.*

- *The point $(a, b)$ on the graph of the exponential function is the reflection across the diagonal line of the point $(b, a)$ on the graph of the logarithm, and the point $(b, a)$ on the graph of the logarithm function is the reflection across the diagonal line of the point $(a, b)$ on the graph of the exponential function.*
- *The graphs of the functions $f(x) = 2^x$ and $g(x) = \log_2(x)$ are reflections of each other across the diagonal line given by $y = x$.*

- Why does this happen? How does the definition of the logarithm tell us that if $(a, b)$ is a point on the exponential graph, then $(b, a)$ is a point on the logarithmic graph? How does the definition of the logarithm tell us that if $(b, a)$ is a point on the logarithmic graph, then $(a, b)$ is a point on the exponential graph?
  - *If $(a, b)$ is a point on the graph of the exponential function $f(x) = 2^x$, then*

$$f(a) = 2^a$$
$$b = 2^a$$
$$\log_2(b) = a$$

  - *So, the point $(b, a)$ is on the graph of the logarithmic function $g(x) = \log_2(x)$.*
    *Likewise, if $(b, a)$ is a point on the graph of the logarithmic function $g(x) = \log_2(x)$, then:*

$$g(b) = \log_2(b)$$
$$\log_2(b) = a$$
$$2^a = b$$

- So, the point $(a, b)$ is on the graph of the exponential function $f(x) = 2^x$.

## Exercise 2 (5 minutes)

**2. Working independently, predict the relation between the graphs of the functions $f(x) = 3^x$ and $g(x) = \log_3(x)$. Test your predictions by sketching the graphs of these two functions. Write your prediction in your notebook, provide justification for your prediction, and compare your prediction with that of your neighbor.**

***The graphs will be reflections of each other about the diagonal.***

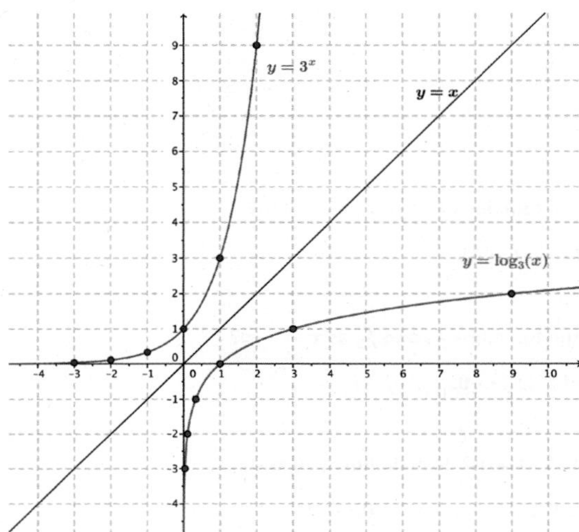

## Exercises 3–4 (10 minutes)

3. **Now let's compare the graphs of the functions $f_2(x) = 2^x$ and $f_3(x) = 3^x$; sketch the graphs of the two exponential functions on the same set of axes; then, answer the questions below.**

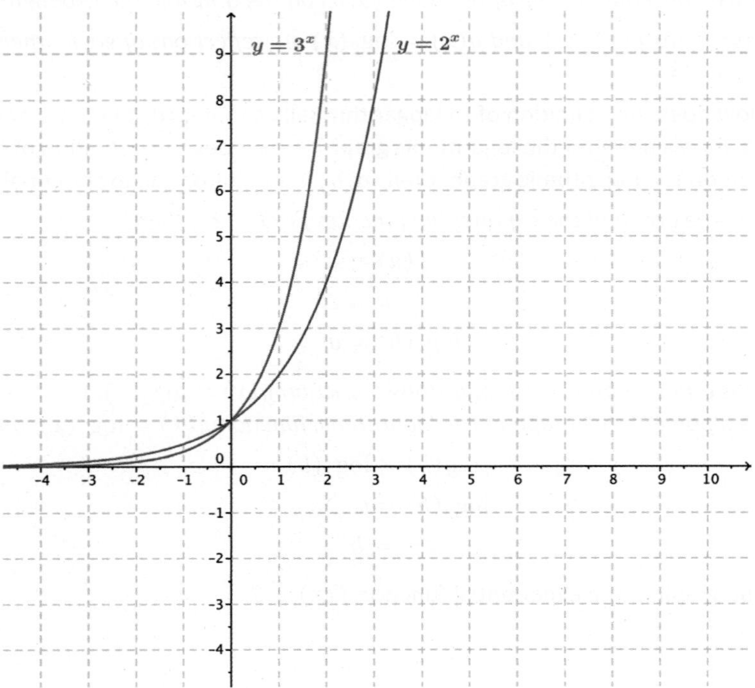


a. **Where do the two graphs intersect?**

*The two graphs intersect at the point* $(0, 1)$.

b. **For which values of $x$ is $2^x < 3^x$?**

*If* $x > 0$, *then* $2^x < 3^x$.

c. **For which values of $x$ is $2^x > 3^x$ ?**

*If* $x < 0$, *then* $2^x > 3^x$.

d. **What happens to the values of the functions $f_2$ and $f_3$ as $x \to \infty$?**

*As* $x \to \infty$, *both* $f_2(x) \to \infty$ *and* $f_3(x) \to \infty$.

e. **What happens to the values of the functions $f_2$ and $f_3$ as $x \to -\infty$?**

*As* $x \to -\infty$, *both* $f_2(x) \to 0$ *and* $f_3(x) \to 0$.

f. **Does either graph ever intersect the $x$-axis? Explain how you know.**

*No. For every value of* $x$, *we know* $2^x \neq 0$ *and* $3^x \neq 0$.

4. **Add sketches of the two logarithmic functions $g_2(x) = \log_2(x)$ and $g_3(x) = \log_3(x)$ to the axes with the graphs of the exponential functions; then, answer the questions below.**

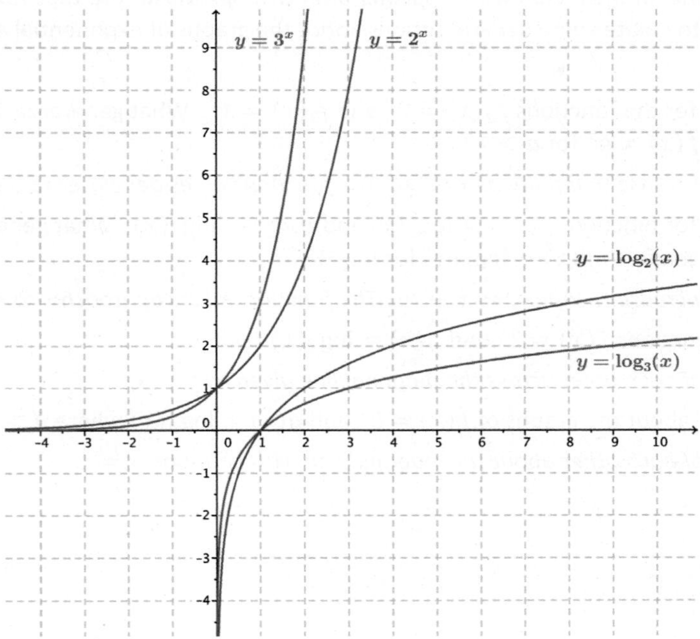


a. **Where do the two logarithmic graphs intersect?**

*The two graphs intersect at the point* $(1, 0)$.

b. **For which values of $x$ is $\log_2(x) < \log_3(x)$?**

*If* $x < 1$, *then* $\log_2(x) < \log_3(x)$.

c. **For which values of $x$ is $g_2(x) > \log_3(x)$?**

*If* $x > 1$, *then* $\log_2(x) > \log_3(x)$.

d. **What happens to the values of the functions $f_2$ and $f_3$ as $x \to \infty$?**

*As* $x \to \infty$, *both* $g_2(x) \to \infty$ *and* $g_3(x) \to \infty$.

e. **What happens to the values of the functions $f_2$ and $f_3$ as $x \to 0$?**

*As* $x \to 0$, *both* $g_2(x) \to -\infty$ *and* $g_3(x) \to -\infty$.

g. **Does either graph ever intersect the $y$-axis? Explain how you know.**

*No. Logarithms are only defined for positive values of* $x$.

h. **Describe the similarities and differences in the behavior of $f_2(x)$ and $g_2(x)$ as $x \to \infty$.**

*As* $x \to \infty$, *both* $f_2(x) \to \infty$ *and* $g_2(x) \to \infty$; *however, the exponential function gets very large very quickly, and the logarithmic function gets large rather slowly.*

## Closing (4 minutes)

Ask students to summarize the key points of the lesson with a partner or in writing. Make sure that students have used the specific examples from the lesson to create some generalizations about the graphs of exponential and logarithmic functions.

MP.8

- Graphical analysis was done for the functions $f_2(x) = 2^x$ and $f_3(x) = 3^x$. What generalizations can we make about functions of the form $f(x) = a^x$ for $a > 1$?
  - *The function values increase to infinity as $x \to \infty$. The function values get closer to $0$ as $x \to -\infty$.*
- Graphical analysis was done for functions $g_2(x) = \log_2(x)$ and $g_3(x) = \log_3(x)$. What generalizations can we make about functions of the form $g(x) = \log_b(x)$ for $b > 1$?
  - *The function values increase to infinity as $x \to \infty$. The function values approach $-\infty$ as $x \to 0$.*
- How are the graphs of the functions $f(x) = 2^x$ and $g(x) = \log_2(x)$ related?
  - *They are reflections of each other across the diagonal line given by $y = x$.*
- What can we say, in general, about the graphs of $f(x) = b^x$ and $g(x) = \log_b(x)$ where $b > 1$?
  - *They are reflections of each other about the diagonal line with equation $y = x$.*

## Exit Ticket (5 minutes)

Name ______________________________ Date ______________

# Lesson 18: Graphs of Exponential Functions and Logarithmic Functions

## Exit Ticket

The graph of a logarithmic function $g(x) = \log_b(x)$ is shown below.

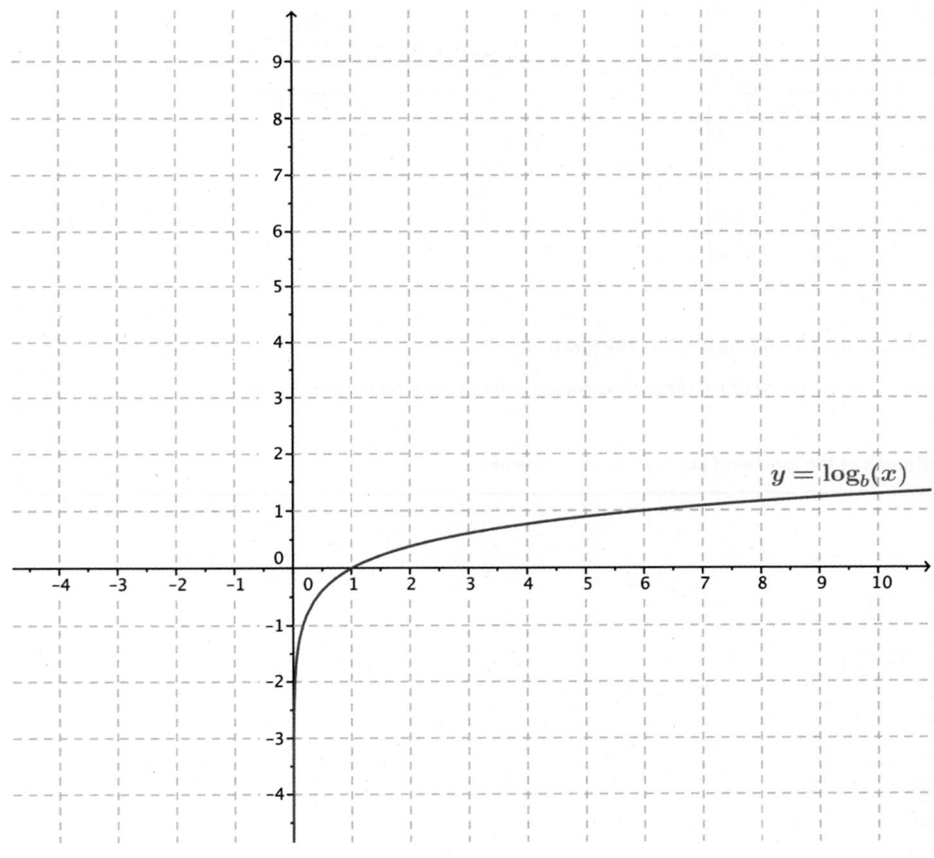


a. Explain how to find points on the graph of the function $f(x) = b^x$.

b. Sketch the graph of the function $f(x) = b^x$ on the same axes.

## Exit Ticket Sample Solutions

**The graph of a logarithmic function $g(x) = \log_b(x)$ is shown below.**

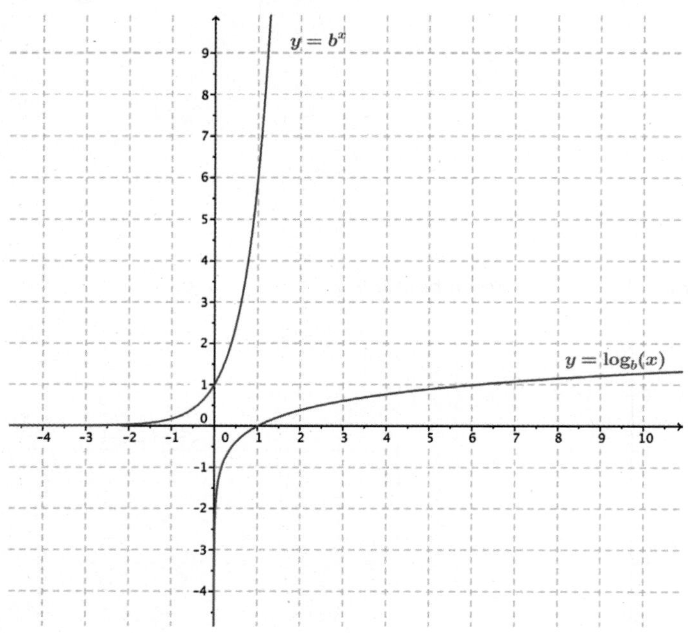


a. **Explain how to find points on the graph of the function $f(x) = b^x$.**

***A point $(x, y)$ is on the graph of $f$ if the corresponding point $(y, x)$ is on the graph of $g$.***

b. **Sketch the graph of the function $f(x) = b^x$ on the same axes.**

## Problem Set Sample Solutions

Problems 5–7 serve to review the process of computing $f(g(x))$ for given functions $f$ and $g$ in preparation for work with inverses of functions in Lesson 19.

1. **Sketch the graphs of the functions $f(x) = 5^x$ and $g(x) = \log_5(x)$.**

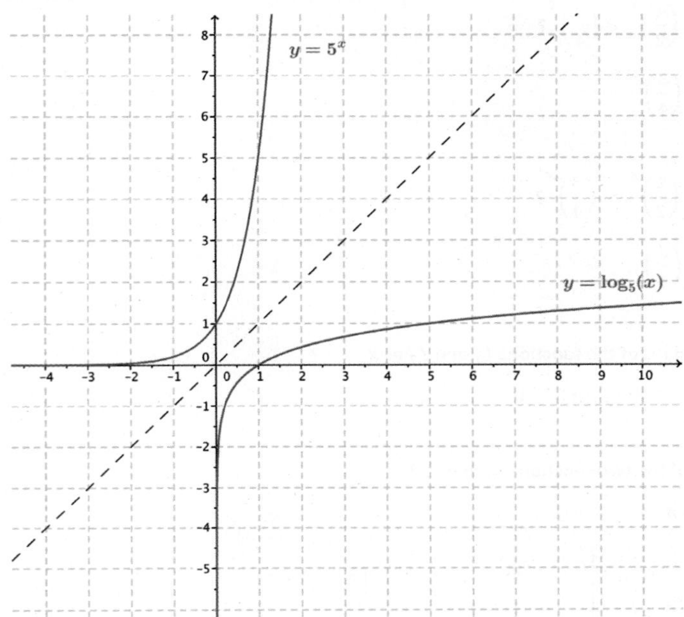


2. **Sketch the graphs of the functions $f(x) = \left(\frac{1}{2}\right)^x$ and $g(x) = \log_{\frac{1}{2}}(x)$.**

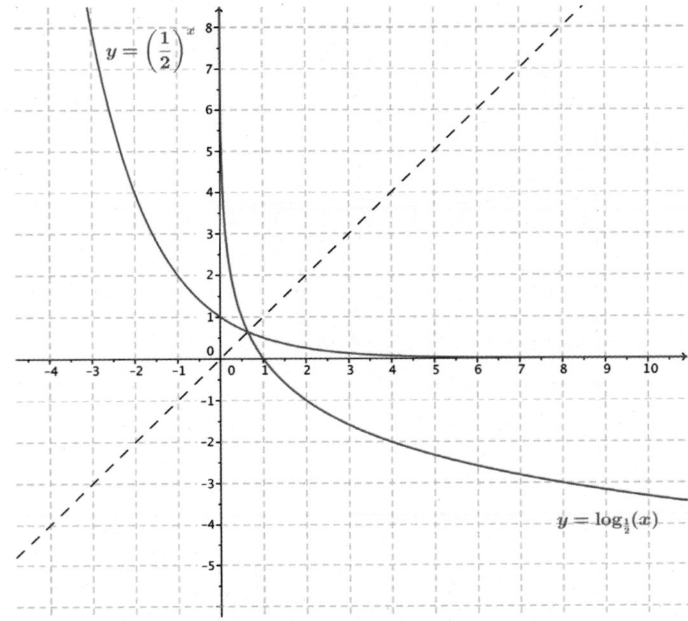

3. **Sketch the graphs of the functions $f_1(x) = \left(\frac{1}{2}\right)^x$ and $f_2(x) = \left(\frac{3}{4}\right)^x$ on the same sheet of graph paper and answer the following questions.**

a. **Where do the two exponential graphs intersect?**

*The graphs intersect at the point* $(0, 1)$.

b. **For which values of $x$ is $\left(\frac{1}{2}\right)^x < \left(\frac{3}{4}\right)^x$?**

*If* $x > 0$, *then* $\left(\frac{1}{2}\right)^x < \left(\frac{3}{4}\right)^x$.

c. **For which values of $x$ is $\left(\frac{1}{2}\right)^x > \left(\frac{3}{4}\right)^x$?**

*If* $x < 0$, *then* $\left(\frac{1}{2}\right)^x > \left(\frac{3}{4}\right)^x$.

d. **What happens to the values of the functions $f_1$ and $f_2$ as $x \to \infty$?**

*As* $x \to \infty$, *both* $f_1(x) \to 0$ *and* $f_2(x) \to 0$.

e. **What are the domains of the two functions $f_1$ and $f_2$?**

*Both functions have domain* $(-\infty, \infty)$.

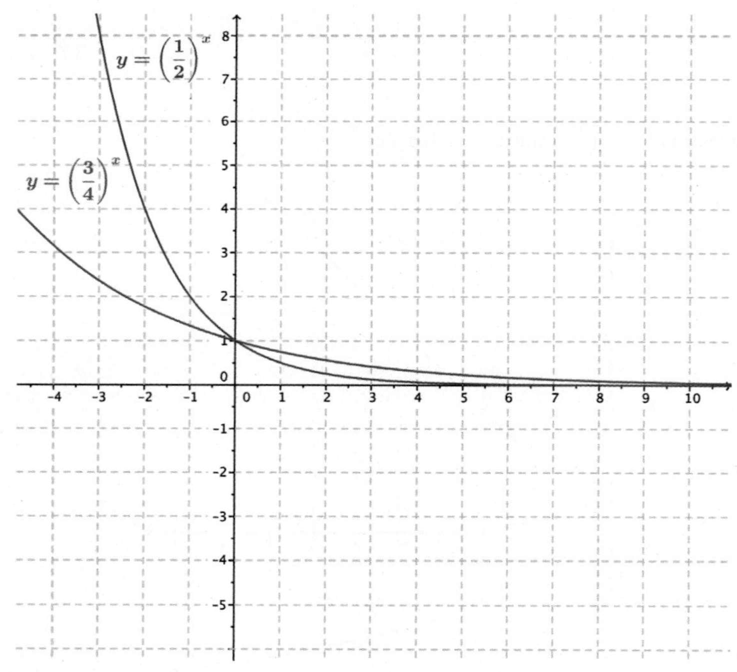

4. **Use the information from Problem 3 together with the relationship between graphs of exponential and logarithmic functions to sketch the graphs of the functions $g_1(x) = \log_{\frac{1}{2}}(x)$ and $g_2(x) = \log_{\frac{3}{4}}(x)$ on the same sheet of graph paper. Then, answer the following questions.**

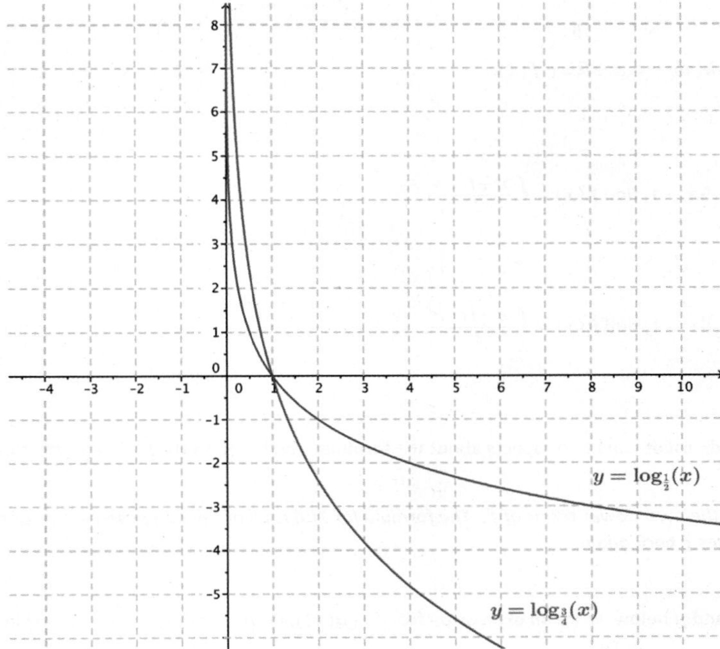


a. **Where do the two logarithmic graphs intersect?**

*The graphs intersect at the point* $(1, 0)$.

b. **For which values of $x$ is $\log_{\frac{1}{2}}(x) < \log_{\frac{3}{4}}(x)$?**

*When* $x < 1$, *we have* $\log_{\frac{1}{2}}(x) < \log_{\frac{3}{4}}(x)$.

c. **For which values of $x$ is $\log_{\frac{1}{2}}(x) > \log_{\frac{3}{4}}(x)$?**

*When* $x > 1$, *we have* $\log_{\frac{1}{2}}(x) > \log_{\frac{3}{4}}(x)$.

d. **What happens to the values of the functions $g_1$ and $g_2$ as $x \to \infty$?**

*As* $x \to \infty$, *both* $g_1(x) \to -\infty$ *and* $g_2(x) \to -\infty$.

e. **What are the domains of the two functions $g_1$ and $g_2$?**

*Both functions have domain* $(0, \infty)$.

5. For each function $f$, find a formula for the function $h$ in terms of $x$.

   a. If $f(x) = x^3$, find $h(x) = 128f\left(\frac{1}{4}x\right) + f(2x)$.

      $h(x) = 10x^3$

   b. If $(x) = x^2 + 1$, find $h(x) = f(x+2) - f(2)$.

      $h(x) = x^2 + 4x$

   c. If $f(x) = x^3 + 2x^2 + 5x + 1$, find $h(x) = \frac{f(x)+f(-x)}{2}$.

      $h(x) = 2x^2 + 1$

   d. If $f(x) = x^3 + 2x^2 + 5x + 1$, find $h(x) = \frac{f(x)-f(-x)}{2}$.

      $h(x) = x^3 + 5x$

6. In Problem 5, parts (c) and (d), list at least two aspects about the formulas you found as they relate to the function $f(x) = x^3 + 2x^2 + 5x + 1$.

   *The formula for 1(c) is all of the even power terms of $f$. The formula for 1(d) is all of the odd power terms of $f$. The sum of the two functions gives $f$ back again.*

7. For each of the functions $f$ and $g$ below, write an expression for (i) $f(g(x))$, (ii) $g(f(x))$, and (iii) $f(f(x))$ in terms of $x$.

   a. $f(x) = x^{\frac{2}{3}}$, $g(x) = x^{12}$

      *i.* $x^8$

      *ii.* $x^8$

      *iii.* $x^{\frac{4}{9}}$

   b. $f(x) = \frac{b}{x-a}$, $g(x) = \frac{b}{x} + a$ for two numbers $a$ and $b$, when $x$ is not $0$ or $a$

      *i.* $x$

      *ii.* $x$

      *iii.* $\frac{b}{\frac{b}{x-a}-a}$, *which is equivalent to* $\frac{b(x-a)}{b+a^2-ax}$

   c. $f(x) = \frac{x+1}{x-1}$, $g(x) = \frac{x+1}{x-1}$, when $x$ is not $1$ or $-1$

      *i.* $x$

      *ii.* $x$

      *iii.* $x$

   d. $f(x) = 2^x$, $g(x) = \log_2(x)$

      *i.* $x$

      *ii.* $x$

      *iii.* $x$

e. $f(x) = \ln(x)$, $g(x) = e^x$

*i.* $x$

*ii.* $x$

*iii.* $\ln(\ln(x))$

f. $f(x) = 2 \cdot 100^x$, $g(x) = \frac{1}{2}\log\left(\frac{1}{2}x\right)$

*i.* $x$

*ii.* $x$

*iii.* $2 \cdot 10000^{100^x}$

# Lesson 19: The Inverse Relationship Between Logarithmic and Exponential Functions

## Student Outcomes

- Students will understand that the logarithm function base $b$ and the exponential function base $b$ are inverse functions.

## Lesson Notes

In the previous lesson, students learned that if they reflected the graph of a logarithmic function across the diagonal line with equation $y = x$, then the reflection is the graph of the corresponding exponential function, and vice-versa. In this lesson, we formalize this graphical observation with the idea of inverse functions. Students have not yet been exposed to the idea of an inverse function, but it is natural for us to have that discussion in this module. In particular, this lesson attends to these standards:

> **F-BF.B.4a**: Solve an equation of the form $f(x) = c$ for a simple function $f$ that has an inverse and write an expression for the inverse.
>
> **F-LE.A.4**: For exponential models, express as a logarithm the solution to $ab^{ct} = d$ where $a$, $c$, and $d$ are numbers and the base $b$ is $2$, $10$, or $e$; evaluate the logarithm using technology.

In order to clarify the procedure for finding an inverse function, we start with algebraic functions before returning to transcendental logarithms and exponential functions.

Note: You might want to consider splitting this lesson over two days.

## Classwork

### Opening Exercise (8 minutes)

Before talking about inverse functions, review the idea of inverse operations. At this point, students have had a lot of practice thinking of division as *undoing* multiplication (in other words, multiplying by $5$, and then dividing by $5$ gives back the original number) and thinking of subtraction as "undoing" addition.

You may also want to recall for your students the compositions of two transformations. For example, in geometry, the image of a *counterclockwise* rotation of a triangle $\triangle ABC$ by $30°$ around a point $P$ is a new triangle congruent to the original. If we apply a $30°$ *clockwise* rotation to this new triangle around $P$ (rotation by $-30°$), the image is the original triangle again. That is, the rotation $R_{P,-30°}$ "undoes" the rotation $R_{P,30°}$: $R_{P,-30°}\left(R_{P,30°}(\triangle ABC)\right) = \triangle ABC$.

In this lesson, we will study functions that "undo" other functions, that is, given a function $f$, sometimes there is another function, $g$, such that if an output value of $f$ is inputted into $g$, the output of $g$ gives back the original number inputted into $f$. Such a function $g$ is called the inverse of the function $f$, just as division is the of multiplication.

**Opening Exercise**

a. **Consider the mapping diagram of the function $f$ below. Fill in the blanks of the mapping diagram of $g$ to construct a function that "undoes" each output value of $f$ by returning the original input value of $f$. (The first one is done for you.)**

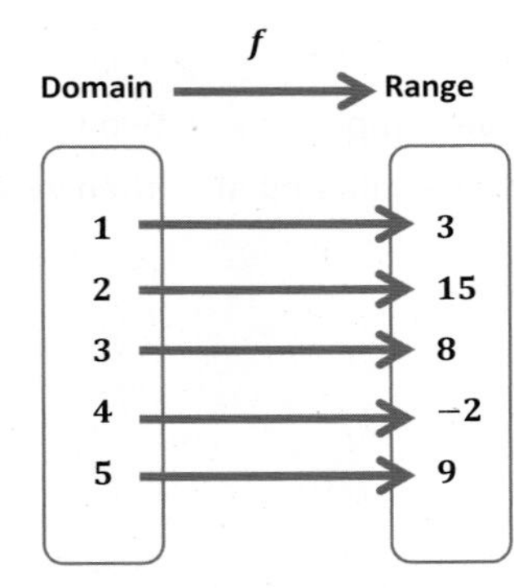


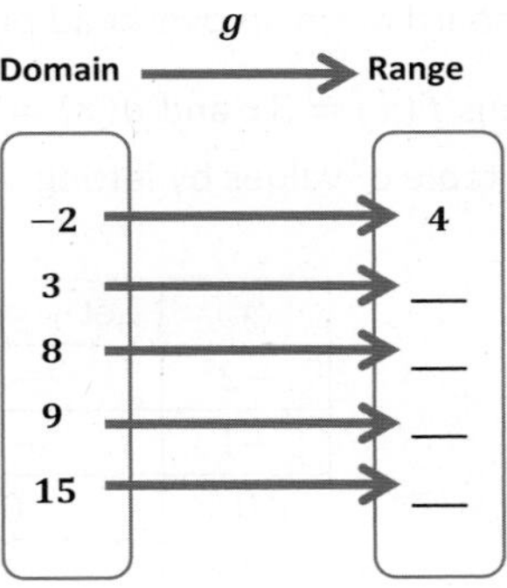


1, 3, 5, 2

As you walk around the room, help struggling students by drawing the analogy of multiplication and division of what students are asked to do above.

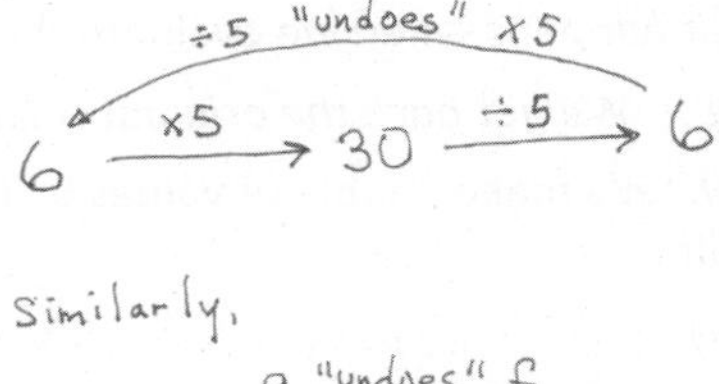


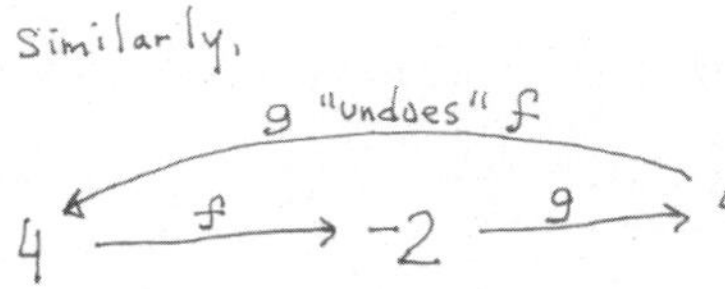


b. **Write the set of input-output pairs for the functions $f$ and $g$ by filling in the blanks below. (The set $F$ for the function $f$ has been done for you.)**

$F = \{(1,3), (2,15), (3,8), (4,-2), (5,9)\}$

$G = \{(-2,4),\quad (3,1),\quad (8,3),\quad (9,5),\quad (15,2)\}$

c. **How can the points in the set $G$ be obtained from the points in $F$?**

*The points in $G$ can be obtained from the points in $F$ by switching the first entry (first coordinate) with the second entry (second coordinate), that is, if $(a, b)$ is a point of $F$, then $(b, a)$ is a point of $G$.*

d. **Peter studied the mapping diagrams of the functions $f$ and $g$ above and exclaimed, "I can get the mapping diagram for $g$ by simply taking the mapping diagram for $f$ and reversing all of the arrows!" Is he correct?**

*He is almost correct. It is true that he can reverse the arrows, but he would also need to switch the domain and range labels to reflect that the range of $f$ is the domain of $g$, and the domain of $f$ is the range of $g$.*

We will explore Problems 1, 3 (as graphs), and 4 in more detail in the examples that follow.

### Discussion (8 minutes)

You may need to point out to students the meaning of Let $y = f(x)$ in this context. Usually, the equation $y = f(x)$ is an equation to be solved for solutions of the form $(x, y)$. However, when we state Let $y = f(x)$, we are using the equal symbol to assign the value $f(x)$ to $y$.

Complete this table either on the board or on an overhead projector.

- Consider the two functions $f(x) = 3x$ and $g(x) = \frac{x}{3}$. What happens if we compose these two functions in sequence? Let's make a table of values by letting $y$ be the value of $f$ when evaluated at $x$, then evaluating $g$ on the result.

| $x$ | Let $y = f(x)$ | $g(y)$ |
|---|---|---|
| $-2$ | $-6$ | $-2$ |
| $-1$ | $-3$ | $-1$ |
| $0$ | $0$ | $0$ |
| $1$ | $3$ | $1$ |
| $2$ | $6$ | $2$ |
| $3$ | $9$ | $3$ |

- What happens when we evaluate the function $f$ on a value of $x$, then the function $g$ on the result?
  - *We get back the original value of $x$.*
- Now, let's make a table of values by letting $y$ be the value of $g$ when evaluated at $x$, then evaluating $f$ on the result.

| $x$ | Let $y = g(x)$ | $f(y)$ |
|---|---|---|
| $-2$ | $-\frac{2}{3}$ | $-2$ |
| $-1$ | $-\frac{1}{3}$ | $-1$ |
| $0$ | $0$ | $0$ |
| $1$ | $\frac{1}{3}$ | $1$ |
| $2$ | $\frac{2}{3}$ | $2$ |
| $3$ | $1$ | $3$ |

- What happens when we evaluate the function $g$ on a value of $x$, then the function $f$ on the result?
  - *We get back the original value of $x$.*
- Does this happen with any two functions? What is special about the functions $f$ and $g$?

  *The formula for the function $f$ multiplies its input by $3$, and the formula for $g$ divides its input by $3$. If we first evaluate $f$ on an input then evaluate $g$ on the result, we are multiplying by $3$ then dividing by $3$, which has a net effect of multiplying by $\frac{3}{3} = 1$, so the result of the composition of $f$ followed by $g$ is the original input. Likewise, if we first evaluate $g$ on an input and then evaluate $f$ on the result, we are dividing a number by $3$ then multiplying the result by $3$ so that the net effect is again multiplication by $\frac{3}{3} = 1$, and the result of the composition is the original input.*

This does not happen with two arbitrarily chosen functions. It is special when it does, and the functions have a special name:

- Functions $f(x) = 3x$ and $g(x) = \frac{x}{3}$ are examples of inverse functions—functions that if you take the output of one function for a given input, and put the output into the other function, you get the original input back.

Inverse functions have a special relationship between their graphs. Let's explore that now and tie it back to what we learned earlier.

- Graph $f(x) = 3x$ and $g(x) = \frac{x}{3}$. (Also, sketch in the graph of the diagonal line $y = x$.)

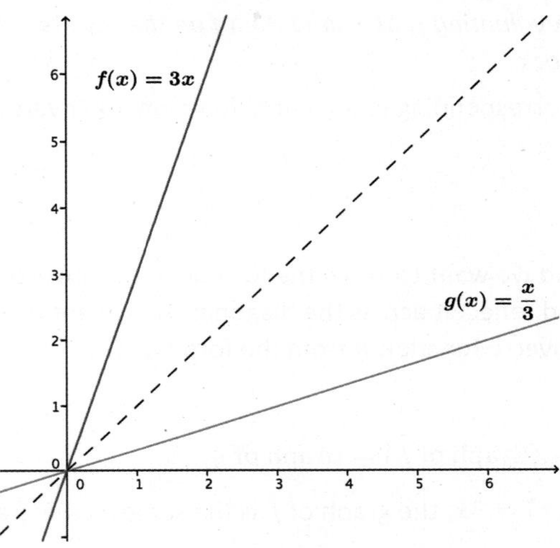


- What do we notice about these two graphs?
  - *They are reflections of each other across the diagonal line given by $y = x$.*
- What is the rule for the transformation that reflects the Cartesian plane across the line given by $y = x$?
  - $r_{y=x}(x, y) = (y, x)$.
- Where have we seen this switching of first and second coordinates before in this lesson? How is that situation similar to this one?
  - *In the Opening Exercise, to obtain the set $G$ from the set $F$, we took each ordered pair of $F$ and switched the first and second coordinates to get a point of $G$. Since plotting the points of $F$ and $G$ produce the graphs of those functions, we see that the graphs are reflections of each other across the diagonal given by $y = x$, that is, $r_{y=x}(F) = G$.*
    *Similarly, for $f(x) = 3x$ and $g(x) = \frac{x}{3}$, we get*
    *$r_{y=x}$(Graph of $f$) = Graph of $g$.*

Finally, let's tie what we learned about graphs of inverse functions to what we learned in the previous lesson.

- What other two functions have we seen whose graphs are reflections of each other across the diagonal line?
  - *The graph of a logarithm and an exponential function with the same base $b$ are reflections of each other across the diagonal line.*
- Make a conjecture about logarithm and exponential functions.
  - *A logarithm and an exponential function with the same base are inverses of each other.*

- Can we verify that using the properties of logarithms and exponents? What happens if we compose the functions by evaluating them one right after another? Let $f(x) = \log(x)$ and $g(x) = 10^x$. What do we know about the result of evaluating $f$ for a number $x$ and then evaluating $g$ on the resulting output? What about evaluating $g$ and then $f$?
  - *Let $y = f(x)$. Then $y = \log(x)$, so $g(y) = 10^y = 10^{\log(x)}$. By logarithmic property 4, $10^{\log(x)} = x$, so evaluating $f$ at $x$, and then $g$ on the results gives us the original input $x$ back.*
  - *Let $y = g(x)$. Then $y = 10^x$, so $f(y) = \log(10^x)$. By logarithmic property 3, $\log(10^x) = x$, so evaluating $g$ at $x$, and then $f$ on the results gives us the original input $x$ back.*
- So, yes, a logarithm function and its corresponding exponential function are inverse functions.

> *Scaffolding:*
>
> Remind students of the identities:
>
> 3. $\log_b(b^x) = x$,
> 4. $b^{\log_b(x)} = x$.
>
> It may also be helpful to include an example next to each property, such as $\log_3(3^x) = x$ and $10^{\log(5)} = 5$.

### Discussion (8 minutes)

What if we have the formula of a function $f$, and we want to know the formula for its inverse function $g$? At this point, all we know is that if we have the graph of $f$ and reflect it across the diagonal line we get the graph of its inverse $g$. We can use this fact to derive the formula for the inverse function $g$ from the formula of $f$.

Above, we saw that

$$r_{x=y}(\text{Graph of } f) = \text{Graph of } g.$$

Let's write out what those sets look like. For $f(x) = 3x$, the graph of $f$ is the same as the graph of the equation $y = f(x)$, that is, $y = 3x$:

$$\text{Graph of } f = \{(x, y) \mid y = 3x\}.$$

For $g(x) = \frac{x}{3}$, the graph of $g$ is the same as the graph of the equation $y = \frac{x}{3}$, which is the same as the graph of the equation $x = 3y$ (Why are they same?):

$$\text{Graph of } g = \{(x, y) \mid x = 3y\}.$$

Thus, the reflection across the diagonal line of the graph of $f$ can be written as follows:

$$\{(x, y) \mid y = 3x\} \xrightarrow{r_{x=y}} \{(x, y) \mid x = 3y\}.$$

- What relationship do you see between the set $\{(x, y) \mid y = 3x\}$ and the set $\{(x, y) \mid x = 3y\}$? How does this relate to the reflection map $r_{x=y}(x, y) = (y, x)$?
  - *To get the second set, we interchange $x$ and $y$ in the equation that defines the first set. This is exactly what the reflection map is telling us to do.*
- Let's double-check your answer with $f(x) = \log(x)$ and its inverse $g(x) = 10^x$. Focusing on $g$, we see that the graph of $g$ is the same as the graph of the equation $y = 10^x$. We can rewrite the equation $y = 10^x$ using logarithms as $x = \log(y)$. (Why are they the same?) Thus, the reflection across the diagonal line of the graph of $f$ can be written as follows:

$$\{(x, y) \mid y = \log(x)\} \xrightarrow{r_{x=y}} \{(x, y) \mid x = \log(y)\}.$$

This pair of sets also has the same relationship.

- How can we use that relationship to obtain the equation for the graph of $g$ from the graph of $f$?
  - *Write the equation $y = f(x)$, and then interchange the symbols to get $x = f(y)$.*
- How can we use the equation $x = f(y)$ to find the formula for the function $g$?
  - *Solve the equation for $y$ to write $y$ as an expression in $x$. The formula for $g$ is the expression in $x$.*
- In general, to find the formula for an inverse function $g$ of a given function $f$:
  i. Write $y = f(x)$ using the formula for $f$.
  ii. Interchange the symbols $x$ and $y$ to get $x = f(y)$.
  iii. Solve the equation for $y$ to write $y$ as an expression in $x$.
  iv. Then, the formula for $g$ is the expression in $x$ found in step (iii).

> *Scaffolding:*
> Use $f(x) = 3x$ to help students discover the steps: $f(x) = 3x \rightarrow y = 3x \rightarrow x = 3y \rightarrow \frac{x}{3} = y \rightarrow y = \frac{x}{3} \rightarrow g(x) = \frac{x}{3}$.

### Exercises 1–7 (8 minutes)

Give students a couple of minutes to use the procedure above on Exercise 1 (in groups of two). They will likely stumble, but that is okay; we want them to think through the procedure on their own and generate questions that they can ask (their partner or you). After giving students a couple of minutes, work through Exercise 1 as a whole class and move on to a selection of the remaining problems.

> *Scaffolding:*
> For Exercises 1–5, it may be useful to ask: What axes setting is required on a calculator to check whether the graphs of the two functions are reflections of each other? (The $x$-axis and $y$-axis must have the same scale.)

**Exercises**

**For each function $f$ in Exercises 1–5, find the formula for the corresponding inverse function $g$. Graph both functions on a calculator to check your work.**

1. $f(x) = 1 - 4x$

$$y = 1 - 4x$$
$$x = 1 - 4y$$
$$4y = 1 - x$$
$$y = \frac{(1 - x)}{4}$$
$$g(x) = \frac{(1 - x)}{4}$$

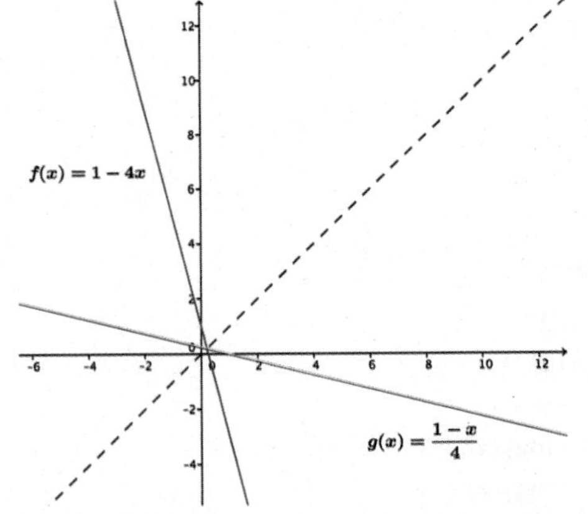

2. $f(x) = x^3 - 3$

$$y = x^3 - 3$$
$$x = y^3 - 3$$
$$y^3 = x + 3$$
$$y = \sqrt[3]{x+3}$$
$$g(x) = \sqrt[3]{x+3}$$

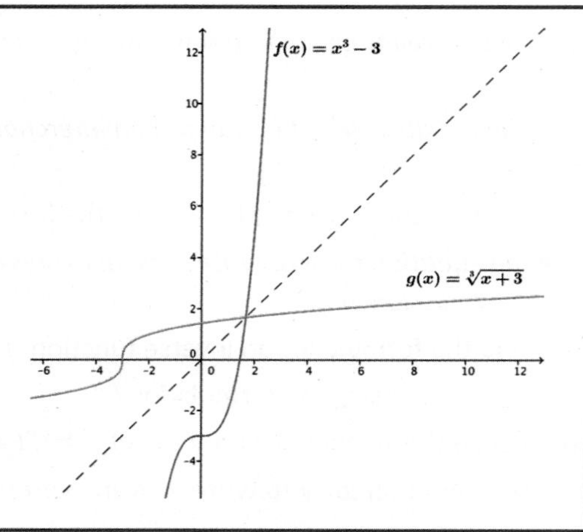


For Exercise 2, you may need to mention that, unlike principal square roots, there are real principal cube roots for negative numbers. This leads to the following identities that hold for all real numbers: $\sqrt[3]{x^3} = x$ and $\left(\sqrt[3]{x}\right)^3 = x$ for any real number $x$. These problems are practiced further in the Problem Set.

3. $f(x) = 3\log(x^2)$ for $x > 0$

$$y = 3\log(x^2)$$
$$x = 3\log(y^2)$$
$$x = 3 \cdot 2\log(y)$$
$$\log(y) = \frac{x}{6}$$
$$y = 10^{\frac{x}{6}}$$
$$g(x) = 10^{\frac{x}{6}}$$

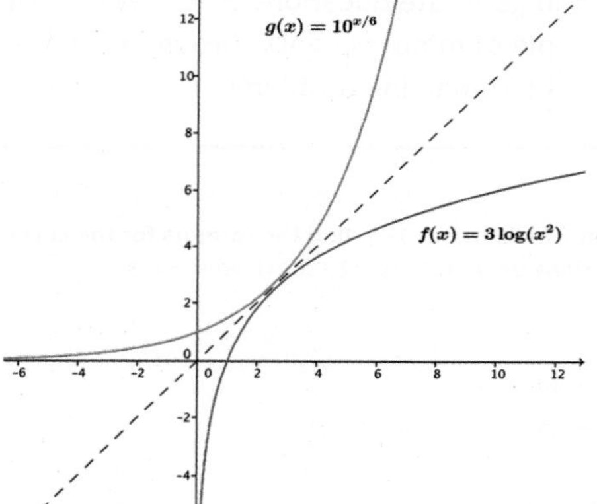


4. $f(x) = 2^{x-3}$

$$y = 2^{x-3}$$
$$x = 2^{y-3}$$
$$\log_2(x) = y - 3$$
$$y = \log_2(x) + 3$$
$$g(x) = \log_2(x) + 3$$

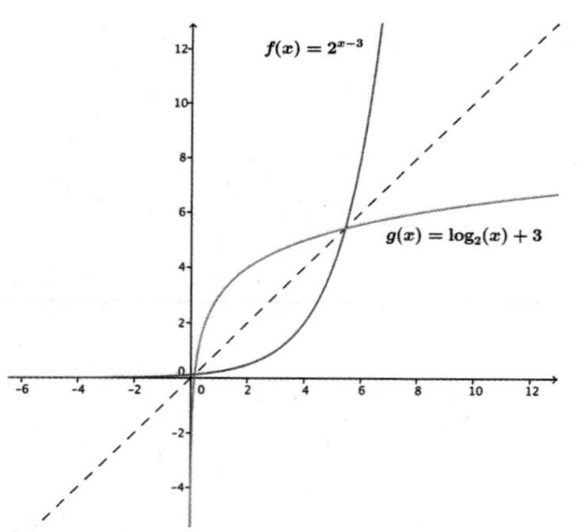

5. $f(x) = \frac{x+1}{x-1}$ for $x \neq 1$

$$y = \frac{x+1}{x-1}$$
$$x = \frac{y+1}{y-1}$$
$$x(y-1) = y+1$$
$$xy - x = y+1$$
$$xy - y = x+1$$
$$y(x-1) = x+1$$
$$y = \frac{x+1}{x-1}$$
$$g(x) = \frac{x+1}{x-1} \textit{ for } x \neq 1$$

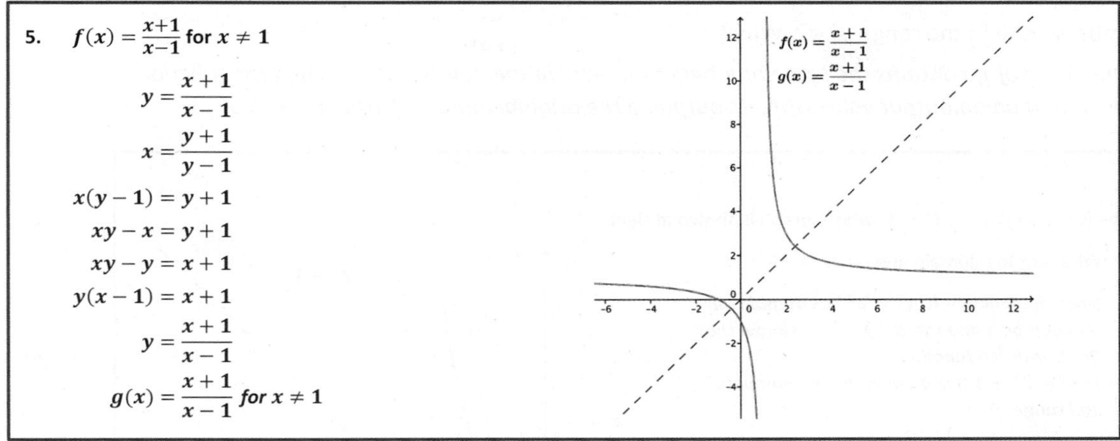


Exercise 5 is quite interesting. You might have students note that the functions $f$ and $g$ are the same. What do students notice about the graph of $f$? (This issue is further explored in the Problem Set.)

6. **Cindy thinks that the inverse of $f(x) = x - 2$ is $g(x) = 2 - x$. To justify her answer, she calculates $f(2) = 0$ and then substitutes the output $0$ into $g$ to get $g(0) = 2$, which gives back the original input. Show that Cindy is incorrect by using other examples from the domain and range of $f$.**

   ***Answers will vary, but any other point other than $2$ will work. For example, $f(3) = 1$, but $g(1) = 1$, not $3$ as needed.***

7. **After finding the inverse for several functions, Henry claims that *every* function must have an inverse. Rihanna says that his statement is not true and came up with the following example: If $f(x) = |x|$ has an inverse, then because $f(3)$ *and* $f(-3)$ both have the same output $3$, the inverse function $g$ would have to map $3$ to *both* $3$ *and* $-3$ simultaneously, which violates the definition of a function. What is another example of a function without an inverse?**

   ***Answers will vary. For example, $f(x) = x^2$ or any even degree polynomial function.***

You might consider showing students graphs of functions without inverses and what the graphs look like after reflecting them along the diagonal line (where it becomes obvious that the reflected figure cannot be a graph of a function).

## Example (5 minutes)

Now we need to address the question of how the domain and range of the function $f$ and its inverse function $g$ relate. You may need to review domain and range of a function with your students first.

- In all exercises we did above, what numbers were in the domain of $g$? Why?
  - *The domain of $g$ contains the same numbers that were in the range of $f$. This is because as the inverse of $f$, the function $g$ takes the output of $f$ (the range) as its input.*

> *Scaffolding:*
>
> For students who are struggling, use concrete examples from the Opening Exercise or from the exercises they just did. For example, "If $f(x) = 2^{x-3}$, what is an example of a number that is in the domain? The range?"

- What numbers were in the range of $g$? Why?
  - *The range of $g$ contains the same numbers that were in the domain of $f$. When the function $g$ is evaluated on an output value of $f$, its output is the original input of $f$ (the domain of $f$).*

**Example**

**Consider the function $f(x) = 2^x + 1$, whose graph is shown at right.**

a. **What are the domain and range of $f$?**

   ***Since the function $h(x) = 2^x$ has domain all real numbers and range $(0, \infty)$, we know that the translated function $f(x) = 2^x + 1$ has domain all real numbers and range $(1, \infty)$.***

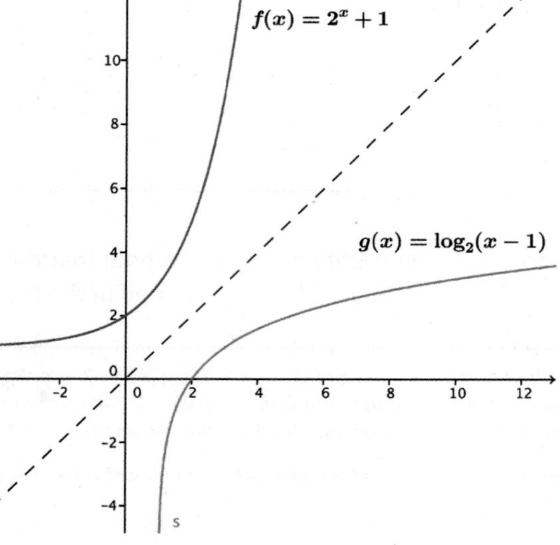


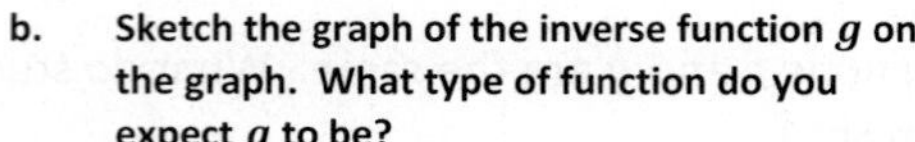

b. **Sketch the graph of the inverse function $g$ on the graph. What type of function do you expect $g$ to be?**

   ***Since logarithm and exponential functions are inverses of each other, $g$ should be some form of a logarithmic function (shown in red).***

c. **What are the domain and range of $g$? How does that relate to your answer in part (a)?**

   ***The range of $g$ is all real numbers, and the domain of $g$ is $(1, \infty)$, which makes sense since the range of $g$ is the domain of $f$, and the domain of $g$ is the range of $f$.***

d. **Find the formula for $g$.**

$$y = 2^x + 1$$
$$x = 2^y + 1$$
$$2^y = x - 1$$
$$y = \log_2(x-1)$$
$$g(x) = \log_2(x-1), \text{ for } x > 1$$

## Closing (3 minutes)

Ask students to summarize the important points of the lesson either in writing, orally with a partner, or as a class. Use this as an opportunity to informally assess understanding of the lesson. In particular, ask students to articulate the process for both graphing and finding the formula for the inverse of a given function. Some important summary elements are contained in the box below.

Lesson Summary

- **INVERTIBLE FUNCTION: Let $f$ be a function whose domain is the set $X$ and whose image is the set $Y$. Then $f$ is *invertible* if there exists a function $g$ with domain $Y$ and image $X$ such that $f$ and $g$ satisfy the property:**

  **For all $x$ in $X$ and $y$ in $Y$, $f(x) = y$ if and only if $g(y) = x$.**

  **The function $g$ is called the *inverse* of $f$.**
- **If two functions whose domain and range are a subset of the real numbers are inverses, then their graphs are reflections of each other across the diagonal line given by $y = x$ in the Cartesian plane.**
- **If $f$ and $g$ are inverses of each other, then**
  - **The domain of $f$ is the same set as the range of $g$.**
  - **The range of $f$ is the same set as the domain of $g$.**
- **In general, to find the formula for an inverse function $g$ of a given function $f$:**
  - **Write $y = f(x)$ using the formula for $f$.**
  - **Interchange the symbols $x$ and $y$ to get $x = f(y)$.**
  - **Solve the equation for $y$ to write $y$ as an expression in $x$.**
  - **Then, the formula for $g$ is the expression in $x$ found in step (iii).**
- **The functions $f(x) = \log_b(x)$ and $g(x) = b^x$ are inverses of each other.**

## Exit Ticket (5 minutes)

Name ______________________________ Date ____________

# Lesson 19: The Inverse Relationship Between Logarithmic and Exponential Functions

## Exit Ticket

1. The graph of a function $f$ is shown below. Sketch the graph of its inverse function $g$ on the same axes.

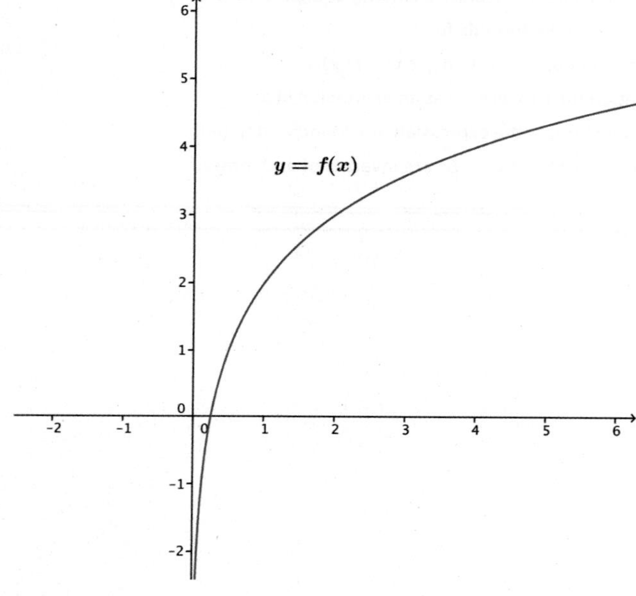


2. Explain how you made your sketch.

3. The function $f$ graphed above is the function $f(x) = \log_2(x) + 2$ for $x > 0$. Find a formula for the inverse of this function.

## Exit Ticket Sample Solutions

1. **The graph of a function $f$ is shown below. Sketch the graph of its inverse function $g$ on the same axes.**

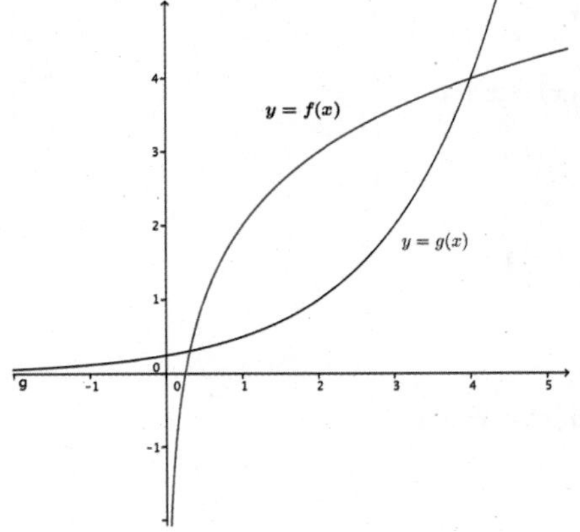


2. **Explain how you made your sketch.**

   *Answers will vary. Example: I drew the line given by $y = x$ and reflected the graph of $f$ across it.*

3. **The graph of the function $f$ above is the function $f(x) = \log_2(x) + 2$ for $x > 0$. Find a formula for the inverse of this function.**

$$y = \log_2(x) + 2$$
$$x = \log_2(y) + 2$$
$$x - 2 = \log_2(y)$$
$$\log_2(y) = x - 2$$
$$y = 2^{x-2}$$
$$g(x) = 2^{x-2}$$

## Problem Set Sample Solutions

1. **For each function $h$ below, find two functions $f$ and $g$ such that $h(x) = f(g(x))$. (There are many correct answers.)**

   a. $h(x) = (3x + 7)^2$

   *Possible answer:* $f(x) = x^2$, $g(x) = 3x + 7$

   b. $h(x) = \sqrt[3]{x^2 - 8}$

   *Possible answer:* $f(x) = \sqrt[3]{x}$, $g(x) = x^2 - 8$

   c. $h(x) = \frac{1}{2x-3}$

   *Possible answer:* $f(x) = \frac{1}{x}$, $g(x) = 2x - 3$

d. $h(x) = \frac{4}{(2x-3)^3}$

*Possible answer:* $f(x) = \frac{4}{x^3}$, $g(x) = 2x - 3$

e. $h(x) = (x+1)^2 + 2(x+1)$

*Possible answer:* $f(x) = x^2 + 2x, g(x) = x + 1$

f. $h(x) = (x+4)^{\frac{4}{5}}$

*Possible answer:* $f(x) = x^{\frac{4}{5}}$, $g(x) = x + 4$

g. $h(x) = \sqrt[3]{\log(x^2+1)}$

*Possible answer:* $f(x) = \sqrt[3]{\log(x)}$, $g(x) = x^2 + 1$

h. $h(x) = \sin(x^2 + 2)$

*Possible answer:* $f(x) = \sin(x)$, $g(x) = x^2 + 2$

i. $h(x) = \ln(\sin(x))$

*Possible answer:* $f(x) = \ln(x)$, $g(x) = \sin(x)$

2. Let $f$ be the function that assigns to each student in your class his or her biological mother.

a. Use the definition of function to explain why $f$ is a function.

*The function has a well-defined domain (students in the class) and range (their mothers), and each student is assigned one and only one biological mother.*

b. In order for $f$ to have an inverse, what condition must be true about the students in your class?

*If a mother has several children in the same classroom, then there would be no way to define an inverse function that picks one and only one student for each mother. The condition that must be true is that there are no siblings in the class.*

c. If we enlarged the domain to include all students in your school, would this larger domain function have an inverse?

*Probably not. Most schools have several students who are siblings.*

3. The table below shows a partially filled-out set of input-output pairs for two functions $f$ and $h$ that have the same finite domain of $\{0, 5, 10, 15, 20, 25, 30, 35, 40\}$.

| $x$ | 0 | 5 | 10 | 15 | 20 | 25 | 30 | 35 | 40 |
|---|---|---|---|---|---|---|---|---|---|
| $f(x)$ | 0 | 0.3 | 1.4 | | 2.1 | | 2.7 | 6 | |
| $h(x)$ | 0 | 0.3 | 1.4 | | 2.1 | | 2.7 | 6 | |

a. Complete the table so that $f$ is invertible but $h$ is definitely not invertible.

*Answers will vary. For $f$, all output values should be different. For $g$, at least two output values for two different inputs should be the same number.*

b. **Graph both functions and use their graphs to explain why $f$ is invertible and $h$ is not.**

***Answers will vary. The graph of $f$ has one unique output for every input, so it is possible to undo $f$ and map each of its outputs to a unique input. The graph of $g$ has at least two input values that map to the same output value. Hence, there is no way to map that output value back to a unique multiple of $5$. Hence, $h$ cannot have an inverse function because such a correspondence is not a function.***

4. **Find the inverse of each of the following functions. In each case, indicate the domain and range of both the original function and its inverse.**

a. $f(x) = \frac{3x-7}{5}$

$$x = \frac{3y-7}{5}$$
$$5x = 3y - 7$$
$$\frac{5x+7}{3} = y$$

***The inverse function is $g(x) = \frac{5x+7}{3}$. Both functions $f$ and $g$ have domain and range all real numbers.***

b. $f(x) = \frac{5+x}{6-2x}$

$$x = \frac{5+y}{6-2y}$$
$$6x - 2yx = 5 + y$$
$$6x - 5 = 2yx + y$$
$$6x - 5 = (2x+1)y$$
$$\frac{6x-5}{2x+1} = y$$

***The inverse function is $g(x) = \frac{6x-5}{2x+1}$.***

***Domain of $f$ and range of $g$: all real numbers $x$ with $x \neq 3$***

***Range of $f$ and domain of $g$: all real numbers $x$ with $x \neq -\frac{1}{2}$***

c. $f(x) = e^{x-5}$

$$x = e^{y-5}$$
$$\ln(x) = y - 5$$
$$\ln(x) + 5 = y$$

***The inverse function is $g(x) = \ln(x) + 5$.***

***Domain of $f$ and range of $g$: all real numbers $x$***

***Range of $f$ and domain of $g$: all real numbers $x$ with $x > 0$***

d. $f(x) = 2^{5-8x}$

$$x = 2^{5-8y}$$
$$\log_2(x) = 5 - 8y$$
$$8y = 5 - \log_2(x)$$
$$y = \frac{1}{8}(5 - \log_2(x))$$

*The inverse function is* $g(x) = \frac{1}{8}(5 - \log_2(x))$.

*Domain of* $f$ *and range of* $g$: *all real numbers* $x$

*Range of* $f$ *and domain of* $g$: *all real numbers* $x$ *with* $x > 0$

e. $f(x) = 7\log(1 + 9x)$

$$x = 7\log(1 + 9y)$$
$$\frac{x}{7} = \log(1 + 9y)$$
$$10^{\frac{x}{7}} = 1 + 9y$$
$$\frac{1}{9}(10^{\frac{x}{7}} - 1) = y$$

*The inverse function is* $g(x) = \frac{1}{9}\left(10^{\frac{x}{7}} - 1\right)$.

*Domain of* $f$ *and range of* $g$: *all real numbers* $x$ *with* $x > -\frac{1}{9}$

*Range of* $f$ *and domain of* $g$: *all real numbers* $x$

f. $f(x) = 8 + \ln(5 + \sqrt[3]{x})$

$$x = 8 + \ln(5 + \sqrt[3]{y})$$
$$x - 8 = \ln(5 + \sqrt[3]{y})$$
$$e^{x-8} = 5 + \sqrt[3]{y}$$
$$e^{x-8} - 5 = \sqrt[3]{y}$$
$$(e^{x-8} - 5)^3 = y$$

*The inverse function is* $g(x) = (e^{x-8} - 5)^3$.

*Domain of* $f$ *and range of* $g$: *all real numbers* $x$ *with* $x > -125$

*Range of* $f$ *and domain of* $g$: *all real numbers* $x$

g. $f(x) = \log\left(\frac{100}{3x+2}\right)$

$$x = \log\left(\frac{100}{3y+2}\right)$$
$$x = \log(100) - \log(3y+2)$$
$$x = 2 - \log(3y+2)$$
$$2 - x = \log(3y+2)$$
$$10^{2-x} = 3y+2$$
$$\frac{1}{3}(10^{2-x} - 2) = y$$

*The inverse function is* $g(x) = \frac{1}{3}(10^{2-x} - 2)$.

*Domain of $f$ and range of $g$: all real numbers $x$ with* $x > -\frac{2}{3}$

*Range of $f$ and domain of $g$: all real numbers $x$*

h. $f(x) = \ln(x) - \ln(x+1)$

$$x = \ln(y) - \ln(y+1)$$
$$x = \ln\left(\frac{y}{y+1}\right)$$
$$e^x = \frac{y}{y+1}$$
$$ye^x + e^x = y$$
$$ye^x - y = -e^x$$
$$y(e^x - 1) = -e^x$$
$$y = \frac{e^x}{1-e^x}$$

*The inverse function is* $g(x) = \frac{e^x}{1-e^x}$.

*Domain of $f$ and range of $g$: all real numbers $x$ with* $x > 0$

*Range of $f$ and domain of $g$: all real numbers* $x < 0$

i. $f(x) = \frac{2^x}{2^x+1}$

$$x = \frac{2^y}{2^y+1}$$
$$x2^y + x = 2^y$$
$$x2^y - 2^y = -x$$
$$2^y(x-1) = -x$$
$$2^y = \frac{-x}{x-1} = \frac{x}{1-x}$$
$$y\ln(2) = \ln\left(\frac{x}{1-x}\right)$$
$$y = \ln\frac{\left(\frac{x}{1-x}\right)}{\ln(2)}$$

*The inverse function is* $g(x) = \frac{\ln\left(\frac{x}{1-x}\right)}{\ln(2)}$ .

*Domain of $f$ and range of $g$: all real numbers $x$*

*Range of $f$ and domain of $g$: all real numbers $x$,* $0 < x < 1$

5. Unlike square roots that do not have any real principal square roots for negative numbers, principal cube roots do exist for negative numbers: $\sqrt[3]{-8}$ is the real number $-2$ since it satisfies $-2 \cdot -2 \cdot -2 = -8$. Use the identities $\sqrt[3]{x^3} = x$ and $\left(\sqrt[3]{x}\right)^3 = x$ for any real number $x$ to find the inverse of each of the functions below. In each case, indicate the domain and range of both the original function and its inverse.

a. $f(x) = \sqrt[3]{2x}$ for any real number $x$.

$$y = \sqrt[3]{2x}$$
$$x = \sqrt[3]{2y}$$
$$x^3 = 2y$$
$$2y = x^3$$
$$y = \frac{1}{2}(x^3)$$
$$g(x) = \frac{1}{2}(x^3)$$

*Domain of $f$ and range of $g$: all real numbers $x$*
*Range of $f$ and domain of $g$: all real numbers $x$*

b. $f(x) = \sqrt[3]{2x-3}$ for any real number $x$.

$$y = \sqrt[3]{2x-3}$$
$$x = \sqrt[3]{2y-3}$$
$$x^3 = 2y - 3$$
$$2y = x^3 + 3$$
$$y = \frac{1}{2}(x^3 + 3)$$
$$g(x) = \frac{1}{2}(x^3 + 3)$$

*Domain of $f$ and range of $g$: all real numbers $x$*
*Range of $f$ and domain of $g$: all real numbers $x$*

c. $f(x) = (x-1)^3 + 3$ for any real number $x$.

$$y = (x-1)^3 + 3$$
$$x = (y-1)^3 + 3$$
$$x - 3 = (y-1)^3$$
$$\sqrt[3]{x-3} = y - 1$$
$$y - 1 = \sqrt[3]{x-3}$$
$$y = \sqrt[3]{x-3} + 1$$
$$g(x) = \sqrt[3]{x-3} + 1$$

*Domain of $f$ and range of $g$: all real numbers $x$*

*Range of $f$ and domain of $g$: all real numbers $x$*

6. Suppose that the inverse of a function is the function itself. For example, the inverse of the function $f(x) = \frac{1}{x}$ (for $x \neq 0$) is just itself again, $g(x) = \frac{1}{x}$ (for $x \neq 0$). What symmetry must the graphs of all such functions have? (Hint: Study the graph of Exercise 5 in the lesson.)

*All graphs of functions that are self-inverses are symmetric with respect to the diagonal line given by the equation $y = x$, i.e., a reflection across the line given by $y = x$ takes the graph back to itself.*

7. When traveling abroad, you will find that daily temperatures in other countries are often reported in Celsius. The sentence, "It will be $25°C$ today in Paris," does not mean it will be freezing in Paris. It will often be necessary for you to convert temperatures reported in degrees Celsius to degrees Fahrenheit, the scale we use in the U.S. for reporting daily temperatures.

Let $f$ be the function that inputs a temperature measure in degrees Celsius and outputs the corresponding temperature measure in degrees Fahrenheit.

a. Assuming that $f$ is linear, we can use two points on the graph of $f$ to determine a formula for $f$. In degrees Celsius, the freezing point of water is $0$, and its boiling point is $100$. In degrees Fahrenheit, the freezing point of water is $32$, and its boiling point is $212$. Use this information to find a formula for the function $f$. (Hint: Plot the points and draw the graph of $f$ first, keeping careful track of the meaning of values on the $x$-axis and $y$-axis.)

$$f(t) = \frac{9}{5}t + 32$$

b. What temperature will Paris be in degrees Fahrenheit if it is reported that it will be $25°C$?

*Since $f(25) = 77$, it will be $77°F$ in Paris that day.*

c. Find the inverse of the function $f$ and explain its meaning in terms of degree scales that its domain and range represent.

$g(t) = \frac{5}{9}(t - 32)$. *Given the measure of a temperature reported in degrees Fahrenheit, the function converts that measure to degrees Celsius.*

d. The graphs of $f$ and its inverse are two lines that intersect in one point. What is that point? What is its significance in terms of degrees Celsius and degrees Fahrenheit?

*The point is $(-40, -40)$.*

*The temperature has the same measure in both degrees Celsius and in degrees Fahrenheit.*

Extension: Use the fact that, for $b > 1$, the functions $f(x) = b^x$ and $g(x) = \log_b(x)$ are increasing to solve the following problems. Recall that an increasing function $f$ has the property that if both $a$ and $b$ are in the domain of $f$ and $a < b$, then $f(a) < f(b)$.

8. For which values of $x$ is $2^x < \frac{1}{1{,}000{,}000}$?

$$2^x < \frac{1}{1{,}000{,}000}$$
$$x < \log_2\left(\frac{1}{1{,}000{,}000}\right) = -\log_2(1{,}000{,}000)$$

9. For which values of $x$ is $\log_2(x) < -1{,}000{,}000$?

$$\log_2(x) < -1{,}000{,}000$$
$$x < 2^{-1{,}000{,}000}$$

# Lesson 20: Transformations of the Graphs of Logarithmic and Exponential Functions

## Student Outcomes

- Students study transformations of the graphs of logarithmic functions.
- Students use the properties of logarithms and exponents to produce equivalent forms of exponential and logarithmic expressions. In particular, they notice that different types of transformations can produce the same graph due to these properties.

## Lesson Notes

Students revisit the use of transformations to produce graphs of exponential and logarithmic functions (**F-BF.B.3**, **F-IF.B.4**, **F-IF.C.7e**). They make and verify conjectures about why certain transformations of the graphs of functions produce the same graph by applying the properties to produce equivalent expressions (MP.3). This work leads to a general form of both logarithmic and exponential functions where the given parameters can be quickly analyzed to determine key features and to sketch graphs of logarithmic and exponential functions (MP.7, MP.8). This lesson reinforces sketching graphs of functions by applying knowledge of transformations and the properties of logarithms and exponents.

## Classwork

### Opening Exercise (8 minutes)

Since much of the work on this lesson will involve the connections between scaling and translating graphs of functions, this Opening Exercise presents students with an opportunity to reflect on what they already know about transformations of graphs of functions using a simple polynomial function and the sine function. Observe students carefully as they work on these exercises to gauge how much re-teaching or additional support may be needed in the Exploratory Challenge that follows. If students struggle to recall their knowledge of transformations, you may need to provide additional guidance and practice throughout the lesson. A grid is provided for students to use when sketching the graphs in Opening Exercise, part (a), but students could also complete this exercise using graphing technology.

**Opening Exercise**

a. **Sketch the graphs of the three functions $f(x) = x^2$, $g(x) = (2x)^2 + 1$, and $h(x) = 4x^2 + 1$.**

   i. **Describe the transformations that will take the graph of $f(x) = x^2$ to the graph of $g(x) = (2x)^2 + 1$.**

   *The graph of $g$ is a horizontal scaling by a factor of $\frac{1}{2}$ and a vertical translation up $1$ unit of the graph of $f$.*

   ii. **Describe the transformations that will take the graph of $f(x) = x^2$ to the graph of $h(x) = 4x^2 + 1$.**

   *The graph of $h$ is a vertical scaling by a factor of $4$ and a vertical translation up $1$ unit of the graph of $f$.*

iii. **Explain why $g$ and $h$ from parts (i) and (ii) are equivalent functions.**

*These functions are equivalent and have the same graph because the expressions $(2x)^2 + 1$ and $4x^2 + 1$ are equivalent. The blue graph shown is the graph of $f$, and the red graph is the graph of $g$ and $h$.*

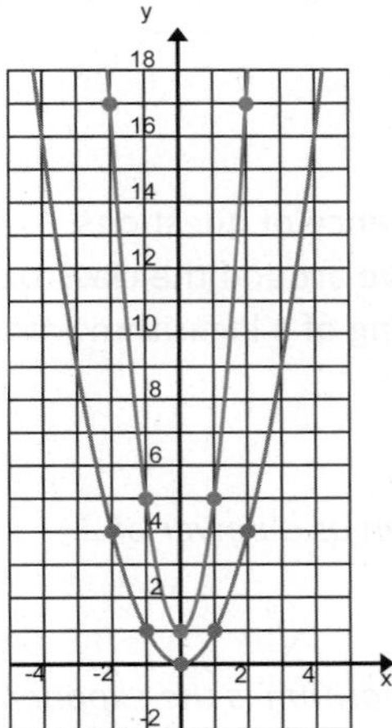

b. **Describe the transformations that will take the graph of $f(x) = \sin(x)$ to the graph of $g(x) = \sin(2x) - 3$.**

*The graph of $g$ is a horizontal scaling by a factor of $\frac{1}{2}$ and a vertical translation down 3 units of the graph of $f$.*

c. **Describe the transformations that will take the graph of $f(x) = \sin(x)$ to the graph of $h(x) = 4\sin(x) - 3$.**

*The graph of $h$ is a vertical scaling by a factor of 4 and a vertical translation down 3 units of the graph of $f$.*

d. **Explain why $g$ and $h$ from parts (b)–(c) are *not* equivalent functions.**

*These functions are not equivalent because they do not have the same graphs, and the two expressions are not equivalent.*

*Scaffolding:*

- For students who struggle with visual processing, provide larger graph paper and/or colored pencils to color code their graphs.
- For struggling students, prominently display the properties of logarithms and exponents on the board or on chart paper in your room for visual reference. Refer students back to these charts during the exploration with questions such as: "Which property could you use to rewrite the expression?"

Have students share responses and revise their work in small groups. Be sure to emphasize that students need to label graphs since many of them appear on the same set of axes. Lead a brief whole-group discussion that focuses on the responses to Opening Exercise part (a-iii) and part (d). Perhaps have one or two students share their written response with the whole class, using either the board or the document camera. Check to make sure students are actually writing responses to these questions.

Conclude this discussion by helping students to understand the following ideas that, as will be demonstrated later, are also true with graphs of logarithmic and exponential functions.

- Certain transformations of the graph of a function can be identical to other transformations depending on the properties of the given function.
- For example, the function $g(x) = x + 1$ is either a horizontal translation of 1 unit to the left or a vertical translation of 1 unit up of the graph of $f(x) = x$. Similarly, $g(x) = |-2x + 2|$ has the same graph as $h(x) = 2|x - 1|$, but could be described using different transformations of the graph of $f(x) = |x|$.

Announce to students that in this lesson, they will explore the properties of logarithms and exponents to understand graphing transformations of those types of functions, and they will explore when two different functions have the same graph in order to reinforce those properties.

## Exploratory Challenge (15 minutes)

Students should work in small groups to complete this sequence of questions. Provide support to individual groups or students as you move around the classroom. As you circulate, keep questioning students as to the meaning of a logarithm with questions like those below.

MP.1

- What does $\log_2(4)$ mean?
  - *The exponent when the number $4$ is written as a power of $2$.*
- Why is $\log_2\left(\frac{1}{4}\right)$ negative?
  - *It is negative because $\frac{1}{4} = 2^{-2}$, and the logarithm is the exponent when the number $\frac{1}{4}$ is written as a power of $2$.*
- What will be the domain and range of the function $f$? Why does this make sense given the definition of a logarithm?
  - *The domain is all real numbers greater than $0$. The range is all real numbers. This makes sense because the range of the exponential function $f(x) = 2^x$ is all real numbers greater than $0$, the domain is all real numbers, and the logarithmic function base $2$ is the inverse of the exponential function base $2$.*

> *Scaffolding:*
>
> Use technology to support learners who are still struggling with arithmetic and need visual reinforcement. Students can investigate using graphing calculators or online graphing programs. Newer calculators and graphing programs have a $\log_b(x)$ function built in. On older models, you may need to coach students to use the change of base property to enter these functions (i.e., to graph $f(x) = \log_2(x)$, you will need to enter the expression $\frac{\log(x)}{\log(2)}$ into the graphing calculator).

You can extend this lesson by using graphing software such as GeoGebra to create parameterized graphs with sliders (variables $a$ and $b$ that can be dynamically changed while viewing graphs). By manipulating the values of $a$ and $b$ in the functions $f(x) = \log_2(ax)$ and $g(x) = \log_2(x) + b$, you can verify that the graphs of $f(x) = \log_2(2x)$ and $g(x) = \log_2(x) + 1$ are the same. This result reinforces properties of logarithms since $\log_2(2x) = \log_2(x) + \log_2(2)$. Students and teachers can similarly confirm the other examples in this lesson as well.

**Exploratory Challenge**

a. **Sketch the graph of $f(x) = \log_2(x)$ by identifying and plotting at least five key points. Use the table below to help you get started.**

***The graph of $f$ is blue, and the graph of $g$ and $h$ is red on the solution graph shown below.***

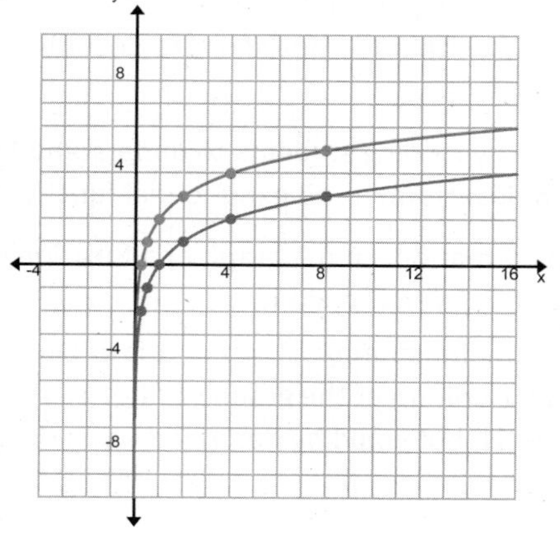

b. **Describe the transformations that will take the graph of $f$ to the graph of $g(x) = \log_2(4x)$.**

*The graph of $g$ is a horizontal scaling by a factor of $\frac{1}{4}$ of the graph of $f$.*

c. **Describe the transformations that will take the graph of $f$ to the graph of $h(x) = 2 + \log_2(x)$.**

*The graph of $h$ is a vertical translation up $2$ units of the graph of $f$.*

d. **Complete the table below for $f$, $g$, and $h$ and describe any patterns that you notice.**

| $x$ | $f(x)$ | $g(x)$ | $h(x)$ |
|---|---|---|---|
| $\frac{1}{4}$ | $-2$ | $0$ | $0$ |
| $\frac{1}{2}$ | $-1$ | $1$ | $1$ |
| $1$ | $0$ | $2$ | $2$ |
| $2$ | $1$ | $3$ | $3$ |
| $4$ | $2$ | $4$ | $4$ |
| $8$ | $3$ | $5$ | $5$ |

*The functions $g$ and $h$ have the same range values at each domain value in the table.*

e. **Graph the three functions on the same coordinate axes and describe any patterns that you notice.**

*See graphs above. The graphs of $g$ and $h$ are identical.*

f. **Use a property of logarithms to show that $g$ and $h$ are equivalent.**

*By the product property and the definition of logarithm, $\log_2(4x) = \log_2(4) + \log_2(x) = 2 + \log_2(x)$.*

Call the entire class together at this point to debrief their work so far. Make sure students understand that by applying the product or quotient property of logarithms, they can rewrite a single logarithmic expression as a sum or difference. In this way, a horizontal scaling of the graph of a logarithmic function will produce the same graph as a vertical translation. The next three parts can be used to informally assess student understanding of the idea that two different transformations can produce the same graph because of the properties of logarithms.

g. **Describe the graph of $g(x) = \log_2\left(\frac{x}{4}\right)$ as a vertical translation of the graph of $f(x) = \log_2(x)$. Justify your response.**

*The graph of $g$ is a vertical translation down $2$ units of the graph of $f$ because $\log_2\left(\frac{x}{4}\right) = \log_2(x) - 2$.*

h. **Describe the graph of $h(x) = \log_2(x) + 3$ as a horizontal scaling of the graph of $f(x) = \log_2(x)$. Justify your response.**

*The graph of $h$ is a horizontal scaling by a factor of $\frac{1}{8}$ of the graph of $f$ because $\log_2(x) + 3 = \log_2(x) + \log_2(8) = \log_2(8x)$.*

i. **Do the functions $f(x) = \log_2(x) + \log_2(4)$ and $g(x) = \log_2(x+4)$ have the same graphs? Justify your reasoning.**

*No, they do not. By substituting $1$ for $x$ in both $f$ and $g$, you can see that the graphs of the two functions will not have the same $y$-coordinate at this point. Therefore, the graphs cannot be the same if at least one point is different.*

Carefully review the answers to the preceding questions to check for student understanding. Students may struggle with expressing $3$ as $\log_2(8)$ in part (h). In the last portion of the Exploratory Challenge, students turn their attention to exponential functions and apply the properties of exponents to explain why graphs of certain exponential functions are identical. Circulate around the room while students are working and encourage them to create the graph of the "parent" function by plotting key points on the graph of the function and then transforming those key points according to the transformation they described.

MP.7

j. **Use properties of exponents to explain why graphs of $f(x) = 4^x$ and $g(x) = 2^{2x}$ are identical.**

*Using the power property of exponents, $2^{2x} = (2^2)^x = 4^x$. Since the expressions are equal, the graphs of the functions would be the same.*

k. **Use the properties of exponents to predict what the graphs of $f(x) = 4 \cdot 2^x$ and $g(x) = 2^{x+2}$ will look like compared to one another. Describe the graphs of $f$ and $g$ as transformations of the graph of $f = 2^x$. Confirm your prediction by graphing $f$ and $g$ on the same coordinate axes.**

*The graphs of these two functions will be the same since $2^{x+2} = 2^x \cdot 2^2 = 4 \cdot 2^x$ by the multiplication property of exponents and the commutative property. The graph of $f$ is the graph of $y = 2^x$ scaled vertically by a factor of $4$. The graph of $g$ is the graph of $y = 2^x$ translated horizontally $2$ units to the left.*

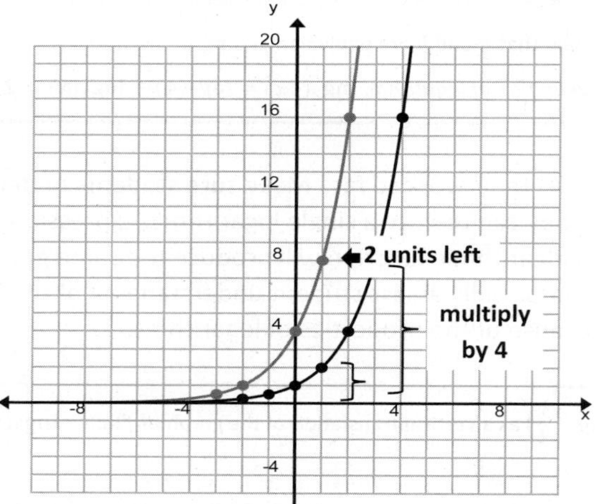

l. **Graph $f(x) = 2^x$, $g(x) = 2^{-x}$, and $h(x) = \left(\frac{1}{2}\right)^x$ on the same coordinate axes. Describe the graphs of $g$ and $h$ as transformations of the graph of $f$. Use the properties of exponents to explain why $g$ and $h$ are equivalent.**

***The graph of $g$ and the graph of $h$ are both reflections about the vertical axis of the graph of $f$. They are equivalent because $\left(\frac{1}{2}\right)^x = (2^{-1})^x = 2^{-x}$ by the definition of a negative exponent and the power property of exponents.***

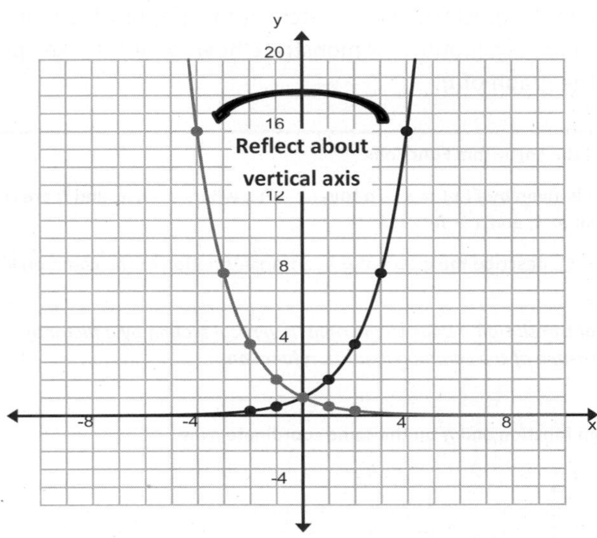


Have groups volunteer to present their findings on the last three parts of the Exploratory Challenge. When debriefing, model both transformations for students by marking the sketch as shown in the solutions above.

Discuss these transformations.

- In part (k), how do you see the transformations that produce the graphs of $f$ and $g$ from the graph of $y = 2^x$?
  - *I see the horizontal translation of 2 units to the left, but others might see the vertical scaling that takes each $y$-value and multiplies it by 4.*
- In part (l), how do the transformations validate the definition of a negative exponent?
  - *Since the graphs of $g$ and $h$ were identical, we have visual confirmation that $\left(\frac{1}{2}\right)^x = 2^{-x}$, which can only be true if $\frac{1}{2} = 2^{-1}$.*

Then, have students respond to the reflection question below in writing or with a partner.

- How do the properties of logarithms and exponents justify the fact that different transformations of the graph of a function can sometimes produce the same graph?
  - *We can use the properties to rewrite logarithmic and exponential expressions in equivalent forms that then represent different transformations of the same original function.*

If time permits, you can also tie these transformations to a simple real-world context. For example, if we rewrite the function $f(x) = 2^{x+3}$ as $f(x) = 8 \cdot 2^x$, students can see that adding 3 to $x$ would be like going three years forward in time. This means the population doubled three times, which is why we are multiplying by 8.

## Example 1 (4 minutes): Graphing Transformations of the Logarithm Functions

MP.7 & MP.8 Introduce the general form of a logarithm function, noting that we do not need a horizontal scaling parameter since a horizontal scaling can always be rewritten as a vertical translation. Continue to reinforce learning from the previous lessons by asking students why the restrictions on $b$ and $x - h$ are necessary. Students should be able to work through part (a) without your assistance, but monitor their work to make sure that all students have the correct answer to refer to when they work the Problem Set. MP.2 Model your expectations for sketching the graphs of logarithm functions in part (b) so students are able to produce accurate and precise graphs. Demonstrate how to plot the key points, and then transform the individual points to produce the graph of $g$.

**Example 1: Graphing Transformations of the Logarithm Functions**

**The general form of a logarithm function is given by $f(x) = k + a\log_b(x - h)$, where $a$, $b$, $k$, and $h$ are real numbers such that $b$ is a positive number not equal to $1$, and $x - h > 0$.**

a. **Given $g(x) = 3 + 2\log(x - 2)$, describe the graph of $g$ as a transformation of the common logarithm function.**

*The graph of $g$ is a horizontal translation $2$ units to the right, a vertical scaling by a factor of $2$, and a vertical translation up $3$ units of the graph of the common logarithm function.*

b. **Graph the common logarithm function and $g$ on the same coordinate axes.**

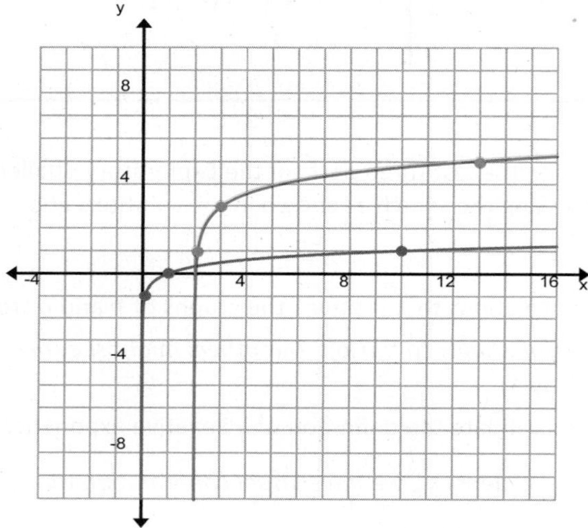


The common logarithm function is shown in blue, and the graph of $f$ is shown in red. Notice the key points that students should include on their hand-drawn sketches.

## Example 2 (4 minutes): Graphing Transformations of Exponential Functions

Introduce the general form of the exponential function, noting that we do not need a horizontal scaling or a horizontal translation since these can always be rewritten using the properties of exponents. Demonstrate, or let students work with a partner on part (a), and make sure students have correct work to refer to when they work on the Problem Set. Since earlier lessons applied transformations to graphing exponential functions, parts (b), (c), and (d) should move along rather quickly. Continue to reinforce your expectations for sketching graphs of functions using transformations.

**Example 2: Graphing Transformations of Exponential Functions**

**The general form of the exponential function is given by $f(x) = a \cdot b^x + k$, where $a$, $b$, and $k$ are real numbers such that $b$ is a positive number not equal to $1$.**

a. **Use the properties of exponents to transform the function $g(x) = 3^{2x+1} - 2$ to the general form, and then graph it. What are the values of $a$, $b$, and $k$?**

*Using the properties of exponents, $3^{2x+1} - 2 = 3^{2x} \cdot 3^1 - 2 = 3 \cdot 9^x - 2$. Thus, $g(x) = 3(9)^x - 2$, so $a = 3$, $b = 9$, and $k = -2$.*

b. **Describe the graph of $g$ as a transformation of the graph of $h(x) = 9^x$.**

*The graph of $g$ is a vertical scaling by a factor of $3$ and a vertical translation down $2$ units of the graph of $h$.*

c. **Describe the graph of $g$ as a transformation of the graph of $h(x) = 3^x$.**

*The graph of $g$ is a horizontal scaling by a factor of $\frac{1}{2}$, a vertical scaling by a factor of $3$, and a vertical translation down $2$ units of the graph of $h$.*

d. **Graph $g$ using transformations.**

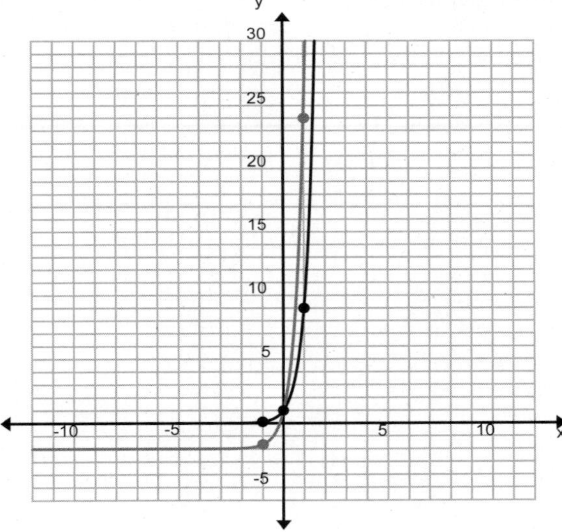


*The graph of $y = 9^x$ is shown in black, and the graph of $f$ is shown in blue.*

## Exercises 1–4 (4 minutes)

Students can work on these exercises independently or with a partner. Monitor their work by circulating around the classroom and checking for accuracy. Encourage students to describe the graph of $g$ as a transformation of the graph of $f$ in more than one way and to justify their answer analytically. In particular, emphasize how rewriting the expression using the properties of logarithms can make sketching the graphs easier because a horizontal scaling is revealed to have the same effect as a vertical translation when graphing logarithm functions.

**Exercises 1–4**

**Graph each pair of functions by first graphing $f$ and then graphing $g$ by applying transformations of the graph of $f$. Describe the graph of $g$ as a transformation of the graph of $f$.**

1. $f(x) = \log_3(x)$ and $g(x) = 2\log_3(x-1)$

   ***The graph of $g$ is the graph of $f$ translated $1$ unit to the right and stretched vertically by a factor of $2$.***

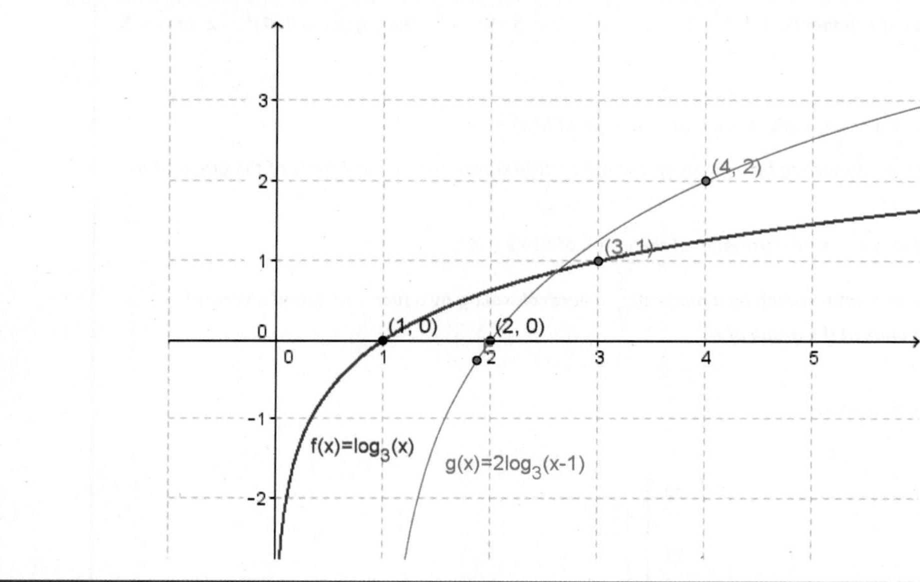

2. $f(x) = \log(x)$ and $g(x) = \log(100x)$

***Because of the product property of logarithms, $g(x) = 2 + \log(x)$. The graph of $g$ is the graph of $f$ translated vertically $2$ units.***

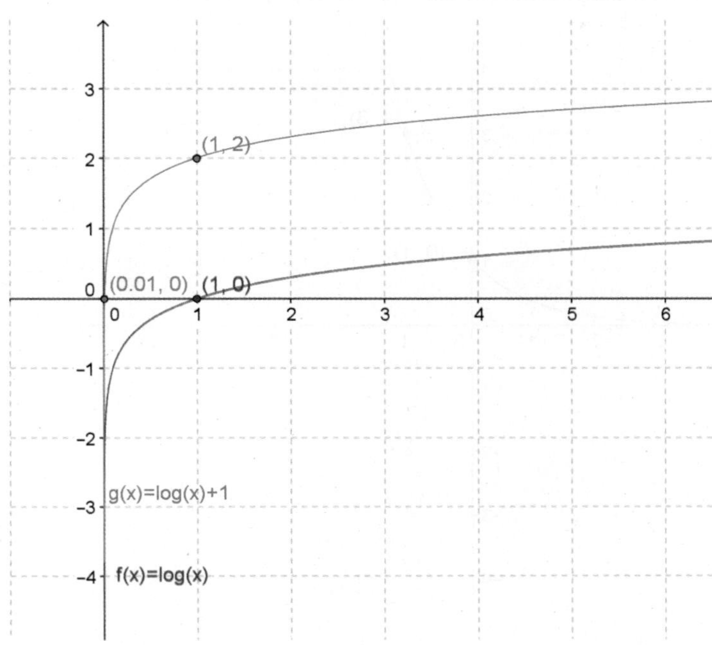


3. $f(x) = \log_5 x$ and $g(x) = -\log_5(5(x+2))$

***Since $-\log_5(5(x+2)) = -1 - \log_5(x+2)$ by the product property of logarithms and the distributive property, the graph of $g$ is the graph of $f$ translated $2$ units to the left, reflected across the horizontal axis, and translated down $1$ unit.***

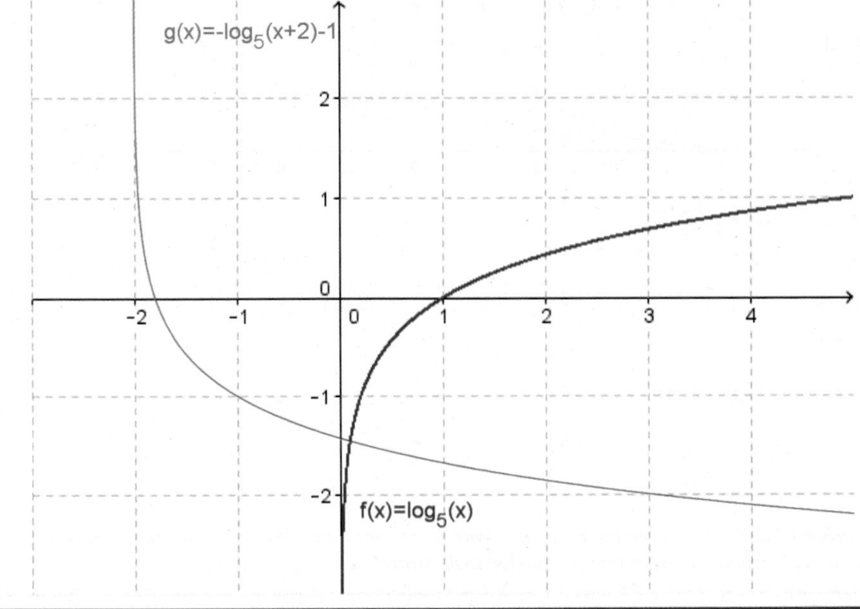

4. $f(x) = 3^x$ and $g(x) = -2 \cdot 3^{x-1}$

***Since $-2 \cdot 3^{x-1} = -2 \cdot 3^x \cdot 3^{-1} = -\frac{2}{3} \cdot 3^x$ by the properties of exponents and the commutative property, the graph of $g$ is the graph of $f$ reflected across the horizontal axis and compressed by a factor of $\frac{2}{3}$.***

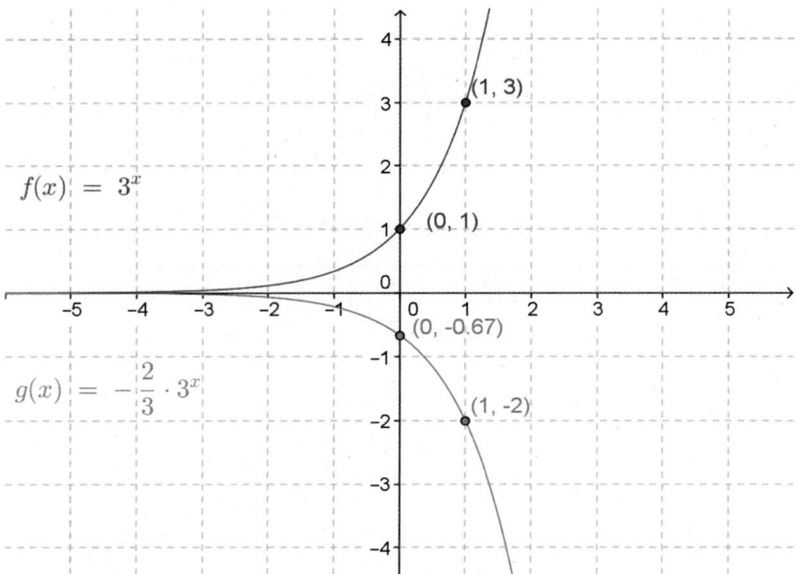


***The graph above shows how to obtain the graph of $g$ from the graph of $f$ by reflecting about the horizontal axis and vertically scaling by a factor of $\frac{2}{3}$. A few points are labeled to illustrate the transformations.***

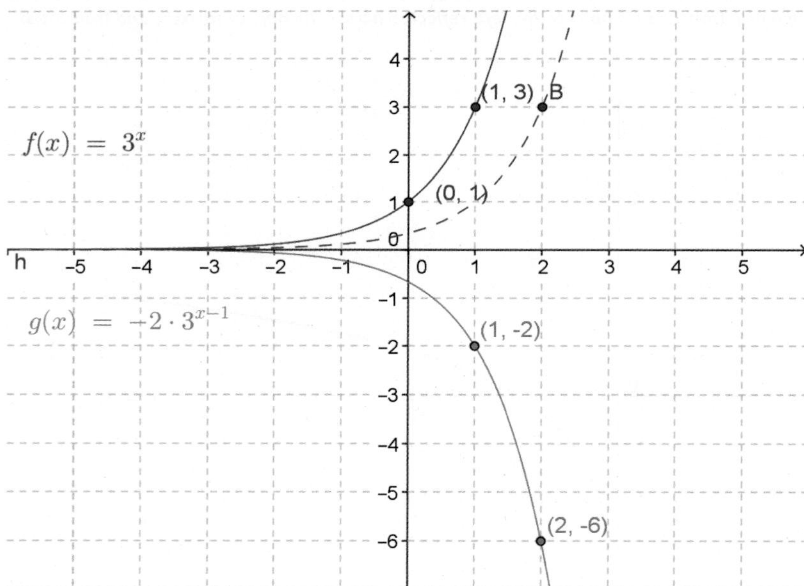


***This graph shows the graph of $g$ obtained from the graph of $f$ with a horizontal translation, a reflection across the horizontal axis, and a vertical scaling. A few points are labeled to illustrate the transformations.***

After a few minutes, have different groups share how they saw the transformations and discuss when it is advantageous to rewrite an expression before graphing and when it is not. For example, it might be easier in Exercise 10 to simply translate the graph 1 unit to the right rather than scale it by a factor of $\frac{2}{3}$. Also, make sure students are including a sketch of the end behavior of the functions.

### Closing (5 minutes)

Provide students with an opportunity to summarize their learning with a partner by responding to the questions below. Their summaries will provide you with additional evidence of their understanding of this lesson.

- How do you apply properties of logarithms or exponents to rewrite $f(x) = \log_2(5x)$ and $g(x) = 3^{x+2} + 2$ in general form?
  - *Using the product properties:* $\log_2(5x) = \log_2(5) + \log_2(x)$, *so* $f(x) = \log_2(5) + \log_2(x)$ *in general form where* $k = \log_2(5)$, $a = 1$, *and* $h = 0$ *in the general form.*
  - *Using the product properties:* $3^{x+2} = 3^x \cdot 3^2$, *so* $g(x) = 9 \cdot 3^x + 2$ *in general form where* $a = 9$ *and* $k = 2$.
- How do transformations help you to sketch quick and accurate graphs of functions?
  - *If you can make a basic logarithm or exponential function graph for a given base, then transformations can be used to quickly sketch a new function that is based on the original function.*

A summary of the key points of this lesson is provided. Review them with your class before they begin the Exit Ticket.

> **Lesson Summary**
>
> **GENERAL FORM OF A LOGARITHMIC FUNCTION:** $f(x) = k + a\log_b(x - h)$ **such that** $a$, $h$, **and** $k$ **are real numbers,** $b$ **is any positive number not equal to** $1$, **and** $x - h > 0$.
>
> **GENERAL FORM OF AN EXPONENTIAL FUNCTION:** $f(x) = a \cdot b^x + k$ **such that** $a$ **and** $k$ **are real numbers, and** $b$ **is any positive number not equal to** $1$.
>
> **The properties of logarithms and exponents can be used to rewrite expressions for functions in equivalent forms that can then be graphed by applying transformations.**

### Exit Ticket (5 minutes)

Name ______________________________ Date ______________

# Lesson 20: Transformations of the Graphs of Logarithmic and Exponential Functions

## Exit Ticket

1. Express $g(x) = -\log_4(2x)$ in the general form of a logarithmic function, $f(x) = k + a\log_b(x - h)$. Identify $a$, $b$, $h$, and $k$.

2. Use the structure of $g$ when written in general from to describe the graph of $g$ as a transformation of the graph of $h(x) = \log_4(x)$.

3. Graph $g$ anad $h$ on the same coordinate axes.

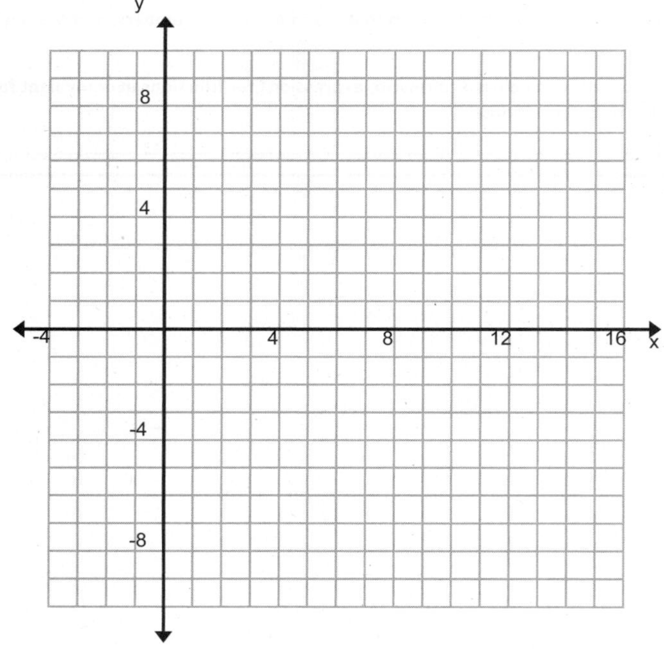

## Exit Ticket Sample Solutions

1. Express $g(x) = -\log_4(2x)$ in the general form of a logarithmic function, $f(x) = k + a\log_b(x - h)$. Identify $a$, $b$, $h$, and $k$.

   *Since* $-\log_4(2x) = -\log_4(2) + \log_4(x) = -\frac{1}{2} - \log_4(x)$, *the function is* $g(x) = -\frac{1}{2} - \log_4(x)$, *and* $a = -1$, $b = 4$, $h = 0$, *and* $k = -\frac{1}{2}$.

2. Use the structure of $g$ when written in general form to describe the graph of $g$ as a transformation of the graph of $h(x) = \log_4(x)$.

   *The graph of* $g$ *is the graph of* $h$ *reflected about the horizontal axis and translated down* $\frac{1}{2}$ *unit.*

3. Graph $g$ and $h$ on the same coordinate axes.

   *The graph of* $h$ *is shown in blue, and the graph of* $g$ *is shown in red.*

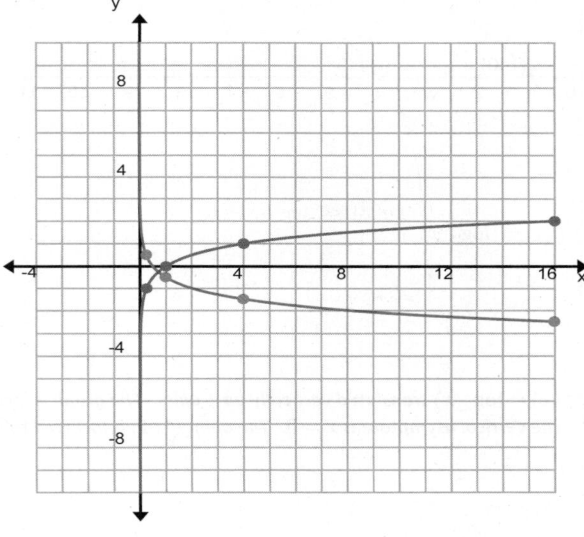

## Problem Set Sample Solutions

1. **Describe each function as a transformation of the graph of a function in the form $f(x) = \log_b(x)$. Sketch the graph of $f$ and the graph of $g$ by hand. Label key features such as intercepts, increasing or decreasing intervals, and the equation of the vertical asymptote.**

   a. $g(x) = \log_2(x - 3)$

   ***The graph of $g$ is the graph of $f(x) = \log_2(x)$ translated horizontally $3$ units to the right. The graph is increasing on $(3, \infty)$. The $x$-intercept is $4$, and the vertical asymptote is $x = 3$.***

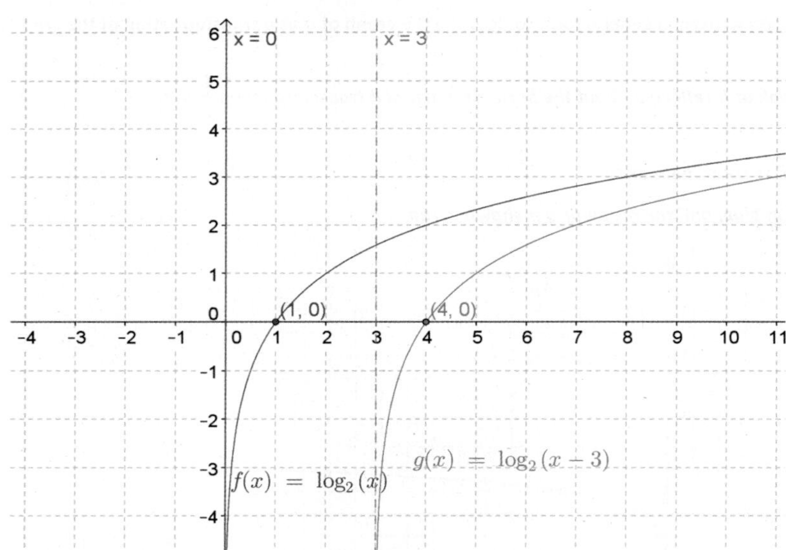


   b. $g(x) = \log_2(16x)$

   ***The graph of $g$ is the graph of $f(x) = \log_2(x)$ translated vertically up $4$ units. The graph is increasing on $(0, \infty)$. The $x$-intercept is $2^{-4}$. The vertical asymptote is $x = 0$. The point $(1, 4)$ is included to illustrate the vertical translation.***

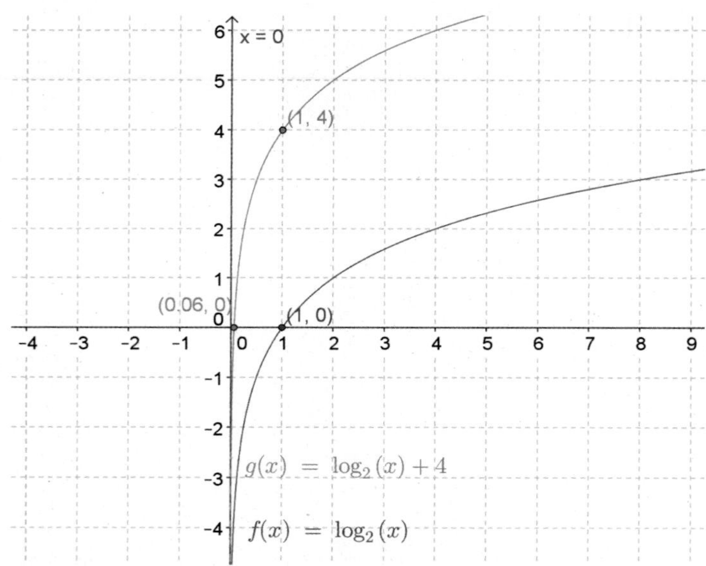

c. $g(x) = \log_2\left(\frac{8}{x}\right)$

***The graph of $g$ is the graph of $f(x) = \log_2(x)$ reflected about the horizontal axis and translated vertically up 3 units. The graph is decreasing on $(0, \infty)$. The $x$-intercept is $2^3$. The vertical asymptote is $x = 0$. The reflected graph and the final graph are both shown in the solution. The point $(1, 3)$ is included to show the vertical translation.***

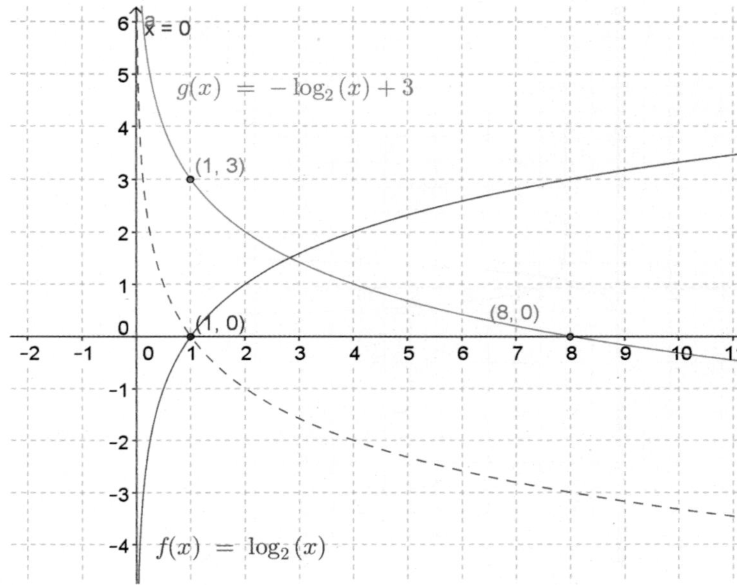


d. $g(x) = \log_2((x-3)^2)$

***The graph of $g$ is the graph of $f(x) = \log_2(x)$ stretched vertically by a factor of 2 and translated horizontally 3 units to the right. The graph is increasing on $(3, \infty)$. The $x$-intercept is 4, and the vertical asymptote is $x = 3$. The points $(2, 1)$ and $(5, 2)$ are labeled to help illustrate the transformations.***

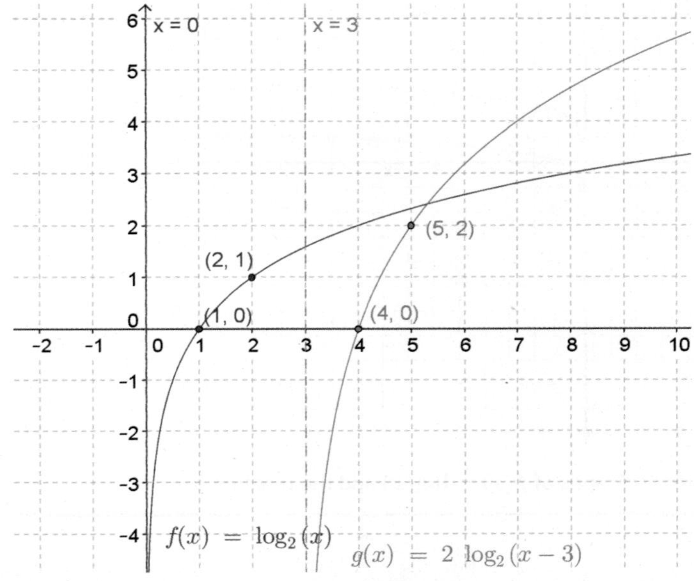

2. **Each function graphed below can be expressed as a transformation of the graph of $f(x) = \log(x)$. Write an algebraic function for $g$ and $h$, and state the domain and range.**

*$g(x) = -\log(x-2)$ for $x > 2$. The domain of $g$ is $x > 2$, and the range of $g$ is all real numbers.*

*$h(x) = 2 + \log(x)$ for $x > 0$. The domain of $h$ is $x > 0$, and the range of $h$ is all real numbers.*

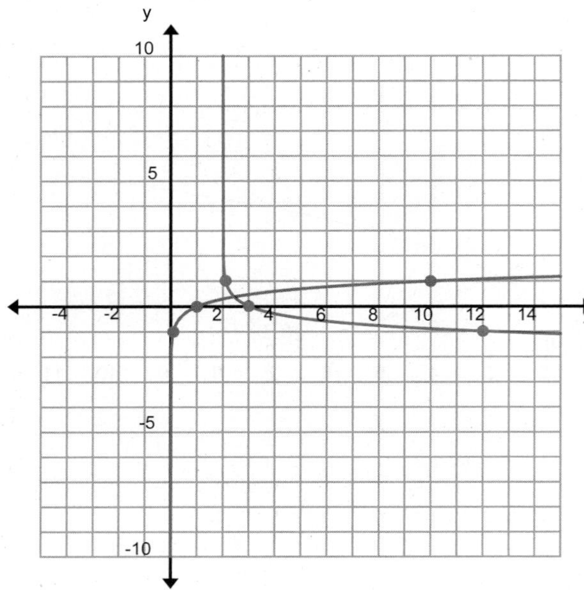

**Figure 1: Graphs of $f(x) = \log(x)$ and the function $g$**

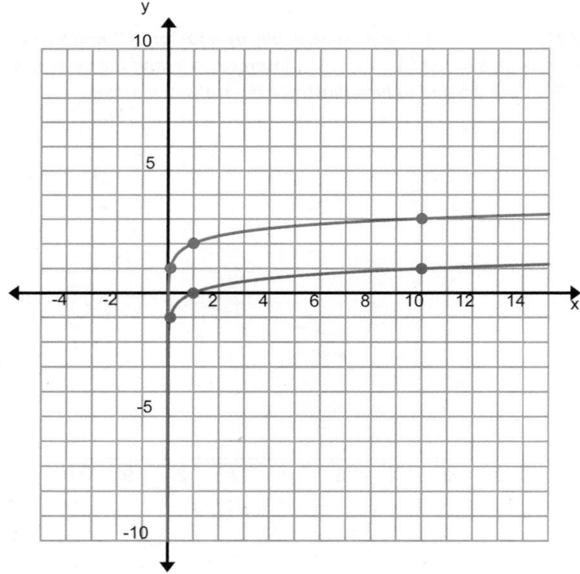

**Figure 2: Graphs of $f(x) = \log(x)$ and the function $h$**

3. Describe each function as a transformation of the graph of a function in the form $f(x) = b^x$. Sketch the graph of $f$ and the graph of $g$ by hand. Label key features such as intercepts, increasing or decreasing intervals, and the horizontal asymptote. (Estimate when needed from the graph.)

   a. $g(x) = 2 \cdot 3^x - 1$

   *The graph of $g$ is the graph of $f(x) = 3^x$ scaled vertically by a factor of $2$ and translated vertically down $1$ unit. The equation of the horizontal asymptote is $y = -1$. The $y$-intercept is $1$, and the $x$-intercept is approximately $-0.631$. The graph of $g$ is increasing for all real numbers.*

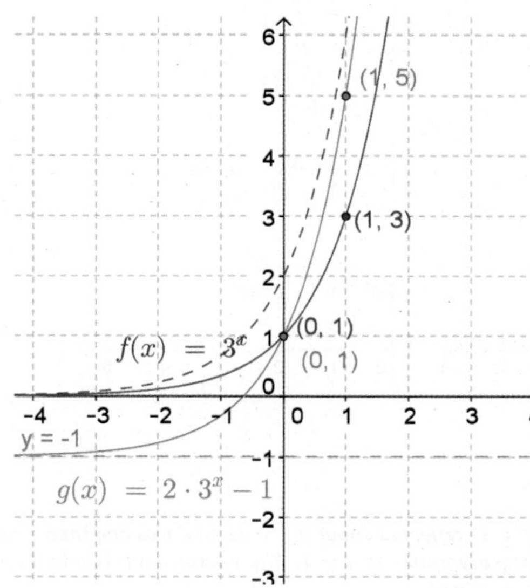


   b. $g(x) = 2^{2x} + 3$

   *The graph of $g$ is the graph of $f(x) = 4^x$ translated vertically up $3$ units. The equation of the horizontal asymptote is $y = 3$. The $y$-intercept is $4$. There is no $x$-intercept. The graph is increasing for all real numbers.*

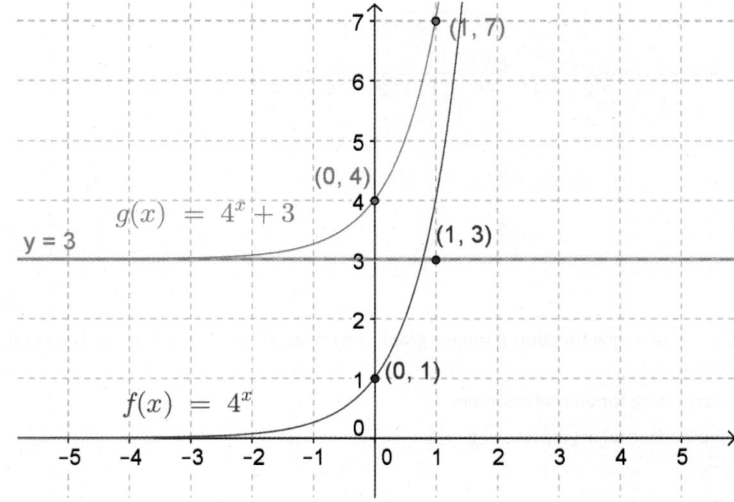

c. $g(x) = 3^{x-2}$

***The graph of $g$ is the graph of $f(x) = 3^x$ translated horizontally $2$ units to the right OR the graph of $f$ scaled vertically by a factor of $\frac{1}{9}$. The equation of the horizontal asymptote is $y = 0$. The $y$-intercept is $\frac{1}{9}$. There is no $x$-intercept, and the graph is increasing for all real numbers.***

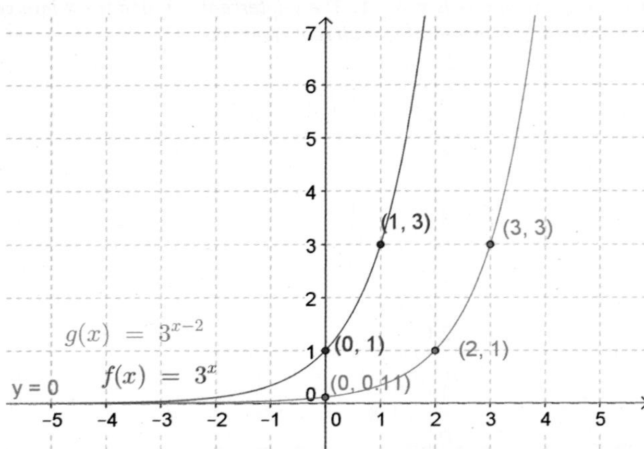


d. $g(x) = -9^{\frac{x}{2}} + 1$

***The graph of $g$ is the graph of $f(x) = 3^x$ reflected about the horizontal axis and then translated vertically up $1$ unit. The equation of the horizontal asymptote is $y = 1$. The $y$-intercept is $0$, and the $x$-intercept is also $0$. The graph of $g$ is decreasing for all real numbers.***

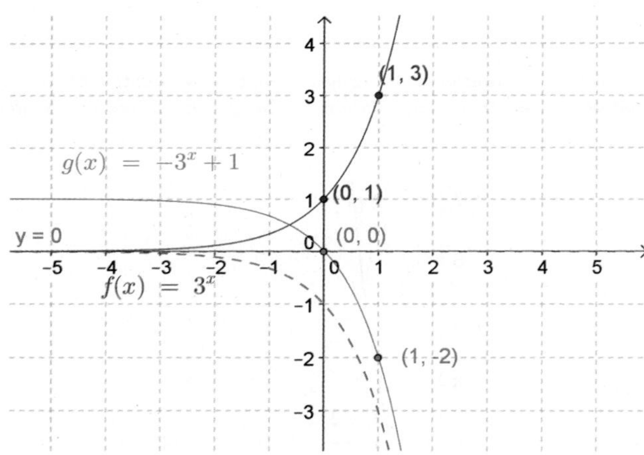


4. **Using the function $f(x) = 2^x$, create a new function $g$ whose graph is a series of transformations of the graph of $f$ with the following characteristics:**
   - **The graph of $g$ is decreasing for all real numbers.**
   - **The equation for the horizontal asymptote is $y = 5$.**
   - **The $y$-intercept is $7$.**

***One possible solution is $g(x) = 2 \cdot 2^{-x} + 5$.***

5. Using the function $f(x) = 2^x$, create a new function $g$ whose graph is a series of transformations of the graph of $f$ with the following characteristics:
   - The graph of $g$ is increasing for all real numbers.
   - The equation for the horizontal asymptote is $y = 5$.
   - The $y$-intercept is $4$.

   *One possible solution is $g(x) = -(2^{-x}) + 5$.*

6. Given the function $g(x) = \left(\frac{1}{4}\right)^{x-3}$:

   a. Write the function $g$ as an exponential function with base $4$. Describe the transformations that would take the graph of $f(x) = 4^x$ to the graph of $g$.

   $$\left(\frac{1}{4}\right)^{x-3} = (4^{-1})^{x-3} = 4^{-x+3} = 4^3 \cdot 4^{-x}$$

   *Thus, $g(x) = 64 \cdot 4^{-x}$. The graph of $g$ is the graph of $f$ reflected about the vertical axis and scaled vertically by a factor of $64$.*

   b. Write the function $g$ as an exponential function with base $2$. Describe two different series of transformations that would take the graph of $f(x) = 2^x$ to the graph of $g$.

   $$\left(\frac{1}{4}\right)^{x-3} = (2^{-2})^{x-3} = 2^{-2(x-3)} = 2^{-2x+6} = 64 \cdot 2^{-2x}$$

   *Thus, $g(x) = 64 \cdot 2^{-2x}$, or $g(x) = 2^{-2(x-3)}$. To obtain the graph of $g$ from the graph of $f$, you can scale the graph horizontally by a factor of $\frac{1}{2}$, reflect the graph about the vertical axis, and scale it vertically by a factor of $64$. Or, you can scale the graph horizontally by a factor of $\frac{1}{2}$, reflect the graph about the vertical axis, and translate the resulting graph horizontally $3$ units to the right.*

7. Explore the graphs of functions in the form $f(x) = \log(x^n)$ for $n > 1$. Explain how the graphs of these functions change as the values of $n$ increase. Use a property of logarithms to support your reasoning.

   *The graphs appear to be a vertical scaling of the common logarithm function by a factor of $n$. This is true because of the property of logarithms that states $\log(x^n) = n\log(x)$.*

8. Use a graphical approach to solve each equation. If the equation has no solution, explain why.

   a. $\log(x) = \log(x - 2)$

   *This equation has no solution because the graphs of $y = \log(x)$ and $y = \log(x - 2)$ are horizontal translations of each other. Thus, their graphs do not intersect, and the corresponding equation has no solution.*

   b. $\log(x) = \log(2x)$

   *This equation has no solution because $\log(2x) = \log(2) + \log(x)$, which means that the graphs of $y = \log(x)$ and $y = \log(2x)$ are a vertical translation of each other. Thus, their graphs do not intersect, and the corresponding equation has no solution.*

c. $\log = \log\left(\frac{2}{x}\right)$

***The solution is the $x$-coordinate of the intersection point of the graphs of $y = \log(x)$ and $y = \log(2) - \log(x)$. Since the graph of the function defined by the right side of the equation is a reflection across the horizontal axis and a vertical translation of the graph of the function defined by the left side of the equation, the graphs of these functions will intersect in exactly one point.***

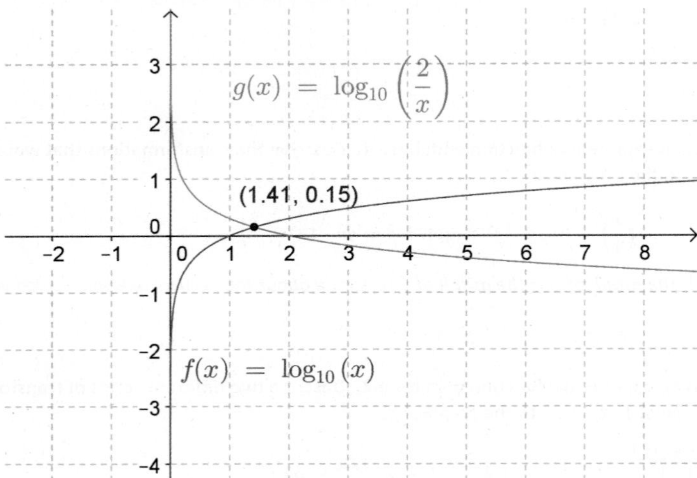


d. **Show algebraically that the exact solution to the equation in part (c) is $\sqrt{2}$.**

$$\log(x) = \log(2) - \log(x)$$
$$2\log(x) = \log(2)$$
$$\log(x) = \frac{1}{2}\log(2)$$
$$\log(x) = \log\left(2^{\frac{1}{2}}\right)$$
$$x = 2^{\frac{1}{2}}$$

***Since $2^{\frac{1}{2}} = \sqrt{2}$, the exact solution is $\sqrt{2}$.***

9. **Make a table of values for $f(x) = x^{\frac{1}{\log(x)}}$ for $x > 1$. Graph this function for $x > 1$. Use properties of logarithms to explain what you see in the graph and the table of values.**

***The table indicates that the function is equal to $10$ for all values of $x$ greater than $1$.***

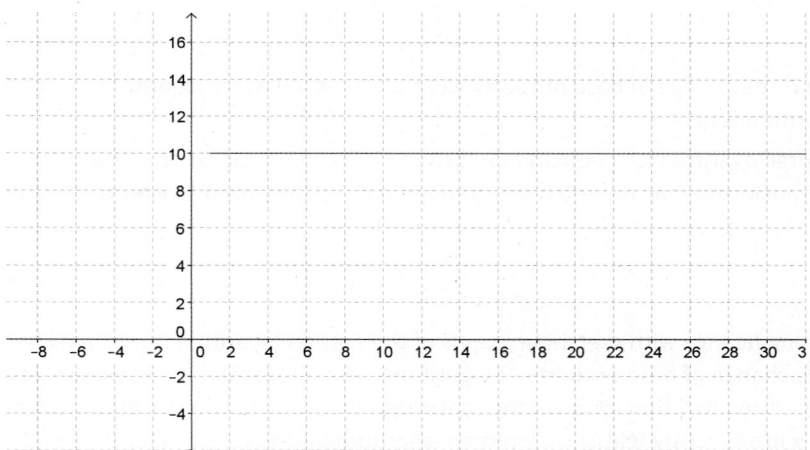

***The expression $x^{\frac{1}{\log(x)}} = 10$ for all $x > 1$ because $\frac{1}{\log(x)} = \frac{\log(10)}{\log(x)} = \log_x(10)$. Therefore, when we substitute $\log_x(10)$ into the expression $x^{\frac{1}{\log(x)}}$, we get $x^{\log_x(10)}$, which is equal to $10$ according to the definition of a logarithm.***

# Lesson 21: The Graph of the Natural Logarithm Function

## Student Outcomes

- Students understand that the change of base property allows us to write every logarithm function as a vertical scaling of a natural logarithm function.
- Students graph the natural logarithm function and understand its relationship to other base $b$ logarithm functions. They apply transformations to sketch the graph of natural logarithm functions by hand.

## Lesson Notes

The focus of this lesson is developing fluency with sketching graphs of the natural logarithm function by hand and understanding that because of the change of base property of logarithms, every logarithm function can be expressed as a vertical scaling of the natural logarithm function (or any other base logarithm we choose, for that matter). This helps to explain why calculators typically prominently feature a common logarithm button and a natural logarithm button. Students may question why we care so much about natural logarithms in mathematics. The importance of the particular base $e$ will become apparent when they study calculus and learn that $\ln(x)$ is equal to the area under the reciprocal function $f(t) = \frac{1}{t}$ from $1$ to $x$; that is, $\ln(x) = \int_1^x \frac{1}{t} dt$.

This lesson begins by challenging students to compare and contrast logarithm functions with different bases in a group exploration. Students complete a graphic organizer to help focus their learning at the end of the exploration. During the exploration, students should be encouraged to use technology. The focus should be on observing the patterns and making generalizations (MP.7, MP.8). A quick set of exercises primes students to explain their observations using the change of base property of logarithms. Once we have established that this property guarantees that graphs of logarithmic functions of one base are a vertical scaling of a graph of a logarithmic function of any other base, we tie the lesson back to the natural logarithm function. The lesson closes with demonstrations and practice with graphing natural logarithm functions to build fluency with creating sketches by hand (**F-IF.B.4**).

## Classwork

### Opening (3 minutes)

Ask students to predict how the graphs of logarithm functions are alike and how they are different when we consider different bases. Post this question on the board, give students a minute or two to think about their response, and then have them share with a partner. Take a few responses from the entire class, but do not really provide any concrete answers at this point. Student responses and the quality of their conversations will help you gauge their understanding of graphs of logarithm functions up to this point and help you decide how to support student learning during the rest of this lesson.

- How are the graphs of $f(x) = \log_2(x)$, $g(x) = \log_3(x)$, and $h(x) = \log(x)$ similar? How are they different?
  - *They are always increasing for $b > 1$ and have one $x$-intercept at $1$. As the base changes, they appear to increase more or less rapidly. They all have the same domain and range.*

## Exploratory Challenge (15 minutes)

Have students work in groups of 4–5. Each group will need access to graphing technology for each student or pair of students, the student materials for this lesson, chart paper or personal white boards, and markers. Have each student select at least one base value from the following list: $b = \left\{\frac{1}{10}, \frac{1}{2}, 2, 5, 20, 100\right\}$. Using a graphing calculator or other graphing technology, students should independently explore how their selected base $b$ logarithm function's graph compares to the graph of the common logarithm function $f(x) = \log(x)$. Next, have them describe what they observed in writing and report it to their group members. As a group, students then categorize their findings based on the value of $b$ and record their observations on chart paper. Have each group present their findings to the entire class. As you debrief this exploration as a whole class, focus on clarifying student language in their descriptions and encourage students to revise their written descriptions to further clarify what they wrote. Student work should be similar to the sample responses shown below.

> *Scaffolding:*
>
> - For English language learners, the graphic organizer provides some scaffolding, but sentence frames may also help students respond to part (c) in this exploration. For example, "Compared to the graph of $f$, the graph of my function was _______."
>
>   OR
>
>   "My function's graph was a ____________ transformation of the graph of $f$."
> - For advanced learners, rather than provide the explicit steps listed in the Exploratory Challenge, present the problem on the board, give them technology and chart paper, and start them on the group presentation.

**Exploratory Challenge**

**Your task is to compare graphs of base $b$ logarithm functions to the graph of the common logarithm function $f(x) = \log(x)$ and summarize your results with your group. Recall that the base of the common logarithm function is 10. A graph of $f$ is provided below.**

a. **Select at least one base value from this list: $\frac{1}{10}, \frac{1}{2}, 2, 5, 20, 100$. Write a function in the form $g(x) = \log_b(x)$ for your selected base value, $b$.**

*Students should use one of the numbers from the list to write their function. For example, $g(x) = \log_5(x)$.*

b. **Graph the functions $f$ and $g$ in the same viewing window using a graphing calculator or other graphing application, and then add a sketch of the graph of $g$ to the graph of $f$ shown below.**

*Several graphs are shown at the end of this exploration.*

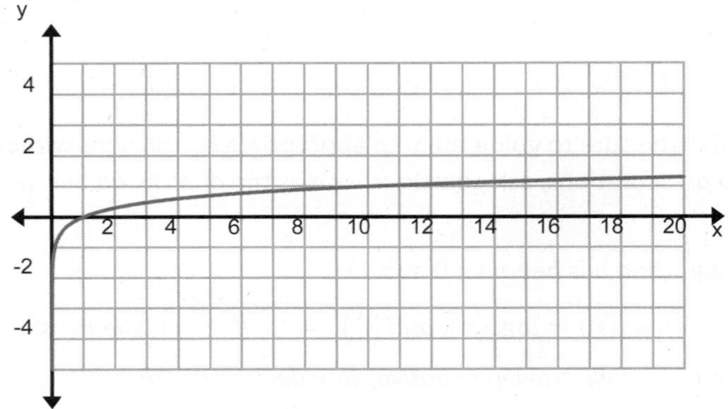


c. **Describe how the graph of $g$ for the base you selected compares to the graph of $f(x) = \log(x)$.**

*Answers will vary depending on the base selected. For example, when the base is 20, the graph of $g$ appears to be a vertical scaling of the common logarithm function by a factor less than 1.*

MP.8

d. **Share your results with your group and record observations on the graphic organizer below. Prepare a group presentation that summarizes the group's findings.**

| How does the graph of $g(x) = \log_b(x)$ compare to the graph of $f(x) = \log(x)$ for various values of $b$? | |
|---|---|
| $0 < b < 1$ | *The function g is decreasing. Its graph is a reflection about the horizontal axis of the graph of a logarithm function whose base is the reciprocal of b.* |
| $1 < b < 10$ | *When b is between 1 and 10, the graph of g appears to be a vertical scaling of the graph of f by a factor greater than 1. As b gets closer to 10, the graph of g gets closer to the graph of f and appears less steep.* |
| $b > 10$ | *When b is greater than 10, the graph of g appears to be a vertical scaling of the graph of f by a factor between 0 and 1. As b grows, the graph of g grows at a slower rate and appears to move closer to the horizontal axis than the graphs of functions whose bases are closest to 10.* |

A graph of logarithm functions with several of the bases listed is shown below.

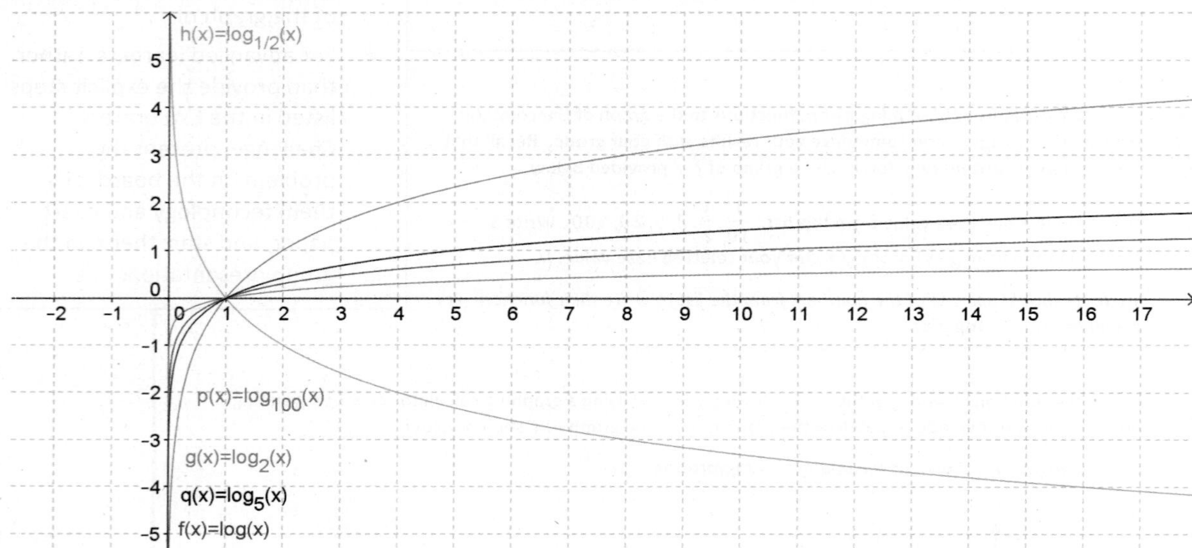


As the groups work through this exploration, be sure to guide them to appropriate conclusions without explicitly telling them answers. After or during the group presentations, ask questions such as the ones listed below to clarify student understanding.

MP.3 & MP.7

- Why are the functions decreasing when $b$ is between 0 and 1?
  - *Consider $b = \frac{1}{2}$. Then, $y = \log_b(x) = \log_{\frac{1}{2}}(x)$, and $\left(\frac{1}{2}\right)^y = x$. If $x > 1$, then $y < 0$. As $x$ increases, $y$ becomes larger in magnitude while staying negative, so $y$ decreases. Thus, the function is decreasing.*
  - *The exponential function for bases between 0 and 1 is a decreasing function, so when we have a logarithm function with these bases, the range values will also decrease as the domain values increase.*

- How does the graph of $y = \log_b(x)$ relate to the graph of $y = \log_{\frac{1}{b}}(x)$? Explain why this relationship exists.
  - *The graphs appear to be reflections of one another about the horizontal axis. The graphs of* $y = b^x$ *and* $y = \left(\frac{1}{b}\right)^x$ *are reflections about the vertical axis because* $\left(\frac{1}{b}\right)^x = b^{-x}$. *Thus, when we exchange the domain and range values to form the related logarithm functions, they will also be reflections of one another, but about the horizontal axis.*
- Why do smaller bases $b > 1$ produce steeper graphs and larger bases produce flatter graphs?
  - *Logarithmic functions with a smaller base grow at a faster rate, making the graph steeper. For example,* $\log_2(64) = 6$, $\log_4(64) = 3$, *and* $\log_8(64) = 2$. *The same input (64) produces a smaller output as the size of the base increases.*
- Where would the graph of $y = \ln(x)$ sit in relation to these graphs? How do you know?
  - *The graph of* $y = \ln(x)$ *would be in between the two graphs of* $y = \log_2(x)$ *and* $y = \log_5(x)$ *because* $e$ *is a number between* $2$ *and* $5$.
- The graphs of these functions appear to be vertical scalings of each other. How could we prove that this is true?
  - *We would have to show that we can rewrite each function as a constant multiple of another logarithm function.*

Check to make sure each student has recorded appropriate information in the graphic organizer in part (d) before moving on. Post the group presentations on the board for reference during the rest of this lesson.

## Exercise 1 (5 minutes)

Announce that now we will explore how all these graphs are related using a property of logarithms. Students should be able to complete this exercise quickly. Some students may already start to understand why the graphs appeared the way they did in the Exploratory Challenge as they work through these exercises.

**Exercise 1**

**Use the change of base property to rewrite each function as a common logarithm.**

| **Base $b$** | **Base $10$ (Common logarithm)** |
|---|---|
| $g(x) = \log_{\frac{1}{4}}(x)$ | $g(x) = \dfrac{\log(x)}{\log\left(\frac{1}{4}\right)}$ |
| $g(x) = \log_{\frac{1}{2}}(x)$ | $g(x) = \dfrac{\log(x)}{\log\left(\frac{1}{2}\right)}$ |
| $g(x) = \log_2(x)$ | $g(x) = \dfrac{\log(x)}{\log(2)}$ |
| $g(x) = \log_5(x)$ | $g(x) = \dfrac{\log(x)}{\log(5)}$ |
| $g(x) = \log_{20}(x)$ | $g(x) = \dfrac{\log(x)}{\log(20)}$ |
| $g(x) = \log_{100}(x)$ | $g(x) = \dfrac{\log(x)}{\log(100)}$ |

## Discussion (5 minutes)

Lead a discussion to help students observe that each function in base 10 is divided by a constant (which is the same as multiplying by the reciprocal of that number). Have students explore the values of the constants using their calculators, and have them make sense of why the graphs appear the way they do compared to the graph of the common logarithm function. For example, $\log(2) \approx 0.69$. When dividing by a number between 0 and 1, you get the same result as multiplying by its reciprocal, which is a number greater than 1. The values of $\log\left(\frac{1}{2}\right)$ and $\log\left(\frac{1}{4}\right)$ are negative, which explains why the graphs of those functions are a vertical scaling and a reflection of the graph of the common logarithm function. When the base is greater than 10, we are dividing by a number greater than 1, which is the same as multiplying by a number between 0 and 1, which compresses the graph vertically.

- How do the functions from Exercise 1 that you wrote in base 10 compare to the function $f(x) = \log(x)$?
  - *They are a constant multiple of the function $f$. For example, $\log(100) = 2$, so the function $g(x) = \frac{\log(x)}{\log(100)}$ could also be written as $g(x) = \frac{1}{2}\log(x)$.*
- Approximate the values of the constants in the functions from Exercise 1. How do those values help to explain why the graphs are a vertical stretch of the common logarithm function when the base is between 1 and 10, and a vertical compression when the base is greater than 10? Why are the functions decreasing when the base is between 0 and 1?
  - *When the base is between 1 and 10, the common logarithms are between 0 and 1. Dividing by a number between 0 and 1 is the same as multiplying by a number larger than 1, which will scale the graph vertically by a factor greater than 1. For bases greater than 10, the common logarithm function is multiplied by a number between 0 and 1. The functions decrease when the base is between 0 and 1 because the common logarithms of those numbers are less than 0.*

Next, revisit the question posed earlier regarding the graph of $y = \ln(x)$, the natural logarithm function, as a way to transition into the last portion of this lesson.

- Where would the graph of $y = \ln(x)$ sit in relation to these graphs? How do you know?
  - *The graph of $y = \ln(x)$ would be in between the two graphs of $y = \log_2(x)$ and $y = \log_5(x)$ because $e$ is a number between 2 and 5.*

### Example 1 (5 minutes): The Graph of the Natural Logarithm Function $f(x) = \ln(x)$

The example that follows demonstrates how to sketch the graph of the natural logarithm function by hand and shows more precisely where it sits in relation to base 2 and base 10 logarithm functions.

**Example 1: The Graph of the Natural Logarithm Function $f(x) = \ln(x)$**

**Graph the natural logarithm function below to demonstrate where it sits in relation to the base 2 and base 10 logarithm functions.**

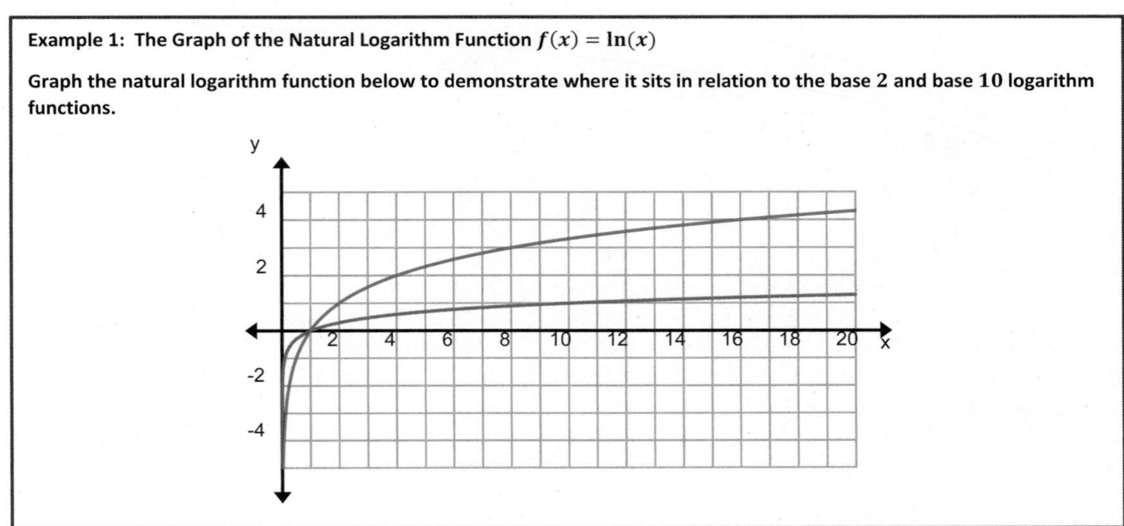


The graphs are not labeled. You can question students about this to informally assess their understanding at this point.

- Which graph is $y = \log_2(x)$, and which one is $y = \log(x)$? How can you tell?
  - *Since the base 2 is smaller, the logarithm function base 2 grows more quickly than the base 10 logarithm function, so the red graph is the graph of $y = \log_2(x)$. You can also verify which graph is which by identifying a few points and substituting them into the equations to see which is true. For example, the blue graph appears to contain the point $(1, 10)$. Since $1 = \log(10)$, the blue graph represents the common logarithm function.*

Remind students that $e \approx 2.718$. Create a table of values like the one shown below and then plot these points. Connect the points with a smooth curve. When students are sketching by hand in the next example, have them plot fewer points, perhaps where the $y$-values are integers only.

| $x$ | $f(x) = \ln(x)$ |
|---|---|
| $\frac{1}{e} \approx 0.369$ | $-1$ |
| $1$ | $0$ |
| $e^{0.5} \approx 1.649$ | $0.5$ |
| $e^1 \approx 2.718$ | $1$ |
| $e^{1.5} \approx 4.482$ | $1.5$ |
| $e^2 \approx 7.389$ | $2$ |
| $e^{2.5} \approx 12.182$ | $2.5$ |

The solution is graphed below with several points labeled on the graph of $f(x) = \ln(x)$.

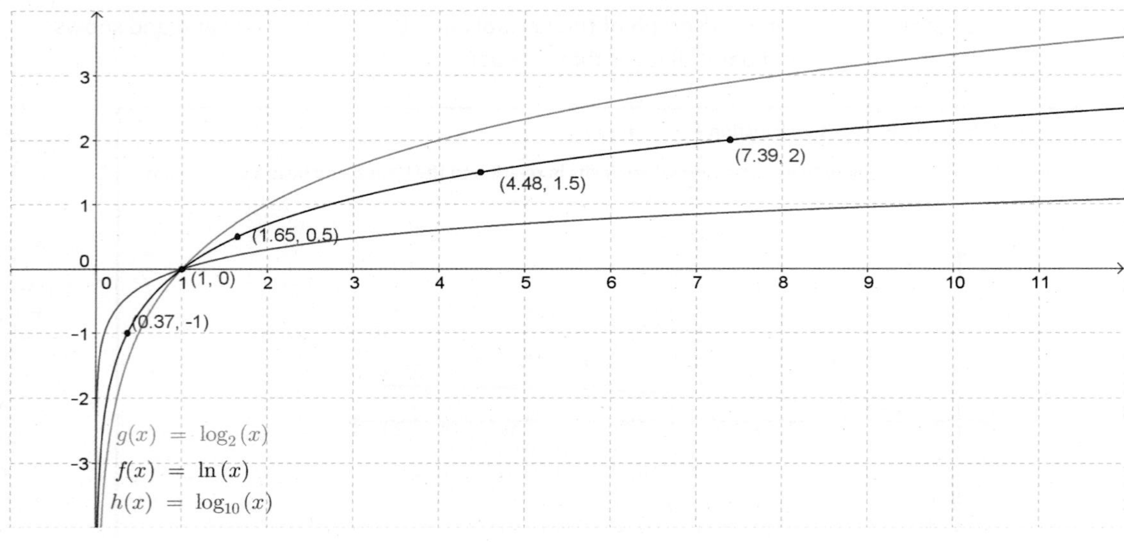


## Example 2 (5 minutes)

In this example, part (a) models how to sketch graphs by applying transformations. Then, show students in part (b) how to rewrite the function as a natural logarithm function, and sketch the graph by applying transformations of the graph of $f(x) = \ln(x)$. Model the transformations in stages. First, sketch the graph of $y = \ln(x)$; next, sketch a second graph applying the first transformation; finally, sketch a graph applying the last transformation to the second graph you made.

> **Example 2**
>
> **Graph each function by applying transformations of the graphs of the natural logarithm function.**
>
> a. $\boldsymbol{f(x) = 3\ln(x-1)}$
>
> ***The graph of*** $\boldsymbol{f}$ ***is the graph of*** $\boldsymbol{y = \ln(x)}$ ***shifted horizontally*** $\boldsymbol{1}$ ***unit to the right, stretched vertically by a factor of*** $\boldsymbol{3}$***.***
>
> 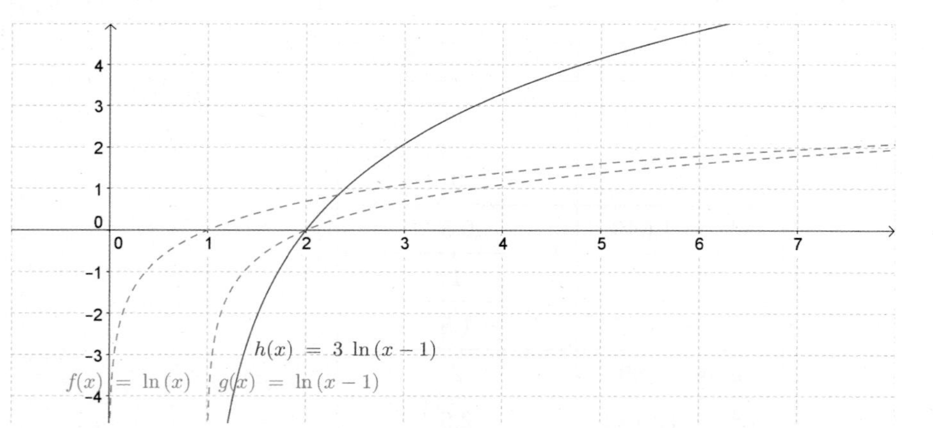
> 

**b.** $g(x) = \log_6(x) - 2$

***First, write g as a natural logarithm function.***

$$g(x) = \frac{\ln(x)}{\ln(6)} - 2$$

***Since*** $\frac{1}{\ln(6)} \approx 0.558$, ***the graph of g will be the graph of*** $y = \ln(x)$ ***scaled vertically by a factor of approximately*** $0.56$ ***and translated down*** $2$ ***units.***

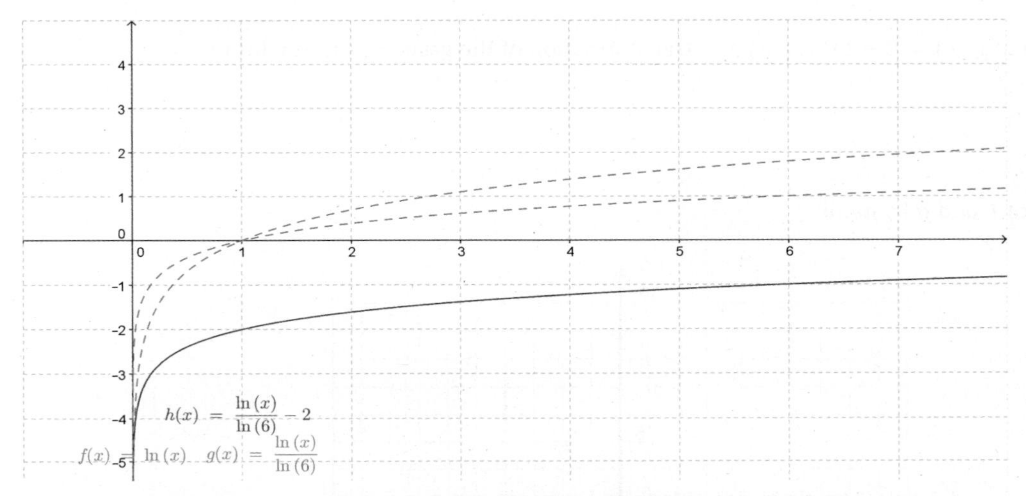


## Closing (2 minutes)

Have students summarize what they have learned in this lesson by revisiting the question from the Opening. Students should revise their initial responses and either discuss their answers with a partner or write a brief individual reflection. The responses should be similar to what is listed in the Lesson Summary.

- How are the graphs of logarithm functions with different bases alike? How are they different?
  - *They have the same* $x$*-intercept* $1$*, and when the base is greater than* $1$*, the functions are increasing. They all have the same domain and range. They are different because as the base changes, the steepness of the graph of the function changes. Larger bases grow at slower rates.*
- How does the change of base property guarantee that every logarithm function could be expressed in the form $f(x) = k + a\ln(x - h)$?
  - *The change of base property guarantees that we can convert any logarithmic expression in base* $b$ *to a natural logarithmic expression where the denominator of the expression is constant.*

## Exit Ticket (5 minutes)

Name ______________________________ Date ______________

# Lesson 21: The Graph of the Natural Logarithm Function

## Exit Ticket

1. Describe the graph of $g(x) = 2 - \ln(x + 3)$ as a transformation of the graph of $f(x) = \ln(x)$.

2. Sketch the graphs of $f$ and $g$ by hand.

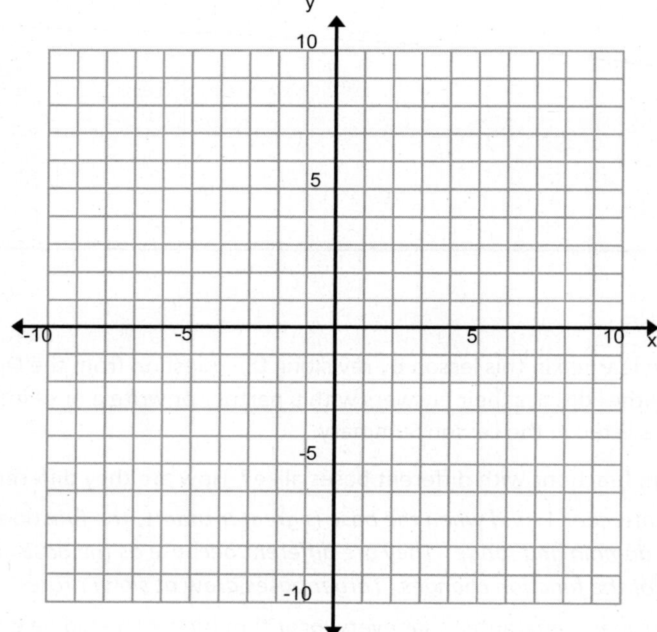

3. Explain where the graph of $g(x) = \log_3(2x)$ would sit in relation to the graph of $f(x) = \ln(x)$. Justify your answer using properties of logarithms and your knowledge of transformations of graph of functions.

## Exit Ticket Sample Solutions

2. **Describe the graph of $g(x) = 2 - \ln(x+3)$ as a transformation of the graph of $f(x) = \ln(x)$.**

   *The graph of $g$ is the graph of $f$ translated $3$ units to the left, reflected about the horizontal axis, and translated up $2$ units.*

3. **Sketch the graphs of $f$ and $g$ by hand.**

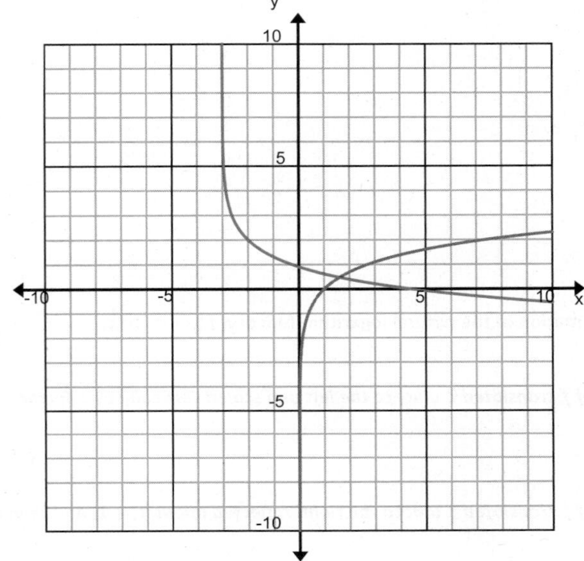


4. **Explain where the graph of $g(x) = \log_3(2x)$ would sit in relation to the graph of $f(x) = \ln(x)$. Justify your answer using properties of logarithms and your knowledge of transformations of graph of functions.**

   *Since $\log_3(2x) = \frac{\ln(2x)}{\ln(3)} = \frac{\ln(2)}{\ln(3)} + \frac{\ln(x)}{\ln(3)}$, the graph of $g$ would be a vertical shift and a vertical scaling by a factor greater than $1$ of the graph of $f$.*

## Problem Set Sample Solutions

5. **Rewrite each logarithm function as a natural logarithm function.**

   a. $f(x) = \log_5(x)$

   $$f(x) = \frac{\ln(x)}{\ln(5)}$$

   b. $f(x) = \log_2(x-3)$

   $$f(x) = \frac{\ln(x-3)}{\ln(2)}$$

c. $f(x) = \log_2\left(\frac{x}{3}\right)$

$$f(x) = \frac{\ln(x)}{\ln(2)} - \frac{\ln(3)}{\ln(2)}$$

d. $f(x) = 3 - \log(x)$

$$f(x) = 3 - \frac{\ln(x)}{\ln(10)}$$

e. $f(x) = 2\log(x+3)$

$$f(x) = \frac{2}{\ln(10)}\ln(x+3)$$

f. $f(x) = \log_5(25x)$

$$f(x) = 2 + \frac{\ln(x)}{\ln(5)}$$

6. **Describe each function as a transformation of the natural logarithm function $f(x) = \ln(x)$.**

a. $g(x) = 3\ln(x+2)$

***The graph of $g$ is the graph of $f$ translated $2$ units to the left and scaled vertically by a factor of $3$.***

b. $g(x) = -\ln(1-x)$

***The graph of $g$ is the graph of $f$ translated $1$ unit to the right, reflected about $x = 1$, and then reflected about the horizontal axis.***

c. $g(x) = 2 + \ln(e^2 x)$

***The graph of $g$ is the graph of $f$ translated up $4$ units.***

d. $g(x) = \log_5(25x)$

***The graph of $g$ is the graph of $f$ translated up $2$ units and scaled vertically by a factor of $\frac{1}{\ln(5)}$.***

7. **Sketch the graphs of each function in Problem 2 and identify the key features including intercepts, decreasing or increasing intervals, and the vertical asymptote.**

   a. ***The equation of the vertical asymptote is $x = -2$. The $x$-intercept is $-1$. The graph is increasing for all $x > -2$. The $y$-intercept is approximately $2.079$.***

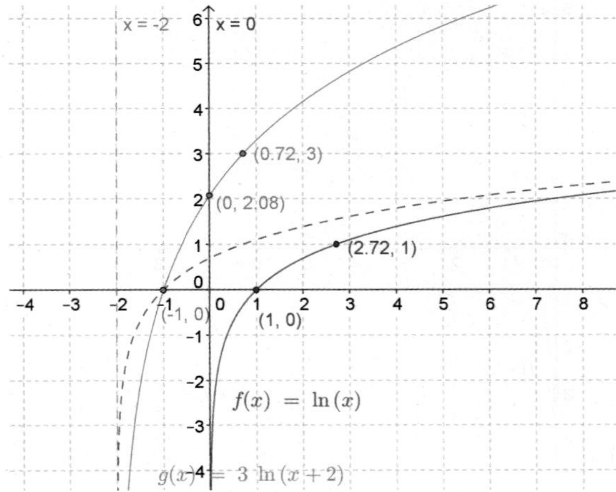


   b. ***The equation of the vertical asymptote is $x = 1$. The $x$-intercept is $0$. The graph is increasing for all $x < 1$. The $y$-intercept is $0$.***

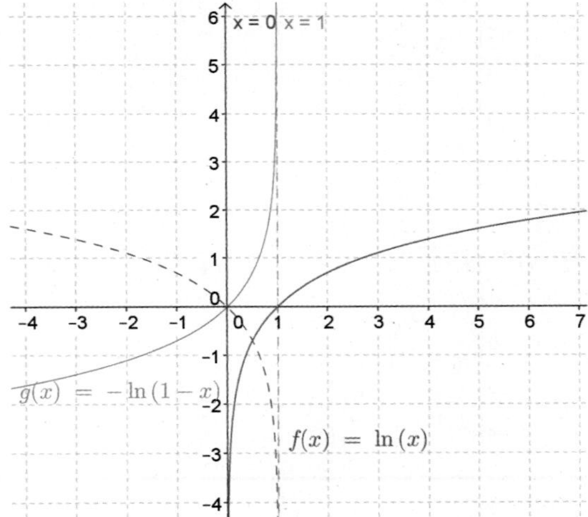

c. ***The equation of the vertical asymptote is $x = 0$. The $x$-intercept is approximately $0.018$. The graph is increasing for all $x > 0$.***

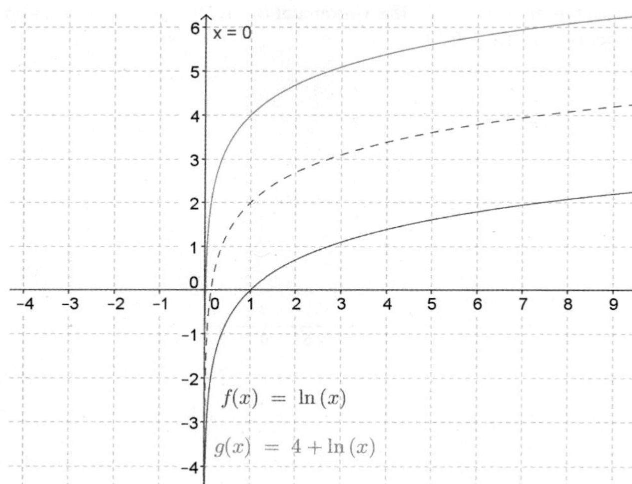


d. ***The equation of the vertical asymptote is $x = 0$. The $x$-intercept is $0.04$. The graph is increasing for all $x > 0$.***

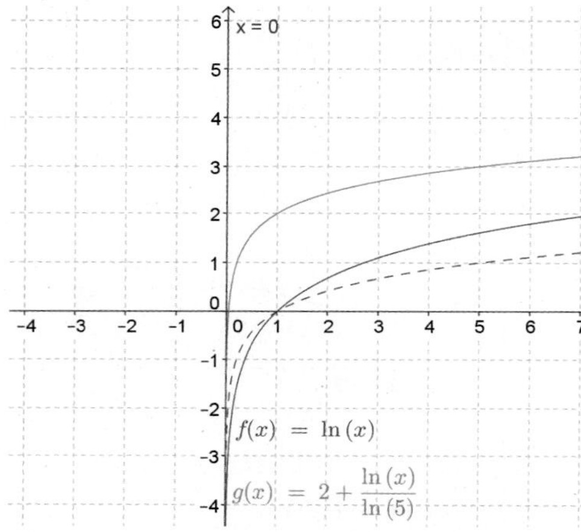

8. Solve the equation $e^{-x} = \ln(x)$ graphically.

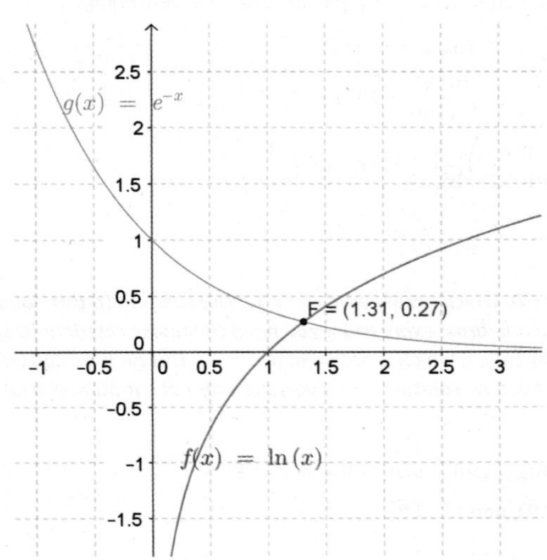


9. Use a graphical approach to explain why the equation $\log(x) = \ln(x)$ has only one solution.

***The graphs intersect in only one point $(1, 0)$, so the equation has only one solution.***

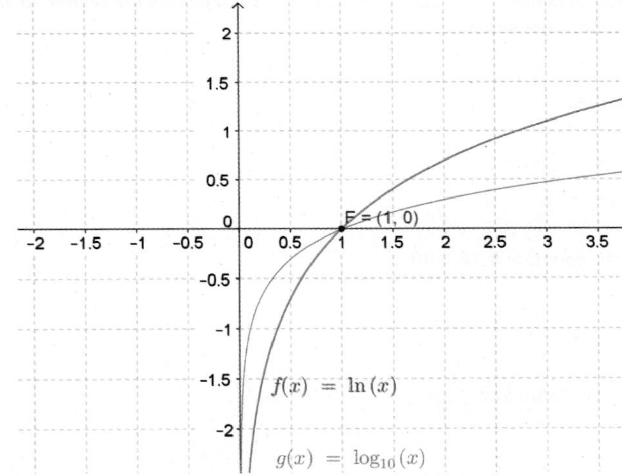

MP.3

10. **Juliet tried to solve this equation as shown below using the change of base property and concluded there is no solution because $\ln(10) \neq 1$. Construct an argument to support or refute her reasoning.**

$$\log(x) = \ln(x)$$
$$\frac{\ln(x)}{\ln(10)} = \ln(x)$$
$$\left(\frac{\ln(x)}{\ln(10)}\right)\frac{1}{\ln(x)} = (\ln(x))\frac{1}{\ln(x)}$$
$$\frac{1}{\ln(10)} = 1$$

***Juliet's approach works as long as $\ln(x) \neq 0$, which occurs when $x = 1$. The solution to this equation is $1$. When you divide both sides of an equation by an algebraic expression, you need to impose restrictions so that you are not dividing by $0$. In this case, Juliet divided by $\ln(x)$, which is not valid if $x = 1$. This division caused the equation in the third and final lines of her solution to have no solution; however, the original equation is true when $x$ is $1$.***

11. **Consider the function $f$ given by $f(x) = \log_x(100)$ for $x > 0$ and $x \neq 1$.**

a. **What are the values of $f(100)$, $f(10)$, and $f(\sqrt{10})$?**

$f(100) = 1,\ f(10) = 2,\ f(\sqrt{10}) = 4$

b. **Why is the value 1 excluded from the domain of this function?**

***The value $1$ is excluded from the domain because $1$ is not a base of an exponential function since it would produce the graph of a constant function. Since logarithm functions by definition are related to exponential functions, we cannot have a logarithm with base $1$.***

c. **Find a value $x$ so that $f(x) = 0.5$.**

$$\log_x(100) = 0.5$$
$$x^{0.5} = 100$$
$$x = 10{,}000$$

***The value of $x$ that satisfies this equation is $10{,}000$.***

d. **Find a value $w$ so that $f(w) = -1$.**

***The value of $w$ that satisfies this equation is $\frac{1}{100}$.***

e. **Sketch a graph of $y = \log_x(100)$ for $x > 0$ and $x \neq 1$.**

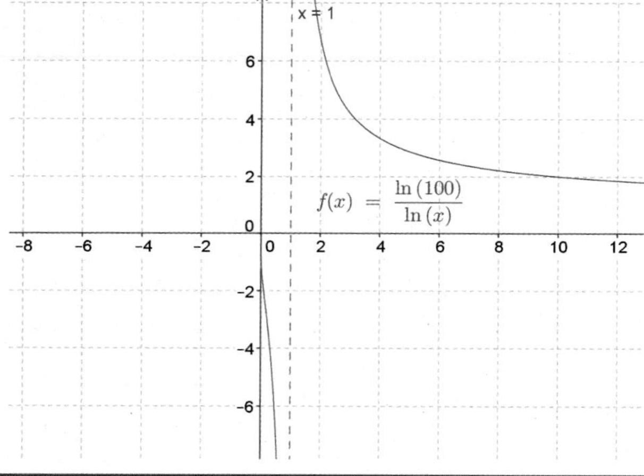

# Lesson 22: Choosing a Model

## Student Outcomes

- Students analyze data and real-world situations and find a function to use as a model.
- Students study properties of linear, quadratic, sinusoidal, and exponential functions.

## Lesson Notes

When modeling authentic data, a mathematician or scientist must apply knowledge of conditions under which the data were gathered before fitting a function to the data. This lesson addresses focus standards **F-BF.A.1a** and **F-LE.A.2**, which ask students to determine an explicit expression for a function from a real-world context. Additionally, the entire lesson focuses on MP.4, as students use and analyze mathematical models for a variety of real-world situations. Students have already studied linear, quadratic, sinusoidal, and exponential functions, so the principal question being asked in this lesson is how to use what we know about the context to choose an appropriate function to model the data. We begin by fitting a curve to existing data points for which the data taken out of context does not clearly suggest the model. We then begin choosing a function type to model various scenarios. This is primarily a summative lesson that begins with a review of properties of linear, polynomial, exponential, and sinusoidal functions.

## Classwork

### Opening Exercise (8 minutes)

In this example, students are given points to plot and a real-world context. The point of this exercise is that either a quadratic model or a sinusoidal model can fit the data, but the real-world context is necessary to determine how to appropriately model the data. Encourage students to use calculators or another graphing utility to produce graphs of the functions in this exercise and to then copy the graph to these axes.

> *Scaffolding:*
>
> Students who struggle with sketching the sinusoidal curve may need to be reminded that its zeros are at $\frac{2\pi n}{3}$ for real $n$.
>
> Those who struggle with sketching the parabola may need the hint that it passes through $(0, 5)$ and $(4.5, 5)$.

**Opening Exercise**

a. **You are working on a team analyzing the following data gathered by your colleagues:**

$$(-1.1, 5), (0, 105), (1.5, 178), (4.3, 120).$$

**Your coworker Alexandra says that the model you should use to fit the data is**

$$k(t) = 100 \cdot \sin(1.5t) + 105.$$

**Sketch Alexandra's model on the axes at left on the next page.**

b. **How does the graph of Alexandra's model $k(t) = 100 \cdot \sin(1.5t) + 105$ relate to the four points? Is her model a good fit to this data?**

***The curve passes through or close to all four of those points, so this model fits the data well.***

c. **Another teammate Randall says that the model you should use to fit the data is**

$$g(t) = -16t^2 + 72t + 105.$$

**Sketch Randall's model on the axes at right on the next page.**

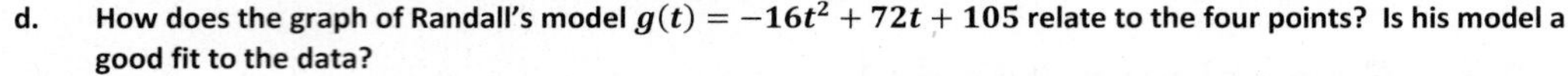

d. **How does the graph of Randall's model $g(t) = -16t^2 + 72t + 105$ relate to the four points? Is his model a good fit to the data?**

*The curve passes through or close to all four of those points, so this model also fits the data well.*

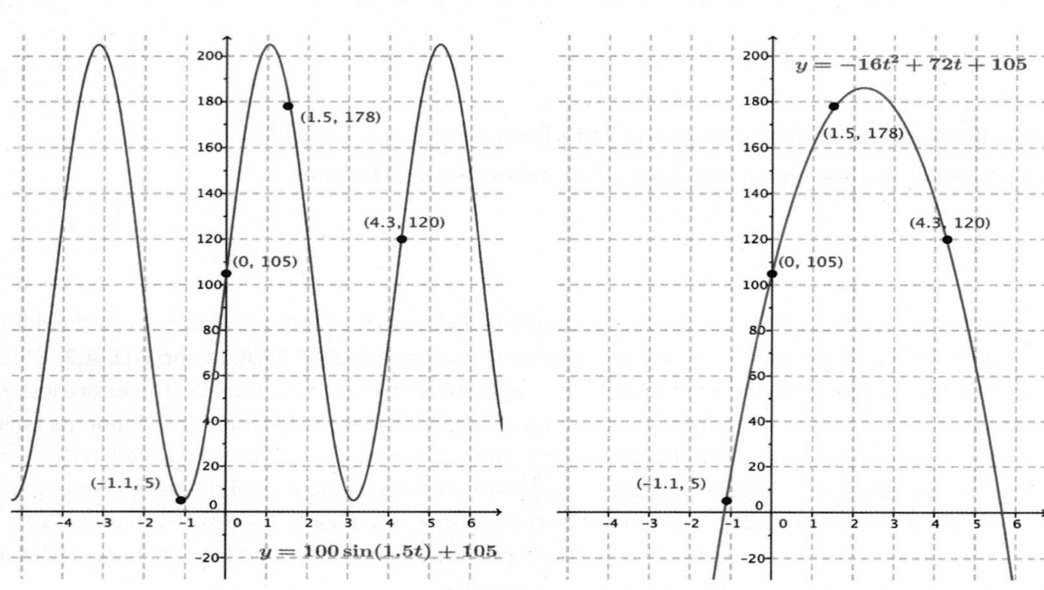


**Alexandra's Model** **Randall's Model**

e. **Suppose the four points represent positions of a projectile fired into the air. Which of the two models is more appropriate in that situation, and why?**

*The quadratic curve of Randall's model makes more sense in this context. Some students may know that the acceleration of the projectile due to gravity warrants a quadratic term in the equation for the function, and all students should understand that the motion of the projectile will not be cyclic—once it hits the ground, it stays there.*

f. **In general, how do we know which model to choose?**

*It entirely depends on the context and what we know about what the data represents.*

## Discussion (8 minutes)

- As the previous exercise showed, just knowing the coordinates of the data points does not tell us which type of function to use to model it. We need to know something about the context in which the data is gathered before we can decide what type of function to use as a model.
- We will focus on linear, quadratic, sinusoidal, and exponential models in this lesson.
- Some things we need to think about: What is the end behavior of this type of function? How does the function change? If the input value ($x$ or $t$) increases by 1 unit, what happens to the output value ($y$)? Is the function increasing or decreasing, or does it do both? Are there any relative maximum or minimum values? What is the range of the function?

- What are the characteristics of a nonconstant linear function? *(Allow students to suggest characteristics, but be sure that the traits listed below have been mentioned before moving on to quadratic models.)*
    - *If $x$ increases by $1$, then $y$ changes by a fixed amount.*
    - *The function is always increasing or always decreasing at the same rate. This rate is the slope of the line when the function is graphed.*
    - *There is no maximum and no minimum value of the function.*
    - *The range is all real numbers.*
    - *The end behavior is that the function increases to $\infty$ in one direction and decreases to $-\infty$ in the other direction.*
- What are the characteristics of a quadratic function?
    - *The second differences of the function are constant, meaning that if $x$ increases by $1$, then $y$ increases linearly with $x$.*
    - *The function increases and decreases, changing direction one time.*
    - *There is either a maximum value or a minimum value.*
    - *The range is either $(-\infty, a)$ or $(a, \infty)$ for some real number $a$.*
    - *The end behavior is that either the function increases to $\infty$ in both directions or the function decreases to $-\infty$ in both directions.*
- What are the characteristics of a sinusoidal function?
    - *The function is periodic; the function values repeat over fixed intervals.*
    - *There is one relative maximum value of the function and one relative minimum value of the function. The function attains these values periodically, alternating between the maximum and minimum value.*
    - *The range of the function is $[a, b]$, for some real numbers $a < b$.*
    - *The end behavior of a sinusoidal function is that it bounces between the relative maximum and relative minimum values as $x \to \infty$ and as $x \to -\infty$.*
- What are the characteristics of an exponential function?
    - *The function increases (or decreases) at a rate proportional to the current value of the function.*
    - *The function is either always increasing or always decreasing.*
    - *Either the function values approach a constant as $x \to -\infty$, and the function values approach $\pm\infty$ as $x \to \infty$, or the function values approach $\pm\infty$ as $x \to -\infty$, and the function values approach a constant as $x \to -\infty$. (The function flattens off in one direction and approaches either $\infty$ or $-\infty$ in the other direction.)*
- What are the clues in the context of a particular situation that suggest the use of a particular type of function as a model?
    - *If we expect from the context that each new term in the sequence of data is a constant added to the previous term, then we try a linear model.*
    - *If we expect from the context that each new term in the sequence of data is a constant multiple of the previous term, then we try an exponential model.*
    - *If we expect from the context that the second differences of the sequence are constant (meaning that the rate of change between terms either grows or shrinks linearly), then we try a quadratic model.*
    - *If we expect from the context that the sequence of terms is periodic, then we try a sinusoidal model.*

> *Scaffolding:*
>
> Have students record this information about when to use each type of function in a chart or graphic organizer.

## Exercise 1 (6 minutes)

This exercise, like the Opening Exercise, provides students with an ambiguous set of data for which it is necessary to understand the context before we can select a model. After students have completed this exercise, go through students' responses as a class to be sure that all students are aware of the ambiguity before setting them to work on the rest of the exercises. Students should work on this exercise in pairs or small groups.

*Scaffolding:*

Tell advanced students that the maximum amount of daylight in Oslo is $18.83$ hours on June $15$, and challenge them to find an appropriate sinusoidal function to model the data.

**Exercises**

1. **The table below contains the number of daylight hours in Oslo, Norway, on the specified dates.**

| Date | Hours and Minutes | Hours |
|---|---|---|
| August 1 | $16:56$ | $16.82$ |
| September 1 | $14:15$ | $14.25$ |
| October 1 | $11:33$ | $11.55$ |
| November 1 | $8:50$ | $8.90$ |

a. **Plot the data on the grid provided. You will need to decide how to best represent it.**

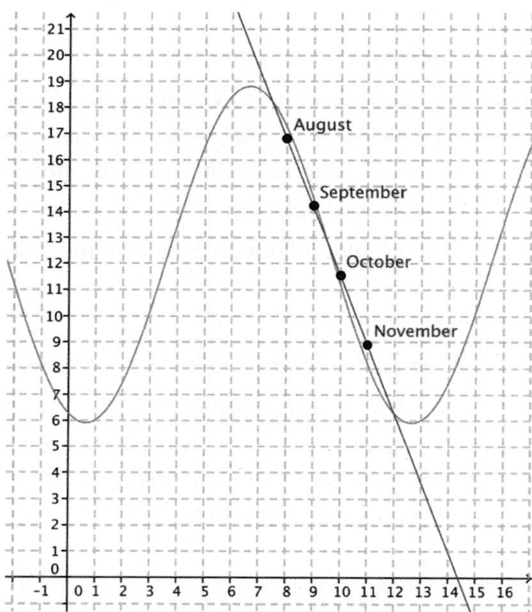


b. **Looking at the data, what type of function appears to be the best fit?**

*The data appears to lie on a straight line.*

c. **Looking at the context in which the data was gathered, what type of function should be used to model the data?**

*Since daylight hours increase and decrease with the season, being shortest in winter and longest in summer and repeating every $12$ months, a linear model is not appropriate. We should model this data with a periodic function.*

d. **Do you have enough information to find a model that is appropriate for this data? Either find a model or explain what other information you would need to do so.**

*We cannot find a complete model for this scenario because we do not know the maximum and minimum number of daylight hours in Oslo. We do know that the maximum is less than $24$ hours and the minimum is more than $0$ hours, so we could find a very rough model using a sinusoidal function; but we do not know the necessary amplitude. (The maximum number of daylight hours in Oslo is $18.83$ hours and occurs in mid-June. With this information, the points can be modeled by the function $d(t) = -6.45\cos\left(\frac{\pi}{6(x-0.65)}\right) + 12.36$, as shown in red.)*

## Exercises 2–6 (12 minutes)

In these exercises, students are asked to determine which type of function should be used to model the given scenario. In some cases, students are given enough information to actually produce a model, and they are expected to do so, and in other cases the students can only specify the type of function that should be used. It is up to them to determine whether or not they have all of the needed information. Students should work on these exercises in pairs or small groups.

2. **The goal of the U.S. Centers for Disease Control and Prevention (CDC) is to protect public health and safety through the control and prevention of disease, injury, and disability. Suppose that $45$ people have been diagnosed with a new strain of the flu virus, and scientists estimate that each person with the virus will infect $5$ people every day with the flu.**

   a. **What type of function should the scientists at the CDC use to model the initial spread of this strain of flu to try to prevent an epidemic? Explain how you know.**

   *Because each person infects $5$ people every day, the number of infected people is multiplied by a factor of $5$ each day. This would be best modeled by an exponential function, at least at the beginning of the outbreak.*

   b. **Do you have enough information to find a model that is appropriate for this situation? Either find a model or explain what other information you would need to do so.**

   *Yes. We know the initial number of infected people is $45$, and we know that the number of infected people is multiplied by $5$ each day. A model for the number of infected people on day $t$ would be $F(t) = 45(5^t)$.*

3. **An artist is designing posters for a new advertising campaign. The first poster takes $10$ hours to design, but each subsequent poster takes roughly $15$ minutes less time than the previous one as he gets more practice.**

   a. **What type of function models the amount of time needed to create $n$ posters, for $n \leq 20$? Explain how you know.**

   *Since the time difference between posters is decreasing linearly, we should model this scenario using a quadratic function.*

*Scaffolding:*

Students may need to be reminded of the methods for finding the coefficients of a quadratic polynomial by solving a linear system as done in Module 1, Lesson 30.

b. **Do you have enough information to find a model that is appropriate for this situation? Either find a model or explain what other information you would need to do so.**

*Yes. The number of hours needed to create $n$ posters can be modeled by a quadratic function*

$$T(n) = an^2 + bn + c,$$

*where we know that $T(0) = 0$, $T(1) = 10$, and $T(2) = 19.75$. This gives us the three linear equations*

$$c = 0$$
$$a + b + c = 10$$
$$4a + 2b + c = 19.75.$$

*We can solve this system of three equations using the methods of Lesson 30 in Module 1, and we find*

$$T(n) = -0.125n^2 + 10.125n.$$

4. **A homeowner notices that her heating bill is the lowest in the month of July and increases until it reaches its highest amount in the month of February. After February, the amount of the heating bill slowly drops back to the level it was in July, when it begins to increase again. The amount of the bill in February is roughly four times the amount of the bill in July.**

a. **What type of function models the amount of the heating bill in a particular month? Explain how you know.**

*Because exterior temperatures repeat fairly periodically, with the coldest temperatures in the winter and the warmest temperatures in the summer, we would expect a periodic use of heating fuel that was highest in the winter and lowest in the summer. Thus, we should use a sinusoidal function to model this scenario.*

b. **Do you have enough information to find a model that is appropriate for this situation? Either find a model or explain what other information you would need to do so.**

*We cannot model this scenario because we do not know the highest or lowest amount of the heating bill. If we knew either of those amounts, we could find the other. Then, we could find the amplitude of the sinusoidal function and create a reasonable model.*

5. **An online merchant sells used books for $\$5.00$ each, and the sales tax rate is $6\%$ of the cost of the books. Shipping charges are a flat rate of $\$4.00$ plus an additional $\$1.00$ per book.**

a. **What type of function models the total cost, including the shipping costs, of a purchase of $x$ books? Explain how you know.**

*We would use a linear function to model this situation because the total cost increases at a constant rate when you increase the number of books purchased.*

b. **Do you have enough information to find a model that is appropriate for this situation? Either find a model or explain what other information you would need to do so.**

*We can model this situation exactly. If we buy $x$ books, then the total cost of the purchase (in dollars) is given by*

$$C(x) = 1.06(5x) + 4 + 1x$$
$$= 6.3x + 4.$$

6. **A stunt woman falls from a tall building in an action-packed movie scene. Her speed increases by $32$ ft/s for every second that she is falling.**

a. **What type of function models her distance from the ground at time $t$ seconds? Explain how you know.**

*Because her speed is increasing by $32$ ft/s every second, her rate at which she gets closer to the ground is increasing linearly; thus, we would model this situation with a quadratic function.*

> **b. Do you have enough information to find a model that is appropriate for this situation? Either find a model or explain what other information you would need to do so.**
>
> ***We cannot create a model for this situation because we do not know the height of the building, so we do not know how far she will fall.***

## Discussion (4 minutes)

Either call on students or ask for volunteers to present their models for the scenarios in Exercises 2–6.

## Closing (6 minutes)

Hold a discussion with the entire class in which students provide responses to the following questions.

- In this lesson, we have looked at four kinds of mathematical models: linear, quadratic, exponential, and sinusoidal. How is a linear model different from a quadratic model?
  - *Responses will vary. Sample responses include: A linear model has a constant rate of change. It has no maximum or minimum. The equation for a quadratic model has one or more squared terms. Its graph is a parabola, and it has a maximum or minimum.*
- How is a quadratic model different from an exponential model?
  - *A quadratic model has a maximum or minimum, whereas an exponential model is unbounded. A quantity increasing exponentially eventually exceeds a quantity increasing quadratically.*
- How is an exponential model different from a linear model?
  - *As $x$ changes by $1$ in a linear model, the $y$-value changes by a fixed amount, but as $x$ changes by $1$ in an exponential model, the $y$-value changes by a multiple of $x$. The range of a linear function (that is not constant) is all real numbers, and the range of an exponential function is either $(-\infty, a)$ or $(a, \infty)$ for some real number $a$.*
- How is a quadratic model different from a sinusoidal model?
  - *A quadratic model has one relative maximum or minimum point and does not repeat, whereas a sinusoidal model has an infinite number of relative maximum and minimum points that repeat periodically.*
- How is a sinusoidal model different from an exponential model?
  - *A sinusoidal model is bounded and cyclic, whereas an exponential model either goes to $\pm\infty$ or to a constant value as $x \to \infty$. A sinusoidal model changes in a periodic fashion, but an exponential model changes at a rate proportional to the current value of the function.*

The four models are summarized in the table below, which can be reproduced and posted in the classroom.

Lesson Summary

- **If we expect from the context that each new term in the sequence of data is a constant added to the previous term, then we try a linear model.**
- **If we expect from the context that the second differences of the sequence are constant (meaning that the rate of change between terms either grows or shrinks linearly), then we try a quadratic model.**
- **If we expect from the context that each new term in the sequence of data is a constant multiple of the previous term, then we try an exponential model.**
- **If we expect from the context that the sequence of terms is periodic, then we try a sinusoidal model.**

| Model | Equation of Function | Rate of Change |
|---|---|---|
| Linear | $f(t) = at + b$ for $a \neq 0$ | Constant |
| Quadratic | $g(t) = at^2 + bt + c$ for $a \neq 0$ | Changing linearly |
| Exponential | $h(t) = ab^{ct}$ for $0 < b < 1$ or $b > 1$ | A multiple of the current value |
| Sinusoidal | $k(t) = A\sin(w(t - h)) + k$ for $A, w \neq 0$ | Periodic |

## Exit Ticket (5 minutes)

Name ______________________________ Date ______________

# Lesson 22: Choosing a Model

## Exit Ticket

The amount of caffeine in a patient's bloodstream decreases by half every 3.5 hours. A latte contains 150 mg of caffeine, which is absorbed into the bloodstream almost immediately.

a. What type of function models the caffeine level in the patient's bloodstream at time $t$ hours after drinking the latte? Explain how you know.

b. Do you have enough information to find a model that is appropriate for this situation? Either find a model or explain what other information you would need to do so.

## Exit Ticket Sample Solutions

The amount of caffeine in a patient's bloodstream decreases by half every $3.5$ hours. A latte contains $150$ mg of caffeine, which is absorbed into the bloodstream almost immediately.

a. What type of function models the caffeine level in the patient's bloodstream at time $t$ hours after drinking the latte? Explain how you know.

*Because the amount of caffeine decreases by half of the current amount in the bloodstream over a fixed time period, we would model this scenario by a decreasing exponential function.*

b. Do you have enough information to find a model that is appropriate for this situation? Either find a model or explain what other information you would need to do so.

*Assuming that the latte was the only source of caffeine for this patient, we can model the amount of caffeine in the bloodstream (in mg) at time t (in hours) by*

$$C(t) = 150\left(\frac{1}{2}\right)^{\frac{t}{3.5}}.$$

## Problem Set Sample Solutions

1. A new car depreciates at a rate of about $20\%$ per year, meaning that its resale value decreases by roughly $20\%$ each year. After hearing this, Brett said that if you buy a new car this year, then after $5$ years the car has a resale value of $\$0.00$. Is his reasoning correct? Explain how you know.

*Brett is not correct. If the car loses $20\%$ of its value each year, then it retains $80\%$ of its resale value each year. In that case, the proper model to use is an exponential function,*

$$V(t) = P(0.80)^t,$$

*where $P$ is the original price paid for the car when it was new, $t$ is the number of years the car has been owned, and $V(t)$ is the resale value of the car in year $t$. Then, when $t = 5$, the value of the car is $V(5) = P(0.80)^5 \approx 0.33P$; so, after $5$ years, the car is worth roughly $33\%$ of its original price.*

2. Alexei just moved to Seattle, and he keeps track of the average rainfall for a few months to see if the city deserves its reputation as the rainiest city in the United States.

| Month | Average rainfall |
|---|---|
| July | $0.93$ in. |
| September | $1.61$ in. |
| October | $3.24$ in. |
| December | $6.06$ in. |

What type of function should Alexei use to model the average rainfall in month $t$?

*Although the data appears to be exponential when plotted, an exponential model does not make sense for a seasonal phenomenon like rainfall. Alexei should use a sinusoidal function to model this data.*

3. **Sunny, who wears her hair long and straight, cuts her hair once per year on January 1, always to the same length. Her hair grows at a constant rate of $2$ cm per month. Is it appropriate to model the length of her hair with a sinusoidal function? Explain how you know.**

   ***No. If we were to use a sinusoidal function to model the length of her hair, then that would imply that her hair grows longer and then slowly shrinks back to its original length. Even though the length of her hair can be represented by a periodic function, it abruptly gets cut off once per year and does not smoothly return to its shortest length. None of our models are appropriate for this situation.***

4. **On average, it takes $2$ minutes for a customer to order and pay for a cup of coffee.**

   a. **What type of function models the amount of time you will wait in line as a function of how many people are in front of you? Explain how you know.**

      ***Because the wait time increases by a constant $2$ minutes for each person in line, we can use a linear function to model this situation.***

   b. **Find a model that is appropriate for this situation.**

      ***If there is no one ahead of you, then your wait time is zero. Thus, the wait time $W$, in minutes, can be modeled by***

      $$W(x) = 2x,$$

      ***where $x$ is the number of people in front of you in line.***

5. **An online ticket-selling service charges $\$50.00$ for each ticket to an upcoming concert. In addition, the buyer must pay $8\%$ sales tax and a convenience fee of $\$6.00$ for the purchase.**

   a. **What type of function models the total cost of the purchase of $n$ tickets in a single transaction?**

      ***The complete price for each ticket is $1.08(\$50.00) = \$54.00$, so the total price of the purchase increases by $\$54.00$ per ticket. Thus, this should be modeled by a linear function.***

   b. **Find a model that is appropriate for this situation.**

      ***The price for buying $n$ tickets, including the convenience fee, is then $T(n) = 54n + 6$ dollars.***

6. **In a video game, the player must earn enough points to pass one level and progress to the next as shown in the table below.**

| To pass this level ... | You need this many total points ... |
|---|---|
| 1 | 5,000 |
| 2 | 15,000 |
| 3 | 35,000 |
| 4 | 65,000 |

   **That is, the increase in the required number of points increases by $10,000$ points at each level.**

   a. **What type of function models the total number of points you need to pass to level $n$? Explain how you know.**

      ***Because the increase in needed points is increasing linearly, we should use a quadratic function to model this situation.***

b. **Find a model that is appropriate for this situation.**

***The amount of points needed to pass level $n$ can be modeled by a quadratic function***

$$P(n) = an^2 + bn + c,$$

***where we know that $P(1) = 5000$, $P(2) = 15000$, and $P(3) = 35000$. This gives us the three linear equations***

$$a + b + c = 5000$$
$$4a + 2b + c = 15000$$
$$9a + 3b + c = 35000.$$

***We can solve this system of three equations using the methods of Lesson 30 in Module 1, and we find***

$$P(n) = 5000n^2 - 5000n + 5000.$$

7. **The southern white rhinoceros reproduces roughly once every 3 years, giving birth to one calf each time. Suppose that a nature preserve houses 100 white rhinoceroses, 50 of which are female. Assume that half of the calves born are female and that females can reproduce as soon as they are 1 year old.**

   a. **What type of function should be used to model the population of female white rhinoceroses in the preserve?**

   ***Because all female rhinoceroses give birth every 3 years and half of those calves are assumed to be female, the population of female rhinoceroses increases by $\frac{1}{6}$ every year. Thus, we should use an exponential function to model the population of female southern white rhinoceroses.***

   b. **Assuming that there is no death in the rhinoceros population, find a function to model the population of female white rhinoceroses in the preserve.**

   ***Since $1 + \frac{1}{6} \approx 1.17$ and the initial population is 50 female southern white rhinoceroses, we can model this by***

   $$R_1(t) = 50(1.17)^t.$$

   c. **Realistically, not all of the rhinoceroses will survive each year, so we will assume a 5% death rate of all rhinoceroses. Now what type of function should be used to model the population of female white rhinoceroses in the preserve?**

   ***We should still use an exponential function, but the growth rate will need to be altered to take the death rate into account.***

   d. **Find a function to model the population of female white rhinoceroses in the preserve, taking into account the births of new calves and the 5% death rate.**

   ***Since 5% of the rhinoceroses die each year, that means that 95% of them survive. The new growth rate is then $0.95(1.17) \approx 1.11$. The new model would be***

   $$R_2(t) = 50(1.11)^t.$$

# Topic D:

# Using Logarithms in Modeling Situations

**A-SSE.B.3c, A-CED.A.1, A-REI.D.11, F-IF.B.3, F-IF.B.6, F-IF.C.8b, F-IF.C.9, F-BF.A.1a, F-BF.1b, F-BF.A.2, F-BF.B.4a, F-LE.A.4, F-LE.B.5**

| | | |
|---|---|---|
| **Focus Standards:** | A-SSE.B.3c | Choose and produce an equivalent form of an expression to reveal and explain properties of the quantity represented by the expression.<br>c. Use the properties of exponents to transform expressions for exponential functions. *For example the expression* $1.15^t$ *can be rewritten as* $\left(1.15^{\frac{1}{12}}\right)^{12t} \approx 1.012^{12t}$ *to reveal the approximate equivalent monthly interest rate if the annual rate is* $15\%$. |
| | A-CED.A.1 | Create equations and inequalities in one variable and use them to solve problems. Include equations arising from linear and quadratic functions, and simple rational and exponential functions. |
| | A-REI.D.11 | Explain why the $x$-coordinates of the points where the graphs of the equations $y = f(x)$ and $y = g(x)$ intersect are the solutions of the equation $f(x) = g(x)$; find the solutions approximately, e.g., using technology to graph the functions, make tables of values, or find successive approximations. Include cases where $f(x)$ and/or $g(x)$ are linear, polynomial, rational, absolute value, exponential, and logarithmic functions.★ |
| | F-IF.B.3 | Recognize that sequences are functions, sometimes defined recursively, whose domain is a subset of the integers. *For example, the Fibonacci sequence is defined recursively by* $f(0) = f(1) = 1$, $f(n+1) = f(n) + f(n-1)$ *for* $n \geq 1$. |
| | F-IF.B.6 | Calculate and interpret the average rate of change of a function (presented symbolically or as a table) over a specified interval. Estimate the rate of change from a graph.★ |

| | |
|---|---|
| F-IF.C.8b | Write a function defined by an expression in different but equivalent forms to reveal and explain different properties of the function.<br>b. Use the properties of exponents to interpret expressions for exponential functions. *For example, identify percent rate of change in functions such as* $y = (1.02)^t$, $y = (0.97)^t$, $y = (1.01)^{12t}$, $y = (1.2)^{\frac{t}{10}}$, *and classify them as representing exponential growth or decay.* |
| F-IF.C.9 | Compare properties of two functions each represented in a different way (algebraically, graphically, numerically in tables, or by verbal descriptions). *For example, given a graph of one quadratic function and an algebraic expression for another, say which has the larger maximum.* |
| F-BF.A.1 | Write a function that describes a relationship between two quantities.★<br>a. Determine an explicit expression, a recursive process, or steps for calculation from a context.<br>b. Combine standard function types using arithmetic operations. *For example, build a function that models the temperature of a cooling body by adding a constant function to a decaying exponential, and relate these functions to the model.* |
| F-BF.A.2 | Write arithmetic and geometric sequences both recursively and with an explicit formula, use them to model situations, and translate between the two forms.★ |
| F-BF.B.4a | Find inverse functions.<br>a. Solve an equation of the form $f(x) = c$ for a simple function f that has an inverse and write an expression for the inverse. *For example,* $f(x) = 2x^3$ *or* $f(x) = (x+1)/(x-1)$ *for* $x \neq 1$. |
| F-LE.A.4 | For exponential models, express as a logarithm the solution to $ab^{ct} = d$ where $a$, $c$, and $d$ are numbers and the base $b$ is 2, 10, or $e$; evaluate the logarithm using technology. |
| F-LE.B.5 | Interpret the parameters in a linear or exponential function in terms of a context. |

**Instructional Days:** 6

**Lesson 23:** Bean Counting (M)[1]
**Lesson 24:** Solving Exponential Equations (P)
**Lesson 25:** Geometric Sequences and Exponential Growth and Decay (P)
**Lesson 26:** Percent Rate of Change (P)
**Lesson 27:** Modeling with Exponential Functions (M)
**Lesson 28:** Newton's Law of Cooling, Revisited (M)

[1] Lesson Structure Key: **P**-Problem Set Lesson, **M**-Modeling Cycle Lesson, **E**-Exploration Lesson, **S**-Socratic Lesson

This topic opens with a simulation and modeling activity where students start with one bean, roll it out of a cup onto the table, and add more beans each time the marked side is up. While clearly an exponential model, the lesson unfolds by having students discover this relationship, without explicitly stating that the results are exponential, by examining patterns when the data is represented numerically and graphically. Students blend what they know about probability and exponential functions to interpret the parameters $a$ and $b$ in the functions $f(t) = a(b^t)$ that they find to model their experimental data (**F-LE.B.5**, **A-CED.A.2**).

In both Algebra I and Lesson 6 in this module, students had to solve exponential equations when modeling real-world situations numerically or graphically. Lesson 24 shows students how to use logarithms to solve these types of equations analytically and makes the connections between numeric, graphical, and analytical approaches explicit, invoking the related standards **F-LE.A.4**, **F-BF.B.4a**, and **A-REI.D.11**. Students will be encouraged to use multiple approaches to solve equations generated in the next several lessons.

In Lessons 25 to 27, a general growth/decay rate formula is presented to students to help construct models from data and descriptions of situations. Students must use properties of exponents to rewrite exponential expressions in order to interpret the properties of the function (**F-IF.C.8b**). For example, in Lesson 27, students compare the initial populations and annual growth rates of population functions given in the forms $E(t) = 281.4(1.0093)^{t-100}$, $f(t) = 81.1(1.0126)^t$, and $g(t) = 76.2(13.6)^{t/10}$. Many of the situations and problems presented here were first encountered in Module 3 of Algebra I; students are now able to solve equations involving exponents that they could only estimate previously, such as finding the time when the population of the United States is expected to surpass a half-billion people. Students answer application questions in the context of the situation. They use technology to evaluate logarithms of base 10 and $e$. Additionally, Lesson 25 begins to develop geometric sequences that will be needed for the financial content in the next Topic (**F-BF.A.2**). Lesson 26 continues developing the skills of distinguishing between situations that require exponential-vs-linear models (**F-LE.A.1**), and Lesson 27 continues the work with geometric sequences that started in Lesson 25 (**F-IF.B.3**, **F-BF.A.1a**).

Lesson 28 closes this topic and addresses **F-BF.A.1b** by revisiting Newton's Law of Cooling, a formula that involves the sum of an exponential function and a constant function. Students first learned about this formula in Algebra I but now that they are armed with logarithms and have more experience understanding how transformations affect the graph of a function, they can find the precise value of the decay constant using logarithms and, thus, can solve problems related to this formula more precisely and with greater depth of understanding.

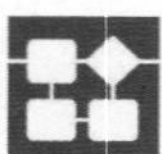

# Lesson 23: Bean Counting

## Student Outcomes

- Students gather experimental data and determine which type of function is best to model the data.
- Students use properties of exponents to interpret expressions for exponential functions.

## Lesson Notes

In the main activity in this lesson, students will work in pairs to gather their own data, plot it (MP.6), and apply the methods of Lesson 22 to decide which type of data to use in modeling the data (MP.4, MP.7). Students should use calculators (or other technological tools) to fit the data with an exponential function (MP.5). Since each group of students will generate its own set of data, each group will find different functions to model the data, but all of those functions should be in the form $f(t) = a(b^t)$, where $a \approx 1$ and $b \approx 1.5$. Take time to discuss why the functions differ between groups and yet are closely related due to probability. If time permits, at the end of the activity average each group's values of the constants $a$ and $b$; the averages should be very close to $1$ and $1.5$, respectively (**F-LE.B.5**).

In the Problem Set, students will investigate the amount of time for quantities to double, triple, or increase by a factor of $10$ using these functions and others like them. For example, students will rewrite functions $f(t) = a(b^t)$ in an equivalent form $f(t) = a(2^{\log_2(b)t})$ and interpret the exponent (**F-IF.C.8b**).

## Materials

Each team of two students will gather data using the following materials:

- At least 50 small, flippable objects marked differently on the two sides. One inexpensive source for these objects would be dried beans, spray painted on one side and left unpainted on the other. Buttons or coins would also work. Throughout this lesson, these objects will be referred to as beans.
- Two paper cups: one to hold the beans, and the other to shake up and dump out the beans onto the paper plate.
- A paper plate (to keep the beans from ending up all over the floor).
- A calculator capable of plotting data and performing exponential regression.

## Classwork

### Opening (3 minutes)

Divide the class into groups of 2–3 students. Smaller groups are better for this exercise, so have students work in pairs if possible. Before distributing the needed materials, model the process for gathering data for two trials. Then, provide each team with a cup containing at least 50 beans, an empty cup, a paper plate, and a calculator.

> *Scaffolding:*
>
> If students are collectively struggling with the concepts of exponential growth and modeling, consider extending this to a 2-day lesson. Devote the entire first day to Exercise 1 on exponential growth, and the second day to Exercise 2 on exponential decay.

- In this lesson, we will gather some data and then decide which type of function to use to model it. In the previous lesson, we studied different types of functions we could use to model data in different situations. What were those types of functions?
  - *Linear, quadratic, sinusoidal, and exponential functions*
- Suppose you are gathering data from an experiment, measuring a quantity on evenly spaced time intervals. How can you recognize from the context that data should be modeled by a linear function?

Accompany the discussion of different model types with visuals showing graphs of each type.

  - *If we expect the data values to increase (or decrease) at an even rate, then the data points should roughly lie on a line.*

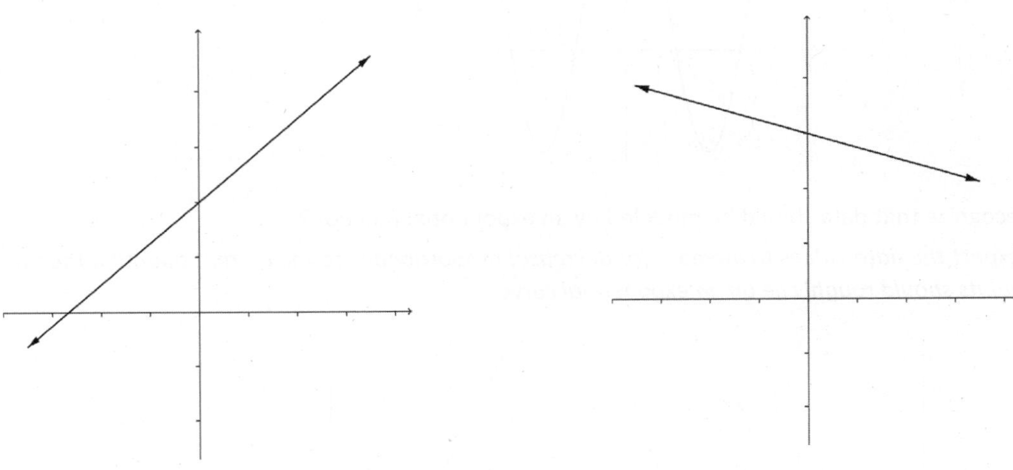

- How do you recognize that data should be modeled by a quadratic function?
  - *If we expect the distance between data values to increase or decrease at a constant rate, then the data points should roughly lie on a parabola.*

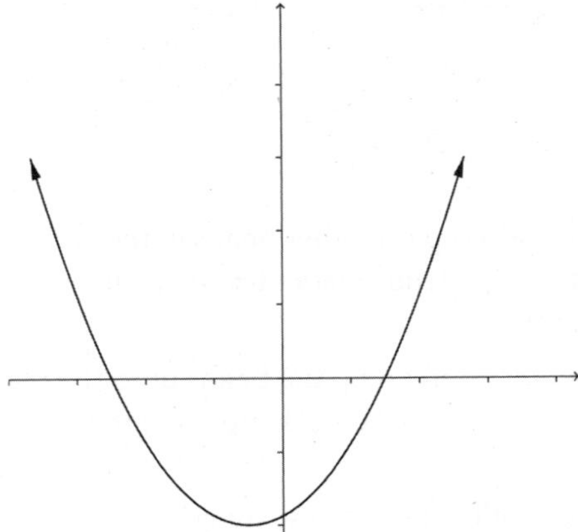

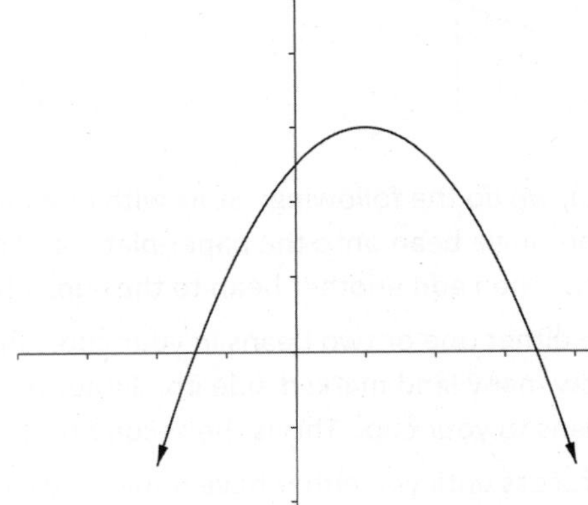

- How do you recognize that data should be modeled by a sinusoidal function?
  - *If we expect the data values to repeat periodically due to a repeating phenomenon, then the data points should roughly lie on a sinusoidal curve.*

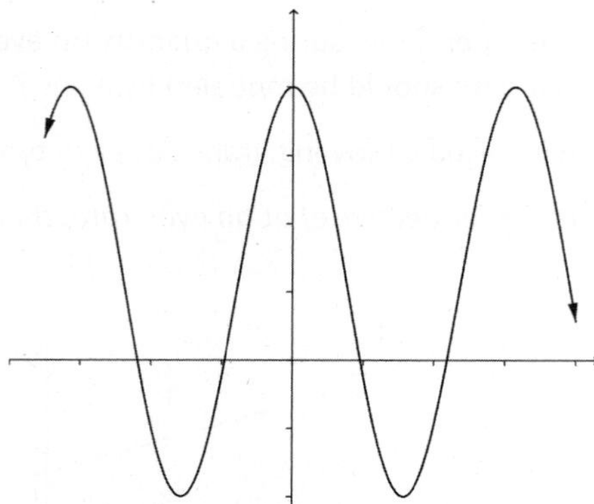

- How do you recognize that data should be modeled by an exponential function?
  - *If we expect the data values to increase (or decrease) proportionally to the current quantity, then the data points should roughly lie on an exponential curve.*

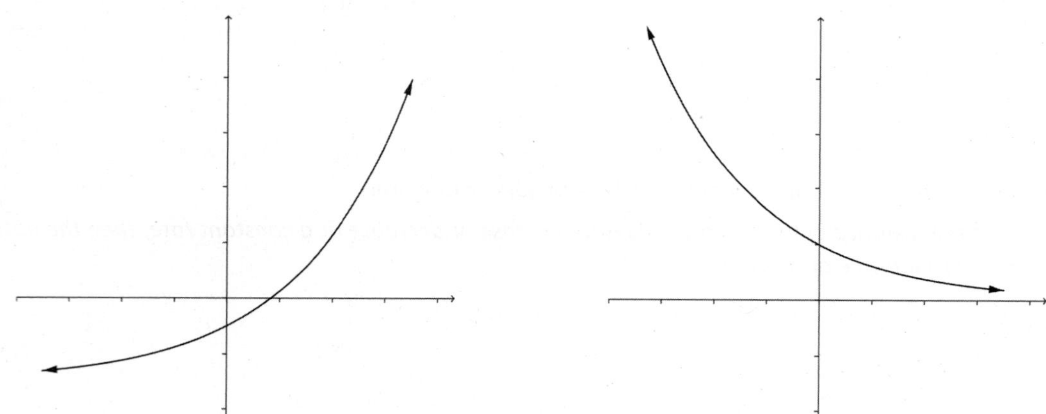

- To gather data, we do the following: Start with one bean in one cup, and keep the other beans in the spare cup. Dump the single bean onto the paper plate, and record in the table if it lands marked-side up. If it lands marked-side up, then add another bean to the cup. This is the first trial.
- You now have either one or two beans in your cup. Shake the cup gently, then dump the beans onto the plate and record how many land marked-side up. Either $0$, $1$, or $2$ beans will land marked-side up. Add that number of beans to your cup. This is the second trial.
- Repeat this process until you either have done $10$ trials or have run out of beans in the spare cup.
- After you have gathered your data, plot the data on the coordinate grid provided. Plot the trial number on the horizontal axis and the number of beans in the cup at the start of the trial on the vertical axis.

## Mathematical Modeling Exercise 1 (15 minutes)

Circulate around the room while the students gather data for this exercise to ensure that they all understand the process for flipping the beans, adding new beans to the cup, and recording data.

**Mathematical Modeling Exercises**

1. **Working with a partner, you are going to gather some data, analyze the data, and find a function to use to model the data. Be prepared to justify your choice of function to the class.**

   a. **Gather your data: For each trial, roll the beans from the cup to the paper plate. Count the number of beans that land marked-side up, and add that many beans to the cup. Record the data in the table below. Continue until you have either completed $10$ trials or the number of beans at the start of the trial exceeds the number that you have.**

| Trial Number, $t$ | Number of Beans at Start of Trial | Number of Beans That Landed Marked-Side Up |
|---|---|---|
| $1$ | $1$ | |
| $2$ | | |
| $3$ | | |
| $4$ | | |
| $5$ | | |
| $6$ | | |
| $7$ | | |
| $8$ | | |
| $9$ | | |
| $10$ | | |

   b. **Based on the context in which you gathered this data, which type of function would best model your data points?**

   ***Since the number of beans we add at each toss is roughly half of the beans we had at the start of that turn, the data should be modeled by an exponential function.***

c. **Plot the data: Plot the trial number on the horizontal axis and the number of beans in the cup at the start of the trial on the vertical axis. Be sure to label the axes appropriately and to choose a reasonable scale for the axes.**

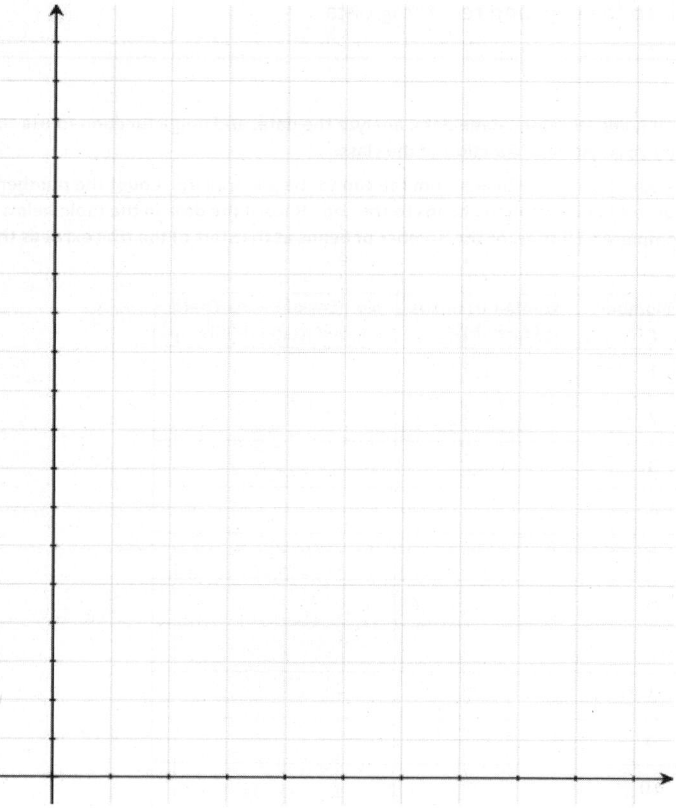

d. **Analyze the data: Which type of function would best fit your data? Explain your reasoning.**

***Students should see that the data follows a clear pattern of exponential growth, and they should decide to model it with an exponential function.***

e. **Model the data: Enter the data into the calculator and use the appropriate type of regression to find an equation that fits this data. Round the constants to two decimal places.**

***Answers will vary but should be of the form $f(t) = a(b^t)$, where $a$ is near $1$ and $b$ is near $1.5$.***

**Discussion (5 minutes)**

After students complete Exercise 1, have them write the equations that they found to model their data on the board, displayed by the document camera, or on poster board visible to all students. After all groups have reported their equations, debrief the class with questions like the following:

- What type of equation did you use to model your data?
  - *We used an exponential function.*

- Why did you choose this type of equation?
  - *The number of beans will always increase, and the amount it increases depends on the current number of beans that we have.*
- Why did all of the teams get different equations?
  - *We all had different data because each roll of the beans is different. Each bean has a 50% chance of landing marked-side up, but that does not mean that half of the beans always land marked-side up. Probability does not guarantee that an event will happen.*
- Look at the function $f(t) = a(b^t)$ that models your data and interpret the value of the constant $a$.
  - *The number $a$ is the number of beans we started with (according to the model).*
- Look at the function $f(t) = a(b^t)$ that models your data and interpret the value of the base $b$.
  - *The base $b$ is the growth factor. This means that the number of beans is multiplied by $b$ with each toss.*
- If the beans landed perfectly with 50% of them marked-side up every time, what would we expect the value of $b$ to be?
  - *If half of the beans always landed marked-side up, then the number of beans would increase by 50% with each trial, so that the new amount would be 150% of the old amount. That is, the number of beans would be multiplied by 1.5 at each trial. Then, the value of $b$ would be 1.5.*
- From the situation we are modeling, what would we expect the value of the coefficient $a$ to be?
  - *For an exponential function of the form $f(t) = a(b^t)$, we have $f(0) = a$, so $a$ represents our initial number of beans. Thus, we should have a value of $a$ near 1.*
- Why didn't your values of $a$ and $b$ turn out to be $a = 1$ and $b = 1.5$?
  - *Even though the probability says that we have a 50% chance of the beans landing marked-side up, this does not mean that half of the beans will always land marked-side up.*
- Let's look at the constants in the equations $f(t) = a(b^t)$ that the groups found to fit their data.
- Have the students calculate the average values of the coefficients $a$ and the base $b$ from each group's equation.
  - *Answers will vary. The average value of $a$ should be near 1, and the average value of $b$ should be near 1.5.*

Create an exponential function $f(t) = a_{\text{avg}}(b_{\text{avg}})^t$, where $a_{\text{avg}}$ and $b_{\text{avg}}$ represent the average values of the coefficient $a$ and base $b$ from each group's equation.

- What would happen to the values of $a_{\text{avg}}$ and $b_{\text{avg}}$ if we had data from 1,000 groups?
  - *The value of $a_{\text{avg}}$ should get very close to 1 and the value of $b_{\text{avg}}$ should get very close to 1.5.*

## Mathematical Modeling Exercise 2 (10 minutes)

2. **This time, we are going to start with $50$ beans in your cup. Roll the beans onto the plate and remove any beans that land marked-side up. Repeat until you have no beans remaining.**

   a. **Gather your data: For each trial, roll the beans from the cup to the paper plate. Count the number of beans that land marked-side up, and remove that many beans from the plate. Record the data in the table below. Repeat until you have no beans remaining.**

| Trial Number, $t$ | Number of Beans at Start of Trial | Number of Beans That Landed Marked-Side Up |
|---|---|---|
| 1 | 50 | |
| 2 | | |
| 3 | | |
| 4 | | |
| 5 | | |
| 6 | | |
| 7 | | |
| 8 | | |
| 9 | | |
| 10 | | |
| | | |
| | | |
| | | |
| | | |

b. **Plot the data: Plot the trial number on the horizontal axis and the number of beans in the cup at the start of the trial on the vertical axis. Be sure to label the axes appropriately and choose a reasonable scale for the axes.**

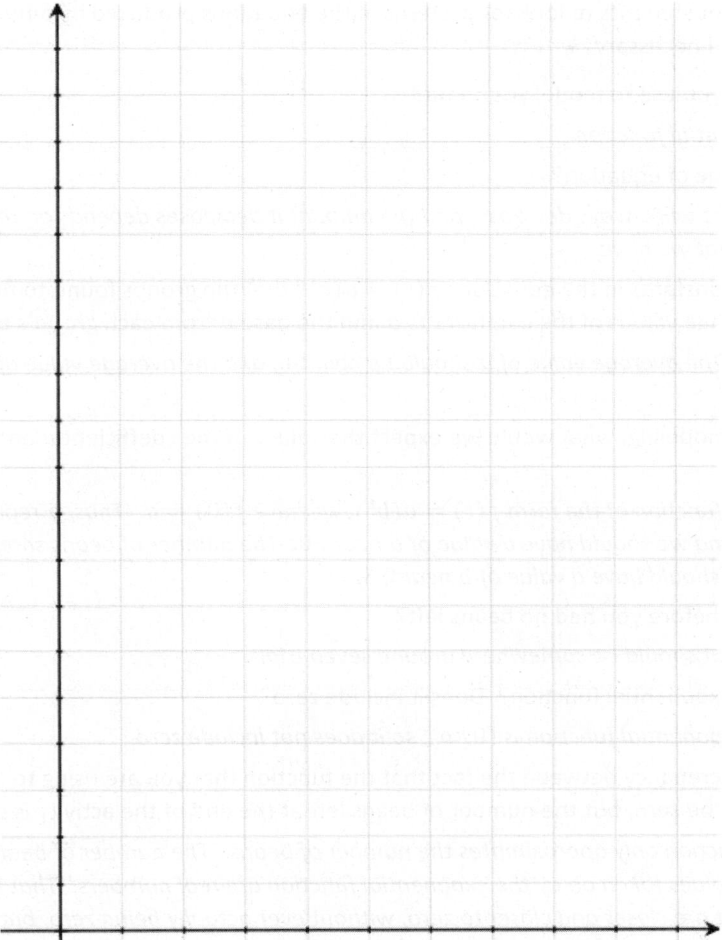

c. **Analyze the data: Which type of function would best fit your data? Explain your reasoning.**

***Students should see that the data follows a clear pattern of exponential decay, and they should decide to model it with an exponential function.***

d. **Make a prediction: What do you expect the values of $a$ and $b$ to be for your function? Explain your reasoning.**

***Using an exponential function $f(t) = a(b^t)$, the value of $a$ is the initial number of beans, so we should expect $a$ to be near $50$. The number of beans decreases by half each time, so we would expect $b = 0.5$.***

e. **Model the data: Enter the data into the calculator. Do not enter your final data point of 0 beans. Use the appropriate type of regression to find an equation that fits this data. Round the constants to two decimal places.**

***Answers will vary but should be of the form $f(t) = a(b^t)$, where $a$ is near $50$ and $b$ is near $0.5$.***

## Discussion (5 minutes)

After students complete Exercise 2, have them write the equations that they found to model their data in a shared location visible to all students. Prompt students to look for patterns in the equations produced by different groups and to revisit their answers from part (d) if necessary.

- What type of equation did you use to model your data?
  - *We used an exponential function.*
- Why did you choose this type of equation?
  - *The number of beans will always decrease, and the amount it decreases depends on the current number of beans that we have.*
- Let's look at the different constants in the equations $f(t) = a(b^t)$ that the groups found to fit the data. Have students calculate the average values of the coefficients $a$ and the base $b$ from each group's equation.
  - *Answers will vary. The average value of $a$ should be near $50$, and the average value of $b$ should be near $0.5$.*

MP.2 & MP.7

- From the situation we are modeling, what would we expect the values of the coefficient $a$ and the base $b$ to be?
  - *For an exponential function of the form $f(t) = a(b^t)$, we have $f(0) = a$. Thus, $a$ represents our initial number of beans, and we should have a value of $a$ near $50$. The number of beans should be cut in half at each trial, so we should have a value of $b$ near $0.5$.*
- How many trials did it take before you had no beans left?
  - *Answers will vary but should be somewhere around seven trials.*
- What is the range of your exponential function? Does it include zero?
  - *The range of the exponential function is $(0, \infty)$, so it does not include zero.*
- How can we explain the discrepancy between the fact that the function that you are using to model the number of beans can never be zero, but the number of beans left at the end of the activity is clearly zero?
  - *The exponential function only approximates the number of beans. The number of beans is always an integer, while the values taken on by the exponential function are real numbers. That is, our function takes on values that are closer and closer to zero, without ever actually being zero, but if we were to round it to integers, then it rounds to zero.*

## Closing (3 minutes)

Have students respond to the following questions individually in writing or orally with a partner.

- What sort of function worked best to model the data we gathered from Mathematical Modeling Exercise 1 where we added beans?
  - *We used an increasing exponential function.*
- Why was this the best type of function?
  - *The number of beans added was roughly a multiple of the current number of beans; if the beans had behaved perfectly and half of them always landed with the marked-side up, we would have added half the number of beans each time.*

- Why could we expect to model this data with a function $f(t) = (1.5)^t$?
  - *When we add half of the beans, that means that the current number of beans is being multiplied by* $1 + 0.5 = 1.5$. *With an exponential function, we get from one data point to the next by multiplying by a constant. For this function, the constant is* $1.5$.
- Why did an exponential function work best to model the data from Mathematical Modeling Exercise 2, in which the number of beans was reduced at each trial?
  - *The number of beans removed at each trial was roughly half the current number of beans. The function that models this is* $f(t) = \left(\frac{1}{2}\right)^t$, *which is a decreasing exponential function.*

**Exit Ticket (4 minutes)**

Name ______________________________ Date ______________

# Lesson 23: Bean Counting

## Exit Ticket

Suppose that you were to repeat the bean activity, but in place of beans, you were to use six-sided dice. Starting with one die, each time a die is rolled with a 6 showing, you add a new die to your cup.

a. Would the number of dice in your cup grow more quickly or more slowly than the number of beans did? Explain how you know.

b. A sketch of one sample of data from the bean activity is shown below. On the same axes, draw a rough sketch of how you would expect the line of best fit from the dice activity to look.

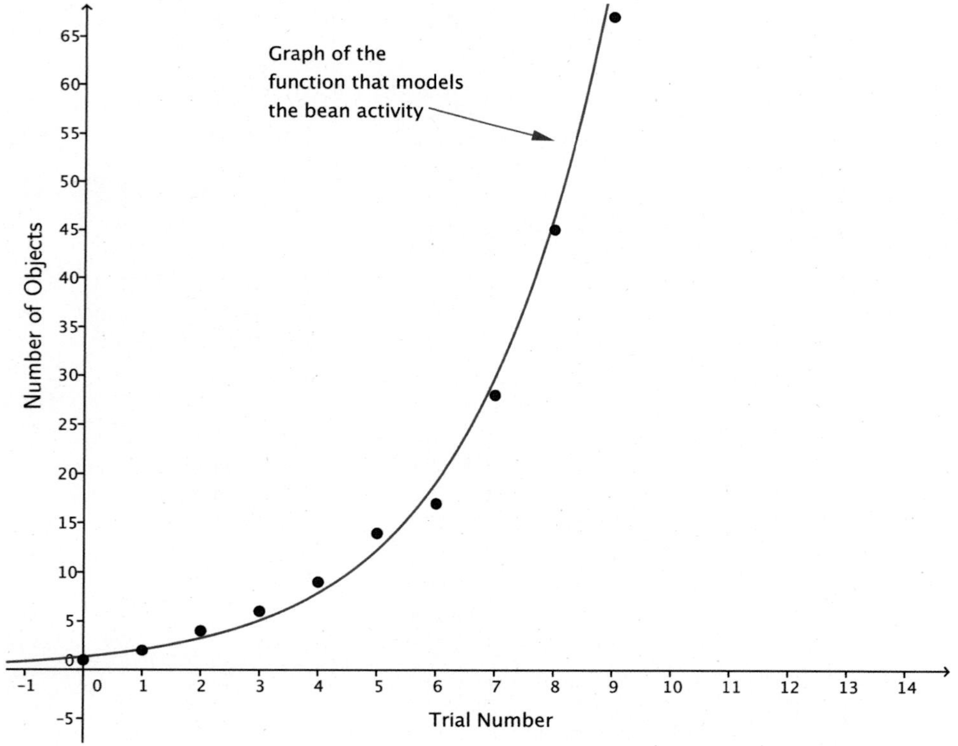

## Exit Ticket Sample Solutions

**Suppose that you were to repeat the bean activity, but in place of beans, you were to use six-sided dice. Starting with one die, each time a die is rolled with a 6 showing, you add a new die to your cup.**

a. **Would the number of dice in your cup grow more quickly or more slowly than the number of beans did? Explain how you know.**

*The number of dice in the cup should grow much more slowly because the probability of rolling a 6 is $\frac{1}{6}$, while the probability of flipping a bean marked-side up is roughly $\frac{1}{2}$. Thus, the beans should land marked-side up, and thus, increase the number of beans in our cup about half of the time, while the dice would show a 6 only $\frac{1}{6}$ of the time. As an example, it could be expected to take one or two flips of the first bean to get a bean to show the marked side, causing us to add one, but it could be expected to take six rolls of the first die to get a 6, causing us to add another die.*

b. **A sketch of one sample of data from the bean activity is shown below. On the same axes, draw a rough sketch of how you would expect the line of best fit from the dice activity to look.**

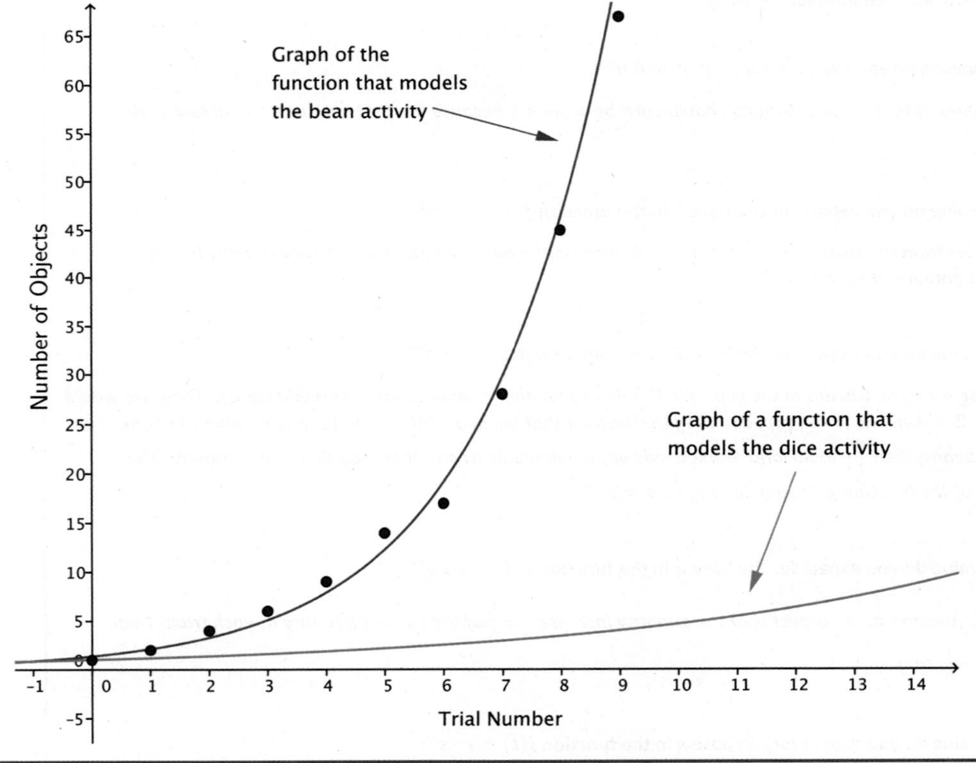

## Problem Set Sample Solutions

1. For this problem, we will consider three scenarios for which data has been collected and functions have been found to model the data, where $a$, $b$, $c$, $d$, $p$, $q$, $r$, $s$, $t$, and $u$ are positive real number constants.

   i. The function $f(t) = a \cdot b^t$ models the original bean activity (Mathematical Modeling Exercise 1). Each bean is painted or marked on one side, and we start with one bean in the cup. A trial consists of throwing the beans in the cup and adding one more bean for each bean that lands marked-side up.

   ii. The function $g(t) = c \cdot d^t$ models a modified bean activity. Each bean is painted or marked on one side, and we start with one bean in the cup. A trial consists of throwing the beans in the cup and adding two more beans for each bean that lands marked-side up.

   iii. The function $h(t) = p \cdot q^t$ models the dice activity from the Exit Ticket. Start with one six-sided die in the cup. A trial consists of rolling the dice in the cup and adding one more die to the cup for each die that lands with a 6 showing.

   iv. The function $j(t) = r \cdot s^t$ models a modified dice activity. Start with one six-sided die in the cup. A trial consists of rolling the dice in the cup and adding one more die to the cup for each die that lands with a 5 or a 6 showing.

   v. The function $k(t) = u \cdot v^t$ models a modified dice activity. Start with one six-sided die in the cup. A trial consists of rolling the dice in the cup and adding one more die to the cup for each die that lands with an even number showing.

   a. What values do you expect for $a$, $c$, $p$, $r$, and $u$?

   *The values of these four constants should each be around $1$ because the first data point in all four cases is $(1, 0)$.*

   b. What value do you expect for the base $b$ in the function $f(t) = a \cdot b^t$?

   *We know from the class activity that $b \approx 1.5$ because the number of beans grows by roughly half of the current amount at each trial.*

   c. What value do you expect for the base $d$ in the function $g(t) = c \cdot d^t$?

   *Suppose we have $4$ beans in the cup. We should expect half of them to land marked-side up. Then, we would add $2 \cdot 2 = 4$ beans to the cup, doubling the amount that we had. This is true for any number of beans; if we had $n$ beans, then $\frac{n}{2}$ should land marked-side up, so we would add $n$ beans, doubling the amount. Thus, values of the function $g$ should double, so $d \approx 2$.*

   d. What value do you expect for the base $q$ in the function $h(t) = p \cdot q^t$?

   *For this function $h$, we expect that the quantity increases by $\frac{1}{6}$ of the current quantity at each trial. Then, $q \approx 1 + \frac{1}{6}$, so $q \approx \frac{7}{6}$.*

   e. What value do you expect for the base $s$ in the function $j(t) = r \cdot s^t$?

   *The probability of rolling a $5$ or $6$ is $\frac{1}{3}$, so we would expect that the number of dice increases by $\frac{1}{3}$ of the current quantity at each trial. Thus, we expect $s \approx 1 + \frac{1}{3}$, which means that $s \approx \frac{4}{3}$.*

f. **What value do you expect for the base $v$ in the function $k(t) = u \cdot v^t$?**

*The probability of rolling an even number on a six-sided die is the same as the probability of getting the marked-side up on a bean, so we would expect that $f$ and $k$ are the same function. Thus, $v \approx 1.5$.*

g. **The following graphs represent the four functions $f$, $g$, $h$, and $j$. Identify which graph represents which function.**

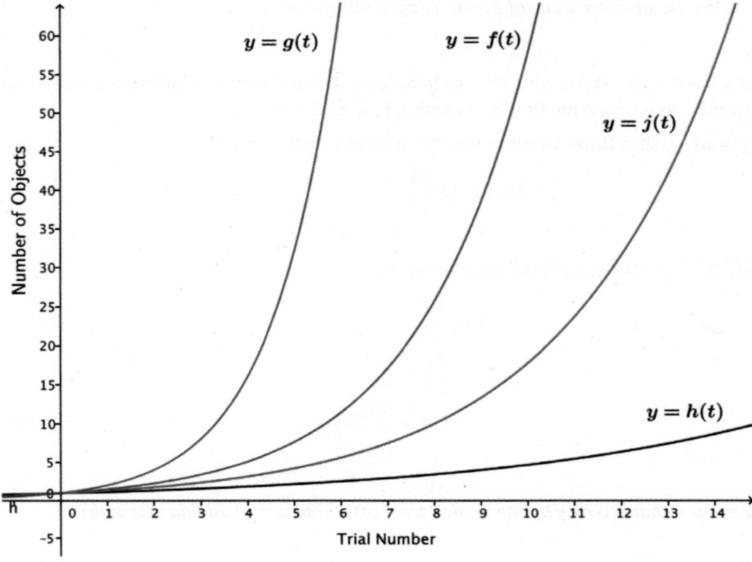


2. **Teams 1, 2, and 3 gathered data as shown in the tables below, and each team modeled the data using an exponential function of the form $f(t) = a \cdot b^t$.**

a. **Which team should have the highest value of $b$? Which team should have the lowest value of $b$? Explain how you know.**

| Team 1 | | Team 2 | | Team 3 | |
|---|---|---|---|---|---|
| Trial Number, $t$ | Number of Beans | Trial Number, $t$ | Number of Beans | Trial Number, $t$ | Number of Beans |
| 0 | 1 | 0 | 1 | 0 | 1 |
| 1 | 1 | 1 | 1 | 1 | 2 |
| 2 | 2 | 2 | 1 | 2 | 3 |
| 3 | 2 | 3 | 2 | 3 | 5 |
| 4 | 4 | 4 | 2 | 4 | 8 |
| 5 | 6 | 5 | 3 | 5 | 14 |
| 6 | 8 | 6 | 5 | 6 | 26 |
| 7 | 14 | 7 | 7 | 7 | 46 |
| 8 | 22 | 8 | 12 | 8 | 76 |
| 9 | 41 | 9 | 18 | 9 | |
| 10 | 59 | 10 | 27 | 10 | |

*The larger the value of the base b, the larger the function will be at $t = 10$. Since team 3 had the most beans at the end (they used their 50 beans first), their equation should have the highest value of the base b. Team 2 has the smallest number of beans after 10 trials, so team 2 should have the smallest value of the base b.*

b. Use a graphing calculator to find the equation that best fits each set of data. Do the equations of the functions provide evidence that your answer in part (a) is correct?

*Team 1's equation:* $f_1(t) = 0.7378(1.5334)^t$

*Team 2's equation:* $f_2(t) = 0.6495(1.4245)^t$

*Team 3's equation:* $f_3(t) = 1.0328(1.7068)^t$

*As predicted, Team 2 has the smallest value of b, and Team 3 has the largest value of b.*

3. Omar has devised an activity in which he starts with 15 dice in his cup. A trial consists of rolling the dice in the cup and adding one more die to the cup for each die that lands with a 1, 2, or 3 showing.

a. Find a function $f(t) = a(b^t)$ that Omar would expect to model his data.

$$f(t) = 15\left(\frac{3}{2}\right)^t$$

b. Solve the equation $f(t) = 30$. What does the solution mean?

$$15\left(\frac{3}{2}\right)^t = 30$$
$$\left(\frac{3}{2}\right)^t = 2$$
$$t = \frac{\log(2)}{\log\left(\frac{3}{2}\right)}$$
$$t \approx 1.71$$

*So, Omar should have more than 30 dice by the second trial, after rolling and adding dice twice.*

c. Omar wants to know in advance how many trials it should take for his initial quantity of 15 dice to double. He uses properties of exponents and logarithms to rewrite the function from part (a) as an exponential function of the form $f(t) = a\left(2^{t\cdot\log_2(b)}\right)$.

$$f(t) = 15\left(2^{t\cdot\log_2\left(\frac{3}{2}\right)}\right) \approx 15(2^{0.5850t})$$

d. Has Omar correctly applied the properties of exponents and logarithms to obtain an equivalent expression for his original equation in part (a)? Explain how you know.

*Yes. The expressions are equal by the property of exponents.*

$$\left(2^{t\cdot\log_2\left(\frac{3}{2}\right)}\right) = \left(2^{\log_2\left(\frac{3}{2}\right)}\right)^t = \left(\frac{3}{2}\right)^t$$

*Thus,*

$$15\left(2^{t\cdot\log_2\left(\frac{3}{2}\right)}\right) = 15\left(\frac{3}{2}\right)^t.$$

e. Explain how the modified formula from part (c) allows Omar to easily find the expected amount of time, $t$, for the initial quantity of dice to double.

*The quantity is doubled at a time t for which* $t \cdot \log_2\left(\frac{3}{2}\right) = 1$. *Thus, we solve the equation* $t \cdot \log_2\left(\frac{3}{2}\right) = 1$ *to find* $t = \frac{1}{\log_2\left(\frac{3}{2}\right)} \approx 1.71$. *This agrees with our answer to part (b).*

4. **Brenna has devised an activity in which she starts with $10$ dice in her cup. A trial consists of rolling the dice in the cup and adding one more die to the cup for each die that lands with a 6 showing.**

   a. **Find a function $f(t) = a(b^t)$ that you would expect to model her data.**

   $$f(t) = 10\left(\frac{7}{6}\right)^t$$

   b. **Solve the equation $f(t) = 30$. What does your solution mean?**

   $$10\left(\frac{7}{6}\right)^t = 30$$
   $$\left(\frac{7}{6}\right)^t = 3$$
   $$t = \frac{\log(3)}{\log\left(\frac{7}{6}\right)}$$
   $$t \approx 7.13$$

   *Brenna's quantity of dice should reach $30$ by the eighth trial.*

   c. **Brenna wants to know in advance how many trials it should take for her initial quantity of $10$ dice to triple. Use properties of exponents and logarithms to rewrite your function from part (a) as an exponential function of the form $f(t) = a(3^{ct})$.**

   *Since* $\frac{7}{6} = 3^{\log_3\left(\frac{7}{6}\right)}$, *we have* $\left(\frac{7}{6}\right)^t = \left(3^{\log_3\left(\frac{7}{6}\right)}\right)^t = 3^{t\cdot\log_3\left(\frac{7}{6}\right)}$. *Then,* $f(t) = 10\left(3^{t\cdot\log_3\left(\frac{7}{6}\right)}\right)$.

   d. **Explain how your formula from part (c) allows you to easily find the expected amount of time, $t$, for the initial quantity of dice to triple.**

   *The quantity triples when* $3^{t\cdot\log_3\left(\frac{7}{6}\right)} = 3$, *so that* $t \cdot \log_3\left(\frac{7}{6}\right) = 1$. *Then, we solve that equation to find the value of* $t$: $t = \log_3\left(\frac{7}{6}\right) \approx 7.13$.

   e. **Rewrite the formula for the function $f$ using a base-$10$ exponential function.**

   *Since* $\frac{7}{6} = 10^{\log\left(\frac{7}{6}\right)}$, *we have* $\left(\frac{7}{6}\right)^t = \left(10^{\log\left(\frac{7}{6}\right)}\right)^t = 10^{t\cdot\log\left(\frac{7}{6}\right)}$. *Then,* $f(t) = 10\left(10^{t\cdot\log\left(\frac{7}{6}\right)}\right)$.

   f. **Use your formula from part (e) to find out how many trials it should take for the quantity of dice to grow to $100$ dice.**

   *The quantity will be $100$ when* $10^{t\cdot\log\left(\frac{7}{6}\right)} = 10$, *so that* $t \cdot \log\left(\frac{7}{6}\right) = 1$. *Then,* $t = \frac{1}{\log\left(\frac{7}{6}\right)} \approx 14.94$, *so that the quantity should exceed $100$ by the $15^{th}$ trial.*

5. **Suppose that one bacteria population can be modeled by the function $P_1(t) = 500(2^t)$ and a second bacteria population can be modeled by the function $P_2(t) = 500(2.83^t)$, where $t$ measures time in hours. Keep four digits of accuracy for decimal approximations of logarithmic values.**

   a. **What does the $500$ mean in each function?**

   *In each function, the $500$ means that each population has $500$ bacteria at the onset of the experiment.*

   b. **Which population should double first? Explain how you know.**

   *Since $2.83 > 2$, the second population is growing at a faster rate than the first, so it should double more quickly.*

c. **How many hours and minutes will it take until the first population doubles?**

*The first population doubles every hour, since the base of the exponential function is* $2$. *Thus, the first population doubles in one hour.*

d. **Rewrite the formula for $P_2(t)$ in the form $P_2(t) = a(2^{ct})$, for some real numbers $a$ and $c$.**

$$\begin{aligned} P_2(t) &= 500(2.83)^t \\ &= 500\left(2^{\log_2(2.83)}\right)^t \\ &= 500\left(2^{t \cdot \log_2(2.83)}\right) \end{aligned}$$

e. **Use your formula in part (d) to find the time, $t$, in hours and minutes until the second population doubles.**

*The second population doubles when* $t \cdot \log_2(2.83) = 1$, *which happens when* $t = \frac{1}{\log_2(2.83)}$. *Thus,* $t \approx 0.6663$ *hours, which is approximately* $40$ *minutes.*

6. **Copper has antibacterial properties, and it has been shown that direct contact with copper alloy C11000 at $20°\text{C}$ kills $99.9\%$ of all methicillin-resistant *Staphylococcus aureus* (MRSA) bacteria in about $75$ minutes. Keep four digits of accuracy for decimal approximations of logarithmic values.**

a. **A function that models a population of $1,000$ MRSA bacteria $t$ minutes after coming in contact with copper alloy C11000 is $P(t) = 1000(0.912)^t$. What does the base $0.912$ mean in this scenario?**

*The base* $0.912$ *means that* $91.2\%$ *of the MRSA bacteria remain at the end of each minute.*

b. **Rewrite the formula for $P$ as an exponential function with base $\frac{1}{2}$.**

*Since* $0.912 = \left(\frac{1}{2}\right)^{\log_{\frac{1}{2}}(0.912)} = \left(\frac{1}{2}\right)^{-\log_2(0.912)}$, *we have*

$$\begin{aligned} P(t) &= 1000(0.912^t) \\ &= 1000\left(\frac{1}{2}\right)^{-t \cdot \log_2(0.912)}. \end{aligned}$$

c. **Explain how your formula from part (b) allows you to easily find the time it takes for the population of MRSA to be reduced by half.**

*The population of MRSA is reduced by half when the exponent is* $1$. *This happens when* $-t \cdot \log_2(0.912) = 1$, *so* $t = -\frac{1}{\log_2(0.912)} \approx 7.52$ *minutes. Thus, half of the MRSA bacteria die every* $7\frac{1}{2}$ *minutes.*

# Lesson 24: Solving Exponential Equations

## Student Outcomes

- Students apply properties of logarithms to solve exponential equations.
- Students relate solutions to $f(x) = g(x)$ to the intersection point(s) on the graphs of $y = f(x)$ and $y = g(x)$ in the case where $f$ and $g$ are constant or exponential functions.

## Lesson Notes

Much of our previous work with logarithms in Topic B provided students with the particular skills needed to manipulate logarithmic expressions and solve exponential equations. Although students have solved exponential equations in earlier lessons in Topic B, this is the first time that they solve such equations in the context of exponential functions. In this lesson, students solve exponential equations of the form $ab^{ct} = d$ using properties of logarithms developed in Lessons 12 and 13 (**F-LE.A.4**). For an exponential function $f$, students solve equations of the form $f(x) = c$ and write a logarithmic expression for the inverse (**F-BF.B.4a**). Additionally, students solve equations of the form $f(x) = g(x)$ where $f$ and $g$ are either constant or exponential functions (**A-REI.D.11**). Examples of exponential functions in this lesson draw from Lesson 7, in which the growth of a bacteria population was modeled by the function $P(t) = 2^t$, and Lesson 23, in which students modeled the growth of an increasing number of beans with a function $f(t) = a(b^t)$, where $a \approx 1$ and $b \approx 1.5$.

Students will need to use technology to calculate logarithmic values and to graph linear and exponential functions.

## Classwork

### Opening Exercise (4 minutes)

The Opening Exercise is a simple example of solving an exponential equation of the form $ab^{ct} = d$. Allow students to work independently or in pairs to solve this problem. Circulate around the room to check that all students know how to apply a logarithm to solve this problem. Students can choose to use either a base 2 or base 10 logarithm.

**Opening Exercise**

**In Lesson 7, we modeled a population of bacteria that doubled every day by the function $P(t) = 2^t$, where $t$ was the time in days. We wanted to know the value of $t$ when there were $10$ bacteria. Since we did not know about logarithms at the time, we approximated the value of $t$ numerically, and we found that $P(t) = 10$ at approximately $t \approx 3.32$ days.**

**Use your knowledge of logarithms to find an exact value for $t$ when $P(t) = 10$, and then use your calculator to approximate that value to $4$ decimal places.**

***Since $P(t) = 2^t$, we need to solve $2^t = 10$.***

$$2^t = 10$$
$$t\log(2) = \log(10)$$
$$t = \frac{1}{\log(2)}$$
$$t \approx 3.3219$$

***Thus, the population will reach $10$ bacteria in approximately $3.3219$ days.***

## Discussion (2 minutes)

Ask students to describe their solution method for the Opening Exercise. Make sure that solutions are discussed using both base 10 and base 2 logarithms. If all students used the common logarithm to solve this problem, then present the following solution using the base 2 logarithm:

$$\begin{aligned} 2^t &= 10 \\ \log_2(2^t) &= \log_2(10) \\ t &= \log_2(10) \\ t &= \frac{\log(10)}{\log(2)} \\ t &= \frac{1}{\log(2)} \\ t &\approx 3.3219 \end{aligned}$$

The remaining exercises ask students to solve equations of the form $f(x) = c$ or $f(x) = g(x)$, where $f$ and $g$ are exponential functions (**F-LE.A.4**, **F-BF.B.4a**, **A-REI.D.11**). For the remainder of the lesson, allow students to work either independently or in pairs or small groups on the exercises. Circulate to ensure students are on task and solving the equations correctly. After completing Exercises 1–4, debrief students to check for understanding, and ensure they are using appropriate strategies to complete problems accurately before moving on to Exercises 5–10.

## Exercises 1–4 (25 minutes)

**Exercises**

1. **Fiona modeled her data from the bean-flipping experiment in Lesson 23 by the function $f(t) = 1.263(1.357)^t$, and Gregor modeled his data with the function $g(t) = 0.972(1.629)^t$.**

   a. **Without doing any calculating, determine which student, Fiona or Gregor, accumulated 100 beans first. Explain how you know.**

      ***Since the base of the exponential function for Gregor's model, $1.629$, is larger than the base of the exponential function for Fiona's model, $1.357$, Gregor's model will grow more quickly than Fiona's, and he will accumulate $100$ beans before Fiona does.***

   b. **Using Fiona's model …**

      i. **How many trials would be needed for her to accumulate 100 beans?**

         ***We need to solve the equation $f(t) = 100$ for t.***

$$\begin{aligned} 1.263(1.357)^t &= 100 \\ 1.357^t &= \frac{100}{1.263} \\ t\log(1.357) &= \log\left(\frac{100}{1.263}\right) \\ t\log(1.357) &= \log(100) - \log(1.263) \\ t &= \frac{2 - \log(1.263)}{\log(1.357)} \\ t &\approx 14.32 \end{aligned}$$

         ***So, it takes $15$ trials for Fiona to accumulate $100$ beans.***

> *Scaffolding:*
>
> Have struggling students begin this exercise with functions $f(t) = 7(2^t)$ and $g(t) = 4(3^t)$.

ii. **How many trials would be needed for her to accumulate $1,000$ beans?**

*We need to solve the equation $f(t) = 1000$ for t.*

$$1.263(1.357)^t = 1000$$
$$1.357^t = \frac{1000}{1.263}$$
$$t\log(1.357) = \log\left(\frac{1000}{1.263}\right)$$
$$t\log(1.357) = \log(1000) - \log(1.263)$$
$$t = \frac{3 - \log(1.263)}{\log(1.357)}$$
$$t \approx 21.86$$

*So, it takes $22$ trials for Fiona to accumulate $1,000$ beans.*

c. **Using Gregor's model …**

i. **How many trials would be needed for him to accumulate $100$ beans?**

*We need to solve the equation $g(t) = 100$ for t.*

$$0.972(1.629)^t = 100$$
$$1.629^t = \frac{100}{0.972}$$
$$t\log(1.629) = \log\left(\frac{100}{0.972}\right)$$
$$t\log(1.629) = \log(100) - \log(0.972)$$
$$t = \frac{2 - \log(0.972)}{\log(1.629)}$$
$$t \approx 9.50$$

*So, it takes $10$ trials for Gregor to accumulate $100$ beans.*

ii. **How many trials would be needed for him to accumulate $1,000$ beans?**

*We need to solve the equation $g(t) = 1000$ for t.*

$$0.972(1.629)^t = 1000$$
$$1.629^t = \frac{1000}{0.972}$$
$$t\log(1.629) = \log\left(\frac{1000}{0.972}\right)$$
$$t\log(1.629) = \log(1000) - \log(0.972)$$
$$t = \frac{3 - \log(0.972)}{\log(1.629)}$$
$$t \approx 14.21$$

*So, it takes $15$ trials for Gregor to accumulate $1,000$ beans.*

d. **Was your prediction in part (a) correct? If not, what was the error in your reasoning?**

***Responses will vary. Either students made the correct prediction, or they did not recognize that the base b determines the growth rate of the exponential function so the larger base $1.629$ causes Gregor's function to grow much more quickly than Fiona's.***

2. **Fiona wants to know when her model $f(t) = 1.263(1.357)^t$ predicts accumulations of $500$, $5,000$, and $50,000$ beans, but she wants to find a way to figure it out without doing the same calculation three times.**

   a. **Let the positive number $c$ represent the number of beans that Fiona wants to have. Then solve the equation $1.263(1.357)^t = c$ for $t$.**

$$1.263(1.357)^t = c$$
$$1.357^t = \frac{c}{1.263}$$
$$t\log(1.357) = \log\left(\frac{c}{1.263}\right)$$
$$t\log(1.357) = \log(c) - \log(1.263)$$
$$t = \frac{\log(c) - \log(1.263)}{\log(1.357)}$$

   b. **Your answer to part (a) can be written as a function $M$ of the number of beans $c$, where $c > 0$. Explain what this function represents.**

   *The function $M(c) = \frac{\log(c)-\log(1.263)}{\log(1.357)}$ calculates the number of trials it will take for Fiona to accumulate $c$ beans.*

   c. **When does Fiona's model predict that she will accumulate …**

   i. **$500$ beans?**

$$M(500) = \frac{\log(500) - \log(1.263)}{\log(1.357)} \approx 19.59$$

   *According to her model, it will take Fiona $20$ trials to accumulate $500$ beans.*

   ii. **$5,000$ beans?**

$$M(5000) = \frac{\log(5000) - \log(1.263)}{\log(1.357)} \approx 27.14$$

   *According to her model, it will take Fiona $28$ trials to accumulate $5,000$ beans.*

   iii. **$50,000$ beans?**

$$M(50000) = \frac{\log(50000) - \log(1.263)}{\log(1.357)} \approx 34.68$$

   *According to her model, it will take Fiona $35$ trials to accumulate $50,000$ beans.*

3. **Gregor states that the function $g$ that he found to model his bean-flipping data can be written in the form $g(t) = 0.972(10^{\log(1.629)t})$. Since $\log(1.629) \approx 0.2119$, he is using $g(t) = 0.972(10^{0.2119t})$ as his new model.**

   a. **Is Gregor correct? Is $g(t) = 0.972\left(10^{\log(1.629)t}\right)$ an equivalent form of his original function? Use properties of exponents and logarithms to explain how you know.**

   *Yes, Gregor is correct. Since $10^{\log(1.629)} = 1.629$, and $10^{\log(1.629)t} = \left(10^{\log(1.629)}\right)^t \approx 10^{0.2119t}$, Gregor is right that $g(t) = 0.972(10^{0.2119t})$ is a reasonable model for his data.*

b. Gregor also wants to find a function that will help him to calculate the number of trials his function $g$ predicts it will take to accumulate $500$, $5,000$, and $50,000$ beans. Let the positive number $c$ represent the number of beans that Gregor wants to have. Solve the equation $0.972(10^{0.2119t}) = c$ for $t$.

$$0.972(10^{0.2119t}) = c$$
$$10^{0.2119t} = \frac{c}{0.972}$$
$$0.2119t = \log\left(\frac{c}{0.972}\right)$$
$$t = \frac{\log(c) - \log(0.972)}{0.2119}$$

c. Your answer to part (b) can be written as a function $N$ of the number of beans $c$, where $c > 0$. Explain what this function represents.

*The function $N(c) = \frac{\log(c)-\log(0.972)}{0.2119}$ calculates the number of trials it will take for Gregor to accumulate $c$ beans.*

d. When does Gregor's model predict that he will accumulate …

i. $500$ beans?

$$N(500) = \frac{\log(500) - \log(0.972)}{0.2119} \approx 12.80$$

*According to his model, it will take Gregor $13$ trials to accumulate $500$ beans.*

ii. $5,000$ beans?

$$N(500) = \frac{\log(5000) - \log(0.972)}{0.2119} \approx 17.51$$

*According to his model, it will take Gregor $18$ trials to accumulate $5,000$ beans.*

iii. $50,000$ beans?

$$N(50000) = \frac{\log(50000) - \log(0.972)}{0.2119} \approx 22.23$$

*According to his model, it will take Gregor $23$ trials to accumulate $50,000$ beans.*

4. Helena and Karl each change the rules for the bean experiment. Helena started with four beans in her cup and added one bean for each that landed marked-side up for each trial. Karl started with one bean in his cup but added two beans for each that landed marked-side up for each trial.

a. Helena modeled her data by the function $h(t) = 4.127(1.468^t)$. Explain why her values of $a = 4.127$ and $b = 1.468$ are reasonable.

*Since Helena starts with four beans, we should expect that $a \approx 4$, so a value $a = 4.127$ is reasonable. Because she is using the same rule for adding beans to the cup as we did in Lesson 23, we should expect that $b \approx 1.5$. Thus, her value of $b = 1.468$ is reasonable.*

b. Karl modeled his data by the function $k(t) = 0.897(1.992^t)$. Explain why his values of $a = 0.897$ and $b = 1.992$ are reasonable.

*Since Karl starts with one bean, we should expect that $a \approx 1$, so a value $a = 0.897$ is reasonable. Because Karl adds two beans to the cup for each that lands marked-side up, he adds two beans instead of one, we should expect that the number of beans roughly doubles with each trial. That is, we should expect $b \approx 2$. Thus, his value of $b = 1.992$ is reasonable.*

c. **At what value of $t$ do Karl and Helena have the same number of beans?**

***We need to solve the equation $h(t) = k(t)$ for $t$.***

$$4.127(1.468^t) = 0.897(1.992^t)$$
$$\log(4.127(1.468^t)) = \log(0.897(1.992^t))$$
$$\log(4.127) + \log(1.468^t) = \log(0.897) + \log(1.992^t)$$
$$\log(4.127) + t\log(1.468) = \log(0.897) + t\log(1.992)$$
$$t\log(1.992) - t\log(1.468) = \log(4.127) - \log(0.897)$$
$$t(\log(1.992) - \log(1.468)) = \log(4.127) - \log(0.897)$$
$$t\left(\log\left(\frac{1.992}{1.468}\right)\right) = \log\left(\frac{4.127}{0.897}\right)$$
$$t(0.13256) \approx 0.66284$$
$$t \approx 5.0003$$

***Thus, after trial number $5$, Karl and Helena have the same number of beans.***

d. **Use a graphing utility to graph $y = h(t)$ and $y = k(t)$ for $0 < t < 10$.**

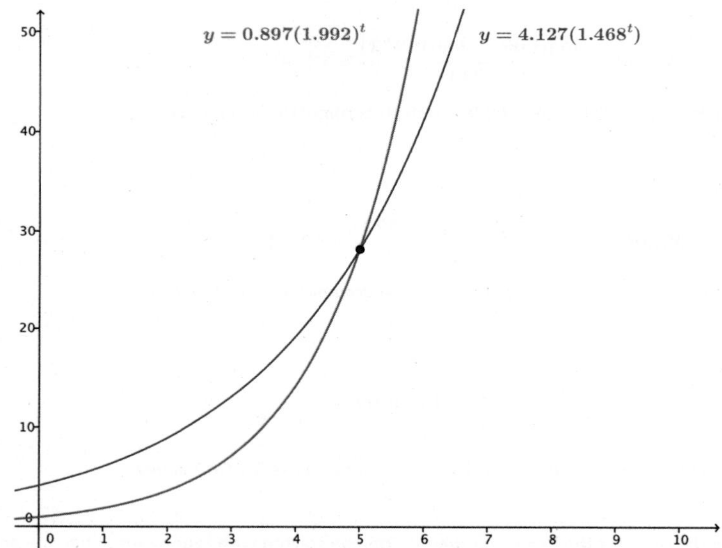


e. **Explain the meaning of the intersection point of the two curves $y = h(t)$ and $y = k(t)$ in the context of this problem.**

***The two curves intersect at the $t$-value where Helena and Karl have the same number of beans. The $y$-value indicates the number of beans they both have after five trials.***

f. **Which student reaches $20$ beans first? Does the reasoning you used with whether Gregor or Fiona would get to $100$ beans first hold true here? Why or why not?**

***Helena reaches $20$ beans first. Although the function modeling Helena's beans has a smaller base, Karl's does not catch up to Helena until after five trials. After five trials, Karl's will always be greater, and he will reach $100$ beans first. The logic we applied to comparing Gregor's model and Fiona's model does not apply here because Helena and Karl do not start with the same initial number of beans.***

Debrief students after they complete Exercises 1–4 to ensure understanding of the exercises and strategies used to solve the exercises before continuing. Exercises 5–10 let students solve exponential functions using what they know about logarithms. After completing Exercises 5–10, debrief students about when it is necessary to use logarithms to solve exponential equations and when it is not. Exercises 7, 8, and 9 are examples of exercises that do not require logarithms to solve but may be appropriate to solve with logarithms depending on the preferences of students.

## Exercise 5–10 (7 minutes)

**For the following functions $f$ and $g$, solve the equation $f(x) = g(x)$. Express your solutions in terms of logarithms.**

5. $f(x) = 10(3.7)^{x+1}$, $g(x) = 5(7.4)^x$

$$10(3.7)^{x+1} = 5(7.4)^x$$
$$2(3.7)^{x+1} = 7.4^x$$
$$\log(2) + \log(3.7^{x+1}) = \log(7.4^x)$$
$$\log(2) + (x+1)\log(3.7) = x\log(7.4)$$
$$\log(2) + x\log(3.7) + \log(3.7) = x\log(7.4)$$
$$\log(2) + \log(3.7) = x(\log(7.4) - \log(3.7))$$
$$\log(7.4) = x\log\left(\frac{7.4}{3.7}\right)$$
$$\log(7.4) = x\log(2)$$
$$x = \frac{\log(7.4)}{\log(2)}$$

6. $f(x) = 135(5)^{3x+1}$, $g(x) = 75(3)^{4-3x}$

$$135(5)^{3x+1} = 75(3)^{4-3x}$$
$$9(5)^{3x+1} = 5(3)^{4-3x}$$
$$\log(9) + (3x+1)\log(5) = \log(5) + (4-3x)\log(3)$$
$$2\log(3) + 3x\log(5) + \log(5) = \log(5) + 4\log(3) - 3x\log(3)$$
$$3x(\log(5) + \log(3)) = 4\log(3) - 2\log(3)$$
$$3x\log(15) = 2\log(3)$$
$$x = \frac{2\log(3)}{3\log(15)}$$

7. $f(x) = 100^{x^3+x^2-4x}$, $g(x) = 10^{2x^2-6x}$

$$100^{x^3+x^2-4x} = 10^{2x^2-6x}$$
$$(10^2)^{x^3+x^2-4x} = 10^{2x^2-6x}$$
$$2(x^3 + x^2 - 4x) = 2x^2 - 6x$$
$$x^3 + x^2 - 4x = x^2 - 3x$$
$$x^3 - x = 0$$
$$x(x^2 - 1) = 0$$
$$x(x+1)(x-1) = 0$$
$$x = 0, x = -1, \text{ or } x = 1$$

*Scaffolding:*

- Challenge advanced students to solve Exercise 6 in more than one way, for example, by using first the logarithm base 5 and then the logarithm base 3, and compare the results.
- Advanced students should be able to solve Exercises 7–9 without logarithms by expressing each function with a common base, but logarithms may be easier and more reliable for students struggling with the exponential properties.

8. $f(x) = 48(4^{x^2+3x})$, $g(x) = 3(8^{x^2+4x+4})$

$$48(4^{x^2+3x}) = 3(8^{x^2+4x+4})$$
$$16(4^{x^2+3x}) = 8^{x^2+4x+4}$$
$$2^4((2^2)^{x^2+3x}) = (2^3)^{x^2+4x+4}$$
$$2^{2x^2+6x+4} = 2^{3x^2+12x+12}$$
$$2x^2 + 6x + 4 = 3x^2 + 12x + 12$$
$$x^2 + 6x + 8 = 0$$
$$(x+4)(x+2) = 0$$
$$x = -4 \text{ or } x = -2$$

9. $f(x) = e^{\sin^2(x)}$, $g(x) = e^{\cos^2(x)}$

$$e^{\sin^2(x)} = e^{\cos^2(x)}$$
$$\sin^2(x) = \cos^2(x)$$
$$\sin(x) = \cos(x) \text{ or } \sin(x) = -\cos(x)$$
$$x = \frac{\pi}{4} + k\pi \text{ or } x = \frac{3\pi}{4} + k\pi \text{ for all integers } k$$

10. $f(x) = (0.49)^{\cos(x)+\sin(x)}$, $g(x) = (0.7)^{2\sin(x)}$

$$(0.49)^{\cos(x)+\sin(x)} = (0.7)^{2\sin(x)}$$
$$\log((0.49)^{\cos(x)+\sin(x)}) = \log(0.7)^{2\sin(x)})$$
$$(\cos(x) + \sin(x))\log(0.49) = 2\sin(x)\log(0.7)$$
$$(\cos(x) + \sin(x))\log(0.7^2) = 2\sin(x)\log(0.7)$$
$$2(\cos(x) + \sin(x))\log(0.7) = 2\sin(x)\log(0.7)$$
$$2\cos(x) + 2\sin(x) = 2\sin(x)$$
$$\cos(x) = 0$$
$$x = \frac{\pi}{2} + k\pi \text{ for all integers } k$$

## Closing (3 minutes)

Ask students to respond to the following prompts either in writing or orally to a partner.

- Describe two different approaches to solving the equation $2^{x+1} = 3^{2x}$. Do not actually solve the equation.
  - *You could begin by taking the logarithm base 10 of both sides, or the logarithm base 2 of both sides. (Or, you could even take the logarithm base 3 of both sides.)*
- Could the graphs of two exponential functions $f(x) = 2^{x+1}$ and $g(x) = 3^{2x}$ ever intersect at more than one point? Explain how you know.
  - *No. The graphs of these functions will always be increasing. They will intersect at one point, but once they cross once they cannot cross again. For large values of $x$, the quantity $3^{2x}$ will always be greater than $2^{x+1}$, so the graph of $g$ will end up above the graph of $f$ after they cross.*

- Discuss how the starting value and base affect the graph of an exponential function and how this can help you compare exponential functions.
  - *The starting value determines the $y$-intercept of an exponential function, so it determines how large or small the function is when $x = 0$. The base is ultimately more important and determines how quickly the function increases (or decreases). When comparing exponential functions, the function with the larger base will always overtake the function with the smaller base no matter how large the value when $x = 0$.*
- If $f(x) = 2^{x+1}$ and $g(x) = 3^{2x}$, is it possible for the equation $f(x) = g(x)$ to have more than one solution?
  - *No. Solutions to the equation $f(x) = g(x)$ correspond to $x$-values of intersection points of the graphs of $y = f(x)$ and $y = g(x)$. Since these graphs can intersect no more than once, the equation can have no more than one solution.*

**Exit Ticket (4 minutes)**

Name ______________________________________________ Date ____________________

# Lesson 24: Solving Exponential Equations

## Exit Ticket

Consider the functions $f(x) = 2^{x+6}$ and $g(x) = 5^{2x}$.

a. Use properties of logarithms to solve the equation $f(x) = g(x)$. Give your answer as a logarithmic expression, and approximate it to two decimal places.

b. Verify your answer by graphing the functions $y = f(x)$ and $y = g(x)$ in the same window on a calculator, and sketch your graphs below. Explain how the graph validates your solution to part (a).

## Exit Ticket Sample Solutions

**Consider the functions $f(x) = 2^{x+6}$ and $g(x) = 5^{2x}$.**

a. **Use properties of logarithms to solve the equation $f(x) = g(x)$. Give your answer as a logarithmic expression, and approximate it to two decimal places.**

$$2^{x+6} = 5^{2x}$$
$$(x+6)\log(2) = 2x\log(5)$$
$$2x\log(5) - x\log(2) = 6\log(2)$$
$$x = \frac{6\log(2)}{2\log(5) - \log(2)}$$
$$x = \frac{\log(64)}{\log(25) - \log(2)}$$
$$x = \frac{\log(64)}{\log\left(\frac{25}{2}\right)}$$
$$x \approx 1.65$$

*Any of the final three forms are acceptable, and other correct forms using logarithms with other bases (such as base 2) are possible.*

b. **Verify your answer by graphing the functions $y = f(x)$ and $y = g(x)$ in the same window on a calculator, and sketch your graphs below. Explain how the graph validates your solution to part (a).**

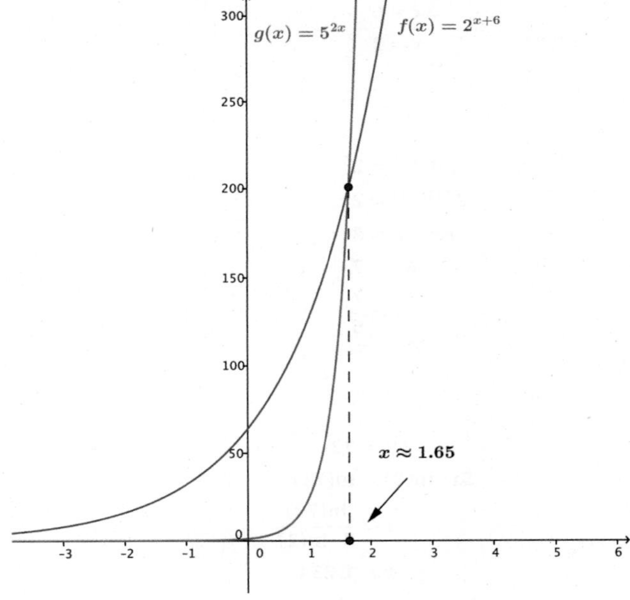


*Because the graphs of $y = f(x)$ and $y = g(x)$ intersect when $x \approx 1.65$, we know that the equation $f(x) = g(x)$ has a solution at approximately $x = 1.65$.*

## Problem Set Sample Solutions

1. **Solve the following equations.**

a. $2 \cdot 5^{x+3} = 6250$

$$5^{x+3} = 3125$$
$$5^{x+3} = 5^5$$
$$x + 3 = 5$$
$$x = 2$$

b. $3 \cdot 6^{2x} = 648$

$$6^{2x} = 216$$
$$6^{2x} = 6^3$$
$$2x = 3$$
$$x = \frac{3}{2}$$

c. $5 \cdot 2^{3x+5} = 10240$

$$2^{3x+5} = 2048$$
$$2^{3x+5} = 2^{11}$$
$$3x + 5 = 11$$
$$3x = 6$$
$$x = 2$$

d. $4^{3x-1} = 32$

$$4^{3x-1} = 2^5$$
$$2^{2\cdot(3x-1)} = 2^5$$
$$6x - 2 = 5$$
$$6x = 7$$
$$x = \frac{7}{6}$$

e. $3 \cdot 2^{5x} = 216$

$$2^{5x} = 72$$
$$5x \cdot \ln(2) = \ln(72)$$
$$x = \frac{\ln(72)}{5 \cdot \ln(2)}$$
$$x \approx 1.234$$

***Note:** Students can also use common logs to solve for the solution.*

f. $5 \cdot 11^{3x} = 120$

$$11^{3x} = 24$$
$$3x \cdot \ln(11) = \ln(24)$$
$$x = \frac{\ln(24)}{3 \cdot \ln(11)}$$
$$x \approx 0.442$$

*Note: Students can also use common logs to solve for the solution.*

g. $7 \cdot 9^x = 5405$

$$9^x = \frac{5405}{7}$$
$$x \cdot \ln(9) = \ln\left(\frac{5405}{7}\right)$$
$$x = \frac{\ln\left(\frac{5405}{7}\right)}{\ln(9)}$$
$$x \approx 3.026$$

*Note: Students can also use common logs to solve for the solution.*

h. $\sqrt{3} \cdot 3^{3x} = 9$

*Solution using properties of exponents:*

$$3^{\frac{1}{2}} \cdot 3^{3x} = 3^2$$
$$3^{\frac{1}{2}+3x} = 3^2$$
$$\frac{1}{2} + 3x = 2$$
$$x = \frac{1}{2}$$

i. $\log(400) \cdot 8^{5x} = \log(160000)$

$$8^{5x} = \frac{\log(160000)}{\log(400)}$$
$$8^{5x} = 2$$
$$8^{5x} = 8^{\frac{1}{3}}$$
$$5x = \frac{1}{3}$$
$$x = \frac{1}{15}$$

2. **Lucy came up with the model $f(t) = 0.701(1.382)^t$ for the first bean activity. When does her model predict that she would have $1,000$ beans?**

$$1000 = 0.701(1.382)^t$$
$$\log(1000) = \log(0.701) + t\log(1.382)$$
$$t = \frac{\log(1000) - \log(0.701)}{\log(1.382)}$$
$$t \approx 22.45$$

*Lucy's model predicts that it will take $23$ trials to have over $1,000$ beans.*

3. Jack came up with the model $g(t) = 1.033(1.707)^t$ for the first bean activity. When does his model predict that he would have $50,000$ beans?

$$50,000 = 1.033(1.707)^t$$
$$\log(50000) = \log(1.033) + t\log(1.707)$$
$$t = \frac{\log(50000) - \log(1.033)}{\log(1.707)}$$
$$t \approx 20.17$$

*Jack's model predicts that it will take $21$ trials to have over $50,000$ beans.*

4. If instead of beans in the first bean activity you were using fair coin, when would you expect to have $\$1,000,000$?

*One million dollars is $10^8$ pennies. Using fair pennies, we can model the situation by $f(t) = 1.5^t$.*

$$10^8 = 1.5^t$$
$$8 = t\log(1.5)$$
$$t = \frac{8}{\log(1.5)}$$
$$t \approx 45.43$$

*We should expect it to take $46$ trials to reach more than $\$1$ million using fair pennies.*

5. Let $f(x) = 2 \cdot 3^x$ and $g(x) = 3 \cdot 2^x$.

   a. Which function is growing faster as $x$ increases? Why?

   *The function $f$ is growing faster due to its larger base, even though $g(0) > f(0)$.*

   b. When will $f(x) = g(x)$?

$$f(x) = g(x)$$
$$2 \cdot 3^x = 3 \cdot 2^x$$
$$\ln(2 \cdot 3^x) = \ln(3 \cdot 2^x)$$
$$\ln(2) + x\ln(3) = \ln(3) + x\ln(2)$$
$$x\ln(3) - x\ln(2) = \ln(3) - \ln(2)$$
$$x\ln\left(\frac{3}{2}\right) = \ln\left(\frac{3}{2}\right)$$
$$x = 1$$

*Note: Students can also use common logs to solve for the solution.*

6. A population of *E. coli* bacteria can be modeled by the function $E(t) = 500(11.547)^t$, and a population of *Salmonella* bacteria can be modeled by the function $S(t) = 4000(3.668)^t$, where $t$ measures time in hours.

   a. Graph these two functions on the same set of axes. At which value of $t$ does it appear that the graphs intersect?

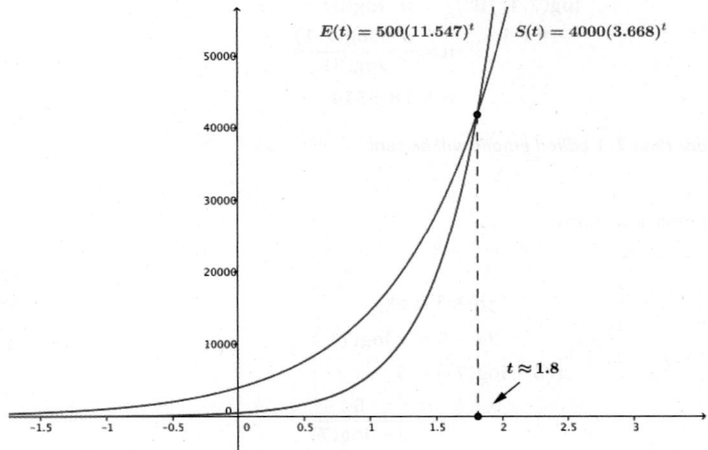


   *From the graph, it appears that the two curves intersect at $t \approx 1.8$ hours.*

   b. Use properties of logarithms to find the time $t$ when these two populations are the same size. Give your answer to two decimal places.

$$E(t) = S(t)$$
$$500(11.547)^t = 4000(3.668)^t$$
$$11.547^t = 8(3.668)^t$$
$$t\log(11.547) = \log(8) + t\log(3.668)$$
$$t\big(\log(11.547) - \log(3.668)\big) = \log(8)$$
$$t = \frac{\log(8)}{\log(11.547) - \log(3.668)}$$
$$t \approx 1.81329$$

   *It takes approximately $1.81$ hours for the populations to be the same size.*

7. Chain emails contain a message suggesting you will have bad luck if you do not forward the email to others. Suppose a student started a chain email by sending the message to $10$ friends and asking those friends to each send the same email to $3$ more friends exactly one day after receiving the message. Assuming that everyone that gets the email participates in the chain, we can model the number of people who will receive the email on the $n^{\text{th}}$ day by the formula $E(n) = 10(3^n)$, where $n = 0$ indicates the day the original email was sent.

   a. If we assume the population of the United States is $318$ million people and everyone who receives the email sends it to $3$ people who have not received it previously, how many days until there are as many emails being sent out as there are people in the United States?

$$318(10^6) = 10 \cdot 3^n$$
$$318(10^5) = 3^n$$
$$\log(318) + \log(10^5) = n \cdot \log(3)$$
$$\log(318) + 5 = n \cdot \log(3)$$
$$n = \frac{5 + \log(318)}{\log(3)}$$
$$n \approx 15.72$$

   *So by the $16^{th}$ day, more than $318$ million emails are being sent out.*

b. **The population of earth is approximately 7.1 billion people. On what day will 7.1 billion emails be sent out?**

$$7.1(10^9) = 10(3^n)$$
$$7.1(10^8) = 3^n$$
$$\log(7.1(10^8)) = n \cdot \log(3)$$
$$n = \frac{8 + \log(7.1)}{\log(3)}$$
$$n \approx 18.5514$$

*By the $19^{th}$ day, more than 7.1 billion emails will be sent.*

8. **Solve the following exponential equations.**

a. $10^{(3x-5)} = 7^x$

$$10^{3x-5} = 7^x$$
$$3x - 5 = x\log(7)$$
$$x(3 - \log(7)) = 5$$
$$x = \frac{5}{3 - \log(7)}$$

b. $3^{\frac{x}{5}} = 2^{4x-2}$

$$3^{\frac{x}{5}} = 2^{4x-2}$$
$$\frac{x}{5}\log(3) = (4x - 2)\log(2)$$
$$4x\log(2) - x\frac{\log(3)}{5} = 2\log(2)$$
$$x\left(4\log(2) - \frac{\log(3)}{5}\right) = 2\log(2)$$
$$x = \frac{2\log(2)}{4\log(2) - \frac{\log(3)}{5}}$$

c. $10^{x^2+5} = 100^{2x^2+x+2}$

$$10^{x^2+5} = 100^{2x^2+x+2}$$
$$x^2 + 5 = (2x^2 + x + 2)\log(100)$$
$$x^2 + 5 = 4x^2 + 2x + 4$$
$$3x^2 + 2x - 1 = 0$$
$$(3x - 1)(x + 1) = 0$$
$$x = \frac{1}{3} \text{ or } x = -1$$

d. $4^{x^2-3x+4} = 2^{5x-4}$

$$4^{x^2-3x+4} = 2^{5x-4}$$
$$(x^2-3x+4)\log_2(4) = (5x-4)\log_2(2)$$
$$2(x^2-3x+4) = 5x-4$$
$$2x^2-6x+8 = 5x-4$$
$$2x^2-11x+12 = 0$$
$$(2x-3)(x-4) = 0$$
$$x = \frac{3}{2} \text{ or } x = 4$$

9. **Solve the following exponential equations.**

a. $(2^x)^x = 8^x$

$$2^{x^2} = 8^x$$
$$x^2\log_2(2) = x\log_2(8)$$
$$x^2 = 3x$$
$$x^2-3x = 0$$
$$x(x-3) = 0$$
$$x = 0 \text{ or } x = 3$$

b. $(3^x)^x = 12$

$$3^{x^2} = 12$$
$$x^2 \log(3) = \log(12)$$
$$x^2 = \frac{\log(12)}{\log(3)}$$
$$x = \sqrt{\frac{\log(12)}{\log(3)}} \text{ or } x = -\sqrt{\frac{\log(12)}{\log(3)}}$$

10. **Solve the following exponential equations.**

a. $10^{x+1} - 10^{x-1} = 1,287$

$$10^{x+1} - 10^{x-1} = 1,287$$
$$100(10^{x-1}) - 10^{x-1} = 1,287$$
$$10^{x-1}(100-1) = 1,287$$
$$99(10^{x-1}) = 1,287$$
$$10^{x-1} = 13$$
$$x - 1 = \log(13)$$
$$x = \log(13) + 1$$

b. $2(4^x) + 4^{x+1} = 342$

$$2(4^x) + 4^{x+1} = 342$$
$$2(4^x) + 4(4^x) = 342$$
$$6(4^x) = 342$$
$$4^x = 57$$
$$x = \log_4(57) = \frac{\log(57)}{\log(4)} = \frac{1}{2}\log_2(57)$$

11. **Solve the following exponential equations.**

a. $(10^x)^2 - 3(10^x) + 2 = 0$ **Hint: Let $u = 10^x$, and solve for $u$ before solving for $x$.**

*Let $u = 10^x$. Then*

$$\begin{aligned} u^2 - 3u + 2 &= 0 \\ (u-2)(u-1) &= 0 \\ u = 2 \text{ or } u &= 1 \end{aligned}$$

*If $u = 2$, we have $2 = 10^x$, and then $x = \log(2)$.*

*If $u = 1$, we have $1 = 10^x$, and then $x = 0$.*

*Thus, the two solutions to this equation are $0$ and $\log(2)$.*

b. $(2^x)^2 - 3(2^x) - 4 = 0$

*Let $u = 2^x$.*

$$\begin{aligned} u^2 - 3u - 4 &= 0 \\ (u-4)(u+1) &= 0 \\ u = 4 \text{ or } u &= -1 \end{aligned}$$

*If $u = 4$, we have $2^x = 4$, and then $x = 2$.*

*If $u = -1$, we have $2^x = -1$, which has no solution.*

*Thus, the only solution to this equation is $2$.*

c. $3(e^x)^2 - 8(e^x) - 3 = 0$

*Let $u = e^x$.*

$$\begin{aligned} 3u^2 - 8u - 3 &= 0 \\ (u-3)(3u+1) &= 0 \\ u = 3 \text{ or } u &= -\frac{1}{3} \end{aligned}$$

*If $u = 3$, we have $e^x = 3$, and then $x = \ln(3)$.*

*If $u = -\frac{1}{3}$, we have $e^x = -\frac{1}{3}$, which has no solution because $e^x > 0$ for every value of $x$.*

*Thus, the only solution to this equation is $\ln(3)$.*

d. $4^x + 7(2^x) + 12 = 0$

*Let $u = 2^x$.*

$$\begin{aligned} (2^x)^2 + 7(2^x) + 12 &= 0 \\ u^2 + 7u + 12 &= 0 \\ (u+3)(u+4) &= 0 \\ u = -3 \text{ or } u &= -4 \end{aligned}$$

*But $2^x > 0$ for every value of $x$, thus there are no solutions to this equation.*

e. $(10^x)^2 - 2(10^x) - 1 = 0$

*Let* $u = 10^x$.

$$u^2 - 2u - 1 = 0$$
$$u = 1 + \sqrt{2} \text{ or } u = 1 - \sqrt{2}$$

*If* $u = 1 + \sqrt{2}$*, we have* $10^x = 1 + \sqrt{2}$*, and then* $x = \log(1 + \sqrt{2})$.

*If* $u = 1 - \sqrt{2}$*, we have* $10^x = 1 - \sqrt{2}$*, which has no solution because* $1 - \sqrt{2} < 0$.

*Thus, the only solution to this equation is* $\log(1 + \sqrt{2})$.

12. Solve the following systems of equations.

a. $2^{x+2y} = 8$
$4^{2x+y} = 1$

$$2^{x+2y} = 2^3$$
$$4^{2x+y} = 4^0$$

$$x + 2y = 3$$
$$2x + y = 0$$

$$x + 2y = 3$$
$$4x + 2y = 0$$

$$y = 2$$
$$x = -1$$

b. $2^{2x+y-1} = 32$
$4^{x-2y} = 2$

$$2^{2x+y-1} = 2^5$$
$$(2^2)^{x-2y} = 2^1$$

$$2x + y - 1 = 5$$
$$2(x - 2y) = 1$$

$$2x + y = 6$$
$$2x - 4y = 1$$

$$y = 1$$
$$x = \frac{5}{2}$$

c. $2^{3x} = 8^{2y+1}$
$9^{2y} = 3^{3x-9}$

$$2^{3x} = (2^3)^{2y+1}$$
$$(3^2)^{2y} = 3^{3x-9}$$

$$3x = 3(2y + 1)$$
$$2(2y) = (3x - 9)$$

$$3x - 6y = 3$$
$$3x - 4y = 9$$

$$y = 3$$
$$x = 7$$

**13.** **Because $f(x) = \log_b(x)$ is an increasing function, we know that if $p < q$, then $\log_b(p) < \log_b(q)$. Thus, if we take logarithms of both sides of an inequality, then the inequality is preserved. Use this property to solve the following inequalities.**

**a.** $4^x > \frac{5}{3}$

$$\begin{aligned} 4^x &> \frac{5}{3} \\ \log(4^x) &> \log\left(\frac{5}{3}\right) \\ x\log(4) &> \log(5) - \log(3) \\ x &> \frac{\log(5) - \log(3)}{\log(4)} \end{aligned}$$

**b.** $\left(\frac{2}{7}\right)^x > 9$

$$\begin{aligned} \left(\frac{2}{7}\right)^x &> 9 \\ x\log\left(\frac{2}{7}\right) &> \log(9) \end{aligned}$$

***But, remember that $\log\left(\frac{2}{7}\right) < 0$, so we need to divide by a negative number. We then have***

$$x < \frac{\log(9)}{\log(2) - \log(7)}$$

**c.** $4^x > 8^{x-1}$

$$\begin{aligned} (2^2)^x &> (2^3)^{x-1} \\ 2^{2x} &> 2^{3x-3} \\ 2x &> 3x - 3 \\ 3 &> x \end{aligned}$$

**d.** $3^{x+2} > 5^{3-2x}$

$$\begin{aligned} 3^{x+2} &> 5^{3-2x} \\ (x+2)\log(3) &> (3-2x)\log(5) \\ 2x\log(5) + x\log(3) &> 3\log(5) - 2\log(3) \\ x &> \frac{3\log(5) - 2\log(3)}{2\log(5) + \log(3)} \\ x &> \frac{\log\left(\frac{125}{9}\right)}{\log(75)} \end{aligned}$$

e. $\left(\frac{3}{4}\right)^x > \left(\frac{4}{3}\right)^{x+1}$

$$\left(\frac{3}{4}\right)^x > \left(\frac{4}{3}\right)^{x+1}$$
$$x\log\left(\frac{3}{4}\right) > (x+1)\log\left(\frac{4}{3}\right)$$
$$x\left(\log\left(\frac{3}{4}\right) - \log\left(\frac{4}{3}\right)\right) > \log\left(\frac{4}{3}\right)$$

***But,*** $\log\left(\frac{3}{4}\right) = -\log\left(\frac{4}{3}\right)$***, so we have***

$$x\left(-\log\left(\frac{4}{3}\right) - \log\left(\frac{4}{3}\right)\right) > \log\left(\frac{4}{3}\right)$$
$$x\left(-2\log\left(\frac{4}{3}\right)\right) > \log\left(\frac{4}{3}\right)$$

***But,*** $-2\log\left(\frac{4}{3}\right) < 0$***, so we need to divide by a negative number, so we have***

$$x < \frac{\log\left(\frac{4}{3}\right)}{-2\log\left(\frac{4}{3}\right)}$$
$$x < -\frac{1}{2}$$

# Lesson 25: Geometric Sequences and Exponential Growth and Decay

## Student Outcomes

- Students use geometric sequences to model situations of exponential growth and decay.
- Students write geometric sequences explicitly and recursively and translate between the two forms.

## Lesson Notes

In Algebra I, students learned to interpret arithmetic sequences as linear functions and geometric sequences as exponential functions but both in simple contexts only. In this lesson, which focuses on exponential growth and decay, students construct exponential functions to solve multi-step problems. In the homework, they do the same with linear functions. The lesson addresses focus standard **F-BF.A.2**, which asks students to write arithmetic and geometric sequences both recursively and with an explicit formula, use them to model situations, and translate between the two forms. These skills are also needed to develop the financial formulas in Topic E.

In general, a *sequence* is defined by a function $f$ from a domain of positive integers to a range of numbers that can be either integers or real numbers depending on the context, or other nonmathematical objects that satisfy the equation $f(n) = a_n$. When that function is expressed as an algebraic function of the index variable $n$, then that expression of the function is called the *explicit form of the sequence (or explicit formula).* For example, the function $f: N \to Z$, which satisfies $f(n) = 3^n$ for all $n \geq 0$ is the explicit form for the sequence $3, 9, 27, 81, \ldots$. If the function is expressed in terms of the previous terms of the sequence and an initial value, then that expression of the function is called the *recursive form of the sequence (or recursive formula).* The recursive formula for the sequence $3, 9, 27, 81, \ldots$ is $a_n = 3a_{n-1}$, with $a_0 = 3$.

It is important to note that sequences can be indexed by starting with any integer. The convention in Algebra I was that the indices usually started at $1$. In Algebra II, we will often—but not always—start our indices at $0$. In this way, we start counting at the zero term, and count $0, 1, 2$ ... instead of $1, 2, 3$ ... . However, we will not explicitly direct students to list the 3rd or 10th term in a sequence to avoid confusion.

## Classwork

### Opening Exercise (8 minutes)

The opening exercise is essentially a reprise of the use in Algebra I of an exponential decay model with a geometric sequence.

*Scaffolding:*

- Students who struggle in calculating the heights, even with a calculator, should have a much easier time getting the terms and seeing the pattern if the rebound is changed to $50\%$ instead of $60\%$.
- Ask advanced students to develop a model without the scaffolded questions presented here.

**Opening Exercise**

**Suppose a ball is dropped from an initial height $h_0$ and that each time it rebounds, its new height is $60\%$ of its previous height.**

a. **What are the first four rebound heights $h_1$, $h_2$, $h_3$, and $h_4$ after being dropped from a height of $h_0 = 10$ ft.?**

*The rebound heights are $h_1 = 6$ ft., $h_2 = 3.6$ ft., $h_3 = 2.16$ ft., and $h_4 = 1.296$ ft.*

b. **Suppose the initial height is $A$ ft. What are the first four rebound heights? Fill in the following table:**

| Rebound | Height (ft.) |
|---|---|
| 1 | $0.6A$ |
| 2 | $0.36A$ |
| 3 | $0.216A$ |
| 4 | $0.1296A$ |

c. **How is each term in the sequence related to the one that came before it?**

*Each term is $0.6$ times the previous term.*

d. **Suppose the initial height is $A$ ft. and that each rebound, rather than being $60\%$ of the previous height, is $r$ times the previous height, where $0 < r < 1$. What are the first four rebound heights? What is the $n^{\text{th}}$ rebound height?**

*The rebound heights are $h_1 = Ar$, $h_2 = Ar^2$, $h_3 = Ar^3$, and $h_4 = Ar^4$ ft. The $n^{\text{th}}$ rebound height is $h_n = ar^n$ ft.*

e. **What kind of sequence is the sequence of rebound heights?**

*The sequence of rebounds is geometric (geometrically decreasing).*

f. **Suppose that we define a function $f$ with domain all real numbers so that $f(1)$ is the first rebound height, $f(2)$ is the second rebound height, and continuing so that $f(k)$ is the $k^{\text{th}}$ rebound height for positive integers $k$. What type of function would you expect $f$ to be?**

*Since each bounce has a rebound height of $r$ times the previous height, the function $f$ should be exponentially decreasing.*

g. **On the coordinate plane below, sketch the height of the bouncing ball when $A = 10$ and $r = 0.60$, assuming that the highest points occur at $x = 1, 2, 3, 4, \ldots$.**

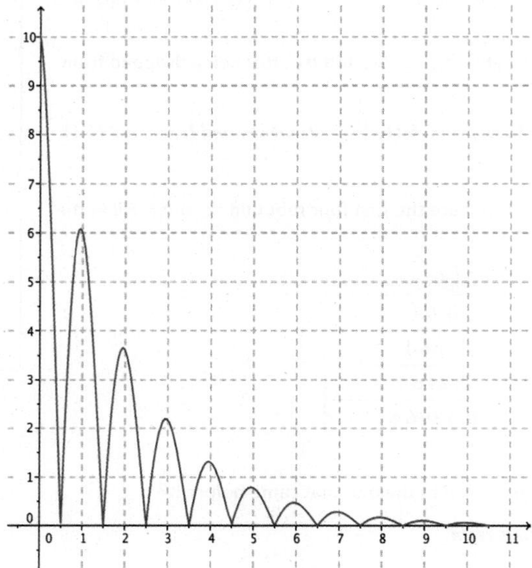

h. **Does the exponential function $f(x) = 10(0.60)^x$ for real numbers $x$ model the height of the bouncing ball? Explain how you know.**

***No. Exponential functions do not have the same behavior as a bouncing ball. The graph of $f$ is the smooth curve that connects the points at the "top" of the rebounds, as shown in the graph at right.***

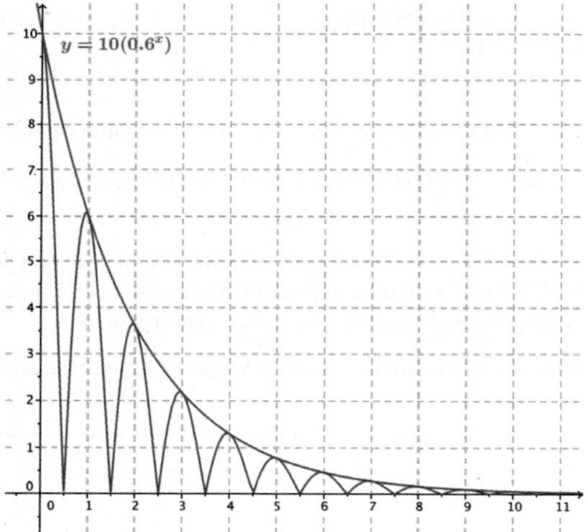


i. **What does the function $f(n) = 10(0.60)^n$ for integers $n \geq 0$ model?**

***The exponential function $f(n) = 10(0.60)^n$ models the height of the rebounds for integer values of $n$.***

## Exercise 1 (4 minutes)

While students are working on Exercise 1, circulate around the classroom to ensure student comprehension. After students complete the exercise, debrief to make sure that everyone understands that the salary model is linear and not exponential.

> *Scaffolding:*
> - If students struggle with calculating the earnings or visualizing the graph, have them calculate the salary for the first five days and make a graph of those earnings.

**Exercises**

**1.**

**a. Jane works for a video game development company that pays her a starting salary of $\$100$ a day, and each day she works, she earns $\$100$ more than the day before. How much does she earn on day $5$?**

***On day $5$, she earns $\$500$.***

**b. If you were to graph the growth of her salary for the first $10$ days she worked, what would the graph look like?**

***The graph would be a set of points lying on a straight line.***

**c. What kind of sequence is the sequence of Jane's earnings each day?**

***The sequence of her earnings is arithmetic (that is, the sequence is arithmetically increasing).***

## Discussion (2 minutes)

Pause here to ask students the following questions:

- What have we learned so far? What is the point of the previous two exercises?
  - *There are two different types of sequences, arithmetic and geometric, that model different ways that quantities can increase or decrease.*
- What do you recall about geometric and arithmetic sequences from Algebra I?
  - *To get from one term of an arithmetic sequence to the next, you add a number $d$, called the common difference. To get from one term of a geometric sequence to the next you multiply by a number $r$, called the common quotient (or common ratio).*

For historical reasons, the number $r$ that we call the *common quotient* is often referred to as the *common ratio*, which is not fully in agreement with our definition of *ratio*. Using the term is acceptable because its use is so standardized in mathematics.

## Exercise 2 (9 minutes)

Students use a geometric sequence to model the following situation and develop closed and recursive formulas for the sequence. Then they find an exponential model first using base 2 and then using base $e$ and solve for *doubling time*. Students should work in pairs on these exercises, using a calculator for calculations. They should be introduced to $P_0$ as the notation for the original number of bacteria (at time $t = 0$) and also the first term of the sequence, which we refer to as the *zero term*. Counting terms starting with $0$ means that if we represent our sequence by a function $f$, then $P_n = f(n)$ for integers $n \geq 0$.

This is an appropriate time to mention to students that we often use a continuous function to model a discrete phenomenon. In this example, the function that we use to represent the bacteria population takes on non-integer values. We need to interpret these function values according to the situation—it is not appropriate to say that the population consists of a non-integer number of bacteria at a certain time, even if the function value is non-integer. In these cases, students should round their answers to an integer that makes sense in the context of the problem.

*Scaffolding:*

- Students may need the hint that in the Opening Exercise, they wrote the terms of a geometric sequence so they can begin with the first three terms of such a sequence and use it to find $r$.

**2. A laboratory culture begins with $1,000$ bacteria at the beginning of the experiment, which we will denote by time $0$ hours. By time $2$ hours, there were $2,890$ bacteria.**

**a. If the number of bacteria is increasing by a common factor each hour, how many bacteria were there at time $1$ hour? At time $3$ hours?**

*If $P_0$ is the original population, the first three terms of the geometric sequence are $P_0, P_0r$, and $P_0r^2$. In this case, $P_0 = 1000$ and $P_2 = P_0r^2 = 2890$, so $r^2 = 2.89$ and $r = 1.7$. Therefore, $P_1 = P_0r = 1700$ and $P_3 = P_0r^3 = 2890 \cdot 1.7 = 4913$.*

**b. Find the explicit formula for term $P_n$ of the sequence in this case.**

$$P_n = P_0r^n = 1000(1.7)^n$$

**c. How would you find term $P_{n+1}$ if you know term $P_n$? Write a recursive formula for $P_{n+1}$ in terms of $P_n$.**

*You would multiply the $n$th term by $r$, which in this case is $1.7$. We have $P_{n+1} = 1.7\, P_n$.*

**d. If $P_0$ is the initial population, the growth of the population $P_n$ at time $n$ hours can be modeled by the sequence $P_n = P(n)$, where $P$ is an exponential function with the following form:**

$$P(n) = P_0 2^{kn}, \text{ where } k > 0.$$

**Find the value of $k$ and write the function $P$ in this form. Approximate $k$ to four decimal places.**

*We know that $P(n) = 1000(1.7)^n$ and $1.7 = 2^{\log_2(1.7)}$, with $k = \log_2(1.7) = \frac{\log(1.7)}{\log(2)} \approx 0.7655$. Thus, we can express $P$ in the form:*

$$P(n) = 1000(2^{0.7655n}).$$

**e. Use the function in part (d) to determine the value of $t$ when the population of bacteria has doubled.**

*We need to solve $2000 = 1000(2^{0.7655t})$, which happens when the exponent is $1$.*

$$0.7655t = 1$$
$$t = \frac{1}{0.7655}$$
$$t \approx 1.306$$

*This population doubles in roughly $1.306$ hours, which is about $1$ hour and $18$ minutes.*

**f. If $P_0$ is the initial population, the growth of the population $P$ at time $t$ can be expressed in the following form:**

$$P(n) = P_0 e^{kn}, \text{ where } k > 0.$$

**Find the value of $k$, and write the function $P$ in this form. Approximate $k$ to four decimal places.**

*Substituting in the formula for $t = 2$, we get $2890 = 1000e^{2k}$. Solving for $k$, we get $k = \ln(1.7) \approx 0.5306$. Thus, we can express $P$ in the form:*

$$P(t) = 1000(e^{0.5306t}).$$

> g. **Use the formula in part (d) to determine the value of $t$ when the population of bacteria has doubled.**
>
> ***Substituting in the formula with $k = 0.5306$, we get $2000 = 1000e^{0.5306t}$. Solving for t, we get $t = \frac{\ln(2)}{0.5306} \approx 1.306$, which is the same value we found in part (e).***

## Discussion (4 minutes)

Students should share their solutions to Exercise 2 with the rest of the class, giving particular attention to parts (b) and (c).

Part (b) of Exercise 2 presents what is called the *explicit formula* (or *closed form*) for a geometric sequence, whereas part (c) introduces the idea of a *recursive formula*. Students need to understand that given any two terms in a geometric (or arithmetic) sequence, they can derive the explicit formula. In working with recursion, they should understand that it provides a way of defining a sequence given one or more initial terms by using the $n^{\text{th}}$ term of the sequence to find the $(n+1)^{\text{st}}$ term (or, by using the $(n-1)^{\text{st}}$ term to find the $n^{\text{th}}$ term).

Discuss with students the distinction between the two functions:

$$P(n) = 1000(2^{0.7655n}) \text{ for integers } n \geq 0, \text{ and}$$

$$P(t) = 1000(2^{0.7655t}) \text{ for real numbers } t \geq 0.$$

In the first case, the function $P$ as a function of an integer $n$ represents the population at discrete times $n = 0, 1, 2, \ldots$, while $P$ as a function of a real number $t$ represents the population at any time $t \geq 0$, regardless of whether that time is an integer. If we graphed these two functions, the first graph would be the points $(0, P(0))$, $(1, P(1))$, $(2, P(2))$, etc., and the second graph would be the smooth curve drawn through the points of the first graph. We can use either statement of the function to define a sequence $P_n = P(n)$ for integers $n$. This was discussed in Opening Exercise part (h), as the difference between the graph of the points at the top of the rebounds of the bouncing ball and the graph of the smooth curve through those points.

Our work earlier in the module that extended the laws of exponents to the set of all real numbers applies here to extend a discretely defined function such as $P(n) = 1000(2^{0.7655n})$ for integers $n \geq 0$ to the continuously-defined function $P(t) = 1000(2^{0.7655t})$ for real numbers $t \geq 0$. Then, we can solve exponential equations involving sequences using our logarithmic tools.

Students may question why we could find two different exponential representations of the function $P$ in parts (d) and (f) of Exercise 2. We can use the properties of exponents to express an exponential function in terms of any base. In Lesson 6 earlier in the module, we saw that the functions $H(t) = ae^t$ for real numbers $a$ have rate of change equal to 1. For this reason, which is important in calculus and beyond, we usually prefer to use base $e$ for exponential functions.

## Exercises 3–4 (5 minutes)

Students should work on these exercises in pairs. They can take turns calculating terms in the sequences. Circulate the room and observe students to call on to share their work with the class before proceeding to the next and final set of exercises.

3. The first term $a_0$ of a geometric sequence is $-5$, and the common ratio $r$ is $-2$.
   a. What are the terms $a_0, a_1$, and $a_2$?

   $$a_0 = -5$$
   $$a_1 = 10$$
   $$a_2 = -20$$

   b. Find a recursive formula for this sequence.

   *The recursive formula is $a_{n+1} = -2a_n$, with $a_0 = -5$.*

   c. Find an explicit formula for this sequence.

   *The explicit formula is $a_n = -5(-2)^n$, for $n \geq 0$.*

   d. What is term $a_9$?

   *Using the explicit formula, we find: $a_9 = (-5) \cdot (-2)^9 = 2560$.*

   e. What is term $a_{10}$?

   *One solution is to use the explicit formula: $a_{10} = (-5) \cdot (-2)^{10} = -5120$.*

   *Another solution is to use the recursive formula: $a_{10} = a_9 \cdot (-2) = -5120$.*

*Scaffolding:*

- Students may need the hint that in the Opening Exercise, they wrote the terms of a geometric sequence so they can begin with the first three terms of such a sequence and use it to find $r$.

4. Term $a_4$ of a geometric sequence is $5.8564$, and term $a_5$ is $-6.44204$.
   a. What is the common ratio $r$?

   *We have $r = \frac{-6.44204}{5.8564} = -1.1$. The common ratio is $-1.1$.*

   b. What is term $a_0$?

   *From the definition of a geometric sequence, $a_4 = a_0r^4 = 5.8564$, so $a_0 = \frac{5.8564}{(-1.1)^4} = \frac{5.8564}{1.4641} = 4$.*

   c. Find a recursive formula for this sequence.

   *The recursive formula is $a_{n+1} = -1.1(a_n)$ with $a_0 = 4$.*

   d. Find an explicit formula for this sequence.

   *The explicit formula is $a_n = 4(-1.1)^n$, for $n \geq 0$.*

## Exercises 5–6 (4 minutes)

This final set of exercises in the lesson attends to **F-BF.A.2**, and asks students to translate between explicit and recursive formulas for geometric sequences. Students should continue to work in pairs on these exercises.

5. The recursive formula for a geometric sequence is $a_{n+1} = 3.92(a_n)$ with $a_0 = 4.05$. Find an explicit formula for this sequence.

   *The common ratio is $3.92$, and the initial value is $4.05$, so the explicit formula is*

   $$a_n = 4.05(3.92)^n \text{ for } n \geq 0.$$

**6. The explicit formula for a geometric sequence is $a_n = 147(2.1)^{3n}$. Find a recursive formula for this sequence.**

***First, we rewrite the sequence as $a_n = 147(2.1^3)^n = 147(9.261)^n$. We then see that the common ratio is $9.261$, and the initial value is $147$, so the recursive formula is***

$$a_{n+1} = (9.261)a_n \text{ with } a_0 = 147.$$

## Closing (4 minutes)

Debrief students by asking the following questions and taking answers as a class:

- If we know that a situation can be described using a geometric series, how can we create the geometric series for that model? How is the geometric series related to an exponential function with base $e$?
  - *The terms of the geometric series are determined by letting $P_n = P(n)$ for an exponential function $P(n) = P_0e^{kn}$, where $P_0$ is the initial amount, $n$ indicates the term of the series, and $e^k$ is the growth rate of the function. Depending on the data given in the situation, we can use either the explicit formula or the recursive formula to find the common ratio $r = e^k$ of the geometric sequence and its initial term $P_0$.*
- Do we need to use an exponential function base $e$?
  - *No. We can choose any base that we want for an exponential function, but mathematicians often choose base e for exponential and logarithm functions.*

Although arithmetic sequences are not emphasized in this lesson, they do make an appearance in the Problem Set. For completeness, the lesson summary includes both kinds of sequences. The two formulas and the function models for each type of sequence are summarized in the box below, which can be reproduced and posted in the classroom:

**Lesson Summary**

**ARITHMETIC SEQUENCE: A sequence is called *arithmetic* if there is a real number $d$ such that each term in the sequence is the sum of the previous term and $d$.**

- ***Explicit formula:*** **Term $a_n$ of an arithmetic sequence with first term $a_0$ and common difference $d$ is given by $a_n = a_0 + nd$, for $n \geq 0$.**
- ***Recursive formula:*** **Term $a_{n+1}$ of an arithmetic sequence with first term $a_0$ and common difference $d$ is given by $a_{n+1} = a_n + d$, for $n \geq 0$.**

**GEOMETRIC SEQUENCE: A sequence is called *geometric* if there is a real number $r$ such that each term in the sequence is a product of the previous term and $r$.**

- ***Explicit formula:*** **Term $a_n$ of a geometric sequence with first term $a_0$ and common ratio $r$ is given by $a_n = a_0r^n$, for $n \geq 0$.**
- ***Recursive formula:*** **Term $a_{n+1}$ of a geometric sequence with first term $a_0$ and common ratio $r$ is given by $a_{n+1} = a_nr$.**

## Exit Ticket (5 minutes)

Name ______________________ Date ____________

# Lesson 25: Geometric Sequences and Exponential Growth and Decay

## Exit Ticket

1. Every year, Mikhail receives a $3\%$ raise in his annual salary. His starting annual salary was $\$40,000$.
   a. Does a geometric or arithmetic sequence best model Mikhail's salary in year $n$? Explain how you know.

   b. Find a recursive formula for a sequence, $S_n$, which represents Mikhail's salary in year $n$.

2. Carmela's annual salary in year $n$ can be modeled by the recursive sequence $C_{n+1} = 1.05\ C_n$, where $C_0 = \$75,000$.
   a. What does the number $1.05$ represent in the context of this problem?

   b. What does the number $\$75,000$ represent in the context of this problem?

   c. Find an explicit formula for a sequence that represents Carmela's salary.

## Exit Ticket Sample Solutions

1. **Every year, Mikhail receives a $3\%$ raise in his annual salary. His starting annual salary was $\$40,000$.**
   a. **Does a geometric or arithmetic sequence best model Mikhail's salary in year $n$? Explain how you know.**

      *Because Mikhail's salary increases by a multiple of itself each year, a geometric series will be an appropriate model.*

   b. **Find a recursive formula for a sequence, $S_n$, which represents Mikhail's salary in year $n$.**

      *Mikhail's annual salary can be represented by the sequence $S_{n+1} = 1.03\ S_n$ with $S_0 = \$40,000$.*

2. **Carmela's annual salary in year $n$ can be modeled by the recursive sequence $C_{n+1} = 1.05\ C_n$, where $C_0 = \$75,000$.**
   a. **What does the number $1.05$ represent in the context of this problem?**

      *The $1.05$ is the growth rate of her salary with time; it indicates that she is receiving a $5\%$ raise each year.*

   b. **What does the number $\$75,000$ represent in the context of this problem?**

      *Carmela's starting annual salary was $\$75,000$, before she earned any raises.*

   c. **Find an explicit formula for a sequence that represents Carmela's salary.**

      *Carmela's salary can be represented by the sequence $C_n = \$75,000\ (1.05)^n$.*

## Problem Set Sample Solutions

1. **Convert the following recursive formulas for sequences to explicit formulas.**
   a. $a_{n+1} = 4.2 + a_n$ **with** $a_0 = 12$

      $a_n = 12 + 4.2n$ *for* $n \geq 0$

   b. $a_{n+1} = 4.2a_n$ **with** $a_0 = 12$

      $a_n = 12(4.2)^n$ *for* $n \geq 0$

   c. $a_{n+1} = \sqrt{5}\ a_n$ **with** $a_0 = 2$

      $a_n = 2(\sqrt{5})^n$ *for* $n \geq 0$

   d. $a_{n+1} = \sqrt{5} + a_n$ **with** $a_0 = 2$

      $a_n = 2 + n\sqrt{5}$ *for* $n \geq 0$

   e. $a_{n+1} = \pi\ a_n$ **with** $a_0 = \pi$

      $a_n = \pi(\pi)^n = \pi^{n+1}$ *for* $n \geq 0$

2. **Convert the following explicit formulas for sequences to recursive formulas.**

a. $a_n = \frac{1}{5}(3^n)$ for $n \geq 0$

$a_{n+1} = 3\,a_n$ *with* $a_0 = \frac{1}{5}$

b. $a_n = 16 - 2n$ for $n \geq 0$

$a_{n+1} = a_n - 2$ *with* $a_0 = 16$

c. $a_n = 16\left(\frac{1}{2}\right)^n$ for $n \geq 0$

$a_{n+1} = \frac{1}{2}a_n$ *with* $a_0 = 16$

d. $a_n = 71 - \frac{6}{7}n$ for $n \geq 0$

$a_{n+1} = a_n - \frac{6}{7}$ *with* $a_0 = 71$

e. $a_n = 190(1.03)^n$ for $n \geq 0$

$a_{n+1} = 1.03\,a_n$ *with* $a_0 = 190$

3. **If a geometric sequence has $a_1 = 256$ and $a_8 = 512$, find the exact value of the common ratio $r$.**

*The recursive formula is $a_{n+1} = a_n \cdot r$, so we have*

$$\begin{aligned} a_8 &= a_7(r) \\ &= a_6(r^2) \\ &= a_5(r^3) \\ \vdots &= a_1(r^7) \\ 512 &= 256(r^7) \\ 2 &= r^7 \\ r &= \sqrt[7]{2}. \end{aligned}$$

4. **If a geometric sequence has $a_2 = 495$ and $a_6 = 311$, approximate the value of the common ratio $r$ to four decimal places.**

*The recursive formula is $a_{n+1} = a_n \cdot r$, so we have*

$$\begin{aligned} a_6 &= a_5(r) \\ &= a_4(r^2) \\ &= a_3(r^3) \\ &= a_2(r^4) \\ 311 &= 495(r^4) \\ r^4 &= \frac{311}{495} \\ r &= \sqrt[4]{\frac{311}{495}} \approx 0.8903. \end{aligned}$$

5. **Find the difference between the terms $a_{10}$ of an arithmetic sequence and a geometric sequence, both of which begin at term $a_0$ and have $a_2 = 4$ and $a_4 = 12$.**

*Arithmetic: The explicit formula has the form $a_n = a_0 + nd$, so $a_2 = a_0 + 2d$ and $a_4 = a_0 + 4d$. Then $a_4 - a_2 = 12 - 4 = 8$ and $a_4 - a_2 = (a_0 + 4d) - (a_0 + 2d)$, so that $8 = 2d$ and $d = 4$. Since $d = 4$, we know that $a_0 = a_2 - 2d = 4 - 8 = -4$. So, the explicit formula for this arithmetic sequence is $a_n = -4 + 4n$. We then know that $a_{10} = -4 + 40 = 36$.*

*Geometric: The explicit formula has the form $a_n = a_0(r^n)$, so $a_2 = a_0(r^2)$ and $a_4 = a_0(r^4)$, so $\frac{a_4}{a_2} r^2$ and $\frac{a_4}{a_2} = \frac{12}{4} = 3$. Thus, $r^2 = 3$, so $r = \pm\sqrt{3}$. Since $r^2 = 3$, we have $a_2 = 4 = a_0(r^2)$, so that $a_0 = \frac{4}{3}$. Then the explicit formula for this geometric sequence is $a_n = \frac{4}{3}(\pm\sqrt{3})^n$. We then know that*

$a_{10} = \frac{4}{3}(\pm\sqrt{3})^{10} = \frac{4}{3}(3^5) = 4(3^4) = 324.$

*Thus, the difference between the terms $a_{10}$ of these two sequences is $324 - 36 = 288$.*

6. **Given the geometric series defined by the following values of $a_0$ and $r$, find the value of $n$ so that $a_n$ has the specified value.**

a. $a_0 = 64, r = \frac{1}{2}, a_n = 2$

*The explicit formula for this geometric series is $a_n = 64\left(\frac{1}{2}\right)^n$ and $a_n = 2$.*

$$2 = 64\left(\frac{1}{2}\right)^n$$
$$\frac{1}{32} = \left(\frac{1}{2}\right)^n$$
$$\left(\frac{1}{2}\right)^5 = \left(\frac{1}{2}\right)^n$$
$$n = 5$$

*Thus, $a_5 = 2$.*

b. $a_0 = 13, r = 3, a_n = 85293$

*The explicit formula for this geometric series is $a_n = 13(3)^n$, and we have $a_n = 85293$.*

$$13(3)^n = 85293$$
$$3^n = 6561$$
$$3^n = 3^8$$
$$n = 8$$

*Thus, $a_8 = 85293$.*

c. $a_0 = 6.7, r = 1.9, a_n = 7804.8$

*The explicit formula for this geometric series is $a_n = 6.7(1.9)^n$, and we have $a_n = 7804.8$.*

$$6.7(1.9)^n = 7804.8$$
$$(1.9)^n = 1164.9$$
$$n\log(1.9) = \log(1164.9)$$
$$n = \frac{\log(1164.9)}{\log(1.9)} = 11$$

*Thus, $a_{11} = 7804.8$.*

d. $a_0 = 10958, r = 0.7, a_n = 25.5$

*The explicit formula for this geometric series is $a_n = 10958(0.7)^n$, and we have $a_n = 25.5$.*

$$10958(0.7)^n = 25.5$$
$$\log(10958) + n\log(0.7) = \log(25.5)$$
$$n = \frac{\log(25.5) - \log(10958)}{\log(0.7)}$$
$$n = 17$$

*Thus, $a_{17} = 25.5$.*

7. Jenny planted a sunflower seedling that started out $5$ cm tall, and she finds that the average daily growth is $3.5$ cm.

   a. Find a recursive formula for the height of the sunflower plant on day $n$.

   $$h_{n+1} = 3.5 + h_n \text{ with } h_0 = 5$$

   b. Find an explicit formula for the height of the sunflower plant on day $n \geq 0$.

   $$h_n = 5 + 3.5n$$

MP.2

8. Kevin modeled the height of his son (in inches) at age $n$ years for $n = 2, 3, \ldots, 8$ by the sequence $h_n = 34 + 3.2(n-2)$. Interpret the meaning of the constants $34$ and $3.2$ in his model.

   *At age $2$, Kevin's son was $34$ in. tall, and between the ages of $2$ and $8$ he grew at a rate of $3.2$ in. per year.*

9. Astrid sells art prints through an online retailer. She charges a flat rate per order for an order processing fee, sales tax, and the same price for each print. The formula for the cost of buying $n$ prints is given by $P_n = 4.5 + 12.6\,n$.

   MP.2

   a. Interpret the number $4.5$ in the context of this problem.

   *The $4.5$ represents a $\$4.50$ order processing fee.*

   b. Interpret the number $12.6$ in the context of this problem.

   *The number $12.6$ represents the cost of each print, including the sales tax. (MP.2)*

   c. Find a recursive formula for the cost of buying $n$ prints.

   $P_n = 12.6 + P_{n-1}$ *with* $P_1 = 17.10$

   *(Notice that it makes no sense to have $P_0$ be the starting value, since that means you need to pay the processing fee when you do not place an order.)*

10. A bouncy ball rebounds to $90\%$ of the height of the preceding bounce. Craig drops a bouncy ball from a height of $20$ ft.

    a. Write out the sequence of the heights $h_1, h_2, h_3$, and $h_4$ of the first four bounces, counting the initial height as $h_0 = 20$.

    $$h_1 = 18$$
    $$h_2 = 16.2$$
    $$h_3 = 14.58$$
    $$h_4 = 13.122$$

b. **Write a recursive formula for the rebound height of a bouncy ball dropped from an initial height of $20$ ft.**

$$h_{n+1} = 0.9\,h_n \text{ with } h_0 = 20$$

c. **Write an explicit formula for the rebound height of a bouncy ball dropped from an initial height of $20$ ft.**

$$h_n = 20(0.9)^n \text{ for } n \geq 0$$

d. **How many bounces will it take until the rebound height is under $6$ ft.?**

$$\begin{aligned} 20(0.9)^n &< 6 \\ n\log(0.9) &< \log(6) - \log(20) \\ n &> \frac{\log(6) - \log(20)}{\log(0.9)} \\ n &> 11.42 \end{aligned}$$

*So, it takes $12$ bounces for the bouncy ball to rebound under $6$ ft.*

e. **Extension: Find a formula for the minimum number of bounces needed for the rebound height to be under $y$ ft., for a real number $0 < y < 20$.**

$$\begin{aligned} 20(0.9)^n &< y \\ n\log(0.9) &< \log(y) - \log(20) \\ n &> \frac{\log(y) - \log(20)}{\log(0.9)} \end{aligned}$$

*Rounding this up to the next integer with the ceiling function, it takes $\left\lceil \frac{\log(y) - \log(20)}{\log(0.9)} \right\rceil$ bounces for the bouncy ball to rebound under $y$ ft.*

11. **Show that when a quantity $a_0 = A$ is increased by $x\%$, its new value is $a_1 = A\left(1 + \frac{x}{100}\right)$. If this quantity is again increased by $x\%$, what is its new value $a_2$? If the operation is performed $n$ times in succession, what is the final value of the quantity $a_n$?**

*We know that $x\%$ of a number $A$ is represented by $\frac{x}{100}A$. Thus, when $a_0 = A$ is increased by $x\%$, the new quantity is*

$$\begin{aligned} a_1 &= A + \frac{x}{100}A \\ &= A\left(1 + \frac{x}{100}\right). \end{aligned}$$

*If we increase it again by $x\%$, we have*

$$\begin{aligned} a_2 &= a_1 + \frac{x}{100}a_1 \\ &= \left(1 + \frac{x}{100}\right)a_1 \\ &= \left(1 + \frac{x}{100}\right)\left(1 + \frac{x}{100}\right)a_0 \\ &= \left(1 + \frac{x}{100}\right)^2 a_0. \end{aligned}$$

*If we repeat this operation $n$ times, we find that*

$$a_n = \left(1 + \frac{x}{100}\right)^n a_0.$$

12. When Eli and Daisy arrive at their cabin in the woods in the middle of winter, the internal temperature is $40°F$.

   a. Eli wants to turn up the thermostat by $2°F$ every $15$ minutes. Find an explicit formula for the sequence that represents the thermostat settings using Eli's plan.

   *Let $n$ represent the number of $15$-minute increments. Then, $E(n) = 40 + 2n$.*

   b. Daisy wants to turn up the thermostat by $4\%$ every $15$ minutes. Find an explicit formula for the sequence that represents the thermostat settings using Daisy's plan.

   *Let $n$ represent the number of $15$-minute increments. Then, $D(n) = 40(1.04)^n$.*

   c. Which plan will get the thermostat to $60°F$ most quickly?

   *Making a table of values, we see that Eli's plan will set the thermostat to $60°F$ first.*

| $n$ | *Elapsed Time* | $E(n)$ | $D(n)$ |
|---|---|---|---|
| $0$ | $0$ *minutes* | $40$ | $40.00$ |
| $1$ | $15$ *minutes* | $42$ | $41.60$ |
| $2$ | $30$ *minutes* | $44$ | $43.26$ |
| $3$ | $45$ *minutes* | $46$ | $45.00$ |
| $4$ | $1$ *hour* | $48$ | $46.79$ |
| $5$ | $1$ *hour* $15$ *minutes* | $50$ | $48.67$ |
| $6$ | $1$ *hour* $30$ *minutes* | $52$ | $50.61$ |
| $7$ | $1$ *hour* $45$ *minutes* | $54$ | $52.64$ |
| $8$ | $2$ *hours* | $56$ | $54.74$ |
| $9$ | $2$ *hours* $15$ *minutes* | $58$ | $56.93$ |
| $10$ | $2$ *hours* $30$ *minutes* | $60$ | $59.21$ |

   d. Which plan will get the thermostat to $72°F$ most quickly?

   *Continuing the table of values from part (c), we see that Daisy's plan will set the thermostat to $72°F$ first.*

| $n$ | *Elapsed Time* | $E(n)$ | $D(n)$ |
|---|---|---|---|
| $11$ | $2$ *hours* $45$ *minutes* | $62$ | $61.58$ |
| $12$ | $3$ *hours* | $64$ | $64.04$ |
| $13$ | $3$ *hours* $15$ *minutes* | $66$ | $66.60$ |
| $14$ | $3$ *hours* $30$ *minutes* | $68$ | $69.27$ |
| $15$ | $3$ *hours* $45$ *minutes* | $70$ | $72.04$ |

13. In nuclear fission, one neutron splits an atom causing the release of two other neutrons, each of which splits an atom and produces the release of two more neutrons, and so on.

   a. Write the first few terms of the sequence showing the numbers of atoms being split at each stage after a single atom splits. Use $a_0 = 1$.

   $a_0 = 1$, $a_1 = 2$, $a_2 = 4$, $a_3 = 8$

   b. Find the explicit formula that represents your sequence in part (a).

   $a_n = 2^n$

c. **If the interval from one stage to the next is one-millionth of a second, write an expression for the number of atoms being split at the end of one second.**

*At the end of one second* $n = 1{,}000{,}000$, *so* $2^{1{,}000{,}000}$ *atoms are being split.*

d. **If the number from part (c) were written out, how many digits would it have?**

*The number of digits in a number* $x$ *is given by rounding up* $\log(x)$ *to the next largest integer; that is, by the ceiling of* $\log(x)$, $\lceil \log(x) \rceil$. *Thus, there are* $\lceil \log(2^{1{,}000{,}000}) \rceil$ *digits.*

*Since* $\log(2^{1{,}000{,}000}) = 1{,}000{,}000 \log(2) \approx 301{,}030$, *there will be* $301{,}030$ *digits in the number* $2^{1{,}000{,}000}$.

# Lesson 26: Percent Rate of Change

## Student Outcomes

- Students develop a *general growth/decay rate formula* in the context of compound interest.
- Students compute *future values* of investments with continually compounding interest rates.

## Lesson Notes

In this lesson, we develop a general growth/decay rate formula by investigating the compound interest formula. In Algebra I, the compound interest formula was described via sequences or functions whose domain is a subset of the integers. We start from this point (**F-IFA.3**) and extend the function to a domain of all real numbers. The function for compound interest is developed first using a recursive process to generate a geometric sequence, which is then rewritten in its explicit form (**F-BF.A.1a**, **F-BF.A.2**). Many of the situations and problems presented here were first encountered in Module 3 of Algebra I, but now students are able to use logarithms to find solutions, using technology appropriately to evaluate the logarithms (MP.5). Students also work on converting between different growth rates and time units (**A-SSE.B.3c**). Students continue to create equations in one variable from the exponential models to solve problems (**A-CED.A.3**).

Note: In this lesson, the letter $r$ stands for *the percent rate of change,* which is different from how the letter $r$ was used in the Lesson 25 where it denoted the common ratio. These two concepts are slightly different (in this lesson, $1 + r$ is *the common ratio*), and this difference might cause confusion for your students. We use the letter $r$ to refer to both, due to historical reasons and because $r$ is the notation most commonly used by adults in both situations. You will need to help your students understand how the context dictates whether $r$ stands for the common ratio or the percent rate of change.

## Classwork

### Example 1 (8 minutes)

MP.4

Present the following situation, which was first seen in Algebra I, to the students. Some trigger questions are presented to help progress student understanding. A general exponential model is presented of the form $F = P(1 + r)^t$, which is appropriate in most applications that can be modeled using exponential functions and was introduced in Module 3, Lesson 4, of Algebra I. It has been a while since the students have seen this formula, so it is developed slowly through this example first using a recursive process before giving the explicit translation (**F-BF.A.1a**, **F-BF.A.2**).

> *Scaffolding:*
> - Either present the following information explicitly or encourage students to write out the first few terms without evaluating to see the structure. Once they see that $P_2 = 800 \cdot 1.03^2$ and that $P_3 = 800 \cdot 1.03^3$, they should be able to see that $P_m = 800 \cdot 1.03^m$.
> - Have advanced learners work on their own to develop the values for years 0–3 and year $m$.

- A youth group has a yard sale to raise money for charity. The group earns $800 but decides to put the money in the bank for a while. Their local bank pays an interest rate of $3\%$ per year, and the group decides to put all of the interest they earn back into the account to earn even more interest.

- We will refer to the time at which the money was deposited into the bank as year $0$. At the end of each year, how can we calculate how much money is in the bank if we know the previous year's balance?
  - *Each year, multiply the previous year's balance by* $1.03$. *For example, since* $3\%$ *can be written* $0.03$, *the amount at the end of the first year is* $800 + 800(0.03) = 800(1 + 0.03) = 800(1.03)$.
- How much money is in the bank at the following times?

| Year | Balance in terms of last year's balance | Balance in terms of the year, $m$ |
|---|---|---|
| $0$ | $\$800$ | $\$800$ |
| $1$ | $\$824 = 800(1.03)$ | $\$824 = 800(1.03)$ |
| $2$ | $\$848.72 = 824(1.03)$ | $\$848.72 = 800(1.03)(1.03)$ |
| $3$ | $\$874.18 \approx 848.72(1.03)$ | $\$874.18 \approx 800(1.03)(1.03)(1.03)$ |
| $m$ | $b_{m-1} \cdot (1.03)$ | $800(1.03)^m$ |

- If instead of evaluating, we write these balances out as mathematical expressions, what pattern do you notice?
  - *For instance, the second year would be* $800(1.03)(1.03) = 800(1.03)^2$. *From there we can see that the balance in the* $m^{th}$ *year would be* $800(1.03)^m$.
- What kind of sequence do these numbers form? Explain how you know.
  - *They form a geometric sequence because each year's balance is* $1.03$ *times the previous year's balance.*
- Write a recursive formula for the balance in the $(m+1)^{st}$ year, denoted by $b_{m+1}$, in terms of the balance of the $n^{th}$ year, denoted by $b_m$.
  - $b_{m+1} = (1.03)b_m$
- What is the explicit formula that gives the amount of money, $F$ (i.e., future value), in the bank account after $m$ years?
  - *The group started with* $\$800$ *and this increases* $3\%$ *each year. After the first year, the group will have* $\$800 \cdot 1.03$ *in the account, and after* $n$ *years, they should have* $800 \cdot 1.03^m$. *Thus, the formula for the amount they have could be represented by* $F = \$800(1.03)^m$.
- Let us examine the base of the exponent in the above problem a little more closely, and write it as $1.03 = 1 + 0.03$. Rewrite the formula for the amount they have in the bank after $n$ years using $1 + 0.03$ instead of $1.03$.
  - $F = 800(1 + 0.03)^m$
- What does the $800$ represent? What does the $1$ represent? What does the $0.03$ represent?
  - *The number* $800$ *represents the starting amount. The* $1$ *represents* $100\%$ *of the previous balance that is maintained every year. The* $0.03$ *represents the* $3\%$ *of the previous balance that is added each year due to interest.*

*Scaffolding:*

For struggling classes, Example 2 may be omitted in lieu of developing fluency with the formula through practice exercises. The ending discussion questions in Example 2 should be discussed throughout the practice. Some practice exercises are presented below:

- Evaluate $300(1 + 0.12)^3$
- Find the future value of an investment of $\$1000$ growing at a rate of $3\%$ per year, compounded monthly.
- Find the growth rate and how many days it would take to grow $\$2$ into $\$2$ million if the amount doubles every day.
- Find the growth rate per year necessary to grow $\$450$ into $\$900$ after ten years.

- Let $P$ be the present or starting value of $800$, let $r$ represent the interest rate of $3\%$, and $t$ be the number of years. Write a formula for the future value $F$ in terms of $P$, $r$, and $t$.
  - $F = P(1+r)^t$

### Discussion (5 minutes)

Make three important points during this discussion: (1) that the formula $F = P(1+r)^t$ can be used in situations far more general than just finance, (2) that $r$ is the percent rate of change expressed as a unit rate, and (3) that the domain of the function given by the formula now includes all real numbers. Note: $r$ is expressed as a unit rate for a unit of time; in finance, that unit of time is typically a year given by the yearly interest rate. In the next examples, we will investigate compounding interest problem with different compounding periods.

- This formula, $F = P(1+r)^t$, can be used in far more situations than just finance, including radioactive decay and population growth. Given a *percent rate of change*, i.e., the percentage increase or decrease of an amount over a unit of time, the number $r$ is the unit rate of that percentage rate of change. For example, if a bank account grows by 2.5% a year, the unit rate is $\frac{2.5}{100}$, which means $r = 0.025$. What is the unit rate if the percent rate of change is a $12\%$ increase? A $100\%$ increase? A $0.2\%$ increase? A $5\%$ decrease?
  - $r = 0.12, r = 1, r = 0.002, r = -0.05.$
- Given the value $P$ and the percent rate of change represented by the unit rate $r$, we can think of the formula as function of time $t$, that is, $F(t) = P(1+r)^t$. In Algebra I, $t$ represented a positive integer, but now we can think of the function as having a domain of all real numbers. Why can we think of the domain of this function as being all real numbers?
  - *Earlier in this module, we learned how to define the value of an exponent when the power is a rational number, and we showed how to use that definition to evaluate exponents when the power is an irrational number. Thus, we can assume that the domain of the function $F$ can be any real number.*

Students can now use the fact that the function has a domain of all real numbers and their knowledge of logarithms to solve equations involving the function.

- In Example 1, the group's goal is to save $1,000 with the money they made from the yard sale. How many years will it take for the amount in the bank to be at least $1,000?
  - *Substitute* $1000$ *for $F$ and solve for $t$ using logarithms.*

$$\begin{aligned} 1000 &= 800 \cdot 1.03^t \\ \frac{1000}{800} &= 1.03^t \\ 1.25 &= 1.03^t \\ \ln(1.25) &= t \cdot \ln(1.03) \\ t &= \frac{\ln(1.25)}{\ln(1.03)} \approx 7.5 \end{aligned}$$

*Since they earn interest every year, it will take them* $8$ *years to save more than* $\$1{,}000$ *with this money.*

- What does the approximation 7.5 mean?
  - *The amount in the bank will reach* $\$1000$ *after roughly* $7$ *years,* $6$ *months.*

The percent rate of change can also be negative, which usually corresponds to a negative unit rate $r$, with $-1 < r < 0$.

- Can you give an example of percent rate of change that we have studied before that has a negative rate of change?
  - *Radioactive decay, populations that are shrinking, etc. An interesting example is the bean counting experiment where they started with lots of beans and removed beans after each trial.*

At this point in the lesson, you may want to work out one problem from the non-financial Problem Set as an example, or have students work one as an exercise.

## Example 2 (8 minutes)

In the function, $F(t) = P(1+r)^t$, the number $r$ is the unit rate of the percent rate of change, and $t$ is time. Frequently, the time units for the percent rate of change and the time unit for $t$ do not agree and some calculation needs to be done so that they do. For instance, if the growth rate is an amount per hour and the time period is a number of days, the formula needs to be altered by factors of 24 and its inverse.

In this example, students learn about compounding periods and percent rates of change that are based upon different units (**A-SSE.B.3c**). Students explore these concepts through some exercises immediately following the example.

- In finance, the interest rates are almost always tied to a specific time period and only accumulate once this has elapsed (called *compounding*). In this context, we refer to the time periods as compounding periods.
- Interest rates for accounts are frequently given in terms of what is called the *nominal annual percentage rate of change* or *nominal APR.* Specifically, the nominal APR is the percent rate of change per compounding period times the number of compounding periods per year. For example, if the percent rate of change is $0.5\%$ per month, then the nominal APR is $6\%$ since there are $12$ months in a year. The nominal APR is an easy way of discussing a monthly or daily percent rate of change in terms of a yearly rate, but as we will see in the examples below, it does not necessarily reflect actual or effective percent rate of change per year.

Note about language: In this lesson and later lessons, we will often use the phrase "an interest rate of $3\%$ per year compounded monthly" to mean, a nominal APR of $3\%$ compounded monthly. Both phrases refer to nominal APR.

- Frequently in financial problems and real-life situations, the nominal APR is given and the percent rate of change per compounding period is deducted from it. The following example shows how to deduce the future value function in this context.
- If the nominal APR is $6\%$ and is compounded monthly, then monthly percent rate of change is $\frac{6\%}{12}$ or $0.5\%$ per month. That means that, if a starting value of $\$800$ was deposited in a bank, after one month there would be $\$800\left(1+\frac{0.06}{12}\right)^1$ in the account, after two months there would be $\$800\left(1+\frac{0.06}{12}\right)^2$, and after 12 months in the bank there would be $\$800\left(1+\frac{0.06}{12}\right)^{12}$ in the account. In fact, since it is compounding 12 times a year, it would compound $12$ times over $1$ year, $24$ times over $2$ years, $36$ times over $3$ years, and $12t$ times over $t$ years. Hence, a function that describes the amount in the account after $t$ years is
$$F(t) = 800\left(1+\frac{0.06}{12}\right)^{12t}.$$
- Describe a function $F$ that describes the amount that would be in an account after $t$ years if $P$ was deposited in an account with a nominal APR given by the unit rate $r$ that is compounded $n$ times a year.
  - $F(t) = P\left(1+\frac{r}{n}\right)^{nt}$

- In this form, $\frac{r}{n}$ is the unit rate per compounding period, and $nt$ is the total number of compounding periods over time $t$.
- However, time $t$ can be any real number; it does not have to be integer valued. For example, if a savings account earns $1\%$ interest per year, compounded monthly, then we would say that the account compounds at a rate of $\frac{0.01}{12}$ per month. How much money would be in the account after $2\frac{1}{2}$ years with an initial deposit of $\$200$?
  - $F(2.5) = 200\left(1+\frac{0.01}{12}\right)^{12(2.5)} \approx \$205.06.$

## Exercise (8 minutes)

Have students work through the following problem to explore the consequences of having different compounding periods. After students finish, debrief them to ensure understanding.

**Exercise**

**Answer the following questions.**

**The youth group from Example 1 is given the option of investing their money at $2.976\%$ interest per year, compounded monthly.**

a. **After two years, how much would be in each account with an initial deposit of $\$800$?**

*The account from the beginning of the lesson would have $\$848.72$, and the new account would have $\$800\left(1+\frac{0.02976}{12}\right)^{12\cdot 2} \approx \$849.00$.*

b. **Compare the total amount from part (a) to how much they would have made using the interest rate of $3\%$ compounded yearly for two years. Which account would you recommend the youth group invest its money in? Why?**

*The $3\%$ compounded yearly yields $\$848.72$, while the $2.976\%$ compounded monthly yields $\$849.00$ after two years. I would recommend either—the difference between both types of investments is only $\$0.28$, hardly an amount to worry over.*

In part (b), the amount from both options is virtually the same: $\$848.72$ versus $\$849.00$. But point out that there is something strange about the numbers; even though the interest rate of 2.975% is less than the interest rate of 3%, the total amount is more. This is due to compounding every month versus every year.

To illustrate this, rewrite the expression $\left(1+\frac{0.02976}{12}\right)^{12t}$ as $((1+0.00248)^{12})^t$, and take the $12^{\text{th}}$ power of $1.00248$ to get approximately $((1.00248)^{12})^t \approx (1.030169)^t$. This shows that when the nominal APR of 2.975% compounded monthly is written as a percent rate of change compounded yearly (the same compounding period as in the Example 1), then the interest rate is approximately 3.0169%, which is more than 3%. In other words, interest rates can be accurately compared when they are both converted to the same compounding period.

**Example 3 (10 minutes)**

7 In this example, students develop the $F(t) = Pe^{rt}$ model using a numerical analysis approach (MP.7, MP.8). Have
students perform the beginning calculations on their own as much as possible before transitioning into continuous
8 compounding.

- Thus far, we have seen that the number of times a quantity compounds per year does have an effect on the future value. For instance, if someone tells you that one savings account gives you a nominal APR $3\%$ per year compounded yearly, and another gives you a nominal APR $3\%$ per year compounded monthly, which account will give you more interest at the end of the year?
  - *The account that compounds monthly gives more interest.*
- How much more interest though? Does it give twelve times as much? How can we find out how much money we will have at the end of the year if we deposit $\$100$?
  - *Calculate using the formula.*

$$F = 100(1 + 0.03)^1 = 103$$

$$F = 100\left(1 + \frac{0.03}{12}\right)^{12} = 103.04$$

    *The account compounding monthly earned* $4$ *cents more.*

- So, even though the second account compounded twelve times as much as the other, it only earned a fraction of a dollar more. Do we think that there is a limit to how much an account can earn through increasing the number of times compounding?
  - *Answers may vary. At this point, although the increase is very small, students have experience with logarithms that grow incredibly slowly but have no upper bound. This could be a situation with an upper bound or not.*
- Let's explore this idea of a limit using our calculators to do the work. Holding the principal, percent rate of change, and number of time units constant, is there a limit to how large the future value can become solely through increasing the number of compounding periods?
- We can simplify this question by setting $P = 1$, $r = 1$, and $t = 1$. What does the expression become for $n$ compounding periods?
  - $F = 1\left(1 + \frac{1}{n}\right)^n$
- Then the question becomes, as $n \to \infty$, does $F$ converge to a specific value, or does it also increase without bound?

- Let's rewrite this expression as something our calculators and computers can evaluate: $y = \left(1 + \frac{1}{x}\right)^x$. For now, go into the table feature of your graphing utilities, and let $x$ start at 1, and go up by 1. Can we populate the following table as a class?

| $x$ | $y = \left(1 + \frac{1}{x}\right)^x$ |
|---|---|
| 1 | 2 |
| 2 | 2.25 |
| 3 | 2.3704 |
| 4 | 2.4414 |
| 5 | 2.4883 |
| 6 | 2.5216 |
| 7 | 2.5465 |

- This demonstrates that although the value of the function is continuing to increase as $x$ increases, it is increasing at a decreasing rate. Still, does this function ever start decreasing? Let's set our table to start at 10,000 and increase by 10,000.

| $x$ | $y = \left(1 + \frac{1}{x}\right)^x$ |
|---|---|
| 10000 | 2.7181459 |
| 20000 | 2.7182139 |
| 30000 | 2.7182365 |
| 40000 | 2.7182478 |
| 50000 | 2.7182546 |
| 60000 | 2.7182592 |
| 70000 | 2.7182624 |

- It turns out that we are rapidly approaching the limit of what our calculators can reliably compute. Much past this point, the rounding that the calculator does to perform its calculations starts to insert horrible errors into the table. However, it is true that the value of the function will increase forever but at a slower and slower rate. In fact, as $x \to \infty$, $y$ does approach a specific value. You may have started to recognize that value from earlier in the module: Euler's number, $e$.
- Unfortunately, a proof that the expression $\left(1 + \frac{1}{x}\right)^x$ approaching is $e$ as $x \to \infty$ requires a lot more mathematics than we have available currently. Using calculus and other advanced mathematics, mathematicians have been able to show not only that as $x \to \infty$, $\left(1 + \frac{1}{x}\right)^x \to e$, but also they have been able to show that as $x \to \infty$, $\left(1 + \frac{r}{x}\right)^x \to e^r$!

Note: As an extension, you can hint at why the expression involving $r$ converges to $e^r$: Rewrite the expression above as $\left(1 + \frac{r}{x}\right)^{\left(\frac{x}{r}\right)\cdot r}$ or $\left(\left(1 + \frac{r}{x}\right)^{\frac{x}{r}}\right)^r$, and substitute $u = \frac{x}{r}$. Then the expression in terms of $u$ becomes $\left(\left(1 + \frac{1}{u}\right)^u\right)^r$. If $x \to \infty$, then $u$ does also, but as $u \to \infty$, the expression $\left(1 + \frac{1}{u}\right)^u \to e$.

- Revisiting our earlier application, what does $x$ represent in our original formula, and what could it mean that $x \to \infty$?
  - The number $x$ *represents the number of compounding periods in a year. If* $x$ *was large (e.g.,* $365$*), it would imply that interest was compounding once a day. If it were very large, say* $365{,}000$*, it would imply that interest was compounding* $1{,}000$ *times a day. As* $x \to \infty$*, the interest would be compounding continuously.*
- Thus, we have a new formula for when interest is compounding continuously: $F = Pe^{rt}$. This is just another representation of the exponential function that we have been using throughout the module.
- The formula is often called the *pert* formula.

## Closing (2 minutes)

Have students summarize the key points of the lesson in writing. A sample is included which you may want to share with the class, or you can guide students to these conclusions on their own.

**Lesson Summary**

- **For application problems involving a percent rate of change represented by the unit rate $r$, we can write $F(t) = P(1 + r)^t$, where $F$ is the future value (or ending amount), $P$ is the present amount, and $t$ is the number of time units. When the percent rate of change is negative, $r$ is negative, and the quantity decreases with time.**
- **The nominal APR is the percent rate of change per compounding period times the number of compounding periods per year. If the nominal APR is given by the unit rate $r$ and is compounded $n$ times a year, then function $F(t) = P\left(1 + \frac{r}{n}\right)^{nt}$ describes the future value at time $t$ of an account given that is given nominal APR and an initial value of $P$.**
- **For continuous compounding, we can write $F = Pe^{rt}$, where $e$ is Euler's number and $r$ is the unit rate associated to the percent rate of change.**

## Exit Ticket (4 minutes)

Name ___________________________________ Date ____________________

# Lesson 26: Percent Rate of Change

## Exit Ticket

April would like to invest $200 in the bank for one year. Three banks all have a nominal APR of 1.5%, but compound the interest differently.

a. Bank A computes interest just once at the end of the year. What would April's balance be after one year with this bank?

b. Bank B compounds interest at the end of each six-month period. What would April's balance be after one year with this bank?

c. Bank C compounds interest continuously. What would April's balance be after one year with this bank?

d. Each bank decides to double the nominal APR it offers for one year. That is, they offer a nominal APR of 3%. Each bank advertises, "DOUBLE THE AMOUNT YOU EARN!" For which of the three banks, if any, is this advertised claim correct?

## Exit Ticket Sample Solutions

**April would like to invest $\$200$ in the bank for one year. Three banks all have a nominal APR of $1.5\%$, but compound the interest differently.**

a. **Bank A computes interest just once at the end of the year. What would April's balance be after one year with this bank?**

$$I = 200 \cdot 0.015 = 3$$

*April would have* $\$203$ *at the end of the year.*

b. **Bank B compounds interest at the end of each six-month period. What would April's balance be after one year with this bank?**

$$F = 200\left(1 + \frac{0.015}{2}\right)^2$$
$$\approx 203.01$$

*April would have* $\$203.01$ *at the end of the year.*

c. **Bank C compounds interest continuously. What would April's balance be after one year with this bank?**

$$F = 200e^{0.015}$$
$$\approx 203.02$$

*April would have* $\$203.02$ *at the end of the year.*

d. **Each bank decides to double the nominal APR it offers for one year. That is, they offer a nominal APR of $3\%$. Each bank advertises, "DOUBLE THE AMOUNT YOU EARN!" For which of the three banks, if any, is this advertised claim correct?**

*Bank A:*

$$I = 200 \cdot 0.03 = 6$$

*Bank B:*

$$F = 200\left(1 + \frac{0.03}{2}\right)^2$$
$$\approx 206.045$$

*Bank C:*

$$F = 200e^{0.015}$$
$$\approx 206.09$$

*All three banks earn at least twice as much with a double interest rate. Bank A earns exactly twice as much, Bank B earns* $2$ *cents more than twice as much, and Bank C earns* $5$ *cents more than twice as much.*

## Problem Set Sample Solutions

1. **Write each recursive sequence in explicit form. Identify each sequence as arithmetic, geometric, or neither.**

   a. $a_1 = 3\,, a_{n+1} = a_n + 5$

      $a_n = 3 + 5(n-1)$, *arithmetic*

   b. $a_1 = -1\,, a_{n+1} = -2a_n$

      $a_n = -(-2)^{n-1}$, *geometric*

   c. $a_1 = 30\,, a_{n+1} = a_n - 3$

      $a_n = 30 - 3(n-1)$, *arithmetic*

   d. $a_1 = \sqrt{2}\,, a_{n+1} = \dfrac{a_n}{\sqrt{2}}$

      $a_n = \sqrt{2}\left(\dfrac{1}{\sqrt{2}}\right)^{n-1}$, *geometric*

   e. $a_1 = 1\,, a_{n+1} = \cos(\pi a_n)$

      $a_1 = 1, a_n = -1$ *for* $n > 1$, *neither.*

2. **Write each sequence in recursive form. Assume the first term is when $n = 1$.**

   a. $a_n = \dfrac{3}{2}n + 3$

      $a_1 = \dfrac{9}{2}, a_{n+1} = a_n + \dfrac{3}{2}$

   b. $a_n = 3\left(\dfrac{3}{2}\right)^n$

      $a_1 = \dfrac{9}{2}, a_{n+1} = \dfrac{3}{2} \cdot a_n$

   c. $a_n = n^2$

      $a_1 = 1, a_{n+1} = a_n + 2n + 1$

   d. $a_n = \cos(2\pi n)$

      $a_1 = 1, a_{n+1} = a_n$

3. **Consider two bank accounts. Bank A gives simple interest on an initial investment in savings accounts at a rate of $3\%$ per year. Bank B gives compound interest on savings accounts at a rate of $2.5\%$ per year. Fill out the following table.**

| Number of Years, $n$ | Bank A Balance, $a_n$ | Bank B Balance, $b_n$ |
|---|---|---|
| $0$ | $\$1000.00$ | $\$1000.00$ |
| $1$ | $\$1030.00$ | $\$1025.00$ |
| $2$ | $\$1060.00$ | $\$1050.63$ |
| $3$ | $\$1090.00$ | $\$1076.89$ |
| $4$ | $\$1120.00$ | $\$1103.81$ |
| $5$ | $\$1150.00$ | $\$1131.41$ |

a. **What type of sequence do the Bank A balances represent?**

***Balances from Bank A represent an arithmetic sequence with constant difference*** $\$30$.

b. **Give both a recursive and an explicit formula for the Bank A balances.**

***Recursive:*** $a_1 = 1000$, $a_n = a_{n-1} + 30$

***Explicit:*** $a_n = 1000 + 30n$ ***or*** $f(n) = 1000 + 30n$

c. **What type of sequence do the Bank B balances represent?**

***Balances from Bank B represent a geometric sequence with common ratio*** $1.025$.

d. **Give both a recursive and an explicit formula for the Bank B balances.**

***Recursive:*** $b_1 = 1000$, $b_n = b_{n-1} \cdot 1.025$

***Explicit:*** $b_n = 1000 \cdot 1.025^n$ ***or*** $f(n) = 1000 \cdot 1.025^n$

e. **Which bank account balance is increasing faster in the first five years?**

***During the first five years, the balance at Bank A is increasing faster at a constant rate of*** $\$30$ ***per year.***

f. **If you were to recommend a bank account for a long-term investment, which would you recommend?**

***The balance at Bank B would eventually outpace the balance at Bank A since the balance at Bank B is increasing geometrically.***

g. **At what point is the balance in Bank B larger than the balance in Bank A?**

***Once the balance in Bank B overtakes the balance in Bank A, it will always be larger, so we just have to find when they are equal. Because of the complication of solving when a linear function is equal to an exponential function, it is probably easiest to graph the two functions and see where they intersect.***

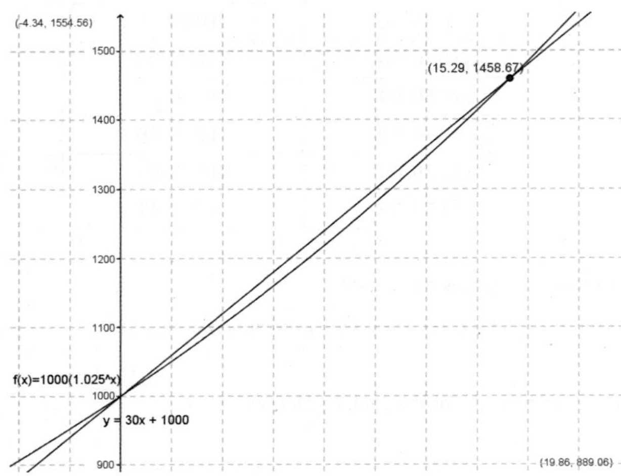


***It appears that the balance in Bank B will overtake the balance in Bank A in the $16^{th}$ year and be larger from then on. Any investment made for $0$ to $15$ years would be better in Bank A than Bank B.***

4. **You decide to invest your money in a bank that uses continuous compounding at $5.5\%$ interest per year. You have $\$500$.**

   a. **Ja'mie decides to invest $\$1,000$ in the same bank for one year. She predicts she will have double the amount in her account than you will have. Is this prediction correct? Explain.**

$$F = 1000 \cdot e^{0.055}$$
$$\approx 1056.54$$

$$F = 500 \cdot e^{0.055}$$
$$\approx 528.27$$

***Her prediction was correct. Evaluating the formula with $1,000$, we can see that $1000e^{0.055} = 2 \cdot 500 \cdot e^{0.055}$.***

   b. **Jonas decides to invest $\$500$ in the same bank as well, but for two years. He predicts that after two years he will have double the amount of cash that you will after one year. Is this prediction correct? Explain.**

***Jonas will earn more than double the amount of interest since the value increasing is in the exponent but will not have more than double the amount of cash.***

5. **Use the properties of exponents to identify the percent rate of change of the functions below, and classify them as representing exponential growth or decay. (The first two problems are done for you.)**

   a. $f(t) = (1.02)^t$

***The percent rate of change is $2\%$ and represents exponential growth.***

b. $f(t) = (1.01)^{12t}$

*Since $(1.01)^{12t} = ((1.01)^{12})^t \approx (1.1268)^t$, the percent rate of change is $12.68\%$ and represents exponential growth.*

c. $f(t) = (0.97)^t$

*Since $(0.97)^t = (1 - 0.03)^t$, the percent rate of change is – $3\%$ and represents exponential decay.*

d. $f(t) = 1000(1.2)^t$

*The percent rate of change is $20\%$ and represents exponential growth.*

e. $f(t) = \frac{(1.07)^t}{1000}$

*The percent rate of change is $7\%$ and represents exponential growth.*

f. $f(t) = 100 \cdot 3^t$

*Since $3^t = (1 + 2)^t$, the percent rate of change is $200\%$ and represents exponential growth.*

g. $f(t) = 1.05 \cdot \left(\frac{1}{2}\right)^t$

*Since $\left(\frac{1}{2}\right)^t = (0.5)^t = (1 - 0.5)^t$, the percent rate of change is $-50\%$ and represents exponential decay.*

h. $f(t) = 80 \cdot \left(\frac{49}{64}\right)^{\frac{1}{2}t}$

*Since $\left(\frac{49}{64}\right)^{\frac{1}{2}t} = \left(\left(\frac{49}{64}\right)^{\frac{1}{2}}\right)^t = \left(\frac{7}{8}\right)^t = \left(1 - \frac{1}{8}\right)^t = (1 - 0.125)^t$, the percent rate of change is $-12.5\%$ and represents exponential decay.*

i. $f(t) = 1.02 \cdot (1.13)^{\pi t}$

*Since $(1.13)^{\pi t} = ((1.13)^{\pi})^t \approx (1.468)^t$, the percent rate of change is $46.8\%$ and represents exponential growth.*

6. The effective rate of an investment is the percent rate of change per year associated with the nominal APR. The effective rate is very useful in comparing accounts with different interest rates and compounding periods. In general, the effective rate can be found with the following formula: $r_E = \left(1 + \frac{r}{k}\right)^k - 1$. The effective rate presented here is the interest rate needed for annual compounding to be equal to compounding $n$ times per year.

a. For investing, which account is better: an account earning a nominal APR of $7\%$ compounded monthly or an account earning a nominal APR of $6.875\%$ compounded daily? Why?

*The $7\%$ account is better. The effective rate for the $7\%$ account is $\left(1 + \frac{0.07}{12}\right)^{12} - 1 \approx 0.07229$ compared to the effective rate for the $6.875\%$ account, which is $0.07116$.*

b. The effective rate formula for an account compounded continuously is $r_E = e^r - 1$. Would an account earning $6.875\%$ interest compounded continuously be better than the accounts in part (a)?

*The effective rate of the account continuously compounded at $6.75\%$ is $e^{0.06875} - 1 \approx 0.07117$, which is less than the effective rate of the $7\%$ account, so the $7\%$ account is the best.*

7. Radioactive decay is the process in which radioactive elements decay into more stable elements. A half-life is the time it takes for half of an amount of an element to decay into a more stable element. For instance, the half-life for half of an amount of uranium-$235$ to transform into lead-$207$ is $704$ million years. Thus, after $704$ million years, only half of any sample of uranium-$235$ will remain, and the rest will have changed into lead-$207$. We will assume that radioactive decay is modeled by exponential decay with a constant decay rate.

   a. Suppose we have a sample of $A$ g of uranium-$235$. Write an exponential formula that gives the amount of uranium-$235$ remaining after $m$ half-lives.

   *The decay rate is constant on average and is $0.5$. If the present value is $A$, then we have $F = A(1 + (-0.50))^m$, which simplifies to $F = A\left(\frac{1}{2}\right)^m$.*

   b. Does the formula that you wrote in part (a) work for any radioactive element? Why?

   *Since $m$ represents the number of half-lives, this should be an appropriate formula for any decaying element.*

   c. Suppose we have a sample of $A$ g of uranium-$235$. What is the decay rate per million years? Write an exponential formula that gives the amount of uranium-$235$ remaining after $t$ million years.

   *The decay rate will be $0.5$ every $704$ million years. If the present value is $A$, then we have $F = A\big(1 + (-0.5)\big)^{\frac{t}{704}} = A(0.5)^{\frac{t}{704}}$. This tells us that the growth rate per million years is $(0.5)^{\frac{1}{704}} \approx 0.9990159005$, and the decay rate is $0.0009840995$ per million years. Written with this decay rate, the formula becomes $F = A(0.9990159005)^t$.*

   d. How would you calculate the number of years it takes to get to a specific percentage of the original amount of material? For example, how many years will it take there to be $80\%$ of the original amount of uranium-$235$ remaining?

   *Set $F = 0.80A$ in our formula and solve for t. For this example, this gives*

$$0.80A = A\left(\frac{1}{2}\right)^{\frac{t}{704}}$$
$$0.80 = \left(\frac{1}{2}\right)^{\frac{t}{704}}$$
$$\ln(0.80) = \frac{t}{704}\left(\ln\left(\frac{1}{2}\right)\right)$$
$$t = 704\frac{\ln(0.80)}{\ln(0.5)}$$
$$t \approx 226.637$$

   *Remember that t represents the number of millions of years. So, it takes approximately $227{,}000{,}000$ years.*

e. **How many millions of years would it take $2.35$ kg of uranium-$235$ to decay to $1$ kg of uranium?**

*For our formula, the future value is $1$ kg, and the present value is $2.35$.*

$$1 = 2.35\left(\frac{1}{2}\right)^{\frac{t}{704}}$$
$$\frac{1}{2.35} = \left(\frac{1}{2}\right)^{\frac{t}{704}}$$
$$\ln\left(\frac{1}{2.35}\right) = \frac{t}{704}\cdot\ln\left(\frac{1}{2}\right)$$
$$t = 704\frac{\ln\left(\frac{1}{2.35}\right)}{\ln\left(\frac{1}{2}\right)}$$
$$t \approx 867.793$$

*Since t is the number of millions of years, it would take approximately $868$ million years for $2.35$ kg of uranium-$235$ to decay to $1$ kg.*

8. **Doug drank a cup of tea with $130$ mg of caffeine. Each hour, the caffeine in Doug's body diminishes by about $12\%$. (This rate varies between $6\%$ and $14\%$ depending on the person.)**

a. **Write a formula to model the amount of caffeine remaining in Doug's system after each hour.**

$$c(t) = 130\cdot(1-0.12)^t$$
$$c(t) = 130\cdot(0.88)^t$$

b. **About how long will it take for the level of caffeine in Doug's system to drop below $30$ mg?**

$$30 = 130\cdot(0.88)^t$$
$$\frac{3}{13} = 0.88^t$$
$$\ln\left(\frac{3}{13}\right) = t\cdot\ln(0.88)$$
$$t = \frac{\ln\left(\frac{3}{13}\right)}{\ln(0.88)}$$
$$t \approx 11.471$$

*The caffeine level is below $30$ mg after about $11$ hours and $28$ minutes.*

c. **The time it takes for the body to metabolize half of a substance is called a *half-life*. To the nearest $5$ minutes, how long is the half-life for Doug to metabolize caffeine?**

$$65 = 130\cdot(0.88)^t$$
$$\frac{1}{2} = 0.88^t$$
$$ln\left(\frac{1}{2}\right) = t\cdot ln(0.88)$$
$$t = \frac{ln\left(\frac{1}{2}\right)}{ln(0.88)}$$
$$t \approx 5.422$$

*The half-life of caffeine in Doug's system is about $5$ hours and $25$ minutes.*

d. **Write a formula to model the amount of caffeine remaining in Doug's system after $m$ half-lives.**

$$c = 130 \cdot \left(\frac{1}{2}\right)^m$$

MP.4

9. **A study done from $1950$ through $2000$ estimated that the world population increased on average by $1.77\%$ each year. In $1950$, the world population was $2.519$ billion.**

a. **Write a function $p$ for the world population $t$ years after $1950$.**

$$p(t) = 2.519 \cdot (1 + 0.0177)^t$$
$$p(t) = 2.519 \cdot (1.0177)^t$$

b. **If this trend continued, when should the world population have reached 7 billion?**

$$7 = 2.519 \cdot (1.0177)^t$$
$$\frac{7}{2.519} = 1.0177^t$$
$$\ln\left(\frac{7}{2.519}\right) = t \cdot ln(1.0177)$$
$$t = \frac{\ln\left(\frac{7}{2.519}\right)}{\ln(1.0177)}$$
$$t \approx 58.252$$

*The model says that the population should reach 7 billion sometime roughly $58\frac{1}{4}$ years after $1950$. This would be around April $2008$.*

c. **The world population reached 7 billion October $31$, $2011$, according to the United Nations. Is the model reasonably accurate?**

*Student responses will vary. The model was accurate to within three years, so, yes, it is reasonably accurate.*

d. **According to the model, when will the world population be greater than $12$ billion people?**

$$12 = 2.519 \cdot (1.0177)^t$$
$$\frac{12}{2.519} = 1.0177^t$$
$$\ln\left(\frac{12}{2.519}\right) = t \cdot \ln(1.0177)$$
$$t = \frac{\ln\left(\frac{12}{2.519}\right)}{\ln(1.0177)}$$
$$t \approx 88.973$$

*According to the model, it will take a little less than $89$ years from $1950$ to get a world population of $12$ billion. This would be the year $2039$.*

10. **A particular mutual fund offers $4.5\%$ nominal APR compounded monthly. Trevor wishes to deposit $\$1,000$.**

a. **What is the percent rate of change per month for this account?**

*There are twelve months in a year, so* $\frac{4.5\%}{12} = 0.375\% = 0.00375$.

b. **Write a formula for the amount Trevor will have in the account after $m$ months.**

$$A = 1000 \cdot (1 + 0.00375)^m$$
$$A = 1000 \cdot (1.00375)^m$$

c. ***Doubling time*** **is the amount of time it takes for an investment to double. What is the doubling time of Trevor's investment?**

$$2000 = 1000 \cdot (1.00375)^m$$
$$2 = 1.00375^m$$
$$\ln(2) = m \cdot \ln(1.00375)$$
$$m = \frac{\ln(2)}{\ln(1.00375)}$$
$$m \approx 185.186$$

*It will take $186$ months for Trevor's investment to double. This is $15$ years and $6$ months.*

11. **When paying off loans, the monthly payment first goes to any interest owed before being applied to the remaining balance. Accountants and bankers use tables to help organize their work.**

a. **Consider the situation that Fred is paying off a loan of $\$125,000$ with an interest rate of $6\%$ per year compounded monthly. Fred pays $\$749.44$ every month. Complete the following table:**

| Payment | Interest Paid | Principal Paid | Remaining Principal |
|---|---|---|---|
| $\$749.44$ | $\$625.00$ | $\$124.44$ | $\$124,875.56$ |
| $\$749.44$ | $\$624.38$ | $\$125.06$ | $\$124,750.50$ |
| $\$749.44$ | $\$623.75$ | $\$125.69$ | $\$124,624.82$ |

b. **Fred's loan is supposed to last for 30 years. How much will Fred end up paying if he pays $\$749.44$ every month for 30 years? How much of this is interest if his loan was originally for $\$125,000$?**

$$\$749.44(30)(12) = \$269,798.40$$

*Fred will pay $\$269,798.40$ for his loan, paying $\$269,798.40 - \$125,000.00 = \$144,793.40$ in interest.*

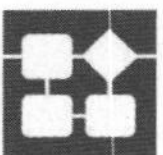

# Lesson 27: Modeling with Exponential Functions

## Student Outcomes

- Students create exponential functions to model real-world situations.
- Students use logarithms to solve equations of the form $f(t) = a \cdot b^{ct}$ for $t$.
- Students decide which type of model is appropriate by analyzing numerical or graphical data, verbal descriptions, and by comparing different data representations.

## Lesson Notes

In this summative lesson, students write exponential functions for different situations to describe the relationships between two quantities (**F-BF.A.1a**). This lesson uses real U.S. Census data to demonstrate how to create a function of the form $f(t) = a \cdot b^{ct}$ that could be used to model quantities that exhibit exponential growth or decay. Students must use properties of exponents to rewrite exponential expressions in order to interpret the properties of the function (**F-IF.C.8b**). They will estimate populations at a given time and determine the time when a population will reach a certain value by writing exponential equations (**A-CED.A.1**) and solving them analytically (**F-LE.A.4**). In Algebra I, students solved these types of problems graphically or numerically, but we have developed the necessary skills in this module to solve these problems algebraically. The data is presented in different forms (**F-IF.C.9**), and students use average rate of change (**F-IF.B.6**) to decide which type of function is most appropriate between linear or exponential functions (**F-LE.A.1**). Students have several different methods for determining the formula for an exponential function from given data: using a calculator's regression feature, solving for the parameters in the function analytically, and estimating the growth rate from a table of data (as covered in this lesson). This lesson ties those methods together and asks students to determine which seem most appropriate (MP.4).

## Classwork

### Opening (1 minute)

Pose this question, which will recall the work students did in Lesson 22:

- If you only have two data points, how should you decide which type of function to use to model the data?
  - *Two data points could be modeled using a linear, quadratic, sinusoidal, or exponential function. You would have to have additional information or know something about the real-world situation to make a decision about which model would be best.*

The Opening Exercise has students review how to find a linear and exponential model given two data points. Later in the lesson, students are then given more information about the data and asked to select and refine a model.

*Scaffolding:*

- If students struggle with the opening question, use this problem to provide a more concrete approach:

  Given the ordered pairs $(0,3)$ and $(3,6)$, we could write the following functions:

  $$f(t) = 3 + t$$

  $$g(t) = 3(2)^{\frac{t}{3}}$$

  Match each function to the appropriate verbal description and explain how you made your choice.

  A: A plant seedling is $3$ ft. tall, and each week the height increases by a fixed amount. After three weeks, the plant is $6$ ft. tall.

  B: Bacteria are dividing in a petri dish. Initially there are $300$ bacteria and three weeks later, there are $600$.

## Opening Exercise (5 minutes)

Give students time to work this Opening Exercise either independently or with a partner. Observe whether they are able to successfully write a linear and an exponential function for this data. If most of your students cannot complete these exercises without your assistance, then you will need to make adjustments during the lesson to help them build fluency with writing a function from given numerical data.

*Scaffolding:*

- For students who struggle with the algebraic manipulations, encourage them to use the statistical features of a graphing calculator to create a linear regression and an exponential regression equation in part (ii) of each Opening Exercise.

**Opening Exercise**

**The following table contains U.S. population data for the two most recent census years, $2000$ and $2010$.**

| Census Year | U.S. Population (in millions) |
|---|---|
| $2000$ | $281.4$ |
| $2010$ | $308.7$ |

a. **Steve thinks the data should be modeled by a linear function.**

i. **What is the average rate of change in population per year according to this data?**

*The average rate of change is* $\dfrac{308.7-281.4}{2010-2000} = 2.73$ *million people per year.*

ii. **Write a formula for a linear function, $L$, that will estimate the population $t$ years since the year $2000$.**

$$L(t) = 2.73t + 281.4$$

b. **Phillip thinks the data should be modeled by an exponential function.**

i. **What is the growth rate of the population per year according to this data?**

*The population will increase by the factor* $\dfrac{308.7}{281.4} = 1.097$ *every* $10$ *years. To determine the yearly rate, we would need to express* $1.097$ *as the product of* $10$ *equal numbers (e.g.,* $1.097^{\frac{1}{10}} \cdot 1.097^{\frac{1}{10}} \cdot \ldots 1.097^{\frac{1}{10}}$ *ten times). The annual rate would be* $1.097^{\frac{1}{10}} \approx 1.0093$.

ii. **Write a formula for an exponential function, $E$, that will estimate the population $t$ years since the year $2000$.**

*Start with* $E(t) = a \cdot b^t$. *Substitute* $(0, 281.4)$ *into the formula to solve for* $a$.

$$281.4 = a \cdot b^0$$

*Thus,* $a = 281.4$.

*Next, substitute the value of* $a$ *and the ordered pair* $(10, 308.7)$ *into the formula to solve for* $b$.

$$308.7 = 281.4b^{10}$$

$$b^{10} = 1.097$$

$$b = \sqrt[10]{1.097}$$

*Thus,* $b = 1.0093$ *when you round to the ten-thousandths place and*

$$E(t) = 281.4(1.0093)^t.$$

c. **Who has the correct model? How do you know?**

***You cannot determine who has the correct model without additional information. However, populations over longer intervals of time tend to grow exponentially if environmental factors do not limit the growth, so Phillip's model is likely to be more appropriate.***

### Discussion (3 minutes)

Before students start working in pairs or small groups on the modeling exercises, debrief the Opening Exercise with the following discussion to ensure that all students are prepared to begin the Modeling Exercise.

MP.7

- What function best modeled the given data? Allow students to debate about whether they chose a linear or an exponential model, and encourage them to provide justification for their decision.
  - $E(t) = 281.4(1.0093)^t$
- What does the number $281.4$ represent?
  - *The initial population in the year* $2000$ *was* $281.4$ *million people.*
- What does the number $1.0093$ represent?
  - *The population is increasing by a factor of* $1.0093$ *each year.*
- How does rewriting the base as $1 + 0.0093$ help us to understand the population growth rate?
  - *We can see the population is increasing by approximately* $0.93\%$ *every year according to our model.*

### Mathematical Modeling Exercises 1–14 (24 minutes)

These problems ask students to compare their model from the Opening Exercise to additional models created when given additional information about the U.S. population, and then ask students to use additional data to find a better model. Students should form small groups and work these exercises collaboratively. Provide time at the end of this portion of the lesson for different groups to share their rationale for the choices that they made. Students are exposed to both tabular and graphical data (**F-IF.C.9**) as they work through these exercises. They must use the properties of exponents to interpret and compare exponential functions (**F-IF.C.8b**).

Exercise 11 requires access to the Internet to look up the current population estimate for the U.S. If students do not have convenient Internet access, you can either display the website http://www.census.gov/popclock, which would be an interesting way to introduce this exercise, or look up the current population estimate at the onset of class and provide this information to the students. The U.S. population clock is updated every 10 or 12 seconds, so it will show a dramatic population increase through a single class period.

**Mathematical Modeling Exercises 1–14**

**In this challenge, you will continue to examine U.S. census data to select and refine a model for the population of the United States over time.**

1. **The following table contains additional U.S. census population data. Would it be more appropriate to model this data with a linear or an exponential function? Explain your reasoning.**

| Census Year | U.S. Population (in millions of people) |
|---|---|
| $1900$ | $76.2$ |
| $1910$ | $92.2$ |
| $1920$ | $106.0$ |
| $1930$ | $122.8$ |
| $1940$ | $132.2$ |
| $1950$ | $150.7$ |
| $1960$ | $179.3$ |
| $1970$ | $203.3$ |
| $1980$ | $226.5$ |
| $1990$ | $248.7$ |
| $2000$ | $281.4$ |
| $2010$ | $308.7$ |

*Scaffolding:*

- For students who are slow to recognize data as linear or exponential, create an additional column that shows the average rate of change and reinforce that unless those values are very close to a constant, a linear function is not the best model.

***It is not clear by looking at a graph of this data whether it lies on an exponential curve or a line. However, from the context, we know that populations tend to grow as a constant factor of the previous population, so we should use an exponential function to model it. The graph below uses $t = 0$ to represent the year $1900$.***

***OR***

***The differences between consecutive population values do not remain constant and in fact get larger as time goes on, but the quotients of consecutive population values are nearly constant around $1.1$. This indicates that a linear model is not appropriate but an exponential model is.***

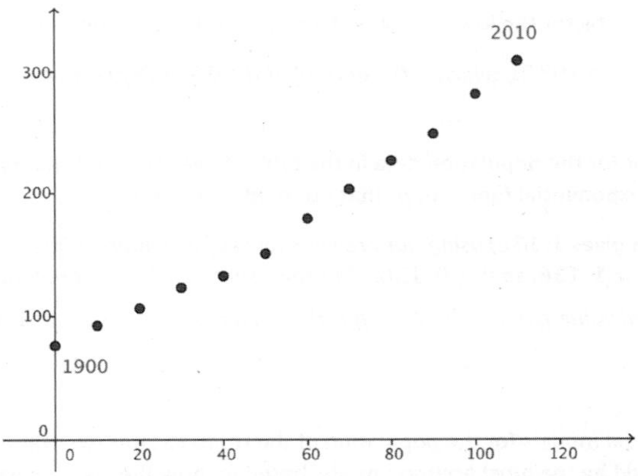


After the work in Lesson 22, students should know that a situation such as this one involving population growth should be modeled by an exponential function. However, the reasoning used by each group of students will vary. Some may plot the data and note the characteristic shape of an exponential curve. Some may calculate the quotients and differences between consecutive population values. If time permits, have students share the reasoning they used to decide which type of function to use.

2. Use a calculator's regression capability to find a function, $f$, that models the U.S. Census Bureau data from $1900$ to $2010$.

*Using a graphing calculator and letting the year $1900$ correspond to $t = 0$ gives the following exponential regression equation.*

$$P(t) = 81.1(1.0126)^t$$

*Scaffolding:*

- Students may need to be shown how to use the calculator to find the exponential regression function.

3. Find the growth factor for each $10$-year period and record it in the table below. What do you observe about these growth factors?

| Census Year | US Population (in millions of people) | Growth Factor ($10$-year period) |
|---|---|---|
| $1900$ | $76.2$ | -- |
| $1910$ | $92.2$ | $1.209974$ |
| $1920$ | $106.0$ | $1.149675$ |
| $1930$ | $122.8$ | $1.158491$ |
| $1940$ | $132.2$ | $1.076547$ |
| $1950$ | $150.7$ | $1.139939$ |
| $1960$ | $179.3$ | $1.189781$ |
| $1970$ | $203.3$ | $1.133854$ |
| $1980$ | $226.5$ | $1.114117$ |
| $1990$ | $248.7$ | $1.098013$ |
| $2000$ | $281.4$ | $1.131484$ |
| $2010$ | $308.7$ | $1.097015$ |

*The growth factors are fairly constant around $1.1$.*

4. For which decade is the $10$-year growth factor the lowest? What factors do you think caused that decrease?

*The $10$-year growth factor is lowest in the $1930$s, which is the decade of the Great Depression.*

5. Find an average $10$-year growth factor for the population data in the table. What does that number represent? Use the average growth factor to find an exponential function, $g$, that can model this data.

*Averaging the $10$-year growth factors gives $1.136$; using our previous form of an exponential function; this means that the growth rate $r$ satisfies $1 + r = 1.136$, so $r = 0.136$. This represents a $13.6\%$ population increase every ten years. The function $g$ has an initial value $g(0) = 76.2$, so $g$ is then given by $g(t) = 76.2(1.136)^{\frac{t}{10}}$, where $t$ represents year since $1900$.*

6. You have now computed three potential models for the population of the United States over time: functions $E$, $f$, and $g$. Which one do you expect would be the most accurate model based on how they were created? Explain your reasoning.

*Student responses will vary. Potential responses:*

- *I expect that function $f$ that we found through exponential regression on the calculator will be the most accurate because it used all of the data points to compute the coefficients of the function.*
- *I expect that the function $E$ will be most accurate because it uses only the most recent population values.*

Students should notice that function $g$ is expressed in terms of a 10-year growth rate (the exponent is $\frac{t}{10}$), while the other two functions are expressed in terms of single-year growth rates (the exponent is $t$). In Exercise 8, encourage students to realize that they will need to use properties of exponents to rewrite the exponential expression in $g$ in the form $g(t) = A(1+r)^t$ with an annual growth rate $r$ so that the three functions can be compared in Exercise 10 (**F-IF.C.8b**). Through questioning, lead students to notice that time $t = 0$ does not have the same meaning for all three functions $E$, $f$, and $g$. In Exercise 9, they will need to transform function $E$ so that $t = 0$ corresponds to the year 1900 instead of 2000. This is the equivalent of translating the graph of $y = E(t)$ horizontally to the right by 100 units.

7. **Summarize the three formulas for exponential models that you have found so far: Write the formula, the initial populations, and the growth rates indicated by each function. What is different between the structures of these three functions?**

   ***We have the three models:***

   - $E(t) = 281.4(1.0093)^t$: ***Population is 281.4 million in the year 2000; annual growth rate is 0.93%.***
   - $f(t) = 81.1(1.0126)^t$: ***Population is 81.1 million in the year 1900; annual growth rate is 1.26%.***
   - $g(t) = 76.2(13.6)^{\frac{t}{10}}$: ***Population is 76.2 million in the year 1900; 10-year growth rate is 13.6%.***

   ***Function $g$ is expressed in terms of a 10-year growth factor instead of an annual growth factor as in functions $E$ and $f$. Function $E$ has the year 2000 corresponding to $t = 0$, while in functions $f$ and $g$ the year $t = 0$ represents the year 1900.***

8. **Rewrite the functions $E$, $f$, and $g$ as needed in terms of an annual growth rate.**

   ***We need to use properties of exponents to rewrite $g$.***

$$\begin{aligned} g(t) &= 76.2(1.136)^{\frac{t}{10}} \\ &= 76.2\left((1.136)^{\frac{1}{10}}\right)^t \\ &\approx 76.2(1.0128)^t \end{aligned}$$

*Scaffolding:*

- Struggling students may need to be explicitly told that they need to re-express $g$ in the form $g(t) = A(1+r)^t$ with an annual growth rate $r$.

9. **Transform the functions as needed so that the time $t = 0$ represents the same year in functions $E$, $f$, and $g$. Then compare the values of the initial populations and annual growth rates indicated by each function.**

   ***In function $E$, $t = 0$ represents the year 2000, and in functions $f$ and $g$, $t = 0$ represents the year 1900.***

   ***Thus, we need to translate function $E$ horizontally to the right by 100 years, giving a new function:***

$$\begin{aligned} E(t) &= 281.4(1.0093)^{t-100} \\ &= 281.4(1.0093)^{-100}(1.0093)^t \\ &\approx 111.5(1.0093)^t. \end{aligned}$$

   ***Then we have the three functions:***

$$\begin{aligned} E(t) &= 111.5(1.0093)^t \\ f(t) &= 81.1(1.0126)^t \\ g(t) &= 76.2(1.0128)^t \end{aligned}$$

   - ***Function $E$ has the largest initial population and the smallest growth rate at 0.93% increase per year.***
   - ***Function $g$ has the smallest initial population and the largest growth rate at 1.28% increase per year.***

*Scaffolding:*

- Struggling students may need to be explicitly told that they need to translate function $E$ so that $t = 0$ represents the year 1900 for all three functions.

10. **Which of the three functions is the best model to use for the U.S. census data from $1900$ to $2010$? Explain your reasoning.**

*Student responses will vary.*

*Possible response: Graphing all three functions together with the data, we see that function $f$ appears to be the closest to all of the data points.*

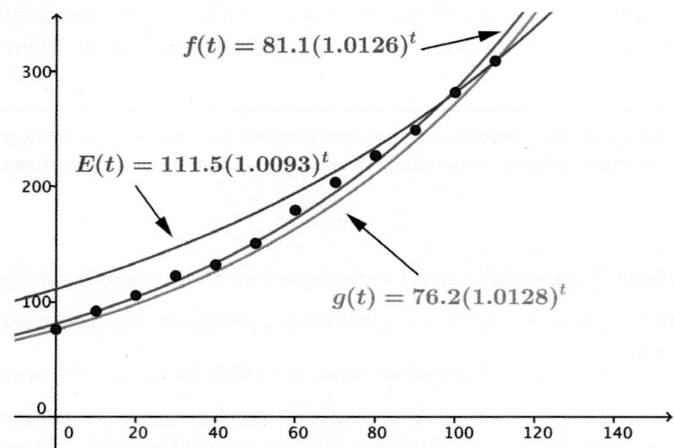


11. **The US Census Bureau website http://www.census.gov/popclock displays the current estimate of both the United States and world populations.**

   a. **What is today's current estimated population of the US?**

   *This will vary by the date. The solution shown here will use the population $318.7$ million and the date August $16, 2014$.*

   b. **If time $t = 0$ represents the year $1900$, what is the value of $t$ for today's date? Give your answer to two decimal places.**

   *August $16$ is the $228^{th}$ day of the year, so the time is $t = 114 + \frac{228}{365}$. We will use $t = 114.62$.*

   c. **Which of the functions $E$, $f$, and $g$ gives the best estimate of today's population? Does that match what you expected? Justify your reasoning.**

$$E(114.62) = 322.2$$
$$f(114.62) = 340.7$$
$$g(114.62) = 327.4$$

   *The function $E$ gives the closest value to today's estimated population, but all three functions produce estimates that are too high. Possible response: I had expected that function $f$, which was obtained through regression, to produce the closest population estimate, so this is a surprise.*

   d. **With your group, discuss some possible reasons for the discrepancy between what you expected in Exercise 8 and the results of part (c) above.**

   *Student responses will vary.*

12. **Use the model that most accurately predicted today's population in Exercise 9, part (c) to predict when the U.S. population will reach half a billion.**

***Half a billion is $500$ million. Set the formula for $E$ equal to $500$ and solve for $t$.***

$$111.5(1.0093)^t = 500$$
$$1.0093^t = \frac{500}{111.5}$$
$$1.0093^t = 4.4843$$
$$\log(1.0093)^t = \log(4.4843)$$
$$t\log(1.0093) = \log(4.4843)$$
$$t = \frac{\log(4.4843)}{\log(1.0093)}$$
$$t \approx 162$$

***Assuming the same rate of growth, the population will reach half a billion people $162$ years from the year $1900$, in the year $2062$.***

13. **Based on your work so far, do you think this is an accurate prediction? Justify your reasoning.**

***Student responses will vary. Possible response: From what we know of population growth, the data should most likely be fit with an exponential function, however the growth rate appears to be decreasing because the models that use all of the census data produce estimates for the current population that are too high. I think the population will reach half a billion sometime after the year $2062$ because the US Census Bureau expects the growth rate to slow down. Perhaps the United States is reaching its capacity and cannot sustain the same exponential rate of growth into the future.***

14. **Here is a graph of the US population since the census began in $1790$. Which type of function would best model this data? Explain your reasoning.**

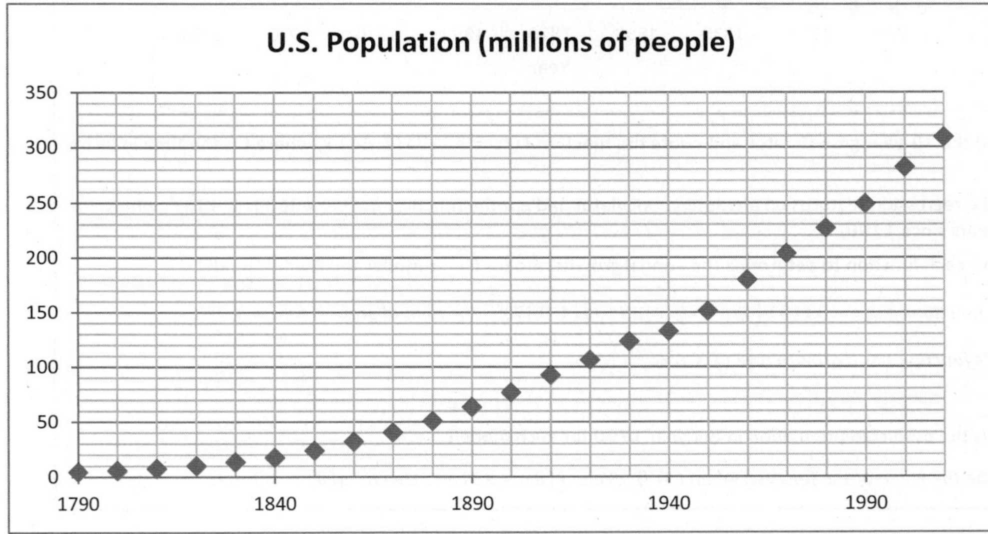


**Figure 1: Source U.S. Census Bureau**

***The shape of the curve indicates that an exponential model would be the best choice. You could model the data for short periods of time using a series of piecewise linear functions, but the average rate of change in the early years is clearly less than that in later years. A linear model would also not make sense because at some point in the past you would have had a negative number of people living in the U.S.***

## Exercises 15–16 (6 minutes)

Exercises 15–16 are provided for students who complete the Modeling Exercises. You might consider assigning these exercises as additional Problem Sets for the rest of the class.

In these two exercises, students are asked to compare different exponential population models. They will need to rewrite them to interpret the parameters when they compare the functions and apply the formula to solve a variety of problems. They are asked to compare the functions that model this data with an actual graph of the data. These problems are examples of **F-IF.C.8b**, **F-LE.A.1**, **F-LE.A.4**, and **F-IF.C.9**.

**Exercises 15–16**

**15. The graph below shows the population of New York City during a time of rapid population growth.**

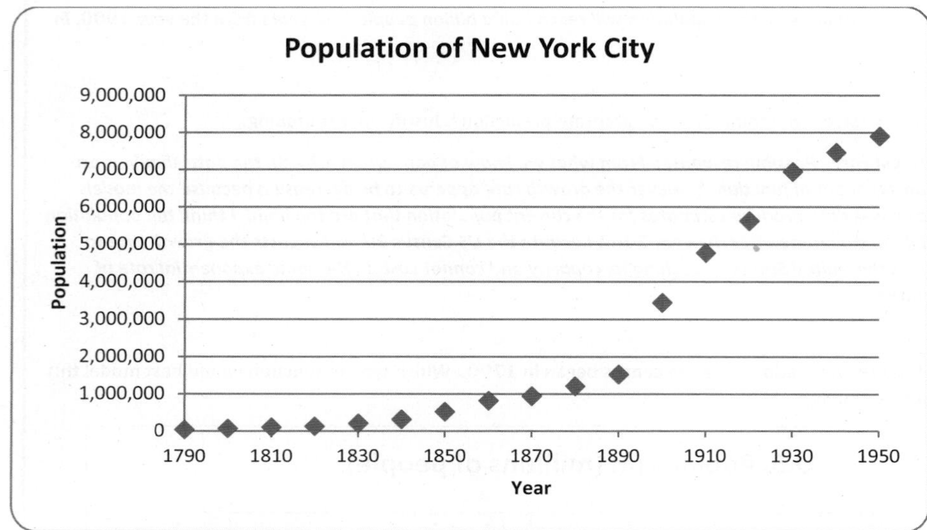


**Finn averaged the $10$-year growth rates and wrote the function $f(t) = 33131(1.44)^{\frac{t}{10}}$, where $t$ is the time in years since $1790$.**

**Gwen used the regression features on a graphing calculator and got the function $g(t) = 48661(1.036)^t$, where $t$ is the time in years since $1790$.**

**a. Rewrite each function to determine the annual growth rate for Finn's model and Gwen's model.**

*Finn's function:* $f(t) = 33131\left(1.44^{\frac{1}{10}}\right)^t = 33131(1.037)^t$*. The annual growth rate is* $3.7\%$.

*Gwen's function has a growth rate of* $3.6\%$.

**b. What is the predicted population in the year $1790$ for each model?**

*It will be the value of the function when* $t = 0$*. Finn:* $f(0) = 33131$*. Gwen:* $g(0) = 48661$.

c. Lenny calculated an exponential regression using his graphing calculator and got the same growth rate as Gwen, but his initial population was very close to $0$. Explain what data Lenny may have used to find his function.

*He may have used the actual year for his time values; where Gwen represented year $1790$ by $t = 0$, Lenny may have represented year $1790$ by $t = 1790$. If you translate Gwen's function $1790$ units to the right write the resulting function in the form $f(t) = a \cdot b^t$, the value of $a$ would be very small.*

$$48661(1.036)^{t-1790} = \frac{48661(1.036)^t}{1.036^{1790}} \text{ and } \frac{48661}{1.036^{1790}} \approx 1.56 \times 10^{-23}$$

d. When does Gwen's function predict the population will reach $1{,}000{,}000$? How does this compare to the graph?

*Solve the equation:* $48661(1.036)^t = 1{,}000{,}000$.

$$1.036^t = \frac{1{,}000{,}000}{48661}$$

$$\log(1.036)^t = \log\left(\frac{1{,}000{,}000}{48{,}661}\right)$$

$$t\log(1.036) = \log\left(\frac{1{,}000{,}000}{48{,}661}\right)$$

$$t = \frac{\log\left(\frac{1{,}000{,}000}{48{,}661}\right)}{\log(1.036)}$$

$$t \approx 85.5$$

*Gwen's model predicts that the population will exceed one million after $86$ years, which would be during the year $1867$. It appears that the population was close to one million around $1870$ so the model does a fairly good job of estimating the population.*

e. Based on the graph, do you think an exponential growth function would be useful for predicting the population of New York in the years after $1950$?

*The graph appears to be increasing but curving downwards, and an exponential model with a base greater than $1$ would always be increasing at an increasing rate, so its graph would curve upwards. The difference between the function and the data would be increasing, so this is probably not an appropriate model.*

16. Suppose each function below represents the population of a different US city since the year $1900$.

a. Complete the table below. Use the properties of exponents to rewrite expressions as needed to help support your answers.

| City Population Function ($t$ is years since $1900$) | Population in the Year $1900$ | Annual Growth/Decay Rate | Predicted in $2000$ | Between Which Years Did the Population Double? |
|---|---|---|---|---|
| $A(t) = 3000(1.1)^{\frac{t}{5}}$ | $3000$ | $1.9\%$ *growth* | $20182$ | *Between* $1936$ *and* $1937$ |
| $B(t) = \frac{(1.5)^{2t}}{2.25}$ | $1$ | $125\%$ *growth* | $7.3 \times 10^{34}$ | *Between* $1901$ *and* $1902$ |
| $C(t) = 10000(1-0.01)^t$ | $10000$ | $1\%$ *decay* | $475$ | *Never* |
| $D(t) = 900(1.02)^t$ | $900$ | $2\%$ *growth* | $6520$ | *Between* $1935$ *and* $1936$ |

b. **Could the function $E(t) = 6520(1.219)^{\frac{t}{10}}$, where $t$ is years since $2000$ also represent the population of one of these cities? Use the properties of exponents to support your answer.**

*Yes, it could represent the population in the city with function $D$. The expression $1.219^{\frac{t}{10}} \approx 1.02^t$ for any real number t. Also, $E(0) \approx D(100)$, which would make sense if the point of reference in time is $100$ years apart.*

c. **Which cities are growing in size, and which are decreasing according to these models?**

*The cities represented by functions $A$, $B$, and $D$ are growing because their base value is greater than $1$. The city represented by function $C$ is shrinking because $1 - 0.01$ is less than $1$.*

d. **Which of these functions might realistically represent city population growth over an extended period of time?**

*Based on the United States and New York City data, it is unlikely that a city in the United States could sustain a $50\%$ growth rate every two years for an extended period of time as indicated by function $B$ and its predicted population in the year $2000$. The other functions seem more realistic, with annual growth or decay rates similar to other city populations we examined.*

## Closing (2 minutes)

Have students respond to this question either in writing or with a partner.

- How do you decide when an exponential function would be an appropriate model for a given situation?
  - *You must consider the real-world situation to determine whether growth or decay by a constant factor is appropriate or not. Analyzing patterns in the graphs or data tables can also help.*
- Which method do you prefer for determining a formula for an exponential function?
  - *Student responses will vary. A graphing calculator provides a statistical regression equation, but you have to type in the data to use that feature.*
- Why did we rewrite the expression for function $g$?
  - *We can more easily compare the properties of functions if they have the same structure.*

Then review the points in the Lesson Summary.

**Lesson Summary**

**To model data with an exponential function:**

- **Examine the data to see if there appears to be a constant growth or decay factor.**
- **Determine a growth factor and a point in time to correspond to $t = 0$.**
- **Create a function to model the situation $f(t) = a \cdot b^{ct}$, where $b$ is the growth factor every $\frac{1}{c}$ years and $a$ is the value of $f$ when $t = 0$.**

**Logarithms can be used to solve for $t$ when you know the value of $f(t)$ in an exponential function model.**

## Exit Ticket (4 minutes)

Name ______________________________________________ Date ____________________

# Lesson 27: Modeling with Exponential Functions

## Exit Ticket

1. The table below gives the average annual cost (e.g., tuition, room, and board) for four-year public colleges and universities. Explain why a linear model might not be appropriate for this situation.

| Year | Average Annual Cost |
|---|---|
| 1981 | $2,550 |
| 1991 | $5,243 |
| 2001 | $8,653 |
| 2011 | $15,918 |

2. Algebraically determine an exponential function to model this situation.

3. Use the properties of exponents to rewrite the function from Problem 2 to determine an annual growth rate.

4. If this trend continues, when will the average annual cost of attendance exceed $35,000?

## Exit Ticket Sample Solutions

1. **The table below gives the average annual cost (e.g., tuition, room, and board) for four-year public colleges and universities. Explain why a linear model might not be appropriate for this situation.**

| Year | Average Annual Cost |
|---|---|
| $1981$ | $\$2,550$ |
| $1991$ | $\$5,243$ |
| $2001$ | $\$8,653$ |
| $2011$ | $\$15,918$ |

*A linear function would not be appropriate because the average rate of change is not constant.*

2. **Write an exponential function to model this situation.**

*If you calculate the growth factor every* $10$ *years, you get the following values.*

$$1981-1991: \frac{5243}{2550} = 2.056$$

$$1991-2001: \frac{8653}{5243} = 1.650$$

$$2001-2011: \frac{15918}{8653} = 1.840$$

*The average of these growth factors is* $1.85$.

*Then the average annual cost in dollars t years after* $1981$ *is* $C(t) = 2550(1.85)^{\frac{t}{10}}$.

3. **Use the properties of exponents to rewrite the function from Problem 2 to determine an annual growth rate.**

*We know that* $2550(1.85)^{\frac{t}{10}} = 2250\left(1.85^{\frac{1}{10}}\right)^t$ *and* $1.85^{\frac{1}{10}} \approx 1.063$. *Thus the annual growth rate is* $6.3\%$.

4. **If this trend continues, when will the average annual cost exceed** $\$35,000$**?**

*We need to solve the equation* $C(t) = 35000$.

$$2550(1.85)^{\frac{t}{10}} = 35000$$

$$(1.85)^{\frac{t}{10}} = 13.725$$

$$\log\left((1.85)^{\frac{t}{10}}\right) = \log(13.725)$$

$$\frac{t}{10} = \frac{\log(13.725)}{\log(1.85)}$$

$$t = 10\left(\frac{\log(13.725)}{\log(1.85)}\right)$$

$$t \approx 42.6$$

*The cost will exceed* $\$35,000$ *after* $43$ *years, which will be in the year* $2024$.

## Problem Set Sample Solutions

1. Does each pair of formulas described below represent the same sequence? Justify your reasoning.

   a. $a_{n+1} = \frac{2}{3}a_n$, $a_0 = -1$ and $b_n = -\left(\frac{2}{3}\right)^n$ for $n \geq 0$.

   *Yes. Checking the first few terms in each sequence gives the same values. Both sequences start with $-1$ and are repeatedly multiplied by $\frac{2}{3}$.*

   b. $a_n = 2a_{n-1} + 3$, $a_0 = 3$ and $b_n = 2(n-1)^3 + 4(n-1) + 3$ for $n \geq 1$.

   *No. The first two terms are the same, but the third term is different.*

   c. $a\_n = \frac{1}{3}(3)^n$ for $n \geq 0$ and $b_n = 3^{n-2}$ for $n \geq 0$.

   *Yes. The first terms are equal $a_0 = \frac{1}{3}$ and $b_0 = 3^{-1} = \frac{1}{3}$, and the next term is found by multiplying the previous term by 3 in both sequences.*

2. Tina is saving her babysitting money. She has $\$500$ in the bank, and each month she deposits another $\$100$. Her account earns $2\%$ interest compounded monthly.

   a. Complete the table showing how much money she has in the bank for the first four months.

| Month | Amount |
|---|---|
| 1 | $500$ |
| 2 | $500(1.00167) + 100 = 600.84$ |
| 3 | $(500(1.00167) + 100)(1.00167) + 100 = 701.84$ |
| 4 | $((500(1.00167) + 100)(1.00167) + 100)1.00167 + 100 = 803.01$ |

   b. Write a recursive sequence for the amount of money she has in her account after $n$ months.

   $a_1 = 500$, $a_{n+1} = a_n\left(1 + \frac{0.02}{12}\right) + 100$

3. Assume each table represents values of an exponential function of the form $f(t) = a(b)^{ct}$ where $b$ is a positive real number and $a$ and $c$ are real numbers. Use the information in each table to write a formula for $f$ in terms of $t$ for parts (a)–(d).

   a.

| $t$ | $f(t)$ |
|---|---|
| 0 | 10 |
| 4 | 50 |

   $f(t) = 10(5)^{\frac{t}{4}}$

   b.

| $t$ | $f(t)$ |
|---|---|
| 0 | 1000 |
| 5 | 750 |

   $f(t) = 1000(0.75)^{\frac{t}{5}}$

   c.

| $t$ | $f(t)$ |
|---|---|
| 6 | 25 |
| 8 | 45 |

   $f(t) = 4.287\left(\frac{9}{5}\right)^{\frac{t}{2}}$

   d.

| $t$ | $f(t)$ |
|---|---|
| 3 | 50 |
| 6 | 40 |

   $f(t) = 62.5\left(\frac{4}{5}\right)^{\frac{t}{3}}$

e. **Rewrite the expressions for each function in parts (a)–(d) to determine the annual growth or decay rate.**

*For part (a),* $5^{\frac{t}{4}} = \left(5^{\frac{1}{4}}\right)^t$ *so the annual growth factor is* $5^{\frac{1}{4}} \approx 1.495$*, and the annual growth rate is* $49.5\%$.

*For part (b),* $0.75^{\frac{t}{5}} = \left(0.75^{\frac{1}{5}}\right)^t$ *so the annual growth factor is* $0.75^{\frac{1}{5}} \approx 0.596$*, so the annual growth rate is* $-40.4\%$*, meaning that the quantity is decaying at a rate of* $40.4\%$.

*For part (c),* $\left(\frac{9}{5}\right)^{\frac{t}{2}} = \left(\left(\frac{9}{5}\right)^{\frac{1}{2}}\right)^t$ *so the annual growth factor is* $\left(\frac{9}{5}\right)^{\frac{1}{2}} \approx 1.312$ *and the annual growth rate is* $31.2\%$.

*For part (a),* $\left(\frac{4}{5}\right)^{\frac{t}{3}} = \left(\left(\frac{4}{5}\right)^{\frac{1}{3}}\right)^t$ *so the annual growth factor is* $\left(\frac{4}{5}\right)^{\frac{1}{3}} \approx 0.928$ *and the annual growth rate is* $-0.072$*, which is a decay rate of* $7.2\%$.

f. **For parts (a) and (c), determine when the value of the function is double its initial amount.**

*For part (a), solve the equation* $2 = 5^{\frac{t}{4}}$ *for* $t$.

$$2 = 5^{\frac{t}{4}}$$
$$\log(2) = \log\left(5^{\frac{t}{4}}\right)$$
$$\frac{t}{4} = \frac{\log(2)}{\log(5)}$$
$$t = 4\left(\frac{\log(2)}{\log(5)}\right)$$
$$t \approx 1.723$$

*For part (c), solve the equation* $2 = \left(\frac{9}{5}\right)^{\frac{t}{2}}$ *for* $t$*. The solution is* $2.358$.

g. **For parts (b) and (d), determine when the value of the function is half its initial amount.**

*For part (b), solve the equation* $\frac{1}{2} = (0.75)^{\frac{t}{5}}$ *for* $t$*. The solution is* $12.047$.

*For part (d), solve the equation* $\frac{1}{2} = \left(\frac{4}{5}\right)^{\frac{t}{3}}$ *for* $t$*. The solution is* $9.319$.

4. **When examining the data in Example 1, Juan noticed the population doubled every five years and wrote the formula** $P(t) = 100(2)^{\frac{t}{5}}$**. Use the properties of exponents to show that both functions grow at the same rate per year.**

*Using properties of exponents,* $100(2)^{\frac{t}{5}} = 100\left(2^{\frac{1}{5}}\right)^t$*. The annual growth is* $2^{\frac{1}{5}}$*. In the other function, the annual growth is* $4^{\frac{1}{10}} = \left(4^{\frac{1}{2}}\right)^{\frac{1}{5}} = 2^{\frac{1}{5}}$.

5. **The growth of a tree seedling over a short period of time can be modeled by an exponential function. Suppose the tree starts out 3 ft. tall and its height increases by** $15\%$ **per year. When will the tree be 25 ft. tall?**

*We model the growth of the seedling by* $h(t) = 3(1.15)^t$*, where* $t$ *is measured in years, and we find that* $3(1.15)^t = 25$ *when* $t = 15.171$ *years. The exact solution is* $t = \frac{\log\left(\frac{25}{3}\right)}{\log(1.15)}$.

6. Loggerhead turtles reproduce every 2–4 years, laying approximately 120 eggs in a clutch. Studying the local population, a biologist records the following data in the second and fourth years of her study:

| Year | Population |
|---|---|
| 2 | 50 |
| 4 | 1250 |

a. Find an exponential model that describes the loggerhead turtle population in year $t$.

*From the table, we see that $P(2) = 50$ and $P(4) = 1250$. So, the growth rate over two years is $\frac{1250}{50} = 25$. Since $P(2) = 50$, and $P(t) = P_0(25)^{\frac{t}{2}}$, we know that $50 = P_0(25)$, so $P_0 = 2$. Then $50\,r^2 = P_0r^4$, so $50r^2 = 1250$. Thus, $r^2 = 25$ and then $r = 5$. Since $50 = P_0r^2$, we see that $P_0 = 2$. Therefore,*

$$P(t) = 2(5^t)$$

b. According to your model, when will the population of loggerhead turtles be over $5,000$? Give your answer in years and months.

$$2(5^t) = 5000$$
$$5^t = 2500$$
$$t\log(5) = \log(2500)$$
$$t = \frac{\log(2500)}{\log(5)}$$
$$t \approx 4.86$$

*The population of loggerhead turtles will be over $5,000$ after year $4.86$, which is roughly $4$ years and $11$ months.*

7. The radioactive isotope seaborgium-266 has a half-life of $30$ seconds, which means that if you have a sample of $A$ g of seaborgium-266, then after $30$ seconds half of the sample has decayed (meaning it has turned into another element), and only $\frac{A}{2}$ g of seaborgium-266 remain. This decay happens continuously.

a. Define a sequence $a_0, a_1, a_2, \ldots$ so that $a_n$ represents the amount of a $100$ g sample that remains after $n$ minutes.

*In one minute, the sample has been reduced by half two times, leaving only $\frac{1}{4}$ of the sample. We can represent this by the sequence $a_n = 100\left(\frac{1}{2}\right)^{2n} = 100\left(\frac{1}{4}\right)^n$. (Either form is acceptable.)*

b. Define a function $a(t)$ that describes the amount of seaborgium-266 that remains of a $100$ g sample after $t$ minutes.

$$a(t) = 100\left(\frac{1}{4}\right)^t = 100\left(\frac{1}{2}\right)^{2t}$$

c. Does your sequence from part (a) and your function from part (b) model the same thing? Explain how you know.

*The function models the amount of seaborgium-266 as it constantly decreases every fraction of a second, and the sequence models the amount of seaborgium-266 that remains only in $30$-second intervals. They model nearly the same thing, but not quite. The function is continuous and the sequence is discrete.*

d. How many minutes does it take for less than $1$ g of seaborgium-$266$ to remain from the original $100$ g-sample? Give your answer to the nearest minute.

*The sequence is $a_0 = 100$, $a_1 = 25$, $a_2 = 6.25$, $a_3 = 1.5625$, $a_4 = 0.390625$, so after $4$ minutes there is less than $1$ g of the original sample remaining.*

8. Compare the data for the amount of substance remaining for each element: strontium-$90$, magnesium-$28$, and bismuth.

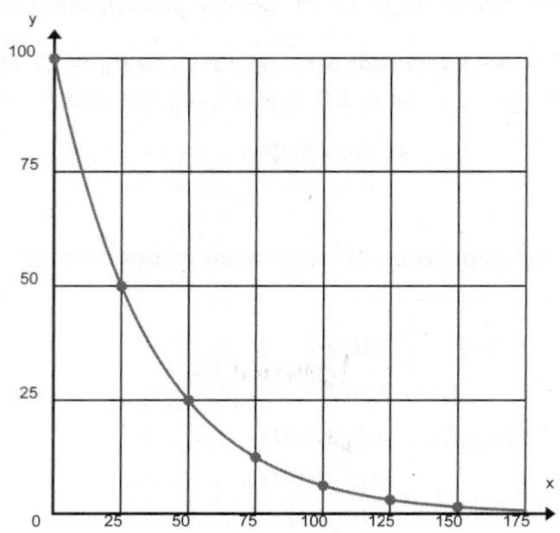


| Radioactive Decay of Magnesium-28 | |
|---|---|
| $R$ | $t$ hours |
| $1$ | $0$ |
| $0.5$ | $21$ |
| $0.25$ | $42$ |
| $0.125$ | $63$ |
| $0.0625$ | $84$ |

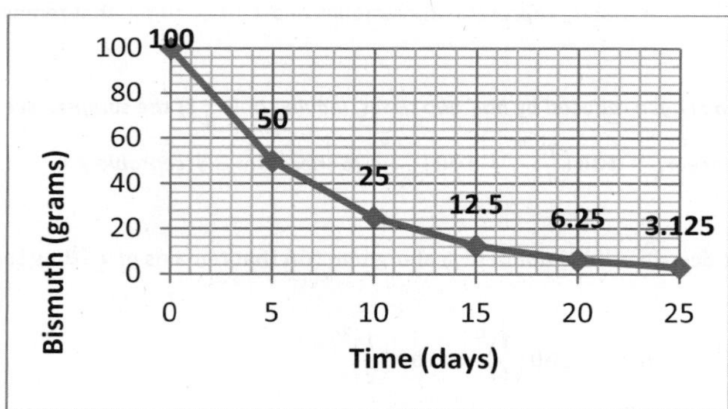


a. Which element decays most rapidly? How do you know?

*Magnesium-$23$ decays most rapidly. It loses half its amount every $21$ hours.*

b. Write an exponential function for each element that shows how much of a $100$ g sample will remain after $t$ days. Rewrite these expressions to show precisely how their exponential decay rates compare to confirm your answer to part (a).

- *Strontium-90: We model the remaining quantity by* $f(t) = 100\left(\frac{1}{2}\right)^{\frac{t}{\frac{25}{24}}}$ *where t is in days.*

  *Rewriting the expression gives a growth factor of* $\left(\frac{1}{2}\right)^{\frac{24}{25}} \approx 0.514$*, so* $f(t) = 100(0.514)^t$.

- *Magnesium-24: We model the remaining quantity by* $f(t) = 100\left(\frac{1}{2}\right)^{\frac{t}{\frac{21}{24}}}$ *where t is in days.*

  *Rewriting the expression give a growth factor of* $\left(\frac{1}{2}\right)^{\frac{24}{21}} \approx 0.453$*, so* $f(t) = 100(0.453)^t$

- *Bismuth: We model the remaining quantity by* $f(t) = 100\left(\frac{1}{2}\right)^{\frac{t}{5}}$ *where t is in days. Rewriting the expression gives a growth factor of* $\left(\frac{1}{2}\right)^{\frac{1}{5}} \approx 0.870$*, so* $f(t) = 100(0.870)^t$.

*The function with the smallest daily growth factor is decaying the fastest, so magnesium-24 decays the fastest.*

9. The growth of two different species of fish in a lake can be modeled by the functions shown below where $t$ is time in months since January $2000$. Assume these models will be valid for at least $5$ years.

Fish A: $f(t) = 5000(1.3)^t$

Fish B: $g(t) = 10,000(1.1)^t$

According to these models, explain why the fish population modeled by function $f$ will eventually catch up to the fish population modeled by function $g$. Determine precisely when this will occur.

*The fish population with the larger growth rate will eventually exceed the population with a smaller growth rate, so eventually Fish A will have a larger population.*

*Solve the equation* $f(t) = g(t)$ *for t to determine when the populations will be equal. After that point in time, the population of Fish A will exceed the population of Fish B.*

*The solution is*

$$5000(1.3)^t = 10,000(1.1)^t$$
$$\frac{(1.3)^t}{(1.1)^t} = 2$$
$$\left(\frac{1.3}{1.1}\right)^t = 2$$
$$t = \frac{\log(2)}{\log\left(\frac{1.3}{1.1}\right)}$$
$$t \approx 4.15$$

*During the fourth year, the population of Fish A will catch up to and then exceed the population of Fish B.*

10. **When looking at U.S. minimum wage data, you can consider the nominal minimum wage, which is the amount paid in dollars for an hour of work in the given year. You can also consider the minimum wage adjusted for inflation. Below are a table showing the nominal minimum wage and a graph of the data when the minimum wage is adjusted for inflation. Do you think an exponential function would be an appropriate model for either situation? Explain your reasoning.**

| Year | Nominal Minimum Wage |
|---|---|
| 1940 | $0.30 |
| 1945 | $0.40 |
| 1950 | $0.75 |
| 1955 | $0.75 |
| 1960 | $1.00 |
| 1965 | $1.25 |
| 1970 | $1.60 |
| 1975 | $2.10 |
| 1980 | $3.10 |
| 1985 | $3.35 |
| 1990 | $3.80 |
| 1995 | $4.25 |
| 2000 | $5.15 |
| 2005 | $5.15 |
| 2010 | $7.25 |

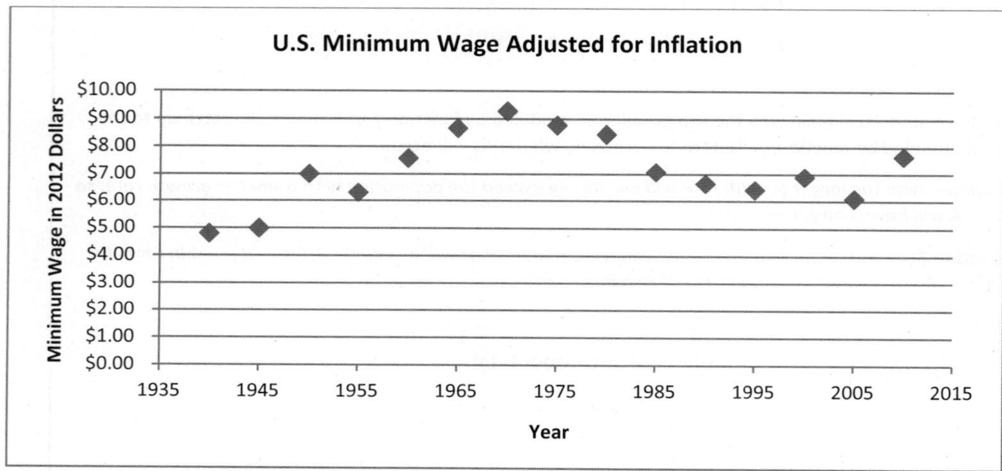


***Student solutions will vary. The inflation-adjusted minimum wage is clearly not exponential because it does not strictly increase or decrease. The other data when graphed does appear roughly exponential, and a good model would be*** $f(t) = 0.40(1.044)^t$.

11. **A dangerous bacterial compound forms in a closed environment but is immediately detected. An initial detection reading suggests the concentration of bacteria in the closed environment is one percent of the fatal exposure level. Two hours later, the concentration has increased to four percent of the fatal exposure level.**

   a. **Develop an exponential model that gives the percent of fatal exposure level in terms of the number of hours passed.**

$$\begin{aligned} P(t) &= 1 \cdot \left(\frac{4}{1}\right)^{\frac{t}{2}} \\ &= 4^{\frac{t}{2}} \\ &= 2^t \end{aligned}$$

   b. **Doctors and toxicology professionals estimate that exposure to two-thirds of the bacteria's fatal concentration level will begin to cause sickness. Provide a rough time limit (to the nearest $15$ minutes) for the inhabitants of the infected environment to evacuate in order to avoid sickness.**

$$\begin{aligned} 66.66 &= 2^t \\ \log(66.66) &= t \cdot \log(2) \\ t &= \frac{\log(66.66)}{\log(2)} \approx 6.0587 \end{aligned}$$

*Inhabitant should evacuate before $6$ hours and $3$ minutes to avoid becoming sick.*

   c. **A prudent and more conservative approach is to evacuate the infected environment before bacteria concentration levels reach $45\%$ of the fatal level. Provide a rough time limit (to the nearest $15$ minutes) for evacuation in this circumstance.**

$$\begin{aligned} 2^t &= 45 \\ t \cdot \log(2) &= \log(45) \\ t &= \frac{\log(45)}{\log(2)} \approx 5.492 \end{aligned}$$

*Inhabitants should evacuate within $5$ hours and $30$ minutes to avoid becoming sick at this conservative level.*

   d. **When will the infected environment reach $100\%$ of the fatal level of bacteria concentration (to the nearest minute)?**

$$\begin{aligned} t \cdot \log(2) &= \log(100) \\ t &= \frac{2}{\log(2)} \approx 6.644 \end{aligned}$$

*Inhabitants should evacuate within $6$ hours and $39$ minutes to avoid fatal levels.*

12. **Data for the number of users at two different social media companies is given below. Assuming an exponential growth rate, which company is adding users at a faster annual rate? Explain how you know.**

| Social Media Company A | |
|---|---|
| Year | Number of Users (Millions) |
| 2010 | 54 |
| 2012 | 185 |

| Social Media Company B | |
|---|---|
| Year | Number of Users (Millions) |
| 2009 | 360 |
| 2012 | 1,056 |

*Company A: The number of users (in millions) can be modeled by* $A(t) = a\left(\frac{185}{54}\right)^{\frac{t}{2}}$ *where* $a$ *is the initial amount and* $t$ *is time in years since* $2010$.

*Company B: The number of users (in millions) can be modeled by* $B(t) = b\left(\frac{1056}{360}\right)^{\frac{t}{3}}$ *where* $b$ *is the initial amount and* $t$ *is time in years since* $2009$.

*Rewriting the expressions, you can see that Company A's annual growth factor is* $\left(\frac{185}{54}\right)^{\frac{1}{2}} \approx 1.851$ *and Company B's annual growth factor is* $\left(\frac{1056}{360}\right)^{\frac{1}{3}} \approx 1.432$. *Thus, Company A is growing at the faster rate of* $85.1\%$ *compared to Company B's* $43.2\%$.

# Lesson 28: Newton's Law of Cooling, Revisited

## Student Outcomes

- Students apply knowledge of exponential and logarithmic functions and transformations of functions to a contextual situation.

## Lesson Notes

*Newton's law of cooling* is a complex topic that appears in physics and calculus; the formula can be derived using differential equations. In Algebra I (Module 3), students completed a modeling lesson in which Newton's law of cooling was simplified to focus on the idea of applying transformations of functions to a contextual situation. In this lesson, students take another look at Newton's law of cooling, this time incorporating their knowledge of the number $e$ and logarithms. Students now have the capability of finding the decay constant, $k$, for a contextual situation through the use of logarithms (**F-LE.A.4**). Students expand their understanding of exponential functions and transformations to build a function that models the temperature of a cooling body by adding a constant function to a decaying exponential and relate these functions to the model (**F-BF.A.1.b**). The entire lesson highlights modeling with mathematics (MP.4) and also provides students with an opportunity to interpret scenarios using Newton's law of cooling when presented with functions represented in various ways (numerically, graphically, algebraically, or verbally) (**F-IF.C.9**).

## Classwork

### Opening (2 minutes)

Review the formula $T(t) = T_a + (T_0 - T_a) \cdot e^{-kt}$ that was first introduced in Algebra I. There is one difference in the current presentation of the formula; in Algebra I, the base was expressed as 2.718 because students had not yet learned about the number $e$. Allow students a minute to examine the given formula. Before they begin working, discuss each parameter in the formula as a class.

- What does $T_a$ represent? $T_0$? $k$? $T(t)$?
  - *The notation $T_a$ represents the temperature surrounding the object, often called the "ambient temperature." The initial temperature of the object is denoted by $T_0$. The constant $k$ is called the decay constant. The temperature of the object after time $t$ has elapsed is denoted by $T(t)$.*
- Is $e$ one of the parameters in the formula?
  - *No; the number $e$ is a constant that is approximately equal to 2.718.*
- Assuming that the temperature of the object is greater than the temperature of the environment, is this formula an example of exponential growth or decay?
  - *It is an example of decay, because the temperature will be decreasing.*
- Why would it be decay when the base $e$ is greater than 1? Shouldn't that be exponential growth?
  - *Because the base is raised to a negative exponent. The negative reflects the graph about the $y$-axis making it decay rather than growth. If we rewrite the exponential expression using properties of exponents, we see that $e^{-kt} = \left(\frac{1}{e}\right)^{kt}$, and $\frac{1}{e} < 1$. In this form, we can clearly identify exponential decay.*

**Newton's law of cooling is used to model the temperature of an object of some temperature placed in an environment of a different temperature. The temperature of the object $t$ hours after being placed in the new environment is modeled by the formula**

$$T(t) = T_a + (T_0 - T_a) \cdot e^{-kt},$$

**where:**

**$T(t)$ is the temperature of the object after a time of $t$ hours has elapsed,**

**$T_a$ is the ambient temperature (the temperature of the surroundings), assumed to be constant and not impacted by the cooling process,**

**$T_0$ is the initial temperature of the object, and**

**$k$ is the decay constant.**

*Scaffolding:*

Use the interactive demonstration on Wolfram Alpha that was used in Algebra I to assist in analyzing the formula.

http://demonstrations.wolfram.com/NewtonsLawOfCooling/

## Mathematical Modeling Exercise 1 (15 minutes)

Have students work in groups on parts (a) and (b) of the exercise. Circulate the room and provide assistance as needed. Stop and debrief to ensure that students set up the equations correctly. Discuss the next scenario as a class before having students continue through the exercise.

**Mathematical Modeling Exercise 1**

**A crime scene investigator is called to the scene of a crime where a dead body has been found. He arrives at the scene and measures the temperature of the dead body at $9{:}30$ p.m. to be $78.3$°F. He checks the thermostat and determines that the temperature of the room has been kept at $74$°F. At $10{:}30$ p.m., the investigator measures the temperature of the body again. It is now $76.8$°F. He assumes that the initial temperature of the body was $98.6$°F (normal body temperature). Using this data, the crime scene investigator proceeds to calculate the time of death. According to the data he collected, what time did the person die?**

a. **Can we find the time of death using only the temperature measured at $9{:}30$ p.m.? Explain.**

***No. There are two parameters that are unknown, $k$ and $t$. We need to know the decay constant, $k$, in order to be able to find the elapsed time.***

b. **Set up a system of two equations using the data.**

***Let $t_1$ represent the elapsed time from the time of death until $9{:}30$ when the first measurement was taken, and let $t_2$ represent the elapsed time between the time of death and $10{:}30$ when the second measurement was taken. Then $t_2 = t_1 + 1$. We have the following equations:***

$$T(t_1) = 74 + (98.6 - 74)e^{-kt_1}$$
$$T(t_2) = 74 + (98.6 - 74)e^{-kt_2}.$$

***Substituting in our known value $T(t_1) = 78.3$ and $T(t_2) = 76.8$, we get the system:***

$$78.3 = 74 + (98.6 - 74)e^{-kt_1}$$
$$76.8 = 74 + (98.6 - 74)e^{-k(t_1+1)}.$$

- Why do we need two equations to solve this problem?
  - *Because there are two unknown parameters.*
- What does $t_1$ represent in the equation? Why does the second equation contain $(t_1 + 1)$ instead of just $t_1$?
  - *The variable $t_1$ represents the elapsed time from time of death to $9{:}30$ p.m. The second equation uses $(t_1 + 1)$ because the time of the second measurement is one hour later, so one additional hour has passed.*

- Joanna set up her equations as follows:

$$78.3 = 74 + (98.6 - 74)e^{-k(t_2-1)}$$
$$76.8 = 74 + (98.6 - 74)e^{-kt_2}$$

- In her equations, what does $t_2$ represent?
  - .2 *Elapsed time from time of death to* $10{:}30$ *p.m.*
- .3 Will she still find the same time of death? Explain why.

If students are unsure, have some groups work through the problem using one set of equations and some using the other. Re-address this question at the end.

  - *Yes, she will still get the same time of death. She will get a value of* $t$ *that is one hour greater since she is measuring elapsed time to* $10{:}30$ *rather than* $9{:}30$*, but she will still get the same time of death.*
- Now that we have this system of equations, how should we go about solving it?

Allow students to struggle with this for a few minutes. They may propose subtracting 74 from both sides or subtracting $98.6 - 74$.

$$4.3 = 24.6e^{-kt_1}$$
$$2.8 = 24.6e^{-k(t_1+1)}$$

- What do we need to do now?
  - *Combine the two equations in some way using the method of substitution or elimination.*
- .1 What is our goal in doing this?
  - *We want to eliminate one of the variables.*
- Would it be helpful to subtract the two equations? If students say yes, have them try it.
  - *No. Subtracting one equation from the other did not eliminate a variable.*
- How else could we combine the equations?
  - *We could use the multiplication property of equality to divide* $4.2$ *by* $2.8$ *and* $24.6e^{-kt_1}$ *by* $24.6e^{-k(t_1+1)}$.

If nobody offers this suggestion, lead students to the idea by reminding them of the properties of exponents. If we divide the exponential expressions, we will subtract the exponents and eliminate the variable $t_1$.

Have students continue the rest of the problem in groups.

c. **Find the value of the decay constant, $k$.**

$$\mathbf{4.3 = 24.6}e^{-kt_1}$$
$$\mathbf{2.8 = 24.6}e^{-k(t_1+1)}$$
$$\frac{\mathbf{4.3}}{\mathbf{2.8}} = e^{-kt_1+k(t_1+1)}$$
$$\frac{\mathbf{4.3}}{\mathbf{2.8}} = e^{k}$$
$$\ln\left(\frac{\mathbf{4.3}}{\mathbf{2.8}}\right) = \ln(e^{k})$$
$$\ln\left(\frac{\mathbf{4.3}}{\mathbf{2.8}}\right) = k \approx \mathbf{0.429}$$

**d. What was the time of death?**

$$4.3 = 24.6e^{-0.429t_1}$$
$$\frac{4.3}{24.6} = e^{-0.429t_1}$$
$$\ln\left(\frac{4.3}{24.6}\right) = \ln(e^{-0.429t_1})$$
$$\ln\left(\frac{4.3}{24.6}\right) = -0.429t_1$$
$$4.0656 = t_1$$

***The person died approximately 4 hours before 9:30 p.m., so the time of death was approximately 5:30 p.m.***

- Would we get the same time of death if we used the set of equations where $t_2$ represents time elapsed from death until 10:30 p.m.?
  - *Yes.*

## Mathematical Modeling Exercise 2 (10 minutes)

Allow students time to work in groups before discussing responses as a class. During the debrief, share and discuss work from different groups.

**Mathematical Modeling Exercise 2**

**A pot of tea is heated to 90°C. A cup of the tea is poured into a mug and taken outside where the temperature is 18°C. After 2 minutes, the temperature of the cup of tea is approximately 65°C.**

**a. Determine the value of the decay constant, $k$.**

$$T(2) = 18 + (90 - 18)e^{-k\cdot 2} = 65$$
$$72e^{-2k} = 47$$
$$e^{-2k} = \frac{47}{72}$$
$$-2k = \ln\left(\frac{47}{72}\right)$$
$$k \approx 0.2133$$

**b. Write a function for the temperature of the tea in the mug, $T$, in °C, as a function of time, $t$, in minutes.**

$$T(t) = 18 + 72\,e^{-0.213t}$$

c. **Graph the function $T$.**

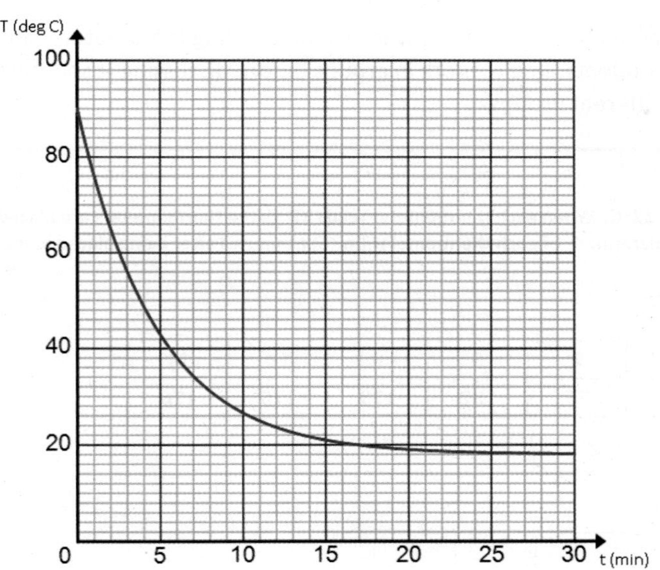


> *Scaffolding:*
>
> Provide students who are struggling with a graphing calculator or other graphing utility so that they can better focus on the key concepts of the lesson.

d. **Use the graph of $T$ to describe how the temperature decreases over time.**

***Because the temperature is decreasing exponentially, the temperature drops rapidly at first and then slows down. After about $25$ minutes, the temperature of the tea levels off.***

e. **Use properties of exponents to rewrite the temperature function in the form $T(t) = 18 + 72(1 + r)^t$.**

$$\begin{aligned} T(t) &= 18 + 72\,e^{-0.213\,t} \\ &= 18 + 72(e^{-0.213})^t \\ &\approx 18 + 72(0.8082)^t \\ &\approx 18 + 72(1 - 0.1918)^t \end{aligned}$$

f. **In Lesson 26, we saw that the value of $r$ represents the percent change of a quantity that is changing according to an exponential function of the form $f(t) = A(1 + r)^t$. Describe what $r$ represents in the context of the cooling tea.**

***The number $r$ represents the percent change in the difference between the temperature of the tea and the temperature of the room. Because $r = -0.1918$, the temperature difference is decreasing by $19.18\%$ each minute.***

g. **As more time elapses, what temperature does the tea approach? Explain using both the context of the problem and the graph of the function $T$.**

***The temperature of the tea approaches $18°\text{C}$. Within the context of the problem, this makes sense because that is the ambient temperature (the outside temperature), so when the tea reaches $18°\text{C}$ it will stop cooling. Looking at the expression of the function $T$, the number $18$ represents a vertical translation so as $t \to \infty$, $T \to 18$.***

## Mathematical Modeling Exercise 3 (10 minutes)

Newton's law of cooling also applies when a cooler object is placed in a warmer surrounding temperature. (In this case, we could call it Newton's Law of Heating.) Allow students time to work in groups before discussing responses as a class. During the debrief, share and discuss work from different groups.

**Mathematical Modeling Exercise 3**

**Two thermometers are sitting in a room that is $22°C$. When each thermometer reads $22°C$, the thermometers are placed in two different ovens. Select data for the temperature $T$ of each thermometer (in $°C$) $t$ minutes after being placed in the oven is provided below.**

**Thermometer 1:**

| $t$ (minutes) | 0 | 2 | 5 | 8 | 10 | 14 |
|---|---|---|---|---|---|---|
| $T$ (°C) | 22 | 75 | 132 | 173 | 175 | 176 |

**Thermometer 2:**

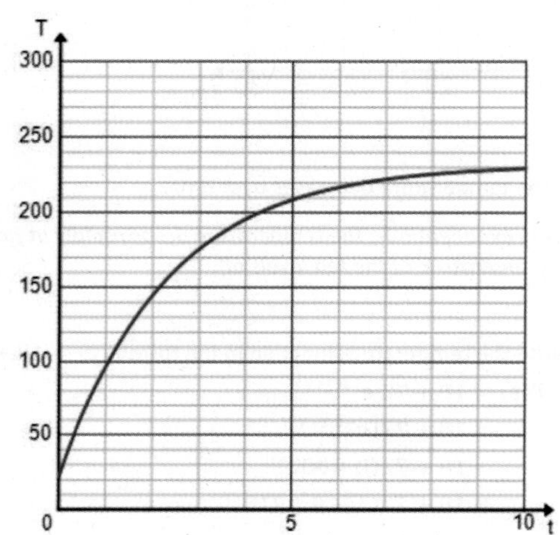


a. **Do the table and graph given for each thermometer support the statement that Newton's law of cooling also applies when the surrounding temperature is warmer? Explain.**

*Yes. The graph shows a reflected exponential curve, which would indicate that a similar formula could be used. From both the table and the graph, it can be seen that the temperature increases rapidly at first and then levels off to the temperature of its surroundings; this coincides with what happens when an object is cooling (that is, the temperature decreases rapidly and then levels off).*

b. **Which thermometer was placed in a hotter oven? Explain.**

*Thermometer 2 was placed in a hotter oven. The graph shows its temperature leveling off at approximately* $230°C$*, while the table indicates that thermometer 1 levels off at approximately* $176°C$.

c. **Using a generic decay constant, $k$, without finding its value, write an equation for each thermometer expressing the temperature as a function of time.**

*Thermometer 1:* $T(t) = 176 + (22 - 176)e^{-kt}$

*Thermometer 2:* $T(t) = 230 + (22 - 230)e^{-kt}$

d. **How do the equations differ when the surrounding temperature is warmer than the object rather than cooler as in previous examples?**

***In the case where we are placing a cool object into a warmer space, the coefficient in front of the exponential expression is negative rather than positive.***

e. **How do the graphs differ when the surrounding temperature is warmer than the object rather than cooler as in previous examples?**

***In the case where we are placing a cool object into a warmer space, the graph increases rather than decreases. The negative coefficient in front of the exponential expression causes the graph to reflect across the $x$-axis.***

## Closing (3 minutes)

Use the closing to highlight how this lesson built on their experiences from Algebra I with exponential decay and transformations of functions as well as the content learned in this module, such as the number $e$ and logarithms.

- For Exercise 2, describe the transformations required to graph $T$ starting from the graph of the natural exponential function $f(t) = e^t$.
  - *The graph is reflected across the $y$-axis, stretched both vertically and horizontally, and translated up.*
- Why were logarithms useful in exploring Newton's law of cooling?
  - *It allowed us to find the decay constant or the amount of time elapsed, both of which involve solving an exponential equation.*
- How do you find the percent rate of change of the temperature difference from the Newton's law of cooling equation?
  - *Rewrite $T(t) = T_0 + (T_a - T_0)e^{-kt}$ as $T(t) = T_0 + (T_a - T_0)(e^{-k})^t$, then express $e^{-k}$ as $e^{-k} = 1 - r$, for some number $r$. Then $r$ represents the percent rate of change of the temperature difference.*

## Exit Ticket (5 minutes)

Name ______________________________ Date ______________

# Lesson 28: Newton's Law of Cooling, Revisited

## Exit Ticket

A pizza, heated to a temperature of $400°\text{F}$, is taken out of an oven and placed in a $75°\text{F}$ room at time $t = 0$ minutes. The temperature of the pizza is changing such that its decay constant, $k$, is $0.325$. At what time is the temperature of the pizza $150°\text{F}$ and, therefore, safe to eat? Give your answer in minutes.

## Exit Ticket Sample Solutions

**A pizza, heated to a temperature of $400°$ Fahrenheit, is taken out of an oven and placed in a $75°\text{F}$ room at time $t = 0$ minutes. The temperature of the pizza is changing such that its decay constant, $k$, is $0.325$. At what time is the temperature of the pizza $150°\text{F}$ and, therefore, safe to eat? Give your answer in minutes.**

$$T(t) = 75 + (400 - 75)e^{-0.325t} = 150$$
$$325e^{-0.325t} = 75$$
$$e^{-0.325t} = \frac{75}{325}$$
$$-0.325t \approx \ln\left(\frac{75}{325}\right)$$
$$t \approx 4.512$$

*The pizza will reach* $150°\text{F}$ *after approximately* $4\frac{1}{2}$ *minutes.*

## Problem Set Sample Solutions

1. **Experiments with a covered cup of coffee show that the temperature (in degrees Fahrenheit) of the coffee can be modeled by the following equation:**

$$f(t) = 112e^{-0.08t} + 68,$$

**where the time is measured in minutes after the coffee was poured into the cup.**

a. **What is the temperature of the coffee at the beginning of the experiment?**

$180°\text{F}$

b. **What is the temperature of the room?**

$68°\text{F}$

c. **After how many minutes is the temperature of the coffee $140°\text{F}$? Give your answer to $3$ decimal places.**

$5.523$ *minutes.*

d. **What is the temperature of the coffee after *how many* minutes have elapsed?**

*The temperature will be slightly above* $68°\text{F}$.

e. **What is the percent rate of change of the difference between the temperature of the room and the temperature of the coffee?**

$$\begin{aligned} f(t) &= 112(e^{-0.08t}) + 68 \\ &= 112(e^{-0.08})^t + 68 \\ &\approx 112(0.9231)^t + 68 \\ &\approx 112(1 - 0.0769)^t + 68 \end{aligned}$$

*Thus, the percent rate of change of the temperature difference is a decrease of* $7.69\%$ *each minute.*

2. Suppose a frozen package of hamburger meat is removed from a freezer that is set at $0°\text{F}$ and placed in a refrigerator that is set at $38°\text{F}$. Six hours after being placed in the refrigerator, the temperature of the meat is $12°\text{F}$.

   a. Determine the decay constant, $k$.

   $$k = 0.063$$

   b. Write a function for the temperature of the meat, $T$ in Fahrenheit, as a function of time, $t$ in hours.

   $$T(t) = 38 - 38e^{-0.063t}$$

   c. Graph the function $T$.

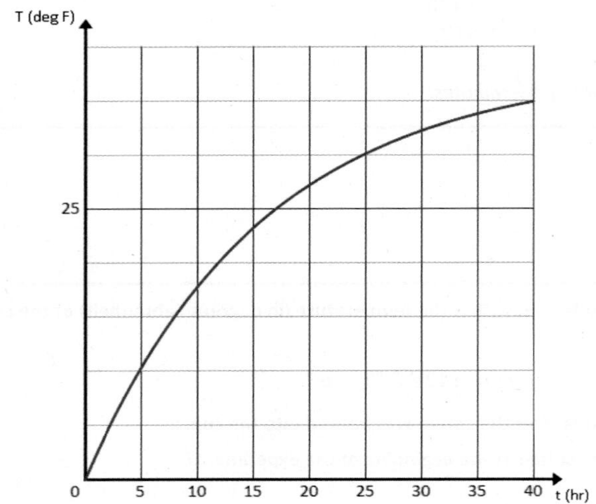


   d. Describe the transformations required to graph the function $T$ beginning with the graph of the natural exponential function $f(t) = e^t$.

   *The graph is stretched horizontally, reflected across the $y$-axis, stretched vertically, reflected across the $x$-axis, and translated up.*

   e. How long will it take the meat to thaw (reach a temperature above $32°\text{F}$)? Give answer to three decimal places.

   $29.299$ *hours*

   f. What is the percent rate of change of the difference between the temperature of the refrigerator and the temperature of the meat?

   $$\begin{aligned} T(t) &= 38 - 38e^{-0.063t} \\ &= 38 - 38(0.9389)^t \\ &= 38 - 38(1 - 0.0611)^t \end{aligned}$$

   *So, the percent rate of change in the difference of temperature is $6.11\%$.*

3. The table below shows the temperature of biscuits that were removed from an oven at time $t = 0$.

| $t$ (min) | 0 | 10 | 20 | 30 | 40 | 50 | 60 |
|---|---|---|---|---|---|---|---|
| $T$ (° C) | 100 | 34.183 | 22.514 | 20.446 | 20.079 | 20.014 | 20.002 |

a. What is the initial temperature of the biscuits?

$100°\text{C}$

b. What does the ambient temperature (room temperature) appear to be?

$20°\text{C}$

c. Use the temperature at $t = 10$ minutes to find the decay constant, $k$.

$k = 0.173$

d. Confirm the value of $k$ by using another data point from the table.

$T(40) = 20 + 80e^{-0.173 \cdot 40} = 20.079$

e. Write a function for the temperature of the biscuits (in Celsius) as a function of time in minutes.

$T(t) = 20 + 80e^{-0.173t}$

f. Graph the function $T$.

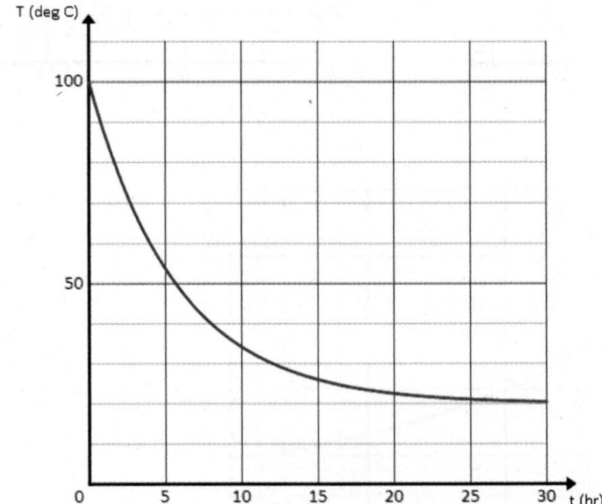

4. **Match each verbal description with its correct graph and write a possible equation expressing temperature as a function of time.**

   a. **A pot of liquid is heated to a boil and then placed on a counter to cool.**

      ***(iii)***, $T(t) = 75 + (212 - 75)e^{-kt}$ ***(Equations will vary.)***

   b. **A frozen dinner is placed in a preheated oven to cook.**

      ***(ii)***, $T(t) = 400 + (32 - 400)e^{-kt}$ ***(Equations will vary.)***

   c. **A can of room-temperature soda is placed in a refrigerator.**

      ***(i)***, $T(t) = 40 + (75 - 40)e^{-kt}$ ***(Equations will vary.)***

(i)

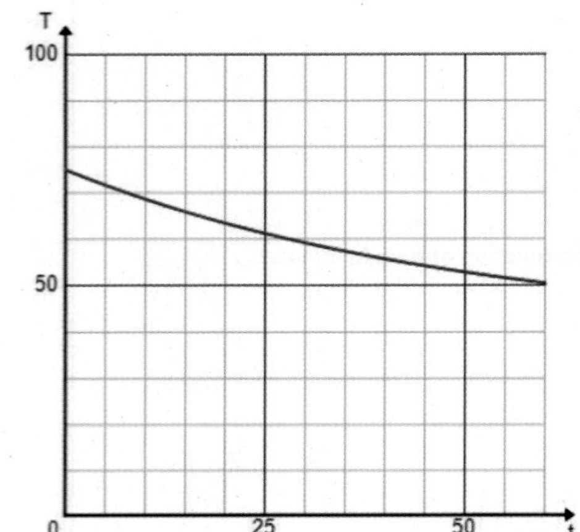


(ii)

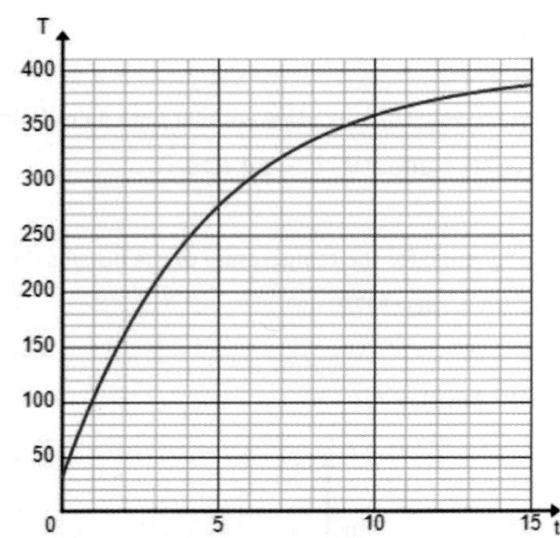


(iii)

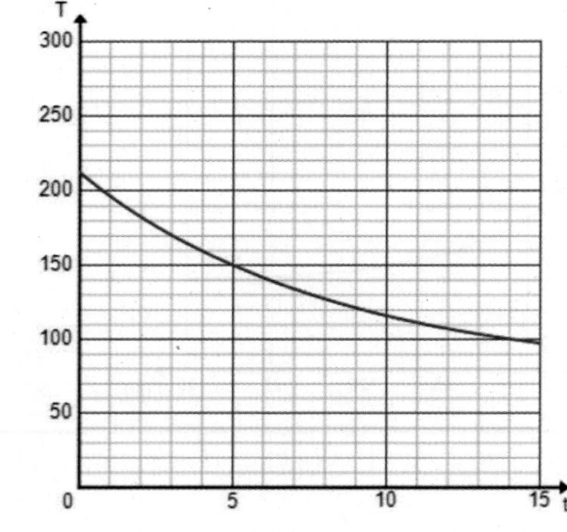

# Topic E:

# Geometric Series and Finance

**A-SSE.B.4, F-IF.C.7e, F-IF.C.8b, F-IF.C.9, F-BF.A.1b, F-BF.A.2, F-LE.B.5**

| **Focus Standards:** | A-SSE.B.4 | Derive the formula for the sum of a finite geometric series (when the common ratio is not 1), and use the formula to solve problems. *For example, calculate mortgage payments.* |
|---|---|---|
| | F-IF.C.7e | Graph functions expressed symbolically and show key features of the graph, by hand in simple cases and using technology for more complicated cases.★<br>e. Graph exponential and logarithmic functions, showing intercepts and end behavior, and trigonometric functions, showing period, midline, and amplitude. |
| | F-IF.C.8b | Write a function defined by an expression in different but equivalent forms to reveal and explain different properties of the function.<br>b. Use the properties of exponents to interpret expressions for exponential functions. *For example, identify percent rate of change in functions such as* $y = (1.02)^t$, $y = (0.97)^t$, $y = (1.01)^{12t}$, $y = (1.2)^{\frac{t}{10}}$, *and classify them as representing exponential growth or decay.* |
| | F-IF.C.9 | Compare properties of two functions each represented in a different way (algebraically, graphically, numerically in tables, or by verbal descriptions). *For example, given a graph of one quadratic function and an algebraic expression for another, say which has the larger maximum.* |
| | F-BF.A.1b | Write a function that describes a relationship between two quantities.★<br>b. Combine standard function types using arithmetic operations. *For example, build a function that models the temperature of a cooling body by adding a constant function to a decaying exponential, and relate these functions to the model.* |
| | F-BF.A.2 | Write arithmetic and geometric sequences both recursively and with an explicit formula, use them to model situations, and translate between the two forms.★ |
| | F-LE.B.5 | Interpret the parameters in a linear or exponential function in terms of a context. |

| **Instructional Days:** | 5 |
|---|---|
| **Lesson 29:** | The Mathematics Behind a Structured Savings Plan (M)[1] |
| **Lesson 30:** | Buying a Car (M) |
| **Lesson 31:** | Credit Cards (M) |
| **Lesson 32:** | Buying a House (M) |
| **Lesson 33:** | The Million Dollar Problem (M) |

Topic E is a culminating series of lessons driven by MP.4, Modeling with Mathematics. Students apply what they have learned about mathematical models and exponential growth to financial literacy, while developing and practicing the formula for the sum of a finite geometric series. Lesson 29 develops the future value formula for a structured savings plan and, in the process, develops the formula for the sum of a finite geometric series (**A-SSE.B.4**). The summation symbol, $\Sigma$, is introduced in this lesson.

Lesson 30 introduces loans through the context of purchasing a car. To develop the formula for the present value of an annuity, students combine two formulas for the future value of the annuity (**F-BF.A.1b**) and apply the sum of a finite geometric series formula. Throughout the remaining lessons, various forms of the present value of an annuity formula are used to calculate monthly payments and loan balances. Students compare the effects of various interest rates and repayment schedules, which requires that students translate between symbolic and numerical representations of functions (**F-IF.C.9**). Lesson 31 addresses the issue of revolving credit such as credit cards, for which the borrower can choose how much of the debt to pay each cycle. Students again sum a geometric series to develop a formula for this scenario, and it turns out to be equivalent to the formula used for car loans. Key features of tables and graphs are used to answer questions about finances (**F-IF.C.7e**).

Lessons 32 and 33 are modeling lessons in which students apply what they have done in earlier lessons to new financial situations (MP.4). Lesson 32 may be extended to an open-ended project in which students research buying a home and justify its affordability. Students graph the present value function and compare that with an amortization table, in accordance with **F-IF.C.9**. Lesson 33, the final lesson of the module, is primarily a summative lesson in which students formulate a plan to have $\$1{,}000{,}000$ in assets within a fixed time frame, using the formulas developed in the prior lessons in the topic. In both of these lessons, students need to combine functions using standard arithmetic operations (**F-IF.A.1b**).

[1] Lesson Structure Key: **P**-Problem Set Lesson, **M**-Modeling Cycle Lesson, **E**-Exploration Lesson, **S**-Socratic Lesson

# Lesson 29: The Mathematics Behind a Structured Savings Plan

## Student Outcomes

- Students derive the sum of a finite geometric series formula.
- Students apply the sum of a finite geometric series formula to a structured savings plan.

## Lesson Notes

Module 3 ends with a series of lessons centered on finance. In prior lessons, students progressed through the mathematics of a structured savings plan, buying a car, borrowing on credit cards, and buying a house. The module ends with an investigation of how to save over one million dollars in assets by the time the students are $40$ years old. Throughout these lessons, students engage in various parts of the modeling cycle: formulating, computing, interpreting, validating, etc.

In Lesson 29, students derive the formula for the sum of a finite geometric series (**A-SSE.B.4**). Once established, students work with and develop fluency with summation notation, sometimes called sigma notation. Students are then presented with the problem of a structured savings plan (known as a sinking fund, but this terminology is avoided). Students use the modeling cycle to identify essential features of structured savings plans and develop a model from the formula for the sum of a finite geometric series. By the end of the lesson, students will have both the formula for a sum of a finite geometric series (where the common ratio is not 1) and the formula for a structured savings plan. The structured savings plan will be modified in the remaining lessons to apply to other types of loans, and additional formulas will be developed based on the structured savings plan.

The formula $A_f = R \cdot \frac{(1+i^n)-1}{i}$ gives the future value of a structured savings plan $A_f$ with recurring payment $R$, interest rate $i$ per compounding period, and number of compounding periods $n$. In the context of loans, sometimes $P$ is used to represent the payment instead of $R$, but $R$ has been chosen because the formula is first presented as a structured savings plan and to avoid conflict with the common practice of using $P$ for principal.

Recall that the definition of a sequence in high school is simply a function whose domain is the natural numbers or non-negative integers. Students have previously worked with arithmetic and geometric sequences. Sequences can be described both recursively and explicitly (**F-BF.A.2**). Although students will primarily be working with geometric sequences in Lessons 29 through 33, arithmetic sequences will be reviewed in the problem sets of the lessons.

A copy of the modeling cycle flowchart is included below to assist with the modeling portions of these lessons. Whenever students consider a modeling problem, they need to first identify variables representing essential features in the situation, formulate a model to describe the relationships between the variables, analyze and perform operations on the relationships, interpret their results in terms of the original situation, validate their conclusions, and either improve the model or report on their conclusions and their reasoning. Both descriptive and analytic modeling will be used. Suggestions are made for the teacher, but the full extent of the modeling portions of the first three lessons in this topic are left to the discretion of the teacher and should be completed as time permits.

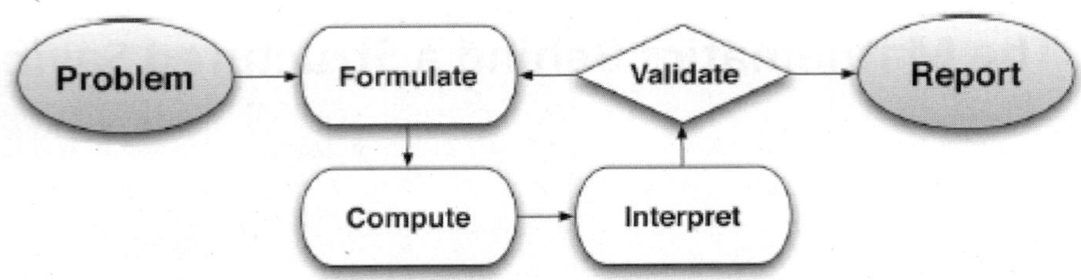


Note: You might consider breaking this lesson up over two days.

## Classwork

### Opening Exercise (3 minutes)

This is a quick exercise designed to remind students of the function $F = P\left(1+\frac{r}{n}\right)^{nt}$ from Lesson 26.

> **Opening Exercise**
>
> **Suppose you invested $\$1000$ in an account that paid an annual interest rate of $3\%$ compounded monthly. How much would you have after $1$ year?**
>
> *Since $F = 1000\left(1+\frac{0.03}{12}\right)^{12}$, we have $F = (1.0025)^{12} = \$1030.42$.*

After students have found the answer, have the following discussion *quickly*. It is important for setting up the notation for Topic E.

- To find the percent rate of change in the problem above, we took the annual interest rate of $3\%$ compounded monthly (called the nominal APR) and divided it by $12$. How do we express that in the future value formula from Lesson 26 using $r$ and $n$?
  - *We can express in the future value formula by using the expression $\frac{r}{n}$, where $r$ is the unit rate associated to the nominal APR and $n$ is the number of compounding periods in a year.*
- In today's lesson and in the next set of lessons, we are going to replace $\frac{r}{n}$ with something simpler: the letter $i$ for unit rate of the percent rate of change per time period (i.e., the interest rate per compounding period). What is the value of $i$ in the problem above?
  - $i = 0.0025$
- Also, since we only need to make calculations based upon the number of time periods, we can use $m$ to stand for the total number of time periods (i.e., total number of compounding periods). What is $m$ in the problem above?
  - $m = 12$
- Thus, the formula we will work with is $F = P(1+i)^m$ where $P$ is the present value and $F$ is the future value. This formula will help make our calculations more transparent, but we need to always remember to find the interest rate per the compounding period first.

**Discussion (7 minutes)**

This discussion sets up the reason for studying geometric series: Finding the future value of a structured savings plan. After this discussion, we will develop the formula for the sum of a series and then return to this discussion to show how to use the formula to find the future value of a structured savings plan. This sum of a geometric series will be brought up again and again throughout the remainder of the finance lessons to calculate information on structured savings plans, loans, annuities, and more. Look for ways you can use the discussion to help students formulate the problem.

- Let us consider the example of depositing $100 at the end of every month into an account for 12 months. If there is no interest earned on these deposits, how much money will you have at the end of 12 months?
    - *We will have* $12 \cdot 100 = \$1{,}200$. *(Some students may be confused about the first and last payment. Have them imagine they start a job where they get paid at the end of each month, and they put $100 from each paycheck into a special account. Then they would have $0 in the account until the end of the first month at which time they deposit $100. Similarly, they would have $200 at the end of month 2, $300 at the end of month 3, and so on.)*
- Now let us make an additional assumption: Suppose that our account is earning an annual interest rate of 3% compounded monthly. We would like to find how much will be in the account at the end of the year. The answer requires several calculations, not just one.
- What is the interest rate per month?
    - $i = 0.025$
- How much will the $100 deposited at the end of month 1 and its interest be worth at the end of the year?
    - *Since the interest on the $100 would be compounded 11 times, it and its interest would be worth* $100(1.025)^{11}$ *using the formula* $F = P(1+i)^m$. *(Some students may calculate this to be $131.21, but tell them to leave their answers in exponential form for now.)*
- How much will the $100 deposited at the end of month 2 and its interest be worth at the end of the year?
    - $100(1.025)^{10}$
- Continue: Find how much the $100 deposited at the each of the remaining months plus its compounded interest will be worth at the end of the year by filling out the table:

| Month deposited | Amount at the end of the year |
|---|---|
| 1 | $100(1.025)^{11}$ |
| 2 | $100(1.025)^{10}$ |
| 3 | $100(1.025)^{9}$ |
| 4 | $100(1.025)^{8}$ |
| 5 | $100(1.025)^{7}$ |
| 6 | $100(1.025)^{6}$ |
| 7 | $100(1.025)^{5}$ |
| 8 | $100(1.025)^{4}$ |
| 9 | $100(1.025)^{3}$ |
| 10 | $100(1.025)^{2}$ |
| 11 | $100(1.025)$ |
| 12 | $100$ |

- Every deposit except for the final deposit will earn interest. If we list the calculations from the final deposit to the first deposit, we get the following sequence:

$$100, 100(1.025)^1, 100(1.025)^2, 100(1.025)^3, \ldots, 1000(1.025)^{10}, 1000(1.02)^{11}$$

- What do you notice about this sequence?
  - *The sequence is geometric with initial term* $100$ *and common ratio* $1.025$.
- Describe how to calculate the amount of money we will have at the end of $12$ months.
  - *Sum the* $12$ *terms in the geometric sequence to get the total amount that will be in the account at the end of* $12$ *months.*
- The answer,

$$100 + 100(1.025) + 100(1.025)^2 + \cdots + 100(1.025)^{11},$$

is an example of a geometric series. In the next example, we will find a formula to find the sum of this series quickly. For now, let's discuss the definition of series. When we add some of the terms of a sequence we get a series:

- **SERIES:** Let $a_1, a_2, a_3, a_4, \ldots$ be a sequence of numbers. A sum of the form

$$a_1 + a_2 + a_3 + \cdots + a_n$$

for some positive integer $n$ is called a *series* (or *finite series*) and is denoted $S_n$.

The $a_i$'s are called the *terms* of the series. The number $S_n$ that the series adds to is called the *sum* of the series.

- **GEOMETRIC SERIES:** A *geometric series* is a series whose terms form a geometric sequence.

Since a geometric sequence is of the form $a, ar, ar^2, ar^3, \ldots$, the general form of a finite geometric sequence is of the form

$$S_n = a + ar + ar^2 + \cdots + ar^{n-1}.$$

- In the geometric series we generated from our structured savings plan, what is the first term $a$? What is the common ratio $r$? What is $n$?
  - $a = 100$, $r = 1.025$, *and* $n = 12$

## Example 1 (10 minutes)

Work through the following example to establish the formula for finding the sum of a finite geometric series. In the example, we calculate the general form of a sum of a generic geometric series using letters $a$ and $r$. We use letters instead of a numerical example for the following reasons:

- The use of letters in this situation actually makes the computations clearer. Students need only keep track of two letters instead of several numbers (as in $3, 6, 12, 24, 48, 96, \ldots$).

- Students have already investigated series several times before in different forms without realizing it. For example, throughout Topic A of Module 1 of Algebra II, they worked with the identity $(1-r)(1+r^2+r^3+\cdots+r^{n-1}) = 1-r^n$ for any real number $r$ and even used the identity to find the sum $1+2+2^2+2^3+\cdots+2^{31}$. In fact, students already derived the geometric series sum formula in the Problem Set of Lesson 1 of this module.

However, depending upon your class's strengths, you might consider doing a "mirror" calculation—on the left half of your board/screen do the calculation below, and on the right half "mirror" a numerical example; that is, for every algebraic line you write on the left, write the corresponding line using numbers on the right. A good numerical series to use is $2+6+18+54+162$.

**Example 1**

**Let $a, ar, ar^2, ar^3, ar^4, \ldots$ be a geometric sequence with first term $a$ and common ratio $r$. Show that the sum $S_n$ of the first $n$ terms of the geometric series**

$$S_n = a + ar + ar^2 + ar^3 + \cdots + ar^{n-1} \qquad (r \neq 1)$$

**is given by the equation**

$$S_n = a\frac{1-r^n}{1-r}.$$

Give students 2–3 minutes to try the problem on their own (or in groups of two). It is completely okay if no one gets an answer—you are giving them structured time to persevere, test conjectures, and grapple with what they are expected to show. Give a hint as you walk around the class: What identity from Module 1 can we use? Or, just ask them to see if the formula works for $a=1, r=2, n=5$. If a student or group of students solves the problem, tell them to hold on to their solution for a couple of minutes. Then go through this discussion:

- Multiply both sides of the equation $S_n = a + ar + ar^2 + \cdots + ar^{n-2} + ar^{n-1}$ by $r$.
  - $r \cdot S_n = ar + ar^2 + ar^3 + \cdots + ar^{n-1} + ar^n$
- Compare the terms on the right-hand side of the old and new equations:

$$\begin{aligned} S_n &= a + ar + ar^2 + ar^3 + \cdots + ar^{n-1} \\ r \cdot S_n &= \phantom{a +{}} ar + ar^2 + ar^3 + \cdots + ar^{n-1} + ar^n \end{aligned}$$

- What are the only terms on the right-hand side of the original equation and the new equation that are not found in both?
  - *The only terms that are not found in both are $a$ and $ar^n$.*

> *Scaffolding:*
> - For struggling students, do this calculation twice, first using a concrete sequence such as $a = 3$ and $r = 2$ before generalizing.

- Therefore, when we subtract $S_n - rS_n$, all the common terms on the right-hand side of the equations subtract to zero leaving only $a - ar^n$ (after applying the associative and commutative properties repeatedly):

$$\begin{aligned} S_n - rS_n &= (a + ar + ar^2 + \cdots + ar^{n-1}) - (ar + ar^2 + \cdots + ar^{n-1} + ar^n) \\ &= a + (ar - ar) + (ar^2 - ar^2) + \cdots + (ar^{n-1} - ar^{n-1}) - ar^n \\ &= a - ar^n, \end{aligned}$$

or $S_n - rS_n = a - ar^n$.

- Isolate $S_n$ in the equation $S_n - rS_n = a - ar^n$ to get the formula (when $r \neq 1$):

$$S_n(1-r) = a(1-r^n) \;\Rightarrow\; S_n = a\frac{1-r^n}{1-r}.$$

Any student who used the identity equation $(1-r)(1+r^2+r^3+\cdots+r^{n-1}) = 1-r^n$ or a different verifiable method (like proof by induction), should be sent to the board to explain the solution to the rest of class as an alternative explanation to your explanation. If no one was able to show the formula, you can wrap up the explanation by linking the formula back to the work they have done with this identity: Divide both sides of the identity equation by $(1-r)$ to isolate the sum $1+r+r^2+\cdots+r^{n-1}$ and then for $r \neq 1$,

$$a+ar+ar^2+\cdots+ar^{n-1} = a(1+r+r^2+\cdots+r^{n-1}) = a\frac{1-r^n}{1-r}.$$

## Exercises 1–3 (5 minutes)

**Exercises 1–3**

1. **Find the sum of the geometric series $3+6+12+24+48+96+192$.**

   *The first term is 3, so $a=3$. The common ratio is $6/3=2$, so $r=2$. Since there are 7 terms, $n=7$. Thus, $S_7 = 3\cdot\frac{1-2^7}{1-2}$, or $S_7 = 381$.*

2. **Find the sum of the geometric series $40+40(1.005)+40(1.005)^2+\cdots+40(1.005)^{11}$.**

   *Reading directly off the sum, $a=40$, $r=1.005$, and $n=12$.*

   *Thus, $S_{12} = 40\cdot\frac{1-1.005^{12}}{1-1.005}$, so $S_{12}\approx 493.42$.*

3. **Describe a situation that might lead to calculating the sum of the geometric series in Exercise 2.**

   *Investing $\$40$ a month into an account with an annual interest rate of $6\%$ compounded monthly (or investing $\$40$ a month into an account with an interest rate of $0.05\%$ per month) can lead to the geometric series in Exercise $2$.*

*Scaffolding:*

Ask advanced students to first find the sum

$$a+ar^2+ar^4+\cdots+ar^{2n-2}$$

and then the sum

$$ar+ar^3+ar^5+\cdots+ar^{2n-1}.$$

## Example 2 (2 minutes)

Let's now return to the Opening Exercise and answer the problem we encountered there.

**Example 2**

**A $\$100$ deposit is made at the end of every month for 12 months in an account that earns interest at an annual interest rate of $3\%$ compounded monthly. How much will be in the account immediately after the last payment?**

*Answer: The total amount is the sum $100+100(1.025)+100(1.025)^2+\cdots+100(1.025)^{11}$. This is a geometric series with $a=100$, $r=1.025$, and $n=12$. Using the formula for the sum of a geometric series, $S_{12} = 100\cdot\frac{1-1.025^{12}}{1-1.025}$, so $S\approx 1{,}379.56$. The account will have $\$1{,}379.56$ in it immediately after the last payment.*

Point out to your students that \$1,379.56 is significantly more money than stuffing $\$100$ in your mattress every month for 12 months. A structured savings plan like this is one way people can build wealth over time. Structured savings plans are examples of annuities. An annuity is commonly thought of as a type of retirement plan, but in this lesson, we are using the term in a much simpler way to refer to any situation where money is transferred into an account in equal amounts on a regular, recurring basis.

## Discussion (5 minutes)

Explain the following text to your students and work with them to answer the questions below it.

> **Discussion**
>
> **An *annuity* is a series of payments made at fixed intervals of time. Examples of annuities include structured savings plans, lease payments, loans, and monthly home mortgage payments. The term annuity sounds like it is only a yearly payment, but annuities are often monthly, quarterly, or semiannually. The *future amount of the annuity,* denoted $A_f$, is the sum of all the individual payments made plus all the interest generated from those payments over the specified period of time.**
>
> **We can generalize the structured savings plan example above to get a generic formula for calculating the future value of an annuity $A_f$ in terms of the recurring payment $R$, interest rate $i$, and number of payment periods $n$. In the example above, we had a recurring payment of $R = 100$, an interest rate per time period of $i = 0.025$, and 12 payments, so $n = 12$. To make things simpler, we always assume that the payments and the time period in which interest is compounded are at the same time. That is, we do not consider plans where deposits are made halfway through the month with interest compounded at the end of the month.**
>
> **In the example, the amount $A_f$ of the structured savings plan annuity was the sum of all payments plus the interest accrued for each payment:**
>
> $$A_f = R + R(1+i)^1 + R(1+i)^2 + \cdots + R(1+i)^{n-1}.$$
>
> **This, of course, is a geometric series with $n$ terms, $a = R$, and $r = 1 + i$, which after substituting into the formula for a geometric series and rearranging is**
>
> $$A_f = R \cdot \frac{(1+i)^n - 1}{i}.$$

- Explain how to get the formula above from the sum of a geometric series formula.
  - *Substitute $R$, $1+i$, and $n$ into the geometric series and rearrange:*

$$A_f = R \cdot \frac{1-(1+i)^n}{1-(1+i)} = R \cdot \frac{1-(1+i)^n}{-i} = R \cdot \frac{(1+i)^n - 1}{i}.$$

- (Optional depending on time.) How much money would need to be invested every month into an account with an annual interest rate of 12% compounded monthly in order to have \$3,000 after 18 months?
  - $3000 = R \cdot \frac{(1.01)^{18}-1}{0.01}$. *Solving for $R$ yields $R \approx 152.95$. For such a savings plan, \$152.95 would need to be invested monthly.*

## Example 3 (5 minutes)

Example 3 develops summation notation, sometimes called sigma notation. Including summation notation at this point is natural to reduce the amount of writing in sums and to develop fluency before relying on summation notation in precalculus and calculus. Plus, it "wraps up" all the important components of describing a series (starting value, index, explicit formula, and ending value) into one convenient notation. Summation notation will be further explored in the problem sets of Lessons 29 through 33 to develop fluency with the notation. Summation notation is specifically used to represent series. Students may be interested to know that a similar notation exists for products; product notation makes use of a capital pi, $\Pi$, instead of a capital sigma but is otherwise identical in form to summation notation.

Mathematicians use notation to reduce the amount of writing and to prevent ambiguity in mathematical sentences. Mathematicians use a special symbol with sequences to indicate that we would like to sum up the sequence. This symbol is a capital sigma, $\Sigma$. The sum of a sequence is called a *series*.

- The first letter of summation is "S," and the Greek letter for "S" is sigma. Capital sigma looks like this: $\Sigma$.
- There is no rigid way to use $\Sigma$ to represent a summation, but all notations generally follow the same rules. We will discuss the most common way it used. Given a sequence $a_1, a_2, a_3, a_4, \ldots$, we can write the sum of the first $n$ terms of the sequence using the expression:

$$\sum_{k=1}^{n} a_k.$$

- It is read, "The sum of $a_k$ from $k = 1$ to $k = n$." The letter $k$ is called the index of the summation. The notation acts like a for-next loop in computer programming: Replace $k$ in the expression right after the sigma by the integers (in this case) $1, 2, 3, \ldots, n$, and add the resulting expressions together. Since

$$\sum_{k=1}^{n} a_k = a_1 + a_2 + a_3 + \cdots + a_n,$$

the notation can be used to greatly simplify writing out the sum of a series.

- When the terms in a sequence are given by an explicit function, like the geometric sequence given by $a_k = 2^k$ for example, we often use the expression defining the explicit function instead of sequence notation. For example, the sum of the first five powers of 2 can be written as

$$\sum_{k=1}^{5} 2^k = 2^1 + 2^2 + 2^3 + 2^4 + 2^5.$$

## Exercises 4–5 (3 minutes)

**Exercises 4–5**

**4. Write the sum without using summation notation, and find the sum.**

a. $\sum_{k=0}^{5} k$

$$\sum_{k=0}^{5} k = 0 + 1 + 2 + 3 + 4 + 5 = 15$$

b. $\sum_{j=5}^{7} j^2$

$$\sum_{j=5}^{7} j^2 = 5^2 + 6^2 + 7^2 = 25 + 36 + 49 = 110$$

c. $\sum_{i=2}^{4} \frac{1}{i}$

$$\sum_{i=2}^{4} \frac{1}{i} = \frac{1}{2} + \frac{1}{3} + \frac{1}{4} = \frac{13}{12}$$

5. **Write each sum using summation notation.**

a. $1^4 + 2^4 + 3^4 + 4^4 + 5^4 + 6^4 + 7^4 + 8^4 + 9^4$

$$\sum_{k=1}^{9} k^4$$

b. $1 + \cos(\pi) + \cos(2\pi) + \cos(3\pi) + \cos(4\pi) + \cos(5\pi)$

$$\sum_{k=0}^{5} \cos(k\pi)$$

c. $2 + 4 + 6 + \cdots + 1000$

$$\sum_{k=1}^{500} 2k$$

## Closing (2 minutes)

- Have students summarize the lesson in writing or with a partner. Circulate and informally assess understanding. Ensure that every student has the formula for the sum of a finite geometric series and the formula for the future value of a structured savings plan. Ask questions to prompt student memory of the definitions and formulas below.

**Lesson Summary**

- **Series: Let $a_1, a_2, a_3, a_4, \ldots$ be a sequence of numbers. A sum of the form**
$$a_1 + a_2 + a_3 + \cdots + a_n$$
**for some positive integer $n$ is called a *series* (or finite series) and is denoted $S_n$. The $a_i$'s are called the terms of the series. The number $S_n$ that the series adds to is called the sum of the series.**
- **GEOMETRIC SERIES: A *geometric series* is a series whose terms form a geometric sequence.**
- **SUM OF A FINITE GEOMETRIC SERIES: The sum $S_n$ of the first $n$ terms of the geometric series $S_n = a + ar + \cdots + ar^{n-1}$ (when $r \neq 1$) is given by**
$$S_n = a\frac{1 - r^n}{1 - r}.$$
- **The sum of a finite geometric series can be written in summation notation as**
$$\sum_{k=0}^{n-1} ar^k = a \cdot \frac{1 - r^n}{1 - r}.$$
- **The generic formula for calculating the future value of an annuity $A_f$ in terms of the recurring payment $R$, interest rate $i$, and number of periods $n$ is given by**
$$A_f = R \cdot \frac{(1 + i)^n - 1}{i}.$$

## Exit Ticket (3 minutes)

Name ______________________________________________ Date ________________

# Lesson 29: The Mathematics Behind a Structured Savings Plan

## Exit Ticket

Martin attends a financial planning conference and creates a budget for himself, realizing that he can afford to put away $\$200$ every month in savings and that he should be able to keep this up for two years. If Martin has the choice between an account earning an interest rate of $2.3\%$ yearly versus an account earning an annual interest rate of $2.125\%$ compounded monthly, which account will give Martin the largest return in two years?

## Exit Ticket Sample Solutions

MP.2

**Martin attends a financial planning conference and creates a budget for himself, realizing that he can afford to put away $\$200$ every month in savings and that he should be able to keep this up for two years. If Martin has the choice between an account earning an interest rate of $2.3\%$ yearly versus an account earning an annual interest rate of $2.125\%$ compounded monthly, which account will give Martin the largest return in two years?**

$$A_f = R \cdot \frac{(1+i)^n - 1}{i}$$

$$A_f = 2400 \cdot \frac{(1.023)^2 - 1}{0.023} = 4855.20$$

$$A_f = 200 \cdot \frac{\left(1 + \frac{0.0215}{12}\right)^{24} - 1}{\frac{0.0215}{12}} \approx 4900.21$$

*The account earning an interest rate of $2.125\%$ compounded monthly will return more than the yearly account.*

## Problem Set Sample Solutions

The Problem Set in this lesson begins with performing the research necessary to personalize Lesson 30 before continuing with more work in sequences and sums and practice using the future value of a structured savings plan formula.

1. **A car loan is one of the first secured loans most Americans obtain. Research used car prices and specifications in your area to find a reasonable used car that you would like to own (under $\$10,000$). If possible, print out a picture of the car you selected.**

   a. **What is the year, make, and model of your vehicle?**

      *Answers will vary. For instance, $2006$ Pontiac G6 GT.*

   b. **What is the selling price for your vehicle?**

      *Answers will vary. For instance, $\$7500$.*

c. The following table gives the monthly cost per $\$1000$ financed on a 5-year auto loan. Assume you can get a $5\%$ annual interest rate. What is the monthly cost of financing the vehicle you selected? (A formula will be developed to find the monthly payment of a loan in Lesson 30.)

| Five-Year (60-month) Loan | |
|---|---|
| Interest Rate | Amount per $1000 Financed |
| $1.0\%$ | $\$17.09$ |
| $1.5\%$ | $\$17.31$ |
| $2.0\%$ | $\$17.53$ |
| $2.5\%$ | $\$17.75$ |
| $3.0\%$ | $\$17.97$ |
| $3.5\%$ | $\$18.19$ |
| $4.0\%$ | $\$18.41$ |
| $4.5\%$ | $\$18.64$ |
| $5.0\%$ | $\$18.87$ |
| $5.5\%$ | $\$19.10$ |
| $6.0\%$ | $\$19.33$ |
| $6.5\%$ | $\$19.56$ |
| $7.0\%$ | $\$19.80$ |
| $7.5\%$ | $\$20.04$ |
| $8.0\%$ | $\$20.28$ |
| $8.5\%$ | $\$20.52$ |
| $9.0\%$ | $\$20.76$ |

*Answers will vary. At $5\%$ interest, financing a $\$7,500$ car for $60$ months will cost $18.87 \cdot 7.5 = 141.525$, which is approximately $\$141.53$ per month.*

d. What is the gas mileage for your vehicle?

*Answers will vary. For instance, the gas mileage might be $29$ miles per gallon.*

e. If you drive $120$ miles per week and gas is $\$4$ per gallon, then how much will gas cost per month?

*Answers will vary but should be based on the answer to part (d). For instance, with gas mileage of $29$ miles per gallon, the gas cost would be $\frac{120}{29} \cdot 4.3 \cdot 4 \approx \$71.17$ per month.*

2. Write the sum without using summation notation, and find the sum.

a. $\sum_{k=1}^{8} k$

$$1 + 2 + 3 + 4 + 5 + 6 + 7 + 8 = 4 \cdot 9 = 36$$

b. $\sum_{k=-8}^{8} k$

$$-8+-7+-6+-5+-4+-3+-2+-1+0+1+2+3+4+5+6+7+8=0$$

c. $\sum_{k=1}^{4} k^3$

$$1^3+2^3+3^3+4^3=1+8+27+64=100$$

d. $\sum_{m=0}^{6} 2m$

$$\begin{aligned}2\cdot 0+2\cdot 1+2\cdot 2+2\cdot 3+2\cdot 4+2\cdot 5+2\cdot 6 &= 0+2+4+6+8+10+12\\ &= 3.5\cdot 12\\ &= 42\end{aligned}$$

e. $\sum_{m=0}^{6} 2m+1$

$$\begin{aligned}(2\cdot 0+1)+(2\cdot 1+1)+(2\cdot 2+1)+(2\cdot 3+1)+(2\cdot 4+1)+(2\cdot 5+1)+(2\cdot 6+1) &= 42+7\\ &= 49\end{aligned}$$

f. $\sum_{k=2}^{5} \frac{1}{k}$

$$\begin{aligned}\frac{1}{2}+\frac{1}{3}+\frac{1}{4}+\frac{1}{5} &= \frac{30}{60}+\frac{20}{60}+\frac{15}{60}+\frac{12}{60}\\ &= \frac{77}{60}\end{aligned}$$

g. $\sum_{j=0}^{3} (-4)^{j-2}$

$$\begin{aligned}(-4)^{-2}+(-4)^{-1}+(-4)^{0}+(-4)^{1} &= \frac{1}{16}+-\frac{1}{4}+1+-4\\ &= \frac{1}{16}-\frac{4}{16}+\frac{16}{16}-\frac{64}{16}\\ &= -\frac{51}{16}\end{aligned}$$

h. $\sum_{m=1}^{4} 16\left(\frac{3}{2}\right)^m$

$$\begin{aligned}\left(16\left(\frac{3}{2}\right)^1\right)+\left(16\left(\frac{3}{2}\right)^2\right)+\left(16\left(\frac{3}{2}\right)^3\right)+\left(16\left(\frac{3}{2}\right)^4\right) &= 16\left(\frac{3}{2}+\frac{9}{4}+\frac{27}{8}+\frac{81}{16}\right)\\ &= 16\left(\frac{24}{16}+\frac{36}{16}+\frac{54}{16}+\frac{81}{16}\right)\\ &= 24+36+54+81\\ &= 195\end{aligned}$$

i. $\sum_{j=0}^{3} \frac{105}{2j+1}$

$$\frac{105}{2\cdot 0+1}+\frac{105}{2\cdot 1+1}+\frac{105}{2\cdot 2+1}+\frac{105}{2\cdot 3+1}=\frac{105}{1}+\frac{105}{3}+\frac{105}{5}+\frac{105}{7}$$
$$=105+35+21+15$$
$$=176$$

j. $\sum_{p=1}^{3} p\cdot 3^p$

$$1\cdot 3^1+2\cdot 3^2+3\cdot 3^3=3+18+81$$
$$=102$$

k. $\sum_{j=1}^{6} 100$

$$100+100+100+100+100+100=600$$

l. $\sum_{k=0}^{4} \sin\left(\frac{k\pi}{2}\right)$

$$\sin\left(\frac{0\pi}{2}\right)+\sin\left(\frac{1\pi}{2}\right)+\sin\left(\frac{2\pi}{2}\right)+\sin\left(\frac{3\pi}{2}\right)+\sin\left(\frac{4\pi}{2}\right)=0+1+0+-1+0$$
$$=0$$

m. $\sum_{k=1}^{9} \log\left(\frac{k}{k+1}\right)$

(Hint: You do not need a calculator to find the sum.)

$$\log\left(\frac{1}{2}\right)+\log\left(\frac{2}{3}\right)+\log\left(\frac{3}{4}\right)+\log\left(\frac{4}{5}\right)+\log\left(\frac{5}{6}\right)+\log\left(\frac{6}{7}\right)+\log\left(\frac{7}{8}\right)+\log\left(\frac{8}{9}\right)+\log\left(\frac{9}{10}\right)$$
$$=\log(1)-\log(2)+\log(2)-\log(3)+\cdots-\log(10)$$
$$=\log(1)-\log(10)$$
$$=0-1$$
$$=-1$$

3. Write the sum without using sigma notation (you do not need to find the sum).

a. $\sum_{k=0}^{4} \sqrt{k+3}$

$$\sqrt{0+3}+\sqrt{1+3}+\sqrt{2+3}+\sqrt{3+3}+\sqrt{4+3}=\sqrt{3}+\sqrt{4}+\sqrt{5}+\sqrt{6}+\sqrt{7}$$

b. $\sum_{i=0}^{8} x^i$

$$1+x^1+x^2+x^3+x^4+x^5+x^6+x^7+x^8$$

c. $\sum_{j=1}^{6} jx^{j-1}$

$$1x^{1-1} + 2x^{2-1} + 3x^{3-1} + 4x^{4-1} + 5x^{5-1} + 6x^{6-1} = 1 + 2x^1 + 3x^2 + 4x^3 + 5x^4 + 6x^5$$

d. $\sum_{k=0}^{9} (-1)^k x^k$

$$(-1)^0x^0 + (-1)^1x^1 + (-1)^2x^2 + (-1)^3x^3 + (-1)^4x^4 + (-1)^5x^5 + (-1)^6x^6 + (-1)^7x^7 + (-1)^8x^8 + (-1)^9x^9$$
$$= 1 - x + x^2 - x^3 + x^4 - x^5 + x^6 - x^7 + x^8 - x^9$$

4. **Write each sum using summation notation.**

a. $1 + 2 + 3 + 4 + \cdots + 1000$

$\sum_{k=1}^{1000} k$

b. $2 + 4 + 6 + 8 + \cdots + 100$

$\sum_{k=1}^{50} 2k$

c. $1 + 3 + 5 + 7 + \cdots + 99$

$\sum_{k=1}^{50} 2k - 1$

d. $\frac{1}{2} + \frac{2}{3} + \frac{3}{4} + \cdots + \frac{99}{100}$

$\sum_{k=1}^{99} \frac{k}{k+1}$

e. $1^2 + 2^2 + 3^2 + 4^2 + \cdots + 10,000^2$

$\sum_{k=1}^{10000} k^2$

f. $1 + x + x^2 + x^3 + \cdots + x^{200}$

$\sum_{k=0}^{200} x^k$

g. $\frac{1}{1\cdot 2} + \frac{1}{2\cdot 3} + \frac{1}{3\cdot 4} + \cdots + \frac{1}{49\cdot 50}$

$\sum_{k=1}^{49} \frac{1}{k(k+1)}$

h. $1\ln(1) + 2\ln(2) + 3\ln(3) + \cdots + 10\ln(10)$

$$\sum_{k=1}^{10} k\ln(k)$$

5. Find the sum of the geometric series.

a. $1 + 3 + 9 + \cdots + 2187$

$$\frac{1-3^8}{1-3} = \frac{6560}{2} = 3280$$

b. $1 + \frac{1}{2} + \frac{1}{4} + \frac{1}{8} + \cdots + \frac{1}{512}$

$$\frac{1-\left(\frac{1}{2}\right)^{10}}{1-\frac{1}{2}} = \frac{\frac{1023}{1024}}{\frac{1}{2}} = \frac{1023}{1024}\cdot 2 = \frac{1023}{512}$$

c. $1 - \frac{1}{2} + \frac{1}{4} - \frac{1}{8} + \cdots - \frac{1}{512}$

$$\frac{1-\left(-\frac{1}{2}\right)^{10}}{1+\frac{1}{2}} = \frac{1023}{1024}\cdot\frac{2}{3} = \frac{341}{512}$$

d. $0.8 + 0.64 + 0.512 + \cdots + 0.32768$

$$0.8\cdot\frac{1-0.8^5}{1-0.8} = 0.8\cdot\frac{0.67232}{0.2} = 2.68928$$

e. $1 + \sqrt{3} + 3 + 3\sqrt{3} + \cdots + 243$

$$\frac{1-\sqrt{3}^{11}}{1-\sqrt{3}} \approx 573.5781477$$

f. $\sum_{k=0}^{5} 2^k$

$$\frac{1-2^6}{1-2} = 63$$

g. $\sum_{m=1}^{4} 5\left(\frac{3}{2}\right)^m$

$$\left(5\left(\frac{3}{2}\right)\right)\left(\frac{1-\left(\frac{3}{2}\right)^4}{1-\frac{3}{2}}\right) = \left(\frac{15}{2}\right)\left(\frac{\frac{81}{16}-1}{\frac{1}{2}}\right)$$
$$= \frac{15}{2}\left(\frac{65}{16}\right)2$$
$$= \frac{975}{16}$$
$$= 60.9375$$

h. $1 - x + x^2 - x^3 + \cdots + x^{30}$ in terms of $x$

$$\frac{1-(-x)^{31}}{1-(-x)} = \frac{x^{31}+1}{x+1}$$

i. $\sum_{m=0}^{11} 4^{\frac{m}{3}}$

$$\frac{1-\left(4^{\frac{1}{3}}\right)^{12}}{1-4^{\frac{1}{3}}} \approx 434.1156679$$

j. $\sum_{n=0}^{14} (\sqrt[5]{6})^n$

$$\frac{1-(\sqrt[5]{6})^{15}}{1-\sqrt[5]{6}} \approx 498.8756953$$

k. $\sum_{k=0}^{6} 2 \cdot (\sqrt{3})^k$

$$2 \cdot \frac{1-\sqrt{3}^7}{1-\sqrt{3}} \approx 125.033321$$

6. Let $a_i$ represent the sequence of even natural numbers $\{2, 4, 6, 8, \dots\}$, and evaluate the following expressions.

a. $\sum_{i=1}^{5} a_i$

$$2 + 4 + 6 + 8 + 10 = 30$$

b. $\sum_{i=1}^{4} a_{2i}$

$$\begin{aligned} a_2 + a_4 + a_6 + a_8 &= 4 + 8 + 12 + 16 \\ &= 40 \end{aligned}$$

c. $\sum_{i=1}^{5} (a_i - 1)$

$$\begin{aligned} (2-1) + (4-1) + (6-1) + (8-1) + (10-1) &= 1 + 3 + 5 + 7 + 9 \\ &= 25 \end{aligned}$$

7. Let $a_i$ represent the sequence of integers giving the yardage gained per rush in a high school football game $\{3, -2, 17, 4, -8, 19, 2, 3, 3, 4, 0, 1, -7\}$.

a. Evaluate $\sum_{i=1}^{13} a_i$. What does this sum represent in the context of the situation?

$$\begin{aligned} \sum_{i=1}^{13} a_i &= 3 + -2 + 17 + 4 + -8 + 19 + 2 + 3 + 3 + 4 + 0 + 1 + -7 \\ &= 56 + -17 \\ &= 39 \end{aligned}$$

*This sum is the total rushing yards.*

b. Evaluate $\frac{\sum_{i=1}^{13} a_i}{13}$. What does this expression represent in the context of the situation?

$$\frac{39}{13} = 3$$

*The average yardage per rush is* 3.

c. In general, if $a_n$ describes any sequence of numbers, what does $\frac{\sum_{i=1}^{n} a_i}{n}$ represent?

*The total divided by the number of numbers is the arithmetic mean or average of the set.*

8. Let $b_n$ represent the sequence given by the following recursive formula: $b_1 = 10, b_n = b_{n-1} \cdot 5$.

a. Write the first 4 terms of this sequence.

$$10, 50, 250, 1250$$

b. Expand the sum $\sum_{i=1}^{4} b_i$. Is it easier to add this series, or is it easier to use the formula for the sum of a finite geometric sequence? Explain your answer. Evaluate $\sum_{i=1}^{4} b_i$.

$$\sum_{i=1}^{4} b_i = 10 + 50 + 250 + 1250$$

*Answers may vary based on personal opinion. Since this series consists of only four terms, it may be easier to simply add the terms together to find the sum. The sum is* 1560.

c. Write an explicit form for $b_n$.

$b_n = 10 \cdot 5^{n-1}$, *where $n$ is a positive integer.*

d. Evaluate $\sum_{i=1}^{10} b_i$.

$$(10) \cdot \left(\frac{1-5^{10}}{1-5}\right) = 10 \cdot \frac{9765624}{4} = 24,414,060$$

9. Consider the sequence given by $a_1 = 20, a_n = \frac{1}{2} \cdot a_{n-1}$.

a. Evaluate $\sum_{i=1}^{10} a_i$, $\sum_{i=1}^{100} a_i$, and $\sum_{i=1}^{1000} a_i$.

$$\sum_{i=1}^{10} a_i = 20 \cdot \frac{1-\left(\frac{1}{2}\right)^{10}}{1-\frac{1}{2}} = 20 \cdot \frac{\frac{1023}{1024}}{\frac{1}{2}} = 39.9609375$$

$$\sum_{i=1}^{100} a_i = 20 \cdot \frac{1-\left(\frac{1}{2}\right)^{100}}{1-\frac{1}{2}} \approx 40$$

$$\sum_{i=1}^{1000} a_i = 20 \cdot \frac{1-\left(\frac{1}{2}\right)^{1000}}{1-\frac{1}{2}} \approx 40$$

b. What value does it appear this series is approaching as $n$ continues to increase? Why might it seem like the series is bounded?

*The series is almost exactly* 40. *In the numerator we are subtracting a number that is incredibly small and gets even smaller the farther we go in the sequence. So, as $n \to \infty$, the sum approaches* $\frac{20}{\frac{1}{2}} = 40$.

10. The sum of a geometric series with 4 terms is 60, and the common ratio is $r = \frac{1}{2}$. Find the first term.

$$60 = a\left(\frac{1-\left(\frac{1}{2}\right)^4}{1-\frac{1}{2}}\right)$$
$$60 = a\left(\frac{1-\frac{1}{16}}{\frac{1}{2}}\right)$$
$$60 = a\left(\frac{15}{16}\cdot 2\right)$$
$$60 = a\left(\frac{15}{8}\right)$$
$$a = 4\cdot 8 = 32$$

11. The sum of the first 4 terms of a geometric series is 203, and the common ratio is 0.4. Find the first term.

$$203 = a\left(\frac{1-0.4^4}{1-0.4}\right)$$
$$a = 203\left(\frac{0.6}{1-0.4^4}\right) = 125$$

12. The third term in a geometric series is $\frac{27}{2}$, and the sixth term is $\frac{729}{16}$. Find the common ratio.

$$ar^2 = \frac{27}{2}$$
$$ar^5 = \frac{729}{16}$$
$$r^3 = \frac{729}{16}\cdot\frac{2}{27} = \frac{27}{8}$$
$$r = \frac{3}{2}$$

13. The second term in a geometric series is 10, and the seventh term is 10240. Find the sum of the first six terms.

$$ar = 10$$
$$ar^6 = 10240$$
$$r^5 = 1024$$
$$r = 4$$
$$a = \frac{10}{4} = \frac{5}{2}$$
$$S_6 = \frac{5}{2}\left(\frac{1-4^6}{1-4}\right)$$
$$= \frac{5}{2}\left(\frac{4095}{3}\right)$$
$$= 3412.5$$

14. **Find the interest earned and the future value of an annuity with monthly payments of $\$200$ for two years into an account that pays $6\%$ interest per year compounded monthly.**

$$A_f = 200\left(\frac{\left(1+\frac{0.06}{12}\right)^{24}-1}{\frac{0.06}{12}}\right)$$
$$\approx 5086.39$$

*The future value is* $\$5086.39$*, and the interest earned is* $\$286.39$.

15. **Find the interest earned and the future value of an annuity with annual payments of $\$1200$ for $15$ years into an account that pays $4\%$ interest per year.**

$$A_f = 1200\left(\frac{(1+0.04)^{15}-1}{0.04}\right)$$
$$\approx 24028.31$$

*The future value is* $\$24028.31$*, and the interest earned is* $\$6028.31$.

16. **Find the interest earned and the future value of an annuity with semiannual payments of $\$1000$ for $20$ years into an account that pays $7\%$ interest per year compounded semiannually.**

$$A_f = 1000\left(\frac{\left(1+\frac{0.07}{2}\right)^{40}-1}{\frac{0.07}{2}}\right)$$
$$\approx 84550.28$$

*The future value is* $\$84550.28$*, and the interest earned is* $\$44550.28$.

17. **Find the interest earned and the future value of an annuity with weekly payments of $\$100$ for three years into an account that pays $5\%$ interest per year compounded weekly.**

$$A_f = 100\left(\frac{\left(1+\frac{0.05}{52}\right)^{156}-1}{\frac{0.05}{52}}\right)$$
$$\approx 16822.05$$

*The future value is* $\$16822.05$*, and the interest earned is* $\$1222.05$.

18. **Find the interest earned and the future value of an annuity with quarterly payments of $\$500$ for $12$ years into an account that pays $3\%$ interest per year compounded quarterly.**

$$A_f = 500\left(\frac{\left(1+\frac{0.03}{4}\right)^{48}-1}{\frac{0.03}{4}}\right)$$
$$\approx 28760.36$$

*The future value is* $\$28760.36$*, and the interest earned is* $\$3760.36$.

19. **How much money should be invested every month with $8\%$ interest per year compounded monthly in order to save up $\$10,000$ in $15$ months?**

$$10000 = R\left(\frac{\left(1+\frac{0.08}{12}\right)^{15}-1}{\frac{0.08}{12}}\right)$$
$$R = 10000\left(\frac{\frac{0.08}{12}}{\left(1+\frac{0.08}{12}\right)^{15}-1}\right)$$
$$\approx 636.11$$

*Invest $\$636.11$ every month for $15$ months at this interest rate to save up $\$10000$.*

20. **How much money should be invested every year with $4\%$ interest per year in order to save up $\$40,000$ in $18$ years?**

$$40000 = R\left(\frac{(1+0.04)^{18}-1}{0.04}\right)$$
$$R = 40000\left(\frac{0.04}{(1.04)^{15}-1}\right)$$
$$\approx 1559.733$$

*Invest $\$1559.74$ every year for $18$ years at $4\%$ interest per year to save up $\$40000$.*

21. **Julian wants to save up to buy a car. He is told that a loan for a car will cost $\$274$ a month for five years, but Julian does not need a car presently. He decides to invest in a structured savings plan for the next three years. Every month Julian invests $\$274$ at an annual interest rate of $2\%$ compounded monthly.**

   a. **How much will Julian have at the end of three years?**

$$A_f = 274\left(\frac{\left(1+\frac{0.02}{12}\right)^{36}-1}{\frac{0.02}{12}}\right) \approx 10157.21$$

   *Julian will have $\$10157.21$ at the end of the three years.*

   b. **What are the benefits of investing in a structured savings plan instead of taking a loan out? What are the drawbacks?**

   *The biggest benefit is that instead of paying interest on a loan, you earn interest on your savings. The drawbacks include that you have to wait to get what you want.*

22. **An *arithmetic series* is a series whose terms form an arithmetic sequence. For example, $2+4+6+\cdots+100$ is an arithmetic series since $2, 4, 6, 8, \ldots, 100$ is an arithmetic sequence with constant difference $2$.**

**The most famous arithmetic series is $1+2+3+4+\cdots+n$ for some positive integer $n$. We studied this series in Algebra I and showed that its sum is $S_n = \frac{n(n+1)}{2}$. It can be shown that the general formula for the sum of an arithmetic series $a+(a+d)+(a+2d)+\cdots+[a+(n-1)d]$ is**

$$S_n = \frac{n}{2}[2a+(n-1)d]$$

**where $a$ is the first term and $d$ is the constant difference.**

   a. **Use the general formula to show that the sum of $1+2+3+\cdots+n$ is $S_n = \frac{n(n+1)}{2}$.**

$$S_n = \frac{n}{2}(2\cdot 1+(n-1)\cdot 1) = \frac{n}{2}(2+n-1) = \frac{n}{2}(n+1)$$

b. **Use the general formula to find the sum of $2 + 4 + 6 + 8 + 10 + \cdots + 100$.**

$$S_n = \frac{50}{2}(4 + (50 - 1)2) = 25(102) = 2550$$

23. **The sum of the first five terms of an arithmetic series is $25$, and the first term is $2$. Find the constant difference.**

$$25 = \frac{5}{2}(2 + a_5)$$
$$10 = 2 + a_5$$
$$a_5 = 8$$

$$8 = 2 + d(4)$$
$$6 = d(4)$$
$$d = \frac{3}{2}$$

24. **The sum of the first nine terms of an arithmetic series is $135$, and the first term is $17$. Find the ninth term.**

$$135 = \frac{9}{2}(17 + a_9)$$
$$30 = 17 + a_9$$
$$13 = a_9$$

$$13 = 17 + d(8)$$
$$-4 = d(8)$$
$$d = -\frac{1}{2}$$

25. **The sum of the first and $100^{\text{th}}$ terms of an arithmetic series is $101$. Find the sum of the first $100$ terms.**

$$S_{100} = \frac{100}{2}(101) = 5050$$

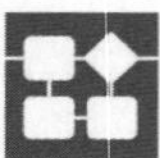

# Lesson 30: Buying a Car

## Student Outcomes

- Students use the sum of a finite geometric series formula to develop a formula to calculate a payment plan for a car loan and use that calculation to derive the present value of an annuity formula.

## Lesson Notes

In this lesson, students will explore the idea of getting a car loan. The lesson extends their knowledge on saving money from the last lesson to the mathematics behind borrowing it. The formula for the monthly payment on a loan is derived using the formula for the sum of a geometric series. Amortization tables are used to help students develop an understanding of borrowing money.

In this lesson, we derive the future amount of an annuity formula again in the context of purchasing a car and use it to understand the present value of an annuity formula,

$$A_p = R \cdot \frac{1-(1+i)^{-n}}{i}.$$

It is helpful to think of the present value of an annuity $A_p$ in the following way: Calculate the future amount of an annuity $A_f$ (as in Lesson 29) to find out the total amount that would be in an account after making all of the payments. Then, use the compound interest formula $F = P(1+i)^n$ from Lesson 26 to compute how much would need to be invested today (i.e., $A_p$) in one single large deposit to equal the amount $A_f$ in the future. More specifically, for an interest rate of $i$ per time period with $n$ payments each of amount $R$, then the present value can be computed (substituting $A_p$ for $P$ and $A_f$ for $F$ in the compound interest formula) to be

$$A_f = A_p(1+i)^n.$$

Using the future amount of an annuity formula and solving for $A_p$ gives

$$A_p = R \cdot \frac{(1+i)^n - 1}{i}(1+i)^{-n},$$

which simplifies to the first formula above. The play between the sum of a geometric series (**A-SSE.B.4**) and the combination of functions to get the new function $A_p$ (**F-BF.A.1b**) constitutes the entirety of the mathematical content of this lesson.

While the mathematics is fairly simple, the context—car loans and the amortization process—is also new to students. To help the car loans process make sense (and loans in general), we have students think about the following situation: Instead of paying the full price of a car immediately, a student asks the dealer to develop a loan payment plan in which the student pays the same amount each month. The car dealer agrees and does the following calculation to determine the amount $R$ that the student should pay each month:

- The car dealer first imagines how much she would have if she took the amount of the loan (i.e., price of the car) and deposited it into an account for 60 months (5 years) at a certain interest rate per month.

- The car dealer then imagines taking the student's payments ($R$ dollars) and depositing them into an account making the same interest rate per month. The final amount is calculated just like calculating the final amount of a structured savings plan from Lesson 29.
- The car dealer then reasons that, to be fair to her and her customer, the two final amounts should be the same—that is, the car dealer should have the same amount in each account at the end of 60 months either way. This sets up the equation above, which can then be solved for $R$.

This lesson is the first lesson where the concept of amortization appears. An example of amortization is the process of decreasing the amount owed on a loan over time, which decreases the amount of interest owed over time as well. This can be thought of as doing an annuity calculation like in Lesson 29 but run backward in time. Whenever possible, use online calculators such as http://www.bankrate.com/calculators/mortgages/amortization-calculator.aspx to generate amortization tables (i.e., tables that show the amount of the principal and interest for each payment). Students have filled in a few amortization tables in Lesson 26 as an application of interest, but the concept was not presented in its entirety.

## Classwork

### Opening Exercise (2 minutes)

The following problem is similar to homework students did in the previous lesson; however, the savings terms are very similar to those found in car loans.

> **Opening Exercise**
>
> **Write a sum to represent the future amount of a structured savings plan (i.e., annuity) if you deposit $\$250$ into an account each month for $5$ years that pays $3.6\%$ interest per year, compounded monthly. Find the future amount of your plan at the end of $3$ years.**
>
> $250(1.003)^{59} + 250(1.003)^{58} + \cdots + 250(1.003) + 250$. *The amount in dollars in the account after $3$ years will be*
>
> $$250 \cdot \frac{(1.003)^{60} - 1}{0.003} \approx 16,407.90.$$

### Example (15 minutes)

Many people take out a loan to purchase a car and then repay the loan on a monthly basis. Announce that we will figure out how banks determine the monthly loan payment for a loan in today's class.

- If you decide to get a car loan, there are many things that you will have to consider. What do you know that goes into getting a loan for a vehicle?
  - *Look for the following: down payment, a monthly payment, interest rates on the loan, number of years of the loan. Explain any of these terms that students may not know.*

For car loans, a down payment is not always required, but a typical down payment is $15\%$ of the total cost of the vehicle. We will assume throughout this example that no down payment is required.

This example is a series of problems to work through with your students that guides students through the process for finding the recurring monthly payment for a car loan described in the teacher notes. After the example, students will be given more information on buying a car and will calculate the monthly payment for a car that they researched on the Internet as part of their homework in Lesson 29.

**Example**

**Jack wanted to buy a $\$9,000$ 2-door sports coupe but could not pay the full price of the car all at once. He asked the car dealer if she could give him a loan where he paid a monthly payment. She told him she could give him a loan for the price of the car at an annual interest rate of $3.6\%$ compounded monthly for $60$ months ($5$ years).**

**The problems below exhibit how Jack's car dealer used the information above to figure out how much his monthly payment of $R$ dollars per month should be.**

a. **First, the car dealer imagined how much she would have in an account if she deposited $\$9,000$ into the account and left it there for $60$ months at an annual interest rate of $3.6\%$ compounded monthly. Use the compound interest formula $F = P(1+i)^n$ to calculate how much she would have in that account after $5$ years. This is the amount she would have in the account after $5$ years if Jack gave her $\$9,000$ for the car, and she immediately deposited it.**

$F = 9000(1 + 0.003)^{60} = 9000(1.003)^{60} \approx 10,772.05$. ***At the end of $60$ months, she would have $\$10,772.05$ in the account.***

b. **Next, she figured out how much would be in an account after $5$ years if she took each of Jack's payments of $R$ dollars and deposited it into a bank that earned $3.6\%$ per year (compounded monthly). Write a sum to represent the future amount of money that would be in the annuity after $5$ years in terms of $R$, and use the sum of a geometric series formula to rewrite that sum as an algebraic expression.**

***This is like the structured savings plan in Lesson 29. The future amount of money in the account after $5$ years can be represented as***

$$R(1.003)^{59} + R(1.003)^{58} + \cdots + R(1.003) + R.$$

***Applying the sum of a geometric series formula***

$$S_n = a \cdot \frac{1-r^n}{1-r}$$

***to the geometric series above using $a = R$, $r = 1.003$, and $n = 60$, one gets***

$$S_n = R \cdot \frac{1-(1.003)^{60}}{1-1.003} = R \cdot \frac{(1.003)^{60}-1}{0.003}.$$

At this point, we have re-derived the future amount of an annuity formula. Point this out to your students! Help them to see the connection between what they are doing in this context with what they did in Lesson 29. The future value formula is

$$A_f = R \cdot \frac{(1+i)^n - 1}{i}.$$

c. **The car dealer then reasoned that, to be fair to her and Jack, the two final amounts in both accounts should be the same—that is, she should have the same amount in each account at the end of $60$ months either way. Write an equation in the variable $R$ that represents this equality.**

$$9000(1.003)^{60} = R \cdot \frac{(1.003)^{60}-1}{0.003}$$

> **d. She then solved her equation to get the amount $R$ that Jack would have to pay monthly. Solve the equation in part (c) to find out how much Jack needed to pay each month.**
>
> ***Solving for $R$ in the equation above, we get***
>
> $$R = 9000 \cdot (1.003)^{60} \cdot \frac{0.003}{(1.003)^{60} - 1} \approx 164.13.$$
>
> ***Thus, Jack will need to make regular payments of $\$164.13$ a month for $60$ months.***

Ask students questions to see if they understand what the \$164.13 means. For example, if Jack decided not to buy the car and instead deposited \$164.13 a month into an account earning 3.6% interest compounded monthly, how much will he have at the end of 60 months? Students should be able to answer \$10,772.05, the final amount of the annuity that the car dealer calculated in part (a) (or (b)). Your goal is to help them see that both ways of calculating the future amount should be equal.

## Discussion (10 minutes)

In this discussion, students are lead to the present value of an annuity formula using the calculations they just did in the example (**F-BF.A.1b**).

- Let's do the calculations in part (a) of the example again but this time using $A_p$ for the loan amount (the present value of an annuity), $i$ for the interest rate per time period, $n$ to be the number of time periods. As in part (a), what is the future value of $A_p$ if it is deposited in an account with an interest rate of $i$ per time period for $n$ compounding periods?
  - $F = A_p(1+i)^n$
- As in part (b) of the example above, what is the future value of an annuity $A_f$ in terms of the recurring payment $R$, interest rate $i$, and number of periods $n$?
  - $A_f = R \cdot \frac{1-(1+i)^n}{i}$
- If we assume (as in the example above) that both methods produce the same future value, we can equate $F = A_f$ and write the following equation:

$$A_p(1+i)^n = R \cdot \frac{(1+i)^n - 1}{i}.$$

- What equation is this in example above?
  - *The equation derived in part (c).*
- We can now solve this equation for $R$ as we did in the example, but it is more common in finance to solve for $A_p$ by multiplying both sides by $(1+i)^{-n}$:

$$A_p = R \cdot \frac{(1+i)^n - 1}{i} \cdot (1+i)^{-n},$$

  and then distributing it through the binomial to get the *present value of an annuity* formula:

$$A_p = R \cdot \frac{1-(1+i)^{-n}}{i}.$$

- When a bank (or a car dealer) makes a loan that is to be repaid with recurring payments $R$, then the payments form an annuity whose present value $A_p$ is the amount of the loan. Thus, we can use this formula to find the payment amount $R$ given the size of the loan $A_p$ (as in Example 1), or we can find the size of the loan $A_p$ if we know the size of the payments $R$.

### Exercise (3 minutes)

**Exercise**

**A college student wants to buy a car and can afford to pay $\$200$ per month. If she plans to take out a loan at $6\%$ interest per year with a recurring payment of $\$200$ per month for four years, what price car can she buy?**

$$A_p = 200 \cdot \frac{1-(1.005)^{-48}}{0.005} \approx 8,516.06$$

*She can afford to take out a* $\$8,516.06$ *loan. If she has no money for a down payment, she can afford a car that is about* $\$8,500$.

You might want to point out to your students that the present value formula can always be easily and quickly derived from the future amount of annuity formula $A_f = R \cdot \frac{1-(1+i)^n}{i}$ and the compound interest formula $A_f = A_p(1+i)^n$ (using the variables $A_f$ and $A_p$ instead of $F$ and $P$).

### Mathematical Modeling Exercise (8 minutes)

The customization and open-endedness of this challenge depends upon how successful students were in researching the price of a potential car in the Problem Set to Lesson 29. For students who did not find a car, you can have them use the list provided below. After the challenge, there are some suggestions for ways to introduce other modeling elements into the challenge. Use the suggestions as you see fit. The solutions throughout this section are based on the 2007 two-door small coupe.

MP.2 & MP.4

**Mathematical Modeling Exercise**

**In the Problem Set of Lesson 29, you researched the price of a car that you might like to own. In this exercise, you will determine how much a car payment would be for that price for different loan options.**

**If you did not find a suitable car, select a car and selling price from the list below:**

| Car | Selling Price |
|---|---|
| 2005 Pickup Truck | $\$9,000$ |
| 2007 Two-Door Small Coupe | $\$7,500$ |
| 2003 Two-Door Luxury Coupe | $\$10,000$ |
| 2006 Small SUV | $\$8,000$ |
| 2008 Four-Door Sedan | $\$8,500$ |

a. **When you buy a car, you must pay sales tax and licensing and other fees. Assume that sales tax is $6\%$ of the selling price and estimated license/title/fees will be $2\%$ of the selling price. If you put a $\$1,000$ down payment on your car, how much money will you need to borrow to pay for the car and taxes and other fees?**

*Answers will vary. For the* $2007$ *two-door small coupe:* $7500 + 7500(0.06) + 7500(0.02) - 1000 = 7100$

*You would have to borrow* $\$7,100$.

*Scaffolding:*

For English Language Learners, provide a visual image of each vehicle type along with a specific make and model.

- Pickup Truck
- 2-Door Small Coupe
- 2-Door Luxury Coupe
- Small SUV
- 4-Door Sedan

b. **Using the loan amount you computed above, calculate the monthly payment for the different loan options shown below:**

| Loan 1 | 36-month loan at $2\%$ |
|---|---|
| Loan 2 | 48-month loan at $3\%$ |
| Loan 3 | 60-month loan at $5\%$ |

*Answers will vary. For the $2007$ two-door small coupe:*

*Loan 1:* $7100 = R \cdot \frac{1-\left(1+\frac{0.02}{12}\right)^{-36}}{0.\frac{02}{12}}$*; therefore,* $R \approx 203.36$*. The monthly payment would be* $\$203.36$*.*

*Loan 2:* $7100 = R \cdot \frac{1-\left(1+\frac{0.03}{12}\right)^{-48}}{0.\frac{03}{12}}$*; therefore,* $R \approx 157.15$*. The monthly payment would be* $\$157.15$*.*

*Loan 3:* $7100 = R \cdot \frac{1-\left(1+\frac{0.05}{12}\right)^{-60}}{0.\frac{05}{12}}$*; therefore,* $R \approx 133.99$*. The monthly payment would be* $\$133.99$*.*

c. **Which plan, if any, will keep your monthly payment under $\$175$? Of the plans under $\$175$ per month, why might you choose a plan with fewer months even though it costs more per month?**

***Answers will vary. Loan 2 and Loan 3 are both under $\$175$ a month. When the monthly payments are close (like Loan 2 and Loan 3), the fewer payments you make with Loan 2 means you pay less overall for that loan.***

If a student found a dealer that offered a loan for the car they were researching, encourage them to do the calculations above for terms of that loan. (Call it loan option 4.)

## Further Modeling Resources

If students are interested in the actual details of purchasing and budgeting for a car, the following websites can be referenced for further exploration.

Vehicle Fees: http://www.dmv.org/ny-new-york/car-registration.php

Inspection: http://dmv.ny.gov/forms/vs77.pdf

Car Maintenance: http://www.edmunds.com/tco.html ($2009$ and newer models only) or http://www.edmunds.com/calculators/

Car Insurance: http://dmv.ny.gov/insurance/looking-insurance-information

Furthermore, you can ask:

- What sort of extra fees go into buying a car?
  - *Answers may vary, but students should be able to come up with sales tax, which is $4\%$ in New York State. Additional sales taxes may also apply for your local jursidiction. Other fees include license plate ($\$25$) and title ($\$50$). An inspection is also required within $10$ days of purchase ($\$10$). Other states may have additional fees.*
- What extra costs go into maintaining a car?
  - *Insurance, repairs, maintenance like oil changes, car washing/detailing, etc. For used and new cars, use an online calculator to estimate car insurance and maintenance costs as well as likely depreciation and interest costs for a loan.*

## Closing (2 minutes)

Close this lesson by asking students to summarize in writing or with a partner what they know so far about borrowing money to buy a car.

- Based on the work you did in this lesson, summarize what you know so far about borrowing money to buy a car.
  - *Making the loan term longer does make the monthly payment go down but causes the total interest paid to go up. Interest rates, down payment, and total length of the loan all affect the monthly payment. In the end, the amount of the loan you get depends on what you can afford to pay per month based on your budget.*

**Lesson Summary**

**The total cost of car ownership includes many different costs in addition to the selling price, such as sales tax, insurance, fees, maintenance, interest on loans, gasoline, etc.**

**The present value of an annuity formula can be used to calculate monthly loan payments given a total amount borrowed, the length of the loan, and the interest rate. The present value $A_p$ (i.e., loan amount) of an annuity consisting of $n$ recurring equal payments of size $R$ and interest rate $i$ per time period is**

$$A_p = R \cdot \frac{1-(1+i)^{-n}}{i}.$$

**Amortization tables and online loan calculators can also help you plan for buying a car.**

**The amount of your monthly payment depends on the interest rate, the down payment, and the length of the loan.**

## Exit Ticket (5 minutes)

Name ______________________________ Date ____________

# Lesson 30: Buying a Car

## Exit Ticket

Fran wants to purchase a new boat. She starts looking for a boat around $6,000. Fran creates a budget and thinks that she can afford $250 every month for 2 years. Her bank charges her 5% interest per year, compounded monthly.

1. What is the actual monthly payment for Fran's loan?

2. If Fran can only pay $250 per month, what is the most expensive boat she can buy without a down payment?

## Exit Ticket Sample Solutions

Fran wants to purchase a new boat. She starts looking for a boat around $\$6,000$. Fran creates a budget and thinks that she can afford $\$250$ every month for $2$ years. Her bank charges her $5\%$ interest per year, compounded monthly.

1. What is the actual monthly payment for Fran's loan?

$$6000 = R\left(\frac{1-\left(1+\frac{0.05}{12}\right)^{-24}}{\frac{0.05}{12}}\right)$$

$$R = 6000\left(\frac{\left(\frac{0.05}{12}\right)}{1-\left(1+\frac{0.05}{12}\right)^{-24}}\right)$$

$$R \approx \$263.23$$

2. If Fran can only pay $\$250$ per month, what is the most expensive boat she can buy without a down payment?

$$P = 250\left(\frac{1-\left(1+\frac{0.05}{12}\right)^{-24}}{\frac{0.05}{12}}\right)$$

$$P \approx \$5698.47$$

*Fran can afford a boat that costs about $\$5,700$ if she does not have a down payment.*

## Problem Set Sample Solutions

1. Benji is $24$ years old and plans to drive his new car about $200$ miles per week. He has qualified for first-time buyer financing, which is a $60$-month loan with $0\%$ down at an interest rate of $4\%$. Use the information below to estimate the monthly cost of each vehicle.

CAR A: $2010$ Pickup Truck for $\$12,000$, $22$ miles per gallon

CAR B: $2006$ Luxury Coupe for $\$11,000$, $25$ miles per gallon

Gasoline: $\$4.00$ per gallon New vehicle fees: $\$80$ Sales Tax: $4.25\%$

Maintenance Costs:

($2010$ model year or newer): $10\%$ of purchase price annually

($2009$ model year or older): $20\%$ of purchase price annually

Insurance:

| Average Rate Ages $25$–$29$ | $\$100$ per month |
|---|---|
| If you are male<br>If you are female | Add $\$10$ per month<br>Subtract $\$10$ per month |
| Type of Car<br>Pickup Truck<br>Small Two-Door Coupe or Four-Door Sedan<br>Luxury Two- or Four-Door Coupe | <br>Subtract $\$10$ per month<br>Subtract $\$10$ per month<br>Add $\$15$ per month |
| Ages $18$–$25$ | Double the monthly cost |

a. **How much money will Benji have to borrow to purchase each car?**

*$\$12,000$ for the truck and $\$11,000$ for the coupe.*

b. **What is the monthly payment for each car?**

$$12000 = R\left(\frac{1-\left(1+\frac{0.04}{12}\right)^{-60}}{\frac{0.04}{12}}\right)$$

$$R = 12000\left(\frac{\left(\frac{0.04}{12}\right)}{1-\left(1+\frac{0.04}{12}\right)^{-60}}\right)$$

$$R \approx 221.00$$

*The truck would cost $\$221.00$ every month.*

$$11000 = R\left(\frac{1-\left(1+\frac{0.04}{12}\right)^{-60}}{\frac{0.04}{12}}\right)$$

$$R = 11000\left(\frac{\left(\frac{0.04}{12}\right)}{1-\left(1+\frac{0.04}{12}\right)^{-60}}\right)$$

$$R \approx 202.58$$

*The coupe would cost $\$202.58$ every month.*

c. **What are the annual maintenance costs and insurance costs for each car?**

*Truck: $10\% \cdot 12000 = 1200$ for the maintenance. Insurance will vary based on the gender of student. Male students will be $\$200$ per month or $\$2400$ per year, while female students will be $\$160$ per month or $\$1,920$ per year.*

*Car: $20\% \cdot 11000 = 2200$ for maintenance. Male students will cost $\$250$ per month or $\$3,000$ per year, while female students will cost $\$210$ per month or $\$2,520$ per year.*

d. **Which car should Benji purchase? Explain your choice.**

*Answers will vary depending on personal preference and experience, as well as financial backgrounds. Answers should be supported using the mathematics of parts (a), (b), and (c).*

2. **Use the total initial cost of buying your car from the lesson to calculate the monthly payment for the following loan options.**

| Option | Number of Months | Down Payment | Interest Rate | Monthly Payment |
|---|---|---|---|---|
| Option A | 48 months | $\$0$ | $2.5\%$ | $\$175.31$ |
| Option B | 60 months | $\$500$ | $3.0\%$ | $\$134.77$ |
| Option C | 60 months | $\$0$ | $4.0\%$ | $\$147.33$ |
| Option D | 36 months | $\$1,000$ | $0.9\%$ | $\$197.15$ |

*Answers will vary. Suggested answers assume an $\$8,000$ car.*

a. **For each option, what is the total amount of money you will pay for your vehicle over the life of the loan?**

*Option A:* $175.31 \cdot 48 = \$8414.88$

*Option B:* $500 + 134.77 \cdot 60 = \$8586.20$

*Option C:* $147.33 \cdot 60 = \$8839.80$

*Option D:* $1000 + 197.15 \cdot 36 = \$8097.40$

b. **Which option would you choose? Justify your reasoning.**

*Answers will vary. Option B is the cheapest per month but requires a down payment. Of the plans without down payments, Option A saves the most money in the end, but Option C is cheaper per month. Option D saves the most money long term but requires the largest down payment and the largest monthly payment.*

3. **Many lending institutions will allow you to pay additional money toward the principal of your loan every month. The table below shows the monthly payment for an $\$8,000$ loan using Option A above if you pay an additional $\$25$ per month.**

| Month/ Year | Payment | Principal Paid | Interest Paid | Total Interest | Balance |
|---|---|---|---|---|---|
| Aug. 2014 | $ 200.31 | $ 183.65 | $ 16.67 | $ 16.67 | $ 7,816.35 |
| Sept. 2014 | $ 200.31 | $ 184.03 | $ 16.28 | $ 32.95 | $ 7,632.33 |
| Oct. 2014 | $ 200.31 | $ 184.41 | $ 15.90 | $ 48.85 | $ 7,447.91 |
| Nov. 2014 | $ 200.31 | $ 184.80 | $ 15.52 | $ 64.37 | $ 7,263.12 |
| Dec. 2014 | $ 200.31 | $ 185.18 | $ 15.13 | $ 79.50 | $ 7,077.94 |
| Jan. 2015 | $ 200.31 | $ 185.57 | $ 14.75 | $ 94.25 | $ 6,892.37 |
| Feb. 2015 | $ 200.31 | $ 185.95 | $ 14.36 | $ 108.60 | $ 6,706.42 |
| Mar. 2015 | $ 200.31 | $ 186.34 | $ 13.97 | $ 122.58 | $ 6,520.08 |
| April 2015 | $ 200.31 | $ 186.73 | $ 13.58 | $ 136.16 | $ 6,333.35 |
| May 2015 | $ 200.31 | $ 187.12 | $ 13.19 | $ 149.35 | $ 6,146.23 |
| June 2015 | $ 200.31 | $ 187.51 | $ 12.80 | $ 162.16 | $ 5,958.72 |
| July 2015 | $ 200.31 | $ 187.90 | $ 12.41 | $ 174.57 | $ 5,770.83 |
| Aug. 2015 | $ 200.31 | $ 188.29 | $ 12.02 | $ 186.60 | $ 5,582.54 |
| Sept. 2015 | $ 200.31 | $ 188.68 | $ 11.63 | $ 198.23 | $ 5,393.85 |
| Oct. 2015 | $ 200.31 | $ 189.08 | $ 11.24 | $ 209.46 | $ 5,204.78 |
| Nov. 2015 | $ 200.31 | $ 189.47 | $ 10.84 | $ 220.31 | $ 5,015.31 |
| Dec. 2015 | $ 200.31 | $ 189.86 | $ 10.45 | $ 230.75 | $ 4,825.45 |

**Note: The months from January 2016 to December 2016 are not shown.**

| | | | | | |
|---|---|---|---|---|---|
| Jan. 2017 | $ 200.31 | $ 195.07 | $ 5.24 | $ 330.29 | $ 2,320.92 |
| Feb. 2017 | $ 200.31 | $ 195.48 | $ 4.84 | $ 335.12 | $ 2,125.44 |
| Mar. 2017 | $ 200.31 | $ 195.88 | $ 4.43 | $ 339.55 | $ 1,929.56 |
| April 2017 | $ 200.31 | $ 196.29 | $ 4.02 | $ 343.57 | $ 1,733.27 |
| May 2017 | $ 200.31 | $ 196.70 | $ 3.61 | $ 347.18 | $ 1,536.57 |
| June 2017 | $ 200.31 | $ 197.11 | $ 3.20 | $ 350.38 | $ 1,339.45 |
| July 2017 | $ 200.31 | $ 197.52 | $ 2.79 | $ 353.17 | $ 1,141.93 |
| Aug. 2017 | $ 200.31 | $ 197.93 | $ 2.38 | $ 355.55 | $ 944.00 |
| Sept. 2017 | $ 200.31 | $ 198.35 | $ 1.97 | $ 357.52 | $ 745.65 |
| Oct. 2017 | $ 200.31 | $ 198.76 | $ 1.55 | $ 359.07 | $ 546.90 |
| Nov. 2017 | $ 200.31 | $ 199.17 | $ 1.14 | $ 360.21 | $ 347.72 |
| Dec. 2017 | $ 200.31 | $ 199.59 | $ 0.72 | $ 360.94 | $ 148.13 |
| Jan. 2018 | $ 148.44 | $ 148.13 | $ 0.31 | $ 361.25 | $ 0.00 |

**How much money would you save over the life of an $\$8,000$ loan using Option A if you paid an extra $\$25$ per month compared to the same loan without the extra payment toward the principal?**

***Using Option A without paying extra toward the principal each month is a monthly payment of*** $\$175.31$***. The total amount you will pay is*** $\$8,414.88$***. If you pay the extra*** $\$25$ ***per month, you make*** $41$ ***payments of*** $\$200.31$ ***and a final payment of*** $\$148.44$ ***for a total amount of*** $\$8,361.15$***. You would save*** $\$53.73$***.***

4. **Suppose you can afford only $\$200$ a month in car payments, and your best loan option is a $60$-month loan at $3\%$. How much money could you spend on a car? That is, calculate the present value of the loan with these conditions.**

$$P = 200\left(\frac{1-\left(1+\frac{0.03}{12}\right)^{-60}}{\frac{0.03}{12}}\right)$$

$$P \approx 11130.47$$

***You can afford a loan of about*** $\$11,000$***. If there is no down payment, then the car would need to cost about*** $\$11,000$***.***

5. **Would it make sense for you to pay an additional amount per month toward your car loan? Use an online loan calculator to support your reasoning.**

***While pre-paying on a loan can save you money for a relatively short-term loan like a vehicle loan, there is usually not a significant cost savings. Most students will probably elect to pocket the extra monthly costs and pay slightly more over the life of the loan. One option is paying off a loan early. That can save you more money and can be explored online as an extension question for advanced learners.***

6. **What is the sum of each series?**

   a. $900 + 900(1.01)^1 + 900(1.01)^2 + \cdots 900(1.01)^{59}$

$$900\left(\frac{1-(1.01)^{60}}{1-1.01}\right) \approx 73502.703$$

b. $\sum_{n=0}^{47} 15{,}000\left(1+\frac{0.04}{12}\right)^n$

$$\sum_{n=0}^{47} 15{,}000\left(1+\frac{0.04}{12}\right)^n = 15000\left(\frac{1-\left(1+\frac{0.04}{12}\right)^{48}}{1-\left(1+\frac{0.04}{12}\right)}\right)$$

$$= 15000\left(\frac{\left(1+\frac{0.04}{12}\right)^{48}-1}{\frac{0.04}{12}}\right)$$

$$\approx \$779394.015$$

7. Gerald wants to borrow $\$12{,}000$ in order to buy an engagement ring. He wants to repay the loan by making monthly installments for two years. If the interest rate on this loan is $9\frac{1}{2}\%$ per year, compounded monthly, what is the amount of each payment?

$$12000 = R\left(\frac{1-\left(1+\frac{0.095}{12}\right)^{-24}}{\frac{0.095}{12}}\right)$$

$$R = 12000\left(\frac{\left(\frac{0.095}{12}\right)}{1-\left(1+\frac{0.95}{12}\right)^{-24}}\right)$$

$$R \approx 550.97$$

*Gerald will need to pay* $\$550.97$ *each month.*

8. Ivan plans to surprise his family with a new pool using his Christmas bonus of $\$4{,}200$ as a down payment. If the price of the pool is $\$9{,}500$ and Ivan can finance it at an interest rate of $2\frac{7}{8}\%$ per year, compounded quarterly, how long is the loan for if he pays $\$285.45$ per quarter?

$$5300 = 285.45\left(\frac{1-\left(1+\frac{0.02875}{4}\right)^{-n}}{\frac{0.02875}{4}}\right)$$

$$\frac{5300}{285.45}\cdot\frac{0.02875}{4} = 1-\left(1+\frac{0.02875}{4}\right)^{-n}$$

$$\left(1+\frac{0.02875}{4}\right)^{-n} = 1-\frac{5300}{285.45}\cdot\frac{0.02875}{4}$$

$$-n\cdot\ln\left(1+\frac{0.02875}{4}\right) = \ln\left(1-\frac{5300}{285.45}\cdot\frac{0.02875}{4}\right)$$

$$n = -\frac{\ln\left(1-\frac{5300}{285.45}\cdot\frac{0.02875}{4}\right)}{\ln\left(1+\frac{0.02875}{4}\right)}$$

$$n \approx 20$$

*It will take Ivan 20 quarters, or five years, to pay off the pool at this rate.*

9. **Jenny wants to buy a car by making payments of $\$120$ per month for three years. The dealer tells her that she will need to put a down payment of $\$3,000$ on the car in order to get a loan with those terms at a $9\%$ interest rate per year, compounded monthly. How much is the car that Jenny wants to buy?**

$$P - 3000 = 120\left(\frac{1-\left(1+\frac{0.09}{12}\right)^{-36}}{\frac{0.09}{12}}\right)$$

$$P \approx 3773.62 + 3000$$

*The car Jenny wants to buy is about* $\$6,773.62$.

10. **Kelsey wants to refinish the floors in her house and estimates that it will cost $\$39,000$ to do so. She plans to finance the entire amount at $3\frac{1}{4}\%$ interest per year, compounded monthly for 10 years. How much is her monthly payment?**

$$39000 = R\left(\frac{1-\left(1+\frac{0.0325}{12}\right)^{-120}}{\frac{0.0325}{12}}\right)$$

$$R = 39000\left(\frac{\left(\frac{0.0325}{12}\right)}{1-\left(1+\frac{0.0325}{12}\right)^{-120}}\right)$$

$$R \approx 381.10$$

*Kelsey will have to pay* $\$381.10$ *every month.*

11. **Lawrence coaches little league baseball and needs to purchase all new equipment for his team. He has $\$489$ in donations, and the team's sponsor will take out a loan at $4\frac{1}{2}\%$ interest per year, compounded monthly for one year, paying up to $\$95$ per month. What is the most that Lawrence can purchase using the donations and loan?**

$$P - 489 = 95\left(\frac{1-\left(1+\frac{0.045}{12}\right)^{-12}}{\frac{0.045}{12}}\right)$$

$$P \approx 489 + 1112.69$$

*The team will have access to* $\$1,601.69$.

# Lesson 31: Credit Cards

## Student Outcomes

- Students will compare payment strategies for a decreasing credit card balance.
- Students will apply the sum of a finite geometric series formula to a decreasing balance on a credit card.

## Lesson Notes

This lesson develops the necessary tools and terminology to analyze the mathematics behind credit cards and other unsecured loans. Credit cards can provide flexibility to budgets, but they must be carefully managed to avoid the pitfalls of bad credit. For young adults, credit card interest rates can be expected to be between $19.99\%$ and $29.99\%$ per year ($29.99\%$ is currently the maximum allowable interest rate by federal law). Adults with established credit can be offered interest rates around $8\%$ to $14\%$. The credit limit for a first credit card is typically around $\$500$, but these limits quickly increase with a history of timely payments.

In this modeling lesson, students explore the mathematics behind calculating the monthly balance on a single credit card purchase and recognize that the decreasing balance can be modeled by the sum of a finite geometric series (**A-SSE.B.4**). We are intentionally keeping the use of rotating credit such as credit cards simple in this lesson. The students make one charge of $\$1,500$ on this hypothetical credit card and pay down the balance without making any additional charges. With this simple example, we can realistically ignore the fact that the interest on a credit card is charged based on the average daily balance of the account; in our example, the daily balance only changes once per month when the payment is made.

The students will need to recall the following definitions from Lesson 29:

- **SERIES:** Let $a_1, a_2, a_3, a_4, \ldots$ be a sequence of numbers. A sum of the form

  $$a_1 + a_2 + a_3 + \cdots + a_n$$

  for some positive integer $n$ is called a *series* (or *finite series*) and is denoted $S_n$. The $a_i$'s are called the *terms* of the series. The number $S_n$ that the series adds to is called the *sum* of the series.

- **GEOMETRIC SERIES:** A *geometric series* is a series whose terms form a geometric sequence.

  The sum $S_n$ of the first $n$ terms of the finite geometric series $S_n = a + ar + \cdots + ar^{n-1}$ (when $r \neq 1$) is given by

  $$S_n = a\frac{1 - r^n}{1 - r}.$$

  The sum formula of a geometric series can be written in summation notation as

  $$\sum_{k=0}^{n-1} ar^k = a \cdot \frac{1 - r^n}{1 - r}.$$

## Classwork

### Opening (3 minutes)

Assign students to small groups, and keep them in the same groups throughout this lesson. In the first mathematical modeling exercise, all groups will be working on the same problem, but in the second mathematical modeling exercise, the groups will be assigned one of three different payment schemes to investigate.

- In the previous lesson, you investigated the mathematics needed for a car loan. What if you have decided to buy a car, but you have not saved up enough money for the down payment? If you are buying through a dealership, it is possible to put the down payment onto a credit card. For today's lesson, you are going to investigate the finances of charging $\$1{,}500$ onto a credit card for the down payment on your car. We will investigate different payment plans and how much you end up paying in total using each plan.
- The annual interest rates on a credit card for people who have not used credit in the past tend to be much higher than for adults with established good credit, ranging between $14.99\%$ and $29.99\%$, which is the maximum interest rate allowed by law. Throughout this lesson, we will use a $19.99\%$ annual interest rate, and you will explore problems with other interest rates in your Problem Set.
- One of the differences between a credit card and a loan is that you can pay as much as you want toward your credit card balance, as long as it is at least the amount of the "minimum payment," which is determined by the lender. In many cases, the minimum payment is the sum of the interest that has accrued over the month and $1\%$ of the outstanding balance, or $\$25$, whichever is greater.
- Another difference between a credit card and a loan is that a loan has a fixed term of repayment—you pay it off over an agreed-upon length of time such as five years—and that there is no fixed term of repayment for a credit card. You can pay it off as quickly as you like by making large payments, or you can pay less and owe money for a longer period of time. In the mathematical modeling exercise, we will investigate the scenario of paying a fixed monthly payment of various sizes toward a credit card balance of $\$1{,}500$.

### Mathematical Modeling Exercise (25 minutes)

In this exercise, students will model the repayment of a single charge of $\$1{,}500$ to a credit card that charges $19.99\%$ annual interest. Before beginning the Mathematical Modeling Exercise, assign students to small groups, and assign groups to be either part of the $50$-team, $100$-team, or $150$-team. The groups in each of the three teams will investigate how long it takes to pay down the $\$1{,}500$ balance making fixed payments of either $\$50$, $\$100$, or $\$150$ each month.

As you circulate the room while students are working, take note of groups that are working well together on this set of problems. Select at least one group on each team to present their work at the end of the exercise period.

> **Mathematical Modeling Exercise**
>
> **You have charged $\$1,500$ for the down payment on your car to a credit card that charges $19.99\%$ annual interest, and you plan to pay a fixed amount toward this debt each month until it is paid off. We will denote the balance owed after the $n^{\text{th}}$ payment has been made as $b_n$.**
>
> a. **What is the monthly interest rate, $i$? Approximate $i$ to $5$ decimal places.**
>
> $$i = \frac{0.1999}{12} \approx 0.01666$$

*Scaffolding:*

- For struggling students, use an interest rate of $24.00\%$ so that $i = 0.02$ and $r = 1.02$.

b. You have been assigned to either the 50-team, the 100-team, or the 150-team, where the number indicates the size of the monthly payment $R$ you will make toward your debt. What is your value of $R$?

*Students will answer* $50$, $100$, *or* $150$ *as appropriate.*

c. Remember that you can make any size payment toward a credit card debt, as long as it is at least the minimum payment specified by the lender. Your lender calculates the minimum payment as the sum of $1\%$ of the outstanding balance and the total interest that has accrued over the month or $\$25$, whichever is greater. Under these stipulations, what is the minimum payment? Is your monthly payment $R$ at least as large as the minimum payment?

*The minimum payment is* $0.01(\$1500) + 0.01666(\$1500) = \$39.99$. *All given values of* $R$ *are greater than the minimum payment.*

d. Complete the following table to show 6 months of payments.

| Month, $n$ | Interest Due | Payment, $R$ | Paid to Principal | Balance, $b_n$ |
|---|---|---|---|---|
| 0 | ——— | ——— | ——— | 1,500.00 |
| 1 | 24.99 | 50 | 25.01 | 1,474.99 |
| 2 | 24.57 | 50 | 25.43 | 1,449.56 |
| 3 | 24.15 | 50 | 25.85 | 1,423.71 |
| 4 | 23.72 | 50 | 26.28 | 1,397.43 |
| 5 | 23.28 | 50 | 26.72 | 1,370.71 |
| 6 | 22.83 | 50 | 27.17 | 1,343.54 |

| Month, $n$ | Interest Due | Payment, $R$ | Paid to Principal | Balance, $b_n$ |
|---|---|---|---|---|
| 0 | ——— | ——— | ——— | 1,500.00 |
| 1 | 24.99 | 100 | 75.01 | 1,424.99 |
| 2 | 23.74 | 100 | 76.26 | 1,348.73 |
| 3 | 22.47 | 100 | 77.53 | 1,271.20 |
| 4 | 21.18 | 100 | 78.82 | 1,192.38 |
| 5 | 19.86 | 100 | 80.14 | 1,112.24 |
| 6 | 18.53 | 100 | 81.47 | 1,030.77 |

| Month, $n$ | Interest Due | Payment, $R$ | Paid to Principal | Balance, $b_n$ |
|---|---|---|---|---|
| 0 | ——— | ——— | ——— | 1,500.00 |
| 1 | 24.99 | 150 | 125.01 | 1,374.99 |
| 2 | 22.91 | 150 | 127.09 | 1,247.90 |
| 3 | 20.79 | 150 | 129.21 | 1,118.69 |
| 4 | 18.64 | 150 | 131.36 | 987.33 |
| 5 | 16.45 | 150 | 133.55 | 853.78 |
| 6 | 14.22 | 150 | 135.78 | 718.00 |

e. Write a recursive formula for the balance $b_n$ in month $n$ in terms of the balance $b_{n-1}$.

*To calculate the new balance,* $b_n$, *we compound interest for one month on the previous balance* $b_{n-1}$ *and then subtract the payment* $R$:

$$b_n = b_{n-1}(1+i) - R, \text{ with } b_0 = 1500.$$

f. **Write an explicit formula for the balance $b_n$ in month $n$, leaving the *expression* $1+i$ in symbolic form.**

***We have the following formulas:***

$$b_1 = b_0(1+i) - R$$
$$b_2 = b_1(1+i) - R$$
$$= [b_0(1+i) - R](1+i) - R$$
$$= b_0(1+i)^2 - R(1+i) - R$$
$$b_3 = b_2(1+i) - R$$
$$= [b_0(1+i)^2 - R(1+i) - R](1+i) - R$$
$$= b_0(1+i)^3 - R(1+i)^2 - R(1+i) - R$$
$$\vdots\ b_n = b_0(1+i)^n - R(1+i)^{n-1} - R(1+i)^{n-2} - \cdots - R(1+i) - R$$

g. **Rewrite your formula in part (f) using $r$ to represent the quantity $(1+i)$.**

$$b_n = b_0r^n - Rr^{n-1} - Rr^{n-2} - \cdots - Rr - R$$
$$= b_0r^n - R(1+r+r^2+\cdots+r^{n-1})$$

h. **What can you say about your formula in part (g)? What term do we use to describe $r$ in this formula?**

***The formula in part (g) contains the sum of a finite geometric series with common ratio $r$.***

i. **Write your formula from part (g) in summation notation using $\Sigma$.**

$$b_n = b_0r^n - R(1+r+r^2+\cdots+r^{n-1})$$
$$= b_0r^n - R\sum_{k=0}^{n-1} r^k$$

*Scaffolding:*
- Ask advanced learners to develop a generic formula for the balance $b_n$ in terms of the payment amount R and the growth factor $r$.

j. **Apply the appropriate formula from Lesson 29 to rewrite your formula from part (g).**

***Using the sum of a finite geometric series formula,***

$$b_n = b_0r^n - R(1+r+r^2+\cdots+r^{n-1})$$
$$= b_0r^n - R\left(\frac{1-r^n}{1-r}\right)$$

k. **Find the month when your balance is paid off.**

***The balance is paid off when $b_n \leq 0$. (The final payment will be less than a full payment so that the debt is not overpaid.)***

***Students will likely do this calculation with the values of $r$, $b_0$, and $R$ substituted in.***

$$b_0 r^n - R\left(\frac{1-r^n}{1-r}\right) \leq 0$$

$$b_0 r^n \leq R\left(\frac{1-r^n}{1-r}\right)$$

$$(1-r)(b_0 r^n) \leq R(1-r^n)$$

$$(1-r)(b_0 r^n) + Rr^n \leq R$$

$$r^n(b_0(1-r)+R) \leq R$$

$$r^n \leq \frac{R}{(b_0(1-r)+R)}$$

$$n\log(r) \leq \log\left(\frac{R}{(b_0(1-r)+R)}\right)$$

$$n \geq \frac{\log\left(\frac{R}{(b_0(1-r)+R)}\right)}{\log(r)}$$

***If $R = 50$, then $n \geq 41.925$. The debt is paid off in 42 months.***

***If $R = 100$, then $n \geq 17.49$. The debt is paid off in 18 months.***

***If $R = 150$, then $n \geq 11.0296$. The debt is paid off in 12 months.***

l. **Calculate the total amount paid over the life of the debt. How much was paid solely to interest?**

***For $R = 50$: The debt is paid in 41 payments of \$50, and the last payment is the amount $b_{41}$ with interest:***

$$50(41) + (1+i)b_{41} = 2050 + r\left(b_0 r^n - R\left(\frac{1-r^n}{1-r}\right)\right)$$
$$= 2050 + r(45.52)$$
$$= 2096.28.$$

***The total amount paid using monthly payments of \$50 is \$2096.28. Of this amount, \$596.28 is interest.***

***For $R = 100$: The debt is paid in 17 payments of \$100, and the last payment is the amount $b_{17}$ with interest.***

$$100(17) + (1+i)b_{17} = 1700 + r\left(b_0 r^{17} - R\left(\frac{1-r^{17}}{1-r}\right)\right)$$
$$= 1700 + r(39.86)$$
$$= 1739.86$$

***The total amount paid using monthly payments of \$100 is \$1,739.86. Of this amount, \$239.86 is interest.***

***For $R = 150$: The debt is paid in 11 payments of \$150, and the last payment is the amount $b_{11}$ with interest.***

$$150(11) + (1+i)b_{11} = 1700 + r\left(b_0 r^n - R\left(\frac{1-r^n}{1-r}\right)\right)$$
$$= 1650 + r(4.42)$$
$$= 1654.42$$

***The total amount paid using monthly payments of \$150 is \$1,654.42. Of this amount, \$154.42 is interest.***

## Discussion (9 minutes)

Have students from each team present their solutions to parts (k) and (l) to the class. After the three teams have made their presentations, lead students through the following discussion, which will help them to make sense of the different results that arise from the different payment values $R$.

- What happens to the number of payments as you increase the amount $R$ of the recurring monthly payment?
  - *As the amount $R$ of the payment increases, the number of payments decreases.*
- What happens to the total amount of interest paid as you increase the amount $R$ of the recurring monthly payment?
  - *As the amount $R$ of the payment increases, the number of payments decreases.*
- What is the largest possible amount of the payment $R$? In that case, how many payments are made?
  - *The largest possible payment would be to pay the entire balance in one payment:* $(1+i)\$1500 = \$1524.99$.

Ask students about the formulas that they developed in the Mathematical Modeling Exercise to calculate the balance of the debt in month $n$. Students will use different notations, but they should have come up with a formula similar to $b_n = b_0 r^n - R\left(\frac{1-r^n}{1-r}\right)$. Depending on what notation the students used, you may need to draw the parallel from this formula to the present value of an annuity formula developed in Lesson 30. If we substitute $b_n = 0$ as the future value of the annuity when it is paid off in $n$ payments, and $A_p = b_0$ as the present value/initial value of the annuity, then we have

$$b_n = b_0 r^n - R\left(\frac{1-r^n}{1-r}\right)$$

$$0 = A_p r^n - R\left(\frac{1-r^n}{1-r}\right)$$

$$A_p r^n = R\left(\frac{1-r^n}{1-r}\right)$$

$$A_p(1+i)^n = R\left(\frac{1-(1+i)^n}{1-(1+i)}\right)$$

$$A_p(1+i)^n = R\left(\frac{1-(1+i)^n}{-i}\right)$$

$$A_p = R\left(\frac{(1+i)^n-1}{i}\right)\cdot(1+i)^{-n}$$

$$A_p = R\left(\frac{1-(1+i)^{-n}}{i}\right).$$

### Closing (3 minutes)

Ask students to summarize the main points of the lesson either in writing or with a partner. Some highlights that should be included are listed below.

- Calculating the balance from a single purchase on a credit card requires that we sum a finite geometric series.
- We have a formula from Lesson 29 that calculates the sum of a finite geometric series:

$$\sum_{k=0}^{n-1} ar^k = a \cdot \frac{1-r^n}{1-r}.$$

- When you have incurred a credit card debt, you need to decide how to pay it off.
  - *If you choose to make a lower payment each month, then both the time required to pay off the debt and the total interest paid over the life of the debt will increase.*
  - *If you choose to make a higher payment each month, then both the time required to pay off the debt and the total interest paid over the life of the debt will decrease.*

### Exit Ticket (5 minutes)

Name ______________________________ Date ______________

# Lesson 31: Credit Cards

## Exit Ticket

Suppose that you currently have one credit card with a balance of \$10,000 at an annual rate of 24.00% interest. You have stopped adding any additional charges to this card and are determined to pay off the balance. You have worked out the formula $b_n = b_0r^n - R(1 + r + r^2 + \cdots + r^{n-1})$, where $b_0$ is the initial balance, $b_n$ is the balance after you have made $n$ payments, $r = 1 + i$, where $i$ is the monthly interest rate, and $R$ is the amount you are planning to pay each month.

a. What is the monthly interest rate $i$? What is the growth rate, $r$?

b. Explain why we can rewrite the given formula as $b_n = b_0r^n - R\left(\frac{1-r^n}{1-r}\right)$.

c. How long will it take you to pay off this debt if you can afford to pay a constant \$250 per month? Give your answer in years and months.

## Exit Ticket Sample Solutions

**Suppose that you currently have one credit card with a balance of $\$10,000$ at an annual rate of $24.00\%$ interest. You have stopped adding any additional charges to this card and are determined to pay off the balance. You have worked out the formula $b_n = b_0 r^n - R(1 + r + r^2 + \cdots + r^{n-1})$, where $b_0$ is the initial balance, $b_n$ is the balance after you have made $n$ payments, $r = 1 + i$, where $i$ is the monthly interest rate, and $R$ is the amount you are planning to pay each month.**

a. **What is the monthly interest rate $i$? What is the growth rate, $r$?**

*The monthly interest rate i is given by* $i = \frac{0.24}{12} = 0.02$, *and* $r = 1 + i = 1.02$.

b. **Explain why we can rewrite the given formula as $b_n = b_0 r^n - R\left(\frac{1-r^n}{1-r}\right)$.**

*Using summation notation and the sum formula for a finite geometric series, we have*

$$\begin{aligned} 1 + r + r^2 + \cdots + r^{n-1} &= \sum_{k=0}^{n-1} r^k \\ &= \frac{1 - r^n}{1 - r}. \end{aligned}$$

*Then the formula becomes*

$$\begin{aligned} b_n &= b_0 r^n - R(1 + r + r^2 + \cdots + r^{n-1}) \\ &= b_0 r^n - R\left(\frac{1 - r^n}{1 - r}\right). \end{aligned}$$

c. **How long will it take you to pay off this debt if you can afford to pay a constant $\$250$ per month? Give your answer in years and months.**

*When the debt is paid off,* $b_n \le 0$. *Then* $b_0 r^n - R\left(\frac{1-r^n}{1-r}\right) = 0$, *and* $b_0 r^n = R\left(\frac{1-r^n}{1-r}\right)$. *Since* $b_0 = 10,000$, $R = 250$, *and* $r = 1.02$, *we have*

$$\begin{aligned} 10000(1.02)^n &\le 250\left(\frac{1 - 1.02^n}{1 - 1.02}\right) \\ 10000(1.02)^n &\le -12500(1 - 1.02^n) \\ 10000(1.02)^n &\le 12500(1.02^n - 1) \\ (1.02)^n &\le 1.25(1.02)^n - 1.25 \\ 1.25 &\le 0.25(1.02)^n \\ 5 &\le 1.02^n \\ \log(5) &\le n\log(1.02) \\ n &\ge \frac{\log(5)}{\log(1.02)} \\ n &\ge 81.27 \end{aligned}$$

*It will take* $82$ *months to pay off this debt, which means it will take* $6$ *years and* $10$ *months.*

## Problem Set Sample Solutions

Problems 1–4 ask students to compare credit card scenarios with the same initial debt and the same monthly payments but different interest rates. Problems 5, 6, and 7 require students to compare properties of functions given by different representations, which aligns with **F-IF.C.9** and **F-LE.B.5**.

The final two problems in this Problem Set require students to do some online research in preparation for Lesson 32, in which they select a career and model the purchase of a house. Have some printouts of real-estate listings ready to hand to students who have not brought their own to class. Feel free to add some additional constraints to the criteria for selecting a house to purchase. The career data in Problem 9 can be found at http://themint.org/teens/starting-salaries.html. For additional jobs and more information, please visit the U.S. Bureau of Labor Statistics at http://www.bls.gov/ooh and http://www.bls.gov/ooh/about/teachers-guide.htm. The salary for the "entry-level full-time" position is based on the projected minimum wage in New York in 2016 and a 2,000-hour work year.

1. **Suppose that you have a $\$2,000$ balance on a credit card with a $29.99\%$ annual interest rate, compounded monthly, and you can afford to pay $\$150$ per month toward this debt.**

   a. **Find the amount of time it will take to pay off this debt. Give your answer in months and years.**

$$2000\left(1+\frac{0.2999}{12}\right)^n - 150\left(\frac{1-\left(1+\frac{0.2999}{12}\right)^n}{-\frac{0.2999}{12}}\right) = 0$$

$$2000\left(1+\frac{0.2999}{12}\right)^n = 150\left(\frac{\left(1+\frac{0.2999}{12}\right)^n - 1}{\frac{0.2999}{12}}\right)$$

$$\frac{2999}{9000}\left(1+\frac{0.2999}{12}\right)^n = \left(1+\frac{0.2999}{12}\right)^n - 1$$

$$\left(1+\frac{0.2999}{12}\right)^n\left(\frac{2999}{9000}-1\right) = -1$$

$$\left(1+\frac{0.2999}{12}\right)^n\left(1-\frac{2999}{9000}\right) = 1$$

$$n\cdot\log\left(1+\frac{0.2999}{12}\right) + \log\left(\frac{6001}{9000}\right) = \log(1)$$

$$n\cdot\log\left(1+\frac{0.2999}{12}\right) = -\log\left(\frac{6001}{9000}\right)$$

$$n = -\frac{\log\left(\frac{6001}{9000}\right)}{\log\left(1+\frac{0.2999}{12}\right)}$$

$$n \approx 16.419$$

   *So it will take $1$ year and $5$ months to pay off the debt.*

   b. **Calculate the total amount paid over the life of the debt.**

   $16.419\cdot 150 = \$2462.85$

   c. **How much money was paid entirely to the interest on this debt?**

   $\$462.85$

2. **Suppose that you have a $\$2,000$ balance on a credit card with a $14.99\%$ annual interest rate, and you can afford to pay $\$150$ per month toward this debt.**

a. **Find the amount of time it will take to pay off this debt. Give your answer in months and years.**

$$2000\left(1+\frac{0.1499}{12}\right)^n - 150\left(\frac{1-\left(1+\frac{0.1499}{12}\right)^n}{-\frac{0.1499}{12}}\right) = 0$$

$$2000\left(1+\frac{0.1499}{12}\right)^n = 150\left(\frac{\left(1+\frac{0.1499}{12}\right)^n - 1}{\frac{0.1499}{12}}\right)$$

$$\frac{1499}{9000}\left(1+\frac{0.1499}{12}\right)^n = \left(1+\frac{0.1499}{12}\right)^n - 1$$

$$\left(1+\frac{0.1499}{12}\right)^n\left(\frac{1499}{9000}-1\right) = -1$$

$$\left(1+\frac{0.1499}{12}\right)^n\left(1-\frac{1499}{9000}\right) = 1$$

$$n\cdot\log\left(1+\frac{0.1499}{12}\right) + \log\left(\frac{7501}{9000}\right) = \log(1)$$

$$n\cdot\log\left(1+\frac{0.1499}{12}\right) = -\log\left(\frac{7501}{9000}\right)$$

$$n = -\frac{\log\left(\frac{7501}{9000}\right)}{\log\left(1+\frac{0.1499}{12}\right)}$$

$$n \approx 14.676$$

*The loan will be paid off in $1$ year and $3$ months.*

b. **Calculate the total amount paid over the life of the debt.**

$14.676\cdot 150 = \$2201.40$

c. **How much money was paid entirely to the interest on this debt?**

$\$201.40$

3. Suppose that you have a $\$2,000$ balance on a credit card with a $7.99\%$ annual interest rate, and you can afford to pay $\$150$ per month toward this debt.

   a. Find the amount of time it will take to pay off this debt. Give your answer in months and years.

$$2000\left(1+\frac{0.0799}{12}\right)^n - 150\left(\frac{1-\left(1+\frac{0.0799}{12}\right)^n}{-\frac{0.0799}{12}}\right) = 0$$

$$2000\left(1+\frac{0.0799}{12}\right)^n = 150\left(\frac{\left(1+\frac{0.0799}{12}\right)^n - 1}{\frac{0.0799}{12}}\right)$$

$$\frac{799}{9000}\left(1+\frac{0.0799}{12}\right)^n = \left(1+\frac{0.0799}{12}\right)^n - 1$$

$$\left(1+\frac{0.0799}{12}\right)^n\left(\frac{799}{9000}-1\right) = -1$$

$$\left(1+\frac{0.0799}{12}\right)^n\left(1-\frac{799}{9000}\right) = 1$$

$$n\cdot\log\left(1+\frac{0.0799}{12}\right) + \log\left(\frac{8201}{9000}\right) = \log(1)$$

$$n\cdot\log\left(1+\frac{0.0799}{12}\right) = -\log\left(\frac{8201}{9000}\right)$$

$$n = -\frac{\log\left(\frac{8201}{9000}\right)}{\log\left(1+\frac{0.0799}{12}\right)}$$

$$n \approx 14.009$$

   *The loan will be paid off in 1 year and 3 months.*

   b. Calculate the total amount paid over the life of the debt.

   $14.009 \cdot 150 = \$2101.35$

   c. How much money was paid entirely to the interest on this debt?

   $\$101.35$

4. Summarize the results of Problems 1, 2, and 3.

   *Answers will vary but should include the fact that the total interest paid in each case dropped by about half with every problem. Lower interest rates meant that the loan was paid off more quickly and that less was paid in total.*

5. **Brendan owes $\$1,500$ on a credit card with an interest rate of $12\%$. He is making payments of $\$100$ every month to pay this debt off. Maggie is also making regular payments to a debt owed on a credit card, and she created the following graph of her projected balance over the next $12$ months.**

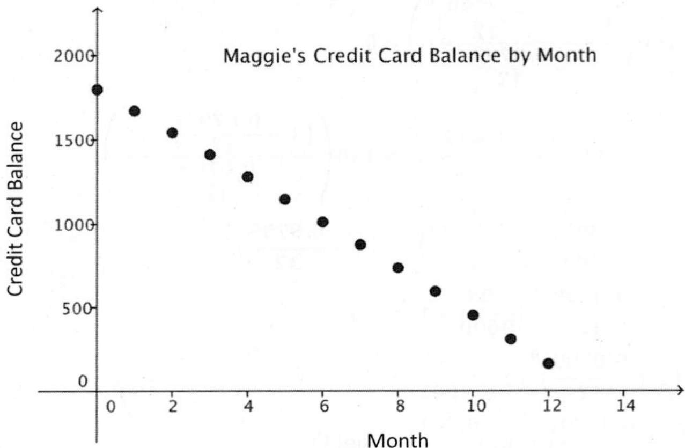


a. **Who has the higher initial balance? Explain how you know.**

*Reading from the graph, Maggie's initial balance is between $\$1,700$ and $\$1,800$, and we are given that Brendan's initial balance is $\$1,500$, so Maggie has the larger initial balance.*

b. **Who will pay their debt off first? Explain how you know.**

*From the graph, it appears that Maggie will pay off her debt between months $12$ and $14$. Brendan's balance in month $n$ can be modeled by the function $b_n = 1500(1.01)^n - 100\left(\frac{1.01^n - 1}{0.01}\right)$, which is equal to zero when $n \approx 16.3$. Thus, Brendan's debt will be paid in month $17$, so Maggie's debt will be paid off first.*

6. **Alan and Emma are both making $\$200$ monthly payments toward balances on credit cards. Alan has prepared a table to represent his projected balances, and Emma has prepared a graph.**

| Alan's Credit Card Balance | | | |
|---|---|---|---|
| Month, $n$ | Interest | Payment | Balance, $b_n$ |
| $0$ | —— | —— | $2,000.00$ |
| $1$ | $41.65$ | $200$ | $1,841.65$ |
| $2$ | $38.35$ | $200$ | $1,680.00$ |
| $3$ | $34.99$ | $200$ | $1,514.99$ |
| $4$ | $31.55$ | $200$ | $1,346.54$ |
| $5$ | $28.04$ | $200$ | $1,174.58$ |
| $6$ | $24.46$ | $200$ | $999.04$ |
| $7$ | $20.81$ | $200$ | $819.85$ |
| $8$ | $17.07$ | $200$ | $636.92$ |
| $9$ | $13.26$ | $200$ | $450.18$ |
| $10$ | $9.37$ | $200$ | $259.55$ |
| $11$ | $5.41$ | $200$ | $64.96$ |

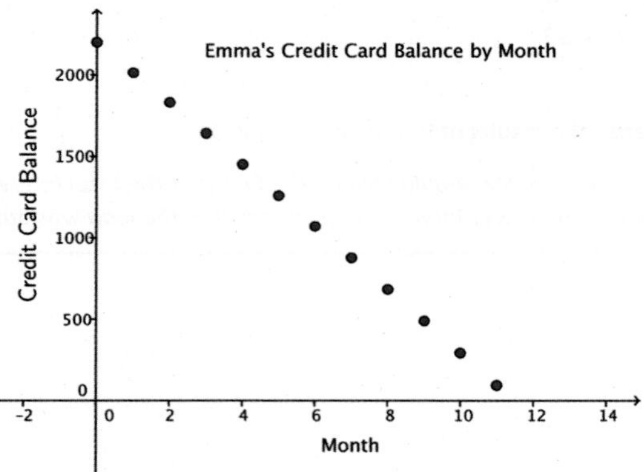


a. **What is the annual interest rate on Alan's debt? Explain how you know.**

*One month's interest on the balance of $\$2,000$ was $\$41.65$, so $41.65 = i(2000)$. Then the monthly interest rate is $i = 0.020825$, and the annual rate is $12i = 0.2499$, so the annual rate on Alan's debt is $24.99\%$.*

b. **Who has the higher initial balance? Explain how you know.**

*From the table, we can see that Alan's initial balance is $\$2,000$, while Emma's initial balance is the $y$-intercept of the graph, which is above $\$2,000$. Thus, Emma's initial balance is higher.*

c. **Who will pay their debt off first? Explain how you know.**

*Both Alan and Emma will pay their debts off in month $12$ because both of their balances in month $11$ are under $\$100$.*

d. **What do your answers to parts (a), (b), and (c) tell you about the interest rate for Emma's debt?**

*Because Emma had the higher initial balance, and they made the same number of payments, Emma must have a lower interest rate on her credit card than Alan does. In fact, since the graph decreases apparently linearly, this implies that Emma has an interest rate of $0\%$.*

7. **Both Gary and Helena are paying regular monthly payments to a credit card balance. The balance on Gary's credit card debt can be modeled by the recursive formula $g_n = g_{n-1}(1.01666) - 200$ with $g_0 = 2500$, and the balance on Helena's credit card debt can be modeled by the explicit formula $h_n = 2000(1.01666)^n - 250\left(\frac{1.01666^n - 1}{0.01666}\right)$ for $n \geq 0$.**

   a. **Who has the higher initial balance? Explain how you know.**

   *Gary has the higher initial balance. Helena's initial balance is $\$2,000$, and Gary's is $\$2,500$.*

   b. **Who has the higher monthly payment? Explain how you know.**

   *Helena has the higher monthly payment. She is paying $\$250$ every month while Gary is paying $\$200$.*

   c. **Who will pay their debt off first? Explain how you know.**

   *Helena will pay her debt off first since she starts at a lower balance and is paying more per month. Additionally, they appear to have the same interest rates.*

8. **In the next lesson, we will apply the mathematics we have learned to the purchase of a house. In preparation for that task, you need to come to class prepared with an idea of the type of house you would like to buy.**

   a. **Research the median housing price in the county where you live or where you wish to relocate.**

   *Answers will vary. A complete list can be found at http://www.tax.ny.gov/research/property/assess/sales/resmedian.htm. For instance, Albany County, New York, had a median sale price of $\$195,000$ in $2013$, and Allegany County, New York, had a median sales price of $\$58,000$.*

   b. **Find the range of prices that are within $25\%$ of the median price from part (a). That is, if the price from part (a) was $P$, then your range is $0.75P$ to $1.25P$.**

   *Answers will vary. Houses in Albany County would range from $\$146,250$ to $\$243,750$, and houses in Allegany County would range from $\$43,500$ to $\$72,500$.*

   c. **Look at online real estate websites, and find a house located in your selected county that falls into the price range specified in part (b). You will be modeling the purchase of this house in Lesson 32, so bring a printout of the real estate listing to class with you.**

   *Answers will vary.*

9. Select a career that interests you from the following list of careers. If the career you are interested in is not on this list, check with your teacher to obtain permission to perform some independent research. Once it has been selected, you will use the career to answer questions in Lesson 32 and Lesson 33.

| Occupation | Median Starting Salary | Education Required |
|---|---|---|
| Entry-level full-time (wait staff, office clerk, lawn care worker, etc.) | $\$18,000$ | High school diploma or GED |
| Accountant | $\$54,630$ | 4-year college degree |
| Athletic Trainer | $\$36,560$ | 4-year college degree |
| Chemical Engineer | $\$78,860$ | 4-year college degree |
| Computer Scientist | $\$93,950$ | 4-year college degree or more |
| Database Administrator | $\$64,600$ | 4-year college degree |
| Dentist | $\$136,960$ | Graduate degree |
| Desktop Publisher | $\$34,130$ | 4-year college degree |
| Electrical Engineer | $\$75,930$ | 4-year college degree |
| Graphic Designer | $\$39,900$ | 2- or 4-year college degree |
| HR Employment Specialist | $\$42,420$ | 4-year college degree |
| HR Compensation Manager | $\$66,530$ | 4-year college degree |
| Industrial Designer | $\$54,560$ | 4-year college degree or more |
| Industrial Engineer | $\$68,620$ | 4-year college degree |
| Landscape Architect | $\$55,140$ | 4-year college degree |
| Lawyer | $\$102,470$ | Law degree |
| Occupational Therapist | $\$60,470$ | Master's degree |
| Optometrist | $\$91,040$ | Master's degree |
| Physical Therapist | $\$66,200$ | Master's degree |
| Physician—Anesthesiology | $\$259,948$ | Medical degree |
| Physician—Family Practice | $\$137,119$ | Medical degree |
| Physician's Assistant | $\$74,980$ | 2 years college plus 2-year program |
| Radiology Technician | $\$47,170$ | 2-year degree |
| Registered Nurse | $\$57,280$ | 2- or 4-year college degree plus |
| Social Worker—Hospital | $\$48,420$ | Master's degree |
| Teacher—Special Education | $\$47,650$ | Master's degree |
| Veterinarian | $\$71,990$ | Veterinary degree |

# Lesson 32: Buying a House

## Student Outcomes

- Students model the scenario of buying a house.
- Students recognize that a mortgage is mathematically equivalent to car loans studied in Lesson 30 and apply the present value of annuity formula to a new situation.

## Lesson Notes

In the Problem Set of Lesson 31, students selected both a future career and a home that they would like to purchase. In this lesson, the students investigate the question of whether or not they can afford the home that they have selected on the salary of the career that they have chosen. We will not develop the standard formulas for mortgage payments, but rather the students will use the concepts from prior lessons on buying a car and paying off a credit card balance to decide for themselves how to model mortgage payments (MP.4). Have students work in pairs or small groups through this lesson, but each student should be working through their own scenario with their own house and their own career. That is, the students will be deciding together how to approach the problem, but they will each be working with their own numbers.

If you teach in a region where the cost of living is particularly high, the median starting salaries given in the list in Problem 9 of Lesson 31 may need to be appropriately adjusted upward in order to make any home purchase feasible in this exercise. Use your professional judgment to make these adjustments.

The students have the necessary mathematical tools to model the payments on a mortgage, but they may not realize it. Allow them to struggle, to debate, and to persevere with the task of deciding how to model this situation (MP.1). It will eventually become apparent that the process of buying a house is only slightly more complicated mathematically than the process of buying a car and that the present value of an annuity formula developed in Lesson 30 applies in this situation (**A-SSE.B.4**). The formula

$$A_p = R\left(\frac{1-(1+i)^{-n}}{i}\right)$$

can be solved for the monthly payment $R$:

$$R = \frac{A_p \cdot i}{1-(1+i)^{-n}},$$

and this formula can be used to answer many of the questions in this lesson. Students may apply the formulas immediately, or they may investigate the balance on the mortgage without using the formulas, which will lead them to develop these formulas on their own. Be sure to ask students to explain their thinking in order to accurately assess their understanding of the mathematics.

## Classwork

### Opening (3 minutes)

- As part of your homework last night, you have selected a potential career that interests you, and you have selected a house that you would like to purchase.

Call on a few students to ask them to share the careers that they have selected, the starting salary, and the price of the home they have chosen.

- Today you will answer the following question: Can you afford the house that you have chosen? There are a few constraints that you need to keep in mind.
  - The total monthly payment for your house cannot exceed $30\%$ of your monthly salary.
  - Your payment includes the payment of the loan for the house and payments into an account called an *escrow account*, which is used to pay for taxes and insurance on your home.
  - Mortgages are usually offered with $30$, $20$, or $15$-year repayment options. You will start with a $30$-year mortgage.
  - You need to make a down payment on the house, meaning that you pay a certain percentage of the price up front and borrow the rest. You will make a $10\%$ down payment for this exercise.

*Scaffolding:*

For struggling students, illustrate the concepts of mortgage, escrow, and down payments using a concrete example with sample values.

### Mathematical Modeling Exercise (25 minutes)

Students may immediately recognize that the previous formulas from Lessons 30 and 31 can be applied to a mortgage, or they may investigate the balance on the mortgage without using the formulas. Both approaches are presented in the sample responses below.

**Mathematical Modeling Exercise**

**Now that you have studied the mathematics of structured savings plans, buying a car, and paying down a credit card debt, it's time to think about the mathematics behind the purchase of a house. In the problem set in Lesson 31, you selected a future career and a home to purchase. The question of the day is this: Can you buy the house you have chosen on the salary of the career you have chosen? You need to adhere to the following constraints:**

- **Mortgages are loans that are usually offered with $30$-, $20$-, or $15$-year repayment options. You will start with a $30$-year mortgage.**
- **The annual interest rate for your mortgage will be $5\%$.**
- **Your payment includes the payment of the loan for the house and payments into an account called an *escrow account*, which is used to pay for taxes and insurance on your home. We will approximate the annual payment to escrow as $1.2\%$ of the home's selling price.**
- **The bank will only approve a mortgage if the total monthly payment for your house, including the payment to the escrow account, does not exceed $30\%$ of your monthly salary.**
- **You have saved up enough money to put a $10\%$ down payment on this house.**

*Scaffolding:*

Struggling students may need to be presented with a set of carefully structured questions:

1. What is the monthly salary for the career you chose?
2. What is $30\%$ of your monthly salary?
3. How much money needs to be paid into the escrow account each year?
4. How much money needs to be paid into the escrow account each month?
5. What is the most expensive house that the bank will allow you to purchase?
6. Is a mortgage like a car loan?
7. What is the formula we used to model a car loan?
8. Which of the values $A_p$, $n$, $i$, and $R$ do we know?
9. Can you rewrite that formula to isolate the $R$?
10. What is the monthly payment according to the formula?
11. Will the bank allow you to purchase the house that you have chosen?

1. **Will the bank approve you for a 30-year mortgage on the house that you have chosen?**

*I chose the career of a graphic designer, with a starting salary of $\$39,900$. My monthly salary is* $\$39,\frac{900}{12} = \$3,325$.

*Thirty percent of my $\$3,325$ monthly salary is $\$997.50$.*

*I found a home that is suitable for $\$190,000$.*

*I need to contribute $0.012(190,000) = 2,280$ to escrow for the year, which means I need to pay $\$190$ to escrow each month.*

*I will make a $\$19,000$ down payment, meaning that I need a mortgage for $\$171,000$.*

*APPROACH 1: We can think of the total owed on the house in two different ways.*

- *If we had placed the original loan amount $A_p = 171,000$ in a savings account earning $5\%$ annual interest, then the future amount in $30$ years would be $A_f = A_p(1+i)^{360}$.*
- *If we deposit a payment of $R$ into an account monthly and let the money in the account accumulate and earn interest for $30$ years, then the future value is*

$$A_f = R + R(1+i) + R(1+i)^2 + \cdots R(1+i)^{359}$$
$$= R\sum_{k=0}^{359}(1+i)^k$$
$$= R\left(\frac{1-(1+i)^{360}}{1-(1+i)}\right)$$
$$= R\left(\frac{(1+i)^{360}-1}{i}\right)$$

*Setting these two expressions for $A_f$ equal to each other, we have*

$$A_p(1+i)^{360} = R\left(\frac{(1+i)^{360}-1}{i}\right),$$

*so*

$$R = \frac{A_p \cdot i \cdot (1+i)^{360}}{(1+i)^{360}-1},$$

*which can also be expressed as*

$$R = \frac{A_p \cdot i}{1-(1+i)^{-360}}.$$

*This is the formula for the present value of an annuity, but rewritten to isolate $R$.*

*Then using my values of $A_p$, $i$ and $n$ we have*

$$R = \frac{171000(0.004167)}{1-(1.004167)^{-360}}$$
$$R = 918.01.$$

*Then, the monthly payment on the house I chose would be $R + 190 = 1,108.01$. The bank will not lend me the money to buy this house because $\$1,108.01$ is higher than $\$997.50$.*

*APPROACH 2: From Lesson 30, we know that the present value of an annuity formula is* $A_p = R\left(\frac{1-(1+i)^{-n}}{i}\right)$, *where* $i$ *is the monthly interest rate,* $R$ *is the monthly payment, and* $n$ *is the number of months in the term. In my example,* $i = \frac{0.05}{12} \approx 0.004167$, $R$ *is unknown,* $n = 12 \cdot 30 = 360$, *and* $A_p = 171,000$. *We can solve the above formula for* $R$, *then we can substitute the known values of the variables and calculate the resulting payment* $R$.

$$A_p = R\left(\frac{1-(1+i)^{-n}}{i}\right)$$

$$A_p \cdot i = R(1-(1+i)^{-n})$$

$$R = \frac{A_p \cdot i}{1-(1+i)^{-n}}$$

*Then using my values of* $A_p$, $i$ *and* $n$ *we have*

$$R = \frac{171000(0.004167)}{1-(1.004167)^{-360}}$$

$$R = 918.01.$$

*Then, the monthly payment on the house I chose would be* $R + 190 = 1,108.01$. *The bank will not lend me the money to buy this house because* $\$1,108.01$ *is higher than* $\$997.50$.

2. Answer either (a) or (b) as appropriate.

   a. If your bank approved you for a $30$-year mortgage, do you meet the criteria for a $20$-year mortgage? If you could get a mortgage for any number of years that you want, what is the shortest term for which you would qualify?

   *(This scenario did not happen in this example.)*

   b. If your bank did not approve you for the $30$-year mortgage, what is the maximum price of a house that fits your budget?

   *The maximum that the bank will allow for my monthly payment is* $30\%$ *of my monthly salary, which is* $\$997.50$. *This includes the payment to the loan and to escrow. If the total price of the house is* $H$ *dollars, then I will make a down payment of* $0.001H$ *and finance* $0.9H$. *Using the present value of an annuity formula, we have*

$$0.9H = R\left(\frac{1-(1+i)^{-n}}{i}\right)$$

$$0.9H = R\left(\frac{1-(1.004167)^{-360}}{0.004167}\right)$$

$$0.9H = R(186.282)$$

   *However,* $R$ *represents just the payment to the loan and not the payment to the escrow account. We know that the escrow portion is one-twelfth of* $1.2\%$ *of the house value. If we denote the total amount paid for the loan and escrow by* $P$, *then* $P = R + 0.001H$, *so* $R = P - 0.001H$. *We know that the largest value for* $P$ *is* $P = 997.50$, *so then*

$$\begin{aligned} 0.9H &= R(186.282) \\ 0.9H &= (997.50 - 0.001H)(186.282) \\ 0.9H &= 185816 - 0.186282H \\ 1.086282H &= 185816 \\ H &= 171,056.87 \end{aligned}$$

   *Then, I can only afford a house that is priced at or below* $\$171,056.87$.

*Scaffolding:*

Mortgage rates can be as low as 3.0%, and in the 1990s rates were often as high as 10%. Ask early finishers to compute the maximum price of a house that they can afford first with an annual interest rate of 5%, then with an annual interest rate of 3%, and then with an annual interest rate of 10%.

### Discussion (9 minutes)

As time permits, ask students to present their results to the class and to explain their thinking. Select students who were approved for their mortgage and those who were not approved to make presentations. Be sure that students who did not immediately recognize that the present value of an annuity formula applies to a mortgage understand that this method is valid. Then, debrief the modeling exercise with the following questions:

- If the bank did not approve your loan, what are your options?
    - *I could wait to purchase the house and save up a larger down payment, I could get a higher-paying job, or I could look for a more reasonably priced house.*
- What would happen if the annual interest rate on your mortgage increased to $8\%$?
    - *If the annual interest rate on the mortgage increased to $8\%$, then the monthly payments would increase dramatically since the loan term is always fixed.*
- Why does the bank limit the amount of the mortgage to $30\%$ of your income?
    - *The bank wants to ensure that you will pay back the loan and that you will not overextend your finances.*

### Closing (3 minutes)

Ask students to summarize the lesson with a partner or in writing by responding to the following questions:

- Which formula from the previous lessons was useful to calculate the monthly payment on the mortgage? Why did that formula apply to this situation?
- How is a mortgage like a car loan? How is it different?
- How is paying a mortgage like paying a credit card balance? How is it different?

### Exit Ticket (5 minutes)

Name ______________________________ Date ______________

# Lesson 32: Buying a House

## Exit Ticket

1. Recall the present value of an annuity formula, where $A_p$ is the present value, $R$ is the monthly payment, $i$ is the monthly interest rate, and $n$ is the number of monthly payments:

$$A_p = R\left(\frac{1-(1+i)^{-n}}{i}\right).$$

Rewrite this formula to isolate $R$.

2. Suppose that you want to buy a house that costs $\$175{,}000$. You can make a $10\%$ down payment, and $1.2\%$ of the house's value is paid into the escrow account each month.

   a. Find the monthly payment for a $30$-year mortgage on this house.

   b. Find the monthly payment for a $15$-year mortgage on this house.

## Exit Ticket Sample Solutions

1. **Recall the present value of an annuity formula, where $A_p$ is the present value, $R$ is the monthly payment, $i$ is the monthly interest rate, and $n$ is the number of monthly payments:**

$$A_p = R\left(\frac{1-(1+i)^{-n}}{i}\right).$$

**Rewrite this formula to isolate $R$.**

$$R = \frac{A_p}{\frac{1-(1+i)^{-n}}{i}}$$

$$R = \frac{A_p \cdot i}{1-(1+i)^{-n}}$$

2. **Suppose that you want to buy a house that costs $\$175,000$. You can make a $10\%$ down payment, and $1.2\%$ of the house's value is paid into the escrow account each month.**

   a. **Find the total monthly payment for a $30$-year mortgage at $4.25\%$ interest on this house.**

   *We have $A_p = 0.9(175,000) = 157,500$, and the monthly escrow payment is $\frac{1}{12}(0.012)(\$175,000) = \$175$. The monthly interest rate $i$ is given by $i = \frac{0.045}{12} = 0.00375$, and $n - 12 \cdot 30 = 360$. Then the formula from Problem 1 gives*

$$\begin{aligned} R &= \frac{A_p \cdot i}{1-(1+i)^{-n}} \\ &= \frac{(157500)(0.00375)}{1-(1.00375)^{-360}} \\ &= 798.03 \end{aligned}$$

   *Thus, the payment to the loan is $\$798.03$ each month. Then the total monthly payment is $\$798.03 + \$175 = \$973.03$.*

   b. **Find the total monthly payment for a $15$-year mortgage at $3.75\%$ interest on this house.**

   *We have $A_p = 0.9(175,000) = 157,500$, and the monthly escrow payment is $\frac{1}{12}(0.012)(\$175,000) = \$175$. The monthly interest rate $i$ is given by $i = \frac{0.0375}{12} = 0.003125$, and $n - 12 \cdot 15 = 180$. Then the formula from Problem 1 gives*

$$\begin{aligned} R &= \frac{A_p \cdot i}{1-(1+i)^{-n}} \\ &= \frac{(157500)(0.003125)}{1-(1.003125)^{-180}} \\ &= 1145.38 \end{aligned}$$

   *Thus, the payment to the loan is $\$1145.38$ each month. Then the total monthly payment is $\$1,145.38 + \$175 = \$1,320.38$.*

## Problem Set Sample Solutions

The results of Exercise 1 are needed for the modeling exercise in Lesson 33, in which students make a plan to save up \$1,000,000 in assets in 15 years, including paying off their home in that time.

1. **Use the house you selected to purchase in the Problem Set from Lesson 31 for this problem.**

   a. **What was the selling price of this house?**

   *Student responses will vary. The sample response will continue to use a house that sold for* $\$190,000$.

   b. **Calculate the total monthly payment, $R$, for a 15-year mortgage at $5\%$ annual interest, paying $10\%$ as a down payment and an annual escrow payment that is $1.2\%$ of the full price of the house.**

   *Using the payment formula with* $A_p = 0.9(190,000) = 171,000$, $i = \frac{0.05}{12} \approx 0.004167$, *and* $n = 15 \cdot 12 = 180$, *we have*

$$\begin{aligned} R &= \frac{A_p \cdot i}{1-(1+i)^{-n}} \\ &= \frac{(171000)(0.004167)}{1-(1.004167)^{-180}} \\ &= 1,352.29 \end{aligned}$$

   *The escrow payment is* $\frac{1}{12}(0.012)(\$190,000) = \$190$. *The total monthly payment is* $\$1352.29 + \$190 = \$1,542.29$.

2. **In the summer of $2014$, the average listing price for homes for sale in the Hollywood Hills was $\$2,663,995$.**

   a. **Suppose you want to buy a home at that price with a $30$-year mortgage at $5.25\%$ annual interest, paying $10\%$ as a down payment and with an annual escrow payment that is $1.2\%$ of the full price of the home. What is your total monthly payment on this house?**

   *Using the payment formula with* $A_p = 0.9(2663995) = 2,397,595.50$, $i = \frac{0.0525}{12} \approx 0.004375$, *and* $n = 360$, *we have*

$$\begin{aligned} R &= \frac{A_p \cdot i}{1-(1+i)^{-n}} \\ &= \frac{(2397595.50)(0.004375)}{1-(1.004375)^{-360}} \\ &= 13,239.60 \end{aligned}$$

   *The escrow payment is* $\frac{1}{12}(0.012)(\$2,663,995) = \$2,664.00$. *The total monthly payment is* $\$13,239.60 + \$2,664 = \$15,903.60$.

   b. **How much is paid in interest over the life of the loan?**

   *The total amount paid is* $\$13,239.60(60) = \$4,766,256$, *and the purchase price was* $\$2,663,995$. *The amount of interest is the difference* $\$4,766,256 - \$2,663,995 = \$2,102,261$.

3. **Suppose that you would like to buy a home priced at $\$200,000$. You will make a payment of $10\%$ of the purchase price and pay $1.2\%$ of the purchase price into an escrow account annually.**

a. **Compute the total monthly payment and the total interest paid over the life of the loan for a 30-year mortgage at $4.8\%$ annual interest.**

*Using the payment formula with* $A_p = 0.9(200,000) = 180,000$, $i = \frac{0.048}{12} = 0.004$, *and* $n = 360$, *we have*

$$R = \frac{A_p \cdot i}{1-(1+i)^{-n}} = \frac{(180,000)(0.004)}{1-(1.004)^{-360}} = 994.40$$

*The escrow payment is* $\frac{1}{12}(0.012)(\$200,000) = \$200.00$. *The total monthly payment is* $\$994.40 + \$200 = \$1,194.40$.

*The total amount of interest is the difference between the total amount paid,* $360(\$994.40) = \$357,984$, *and the selling price* $\$200,000$, *so the total interest paid is* $\$157,984$.

b. **Compute the total monthly payment and the total interest paid over the life of the loan for a 20-year mortgage at $4.8\%$ annual interest.**

*Using the payment formula with* $A_p = 0.9(200,000) = 180,000$, $i = \frac{0.048}{12} = 0.004$, *and* $n = 240$, *we have*

$$R = \frac{A_p \cdot i}{1-(1+i)^{-n}} = \frac{(180,000)(0.004)}{1-(1.004)^{-240}} = 1,168.12$$

*The escrow payment is* $\frac{1}{12}(0.012)(\$200,000) = \$200.00$. *The total monthly payment is* $\$1,168.12 + \$200 = \$1,368.12$.

*The total amount of interest is the difference between the total amount paid,* $240(\$1,168.12) = \$280,348.80$, *and the selling price,* $\$200,000$, *so the total interest paid is* $\$80,348.80$.

c. **Compute the total monthly payment and the total interest paid over the life of the loan for a 15-year mortgage at $4.8\%$ annual interest.**

*Using the payment formula with* $A_p = 0.9(200,000) = 180,000$, $i = \frac{0.048}{12} = 0.004$, *and* $n = 180$, *we have*

$$R = \frac{A_p \cdot i}{1-(1+i)^{-n}} = \frac{(180,000)(0.004)}{1-(1.004)^{-180}} = 1,404.75$$

*The escrow payment is* $\frac{1}{12}(0.012)(\$200,000) = \$200.00$. *The total monthly payment is* $\$1,404.75 + \$200 = \$1,604.75$.

*The total amount of interest is the difference between the total amount paid,* $180(\$1,404.75) = \$252,855$, *and the selling price,* $\$200,000$, *so the total interest paid is* $\$52,855$.

4. Suppose that you would like to buy a home priced at $\$180,000$. You will qualify for a $30$-year mortgage at $4.5\%$ annual interest and pay $1.2\%$ of the purchase price into an escrow account annually.

a. Calculate the total monthly payment and the total interest paid over the life of the loan if you make a $3\%$ down payment.

*With a three percent down payment, you need to borrow* $A_p = 0.97(\$180,000) = \$174,600$. *We have* $i = \frac{0.045}{12} = 0.00375$, *and* $n = 360$, *so*

$$\begin{aligned} R &= \frac{A_p \cdot i}{1-(1+i)^{-n}} \\ &= \frac{(174,600)(0.00375)}{1-(1.00375)^{-360}} \\ &= 884.67 \end{aligned}$$

*The escrow payment is* $\frac{1}{12}(0.012)(\$180,000) = \$180.00$. *The total monthly payment is* $\$884.67 + \$180 = \$1,064.67$.

*The total amount of interest is the difference between the total amount paid,* $360(\$884.67) = \$318,481.20$, *and the selling price,* $\$180,000$, *so the total interest paid is* $\$138,481.20$.

b. Calculate the total monthly payment and the total interest paid over the life of the loan if you make a $10\%$ down payment.

*With a ten percent down payment, you need to borrow* $A_p = 0.9(\$180,000) = \$162,000$. *We have* $i = \frac{0.045}{12} = 0.00375$, *and* $n = 360$, *so*

$$\begin{aligned} R &= \frac{A_p \cdot i}{1-(1+i)^{-n}} \\ &= \frac{(162,000)(0.00375)}{1-(1.00375)^{-360}} \\ &= 820.83 \end{aligned}$$

*The escrow payment is* $\frac{1}{12}(0.012)(\$180,000) = \$180.00$. *The total monthly payment is* $\$820.83 + \$180 = \$1,000.83$.

*The total amount of interest is the difference between the total amount paid,* $360(\$820.83) = \$295,498.80$, *and the selling price,* $\$180,000$, *so the total interest paid is* $\$115,498.80$.

c. Calculate the total monthly payment and the total interest paid over the life of the loan if you make a $20\%$ down payment.

*With a twenty percent down payment, you need to borrow* $A_p = 0.8(\$180,000) = \$144,000$. *We have* $i = \frac{0.045}{12} = 0.00375$, *and* $n = 360$, *so*

$$\begin{aligned} R &= \frac{A_p \cdot i}{1-(1+i)^{-n}} \\ &= \frac{(144,000)(0.00375)}{1-(1.00375)^{-360}} \\ &= 712.56 \end{aligned}$$

*The escrow payment is* $\frac{1}{12}(0.012)(\$180,000) = \$180.00$. *The total monthly payment is* $\$712.56 + \$180 = \$892.56$.

*The total amount of interest is the difference between the total amount paid,* $360(\$712.56) = \$256,521.60$, *and the selling price,* $\$180,000$, *so the total interest paid is* $\$76,521.60$.

d. **Summarize the results of parts (a), (b), and (c) in the chart below.**

| Percent down payment | Amount of down payment | Total interest paid |
|---|---|---|
| $3\%$ | $\$5,400$ | $\$138,481.20$ |
| $10\%$ | $\$18,000$ | $\$115,498.80$ |
| $20\%$ | $\$36,000$ | $\$76,521.60$ |

5. **The following amortization table shows the amount of payments to principal and interest on a $\$100,000$ mortgage at the beginning and the end of a $30$-year loan. These payments do not include payments to the escrow account.**

| Month/ Year | Payment | Principal Paid | Interest Paid | Total Interest | Balance |
|---|---|---|---|---|---|
| Sept. 2014 | $ 477.42 | $ 144.08 | $ 333.33 | $ 333.33 | $ 99,855.92 |
| Oct. 2014 | $ 477.42 | $ 144.56 | $ 332.85 | $ 666.19 | $ 99,711.36 |
| Nov. 2014 | $ 477.42 | $ 145.04 | $ 332.37 | $ 998.56 | $ 99,566.31 |
| Dec. 2014 | $ 477.42 | $ 145.53 | $ 331.89 | $ 1,330.45 | $ 99,420.78 |
| Jan. 2015 | $ 477.42 | $ 146.01 | $ 331.40 | $ 1,661.85 | $ 99,274.77 |
| Mar. 2044 | $ 477.42 | $ 467.98 | $ 9.44 | $ 71,845.82 | $ 2,363.39 |
| April 2044 | $ 477.42 | $ 469.54 | $ 7.88 | $ 71,853.70 | $ 1,893.85 |
| May 2044 | $ 477.42 | $ 471.10 | $ 6.31 | $ 71,860.01 | $ 1,422.75 |
| June 2044 | $ 477.42 | $ 472.67 | $ 4.74 | $ 71,864.75 | $ 950.08 |
| July 2044 | $ 477.42 | $ 474.25 | $ 3.17 | $ 71,867.92 | $ 475.83 |
| Aug. 2044 | $ 477.42 | $ 475.83 | $ 1.59 | $ 71,869.51 | $ 0.00 |

a. **What is the annual interest rate for this loan? Explain how you know.**

*Since the first interest payment is $i \cdot \$100,000 = \$333.33$, the monthly interest rate is $i = 0.0033333$, and the annual interest rate is then $12i \approx 0.044$, so $i = 4\%$.*

b. **Describe the changes in the amount of principal paid each month as the month $n$ gets closer to $360$.**

*As $n$ gets closer to $360$, the amount of the payment that is allocated to principal increases to nearly the entire payment.*

c. **Describe the changes in the amount of interest paid each month as the month $n$ gets closer to $360$.**

*As $n$ gets closer to $360$, the amount of the payment that is allocated to interest decreases to nearly zero.*

6. Suppose you want to buy a $\$200,000$ home with a $30$-year mortgage at $4.5\%$ annual interest paying $10\%$ down with an annual escrow payment that is $1.2\%$ of the price of the home.

   a. Disregarding the payment to escrow, how much do you pay toward the loan on the house each month?

$$\begin{aligned} R &= \frac{A_p \cdot i}{1-(1+i)^{-n}} \\ &= \frac{(180000)(0.00375)}{1-(1.00375)^{-360}} \\ &= 912.03 \end{aligned}$$

   b. What is the total monthly payment on this house?

   *The monthly escrow payment is* $\frac{1}{12}(0.012)(200,000) = 200$, *so the total monthly payment is* $\$1,112.03$.

   c. The graph below depicts the amount of your payment from part (b) that goes to the interest on the loan and the amount that goes to the principal on the loan. Explain how you can tell which graph is which.

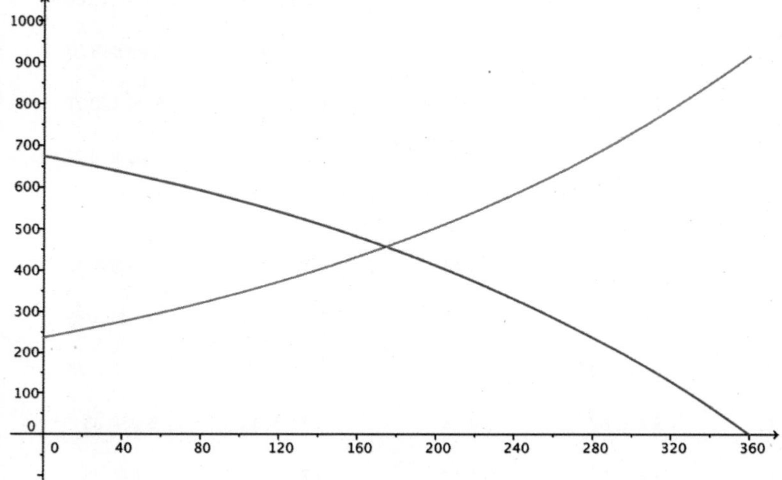


   *The amount paid to interest starts high and decreases, while the amount paid to principal starts low and then increases over the life of the loan. Thus, the blue curve that starts around* $675$ *and decreases represents the amount paid to interest, and the red curve that starts around* $240$ *and increases represents the amount paid to principal.*

7. Student loans are very similar to both car loans and mortgages. The same techniques used for car loans and mortgages can be used for student loans. The difference between student loans and other types of loans is that usually students are not required to pay anything until 6 months after they stop being full-time students.

   a. An unsubsidized student loan will accumulate interest while a student remains in school. Sal borrows $\$9,000$ his first term in school at an interest rate of $5.95\%$ per year compounded monthly and never makes a payment. How much will he owe $4\frac{1}{2}$ years later? How much of that amount is due to compounded interest?

   *This is a compound interest problem without amortization since Sal does not make any payments.*

$$9,000 \cdot \left(1 + \frac{0.0595}{12}\right)^{54} \approx 11,755.40$$

   *Sal will owe* $\$11,755.40$ *at the end of* $4\frac{1}{2}$ *years. Since he borrowed* $\$9,000$, *he owes* $\$2,755.40$ *in interest.*

b. **If Sal pays the interest on his student loan every month while he is in school, how much money has he paid?**

*Since Sal pays the interest on his loan every month, the principal never grows. Every month, the interest is calculated by*

$$9{,}000 \cdot \frac{0.0595}{12} \approx \$44.63.$$

*If Sal pays $\$44.63$ every month for $4\frac{1}{2}$ years, he will have paid $\$2{,}410.02$.*

c. **Explain why the answer to part (a) is different than the answer to part (b).**

*If Sal pays the interest each month, as in part (b), then no interest ever compounds. If he skips the interest payments while he is in school, then the compounding process charges interest on top of interest, increasing the total amount of interest owed on the loan.*

8. **Consider the sequence $a_0 = 10{,}000$, $a_n = a_{n-1} \cdot \frac{1}{10}$ for $n \geq 1$.**

a. **Write the explicit form for the $n^{\text{th}}$ term of the sequence.**

$$a_1 = 10{,}000\left(\frac{1}{10}\right) = 1000$$
$$a_2 = \frac{1}{10}(a_1) = 10{,}000\left(\frac{1}{10}\right)^2$$
$$a_3 = \frac{1}{10}(a_2) = 10{,}000\left(\frac{1}{10}\right)^3$$
$$\vdots$$
$$a_n = 10{,}000\left(\frac{1}{10}\right)^n$$

b. **Evaluate $\sum_{k=0}^{4} a_k$.**

$$\sum_{k=0}^{4} a_k = 10{,}000 + 1{,}000 + 100 + 10 + 1 = 11{,}111$$

c. **Evaluate $\sum_{k=0}^{6} a_k$.**

$$\sum_{k=0}^{6} a_k = 10{,}000 + 1{,}000 + 100 + 10 + 1 + 0.1 + 0.01 = 11{,}111.11$$

d. **Evaluate $\sum_{k=0}^{8} a_k$ using the sum of a geometric series formula.**

$$\sum_{k=0}^{8} a_k = 10{,}000\frac{(1 - r^8)}{1 - r}$$
$$= 10{,}000\frac{\left(1 - \left(\frac{1}{10}\right)^8\right)}{1 - \frac{1}{10}}$$
$$= 11{,}111.1111$$

e. **Evaluate $\sum_{k=0}^{10} a_k$ using the sum of a geometric series formula.**

$$\sum_{k=0}^{10} a_k = 10{,}000\frac{(1-r^{10})}{1-r}$$
$$= 10{,}000\frac{\left(1-\left(\frac{1}{10}\right)^{10}\right)}{1-\frac{1}{10}}$$
$$= 11{,}111.111111$$

f. **Describe the value of $\sum_{k=0}^{n} a_k$ for any value of $n \geq 4$.**

*The value of $\sum_{k=0}^{n} a_k$ for any $n \geq 4$ is $11{,}111.\underbrace{111\cdots1}_{n-4\text{ ones}}$.*

# Lesson 33: The Million Dollar Problem

## Student Outcomes

- Students use geometric series to calculate how much money should be saved each month to have 1 million in assets within a specified amount of time.

## Lesson Notes

In Lesson 33, amortization calculators and other online calculators are not advanced enough to easily develop a savings plan that will result in earning $1 million in assets by the time the student reaches the age of $40$. Students continue their exploration of the formula for the future value of a structured savings plan from Lesson 29 (**A-SSE.B.4**).

We continue to use the formula $A_f = R\left(\frac{(1+i)^n-1}{i}\right)$ from Lesson 29. The formula is discussed more extensively in Lesson 29, but it is always good to remember: The finance formulas in these lessons are direct applications of the sum for a geometric sequence and compound interest formula. Throughout these lessons, we have re-derived these formulas in different contexts for two reasons: firstly, so that students recognize the usefulness of geometric series, and secondly, so that students can make the realization that the types of financial activities (savings plans, car loans, credit cards, etc.) initially appear to be different, but in the end all require the same calculation. The goal is for students to continue to build the formulas from geometric series until they are proficient with the meaning and uses of the formulas.

In the future amount of the annuity formula stated above, the amount of money that somebody wants to have in the future is $A_f$. The amount deposited (which is generically called the payment) per compounding period is $R$, and the interest per compounding period is $i$. The total number of compounding periods is $n$. The amount of money it will take the person to save this amount is the monthly payment times the number of payments, $nR$. When the annuity is representing a loan, the monthly payment times the number of payments is called the total cost of a loan.

For loans, we rewrite the basic compound interest formula as $A_f = A_p(1+i)^n$, set it equal to the formula above from Lesson 29, and solve for the annuity's principal $A_p$. In this context, the present value of the annuity is the same as the principal of the loan (i.e., the loan amount). The present value of an annuity can be thought of as the lump sum amount one would need to invest now in order to earn the future value of an annuity through compound interest alone.

In this modeling lesson, you have the option of including the appreciation (or depreciation) rate of the property value. An interactive map of appreciation rates can be found at http://www.neighborhoodscout.com/ny/rates, and this source claims that the statewide average since 1990 is $2.95\%$. We will use this rate in the sample answers throughout the lesson. The example house used in this lesson will continue to be the $\$190{,}000$ house from Lesson 32.

Some fundamental budgeting is included to provide the framework for the house purchase. The material students developed in Lesson 31 is used to provide flexibility to budgets and to discuss paying off debts in the context of budgets. Students develop and combine functions for the appreciation of their home, the balance in a savings account, and the value of their car to answer the question of the lesson: How can I accumulate $\$1{,}000{,}000$ in assets (**F-BF.A.1b**)?

A copy of the modeling cycle flowchart is included below to assist with the modeling portions of these lessons. Whenever students consider a modeling problem, they need first to identify variables representing essential features in the situation, formulate a model to describe the relationships between the variables, analyze and perform operations on the relationships, interpret their results in terms of the original situation, validate their conclusions, and either improve the model or report their conclusions and their reasoning. The exercises provided in this lesson suggest and provide a road map for you to structure the lesson around the modeling cycle flowchart, but they are only a road map: How much you use the exercises is left to your discretion and as time permits. For example, you may wish to start the class with just the opening question on saving $1 million in 15 years and let the students decide how to move through the modeling cycle on their own without using the exercise questions as prompts. Regardless, each student's report should take into consideration the ideas discussed in the exercises.

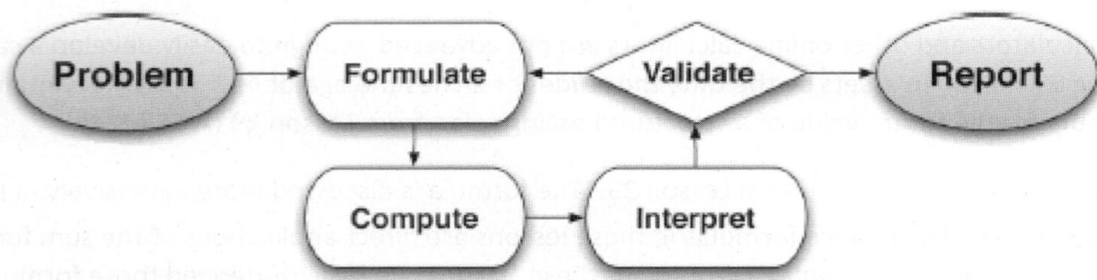


## Classwork

### Opening (4 minutes)

- Now that you are in your mid-twenties, own a car, a house, and have a career, the question remains: What savings plan would you need to generate a $\$1{,}000{,}000$ in assets over the next $15$ years?
- What assets do we have that we can include?
  - *House, car, and savings.*
- Over the long run, property values generally appreciate, but most cars will depreciate. Assume the used car you bought back in Lesson 30 is depreciating and not an asset. But the house you bought does hold value (called equity), and that equity is an asset. Let's focus specifically on the value of your house from Lesson 32. What formula can we use to calculate the value of your house in $15$ years?
  - *The formula is $F = P(1+r)^{15}$, where $r$ is the appreciation rate per year (this can be researched on the internet to find your local appreciation rate). The average appreciation rate for New York is $2.95\%$.*

Inform students what appreciation rate they should use for their house.

- After finding this value, the problem then becomes, "How much do you need to deposit monthly to add up to $\$1{,}000{,}000$ after $15$ years?" What type of problem does this sound like?
  - *This sounds like a structured savings plan like we studied in Lesson 29.*

## Opening Exercise (15 minutes)

Take the time for students to calculate the estimated value of their home and record the results. Have students plot their appreciation curve over 15 years and compare both the appreciated values of their homes and the graphs they produce with each other (**F-IF.C.7e, F-IF.C.8, F-IF.C.9**).

**Opening Exercise**

**In Problem 1 of the Problem Set of Lesson 32, you calculated the monthly payment for a 15-year mortgage at a $5\%$ annual interest rate for the house you chose. You will need that monthly payment for these questions.**

a. **About how much do you expect your home to be worth in 15 years?**

***Answers will vary, but should follow similar steps. For example:***

***Step 1:*** $F = 190,000(1.0295)^{15}$
***Step 2:*** $F \approx 293,866$

***My home will be worth about*** $\$294,000$ ***if it continues to appreciate at an average rate of*** $2.95\%$ ***every year.***

b. **For $0 \leq x \leq 15$, plot the graph of the function $f(x) = P(1+r)^x$ where $r$ is the appreciation rate and $P$ is the initial value of your home.**

c. **Compare the image of the graph you plotted in part (b) with a partner, and write your observations of the differences and similarities. What do you think is causing the differences that you see in the graphs? Share your observations with another group to see if your conclusions are correct.**

***Answers will vary. Although the growth rate is the same for all students in the class, depending on the initial value of the home, different homes will appreciate more quickly than others. For instance, a house that is*** $\$100,000$ ***will increase to*** $\$154,666.15$ ***in 15 years. This increase is about*** $\$49,000$ ***less than the increase for a*** $\$190,000$ ***initial value. The differences are caused by the differences in initial value. Since the rate of increase is a percentage and exponential, the increase is markedly different for more expensive homes and will be even more significant as years increase.***

Your friend Julia bought a home at the same time as you but chose to finance the loan over 30 years. Julia also was able to avoid a down payment and financed the entire value of her home. This allowed her to purchase a more expensive home, but 15 years later she still has not paid off the loan. Consider the following amortization table representing Julia's mortgage, and answer the following questions by comparing the table with your graph.

| Payment # | Beginning Balance | Payment on Interest | Payment on Principal |
|---|---|---|---|
| $1$ | $\$145,000$ | $\$543.75$ | $\$190.94$ |
| $\vdots$ | $\vdots$ | $\vdots$ | $\vdots$ |
| $178$ | $\$96,784.14$ | $\$362.94$ | $\$371.75$ |
| $179$ | $\$96,412.38$ | $\$361.55$ | $\$373.15$ |
| $180$ | $\$96,039.23$ | $\$360.15$ | $\$374.55$ |

d. **In Julia's neighborhood, her home has grown in value at around $2.95\%$ per year. Considering how much she still owes the bank, how much of her home does she own after 15 years (the equity in her home)? Express your answer in dollars to the nearest thousand and as a percent of the value of her home.**

*Julia's home is worth* $145,000(1.0295)^{15} \approx 224,265.91$*; about* $\$224,000$. *She owes* $\$96,039.23$*, which still leaves* $\$224,000 - \$96,000 = \$128,000$. *This means she owns about* $\$128,000$ *of her home or* $\frac{128}{224} = \frac{4}{7} \approx 57\%$ *of her home.*

e. **Reasoning from your graph in part (b) and the table above, if both you and Julia sell your homes in 15 years at the homes' appreciated values, who would have more equity?**

*Answers will vary. Any student whose home is worth more than* $\$83,000$ *initially will have more equity than Julia after* 15 *years, which should be clear from the graph.*

f. **How much more do you need to save over 15 years to have assets over $\$1,000,000$?**

*Answers will vary. For example:*

$$\$1,000,000 - \$294,000 = \$706,000$$

*I will need to save* $\$706,000$ *in the next* 15 *years.*

## Mathematical Modeling Exercises (17 minutes)

Have students work individually or in pairs to figure out the monthly payment they need to save up to $1 million in assets over 15 years. Although 7% compounded quarterly is used as the interest rate on the account, if time permits or with more advanced students, multiple interest rates may be given to different groups of students and the differences between the accounts analyzed and discussed.

The question of what type of account is being used for savings is left to the discretion of the teacher and may be omitted from the discussion. Possibilities include bonds, CDs, and stocks. Bonds and CDs are relatively secure and safe investments but have maximum interest rates around 2–3% annually. The stock market may seem like a risky place to invest, but mutual funds based upon stocks over the long run can provide relatively stable growth. For a list of stock market annual growth rates as well as a compound annual growth rate (CAGR) calculator, please visit http://www.moneychimp.com/features/market_cagr.htm. The data suggests that the CAGR is around 6.86% per year adjusted for inflation, which we have rounded to 7% to give the most optimistic calculations—an annual interest rate of 7% compounded quarterly will double about every 10 years.

Throughout these exercises the modeling cycle should be emphasized so that students use the process correctly. If necessary, draw on the board the modeling flowchart included at the beginning of the lesson to keep students on task.

**Mathematical Modeling Exercises**

**Assume you can earn $7\%$ interest annually, compounded monthly, in an investment account. Develop a savings plan so that you will have $\$1$ million in assets in $15$ years (including the equity in your paid-off house).**

1. **Use your answer to Opening Exercise, part (g) as the future value of your savings plan.**
   a. **How much will you have to save every month to save up $\$1$ million in assets?**

   *Answers will vary. For example, since* $i = \frac{0.07}{12} \approx 0.00583$, $A_f = 706{,}000$, *and* $n = 180$,

$$706000 = R\left(\frac{(1+0.00583)^{180}-1}{0.00583}\right)$$
$$R = 706000 \cdot \frac{0.00583}{(1.00583)^{180}-1}$$
$$R \approx 2{,}227.39$$

   *The monthly payment to save* $\$706{,}000$ *in* $15$ *years at* $7\%$ *interest compounded quarterly would be* $\$2{,}227.39$.

You should not expect students to answer this problem as easily as the answer above implies. Walk around the room encouraging students to try a simpler problem first—maybe one where they earn $5,000 after making four payments. Also, the answer above is the shortest answer possible. Many of your students may need to write out a geometric series inductively to get what the deposits and interest earned will look like. Above all, these last few modeling lessons are meant to let students figure out the solution on their own, so please give them the time to do so. Challenge students who get the answer quickly with the following questions: How is the formula derived? What does it mean?

   b. **Recall the monthly payment to pay off your home in $15$ years (from Problem 1 of the Problem Set of Lesson 32). How much are the two together? What percentage of your monthly income is this for the profession you chose?**

   *The monthly payment on a* $15$*-year loan was about* $\$1{,}352$ *(a* $5\%$ *annual interest loan on* $\$171{,}000$ *for a* $\$190{,}000$ *house with* $\$19{,}000$ *down for* $15$ *years). The savings payment coupled with the monthly mortgage comes to about* $\$3{,}579$.

   *Answers will vary on the percentage of monthly income.*

It is very likely that the total amount of the two may exceed $50\%$ of the monthly income. If so, you can lead your students to recalculate a more reasonable scenario like taking $20$ years to generate $\$1$ million in assets. Have them come up with the plan.

2. **Write a report supported by the calculations you did above on how to save $\$1$ million (or more) in your lifetime.**

   *Answers will vary.*

   *You may wish to assign this as homework so students can type up their plan, make a slide presentation, blog about it, write it in their journal, etc.*

## Closing (4 minutes)

Debrief students on their understanding of the mathematics of finance. Suggested questions are listed below with likely responses. Have students answer on their own or with a partner in writing.

- What formula made all of our work with structured savings plans, credit cards, and loans possible?
  - *The formula for the sum of a finite geometric sequence is* $S_n = a \cdot \frac{1-r^n}{1-r}$.
- What does each part of the formula for the sum of a finite geometric sequence represent?
  - *The* $n^{th}$ *partial sum is* $S_n$*,* $a$ *is the first term,* $r$ *is the common ratio, and* $n$ *is the number of terms.*
- How does this translate to the formula for the future value of a structured savings plan?
  - *Structured savings plans are geometric series with initial terms* $R$ *standing for recurring payment,* $1+i$ *is the common ratio, and* $n$ *is the total number of payments. The sum of all the payments and the interest they earn is the future value of the structured savings plan,* $A_f$*. We get* $A_f = R \cdot \left(\frac{1-(1+i)^n}{1-(1+i)}\right)$ *which simplifies to* $A_f = R \cdot \left(\frac{(1+i)^n-1}{i}\right)$.

The next question is included as a reminder to students to reconnect the work they did in Lessons 30, 31, and 32 with Lesson 33.

- For loans and credit cards, we set the future value of a savings plan equal to the future value of a compound interest account to find the present value, or balance of the loan. State the formula for the present value of a loan, and identify its parts.
  - *The present value of a loan is derived from* $A_p(1+i)^n = R \cdot \left(\frac{(1+i)^n-1}{i}\right)$ *which simplifies to* $A_p = R \cdot \left(\frac{1-(1+i)^{-n}}{i}\right)$*. The present value or balance of the loan is* $A_p$*,* $R$ *is the recurring payment,* $i$ *is the interest rate, and* $n$ *is the number of payments.*

## Exit Ticket (5 minutes)

Name ______________________________ Date ______________

# Lesson 33: The Million Dollar Problem

## Exit Ticket

1. At age $25$, you begin planning for retirement at $65$. Knowing that you have $40$ years to save up for retirement and expecting an interest rate of $4\%$ per year compounded monthly throughout the $40$ years, how much do you need to deposit every month to save up $2 million for retirement?

2. Currently, your savings for each month is capped at $\$400$. If you start investing all of this into a savings plan earning $1\%$ interest annually, compounded monthly, then how long will it take to save $\$160{,}000$? (Hint: Use logarithms.)

## Exit Ticket Sample Solutions

1. **At age $25$, you begin planning for retirement at $65$. Knowing that you have $40$ years to save up for retirement and expecting an interest rate of $4\%$ per year, compounded monthly, throughout the $40$ years, how much do you need to deposit every month to save up $\$2$ million for retirement?**

$$A_f = R\left(\frac{(1+i)^n - 1}{i}\right)$$

$$2 \times 10^6 = R\left(\frac{\left(1+\frac{0.04}{12}\right)^{12\cdot 40} - 1}{\frac{0.04}{12}}\right)$$

$$R = 2 \times 10^6 \cdot \left(\frac{0.04}{12}\right) \div \left(\left(1+\frac{0.04}{12}\right)^{480} - 1\right)$$

$$R \approx 1692.10$$

*You need to deposit $\$1,692.10$ every month for $40$ years to save $\$2$ million at $4\%$ interest.*

2. **Currently, your savings for each month is capped at $\$400$. If you start investing all of this into a savings plan earning $1\%$ interest annually, compounded monthly, then how long will it take to save $\$160,000$? (Hint: Use logarithms.)**

$$160000 = 400\left(\frac{\left(1+\frac{0.01}{12}\right)^{12t} - 1}{\frac{0.01}{12}}\right)$$

$$400 = \frac{\left(1+\frac{0.01}{12}\right)^{12t} - 1}{\frac{0.01}{12}}$$

$$400 \cdot \left(\frac{0.01}{12}\right) = \left(1+\frac{0.01}{12}\right)^{12t} - 1$$

$$\frac{4}{12} + 1 = \left(1+\frac{0.01}{12}\right)^{12t}$$

$$\frac{4}{3} = \left(1+\frac{0.01}{12}\right)^{12t}$$

$$\ln\left(\frac{4}{3}\right) = \ln\left(\left(1+\frac{0.01}{12}\right)^{12t}\right)$$

$$\ln\left(\frac{4}{3}\right) = 12t \cdot \ln\left(1+\frac{0.01}{12}\right)$$

$$t = \frac{\ln\left(\frac{4}{3}\right)}{12\ln\left(1+\frac{0.01}{12}\right)}$$

$$\approx 28.7802$$

*It would take $28$ years and $10$ months to save up $\$160,000$ with only $\$400$ deposited every month.*

## Problem Set Sample Solutions

1. Consider the following scenario: You would like to save up $\$50,000$ after $10$ years and plan to set up a structured savings plan to make monthly payments at $4.125\%$ interest annually, compounded monthly.

   a. What lump sum amount would you need to invest at this interest rate in order to have $\$50,000$ after $10$ years?

$$50000 = P\left(1 + \frac{0.04125}{12}\right)^{120}$$

$$P = 50000 \div \left(1 + \frac{0.04125}{12}\right)^{120}$$

$$P \approx 33123.08$$

   *You would need to deposit* $\$33,123.09$ *now to save up to* $\$50,000$.

   b. Use an online amortization calculator to find the monthly payment necessary to take a loan for the amount in part (a) at this interest rate and for this time period.

   $\$337.33$

   c. Use $A_f = R\left(\frac{(1+i)^n - 1}{i}\right)$ to solve for $R$.

$$50000 = R\left(\frac{\left(1 + \frac{0.04125}{12}\right)^{120} - 1}{\frac{0.04125}{12}}\right)$$

$$R \approx 337.33$$

   *The monthly payment would be* $\$337.33$.

   d. Compare your answers to part (b) and part (c). What do you notice? Why did this happen?

   *The answers are the same. The present value of an annuity is the cost of a loan and can be found by setting the loan equal to the compound interest formula, which is what we did originally. Once we had the cost of a loan, the amortization calculator was able to find the monthly payment. In part (c) we used the future value of an annuity to find the same quantity.*

2. For structured savings plans, the future value of the savings plan as a function of the number of payments made at that point is an interesting function to examine. Consider a structured savings plan with a recurring payment of $\$450$ made monthly and an annual interest rate of $5.875\%$ compounded monthly.

   a. State the formula for the future value of this structured savings plan as a function of the number of payments made. Use $f$ for the function name.

$$f(x) = 450\left(\frac{\left(1 + \frac{0.05875}{12}\right)^x - 1}{\left(\frac{0.05875}{12}\right)}\right)$$

b. Graph the function you wrote in part (a) for $0 \leq x \leq 216$.

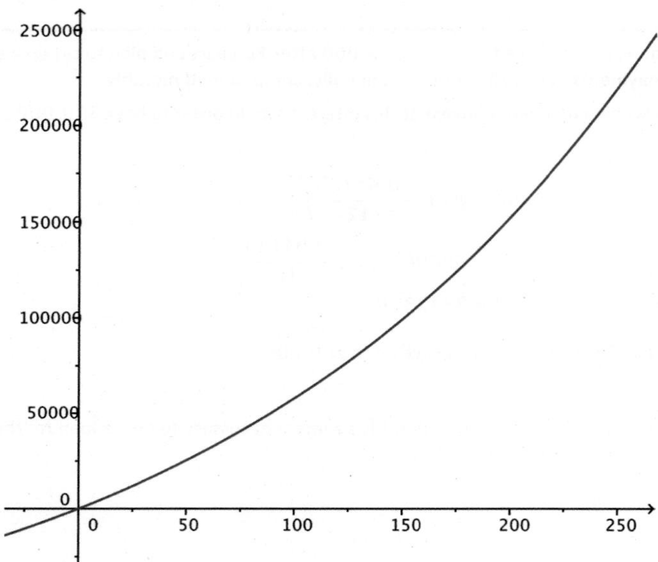


c. State any trends that you notice for this function.

*The function appears to be exponential in nature. It is increasing at an increasing rate.*

d. What is the approximate value of the function $f$ for $x = 216$?

$$f(216) = 450\left(\frac{\left(1 + \frac{0.05875}{12}\right)^{216} - 1}{\left(\frac{0.05875}{12}\right)}\right)$$
$$\approx 172{,}041.21$$

e. What is the domain of $f$? Explain.

*Since the compounding is monthly, the domain of $f$ is normally considered to be a positive integer (i.e., the number of periods).*

f. If the domain of the function is restricted to natural numbers, is the function a geometric sequence? Why or why not?

*No, the function is not a geometric sequence, since it does not have a common ratio. For instance, from $x = 1$ to $x = 2$, the value of the ratio is approximately $2.004896$, but from $x = 2$ to $x = 3$, the value of the ratio is $1.50367$.*

g. Recall that the $n^{\text{th}}$ partial sums of a geometric sequence can be represented with $S_n$. It is true that $f(x) = S_x$ for positive integers $x$, since it is a geometric sequence; that is, $S_x = \sum_{i=1}^{x} ar^i$. State the geometric sequence whose sums of the first $x$ terms are represented by this function.

*The geometric sequence has first term* $450$ *and common ratio* $\left(1 + \frac{0.05875}{12}\right)$. *It can be written as* $a_n = 450 \cdot \left(1 + \frac{0.05875}{12}\right)^{n-1}$.

h. April has been following this structured savings plan for $18$ years. April says that taking out the money and starting over will not affect the total money earned because the interest rate does not change. Explain why April is incorrect in her reasoning.

*The function is increasing exponentially, so the larger the balance, the more it grows. If the money is taken out, then the growth would be reset back to the beginning, although you would have that money.*

3. Henry plans to have $\$195,000$ in property in $14$ years and would like to save up to $\$1$ million by depositing $\$3,068.95$ each month at $6\%$ interest per year, compounded monthly. Tina's structured savings plan over the same time span is described in the following table:

| Deposit # | Amount Saved |
|---|---|
| $30$ | $\$110,574.77$ |
| $31$ | $\$114,466.39$ |
| $32$ | $\$118,371.79$ |
| $33$ | $\$122,291.02$ |
| $34$ | $\$126,224.14$ |
| ⋮ | ⋮ |
| $167$ | $\$795,266.92$ |
| $168$ | $\$801,583.49$ |

a. Who has the higher interest rate? Who pays more every month?

*From the table it looks like Tina pays more every month, but Henry has the higher interest rate. Henry would have about* $\$99,000$ *after* $30$ *payments, but Tina has more at that point. After* $168$ *payments, Henry will save up* $\$805,000$, *while Tina has only saved* $\$801,583$. *Over long periods of time, a higher interest rate will eventually beat larger payments.*

b. At the end of $14$ years, who has more money from their structured savings plan? Does this agree with what you expected? Why or why not?

*Henry has more, but just barely. The larger payment Tina was making was not enough to stay ahead of Henry for* $14$ *years, but he looks to have just passed her recently.*

c. At the end of $40$ years, who has more money from their structured savings plan?

*Henry will have extended his lead significantly by this point. Once he overtakes Tina, his savings will continue to grow at a faster rate than Tina's savings.*

**4. Edgar and Paul are two brothers that both get an inheritance of $\$150,000$. Both plan to save up over $\$1,000,000$ in $25$ years. Edgar takes his inheritance and deposits the money into an investment account earning $8\%$ interest annually, compounded monthly, payable at the end of $25$ years. Paul spends his inheritance but uses a structured savings plan that is represented by the sequence $b_n = 1275 + b_{n-1} \cdot \left(1 + \frac{0.0775}{12}\right)$ with $b_0 = 1275$ in order to save up more than $\$1,000,000$.**

**a. Which of the two has more money at the end of $25$ years?**

*Let $E(x)$ represent Edgar's savings and $P(x)$ represent Paul's savings.*

*Then* $E(x) = 150,000\left(1 + \frac{0.08}{12}\right)^{300} \approx 1,101,026.40.$

$$\begin{aligned} P(x) &= \sum_{i=0}^{299} 1275\left(1 + \frac{0.0775}{12}\right)^i \\ &= 1275\left(\frac{1 - \left(1 + \frac{0.0775}{12}\right)^{300}}{1 - \left(1 + \frac{0.0775}{12}\right)}\right) \\ &= 1275\left(\frac{\left(1 + \frac{0.0775}{12}\right)^{300} - 1}{\frac{0.0775}{12}}\right) \\ &\approx \$1,164,432.17 \end{aligned}$$

**b. What are the pros and cons of both brothers' plans? Which would you rather do? Why?**

*Edgar only has to make a single payment, and he inherited the money, so it does not come out of his normal budget. He does not have to worry about the account again, but he makes less money than Paul overall and cannot access the money until the end of the $25$ years. Edgar also pays much less than Paul does.*

*Paul ends up paying $\$382,500$ in order to save up his million, but he does this slowly over the $25$ years, so he does not have a huge pinch at any point in time. He spends his inheritance, so he will need to pay a lot more taxes up front on the amount. Paul ends up saving more money in the long run.*

*Answers will vary between the two plans and may include a combination of both.*

Name ______________________________ Date ____________________

1. For parts (a) to (c),
   - Sketch the graph of each pair of functions on the same coordinate axes showing end behavior and intercepts, and
   - Describe the graph of $g$ as a series of transformations of the graph of $f$.

   a. $f(x) = 2^x$, and $g(x) = 2^{-x} + 3$

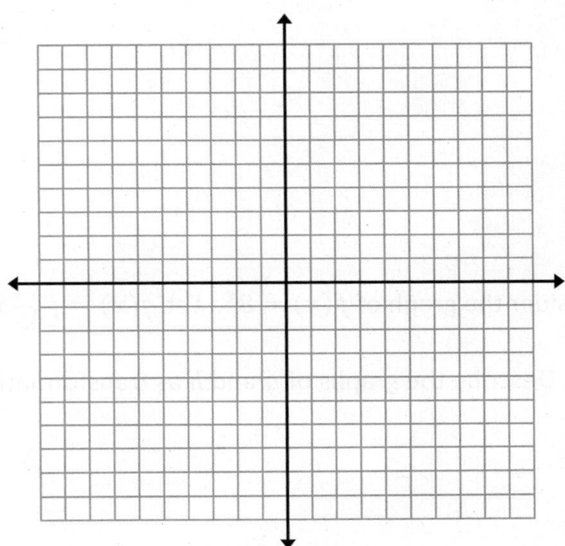

   b. $f(x) = 3^x$, and $g(x) = 9^{x-2}$

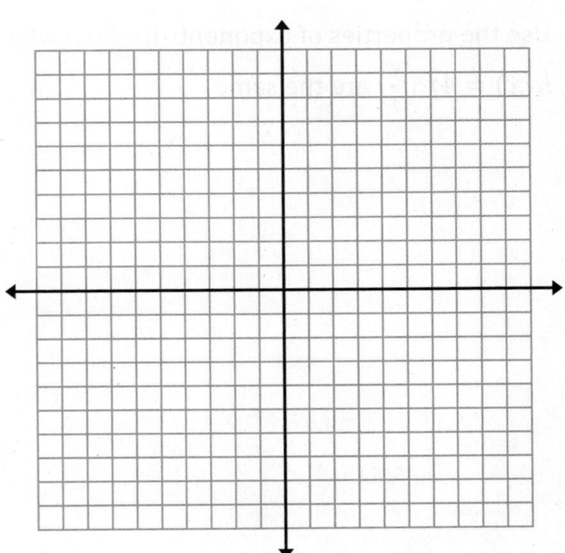

c. $f(x) = \log_2(x)$, and $g(x) = \log_2(x-1)^2$

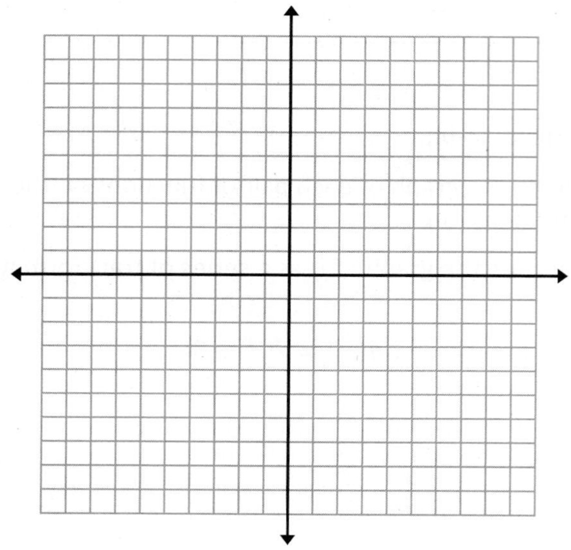

2. Consider the graph of $f(x) = 8^x$. Let $g(x) = f\left(\frac{1}{3}x + \frac{2}{3}\right)$ and $h(x) = 4f\left(\frac{x}{3}\right)$.

   a. Describe the graphs of $g$ and $h$ as transformations of the graph of $f$.

   b. Use the properties of exponents to show why the graphs of the functions $g(x) = f\left(\frac{1}{3}x + \frac{2}{3}\right)$ and $h(x) = 4f\left(\frac{x}{3}\right)$ are the same.

3. The graphs of the functions $f(x) = \ln(x)$ and $g(x) = \log_2(x)$ are shown to the right.

   a. Which curve is the graph of $f$, and which curve is the graph of $g$? Explain.

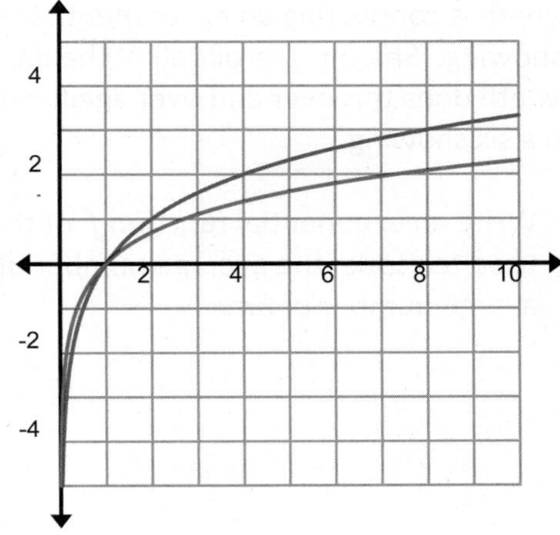

   b. Describe the graph of $g$ as a transformation of the graph of $f$.

   c. By what factor has the graph of $f$ been scaled vertically to produce the graph of $g$? Explain how you know.

4. Gwyneth is conducting an experiment. She rolls $1{,}000$ dice simultaneously and removes any that have a six showing. She then rerolls all of the dice that remain and again removes any that have a six showing. Gwyneth does this over and over again—rerolling the remaining dice and then removing those that land with a six showing.

   a. Write an exponential function $f$ of the form $f(n) = a \cdot b^{cn}$ for any real number $n \geq 0$ that could be used to model the average number of dice she could expect on the $n^{\text{th}}$ roll if she ran her experiment a large number of times.

   b. Gwyneth computed $f(12) = 112.15\ldots$ using the function $f$. How should she interpret the number $112.15\ldots$ in the context of the experiment?

   c. Explain the meaning of the parameters in your function $f$ in terms of this experiment.

   d. Describe in words the key features of the graph of the function $f$ for $n \geq 0$. Be sure to describe where the function is increasing or decreasing, where it has maximums and minimums (if they exist), and the end behavior.

e. According to the model, on which roll does Gwyneth expect, on average, to find herself with only one die remaining? Write and solve an equation to support your answer to this question.

f. For all of the values in the domain of $f$, is there any value for which $f$ will predict an average number of $0$ dice remaining? Explain why or why not. Be sure to use the domain of the function and the graph to support your reasoning.

Suppose the table below represents the results of one trial of Gwyneth's experiment.

| Roll | Number of Dice Left | Roll | Number of Dice Left | Roll | Number of Dice Left |
|---|---|---|---|---|---|
| 0 | 1000 | 10 | 157 | 20 | 26 |
| 1 | 840 | 11 | 139 | 21 | 22 |
| 2 | 692 | 12 | 115 | 22 | 15 |
| 3 | 581 | 13 | 90 | 23 | 13 |
| 4 | 475 | 14 | 78 | 24 | 10 |
| 5 | 400 | 15 | 63 | 25 | 6 |
| 6 | 341 | 16 | 55 | 26 | 2 |
| 7 | 282 | 17 | 43 | 27 | 1 |
| 8 | 232 | 18 | 40 | 28 | 0 |
| 9 | 190 | 19 | 33 | | |

g. Let $g$ be the function that is defined exactly by the data in the table, i.e., $g(0) = 1000$, $g(1) = 840$, $g(2) = 692$, and so forth, up to $g(28) = 0$. Describe in words how the graph of $g$ looks different from the graph of $f$. Be sure to use the domain of $g$ and the domain of $f$ to justify your description.

h. Gwyneth runs her experiment hundreds of times, and each time she generates a table like the one above. How are these tables similar to the function $f$? How are they different?

5. Find the inverse $g$ for each function $f$.

a. $f(x) = \frac{1}{2}x - 3$

b. $f(x) = \frac{x+3}{x-2}$

c. $f(x) = 2^{3x} + 1$

d. $f(x) = e^{x-3}$

e. $f(x) = \log(2x+3)$

6. Dani has $\$1,000$ in an investment account that earns $3\%$ per year, compounded monthly.

   a. Write a recursive sequence for the amount of money in her account after $n$ months.

   b. Write an explicit formula for the amount of money in the account after $n$ months.

   c. Write an explicit formula for the amount of money in her account after $t$ years.

   d. Boris also has $\$1,000$, but in an account that earns $3\%$ per year, compounded yearly. Write an explicit formula for the amount of money in his account after $t$ years.

   e. Boris claims that the equivalent monthly interest rate for his account would be the same as Dani's. Use the expression you wrote in part (d) and the properties of exponents to show why Boris is incorrect.

7. Show that

$$\sum_{k=0}^{n} a \cdot r^k = a\left(\frac{1-r^n}{1-r}\right)$$

where $r \neq 1$.

8. Sami opens an account and deposits $\$100$ into it at the end of each month. The account earns $2\%$ per year compounded monthly. Let $S(n)$ denote the amount of money in her account at the end of $n$ months (just after she makes a deposit). For example, $S(1) = 100$ and $S(2) = 100\left(1 + \frac{0.02}{12}\right) + 100$.

   a. Write a geometric series for the amount of money in the account after 3, 4, and 5 months.

   b. Find a recursive description for $S(n)$.

c. Find an explicit function for $S(n)$, and use it to find $S(12)$.

d. When will Sami have at least $5,000 in her account? Show work to support your answer.

9. Beatrice decides to deposit $100 per month at the end of every month in a bank with an annual interest rate of 5.5% compounded monthly.

   a. Write a geometric series to show how much she will accumulate in her account after one year.

b. Use the formula for the sum of a geometric series to calculate how much she will have in the bank after five years if she keeps on investing $\$100$ per month.

10. Nina has just taken out a car loan for $\$12{,}000$. She will pay an annual interest rate of $3\%$ through a series of monthly payments for $60$ months, which she pays at the end of each month. The amount of money she has left to pay on the loan at the end of the $n^{\text{th}}$ month can be modeled by the function $f(n) = 86248 - 74248(1.0025)^n$ for $0 \leq n \leq 60$.

    At the same time as her first payment (at the end of the first month), Nina placed $\$100$ into a separate investment account that earns $6\%$ per year compounded monthly. She placed $\$100$ into the account at the end of each month thereafter. The amount of money in her savings account at the end of the $n^{\text{th}}$ month can be modeled by the function $g(n) = 20000(1.005)^n - 20000$ for $n \geq 0$.

    a. Use the functions $f$ and $g$ to write an equation whose solution could be used to determine when Nina will have saved enough money to pay off the remaining balance on her car loan.

    b. Use a calculator or computer to graph $f$ and $g$ on the same coordinate plane. Sketch the graphs below, labeling intercepts and indicating end behavior on the sketch. Include the coordinates of any intersection points.

c. How would you interpret the end behavior of each function in the context of this situation?

d. What does the intersection point mean in the context of this situation? Explain how you know.

e. After how many months will Nina have enough money saved to pay off her car loan? Explain how you know.

11. Each function below models the growth of three different trees of different ages over a fixed time interval.

**Tree A:**

$f(t) = 15(1.69)^{\frac{t}{2}}$, where $t$ is time in years since the tree was 15 feet tall, $f(t)$ is the height of the tree in feet, and $0 \le t \le 4$.

**Tree B:**

| Years since the tree was 5 feet tall, $t$ | Height in feet after $t$ years, $g(t)$ |
|---|---|
| 0 | 5 |
| 1 | 6.3 |
| 2 | 7.6 |
| 3 | 8.9 |
| 4 | 10.2 |

**Tree C:** The graph of $h$ is shown where $t$ is years since the tree was 5 feet tall, and $h(t)$ is the height in feet after $t$ years.

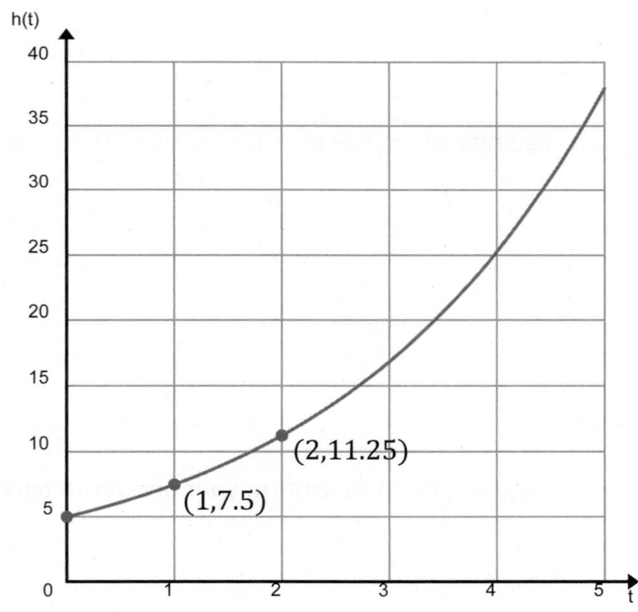

a. Classify each function $f$ and $g$ as representing a linear or nonlinear function. Justify your answers.

b. Use the properties of exponents to show that Tree A has a percent rate of change of $30\%$ per year.

c. Which tree, A or C, has the greatest percent rate of change? Justify your answer.

d. Which function has the greatest average rate of change over the interval $[0,4]$, and what does that mean in terms of tree heights?

e. Write formulas for functions $g$ and $h$, and use them to confirm your answer to part (c).

f. For the exponential models, if the average rate of change of one function over the interval $[0,4]$ is greater than the average rate of change of another function on the same interval, is the percent rate of change also greater? Why or why not?

12. Identify which functions are exponential. For the exponential functions, use the properties of exponents to identify the percent rate of change, and classify them as representing exponential growth or decay.

a. $f(x) = 3(1 - 0.4)^{-x}$

b. $g(x) = \frac{3}{4^x}$

c. $k(x) = 3x^{0.4}$

d. $h(x) = 3^{\frac{x}{4}}$

13. A patient in a hospital needs to maintain a certain amount of a medication in her bloodstream to fight an infection. Suppose the initial dosage is $10$ mg, and the patient is given an additional maintenance dosage of $4$ mg every hour. Assume that the amount of medication in the bloodstream is reduced by $25\%$ every hour.

a. Write a function for the amount of the initial dosage that is in the bloodstream after $n$ hours.

b. Complete the table below to track the amount of medication from the maintenance dosage in the patient's bloodstream for the first five hours.

| Hours since initial dose, $n$ | Amount of the medication in the bloodstream from the maintenance dosage at the beginning of each hour |
|---|---|
| $0$ | $0$ |
| $1$ | $4$ |
| $2$ | $4(1 + 0.75)$ |
| $3$ | |
| $4$ | |
| $5$ | |

c. Write a function that models the total amount of medication in the bloodstream after $n$ hours.

d. Use a calculator to graph the function you wrote in part (c). According to the graph, will there ever be more than 16 mg of the medication present in the patient's bloodstream after each dose is administered?

e. Rewrite this function as the difference of two functions (one a constant function and the other an exponential function), and use that difference to justify why the amount of medication in the patient's bloodstream will not exceed 16 mg after each dose is administered.

| A Progression Toward Mastery | | | | | |
|---|---|---|---|---|---|
| **Assessment Task Item** | | **STEP 1 Missing or incorrect answer and little evidence of reasoning or application of mathematics to solve the problem.** | **STEP 2 Missing or incorrect answer but evidence of some reasoning or application of mathematics to solve the problem.** | **STEP 3 A correct answer with some evidence of reasoning or application of mathematics to solve the problem, or an incorrect answer with substantial evidence of solid reasoning or application of mathematics to solve the problem.** | **STEP 4 A correct answer supported by substantial evidence of solid reasoning or application of mathematics to solve the problem.** |
| **1** | **a–c**<br>**A-SSE.B.3c**<br>**F-IF.C.7e**<br>**F-BF.B.3** | Student provides graphs that contain major errors. Labeled intercepts and end behavior are inconsistent with the functions. | Student provides a graph for $f$ that would be correct for $g$ and vice versa. Intercepts and end behavior are correctly shown in at least one graph. | Student provides correct graphs that contain one minor error. Descriptions of transformations taking the graph of $f$ to the graph of $g$ are wrong or incomplete. | Student provides correct graphs that contain one minor error. Descriptions of transformations taking the graph of $f$ to the graph of $g$ are correct. |
| **2** | **a**<br>**F-BF.B.3** | Student draws one or more of the graphs of $f$, $g$, and $h$ incorrectly. (While not needed for a correct answer, drawing a graph shows evidence of reasoning.) | Student draws the graphs correctly but does not describe, or gives an incomplete description of, the graphs of $g$ and $h$ as transformations of the graph of $f$. | Student describes the graphs of $g$ and $h$ as transformations of $f$, but makes one minor error. | Student describes the graphs of $g$ and $h$ as transformations of $f$. |
| | **b**<br>**F-BF.B.3** | Student provides little or no evidence of understanding why $g$ and $h$ are equivalent functions. | Student claims that the functions are equivalent because the graphs they drew in part (a) are the same. | Student does the operations to show how the expressions are equal but fails to connect it to the graphs of the functions. | Student performs a sequence of valid operations to show how the expressions are equal, and the student uses the calculation to show that $g$ and $h$ are equivalent functions. |

| | | | | | |
|---|---|---|---|---|---|
| 3 | a–c<br><br>F-BF.B.3 | Student incorrectly identifies the blue curve as the graph of $g$ and fails to recognize that the functions are vertical scalings of each other. | Student correctly identifies the graphs (perhaps by noticing $g(2) = 1$) but fails to recognize that the functions are vertical scalings of each other. | Student correctly identifies the graphs and expresses that the graphs are vertical scalings of each other, but fails to find the correct scale factor. | Student correctly identifies the graphs, expresses that the graphs are vertical scalings of each other, and finds the correct scale factor. |
| 4 | a–c<br><br>F-LE.A.2<br>F-LE.B.5 | Student does not provide a correct exponential function or function is missing.<br><br>Student explanations of the meaning of $f(12) = 112.5$ and the meaning of the parameters are missing or contain multiple errors. | Student provides a correct exponential function, but at least one parameter is incorrect given the situation. Explanations contain more than one or two minor errors or inconsistencies. | Student provides a correct exponential function, but explanations of the meaning of $f(12) = 112.5$ or one of the two parameters contain minor errors or are incomplete.<br>OR<br>Student explanations of the point $f(12) = 112.5$ and the meaning of the parameters are correct and consistent with the situation, but one parameter in the function is incorrect. | Student correctly writes an exponential function and provides correct explanations of the meaning of $f(12) = 112.5$ and the meaning of the initial amount ($a$) and growth rate ($b$) in a function of the form $f(n) = a \cdot b^n$. |
| | d<br><br>F-IF.B.4 | Student description is incomplete and contains major errors in vocabulary or notation. | Student description is missing two or three essential features and may contain minor errors in vocabulary or notation. | Student description is fairly accurate and complete with no more than one missing component from those listed in Step 4 of the rubric. | Student description is accurate and complete based on the graph of the model written in part (a). Solutions include information about increasing/decreasing intervals, extrema, intercepts, and end behavior. The appropriate domain is also considered in the solution. |
| | e<br><br>F-LE.A.4<br>F-BF.B.4a | Student provides little or no correct work. Equation is not of the form $f(n) = 1$. | Student provides an equation of the form $f(n) = 1$, but the solutions are incomplete or contains major errors. Little or no attempt is made to interpret the irrational solution in terms of a rolled number. | Student provides an equation of the form $f(n) = 1$. Solutions contain no more than one minor error, but student correctly interprets the irrational solution in terms of a rolled number.<br>OR<br>Student writes and solves equation | Student provides an equation of the form $f(n) = 1$. Solutions are correct and solved using logarithms. Student correctly interprets the irrational solution in terms of a rolled number in the context of the situation. |

| | | | | | |
|---|---|---|---|---|---|
| | | | | correctly but does not interpret the solution in the context of the situation. | |
| | **f–g**<br><br>**F-IF.B.5** | Student solutions are incomplete and contain multiple errors relating to interpretation of the domain. Little or no attempt is made to use the graph to support reasoning. | Student struggles to connect the function model to the real world situation and the solutions have one or more major misconceptions or errors, such as not identifying the correct domain of the table. Student does not use the graph to support reasoning. | Student solutions recognize the limitations of the function model to represent this situation perfectly. Explanations are fairly complete and accurate or contain no more than one or two minor errors. Student uses the graph to support reasoning. | Student recognizes that the function will not perfectly model this situation due to differences in the end behavior of the function and the nature of simulation. Student recognizes the model's limitations and compares and contrasts tables and graphs to justify solution. |
| | **h**<br><br>**F-IF.C.9** | Student fails to provide a valid similarity or difference. Little or no explanation to support reasoning is given. | Student provides one valid similarity or one valid difference. Explanation to support reasoning is limited. | Student provides at least one valid similarity and one valid difference and supports the answer with tables, graphs or both.<br><u>OR</u><br>Student provides at least one valid similarity and two valid differences with limited explanation to support reasoning. | Student provides at least one valid similarity and two valid differences and supports the answer with tables, graph, or both. |
| **5** | **a–e**<br><br>**F-BF.B.4a** | Student finds the correct inverse for one or zero problems. Work shows a limited understanding of the process of finding the inverse of a function. | Student finds the correct inverse for two parts. Solutions to some parts may contain major mathematical errors or significant problems with using proper notation. | Student finds the correct inverse for at least three of the five parts. Solutions contain no more than one minor error. Work demonstrates that student mostly understands the process and is able to use proper notation. | Student finds the correct inverse for each function, showing sufficient work and using proper notation. |
| **6** | **a–e**<br><br>**F-BF.A.2**<br>**A-SSE.B.3c**<br>**A-CED.A.1**<br>**F-BF.A.1a** | Student solutions are incomplete and inaccurate with major mathematical errors throughout. | Student correctly writes one of the three requested formulas. Solution contains multiple minor errors or one or two major errors. | Student solutions to each part are mostly correct with no more than one minor mathematical error and/or no more than one minor difficulty with the use of proper notation. | Student solutions include correct recursive and explicit formulas and an accurate justification in part (e) using the formulas from parts (c) and (d). Student uses proper notation throughout the solution. |

| | | | | | |
|---|---|---|---|---|---|
| 7 | A-SSE.B.4 | Student makes little or no attempt to derive the formula. | Student derivation starts with appropriate identity but is limited and may contain major mathematical errors. | Student derivation is fairly accurate and may contain a minor error in the use of notation or completeness. | Student derivation is accurate, complete, and shows the use of proper notation. Essential steps are shown in the solution. |
| 8 | a–b<br><br>F-IF.A.3<br>F-BF.A.2<br>F-BF.A.1a | Student fails to calculate the amount for more than the second month correctly. Work shown is incomplete with major errors or is missing all together. | Student correctly calculates the recursive description or the amounts after two, three, and four months, but not both. There are more than a few mathematical or notation errors. | Student calculations are mostly correct but may contain a minor mathematical or notation error. | Student correctly calculates the amount after two, three, and four months and uses this information to write a correct recursive description for the function. Solution shows the use of proper notation. |
| | c–d<br><br>A-SSE.B.4<br>F-LE.A.4<br>F-BF.B.4a<br>F-BF.A.1a | Student writes an explicit formula that contains major mathematical errors and is not based on the formula for a geometric series. There is little or no attempt to answer part (d). | Student writes an explicit formula that contains a major mathematical error and the solution to part (d) shows little evidence that student is attempting to solve $S(n) = 5{,}000$. | Student writes a correct explicit formula but struggles to determine successfully when the account will contain $\$5{,}000$ with proper justification and work shown.<br>OR<br>Student writes an explicit formula with a minor error and uses it and the function to determine when the account will contain $\$5{,}000$, giving sufficient justification for the solution. | Student writes a correct explicit formula for the function and successfully uses it to determine when the account will contain $\$5{,}000$. Solution to part (d) is justified by solving the equation algebraically or graphically. |
| 9 | a–b<br><br>A-SSE.B.4<br>A-CED.A.1 | Student solutions to both parts are incomplete and contain two or more major mathematical errors. | Student solutions to both parts contain minor mathematical errors.<br>OR<br>Student solution to one part is mostly correct but the other solution contains major errors. | Student solutions to parts (a) and (b) contain no more than one minor error. Solutions may show limited work, but that work shown is correct. | Student writes a correct series in part (a) and uses the formula correctly in part (b) to find the amount after five years. Work is shown to clearly indicate understanding of the application of a geometric series to structured savings accounts. |

| | | | | | |
|---|---|---|---|---|---|
| 10 | a–e<br><br>A-REI.D.11<br>F-IF.B.4<br>F-IF.C.7e<br>A-CED.A.1 | Student solutions contain many errors and omissions.<br>Equation in part (a) is missing or incorrect and graphs lack sufficient detail or do not reflect the actual graphs of the equations.<br>Little or no attempt is made to relate the functions to the situation. | Student equates $f$ and $g$ and finds the intersection point of the graphs of $f$ and $g$.<br>Student struggles to sketch and label key features of the graph, providing incomplete responses and incorrect or missing interpretations of the features of the graphs in terms of the situation. | Student solutions to all five parts are mostly correct and contain no more than one or two minor errors. | Student equates $f$ and $g$ to write the equation, solve it graphically, and correctly interpret the meaning of the $x$-coordinate of the intersection point of the graphs of $f$ and $g$.<br>Student accurately sketches, labels, and describes at least three key features of each graph.<br>Student correctly interprets features of the graph in terms of the situation. |
| 11 | a–c<br><br>F-IF.C.9<br>F-IF.C.8b | Student correctly classifies one or no functions. Little or no correct work is shown. | Student correctly classifies two out of three functions as linear or exponential. Percent rate of change identification is missing or incorrect. | Student correctly classifies each function as linear or exponential and calculates the percent rate of change with no more than one minor error. Evidence of using rate of change to help classify functions may be limited. | Student correctly classifies each function as linear or exponential showing evidence of using rate of change to do so.<br>Student correctly identifies a growth factor and interprets it using a percent rate of change for the two exponential functions. |
| | d<br><br>F-IF.B.6 | Student shows little or no correct work. There is little or no attempt to interpret rate of change in terms of the situation. | Student makes a major mathematical error or is only able to calculate the rate of change correctly for one of the three functions.<br>Interpretation of rate of change is limited or incorrect. | Student computes and interprets each rate of change with no more than one minor mathematical error. | Student computes and interprets each rate of change appropriately and uses correct notation and vocabulary in the solution. |
| | e<br><br>F-LE.A.2<br>F-IF.B.6 | Student solutions are incorrect or missing. | Student writes either Tree C or Tree B function correctly but has major mathematical errors on the other function.<br>OR<br>Student writes both functions partially correct (e.g., correct slope but not the correct intercept for the linear function). | Student writes a linear function for Tree B and an exponential function for Tree C with no more than one minor mathematical error. Student's solution shows little or no direct evidence of using rate of change to identify the slope. | Student writes a correct linear function for Tree B and a correct exponential function for Tree C by analyzing the information in the respective table and graph. Student uses rate of change to identify the slope of Tree B. The work shown uses proper notation. |

| | | | | | |
|---|---|---|---|---|---|
| | **f**<br><br>**F-IF.C.9** | Student solution is not correct and offers little or no explanation.<br>OR<br>Student solution and explanation are not supported by the given information or are off-task. | Student solution is correct and offers little or no explanation. | Student solution is mostly correct but explanation is limited or contains minor errors. | Student solution is correct and supported using average rate of change based on the functions provided in the problem. |
| **12** | **a–d**<br><br>**F-IF.C.8b** | Student solution is missing or incorrect. | Student identifies two exponential functions correctly including percent rate of change and whether it is growth or decay. Explanation may be missing.<br>OR<br>Student correctly identifies three exponential functions, but the rest of the solution is limited or incorrect. | Student correctly classifies the functions and identifies the percent growth rate in (a), (b), and (d) but explanation is limited or solutions contain minor errors. | Student classifies (a), (b), and (d) as exponential and recognizes that (c) is not exponential.<br>Student rewrites the expression to classify exponential functions as growth or decay and identifies the percent growth rate.<br>Explanations are clear and accurate. |
| **13** | **a–b**<br><br>**A-SSE.B.4** | Student solution is incomplete, inaccurate, or missing. | Student table and/or explicit formula are partially correct with at least one significant error or several minor errors. | Student table and explicit formula are mostly correct, but solution may contain minor mathematical errors. | Student table and explicit formula are correct and clearly show how the maintenance dosage is related to a geometric series. |
| | **c–e**<br><br>**A-SSE.B.4**<br>**F-BF.A.1b** | Student work is incorrect or incomplete with little or no attempt made to accurately graph or correctly rewrite the function. | Student struggles to produce a function that is correct and/or struggles to rewrite the function, which includes one or more significant errors and does not correctly relate the end behavior to the context or to the graph. | Student writes the total amount of the medication as the sum of two functions.<br>Student rewrites the function as the sum of an exponential function and constant function, but student explanation connecting the graph and the situation may be incomplete or contain minor errors. | Student writes the total amount of the medication as the sum of two functions. Student rewrites the function as a sum of an exponential function and a constant function and connects that representation to the graph and to the meaning of the end behavior in the context of the situation. |

Name ______________________________ Date ________________

1. For parts (a) to (c),
   - Sketch the graph of each pair of functions on the same coordinate axes showing end behavior and intercepts, and
   - Describe the graph of $g$ as a series of transformations of the graph of $f$.

a. $f(x) = 2^x$, and $g(x) = 2^{-x} + 3$

*The graph of g is the graph of f reflected across the y-axis and translated vertically 3 units.*

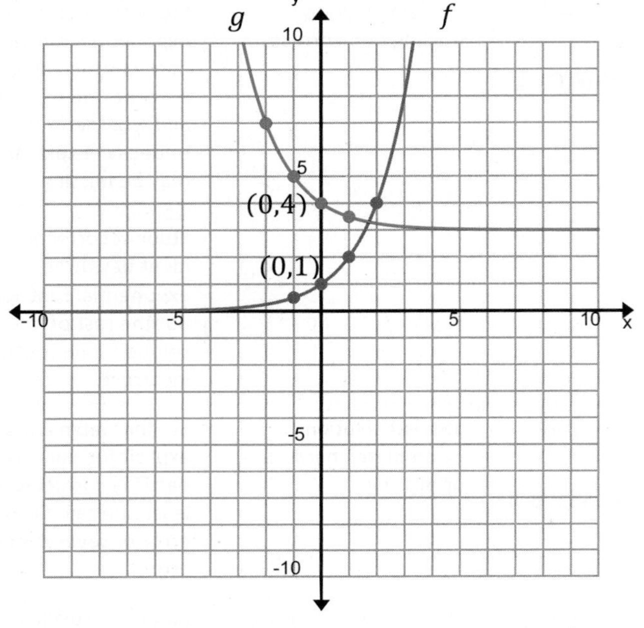


b. $f(x) = 3^x$, and $g(x) = 9^{x-2}$

*Answers will vary. For example, because $9^{x-2} = 3^{2(x-2)}$, the graph of g is the graph of f scaled horizontally by a factor of $\frac{1}{2}$ and translated horizontally by 2 units to the right.*

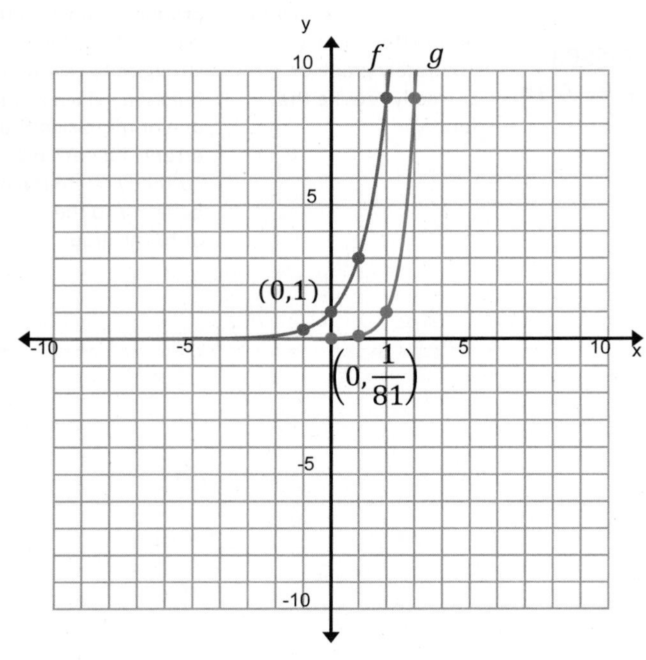

c. $f(x) = \log_2(x)$, and $g(x) = \log_2(x-1)^2$

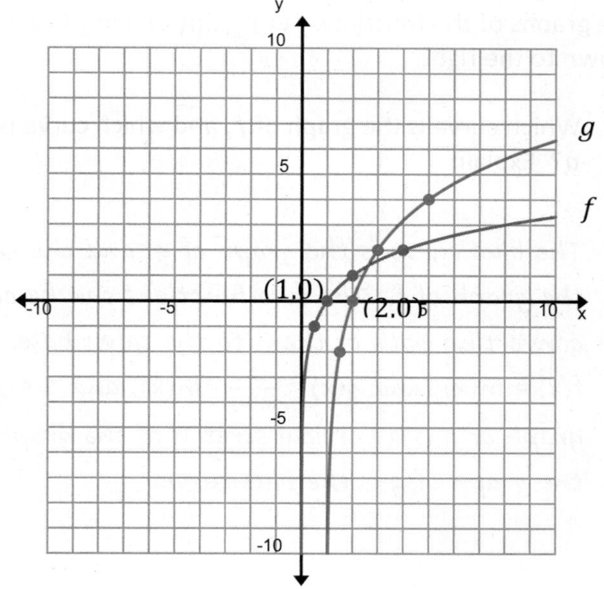


*Most likely answer: Because $\log_2(x-1)^2 = 2\log_2(x-1)$, the graph of g is the graph of f scaled vertically by a factor of 2 and translated horizontally by 1 unit to the right.*

2. Consider the graph of $f(x) = 8^x$. Let $g(x) = f\left(\frac{1}{3}x + \frac{2}{3}\right)$ and $h(x) = 4f\left(\frac{x}{3}\right)$.

   a. Describe the graphs of $g$ and $h$ as transformations of the graph of $f$.

   *The graph of g is the graph of f with a horizontal scaling by a factor of 3 and a horizontal translation 2 units to the left. The graph of h is the graph of f scaled vertically by a factor of 4 and horizontally by a factor of 3.*

   b. Use the properties of exponents to show why the graphs of the functions $g(x) = f\left(\frac{1}{3}x + \frac{2}{3}\right)$ and $h(x) = 4f\left(\frac{x}{3}\right)$ are the same.

   *Since $g(x) = 8^{\frac{x}{3}+\frac{2}{3}}$, $h(x) = 4 \cdot 8^{\frac{x}{3}}$, $8^{\frac{x}{3}+\frac{2}{3}} = 8^{\frac{x}{3}} \cdot 8^{\frac{2}{3}} = 4 \cdot 8^{\frac{x}{3}}$ for any real number x, and their domains are the same, the functions g and h are equivalent. Therefore, they have identical graphs.*

3. The graphs of the functions $f(x) = \ln(x)$ and $g(x) = \log_2(x)$ are shown to the right.

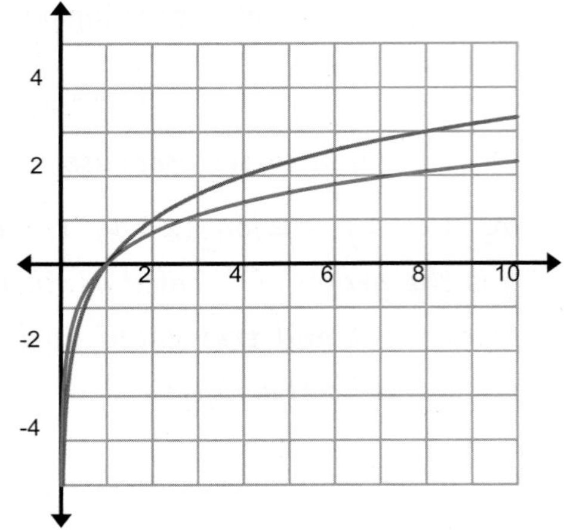


a. Which curve is the graph of $f$, and which curve is the graph of $g$? Explain.

*The blue curve is the graph of g, and the red curve is the graph of f. The two functions can be compared by converting both of them to the same base. Since $f(x) = \ln(x)$, and $g(x) = \frac{1}{\ln(2)} \cdot \ln(x)$, and $1 < \frac{1}{\ln(2)}$, the graph of g is a vertical stretch of the graph of f; thus, the graph of g is the blue curve.*

b. Describe the graph of $g$ as a transformation of the graph of $f$.

*The graph of g is a vertical scaling of the graph of f by a scale factor greater than one.*

c. By what factor has the graph of $f$ been scaled vertically to produce the graph of $g$? Explain how you know.

*By the change of base formula, $g(x) = \frac{1}{\ln(2)} \cdot \ln(x)$; thus, the graph of g is a vertical scaling of the graph of f by a scale factor of $\frac{1}{\ln(2)}$.*

4. Gwyneth is conducting an experiment. She rolls 1,000 dice simultaneously and removes any that have a six showing. She then rerolls all of the dice that remain and again removes any that have a six showing. Gwyneth does this over and over again—rerolling the remaining dice and then removing those that land with a six showing.

   a. Write an exponential function $f$ of the form $f(n) = a \cdot b^{cn}$ for any real number $n \geq 0$ that could be used to model the average number of dice remaining after the $n^{\text{th}}$ roll if she ran her experiment a large number of times.

   $$f(n) = 1000\left(\frac{5}{6}\right)^n$$

   b. Gwyneth computed $f(12) = 112.15\ldots$ using the function $f$. How should she interpret the number $112.15\ldots$ in the context of the experiment?

   *The value $f(12) = 112.15$ means that if she was to run her experiment over and over again, the model predicts that the average number of dice left after the $12^{th}$ roll would be approximately 112.15.*

   c. Explain the meaning of the parameters in your function $f$ in terms of this experiment.

   *The 1,000 represents the initial amount of dice. The $\frac{5}{6}$ represents the fraction of dice remaining from the previous roll each time she rolls the dice. It represents the fact that there is a 1 in 6 chance of any die landing on a 6, so we would predict that after rolling the dice, about $\frac{5}{6}$ of them would not be removed.*

   d. Describe in words the key features of the graph of the function $f$ for $n \geq 0$. Be sure to describe where the function is increasing or decreasing, where it has maximum and minimum (if they exist), and the end behavior.

   *This function is decreasing for all $n \geq 0$. The maximum is the starting number of dice, 1000. There is no minimum. The graph decreases at a decreasing rate with $f(x) \to 0$ as $x \to \infty$.*

e. According to the model, on which roll does Gwyneth expect, on average, to find herself with only one die remaining? Write and solve an equation to support your answer to this question.

$$1000\left(\frac{5}{6}\right)^n = 1$$

$$\left(\frac{5}{6}\right)^n = \frac{1}{1000}$$

$$\log\left(\frac{5}{6}\right)^n = \log\left(\frac{1}{1000}\right)$$

$$n\log\left(\frac{5}{6}\right) = -3$$

$$n = -\frac{3}{\log\left(\frac{5}{6}\right)}$$

$$n \approx 37.88$$

*According to the model, Gwyneth should have 1 die remaining, on average, on the 38th roll.*

f. For all of the values in the domain of $f$, is there any value for which $f$ will predict an average number of 0 dice remaining? Explain why or why not. Be sure to use the domain of the function and the graph to support your reasoning.

*The graph of this function shows the end behavior approaching 0 as the number of rolls increases. This function will never predict an average number of 0 dice remaining because a function of the form $f(n) = ab^{cn}$ never takes on the value of zero. Even as the number of trials becomes very large, the value of $f$ will be a positive number.*

Suppose the table below represents the results of one trial of Gwyneth's experiment.

| Roll | Number of Dice Left | Roll | Number of Dice Left | Roll | Number of Dice Left |
|---|---|---|---|---|---|
| 0 | 1000 | 10 | 157 | 20 | 26 |
| 1 | 840 | 11 | 139 | 21 | 22 |
| 2 | 692 | 12 | 115 | 22 | 15 |
| 3 | 581 | 13 | 90 | 23 | 13 |
| 4 | 475 | 14 | 78 | 24 | 10 |
| 5 | 400 | 15 | 63 | 25 | 6 |
| 6 | 341 | 16 | 55 | 26 | 2 |
| 7 | 282 | 17 | 43 | 27 | 1 |
| 8 | 232 | 18 | 40 | 28 | 0 |
| 9 | 190 | 19 | 33 | | |

g. Let $g$ be the function that is defined exactly by the data in the table, i.e., $g(0) = 1000$, $g(1) = 840$, $g(2) = 692$, and so forth up to $g(28) = 0$. Describe in words how the graph of $g$ looks different from the graph of $f$. Be sure to use the domain of $g$ and the domain of $f$ to justify your description.

*The domain of g is the set of integers from 0 to 28. The graph of g is discrete, while the graph of f is continuous (a curve). The graph of g has an x-intercept at (28, 0), and the graph of f has no x-intercept. The graphs of both functions decrease, but the graph of f is exponential, which means that the ratio $\frac{f(x+1)}{f(x)} = \frac{5}{6}$ for all x in the domain of f. The function values for g do not have this property.*

h. Gwyneth runs her experiment hundreds of times, and each time she generates a table like the one above. How are these tables similar to the function $f$? How are they different?

*Similar: Each table will have the same basic shape as f (i.e., decreasing from 1,000 eventually to 0 with a common ratio between rolls of about $\frac{5}{6}$). Different: (1) The tables are always discrete domains with integer range. (2) Each table will eventually reach 0 after some finite number of rolls, whereas the function f can never take on the value of 0.*

5. Find the inverse $g$ for each function $f$.

a. $f(x) = \frac{1}{2}x - 3$

$$y = \frac{1}{2}x - 3$$

$$x = \frac{1}{2}y - 3$$

$$x + 3 = \frac{1}{2}y$$

$$y = 2x + 6$$

$$g(x) = 2x + 6$$

b. $f(x) = \frac{x+3}{x-2}$

$$y = \frac{x+3}{x-2}$$
$$x = \frac{y+3}{y-2}$$
$$xy - 2x = y + 3$$
$$y(x-1) = 2x + 3$$
$$y = \frac{2x+3}{x-1}$$
$$g(x) = \frac{2x+3}{x-1}$$

c. $f(x) = 2^{3x} + 1$

$$y = 2^{3x} + 1$$
$$x = 2^{3y} + 1$$
$$x - 1 = 2^{3y}$$
$$\log_2(x-1) = 3y$$
$$y = \frac{1}{3}\log_2(x-1)$$
$$g(x) = \frac{1}{3}\log_2(x-1)$$

d. $f(x) = e^{x-3}$

$$y = e^{x-3}$$
$$x = e^{y-3}$$
$$\ln(x) = y - 3$$
$$y = \ln(x) + 3$$
$$g(x) = \ln(x) + 3$$

e. $f(x) = \log(2x+3)$

$$y = \log(2x+3)$$
$$x = \log(2y+3)$$
$$2y + 3 = 10^x$$
$$y = \frac{1}{2} \cdot 10^x - \frac{3}{2}$$
$$g(x) = \frac{1}{2} \cdot 10^x - \frac{3}{2}$$

6. Dani has $1,000 in an investment account that earns 3% per year, compounded monthly.

   a. Write a recursive sequence for the amount of money in her account after $n$ months.

   $$a_1 = 1000, \; a_{n+1} = a_n(1.0025)$$

   b. Write an explicit formula for the amount of money in the account after $n$ months.

   $$a_n = 1000(1.0025)^n$$

   c. Write an explicit formula for the amount of money in her account after $t$ years.

   $$b_n = 1000(1.0025)^{12t}$$

   d. Boris also has $1000, but in an account that earns 3% per year, compounded yearly. Write an explicit formula for the amount of money in his account after $t$ years.

   $$V(t) = 1000(1.03)^t$$

   e. Boris claims that the equivalent monthly interest rate for his account would be the same as Dani's. Use the expression you wrote in part (d) and the properties of exponents to show why Boris is incorrect.

   *Boris is incorrect because the formula for the amount of money in the account, based on a monthly rate, is given by $V(t) = 1000(1+i)^{12t}$, where $i$ is the monthly interest rate. The expression from part d, $1000(1.03)^t$, is equivalent to $1000(1.03)^{\frac{12}{12}t} = 1000(1.03)^{\frac{1}{12}\cdot 12t} = 1000\left((1.03)^{\frac{1}{12}}\right)^{12t}$ by properties of exponents. Therefore, if the expressions must be equivalent, the quantity given by $1.03^{\frac{1}{12}} \approx 1.00247$ means that his monthly rate is about 0.247%, which is less than 0.25%, given by Dani's formula.*

7. Show that

$$\sum_{k=0}^{n} a \cdot r^k = a\left(\frac{1-r^n}{1-r}\right)$$

where $r \neq 1$.

*We know that the statement,* $1 - r^n = (1 - r)(1 + r + r^2 + r^3 + \ldots + r^{n-1})$ *for any real number* $r$ *such that* $r \neq 1$ *and positive integers* $n$*, is an identity. Dividing both sides of the identity by* $1 - r$ *gives* $\frac{1-r^n}{1-r} = 1 + r^2 + r^3 + \ldots + r^{n-1}$.

*Therefore, by factoring the common factor* $a$ *and substituting, we get the formula,*

$$a + ar + ar^2 + \ldots + ar^{n-1} = a(1 + r + r^2 + \ldots + r^{n-1})$$
$$= a\left(\frac{1-r^n}{1-r}\right).$$

8. Sami opens an account and deposits \$100 into it at the end of each month. The account earns 2% per year compounded monthly. Let $S(n)$ denote the amount of money in her account at the end of $n$ months (just after she makes a deposit). For example, $S(1) = 100$, and $S(2) = 100\left(1 + \frac{0.02}{12}\right) + 100$.

a. Write a series for the amount of money in the account after 3, 4, and 5 months.

| Month | Amount |
|---|---|
| 3 | $100\left(1+\frac{0.02}{12}\right)^2 + 100\left(1+\frac{0.02}{12}\right) + 100$ |
| 4 | $100\left(1+\frac{0.02}{12}\right)^3 + 100\left(1+\frac{0.02}{12}\right)^2 + 100\left(1+\frac{0.02}{12}\right) + 100$ |
| 5 | $100\left(1+\frac{0.02}{12}\right)^4 + 100\left(1+\frac{0.02}{12}\right)^3 + 100\left(1+\frac{0.02}{12}\right)^2 + 100\left(1+\frac{0.02}{12}\right) + 100$ |

b. Find a recursive description for $S(n)$.

$S_1 = 100,\ S_{n+1} = S_n\left(1 + \frac{0.02}{12}\right) + 100$

c. Find an explicit formula for $S(n)$, and use it to find $S(12)$.

$$S(n) = \frac{100\left(1-\left(1+\frac{0.02}{12}\right)^n\right)}{1-\left(1+\frac{0.02}{12}\right)}$$

$$S(12) \approx 1211.06$$

d. When will Sami have at least $5,000 in her account? Show work to support your answer.

*An algebraic solution is shown, but students could also solve this equation numerically or algebraically.*

$$5000 = \frac{100\left(1-\left(1+\frac{0.02}{12}\right)^n\right)}{1-\left(1+\frac{0.02}{12}\right)}$$

$$50\left(-\frac{0.02}{12}\right) = 1-\left(1+\frac{0.02}{12}\right)^n$$

$$\frac{13}{12} = \left(1+\frac{0.02}{12}\right)^n$$

$$\log\left(\frac{13}{12}\right) = \log\left(1+\frac{0.02}{12}\right)^n$$

$$n = \frac{\log\left(\frac{13}{12}\right)}{\log\left(1+\frac{0.02}{12}\right)}$$

9. Beatrice decides to deposit $100 per month at the end of every month in a bank with an annual interest rate of 5.5% compounded monthly.

a. Write a geometric series to show how much she will accumulate in her account after one year.

$$100\left(1+\frac{0.055}{12}\right)^{12} + 100\left(1+\frac{0.055}{12}\right)^{11} + \ldots + 100\left(1+\frac{0.055}{12}\right)^{1} + 100$$

b. Use the formula for the sum of a geometric series to calculate how much she will have in the bank after five years if she keeps on investing $100 per month.

$$100\left(\frac{1-\left(1+\frac{0.055}{12}\right)^{60}}{1-\left(1+\frac{0.055}{12}\right)}\right)\approx 6888.08$$

*She will have approximately $6,888 in her account.*

10. Nina has just taken out a car loan for $12,000. She will pay an annual interest rate of 3% through a series of monthly payments for 60 months, which she pays at the end of each month. The amount of money she has left to pay on the loan at the end of the $n^{\text{th}}$ month can be modeled by the function $f(n) = 86248 - 74248(1.0025)^n$ for $0 \leq n \leq 60$.

At the same time as her first payment (at the end of the first month), Nina placed $100 into a separate investment account that earns 6% per year compounded monthly. She placed $100 into the account at the end of each month thereafter. The amount of money in her savings account at the end of the $n^{\text{th}}$ month can be modeled by the function $g(n) = 20000(1.005)^n - 20000$ for $n \geq 0$.

a. Use the functions $f$ and $g$ to write an equation whose solution could be used to determine when Nina will have saved enough money to pay off the remaining balance on her car loan.

$$86248 - 74248(1.0025)^n = 20000(1.005)^n - 20000$$

b. Use a calculator or computer to graph $f$ and $g$ on the same coordinate plane. Sketch the graphs below labeling intercepts and indicating end behavior on the sketch. Include the coordinates of any intersection points.

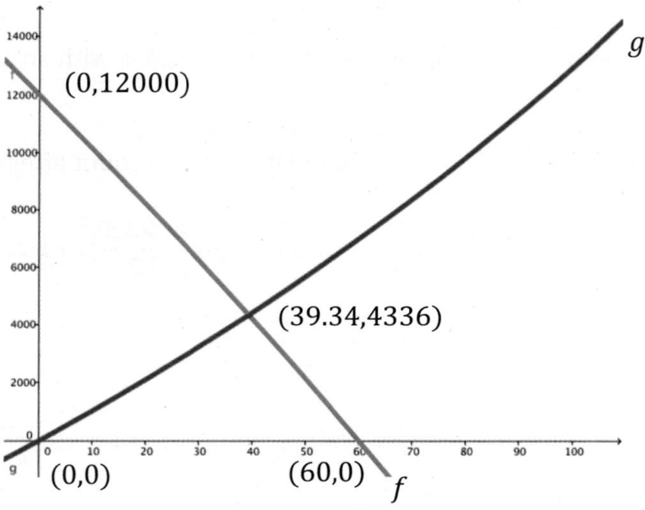

c. How would you interpret the end behavior of each function in the context of this situation?

*For $g$, the end behavior indicates that the amount in the account will continue to grow over time at an exponential rate. For $f$, the end behavior as $x \to \infty$ does not have any real meaning in this situation since the loan will be paid off after 60 months.*

d. What does the intersection point mean in the context of this situation? Explain how you know.

*The intersection point describes the time and amount when approximately the amount left on Nina's loan is equal to the amount she has saved. It is only approximate because the actual amounts are compounded/paid/deposited at the end of the month.*

e. After how many months will Nina have enough money saved to pay off her car loan? Explain how you know.

*After 39.34 months, or at the end of the 40th month, she will have enough money saved to pay off the loan completely. The amount she has at the end of the 40th month in her savings account will be slightly more than \$4,336, and the amount she has left on her loan will be slightly less than \$4,336.*

11. Each function below models the growth of three different trees of different ages over a fixed time interval.

Tree A:

$f(t) = 15(1.69)^{\frac{t}{2}}$, where $t$ is time in years since the tree was 15 feet tall, $f(t)$ is the height of the tree in feet, and $0 \leq t \leq 4$.

Tree B:

| Years since the tree was 5 feet tall, $t$ | Height in feet after $t$ years, $g(t)$ |
|---|---|
| 0 | 5 |
| 1 | 6.3 |
| 2 | 7.6 |
| 3 | 8.9 |
| 4 | 10.2 |

Tree C: The graph of $h$ is shown where $t$ is years since the tree was 5 feet tall, and $h(t)$ is the height in feet after $t$ years.

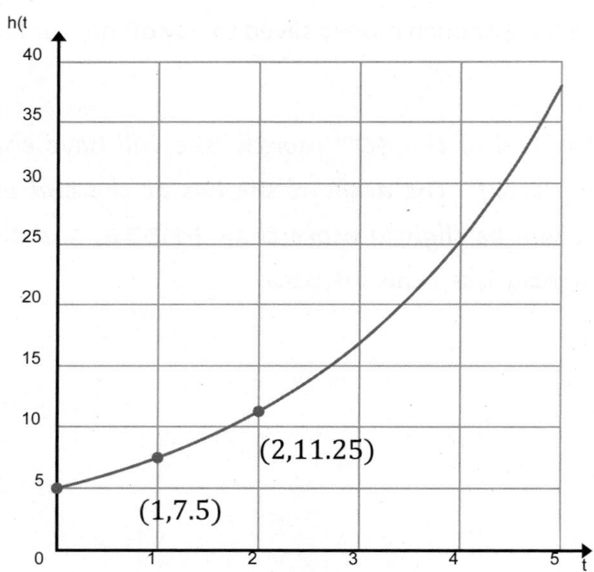

a. Classify each function $f$ and $g$ as representing a linear or non-linear function. Justify your answers.

*The function f is exponential because it is of the form $f(t) = ab^{ct}$, the function g could represent a linear function because its first difference is constant (or it has a constant average rate of change), and the function h is potentially an exponential function because it has a common ratio of 1.5.*

b. Use the properties of exponents to show that Tree A has a percent rate of change of 30% per year.

*The percent rate of change for f is 30% because $(1.69)^{\frac{t}{2}} = \left((1.69)^{\frac{1}{2}}\right)^t = (1.3)^t = (1+0.3)^t$.*

c. Which tree, A or C, has the greatest percent rate of change? Justify your answer?

*The percent rate of change for f is 30% by part (b).*

*The percent rate of change for h is 50% because the common ratio is $\frac{7.5}{5} = \frac{11.25}{7.5} = 1.5$. Tree C has the greatest percent rate of change.*

d. Which function has the greatest average rate of change over the interval [0,4], and what does that mean in terms of tree heights?

*The average rate of change of f on [0,4] is $\frac{15(1.69)^{\frac{4}{2}} - 15}{4} \approx 7$. The average rate of change of g on the interval [0,4] is $\frac{10.2-5}{4} = 1.3$. The average rate of change of h, by estimating from the graph, is approximately $\frac{25.5-5}{4} = 5.125$. The function with the greatest average rate of change is f, which means that over that four-year period, Tree A grew more feet per year on average than the other two trees.*

e. Write formulas for functions $g$ and $h$, and use them to confirm your answer to part (c).

*The functions are given by $g(t) = 5 + 1.3t$ and $h(t) = 5\left(\frac{3}{2}\right)^t$. The average rate of change for g can be computed as $\frac{g(4)-g(0)}{4} = \frac{10.2-5}{4} = 1.3$, and the average rate of change for h on the interval [0,4] is $\frac{h(4)-h(0)}{4} = \frac{25.3125-5}{4} = 5.078125$.*

f. For the exponential models, if the average rate of change of one function over the interval $[0,4]$ is greater than the average rate of change of another function on the same interval, is the percent rate of change also greater? Why or why not?

*No. The average rate of change of f is greater than h over the interval of [0,4], but the percent rate of change of h is greater. The average rate of change is the rate of change over a specific interval, which varies with the interval chosen. The percent rate of change of an exponential function is the percent increase or decrease between the value of the function at x and the value of the function at x + 1 for any real number x; the percent rate of change is constant for an exponential function.*

12. Identify which functions are exponential. For the exponential functions, use the properties of exponents to identify the percent rate of change, and classify them as representing exponential growth or decay.

a. $f(x) = 3(1-0.4)^{-x}$

*Since* $(1-0.4)^{-x} = ((0.6)^{-1})^x = \left(\left(\frac{6}{10}\right)^{-1}\right)^x = \left(\frac{10}{6}\right)^x \approx (1.666)^x$, *the percent rate of change for f is approximately 66%, which represents an exponential growth.*

b. $g(x) = \frac{3}{4^x}$

*Since* $\frac{3}{4^x} = 3\left(\frac{1}{4}\right)^x = 3(0.25)^x = 3(1-0.75)^x$, *the percent rate of change for g is −75%, which represents an exponential decay.*

c. $k(x) = 3x^{0.4}$

*Since* $3x^{0.4} = 3x^{\frac{4}{10}} = 3\sqrt[10]{x^4}$, *this is not an exponential function (it does not have a common ratio, for example).*

d. $h(x) = 3^{\frac{x}{4}}$

$3^{\frac{x}{4}} = \left(3^{\frac{1}{4}}\right)^x \approx 1.316^x$. *The percent growth is approximately 31.6%, which represents an exponential growth.*

13. A patient in a hospital needs to maintain a certain amount of a medication in her bloodstream to fight infection. Suppose the initial dosage is 10 mg, and the patient is given an additional maintenance dosage of 4 mg every hour. Assume that the amount of medication in the bloodstream is reduced by 25% every hour.

a. Write a function for the amount of the initial dosage that is in the bloodstream after $n$ hours.

$$b(t) = 10(0.75)^n$$

b. Complete the table below to track the amount of medication from the maintenance dosage in the patient's bloodstream for the first five hours.

| Hours since initial dose, $n$ | Amount of the medication in the bloodstream from the maintenance dosage at the beginning of each hour |
|---|---|
| 0 | 0 |
| 1 | 4 |
| 2 | $4(1+0.75)$ |
| 3 | $4(1+0.75+0.75^2)$ |
| 4 | $4(1+0.75+0.75^2+0.75^3)$ |
| 5 | $4(1+0.75+0.75^2+\dots+0.75^4)$ |

c. Write a function that models the total amount of medication in the bloodstream after $n$ hours.

$$d(n) = 10(0.75)^n + 4\left(\frac{1-0.75^n}{1-0.75}\right)$$

d. Use a calculator to graph the function you wrote in part (c). According to the graph, will there ever be more than 16 mg of the medication present in the patient's bloodstream after each dose is administered?

According to the graph of the function, the amount of the medication approaches 16 mg as $n \to \infty$.

e. Rewrite this function as the difference of two functions (one a constant function and the other an exponential function), and use that difference to justify why the amount of medication in the patient's bloodstream will not exceed 16 mg after each dose is administered.

*Rewriting this function gives $d(n) = 16 - 6(0.75)^n$. Thus, the function is the difference between a constant function and an exponential function that is approaching a value of 0 as $n$ increases. The amount of the medication will always be 16 reduced by an ever-decreasing quantity.*

# Mathematics Curriculum

## Student Material

# Lesson 1: Integer Exponents

## Classwork

### Opening Exercise

Can you fold a piece of notebook paper in half 10 times?

How thick will the folded paper be?

Will the area of the paper on the top of the folded stack be larger or smaller than a postage stamp?

### Exploratory Challenge

a. What are the dimensions of your paper?

b. How thick is one sheet of paper? Explain how you decided on your answer.

c. Describe how you folded the paper.

d. Record data in the following table based on the size and thickness of your paper.

| Number of Folds | 0 | 1 | 2 | 3 | 4 | 5 | 6 | 7 | 8 | 9 | 10 |
|---|---|---|---|---|---|---|---|---|---|---|---|
| Thickness of the Stack (in.) | | | | | | | | | | | |
| Area of the Top of the Stack (sq. in.) | | | | | | | | | | | |

e. Were you able to fold a piece of notebook paper in half 10 times? Why or why not?

f. Create a formula that approximates the height of the stack after $n$ folds.

g. Create a formula that will give you the approximate area of the top after $n$ folds.

h. Answer the original questions from the Opening Exercise. How do the actual answers compare to your original predictions?

## Example 1: Using the Properties of Exponents to Rewrite Expressions

The table below displays the thickness and area of a folded square sheet of gold foil. In 2001, Britney Gallivan, a California high school junior, successfully folded this size sheet of gold foil in half 12 times to earn extra credit in her mathematics class.

Rewrite each of the table entries as a multiple of a power of 2.

| Number of Folds | Thickness of the Stack (Millionths of a Meter) | Thickness Using a Power of 2 | Area of the Top (Square Inches) | Area Using a Power of 2 |
|---|---|---|---|---|
| 0 | 0.28 | $0.28 \cdot 2^0$ | 100 | $100 \cdot 2^0$ |
| 1 | 0.56 | $0.28 \cdot 2^1$ | 50 | $100 \cdot 2^{-1}$ |
| 2 | 1.12 | | 25 | |
| 3 | 2.24 | | 12.5 | |
| 4 | 4.48 | | 6.25 | |
| 5 | 8.96 | | 3.125 | |
| 6 | 17.92 | | 1.5625 | |

## Example 2: Applying the Properties of Exponents to Rewrite Expressions

Rewrite each expression in the form of $kx^n$ where $k$ is a real number, $n$ is an integer, and $x$ is a nonzero real number.

a. $5x^5 \cdot -3x^2$

b. $\frac{3x^5}{(2x)^4}$

c. $\frac{3}{(x^2)^{-3}}$

d. $\frac{x^{-3}x^{4}}{x^{8}}$

**Exercises 1–5**

Rewrite each expression in the form of $kx^n$ where $k$ is a real number and $n$ is an integer. Assume $x \neq 0$.

1. $2x^5 \cdot x^{10}$

2. $\frac{1}{3x^8}$

3. $\frac{6x^{-5}}{x^{-3}}$

4. $\left(\frac{3}{x^{-22}}\right)^{-3}$

5. $(x^2)^n \cdot x^3$

**Lesson Summary**

The Properties of Exponents

For real numbers $x$ and $y$ with $x \neq 0$, $y \neq 0$, and all integers $a$ and $b$, the following properties hold:

$$x^a \cdot x^b = x^{a+b}$$

$$(x^a)^b = x^{ab}$$

$$(xy)^a = x^a y^a$$

$$\frac{1}{x^a} = x^{-a}$$

$$\frac{x^a}{x^b} = x^{a-b}$$

$$\left(\frac{x}{y}\right)^a = \frac{x^a}{y^a}$$

$$x^0 = 1$$

## Problem Set

1. Suppose your class tried to fold an unrolled roll of toilet paper. It was originally $4$ in. wide and $30$ ft. long. Toilet paper is approximately $0.002$ in. thick.

   a. Complete each table and represent the area and thickness using powers of 2.

| Number of Folds $n$ | Thickness After $n$ Folds (inches) |
|---|---|
| 0 | |
| 1 | |
| 2 | |
| 3 | |
| 4 | |
| 5 | |
| 6 | |

| Number of Folds $n$ | Area on Top After $n$ Folds (square inches) |
|---|---|
| 0 | |
| 1 | |
| 2 | |
| 3 | |
| 4 | |
| 5 | |
| 6 | |

   b. Create an algebraic function that describes the area in square inches after $n$ folds.

   c. Create an algebraic function that describes the thickness in inches after $n$ folds.

2. In the Exit Ticket, we saw the formulas below. The first formula determines the minimum width, $W$, of a square piece of paper of thickness $T$ needed to fold it in half $n$ times, alternating horizontal and vertical folds. The second formula determines the minimum length, $L$, of a long rectangular piece of paper of thickness $T$ needed to fold it in half $n$ times, always folding perpendicular to the long side.

$$W = \pi \cdot T \cdot 2^{\frac{3(n-1)}{2}} \qquad L = \frac{\pi T}{6}(2^n + 4)(2^n - 1)$$

Use the appropriate formula to verify why it is possible to fold a $10$ inch by $10$ inch sheet of gold foil in half $13$ times. Use $0.28$ millionths of a meter for the thickness of gold foil.

3. Use the formula from the Exit Ticket to determine if you can fold an unrolled roll of toilet paper in half more than $10$ times. Assume that the thickness of a sheet of toilet paper is approximately $0.002$ in. and that one roll is $102$ ft. long.

4. Apply the properties of exponents to rewrite expressions in the form $kx^n$, where $n$ is an integer and $x \neq 0$.

   a. $(2x^3)(3x^5)(6x)^2$

   b. $\frac{3x^4}{(-6x)^{-2}}$

   c. $\frac{x^{-3}x^5}{3x^4}$

   d. $5(x^3)^{-3}(2x)^{-4}$

   e. $\left(\frac{x^2}{4x^{-1}}\right)^{-3}$

5. Apply the properties of exponents to verify that each statement is an identity.

   a. $\frac{2^{n+1}}{3^n} = 2\left(\frac{2}{3}\right)^n$ for integer values of $n$.

   b. $3^{n+1} - 3^n = 2 \cdot 3^n$ for integer values of $n$.

   c. $\frac{1}{(3^n)^2} \cdot \frac{4^n}{3} = \frac{1}{3}\left(\frac{2}{3}\right)^{2n}$ for integer values of $n$.

6. Jonah was trying to rewrite expressions using the properties of exponents and properties of algebra for nonzero values of $x$. In each problem, he made a mistake. Explain where he made a mistake in each part and provide a correct solution.

| Jonah's Incorrect Work |
|---|
| a. $(3x^2)^{-3} = -9x^{-6}$ |
| b. $\frac{2}{3x^5} = 6x^{-5}$ |
| c. $\frac{2x-x^3}{3x} = \frac{2}{3} - x^3$ |

7. If $x = 5a^4$, and $a = 2b^3$, express $x$ in terms of $b$.

8. If $a = 2b^3$, and $b = -\frac{1}{2}c^{-2}$, express $a$ in terms of $c$.

9. If $x = 3y^4$, and $y = \frac{s}{2x^3}$, show that $s = 54y^{13}$.

10. Do the following without a calculator:
    a. Express $8^3$ as a power of 2.
    b. Divide $4^{15}$ by $2^{10}$.

11. Use powers of 2 to help you perform each calculation.
    a. $\frac{2^7 \cdot 2^5}{16}$
    b. $\frac{512000}{320}$

12. Write the first five terms of each of the following recursively-defined sequences:
    a. $a_{n+1} = 2a_n$, $a_1 = 3$
    b. $a_{n+1} = (a_n)^2$, $a_1 = 3$
    c. $a_{n+1} = 2(a_n)^2$, $a_1 = x$, where $x$ is a real number. Write each term in the form $kx^n$.
    d. $a_{n+1} = 2(a_n)^{-1}$, $a_1 = y$, $(y \neq 0)$. Write each term in the form $kx^n$.

13. In Module 1, you established the identity
    $(1 - r)(1 + r + r^2 + \cdots + r^{n-1}) = 1 - r^n$, where $r$ is a real number and $n$ is a positive integer.
    Use this identity to find explicit formulas as specified below.
    a. Rewrite the given identity to isolate the sum $1 + r + r^2 + \cdots r^{n-1}$ for $r \neq 1$.
    b. Find an explicit formula for $1 + 2 + 2^2 + 2^3 + \cdots + 2^{10}$.
    c. Find an explicit formula for $1 + a + a^2 + a^3 + \cdots + a^{10}$ in terms of powers of $a$.
    d. Jerry simplified the sum $1 + a + a^2 + a^3 + a^4 + a^5$ by writing $1 + a^{15}$. What did he do wrong?
    e. Find an explicit formula for $1 + 2a + (2a)^2 + (2a)^3 + \cdots + (2a)^{12}$ in terms of powers of $a$.
    f. Find an explicit formula for $3 + 3(2a) + 3(2a)^2 + 3(2a)^3 + \ldots + 3(2a)^{12}$ in terms of powers of $a$. Hint: Use part (e).
    g. Find an explicit formula for $P + P(1 + r) + P(1 + r)^2 + P(1 + r)^3 + \cdots + P(1 + r)^{n-1}$ in terms of powers $(1 + r)$.

# Lesson 2: Base 10 and Scientific Notation

## Classwork

### Opening Exercise

In the last lesson, you worked with the thickness of a sheet of gold foil (a very small number) and some very large numbers that gave the size of a piece of paper that actually could be folded in half more than 13 times.

a. Convert $0.28$ millionths of a meter to centimeters and express your answer as a decimal number.

b. The length of a piece of square notebook paper that could be folded in half 13 times was $3294.2$ in. Use this number to calculate the area of a square piece of paper that could be folded in half 14 times. Round your answer to the nearest million.

c. Sort the following numbers into products and single numeric expressions. Then match the equivalent expressions without using a calculator.

| $3.5 \times 10^5$ | $-6$ | $-6 \times 10^0$ | $0.6$ | $3.5 \times 10^{-6}$ |
|---|---|---|---|---|
| $3{,}500{,}000$ | $350{,}000$ | $6 \times 10^{-1}$ | $0.0000035$ | $3.5 \times 10^6$ |

## Example 1

Write each number as a product of a decimal number between 1 and 10 and a power of 10.

a. $234{,}000$

b. $0.0035$

c. $532{,}100{,}000$

d. $0.0000000012$

e. $3.331$

## Exercises 1–6

For Exercises 1–6, write each decimal in scientific notation.

1. $532{,}000{,}000$

2. $0.0000000000000000123$ (16 zeros after the decimal place)

3. $8{,}900{,}000{,}000{,}000{,}000$ (14 zeros after the 9)

4. $0.00003382$

5. $34{,}000{,}000{,}000{,}000{,}000{,}000{,}000{,}000$ (24 zeros after the 4)

6. $0.0000000000000000000004$ (21 zeros after the decimal place)

## Exercises 7–8

7. Approximate the average distances between the Sun and Earth, Jupiter, and Pluto. Express your answers in scientific notation ($d \times 10^n$), where $d$ is rounded to the nearest tenth.

   a. Sun to Earth:

   b. Sun to Jupiter:

   c. Sun to Pluto:

   d. Earth to Jupiter:

   e. Jupiter to Pluto:

8. Order the numbers in Exercise 7 from smallest to largest. Explain how writing the numbers in scientific notation helps you to quickly compare and order them.

## Example 2: Arithmetic Operations with Numbers Written Using Scientific Notation

a. $(2.4 \times 10^{20}) + (4.5 \times 10^{21})$

b. $(7 \times 10^{-9})(5 \times 10^{5})$

c. $\frac{1.2 \times 10^{15}}{3 \times 10^{7}}$

## Exercises 9–11

9. Perform the following calculations without rewriting the numbers in decimal form.

   a. $(1.42 \times 10^{15}) - (2 \times 10^{13})$

   b. $(1.42 \times 10^{15})(2.4 \times 10^{13})$

   c. $\frac{1.42 \times 10^{-5}}{2 \times 10^{13}}$

10. Estimate how many times farther Jupiter is from the Sun than Earth is from the Sun. Estimate how many times farther Pluto is from the Sun than Earth is from the Sun.

11. Estimate the distance between Earth and Jupiter and between Jupiter and Pluto.

## Problem Set

1. Write the following numbers used in these statements in scientific notation. (Note: Some of these numbers have been rounded.)
   a. The density of helium is $0.0001785$ grams per cubic centimeter.
   b. The boiling point of gold is $5200°\text{F}$.
   c. The speed of light is $186{,}000$ miles per second.
   d. One second is $0.000278$ hours.
   e. The acceleration due to gravity on the Sun is $900\ \text{ft/s}^2$.
   f. One cubic inch is $0.0000214$ cubic yards.
   g. Earth's population in 2012 was $7{,}046{,}000{,}000$ people.
   h. Earth's distance from the sun is $93{,}000{,}000$ miles.
   i. Earth's radius is $4000$ miles.
   j. The diameter of a water molecule is $0.000000028\ \text{cm}$.

2. Write the following numbers in decimal form. (Note: Some of these numbers have been rounded.)
   a. A light year is $9.46 \times 10^{15}\ \text{m}$.
   b. Avogadro's number is $6.02 \times 10^{23}\ \text{mol}^{-1}$.
   c. The universal gravitational constant is $6.674 \times 10^{-11}$ N(m/kg)$^2$.
   d. Earth's age is $4.54 \times 10^{9}$ years.
   e. Earth's mass is $5.97 \times 10^{24}\ \text{kg}$.
   f. A foot is $1.9 \times 10^{-4}$ miles.
   g. The population of China in 2014 was $1.354 \times 10^{9}$ people.
   h. The density of oxygen is $1.429 \times 10^{-4}$ grams per liter.
   i. The width of a pixel on a smartphone is $7.8 \times 10^{-2}\ \text{mm}$.
   j. The wavelength of light used in optic fibers is $1.55 \times 10^{-6}\ \text{m}$.

3. State the necessary value of $n$ that will make each statement true.
   a. $0.000027 = 2.7 \times 10^{n}$
   b. $-3.125 = -3.125 \times 10^{n}$
   c. $7{,}540{,}000{,}000 = 7.54 \times 10^{n}$
   d. $0.033 = 3.3 \times 10^{n}$
   e. $15 = 1.5 \times 10^{n}$
   f. $26{,}000 \times 200 = 5.2 \times 10^{n}$
   g. $3000 \times 0.0003 = 9 \times 10^{n}$
   h. $0.0004 \times 0.002 = 8 \times 10^{n}$
   i. $\frac{16000}{80} = 2 \times 10^{n}$
   j. $\frac{500}{0.002} = 2.5 \times 10^{n}$

Perform the following calculations without rewriting the numbers in decimal form.

k. $(2.5 \times 10^4) + (3.7 \times 10^3)$

l. $(6.9 \times 10^{-3}) - (8.1 \times 10^{-3})$

m. $(6 \times 10^{11})(2.5 \times 10^{-5})$

n. $\frac{4.5 \times 10^8}{2 \times 10^{10}}$

4. The wavelength of visible light ranges from $650$ nanometers to $850$ nanometers, where $1\text{ nm} = 1 \times 10^{-7}\text{ cm}$. Express the wavelength range of visible light in centimeters.

5. In 1694, the Dutch scientist Antonie van Leeuwenhoek was one of the first scientists to see a red blood cell in a microscope. He approximated that a red blood cell was "$25{,}000$ times as small as a grain of sand." Assume a grain of sand is $\frac{1}{2}$ millimeter wide and a red blood cell is approximately $7$ micrometers wide. One micrometer is $1 \times 10^{-6}$ meters. Support or refute Leeuwenhoek's claim. Use scientific notation in your calculations.

6. When the Mars Curiosity Rover entered the atmosphere of Mars on its descent in 2012, it was traveling roughly $13{,}200$ mph. On the surface of Mars, its speed averaged $0.00073$ mph. How many times faster was the speed when it entered the atmosphere than its typical speed on the planet's surface? Use scientific notation in your calculations.

7. Earth's surface is approximately $70\%$ water. There is no water on the surface of Mars, and its diameter is roughly half of Earth's diameter. Assume both planets are spherical. The surface area of a sphere is given by the formula $SA = 4\pi r^2$ where $r$ is the radius of the sphere. Which has more land mass, Earth or Mars? Use scientific notation in your calculations.

8. There are approximately $25$ trillion $(2.5 \times 10^{13})$ red blood cells in the human body at any one time. A red blood cell is approximately $7 \times 10^{-6}$ m wide. Imagine if you could line up all your red blood cells end to end. How long would the line of cells be? Use scientific notation in your calculations.

9. Assume each person needs approximately $100$ sq. ft. of living space. Now imagine that we are going to build a giant apartment building that will be $1$ mile wide and $1$ mile long to house all the people in the United States, estimated to be $313.9$ million people in 2012. If each floor of the apartment building is $10$ ft. high, how tall will the apartment building be?

# Lesson 3: Rational Exponents—What Are $2^{\frac{1}{2}}$ and $2^{\frac{1}{3}}$?

## Classwork

### Opening Exercise

a. What is the value of $2^{\frac{1}{2}}$? Justify your answer.

b. Graph $f(x) = 2^x$ for each integer $x$ from $x = -2$ to $x = 5$. Connect the points on your graph with a smooth curve.

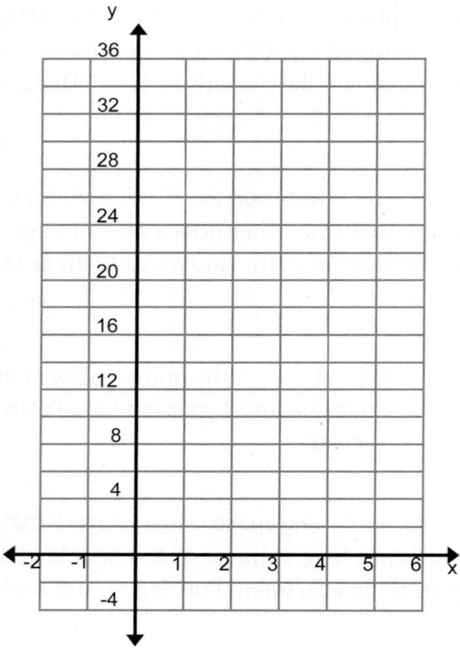

The graph at right shows a close-up view of $f(x) = 2^x$ for $-0.5 < x < 1.5$.

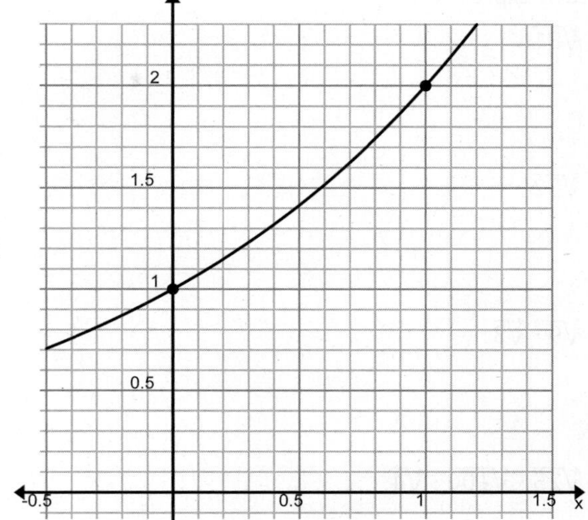


c. Find two consecutive integers that are over and under estimates of the value of $2^{\frac{1}{2}}$.

d. Does it appear that $2^{\frac{1}{2}}$ is halfway between the integers you specified in Exercise 1?

e. Use the graph of $f(x) = 2^x$ to estimate the value of $2^{\frac{1}{2}}$.

f. Use the graph of $f(x) = 2^x$ to estimate the value of $2^{\frac{1}{3}}$.

**Example 1**

a. What is the $4^{\text{th}}$ root of $16$?

b. What is the cube root of $125$?

c. What is the $5^{\text{th}}$ root of $100{,}000$?

## Exercise 1

Evaluate each expression.

a. $\sqrt[4]{81}$

b. $\sqrt[5]{32}$

c. $\sqrt[3]{9} \cdot \sqrt[3]{3}$

d. $\sqrt[4]{25} \cdot \sqrt[4]{100} \cdot \sqrt[4]{4}$

## Discussion

If $2^{\frac{1}{2}} = \sqrt{2}$ and $2^{\frac{1}{3}} = \sqrt[3]{2}$, what does $2^{\frac{3}{4}}$ equal? Explain your reasoning.

## Exercises 2–8

Rewrite each exponential expression as an $n^{\text{th}}$ root.

2. $3^{\frac{1}{2}}$

3. $11^{\frac{1}{5}}$

4. $\left(\frac{1}{4}\right)^{\frac{1}{5}}$

5. $6^{\frac{1}{10}}$

Rewrite the following exponential expressions as equivalent radical expressions. If the number is rational, write it without radicals or exponents.

6. $2^{\frac{3}{2}}$

7. $4^{\frac{5}{2}}$

8. $\left(\frac{1}{8}\right)^{\frac{5}{3}}$

9. Show why the following statement is true:

$$2^{-\frac{1}{2}} = \frac{1}{2^{\frac{1}{2}}}$$

Rewrite the following exponential expressions as equivalent radical expressions. If the number is rational, write it without radicals or exponents.

10. $4^{-\frac{3}{2}}$

11. $27^{-\frac{2}{3}}$

12. $\left(\frac{1}{4}\right)^{-\frac{1}{2}}$

### Lesson Summary

**$n^{\text{TH}}$ ROOT OF A NUMBER:** Let $a$ and $b$ be numbers, and let $n \geq 2$ be a positive integer. If $b = a^n$, then $a$ is an $n^{th}$ *root of* $b$. If $n = 2$, then the root is a called a *square root*. If $n = 3$, then the root is called a *cube root*.

**PRINCIPAL $n^{\text{TH}}$ ROOT OF A NUMBER:** Let $b$ be a real number that has at least one real $n^{\text{th}}$ root. The *principal $n^{th}$ root of $b$* is the real $n^{\text{th}}$ root that has the same sign as $b$ and is denoted by a radical symbol: $\sqrt[n]{b}$.

Every positive number has a unique principal $n^{\text{th}}$ root. We often refer to the principal $n^{\text{th}}$ root of $b$ as just the *$n^{th}$ root of $b$*. The $n^{\text{th}}$ root of $0$ is $0$.

For any positive integers $m$ and $n$, and any real number $b$ for which the principal $n^{\text{th}}$ root of $b$ exists, we have

$$b^{\frac{1}{n}} = \sqrt[n]{b}$$

$$b^{\frac{m}{n}} = \sqrt[n]{b^m} = \left(\sqrt[n]{b}\right)^m$$

$$b^{-\frac{m}{n}} = \frac{1}{\sqrt[n]{b^m}}.$$

## Problem Set

1. Select the expression from (A), (B), and (C) that correctly completes the statement.

| | | (A) | (B) | (C) |
|---|---|---|---|---|
| a. | $x^{\frac{1}{3}}$ is equivalent to ________. | $\frac{1}{3}x$ | $\sqrt[3]{x}$ | $\frac{3}{x}$ |
| b. | $x^{\frac{2}{3}}$ is equivalent to ________. | $\frac{2}{3}x$ | $\sqrt[3]{x^2}$ | $\left(\sqrt{x}\right)^3$ |
| c. | $x^{-\frac{1}{4}}$ is equivalent to ________. | $-\frac{1}{4}x$ | $\frac{4}{x}$ | $\frac{1}{\sqrt[4]{x}}$ |
| d. | $\left(\frac{4}{x}\right)^{\frac{1}{2}}$ is equivalent to ________. | $\frac{2}{x}$ | $\frac{4}{x^2}$ | $\frac{2}{\sqrt{x}}$ |

2. Identify which of the expressions (A), (B), and (C) are equivalent to the given expression.

| | | (A) | (B) | (C) |
|---|---|---|---|---|
| a. | $16^{\frac{1}{2}}$ | $\left(\frac{1}{16}\right)^{-\frac{1}{2}}$ | $8^{\frac{2}{3}}$ | $64^{\frac{3}{2}}$ |
| b. | $\left(\frac{2}{3}\right)^{-1}$ | $-\frac{3}{2}$ | $\left(\frac{9}{4}\right)^{\frac{1}{2}}$ | $\frac{27^{\frac{1}{3}}}{6}$ |

3. Rewrite in radical form. If the number is rational, write it without using radicals.

   a. $6^{\frac{3}{2}}$

   b. $\left(\frac{1}{2}\right)^{\frac{1}{4}}$

   c. $3(8)^{\frac{1}{3}}$

   d. $\left(\frac{64}{125}\right)^{-\frac{2}{3}}$

   e. $81^{-\frac{1}{4}}$

4. Rewrite the following expressions in exponent form.

   a. $\sqrt{5}$

   b. $\sqrt[3]{5^2}$

   c. $\sqrt{5^3}$

   d. $\left(\sqrt[3]{5}\right)^2$

5. Use the graph of $f(x) = 2^x$ shown to the right to estimate the following powers of 2.

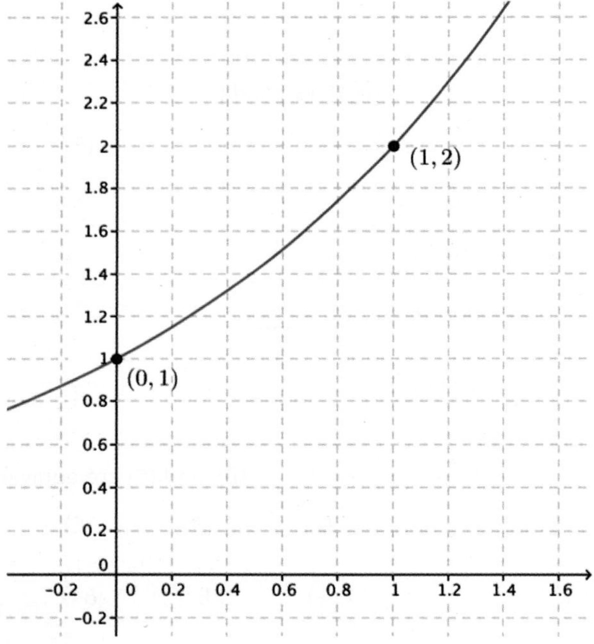


   a. $2^{\frac{1}{4}}$

   b. $2^{\frac{2}{3}}$

   c. $2^{\frac{3}{4}}$

   d. $2^{0.2}$

   e. $2^{1.2}$

   f. $2^{-\frac{1}{5}}$

6. Rewrite each expression in the form $kx^n$, where $k$ is a real number, $x$ is a positive real number, and $n$ is rational.

   a. $\sqrt[4]{16x^3}$
   b. $\frac{5}{\sqrt{x}}$
   c. $\sqrt[3]{1/x^4}$
   d. $\frac{4}{\sqrt[3]{8x^3}}$
   e. $\frac{27}{\sqrt{9x^4}}$
   f. $\left(\frac{125}{x^2}\right)^{-\frac{1}{3}}$

7. Find a value of $x$ for which $2x^{\frac{1}{2}} = 32$.

8. Find a value of $x$ for which $x^{\frac{4}{3}} = 81$.

9. If $x^{\frac{3}{2}} = 64$, find the value of $4x^{-\frac{3}{4}}$.

10. If $b = \frac{1}{9}$, evaluate the following expressions.

    a. $b^{-\frac{1}{2}}$
    b. $b^{\frac{5}{2}}$
    c. $\sqrt[3]{3b^{-1}}$

11. Show that each expression is equivalent to $2x$. Assume $x$ is a positive real number.

    a. $\sqrt[4]{16x^4}$
    b. $\frac{\left(\sqrt[3]{8x^3}\right)^2}{\sqrt{4x^2}}$
    c. $\frac{6x^3}{\sqrt[3]{27x^6}}$

12. Yoshiko said that $16^{\frac{1}{4}} = 4$ because $4$ is one-fourth of $16$. Use properties of exponents to explain why she is or is not correct.

13. Jefferson said that $8^{\frac{4}{3}} = 16$ because $8^{\frac{1}{3}} = 2$ and $2^4 = 16$. Use properties of exponents to explain why he is or is not correct.

14. Rita said that $8^{\frac{2}{3}} = 128$ because $8^{\frac{2}{3}} = 8^2 \cdot 8^{\frac{1}{3}}$, so $8^{\frac{2}{3}} = 64 \cdot 2$, and then $8^{\frac{2}{3}} = 128$. Use properties of exponents to explain why she is or is not correct.

15. Suppose for some positive real number $a$ that $\left(a^{\frac{1}{4}} \cdot a^{\frac{1}{2}} \cdot a^{\frac{1}{4}}\right)^2 = 3$.
    a. What is the value of $a$?
    b. Which exponent properties did you use to find your answer to part (a)?

16. In the lesson, you made the following argument:

$$\begin{aligned}\left(2^{\frac{1}{3}}\right)^3 &= 2^{\frac{1}{3}} \cdot 2^{\frac{1}{3}} \cdot 2^{\frac{1}{3}} \\ &= 2^{\frac{1}{3}+\frac{1}{3}+\frac{1}{3}} \\ &= 2^1 \\ &= 2.\end{aligned}$$

Since $\sqrt[3]{2}$ is a number so that $\left(\sqrt[3]{2}\right)^3 = 2$ and $2^{\frac{1}{3}}$ is a number so that $\left(2^{\frac{1}{3}}\right)^3 = 2$, you concluded that $2^{\frac{1}{3}} = \sqrt[3]{2}$. Which exponent property was used to make this argument?

# Lesson 4: Properties of Exponents and Radicals

## Classwork

### Opening Exercise

Write each exponent as a radical, and then use the definition and properties of radicals to write that expression as an integer.

a. $7^{\frac{1}{2}} \cdot 7^{\frac{1}{2}}$

b. $3^{\frac{1}{3}} \cdot 3^{\frac{1}{3}} \cdot 3^{\frac{1}{3}}$

c. $12^{\frac{1}{2}} \cdot 3^{\frac{1}{2}}$

d. $\left(64^{\frac{1}{3}}\right)^{\frac{1}{2}}$

### Examples 1–3

Write each expression in the form $b^{\frac{m}{n}}$ for positive real numbers $b$ and integers $m$ and $n$ with $n > 0$ by applying the properties of radicals and the definition of $n^{\text{th}}$ root.

1. $b^{\frac{1}{4}} \cdot b^{\frac{1}{4}}$

2. $b^{\frac{1}{3}} \cdot b^{\frac{4}{3}}$

3. $b^{\frac{1}{5}} \cdot b^{\frac{3}{4}}$

## Exercises 1–4

Write each expression in the form $b^{\frac{m}{n}}$. If a numeric expression is a rational number, then write your answer without exponents.

1. $b^{\frac{2}{3}} \cdot b^{\frac{1}{2}}$

2. $\left(b^{-\frac{1}{5}}\right)^{\frac{2}{3}}$

3. $64^{\frac{1}{3}} \cdot 64^{\frac{3}{2}}$

4. $\left(\frac{9^3}{4^2}\right)^{\frac{3}{2}}$

## Example 4

Rewrite the radical expression $\sqrt{48x^5y^4z^2}$ so that no perfect square factors remain inside the radical.

## Exercise 5

5. If $x = 50$, $y = 12$, and $z = 3$, the following expressions are difficult to evaluate without using properties of radicals or exponents (or a calculator). Use the definition of rational exponents and properties of exponents to rewrite each expression in a form where it can be easily evaluated, and then use that exponential expression to find the value.

   a. $\sqrt{8x^3y^2}$

   b. $\sqrt[3]{54y^7z^2}$

## Exercise 6

6. Order these numbers from smallest to largest. Explain your reasoning.

   $16^{2.5}$ $9^{3.1}$ $32^{1.2}$

## Lesson Summary

The properties of exponents developed in Grade 8 for integer exponents extend to rational exponents.

That is, for any integers $m$, $n$, $p$, and $q$, with $n > 0$ and $q > 0$ and any real numbers $a$ and $b$ so that $a^{\frac{1}{n}}$, $b^{\frac{1}{n}}$, and $b^{\frac{1}{q}}$ are defined, we have the following properties of exponents:

$$b^{\frac{m}{n}} \cdot b^{\frac{p}{q}} = b^{\frac{m}{n}+\frac{p}{q}}$$

$$b^{\frac{m}{n}} = \sqrt[n]{b^m}$$

$$\left(b^{\frac{1}{n}}\right)^n = b$$

$$(b^n)^{\frac{1}{n}} = b$$

$$(ab)^{\frac{m}{n}} = a^{\frac{m}{n}} \cdot b^{\frac{m}{n}}$$

$$\left(b^{\frac{m}{n}}\right)^{\frac{p}{q}} = b^{\frac{mp}{nq}}$$

$$b^{-\frac{m}{n}} = \frac{1}{b^{\frac{m}{n}}}$$

## Problem Set

1. Evaluate each expression if $a = 27$ and $b = 64$.

   a. $\sqrt[3]{a}\sqrt{b}$

   b. $\left(3\sqrt[3]{a}\sqrt{b}\right)^2$

   c. $\left(\sqrt[3]{a} + 2\sqrt{b}\right)^2$

   d. $a^{-\frac{2}{3}} + b^{\frac{3}{2}}$

   e. $\left(a^{-\frac{2}{3}} \cdot b^{\frac{3}{2}}\right)^{-1}$

   f. $\left(a^{-\frac{2}{3}} - \frac{1}{8}b^{\frac{3}{2}}\right)^{-1}$

2. Rewrite each expression so that each term is in the form $kx^n$, where $k$ is a real number, $x$ is a positive real number, and $n$ is a rational number.

   a. $x^{-\frac{2}{3}} \cdot x^{\frac{1}{3}}$

   b. $2x^{\frac{1}{2}} \cdot 4x^{-\frac{5}{2}}$

   c. $\dfrac{10x^{\frac{1}{3}}}{2x^2}$

   d. $\left(3x^{\frac{1}{4}}\right)^{-2}$

   e. $x^{\frac{1}{2}}\left(2x^2 - \frac{4}{x}\right)$

   f. $\sqrt[3]{\dfrac{27}{x^6}}$

   g. $\sqrt[3]{x} \cdot \sqrt[3]{-8x^2} \cdot \sqrt[3]{27x^4}$

   h. $\dfrac{2x^4 - x^2 - 3x}{\sqrt{x}}$

   i. $\dfrac{\sqrt{x} - 2x^{-3}}{4x^2}$

3. Show that $(\sqrt{x}+\sqrt{y})^2$ is not equal to $x^1+y^1$ when $x=9$ and $y=16$.

4. Show that $\left(x^{\frac{1}{2}}+y^{\frac{1}{2}}\right)^{-1}$ is not equal to $\frac{1}{x^{\frac{1}{2}}}+\frac{1}{y^{\frac{1}{2}}}$ when $x=9$ and $y=16$.

5. From these numbers, select (a) one that is negative, (b) one that is irrational, (c) one that is not a real number, and (d) one that is a perfect square:

$$3^{\frac{1}{2}}\cdot 9^{\frac{1}{2}},\ 27^{\frac{1}{3}}\cdot 144^{\frac{1}{2}},\ 64^{\frac{1}{3}}-64^{\frac{2}{3}},\text{ and }\left(4^{-\frac{1}{2}}-4^{\frac{1}{2}}\right)^{\frac{1}{2}}$$

6. Show that the expression $2^n\cdot 4^{n+1}\cdot\left(\frac{1}{8}\right)^n$ is equal to $4$.

7. Express each answer as a power of $10$.

   a. Multiply $10^n$ by $10$.

   b. Multiply $\sqrt{10}$ by $10^n$.

   c. Square $10^n$.

   d. Divide $100\cdot 10^n$ by $10^{2n}$.

   e. Show that $10^n=11\cdot 10^n-10^{n+1}$

8. Rewrite each of the following radical expressions as an equivalent exponential expression in which each variable occurs no more than once.

   a. $\sqrt{8x^2y}$

   b. $\sqrt[5]{96x^3y^{15}z^6}$

9. Use properties of exponents to find two integers that are upper and lower estimates of the value of $4^{1.6}$.

10. Use properties of exponents to find two integers that are upper and lower estimates of the value of $8^{2.3}$.

11. Kepler's third law of planetary motion relates the average distance, $a$, of a planet from the Sun to the time $t$ it takes the planet to complete one full orbit around the Sun according to the equation $t^2=a^3$. When the time, $t$, is measured in Earth years, the distance, $a$, is measured in astronomical units (AU). (One AU is equal to the average distance from Earth to the Sun.)

   a. Find an equation for $t$ in terms of $a$ and an equation for $a$ in terms of $t$.

   b. Venus takes about $0.616$ Earth years to orbit the Sun. What is its average distance from the Sun?

   c. Mercury is an average distance of $0.387$ AU from the Sun. About how long is its orbit in Earth years?

# Lesson 5: Irrational Exponents—What Are $2^{\sqrt{2}}$ and $2^{\pi}$?

## Classwork

### Exercise 1

a. Write the following finite decimals as fractions (you do not need to reduce to lowest terms).

$$1, \quad 1.4, \quad 1.41, \quad 1.414, \quad 1.4142, \quad 1.41421$$

b. Write $2^{1.4}$, $2^{1.41}$, $2^{1.414}$, and $2^{1.4142}$ in radical form ($\sqrt[n]{2^m}$).

c. Compute a decimal approximation to 5 decimal places of the radicals you found in part (b) using your calculator. For each approximation, underline the digits that are also in the previous approximation, starting with 2.00000 done for you below. What do you notice?

$$2^1 = 2 = 2.00000$$

**Exercise 2**

a. Write six terms of a sequence that a calculator can use to approximate $2^{\pi}$.
(Hint: $\pi = 3.14159\ldots$)

b. Compute $2^{3.14} = \sqrt[100]{2^{314}}$ and $2^{\pi}$ on your calculator. In which digit do they start to differ?

c. How could you improve the accuracy of your estimate of $2^{\pi}$?

## Problem Set

1. Is it possible for a number to be both rational and irrational?

2. Use properties of exponents to rewrite the following expressions as a number or an exponential expression with only one exponent.

   a. $\left(2^{\sqrt{3}}\right)^{\sqrt{3}}$

   b. $\left(\sqrt{2}^{\sqrt{2}}\right)^{\sqrt{2}}$

   c. $\left(3^{1+\sqrt{5}}\right)^{1-\sqrt{5}}$

   d. $3^{\frac{1+\sqrt{5}}{2}} \cdot 3^{\frac{1-\sqrt{5}}{2}}$

   e. $3^{\frac{1+\sqrt{5}}{2}} \div 3^{\frac{1-\sqrt{5}}{2}}$

   f. $3^{2\cos^2(x)} \cdot 3^{2\sin^2(x)}$

3.

   a. Between what two integer powers of 2 does $2^{\sqrt{5}}$ lie?

   b. Between what two integer powers of 3 does $3^{\sqrt{10}}$ lie?

   c. Between what two integer powers of 5 does $5^{\sqrt{3}}$ lie?

4. Use the process outlined in the lesson to approximate the number $2^{\sqrt{5}}$. Use the approximation $\sqrt{5} \approx 2.23606798$.

   a. Find a sequence of five intervals that contain $\sqrt{5}$ whose endpoints get successively closer to $\sqrt{5}$.

   b. Find a sequence of five intervals that contain $2^{\sqrt{5}}$ whose endpoints get successively closer to $2^{\sqrt{5}}$. Write your intervals in the form $2^r < 2^{\sqrt{5}} < 2^s$ for rational numbers $r$ and $s$.

   c. Use your calculator to find approximations to four decimal places of the endpoints of the intervals in part (b).

   d. Based on your work in part (c), what is your best estimate of the value of $2^{\sqrt{5}}$?

   e. Can we tell if $2^{\sqrt{5}}$ is rational or irrational? Why or why not?

5. Use the process outlined in the lesson to approximate the number $3^{\sqrt{10}}$. Use the approximation $\sqrt{10} \approx 3.1622777$.

   a. Find a sequence of five intervals that contain $3^{\sqrt{10}}$ whose endpoints get successively closer to $3^{\sqrt{10}}$. Write your intervals in the form $3^r < 3^{\sqrt{10}} < 3^s$ for rational numbers $r$ and $s$.

   b. Use your calculator to find approximations to four decimal places of the endpoints of the intervals in part (a).

   c. Based on your work in part (b), what is your best estimate of the value of $3^{\sqrt{10}}$?

6. Use the process outlined in the lesson to approximate the number $5^{\sqrt{7}}$. Use the approximation $\sqrt{7} \approx 2.64575131$.
   a. Find a sequence of seven intervals that contain $5^{\sqrt{7}}$ whose endpoints get successively closer to $5^{\sqrt{7}}$. Write your intervals in the form $5^r < 5^{\sqrt{7}} < 5^s$ for rational numbers $r$ and $s$.
   b. Use your calculator to find approximations to four decimal places of the endpoints of the intervals in part (a).
   c. Based on your work in part (b), what is your best estimate of the value of $5^{\sqrt{7}}$?

7. Can the value of an irrational number raised to an irrational power ever be rational?

# Lesson 6: Euler's Number, $e$

## Classwork

### Exercises 1–3

1. Assume that there is initially $1 \text{ cm}$ of water in the tank and the height of the water doubles every $10$ seconds. Write an equation that could be used to calculate the height $H(t)$ of the water in the tank at any time $t$.

2. How would the equation in Exercise 1 change if ...
   a. The initial depth of water in the tank was $2 \text{ cm}$?

   b. The initial depth of water in the tank was $\frac{1}{2} \text{cm}$?

   c. The initial depth of water in the tank was $10 \text{ cm}$?

   d. The initial depth of water in the tank was $A$ cm, for some positive real number $A$?

3. How would the equation in Exercise 2, part (d) change if ...
   a. The height tripled every ten seconds?

   b. The height doubled every five seconds?

c. The height quadrupled every second?

d. The height halved every ten seconds?

## Example 1

1. Consider two identical water tanks, each of which begins with a height of water 1 cm and fills with water at a different rate. Which equations can be used to calculate the height of water in each tank at time $t$? Use $H_1$ for tank 1 and $H_2$ for tank 2.

| The height of the water in TANK 1 doubles every second. | The height of the water in TANK 2 triples every second. |
|---|---|

a. If both tanks start filling at the same time, which one fills first?

b. We want to know the average rate of change of the height of the water in these tanks over an interval that starts at a fixed time $T$ as they are filling up. What is the formula for the average rate of change of a function $f$ on an interval $[a, b]$?

c. What is the formula for the average rate of change of the function $H_1$ on an interval $[a, b]$?

d. Let's calculate the average rate of change of the function $H_1$ on the interval $[T, T + 0.1]$, which is an interval one-tenth of a second long starting at an unknown time $T$.

## Exercises 4–8

4. For the second tank, calculate the average change in the height, $H_2$, from time $T$ seconds to $T + 0.1$ seconds. Express the answer as a number times the value of the original function at time $T$. Explain the meaning of these findings.

5. For each tank, calculate the average change in height from time $T$ seconds to $\mathrm{T} + 0.001$ seconds. Express the answer as a number times the value of the original function at time $T$. Explain the meaning of these findings.

6. In Exercise 5, the average rate of change of the height of the water in tank 1 on the interval $[T, T + 0.01]$ can be described by the expression $c_1 \cdot 2^T$, and the average rate of change of the height of the water in tank 2 on the interval $[T, T + 0.01]$ can be described by the expression $c_2 \cdot 3^T$. What are approximate values of $c_1$ and $c_2$?

7. As an experiment, let's look for a value of $b$ so that if the height of the water can be described by $H(t) = b^t$, then the expression for the average of change on the interval $[T, T + 0.01]$ is $1 \cdot H(T)$.

   a. Write out the expression for the average rate of change of $H(t) = b^t$ on the interval $[T, T + 0.01]$.

   b. Set your expression in part (a) equal to $1 \cdot H(T)$ and reduce to an expression involving a single $b$.

   c. Now we want to find the value of $b$ that satisfies the equation you found in part (b), but we do not have a way to explicitly solve this equation. Look back at Exercise 6; which two consecutive integers have $b$ between them?

   d. Use your calculator and a guess-and-check method to find an approximate value of $b$ to 2 decimal places.

8. Verify that for the value of $b$ found in Exercise 7, $\frac{H_b(T + 0.001) - H_b(T)}{0.001} = H_b(T)$, where $H_b(T) = b^T$.

**Lesson Summary**

- Euler's number, $e$, is an irrational number that is approximately equal to $e \approx 2.7182818284590$.
- **AVERAGE RATE OF CHANGE:** Given a function $f$ whose domain contains the interval of real numbers $[a, b]$ and whose range is a subset of the real numbers, the average rate of change on the interval $[a, b]$ is defined by the number

$$\frac{f(b) - b(a)}{b - a}.$$

## Problem Set

1. The product $4 \cdot 3 \cdot 2 \cdot 1$ is called 4 factorial and is denoted by $4!$. Then $10! = 10 \cdot 9 \cdot 8 \cdot 7 \cdot 6 \cdot 5 \cdot 4 \cdot 3 \cdot 2 \cdot 1$, and for any positive integer $n$, $n! = n(n-1)(n-2)\cdots 3 \cdot 2 \cdot 1$.

   a. Complete the following table of factorial values:

| $n$ | 1 | 2 | 3 | 4 | 5 | 6 | 7 | 8 |
|---|---|---|---|---|---|---|---|---|
| $n!$ | | | | | | | | |

   b. Evaluate the sum $1 + \frac{1}{1!}$.

   c. Evaluate the sum $1 + \frac{1}{1!} + \frac{1}{2!}$.

   d. Use a calculator to approximate the sum $1 + \frac{1}{1!} + \frac{1}{2!} + \frac{1}{3!}$ to 7 decimal places. Do not round the fractions before evaluating the sum.

   e. Use a calculator to approximate the sum $1 + \frac{1}{1!} + \frac{1}{2!} + \frac{1}{3!} + \frac{1}{4!}$ to 7 decimal places. Do not round the fractions before evaluating the sum.

   f. Use a calculator to approximate sums of the form $1 + \frac{1}{1!} + \frac{1}{2!} + \cdots + \frac{1}{k!}$ to 7 decimal places for $k = 5, 6, 7, 8, 9, 10$. Do not round the fractions before evaluating the sums with a calculator.

   g. Make a conjecture about the sums $1 + \frac{1}{1!} + \frac{1}{2!} + \cdots + \frac{1}{k!}$ for positive integers $k$ as $k$ increases in size.

   h. Would calculating terms of this sequence ever yield an exact value of $e$? Why or why not?

2. Consider the sequence given by the function $a_n = \left(1 + \frac{1}{n}\right)^n$, where $n \geq 1$ is an integer.

   a. Use your calculator to approximate the first 5 terms of this sequence to 7 decimal places.

   b. Does it appear that this sequence settles near a particular value?

   c. Use a calculator to approximate the following terms of this sequence to 7 decimal places.

      i. $a_{100}$

      ii. $a_{1000}$

      iii. $a_{10,000}$

      iv. $a_{100,000}$

      v. $a_{1,000,000}$

      vi. $a_{10,000,000}$

      vii. $a_{100,000,000}$

   d. Does it appear that this sequence settles near a particular value?

   e. Compare the results of this exercise with the results of Problem 1. What do you observe?

3. If $x = 5a^4$ and $a = 2e^3$, express $x$ in terms of $e$ and approximate to the nearest whole number.

4. If $a = 2b^3$ and $b = -\frac{1}{2}e^{-2}$, express $a$ in terms of $e$ and approximate to four decimal places.

5. If $x = 3e^4$ and $= \frac{s}{2x^3}$, show that $s = 54e^{13}$ and approximate $s$ to the nearest whole number.

6. The following graph shows the number of barrels of oil produced by the Glenn Pool well in Oklahoma from 1910 to 1916.

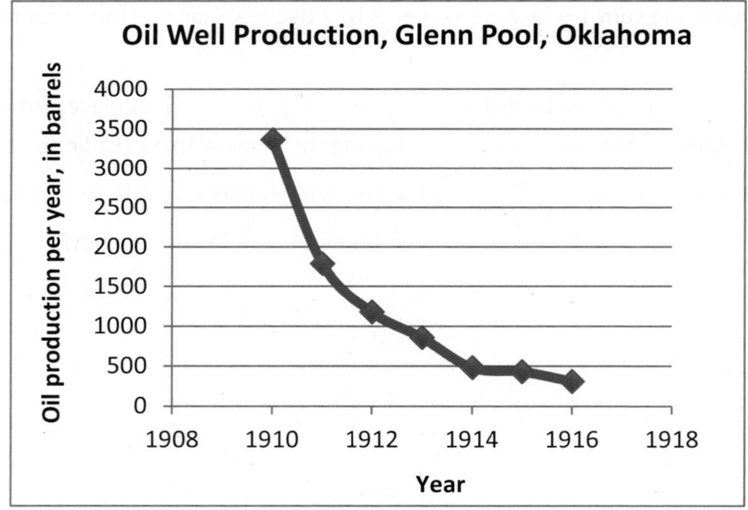


Source: Cutler, Willard W., Jr. Estimation of Underground Oil Reserves by Oil-Well Production Curves, U.S. Department of the Interior, 1924.

a. Estimate the average rate of change of the amount of oil produced by the well on the interval $[1910, 1916]$ and explain what that number represents.

b. Estimate the average rate of change of the amount of oil produced by the well on the interval $[1910, 1913]$ and explain what that number represents.

c. Estimate the average rate of change of the amount of oil produced by the well on the interval $[1913, 1916]$ and explain what that number represents.

d. Compare your results for the rates of change in oil production in the first half and the second half of the time period in question in parts (b) and (c). What do those numbers say about the production of oil from the well?

e. Notice that the average rate of change of the amount of oil produced by the well on any interval starting and ending in two consecutive years is always negative. Explain what that means in the context of oil production.

7. The following table lists the number of hybrid electric vehicles (HEVs) sold in the United States between 1999 and 2013.

| Year | Number of HEVs sold in U.S. |
|---|---|
| 1999 | 17 |
| 2000 | 9350 |
| 2001 | 20,282 |
| 2002 | 36,035 |
| 2003 | 47,600 |
| 2004 | 84,199 |
| 2005 | 209,711 |
| 2006 | 252,636 |

| Year | Number of HEVs sold in U.S. |
|---|---|
| 2007 | 352,274 |
| 2008 | 312,386 |
| 2009 | 290,271 |
| 2010 | 274,210 |
| 2011 | 268,752 |
| 2012 | 434,498 |
| 2013 | 495,685 |

Source: U.S. Department of Energy, Alternative Fuels and Advanced Vehicle Data Center, 2013.

a. During which one-year interval is the average rate of change of the number of HEVs sold the largest? Explain how you know.

b. Calculate the average rate of change of the number of HEVs sold on the interval $[2003, 2004]$ and explain what that number represents.

c. Calculate the average rate of change of the number of HEVs sold on the interval $[2003, 2008]$ and explain what that number represents.

d. What does it mean if the average rate of change of the number of HEVs sold is negative?

**Extension:**

8. The formula for the area of a circle of radius $r$ can be expressed as a function $A(r) = \pi r^2$.

a. Find the average rate of change of the area of a circle on the interval $[4, 5]$.

b. Find the average rate of change of the area of a circle on the interval $[4, 4.1]$.

c. Find the average rate of change of the area of a circle on the interval $[4, 4.01]$.

d. Find the average rate of change of the area of a circle on the interval $[4, 4.001]$.

e. What is happening to the average rate of change of the area of the circle as the interval gets smaller and smaller?

f. Find the average rate of change of the area of a circle on the interval $[4, 4+h]$ for some small positive number $h$.

g. What happens to the average rate of change of the area of the circle on the interval $[4, 4+h]$ as $h \to 0$? Does this agree with your answer to part (d)? Should it agree with your answer to part (e)?

h. Find the average rate of change of the area of a circle on the interval $[r_0, r_0+h]$ for some positive number $r_0$ and some small positive number $h$.

i. What happens to the average rate of change of the area of the circle on the interval $[r_0, r_0+h]$ as $h \to 0$? Do you recognize the resulting formula?

9. The formula for the volume of a sphere of radius $r$ can be expressed as a function $V(r) = \frac{4}{3}\pi r^3$. As you work through these questions, you will see the pattern develop more clearly if you leave your answers in the form of a coefficient times $\pi$. Approximate the coefficient to five decimal places.

a. Find the average rate of change of the volume of a sphere on the interval $[2, 3]$.

b. Find the average rate of change of the volume of a sphere on the interval $[2, 2.1]$.

c. Find the average rate of change of the volume of a sphere on the interval $[2, 2.01]$.

d. Find the average rate of change of the volume of a sphere on the interval $[2, 2.001]$.

e. What is happening to the average rate of change of the volume of a sphere as the interval gets smaller and smaller?

f. Find the average rate of change of the volume of a sphere on the interval $[2, 2+h]$ for some small positive number $h$.

g. What happens to the average rate of change of the volume of a sphere on the interval $[2, 2+h]$ as $h \to 0$? Does this agree with your answer to part (e)? Should it agree with your answer to part (e)?

h. Find the average rate of change of the volume of a sphere on the interval $[r_0, r_0+h]$ for some positive number $r_0$ and some small positive number $h$.

i. What happens to the average rate of change of the volume of a sphere on the interval $[r_0, r_0+h]$ as $h \to 0$? Do you recognize the resulting formula?

# Lesson 7: Bacteria and Exponential Growth

## Classwork

### Opening Exercise

Work with your partner or group to solve each of the following equations for $x$.

a. $2^x = 2$

b. $2^x = 2^3$

c. $2^x = 16$

d. $2^x - 64 = 0$

e. $2^x - 1 = 0$

f. $2^{3x} = 64$

g. $2^{x+1} = 32$

## Example

The *Escherichia coli* bacteria (commonly known as *E. coli*), reproduces once every $30$ minutes, meaning that a colony of *E. coli* can double every half hour. *Mycobacterium tuberculosis* has a generation time in the range of $12$ to $16$ hours. Researchers have found evidence that suggests certain bacteria populations living deep below the surface of the earth may grow at extremely slow rates, reproducing once every several thousand years. With this variation in bacterial growth rates, it is reasonable that we assume a $24$-hour reproduction time for a hypothetical bacteria colony in the next example.

Suppose we have a bacteria colony that starts with $1$ bacterium, and the population of bacteria doubles every day.

What function $P$ can we use to model the bacteria population on day $t$?

| $t$ | $P(t)$ |
|---|---|
| | |
| | |
| | |
| | |
| | |

How many days will it take for the bacteria population to reach $8$?

How many days will it take for the bacteria population to reach $16$?

Roughly how long will it take for the population to reach $10$?

We already know from our previous discussion that if $2^d = 10$, then $3 < d < 4$, and the table confirms that. At this point, we have an underestimate of $3$ and an overestimate of $4$ for $d$. How can we find better under and over estimates for $d$?

| $t$ | $P(t)$ |
|---|---|
| | |
| | |
| | |
| | |
| | |

From our table, we now know another set of under and over estimates for the number $d$ that we seek. What are they?

Continue this process of "squeezing" the number $d$ between two numbers until you are confident you know the value of $d$ to two decimal places.

| $t$ | $P(t)$ |
|---|---|
| | |
| | |
| | |
| | |
| | |

| $t$ | $P(t)$ |
|---|---|
| | |
| | |
| | |
| | |
| | |

What if we had wanted to find $d$ to 5 decimal places?

To the nearest minute, when does the population of bacteria become $10$?

| $t$ | $P(t)$ |
|---|---|
| | |
| | |
| | |
| | |
| | |

| $t$ | $P(t)$ |
|---|---|
| | |
| | |
| | |
| | |
| | |

| $t$ | $P(t)$ |
|---|---|
| | |
| | |
| | |
| | |
| | |

| $t$ | $P(t)$ |
|---|---|
| | |
| | |
| | |
| | |
| | |

## Exercise

Use the method from the Example to approximate the solution to the equations below to two decimal places.

a. $2^x = 1000$

b. $3^x = 1000$

c. $4^x = 1000$

d. $5^x = 1000$

e. $6^x = 1000$

f. $7^x = 1000$

g. $8^x = 1000$

h. $9^x = 1000$

i. $11^x = 1000$

j. $12^x = 1000$

k. $13^x = 1000$

l. $14^x = 1000$

m. $15^x = 1000$

n. $16^x = 1000$

## Problem Set

1. Solve each of the following equations for $x$ using the same technique as was used in the Opening Exercise.
   a. $2^x = 32$
   b. $2^{x-3} = 2^{2x+5}$
   c. $2^{x^2-3x} = 2^{-2}$
   d. $2^x - 2^{4x-3} = 0$
   e. $2^{3x} \cdot 2^5 = 2^7$
   f. $2^{x^2-16} = 1$
   g. $3^{2x} = 27$
   h. $3^{\frac{2}{x}} = 81$
   i. $\frac{3^{x^2}}{3^{5x}} = 3^6$

2. Solve the equation $\frac{2^{2x}}{2^{x+5}} = 1$ algebraically using two different initial steps as directed below.
   a. Write each side as a power of 2.
   b. Multiply both sides by $2^{x+5}$.

3. Find consecutive integers that are under and over estimates of the solutions to the following exponential equations.
   a. $2^x = 20$
   b. $2^x = 100$
   c. $3^x = 50$
   d. $10^x = 432{,}901$
   e. $2^{x-2} = 750$
   f. $2^x = 1.35$

4. Complete the following table to approximate the solution to $10^x = 34{,}198$ to two decimal places.

| $t$ | $P(t)$ |
|---|---|
| 1 | 10 |
| 2 | 100 |
| 3 | 1,000 |
| 4 | 10,000 |
| 5 | 100,000 |
| | |

| $t$ | $P(t)$ |
|---|---|
| 4.1 | |
| 4.2 | |
| 4.3 | |
| 4.4 | |
| 4.5 | |
| 4.6 | |

| $t$ | $P(t)$ |
|---|---|
| 4.51 | |
| 4.52 | |
| 4.53 | |
| 4.54 | |
| | |
| | |

| $t$ | $P(t)$ |
|---|---|
| 4.531 | |
| 4.532 | |
| 4.533 | |
| 4.534 | |
| 4.535 | |
| | |

5. Complete the following table to approximate the solution to $2^x = 18$ to two decimal places.

| $t$ | $P(t)$ |
|---|---|
| | |
| | |
| | |
| | |
| | |
| | |
| | |
| | |
| | |
| | |

| $t$ | $P(t)$ |
|---|---|
| | |
| | |
| | |
| | |
| | |
| | |
| | |
| | |
| | |
| | |

| $t$ | $P(t)$ |
|---|---|
| | |
| | |
| | |
| | |
| | |
| | |
| | |
| | |
| | |
| | |

| $t$ | $P(t)$ |
|---|---|
| | |
| | |
| | |
| | |
| | |
| | |
| | |
| | |
| | |
| | |

6. Approximate the solution to $5^x = 5555$ to four decimal places.

7. A dangerous bacterial compound forms in a closed environment but is immediately detected. An initial detection reading suggests the concentration of bacteria in the closed environment is one percent of the fatal exposure level. This bacteria is known to double in growth (double in concentration in a closed environment) every hour and canbe modeled by the function $P(t) = 100 \cdot 2^t$, where $t$ is measured in hours.

   a. In the function $P(t) = 100 \cdot 2^t$, what does the $100$ mean? What does the $2$ mean?

   b. Doctors and toxicology professionals estimate that exposure to two-thirds of the bacteria's fatal concentration level will begin to cause sickness. Without consulting a calculator or other technology, offer a rough time limit for the inhabitants of the infected environment to evacuate in order to avoid sickness in the doctors' estimation. Note that immediate evacuation is not always practical, so offer extra evacuation time if it is affordable.

   c. A more conservative approach is to evacuate the infected environment before bacteria concentration levels reach one-third of fatal levels. Without consulting a calculator or other technology, offer a rough time limit for evacuation in this circumstance.

d. Use the method of the Example to approximate when the infected environment will reach fatal levels ($100\%$) of bacteria concentration, to the nearest minute.

| $t$ | $2^t$ |
|---|---|
| | |
| | |
| | |
| | |
| | |
| | |
| | |
| | |
| | |

| $t$ | $2^t$ |
|---|---|
| | |
| | |
| | |
| | |
| | |
| | |
| | |
| | |
| | |

| $t$ | $2^t$ |
|---|---|
| | |
| | |
| | |
| | |
| | |
| | |
| | |
| | |
| | |

| $t$ | $2^t$ |
|---|---|
| | |
| | |
| | |
| | |
| | |
| | |
| | |
| | |
| | |

| $t$ | $2^t$ |
|---|---|
| | |
| | |
| | |
| | |
| | |
| | |
| | |
| | |
| | |

# Lesson 8: The "WhatPower" Function

## Classwork

### Opening Exercise

Evaluate each expression. The first two have been completed for you.

a. $\text{WhatPower}_2(8) = 3$

b. $\text{WhatPower}_3(9) = 2$

c. $\text{WhatPower}_6(36) =$ ______

d. $\text{WhatPower}_2(32) =$ ______

e. $\text{WhatPower}_{10}(1000) =$ ______

f. $\text{WhatPower}_{10}(1{,}000{,}000) =$ ______

g. $\text{WhatPower}_{100}(1{,}000{,}000) =$ ______

h. $\text{WhatPower}_4(64) =$ ______

i. $\text{WhatPower}_2(64) =$ ______

j. $\text{WhatPower}_9(3) =$ ______

k. $\text{WhatPower}_5(\sqrt{5}) =$ ______

l. $\text{WhatPower}_{\frac{1}{2}}\left(\frac{1}{8}\right) =$ ______

m. $\text{WhatPower}_{42}(1) =$ ______

n. $\text{WhatPower}_{100}(0.01) =$ ______

o. $\text{WhatPower}_2\left(\frac{1}{4}\right) =$ ______

p. $\text{WhatPower}_{\frac{1}{4}}(2) =$ ______

q. With your group members, write a definition for the function $\text{WhatPower}_b$, where $b$ is a number.

**Exercises 1–9**

Evaluate the following expressions and justify your answers.

2. $\text{WhatPower}_7(49)$

3. $\text{WhatPower}_0(7)$

4. $\text{WhatPower}_5(1)$

5. $\text{WhatPower}_1(5)$

6. $\text{WhatPower}_2(16)$

7. $\text{WhatPower}_{-2}(32)$

8. $\text{WhatPower}_{\frac{1}{3}}(9)$

9. $\text{WhatPower}_{-\frac{1}{3}}(27)$

10. Describe the allowable values of $b$ in the expression $\text{WhatPower}_b(x)$. When can we define a function $f(x) = \text{WhatPower}_b(x)$? Explain how you know.

## Examples

1. $\log_2(8) = 3$
2. $\log_3(9) = 2$
3. $\log_6(36) =$ ______
4. $\log_2(32) =$ ______
5. $\log_{10}(1000) =$ ______
6. $\log_{42}(1) =$ ______
7. $\log_{100}(0.01) =$ ______
8. $\log_2\left(\frac{1}{4}\right) =$ ______

### Exercise 10

10. Compute the value of each logarithm. Verify your answers using an exponential statement.
    a. $\log_2(32)$

b. $\log_3(81)$

c. $\log_9(81)$

d. $\log_5(625)$

e. $\log_{10}(1{,}000{,}000{,}000)$

f. $\log_{1000}(1{,}000{,}000{,}000)$

g. $\log_{13}(13)$

h. $\log_{13}(1)$

i. $\log_9(27)$

j. $\log_7(\sqrt{7})$

k. $\log_{\sqrt{7}}(7)$

l. $\log_{\sqrt{7}}\left(\frac{1}{49}\right)$

m. $\log_x(x^2)$

**Lesson Summary**

- If three numbers, $L$, $b$, and $x$ are related by $x = b^L$, then $L$ is the *logarithm base* $b$ of $x$ and we write $\log_b(x)$. That is, the value of the expression $L = \log_b(x)$ is the power of $b$ needed to obtain $x$.
- Valid values of $b$ as a base for a logarithm are $0 < b < 1$ and $b > 1$.

## Problem Set

1. Rewrite each of the following in the form $\text{WhatPower}_b(x) = L$.

   a. $3^5 = 243$  b. $6^{-3} = \frac{1}{216}$  c. $9^0 = 1$

2. Rewrite each of the following in the form $\log_b(x) = L$.

   a. $16^{\frac{1}{4}} = 2$  b. $10^3 = 1{,}000$  c. $b^k = r$

3. Rewrite each of the following in the form $b^L = x$.

   a. $\log_5(625) = 4$  b. $\log_{10}(0.1) = -1$  c. $\log_{27}9 = \frac{2}{3}$

4. Consider the logarithms base 2. For each logarithmic expression below, either calculate the value of the expression or explain why the expression does not make sense.
   a. $\log_2(1024)$
   b. $\log_2(128)$
   c. $\log_2(\sqrt{8})$
   d. $\log_2\left(\frac{1}{16}\right)$
   e. $\log_2(0)$
   f. $\log_2\left(-\frac{1}{32}\right)$

5. Consider the logarithms base 3. For each logarithmic expression below, either calculate the value of the expression or explain why the expression does not make sense.
   a. $\log_3(243)$
   b. $\log_3(27)$
   c. $\log_3(1)$
   d. $\log_3\left(\frac{1}{3}\right)$
   e. $\log_3(0)$
   f. $\log_3\left(-\frac{1}{3}\right)$

6. Consider the logarithms base 5. For each logarithmic expression below, either calculate the value of the expression or explain why the expression does not make sense.
   a. $\log_5(3125)$
   b. $\log_5(25)$
   c. $\log_5(1)$
   d. $\log_5\left(\frac{1}{25}\right)$
   e. $\log_5(0)$
   f. $\log_5\left(-\frac{1}{25}\right)$

7. Is there any positive number $b$ so that the expression $\log_b(0)$ makes sense? Explain how you know.

8. Is there any positive number $b$ so that the expression $\log_b(-1)$ makes sense? Explain how you know.

9. Verify each of the following by evaluating the logarithms.
   a. $\log_2(8) + \log_2(4) = \log_2(32)$
   b. $\log_3(9) + \log_3(9) = \log_3(81)$
   c. $\log_4(4) + \log_4(16) = \log_4(64)$
   d. $\log_{10}(10^3) + \log_{10}(10^4) = \log_{10}(10^7)$

10. Looking at the results from Problem 9, do you notice a trend or pattern? Can you make a general statement about the value of $\log_b(x) + \log_b(y)$?

11. To evaluate $\log_2(3)$, Autumn reasoned that since $\log_2(2) = 1$ and $\log_2(4) = 2$, $\log_2(3)$ must be the average of 1 and 2 and therefore $\log_2(3) = 1.5$. Use the definition of logarithm to show that $\log_2(3)$ cannot be 1.5. Why is her thinking not valid?

12. Find the value of each of the following.
    a. If $x = \log_2(8)$ and $y = 2^x$, find the value of $y$.
    b. If $\log_2(x) = 6$, find the value of $x$.
    c. If $r = 2^6$ and $s = \log_2(r)$, find the value of $s$.

# Lesson 9: Logarithms—How Many Digits Do You Need?

## Classwork

### Opening Exercise

a. Evaluate $\text{WhatPower}_2(8)$. State your answer as a logarithm and evaluate it.

b. Evaluate $\text{WhatPower}_5(625)$. State your answer as a logarithm and evaluate it.

### Exploratory Challenge

Autumn is starting a new club with eight members including herself. She wants everyone to have a secret identification code made up of only A's and B's. For example, using two characters, her ID code could be AB, which also happens to be her initials.

a. Using A's and B's, can Autumn assign each club member a unique two-character ID using only A's and B's? Justify your answer. Here's what Autumn has so far.

| Club Member Name | Secret ID |
|---|---|
| Autumn | AA |
| Kris | |
| Tia | |
| Jimmy | |

| Club Member Name | Secret ID |
|---|---|
| Robert | |
| Jillian | |
| Benjamin | |
| Scott | |

b. Using A's and B's, how many characters would be needed to assign each club member a unique ID code? Justify your answer by showing the IDs you would assign to each club member by completing the table above (adjust Autumn's ID if needed).

When the club grew to 16 members, Autumn started noticing a pattern.

Using A's and B's:

i. Two people could be given a secret ID with 1 letter: A and B.

ii. Four people could be given a secret ID with 2 letters: AA, BA, AB, BB.

iii. Eight people could be given a secret ID with 3 letters: AAA, BAA, ABA, BBA, AAB, BAB, ABB, BBB.

c. Complete the following statement and list the secret IDs for the 16 people.

16 people could be given a secret ID with _________ letters using A's and B's.

| Club Member Name | Secret ID |
|---|---|
| Autumn | |
| Kris | |
| Tia | |
| Jimmy | |
| Robert | |
| Jillian | |
| Benjamin | |
| Scott | |

| Club Member Name | Secret ID |
|---|---|
| Gwen | |
| Jerrod | |
| Mykel | |
| Janette | |
| Nellie | |
| Serena | |
| Ricky | |
| Mia | |

d. Describe the pattern in words. What type of function could be used to model this pattern?

## Exercises 1–2

In the previous problems, the letters A and B were like the digits in a number. A 4-digit ID for Autumn's club could be any 4-letter arrangement of A's and B's because in her ID system, the only digits are the letters A and B.

1. When Autumn's club grows to include more than $16$ people, she will need $5$ digits to assign a unique ID to each club member. What is the maximum number of people that could be in the club before she needs to switch to a 6-digit ID? Explain your reasoning.

2. If Autumn has $256$ members in her club, how many digits would she need to assign each club member a unique ID using only A's and B's? Show how you got your answers.

## Example 1

A thousand people are given unique identifiers made up of the digits $0, 1, 2, \ldots, 9$. How many digits would be needed for each ID number?

## Exercises 3–4

3. There are approximately $317$ million people in the United States. Compute and use $\log(100{,}000{,}000)$ and $\log(1{,}000{,}000{,}000)$ to explain why Social Security numbers are $9$ digits long.

4. There are many more telephones than the number of people in the United States because of people having home phones, cell phones, business phones, fax numbers, etc. Assuming we need at most 10 billion phone numbers in the United States, how many digits would be needed so that each phone number is unique? Is this reasonable? Explain.

## Problem Set

1. The student body president needs to assign each officially sanctioned club on campus a unique ID number for purposes of tracking expenses and activities. She decides to use the letters A, B, and C to create a unique three-character code for each club.

   a. How many clubs can be assigned a unique ID according to this proposal?

   b. There are actually over $500$ clubs on campus. Assuming the student body president still wants to use the letters A, B, and C, how many characters would be needed to generate a unique ID for each club?

2. Can you use the numbers $1, 2, 3$, and $4$ in a combination of $4$ digits to assign a unique ID to each of $500$ people? Explain your reasoning.

3. Automobile license plates typically have a combination of letters $(26)$ and numbers $(10)$. Over time, the state of New York has used different criteria to assign vehicle license plate numbers.

   a. From $1973$ to $1986$, the state used a 3-letter and 4-number code where the three letters indicated the county where the vehicle was registered. Essex County had $13$ different 3-letter codes in use. How many cars could be registered to this county?

   b. Since $2001$, the state has used a 3-letter and 4-number code but no longer assigns letters by county. Is this coding scheme enough to register $10$ million vehicles?

4. The Richter scale uses common (base $10$) logarithms to assign a magnitude to an earthquake that is based on the amount of force released at the earthquake's source as measured by seismographs in various locations.

   a. Explain the difference between an earthquake that is assigned a magnitude of $5$ versus one assigned a magnitude of $7$.

   b. A magnitude $2$ earthquake can usually be felt by multiple people who are located near the earthquake's origin. The largest recorded earthquake was magnitude $9.5$ in Chile in $1960$. How many times greater force than a magnitude $2$ earthquake was the largest recorded earthquake?

   c. What would be the magnitude of an earthquake whose force was $1000$ times greater than a magnitude 4.3 quake?

5. Sound pressure level is measured in decibels (dB) according to the formula $L = 10\log\left(\frac{I}{I_0}\right)$, where $I$ is the intensity of the sound and $I_0$ is a reference intensity that corresponds to a barely perceptible sound.

   a. Explain why this formula would assign $0$ decibels to a barely perceptible sound.

   b. Decibel levels above $120$ dB can be painful to humans. What would be the intensity that corresponds to this level?

# Lesson 10: Building Logarithmic Tables

## Classwork

### Opening Exercise

Find the value of the following expressions without using a calculator.

$\text{WhatPower}_{10}(1000)$ $\log_{10}(1000)$

$\text{WhatPower}_{10}(100)$ $\log_{10}(100)$

$\text{WhatPower}_{10}(10)$ $\log_{10}(10)$

$\text{WhatPower}_{10}(1)$ $\log_{10}(1)$

$\text{WhatPower}_{10}\left(\frac{1}{10}\right)$ $\log_{10}\left(\frac{1}{10}\right)$

$\text{WhatPower}_{10}\left(\frac{1}{100}\right)$ $\log_{10}\left(\frac{1}{100}\right)$

Formulate a rule based on your results above: If $k$ is an integer, then $\log_{10}(10^k) =$ ______.

Example 1

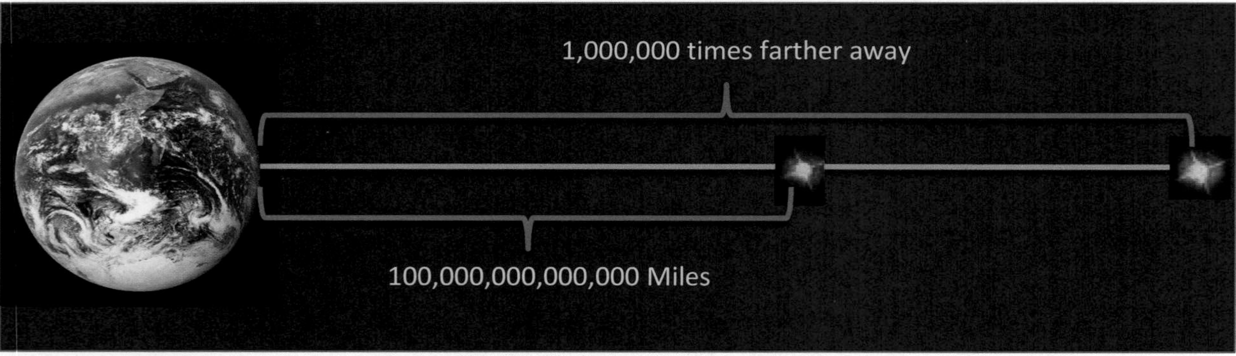


## Exercises

1. Find two consecutive powers of $10$ so that $30$ is between them. That is, find an integer exponent $k$ so that $10^k < 30 < 10^{k+1}$.

2. From your result in Exercise 1, $\log(30)$ is between which two integers?

3. Find a number $k$ to one decimal place so that $10^k < 30 < 10^{k+0.1}$, and use that to find under and over estimates for $\log(30)$.

4. Find a number $k$ to two decimal places so that $10^k < 30 < 10^{k+0.01}$, and use that to find under and over estimates for $\log(30)$.

5. Repeat this process to approximate the value of $\log(30)$ to 4 decimal places.

6. Verify your result on your calculator, using the LOG button.

7. Use your calculator to complete the following table. Round the logarithms to 4 decimal places.

| $x$ | $\log(x)$ |
|---|---|
| 1 | |
| 2 | |
| 3 | |
| 4 | |
| 5 | |
| 6 | |
| 7 | |
| 8 | |
| 9 | |

| $x$ | $\log(x)$ |
|---|---|
| 10 | |
| 20 | |
| 30 | |
| 40 | |
| 50 | |
| 60 | |
| 70 | |
| 80 | |
| 90 | |

| $x$ | $\log(x)$ |
|---|---|
| 100 | |
| 200 | |
| 300 | |
| 400 | |
| 500 | |
| 600 | |
| 700 | |
| 800 | |
| 900 | |

8. What pattern(s) can you see in the table from Exercise 7 as $x$ is multiplied by 10? Write the pattern(s) using logarithmic notation.

9. What pattern would you expect to find for $\log(1000x)$? Make a conjecture and test it to see whether or not it appears to be valid.

10. Use your results from Exercises 8 and 9 to make a conjecture about the value of $\log(10^k \cdot x)$ for any positive integer $k$.

11. Use your calculator to complete the following table. Round the logarithms to 4 decimal places.

| $x$ | $\log(x)$ |
|---|---|
| 1 | |
| 2 | |
| 3 | |
| 4 | |
| 5 | |
| 6 | |
| 7 | |
| 8 | |
| 9 | |

| $x$ | $\log(x)$ |
|---|---|
| 0.1 | |
| 0.2 | |
| 0.3 | |
| 0.4 | |
| 0.5 | |
| 0.6 | |
| 0.7 | |
| 0.8 | |
| 0.9 | |

| $x$ | $\log(x)$ |
|---|---|
| 0.01 | |
| 0.02 | |
| 0.03 | |
| 0.04 | |
| 0.05 | |
| 0.06 | |
| 0.07 | |
| 0.08 | |
| 0.09 | |

12. What pattern(s) can you see in the table from Exercise 11? Write them using logarithmic notation.

13. What pattern would you expect to find for $\log\left(\frac{x}{1000}\right)$? Make a conjecture and test it to see whether or not it appears to be valid.

14. Combine your results from Exercises 10 and 12 to make a conjecture about the value of the logarithm for a multiple of a power of 10; that is, find a formula for $\log(10^k \cdot x)$ for any integer $k$.

**Lesson Summary**

- The notation $\log(x)$ is used to represent $\log_{10}(x)$.
- For integers $k$, $\log(10^k) = k$.
- For integers $m$ and $n$, $\log(10^m \cdot 10^n) = \log(10^m) + \log(10^n)$.
- For integers k and positive real numbers $x$, $\log(10^k \cdot x) = k + \log(x)$.

## Problem Set

1. Complete the following table of logarithms without using a calculator; then, answer the questions that follow.

| $x$ | $\log(x)$ |
|---|---|
| 1,000,000 | |
| 100,000 | |
| 10,000 | |
| 1000 | |
| 100 | |
| 10 | |

| $x$ | $\log(x)$ |
|---|---|
| 0.1 | |
| 0.01 | |
| 0.001 | |
| 0.0001 | |
| 0.00001 | |
| 0.000001 | |

   a. What is $\log(1)$? How does that follow from the definition of a base-10 logarithm?
   b. What is $\log(10^k)$ for an integer $k$? How does that follow from the definition of a base-10 logarithm?
   c. What happens to the value of $\log(x)$ as $x$ gets really large?
   d. For $x > 0$, what happens to the value of $\log(x)$ as $x$ gets really close to zero?

2. Use the table of logarithms below to estimate the values of the logarithms in parts (a)–(h).

| $x$ | $\log(x)$ |
|---|---|
| 2 | 0.3010 |
| 3 | 0.4771 |
| 5 | 0.6990 |
| 7 | 0.8451 |
| 11 | 1.0414 |
| 13 | 1.1139 |

   a. $\log(70{,}000)$
   b. $\log(0.0011)$
   c. $\log(20)$
   d. $\log(0.00005)$
   e. $\log(130{,}000)$
   f. $\log(3000)$
   g. $\log(0.07)$
   h. $\log(11{,}000{,}000)$

3. If $\log(n) = 0.6$, find the value of $\log(10n)$.

4. If $m$ is a positive integer and $\log(m) \approx 3.8$, how many digits are there in $m$? Explain how you know.

5. If $m$ is a positive integer and $\log(m) \approx 9.6$, how many digits are there in $m$? Explain how you know.

6. Vivian says $\log(452{,}000) = 5 + \log(4.52)$, while her sister Lillian says that $\log(452{,}000) = 6 + \log(0.452)$. Which sister is correct? Explain how you know.

7. Write the logarithm base 10 of each number in the form $k + \log(x)$, where $k$ is the exponent from the scientific notation, and $x$ is a positive real number.

   a. $2.4902 \times 10^4$

   b. $2.58 \times 10^{13}$

   c. $9.109 \times 10^{-31}$

8. For each of the following statements, write the number in scientific notation and then write the logarithm base 10 of that number in the form $k + \log(x)$, where $k$ is the exponent from the scientific notation, and $x$ is a positive real number.

   a. The speed of sound is $1116$ ft/s.

   b. The distance from Earth to the Sun is $93$ million miles.

   c. The speed of light is $29{,}980{,}000{,}000$ cm/s.

   d. The weight of the earth is $5{,}972{,}000{,}000{,}000{,}000{,}000{,}000{,}000$ kg.

   e. The diameter of the nucleus of a hydrogen atom is $0.00000000000000175$ m.

   f. For each part (a)–(e), you have written each logarithm in the form $k + \log(x)$, for integers $k$ and positive real numbers $x$. Use a calculator to find the values of the expressions $\log(x)$. Why are all of these values between $0$ and $1$?

# Lesson 11: The Most Important Property of Logarithms

## Classwork

### Opening Exercise

Use the logarithm table below to calculate the specified logarithms.

| $x$ | $\log(x)$ |
|---|---|
| 1 | 0 |
| 2 | 0.3010 |
| 3 | 0.4771 |
| 4 | 0.6021 |
| 5 | 0.6990 |
| 6 | 0.7782 |
| 7 | 0.8451 |
| 8 | 0.9031 |
| 9 | 0.9542 |

a. $\log(80)$

b. $\log(7000)$

c. $\log(0.00006)$

d. $\log(3.0 \times 10^{27})$

e. $\log(9.0 \times 10^{k})$ for an integer $k$

**Exercises 1–5**

1. Use your calculator to complete the following table. Round the logarithms to four decimal places.

| $x$ | $\log(x)$ |
|---|---|
| 1 | 0 |
| 2 | 0.3010 |
| 3 | 0.4771 |
| 4 | 0.6021 |
| 5 | 0.6990 |
| 6 | 0.7782 |
| 7 | 0.8451 |
| 8 | 0.9031 |
| 9 | 0.9542 |

| $x$ | $\log(x)$ |
|---|---|
| 10 | |
| 12 | |
| 16 | |
| 18 | |
| 20 | |
| 25 | |
| 30 | |
| 36 | |
| 100 | |

2. Calculate the following values. Do they appear anywhere else in the table?

a. $\log(2) + \log(4)$

b. $\log(2) + \log(6)$

c. $\log(3) + \log(4)$

d. $\log(6) + \log(6)$

e. $\log(2) + \log(18)$

f. $\log(3) + \log(12)$

3. What pattern(s) can you see in Exercise 2 and the table from Exercise 1? Write them using logarithmic notation.

4. What pattern would you expect to find for $\log(x^2)$? Make a conjecture, and test it to see whether or not it appears to be valid.

5. Make a conjecture for a logarithm of the form $\log(xyz)$, where $x$, $y$, and $z$ are positive real numbers. Provide evidence that your conjecture is valid.

## Example 1

Use the logarithm table from Exercise 1 to approximate the following logarithms:

a. $\log(14)$

b. $\log(35)$

c. $\log(72)$

d. $\log(121)$

**Exercises 6–8**

6. Use your calculator to complete the following table. Round the logarithms to four decimal places.

| $x$ | $\log(x)$ |
|---|---|
| 2 | |
| 4 | |
| 5 | |
| 8 | |
| 10 | |
| 16 | |
| 20 | |
| 50 | |
| 100 | |

| $x$ | $\log(x)$ |
|---|---|
| 0.5 | |
| 0.25 | |
| 0.2 | |
| 0.125 | |
| 0.1 | |
| 0.0625 | |
| 0.05 | |
| 0.02 | |
| 0.01 | |

7. What pattern(s) can you see in the table from Exercise 6? Write a conjecture using logarithmic notation.

8. Use the definition of logarithm to justify the conjecture you found in Exercise 7.

## Example 2

Use the logarithm tables and the rules we discovered to estimate the following logarithms to four decimal places.

a. $\log(2100)$

b. $\log(0.00049)$

c. $\log(42{,}000{,}000)$

d. $\log\left(\frac{1}{640}\right)$

**Lesson Summary**

- The notation $\log(x)$ is used to represent $\log_{10}(x)$.
- The most important property of logarithms base 10 is that for positive real numbers $x$ and $y$,
$$\log(xy) = \log(x) + \log(y).$$
- For positive real numbers $x$,
$$\log\left(\frac{1}{x}\right) = -\log(x).$$

## Problem Set

1. Use the table of logarithms at right to estimate the value of the logarithms in parts (a)–(h).
   a. $\log(25)$
   b. $\log(27)$
   c. $\log(33)$
   d. $\log(55)$
   e. $\log(63)$
   f. $\log(75)$
   g. $\log(81)$
   h. $\log(99)$

| $x$ | $\log(x)$ |
|---|---|
| 2 | 0.30 |
| 3 | 0.48 |
| 5 | 0.70 |
| 7 | 0.85 |
| 11 | 1.04 |
| 13 | 1.11 |

2. Use the table of logarithms at right to estimate the value of the logarithms in parts (a)–(f).
   a. $\log(350)$
   b. $\log(0.0014)$
   c. $\log(0.077)$
   d. $\log(49{,}000)$
   e. $\log(1.69)$
   f. $\log(6.5)$

3. Use the table of logarithms at right to estimate the value of the logarithms in parts (a)–(f).
   a. $\log\left(\frac{1}{30}\right)$
   b. $\log\left(\frac{1}{35}\right)$
   c. $\log\left(\frac{1}{40}\right)$
   d. $\log\left(\frac{1}{42}\right)$
   e. $\log\left(\frac{1}{50}\right)$
   f. $\log\left(\frac{1}{64}\right)$

4. Reduce each expression to a single logarithm of the form $\log(x)$.
   a. $\log(5) + \log(7)$
   b. $\log(3) + \log(9)$
   c. $\log(15) - \log(5)$
   d. $\log(8) + \log\left(\frac{1}{4}\right)$

5. Use properties of logarithms to write the following expressions involving logarithms of only prime numbers.
   a. $\log(2500)$
   b. $\log(0.00063)$
   c. $\log(1250)$
   d. $\log(26{,}000{,}000)$

6. Use properties of logarithms to show that $\log(26) = \log(2) - \log\left(\frac{1}{13}\right)$.

7. Use properties of logarithms to show that $\log(3) + \log(4) + \log(5) - \log(6) = 1$.

8. Use properties of logarithms to show that $-\log(3) = \log\left(\frac{1}{2} - \frac{1}{3}\right) + \log(2)$.

9. Use properties of logarithms to show that $\log\left(\frac{1}{3} - \frac{1}{4}\right) + \left(\log\left(\frac{1}{3}\right) - \log\left(\frac{1}{4}\right)\right) = -2\log(3)$.

# Lesson 12: Properties of Logarithms

## Classwork

### Opening Exercise

Use the approximation $\log(2) \approx 0.3010$ to approximate the values of each of the following logarithmic expressions.

a. $\log(20)$

b. $\log(0.2)$

c. $\log(2^4)$

### Exercises 1–10

For Exercises 1–6, explain why each statement below is a property of base-10 logarithms.

1. Property 1: $\log(1) = 0$.

2. Property 2: $\log(10) = 1$.

3. Property 3: For all real numbers $r$, $\log(10^r) = r$.

4. Property 4: For any $x > 0$, $10^{\log(x)} = x$.

5. Property 5: For any positive real numbers $x$ and $y$, $\log(x \cdot y) = \log(x) + \log(y)$.
Hint: Use an exponent rule as well as Property 4.

6. Property 6: For any positive real number $x$ and any real number $r$, $\log(x^r) = r \cdot \log(x)$.
Hint: Again, use an exponent rule as well as Property 4.

7. Apply properties of logarithms to rewrite the following expressions as a single logarithm or number.

a. $\frac{1}{2}\log(25) + \log(4)$

b. $\frac{1}{3}\log(8) + \log(16)$

c. $3\log(5) + \log(0.8)$

8. Apply properties of logarithms to rewrite each expression as a sum of terms involving numbers, $\log(x)$, and $\log(y)$.

a. $\log(3x^2y^5)$

b. $\log\left(\sqrt{x^7y^3}\right)$

9. In mathematical terminology, logarithms are well defined because if $X = Y$, then $\log(X) = \log(Y)$ for $X, Y > 0$. This means that if you want to solve an equation involving exponents, you can apply a logarithm to both sides of the equation, just as you can take the square root of both sides when solving a quadratic equation. You do need to be careful not to take the logarithm of a negative number or zero.

   Use the property stated above to solve the following equations.

   a. $10^{10x} = 100$

   b. $10^{x-1} = \frac{1}{10^{x+1}}$

   c. $100^{2x} = 10^{3x-1}$

10. Solve the following equations.

   a. $10^x = 2^7$

   b. $10^{x^2+1} = 15$

   c. $4^x = 5^3$

**Lesson Summary**

We have established the following properties for base-10 logarithms, where $x$ and $y$ are positive real numbers and $r$ is any real number:

1. $\log(1) = 0$
2. $\log(10) = 1$
3. $\log(10^r) = r$
4. $10^{\log(x)} = x$
5. $\log(x \cdot y) = \log(x) + \log(y)$
6. $\log(x^r) = r \cdot \log(x)$

Additional properties not yet established are the following:

7. $\log\left(\frac{1}{x}\right) = -\log(x)$
8. $\log\left(\frac{x}{y}\right) = \log(x) - \log(y)$

Also, logarithms are well defined, meaning that for $X, Y > 0$, if $X = Y$, then $\log(X) = \log(Y)$.

## Problem Set

1. Use the approximate logarithm values below to estimate each of the following logarithms. Indicate which properties you used.

$\log(2) = 0.3010$ $\log(3) = 0.4771$

$\log(5) = 0.6990$ $\log(7) = 0.8451$

a. $\log(6)$
b. $\log(15)$
c. $\log(12)$
d. $\log(10^7)$
e. $\log\left(\frac{1}{5}\right)$
f. $\log\left(\frac{3}{7}\right)$
g. $\log(\sqrt[4]{2})$

2. Let $\log(X) = r$, $\log(Y) = s$, and $\log(Z) = t$. Express each of the following in terms of $r$, $s$, and $t$.
   a. $\log\left(\frac{X}{Y}\right)$
   b. $\log(YZ)$
   c. $\log(X^r)$
   d. $\log(\sqrt[3]{Z})$
   e. $\log\left(\sqrt[4]{\frac{Y}{Z}}\right)$
   f. $\log(XY^2Z^3)$

3. Use the properties of logarithms to rewrite each expression in an equivalent form containing a single logarithm.
   a. $\log\left(\frac{13}{5}\right) + \log\left(\frac{5}{4}\right)$
   b. $\log\left(\frac{5}{6}\right) - \log\left(\frac{2}{3}\right)$
   c. $\frac{1}{2}\log(16) + \log(3) + \log\left(\frac{1}{4}\right)$

4. Use the properties of logarithms to rewrite each expression in an equivalent form containing a single logarithm.
   a. $\log(\sqrt{x}) + \frac{1}{2}\log\left(\frac{1}{x}\right) + 2\log(x)$
   b. $\log(\sqrt[5]{x}) + \log(\sqrt[5]{x^4})$
   c. $\log(x) + 2\log(y) - \frac{1}{2}\log(z)$
   d. $\frac{1}{3}(\log(x) - 3\log(y) + \log(z))$
   e. $2(\log(x) - \log(3y)) + 3(\log(z) - 2\log(x))$

5. Use properties of logarithms to rewrite the following expressions in an equivalent form containing only $\log(x)$, $\log(y)$, $\log(z)$, and numbers.
   a. $\log\left(\frac{3x^2y^4}{\sqrt{z}}\right)$
   b. $\log\left(\frac{42\sqrt[3]{xy^7}}{x^2z}\right)$
   c. $\log\left(\frac{100x^2}{y^3}\right)$
   d. $\log\left(\sqrt{\frac{x^3y^2}{10z}}\right)$
   e. $\log\left(\frac{1}{10x^2z}\right)$

6. Express $\log\left(\frac{1}{x} - \frac{1}{x+1}\right) + \left(\log\left(\frac{1}{x}\right) - \log\left(\frac{1}{x+1}\right)\right)$ as a single logarithm for positive numbers $x$.

7. Show that $\log(x + \sqrt{x^2 - 1}) + \log(x - \sqrt{x^2 - 1}) = 0$ for $x \geq 1$.

8. If $xy = 10^{3.67}$, find the value of $\log(x) + \log(y)$.

9. Solve the following exponential equations by taking the logarithm base 10 of both sides. Leave your answers stated in terms of logarithmic expressions.
   a. $10^{x^2} = 320$
   b. $10^{\frac{x}{8}} = 300$
   c. $10^{3x} = 400$
   d. $5^{2x} = 200$
   e. $3^x = 7^{-3x+2}$

10. Solve the following exponential equations.
    a. $10^x = 3$
    b. $10^y = 30$
    c. $10^z = 300$
    d. Use the properties of logarithms to justify why $x$, $y$, and $z$ form an arithmetic sequence whose constant difference is 1.

11. Without using a calculator, explain why the solution to each equation must be a real number between 1 and 2.
    a. $11^x = 12$
    b. $21^x = 30$
    c. $100^x = 2000$
    d. $\left(\frac{1}{11}\right)^x = 0.01$
    e. $\left(\frac{2}{3}\right)^x = \frac{1}{2}$
    f. $99^x = 9000$

12. Express the exact solution to each equation as a base-10 logarithm. Use a calculator to approximate the solution to the nearest $1000^{\text{th}}$.
    a. $11^x = 12$
    b. $21^x = 30$
    c. $100^x = 2000$
    d. $\left(\frac{1}{11}\right)^x = 0.01$
    e. $\left(\frac{2}{3}\right)^x = \frac{1}{2}$
    f. $99^x = 9000$

13. Show that the value of $x$ that satisfies the equation $10^x = 3 \cdot 10^n$ is $\log(3) + n$.

14. Solve each equation. If there is no solution, explain why.
    a. $3 \cdot 5^x = 21$
    b. $10^{x-3} = 25$
    c. $10^x + 10^{x+1} = 11$
    d. $8 - 2^x = 10$

15. Solve the following equation for $n$: $A = P(1+r)^n$.

16. In this exercise, we will establish a formula for the logarithm of a sum. Let $L = \log(x+y)$, where $x, y > 0$.
    a. Show $\log(x) + \log\left(1 + \frac{y}{x}\right) = L$. State as a property of logarithms after showing this is a true statement.
    b. Use part (a) and the fact that $\log(100) = 2$ to rewrite $\log(365)$ as a sum.
    c. Rewrite $365$ in scientific notation, and use properties of logarithms to express $\log(365)$ as a sum of an integer and a logarithm of a number between $0$ and $10$.
    d. What do you notice about your answers to (b) and (c)?
    e. Find two integers that are upper and lower estimates of $\log(365)$.

# Lesson 13: Changing the Base

## Classwork

### Exercises

1. Assume that $x$, $a$, and $b$ are all positive real numbers, so that $a \neq 1$ and $b \neq 1$. What is $\log_b(x)$ in terms of $\log_a(x)$? The resulting equation allows us to change the base of a logarithm from $a$ to $b$.

2. Approximate each of the following logarithms to four decimal places. Use the LOG key on your calculator rather than logarithm tables, first changing the base of the logarithm to 10 if necessary.

   a. $\log(3^2)$

   b. $\log_3(3^2)$

   c. $\log_2(3^2)$

3. In Lesson 12, we justified a number of properties of base 10 logarithms. Working in pairs, justify the following properties of base $b$ logarithms.

   a. $\log_b(1) = 0$

   b. $\log_b(b) = 1$

   c. $\log_b(\mathrm{b}^r) = r$

   d. $b^{\log_b(x)} = x$

   e. $\log_b(x \cdot y) = \log_b(x) + \log_b(y)$

f. $\log_b(x^r) = r \cdot \log_b(x)$

g. $\log_b\left(\frac{1}{x}\right) = -\log_b(x)$

h. $\log_b\left(\frac{x}{y}\right) = \log_b(x) - \log_b(y)$

4. Find each of the following to four decimal places. Use the LN key on your calculator rather than a table.

a. $\ln(3^2)$

b. $\ln(2^4)$

5. Write as a single logarithm:

   a. $\ln(4) - 3\ln\left(\frac{1}{3}\right) + \ln(2)$.

   b. $\ln(5) + \frac{3}{5}\ln(32) - \ln(4)$.

6. Write each expression as a sum or difference of constants and logarithms of simpler terms.

   a. $\ln\left(\frac{\sqrt{5x^3}}{e^2}\right)$

   b. $\ln\left(\frac{(x+y)^2}{x^2+y^2}\right)$

### Lesson Summary

We have established a formula for changing the base of logarithms from $b$ to $a$:

$$\log_b(x) = \frac{\log_a(x)}{\log_a(b)}.$$

In particular, the formula allows us to change logarithms base $b$ to common or natural logarithms, which are the only two kinds of logarithms that calculators compute:

$$\log_b(x) = \frac{\log(x)}{\log(b)} = \frac{\ln(x)}{\ln(b)}.$$

We have also established the following properties for base $b$ logarithms. If $x$, $y$, $a$, and $b$ are all positive real numbers with $a \neq 1$ and $b \neq 1$ and $r$ is any real number, then:

1. $\log_b(1) = 0$
2. $\log_b(b) = 1$
3. $\log_b(b^r) = r$
4. $b^{\log_b(x)} = x$
5. $\log_b(x \cdot y) = \log_b(x) + \log_b(y)$
6. $\log_b(x^r) = r \cdot \log_b(x)$
7. $\log_b\left(\frac{1}{x}\right) = -\log_b(x)$
8. $\log_b\left(\frac{x}{y}\right) = \log_b(x) - \log_b(y)$

## Problem Set

1. Evaluate each of the following logarithmic expressions, approximating to four decimal places if necessary. Use the LN or LOG key on your calculator rather than a table.
   a. $\log_8(16)$
   b. $\log_7(11)$
   c. $\log_3(2) + \log_2(3)$

2. Use logarithmic properties and the fact that $\ln(2) \approx 0.69$ and $\ln(3) \approx 1.10$ to approximate the value of each of the following logarithmic expressions. Do not use a calculator.
   a. $\ln(e^4)$
   b. $\ln(6)$
   c. $\ln(108)$
   d. $\ln\left(\frac{8}{3}\right)$

3. Compare the values of $\log_{\frac{1}{9}}(10)$ and $\log_9\left(\frac{1}{10}\right)$ without using a calculator.

4. Show that for any positive numbers $a$ and $b$ with $a \neq 1$ and $b \neq 1$, $\log_a(b) \cdot \log_b(a) = 1$.

5. Express $x$ in terms of $a$, $e$, and $y$ if $\ln(x) - \ln(y) = 2a$.

6. Rewrite each expression in an equivalent form that only contains one base 10 logarithm.
   a. $\log_2(800)$
   b. $\log_x\left(\frac{1}{10}\right)$, for positive real values of $x \neq 1$
   c. $\log_5(12{,}500)$
   d. $\log_3(0.81)$

7. Write each number in terms of natural logarithms, and then use the properties of logarithms to show that it is a rational number.
   a. $\log_9(\sqrt{27})$
   b. $\log_8(32)$
   c. $\log_4\left(\frac{1}{8}\right)$

8. Write each expression as an equivalent expression with a single logarithm. Assume $x$, $y$, and $z$ are positive real numbers.
   a. $\ln(x) + 2\ln(y) - 3\ln(z)$
   b. $\frac{1}{2}(\ln(x + y) - \ln(z))$
   c. $(x + y) + \ln(z)$

9. Rewrite each expression as sums and differences in terms of $\ln(x)$, $\ln(y)$, and $\ln(z)$.
   a. $\ln(xyz^3)$
   b. $\ln\left(\frac{e^3}{xyz}\right)$
   c. $\ln\left(\sqrt{\frac{x}{y}}\right)$

10. Solve the following equations in terms of base 5 logarithms. Then, use the change of base properties and a calculator to estimate the solution to the nearest $1000^{\text{th}}$. If the equation has no solution, explain why.
   a. $5^{2x} = 20$
   b. $75 = 10 \cdot 5^{x-1}$
   c. $5^{2+x} - 5^x = 10$
   d. $5^{x^2} = 0.25$

11. In Lesson 6, you discovered that $\log(x \cdot 10^k) = k + \log(x)$ by looking at a table of logarithms. Use the properties of logarithms to justify this property for an arbitrary base $b > 0$ with $b \neq 1$. That is, show that $\log_b(x \cdot b^k) = k + \log_b(x)$.

12. Larissa argued that since $\log_2(2) = 1$ and $\log_2(4) = 2$, then it must be true that $\log_2(3) = 1.5$. Is she correct? Explain how you know.

13. Extension: Suppose that there is some positive number $b$ so that

$$\log_b(2) = 0.36$$
$$\log_b(3) = 0.57$$
$$\log_b(5) = 0.84.$$

   a. Use the given values of $\log_b(2)$, $\log_b(3)$, and $\log_b(5)$ to evaluate the following logarithms.
      i. $\log_b(6)$
      ii. $\log_b(8)$
      iii. $\log_b(10)$
      iv. $\log_b(600)$
   b. Use the change of base formula to convert $\log_b(10)$ to base 10, and solve for $b$. Give your answer to four decimal places.

14. Solve the following exponential equations.
   a. $2^{3x} = 16$
   b. $2^{x+3} = 4^{3x}$
   c. $3^{4x-2} = 27^{x+2}$
   d. $4^{2x} = \left(\frac{1}{4}\right)^{3x}$
   e. $5^{0.2x+3} = 625$

15. Solve each exponential equation.
   a. $3^{2x} = 81$
   b. $6^{3x} = 36^{x+1}$
   c. $625 = 5^{3x}$
   d. $25^{4-x} = 5^{3x}$
   e. $32^{x-1} = \frac{1}{2}$
   f. $\frac{4^{2x}}{2^{x-3}} = 1$
   g. $\frac{1}{8^{2x-4}} = 1$
   h. $2^x = 81$
   i. $8 = 3^x$
   j. $6^{x+2} = 12$
   k. $10^{x+4} = 27$
   l. $2^{x+1} = 3^{1-x}$
   m. $3^{2x-3} = 2^{x+4}$
   n. $e^{2x} = 5$
   o. $e^{x-1} = 6$

16. In Problem 9(e) of Lesson 12, you solved the equation $3^x = 7^{-3x+2}$ using the logarithm base $10$.

    a. Solve $3^x = 7^{-3x+2}$ using the logarithm base $3$.

    b. Apply the change of base formula to show that your answer to part (a) agrees with your answer to Problem 9(e) of Lesson 12.

    c. Solve $3^x = 7^{-3x+2}$ using the logarithm base $7$.

    d. Apply the change of base formula to show that your answer to part (c) also agrees with your answer to Problem 9(e) of Lesson 12.

17. Pearl solved the equation $2^x = 10$ as follows:

$$\log(2^x) = \log(10)$$
$$x\log(2) = 1$$
$$x = \frac{1}{\log(2)}.$$

Jess solved the equation $2^x = 10$ as follows:

$$\log_2(2^x) = \log_2(10)$$
$$x\log_2(2) = \log_2(10)$$
$$x = \log_2(10).$$

Is Pearl correct? Is Jess correct? Explain how you know.

# Lesson 14: Solving Logarithmic Equations

## Classwork

### Opening Exercises

Convert the following logarithmic equations to exponential form:

a. $\log(10{,}000) = 4$

b. $\log(\sqrt{10}) = \frac{1}{2}$

c. $\log_2(256) = 8$

d. $\log_4(256) = 4$

e. $\ln(1) = 0$

f. $\log(x + 2) = 3$

**Examples 1–3**

Write each of the following equations as an equivalent exponential equation, and solve for $x$.

1. $\log(3x + 7) = 0$

2. $\log_2(x + 5) = 4$

3. $\log(x + 2) + \log(x + 5) = 1$

## Exercises

1. Drew said that the equation $\log_2[(x+1)^4] = 8$ cannot be solved because he expanded $(x+1)^4 = x^4 + 4x^3 + 6x^2 + 4x + 1$ and realized that he cannot solve the equation $x^4 + 4x^3 + 6x^2 + 4x + 1 = 2^8$. Is he correct? Explain how you know.

Solve the equations in Exercises 2–4 for $x$.

2. $\ln((4x)^5) = 15$

3. $\log((2x+5)^2) = 4$

4. $\log_2((5x+7)^{19}) = 57$

Solve the logarithmic equations in Exercises 5–9, and identify any extraneous solutions.

5. $\log(x^2+7x+12) - \log(x+4) = 0$

6. $\log_2(3x) + \log_2(4) = 4$

7. $2\ln(x+2) - \ln(-x) = 0$

8. $\log(x) = 2 - \log(x)$

9. $\ln(x+2) = \ln(12) - \ln(x+3)$

## Problem Set

1. Solve the following logarithmic equations.
   a. $\log(x) = \frac{5}{2}$
   b. $5\log(x+4) = 10$
   c. $\log_2(1-x) = 4$
   d. $\log_2(49x^2) = 4$
   e. $\log_2(9x^2 + 30x + 25) = 8$

2. Solve the following logarithmic equations.
   a. $\ln(x^6) = 36$
   b. $\log[(2x^2 + 45x - 25)^5] = 10$
   c. $\log[(x^2 + 2x - 3)^4] = 0$

3. Solve the following logarithmic equations.
   a. $\log(x) + \log(x-1) = \log(3x+12)$
   b. $\ln(32x^2) - 3\ln(2) = 3$
   c. $\log(x) + \log(-x) = 0$
   d. $\log(x+3) + \log(x+5) = 2$
   e. $\log(10x+5) - 3 = \log(x-5)$
   f. $\log_2(x) + \log_2(2x) + \log_2(3x) + \log_2(36) = 6$

4. Solve the following equations.
   a. $\log_2(x) = 4$
   b. $\log_6(x) = 1$
   c. $\log_3(x) = -4$
   d. $\log_{\sqrt{2}}(x) = 4$
   e. $\log_{\sqrt{5}}(x) = 3$
   f. $\log_3(x^2) = 4$
   g. $\log_2(y^{-3}) = 12$
   h. $\log_3(8x+9) = 4$
   i. $2 = \log_4(3x-2)$
   j. $\log_5(3-2x) = 0$
   k. $\ln(2x) = 3$
   l. $\log_3(x^2 - 3x + 5) = 2$
   m. $\log((x^2+4)^5) = 10$
   n. $\log(x) + \log(x+21) = 2$
   o. $\log_4(x-2) + \log_4(2x) = 2$
   p. $\log(x) - \log(x+3) = -1$
   q. $\log_4(x+3) - \log_4(x-5) = 2$
   r. $\log(x) + 1 = \log(x+9)$
   s. $\log_3(x^2-9) - \log_3(x+3) = 1$
   t. $1 - \log_8(x-3) = \log_8(2x)$
   u. $\log_2(x^2-16) - \log_2(x-4) = 1$
   v. $\log\left(\sqrt{(x+3)^3}\right) = \frac{3}{2}$
   w. $\ln(4x^2 - 1) = 0$
   x. $\ln(x+1) - \ln(2) = 1$

# Lesson 15: Why Were Logarithms Developed?

## Classwork

### Exercises

1. Solve the following equations. Remember to check for extraneous solutions because logarithms are only defined for positive real numbers.
   a. $\log(x^2) = \log(49)$

   b. $\log(x+1) + \log(x-2) = \log(7x-17)$

   c. $\log(x^2+1) = \log\big(x(x-2)\big)$

d. $\log(x+4) + \log(x-1) = \log(3x)$

e. $\log(x^2 - x) - \log(x-2) = \log(x-3)$

f. $\log(x) + \log(x-1) + \log(x+1) = 3\log(x)$

g. $\log(x-4) = -\log(x-2)$

2. How do you know if you need to use the definition of logarithm to solve an equation involving logarithms as we did in Lesson 15 or if you can use the methods of this lesson?

### Lesson Summary

A table of base 10 logarithms can be used to simplify multiplication of multi-digit numbers:

1. To compute $A \times B$ for positive real numbers $A$ and $B$, look up the values $\log(A)$ and $\log(B)$ in the logarithm table.
2. Add $\log(A)$ and $\log(B)$. The sum can be written as $k + d$, where $k$ is an integer and $0 \leq d < 1$ is the decimal part.
3. Look back at the table and find the entry closest to the decimal part, $d$.
4. The product of that entry and $10^k$ is an approximation to $A \times B$.

A similar process simplifies division of multi-digit numbers:

1. To compute $A \div B$ for positive real numbers $A$ and $B$, look up the values $\log(A)$ and $\log(B)$ in the logarithm table.
2. Calculate $\log(A) - \log(B)$. The difference can be written as $k + d$, where $k$ is an integer and $0 \leq d < 1$ is the decimal part.
3. Look back at the table to find the entry closest to the decimal part, $d$.
4. The product of that entry and $10^k$ is an approximation to $A \div B$.

For any positive values $X$ and $Y$, if $\log_b(X) = \log_b(Y)$, we can conclude that $X = Y$. This property is the essence of how a logarithm table works, and it allows us to solve equations with logarithmic expressions on both sides of the equation.

## Problem Set

1. Use the table of logarithms to approximate solutions to the following logarithmic equations.
   a. $\log(x) = 0.5044$
   b. $\log(x) = -0.5044$ [Hint: Begin by writing $-0.5044$ as $[(-0.5044) + 1] - 1$.]
   c. $\log(x) = 35.5044$
   d. $\log(x) = 4.9201$

2. Use logarithms and the logarithm table to evaluate each expression.
   a. $\sqrt{2.33}$
   b. $13{,}500 \cdot 3{,}600$
   c. $\frac{7.2 \times 10^9}{1.3 \times 10^5}$

3. Solve for $x$: $\log(3) + 2\log(x) = \log(27)$.

4. Solve for $x$: $\log(3x) + \log(x + 4) = \log(15)$.

5. Solve for $x$.
   a. $\log(x) = \log(y + z) + \log(y - z)$
   b. $\log(x) = \left(\log(y) + \log(z)\right) + \left(\log(y) - \log(z)\right)$

6. If $x$ and $y$ are positive real numbers, and $\log(y) = 1 + \log(x)$, express $y$ in terms of $x$.

7. If $x$, $y$, and $z$ are positive real numbers, and $\log(x) - \log(y) = \log(y) - \log(z)$, express $y$ in terms of $x$ and $z$.

8. If $x$ and $y$ are positive real numbers, and $\log(x) = y\left(\log(y + 1) - \log(y)\right)$, express $x$ in terms of $y$.

9. If $x$ and $y$ are positive real numbers, and $\log(y) = 3 + 2\log(x)$, express $y$ in terms of $x$.

10. If $x$, $y$, and $z$ are positive real numbers, and $\log(z) = \log(y) + 2\log(x) - 1$, express $z$ in terms of $x$ and $y$.

11. Solve the following equations.
    a. $\ln(10) - \ln(7 - x) = \ln(x)$
    b. $\ln(x + 2) + \ln(x - 2) = \ln(9x - 24)$
    c. $\ln(x + 2) + \ln(x - 2) = \ln(-2x - 1)$

12. Suppose the formula $P = P_0(1 + r)^t$ gives the population of a city $P$ growing at an annual percent rate $r$, where $P_0$ is the population $t$ years ago.
    a. Find the time $t$ it takes this population to double.
    b. Use the structure of the expression to explain why populations with lower growth rates take a longer time to double.
    c. Use the structure of the expression to explain why the only way to double the population in one year is if there is a $100$ percent growth rate.

13. If $x > 0$, $a + b > 0$, $a > b$, and $\log(x) = \log(a + b) + \log(a - b)$, find $x$ in terms of $a$ and $b$.

14. Jenn claims that because $\log(1) + \log(2) + \log(3) = \log(6)$, then $\log(2) + \log(3) + \log(4) = \log(9)$.
    a. Is she correct? Explain how you know.
    b. If $\log(a) + \log(b) + \log(c) = \log(a + b + c)$, express $c$ in terms of $a$ and $b$. Explain how this result relates to your answer to part (a).
    c. Find other values of $a$, $b$, and $c$ so that $\log(a) + \log(b) + \log(c) = \log(a + b + c)$.

15. In Problem 7 of the Lesson 12 Problem Set, you showed that for $x \geq 1$, $\log\left(x + \sqrt{x^2 - 1}\right) + \log\left(x - \sqrt{x^2 - 1}\right) = 0$. It follows that $\log\left(x + \sqrt{x^2 - 1}\right) = -\log\left(x - \sqrt{x^2 - 1}\right)$. What does this tell us about the relationship between the expressions $x + \sqrt{x^2 - 1}$ and $x - \sqrt{x^2 - 1}$?

16. Use the change of base formula to solve the following equations.

    a. $\log(x) = \log_{100}(x^2 - 2x + 6)$
    b. $\log(x - 2) = \log_{100}(14 - x)$
    c. $\log_2(x + 1) = \log_4(x^2 + 3x + 4)$
    d. $\log_2(x - 1) = \log_8(x^3 - 2x^2 - 2x + 5)$

17. Solve the following equation:

$$\log(9x) = \frac{2\ln(3) + \ln(x)}{\ln(10)}.$$

# Lesson 16: Rational and Irrational Numbers

## Classwork

### Opening Exercise

a. Explain how to use a number line to add the fractions $\frac{7}{5} + \frac{9}{4}$.

b. Convert $\frac{7}{5}$ and $\frac{9}{4}$ to decimals, and explain the process for adding them together.

## Exercises

1. According to the calculator, $\log(4) = 0.6020599913...$ and $\log(25) = 1.3979400087....$ Find an approximation of $\log(4) + \log(25)$ to one decimal place, that is, to an accuracy of $10^{-1}$.

2. Find the value of $\log(4) + \log(25)$ to an accuracy of $10^{-2}$.

3. Find the value of $\log(4) + \log(25)$ to an accuracy of $10^{-8}$.

4. Make a conjecture: Is $\log(4) + \log(25)$ a rational or an irrational number?

5. Why is your conjecture in Exercise 4 true?

Remember that the calculator gives the following values: $\log(4) = 0.6020599913...$ and $\log(25) = 1.3979400087....$

6. Find the value of $\log(4) \cdot \log(25)$ to three decimal places.

7. Find the value of $\log(4) \cdot \log(25)$ to five decimal places.

8. Does your conjecture from the above discussion appear to be true?

**Lesson Summary**

- Irrational numbers occur naturally and frequently.
- The $n^{\text{th}}$ roots of most integers and rational numbers are irrational.
- Logarithms of most positive integers or positive rational numbers are irrational.
- We can locate an irrational number on the number line by trapping it between lower and upper approximations. The infinite process of squeezing the irrational number in smaller and smaller intervals locates exactly where the irrational number is on the number line.
- We can perform arithmetic operations such as addition and multiplication with irrational numbers using lower and upper approximations and squeezing the result of the operation in smaller and smaller intervals between two rational approximations to the result.

## Problem Set

1. Given that $\sqrt{5} \approx 2.2360679775$ and $\pi \approx 3.1415926535$, find the sum $\sqrt{5} + \pi$ to an accuracy of $10^{-8}$, without using a calculator.

2. Put the following numbers in order from least to greatest.

$$\sqrt{2}, \pi, 0, e, \frac{22}{7}, \frac{\pi^2}{3}, 3.14, \sqrt{10}$$

3. Find a rational number between the specified two numbers.
   a. $\frac{4}{13}$ and $\frac{5}{13}$
   b. $\frac{3}{8}$ and $\frac{5}{9}$
   c. $1.7299999$ and $1.73$
   d. $\frac{\sqrt{2}}{7}$ and $\frac{\sqrt{2}}{9}$
   e. $\pi$ and $\sqrt{10}$

4. Knowing that $\sqrt{2}$ is irrational, find an irrational number between $\frac{1}{2}$ and $\frac{5}{9}$.

5. Give an example of an irrational number between $e$ and $\pi$.

6. Given that $\sqrt{2}$ is irrational, which of the following numbers are irrational?

$$\frac{\sqrt{2}}{2}, 2+\sqrt{2}, \frac{\sqrt{2}}{2\sqrt{2}}, \frac{2}{\sqrt{2}}, \left(\sqrt{2}\right)^2$$

7. Given that $\pi$ is irrational, which of the following numbers are irrational?

$$\frac{\pi}{2}, \frac{\pi}{2\pi}, \sqrt{\pi}, \pi^2$$

8. Which of the following numbers are irrational?

$$1, 0, \sqrt{5}, \sqrt[3]{64}, e, \pi, \frac{\sqrt{2}}{2}, \frac{\sqrt{8}}{\sqrt{2}}, \cos\left(\frac{\pi}{3}\right), \sin\left(\frac{\pi}{3}\right)$$

9. Find two irrational numbers $x$ and $y$ so that their average is rational.

10. Suppose that $\frac{2}{3}x$ is an irrational number. Explain how you know that $x$ must be an irrational number. (Hint: What would happen if there were integers $a$ and $b$ so that $x = \frac{a}{b}$?)

11. If $r$ and $s$ are rational numbers, prove that $r + s$ and $r - s$ are also rational numbers.

12. If $r$ is a rational number and $x$ is an irrational number, determine whether the following numbers are always rational, sometimes rational, or never rational. Explain how you know.
    a. $r + x$
    b. $r - x$
    c. $rx$
    d. $x^r$

13. If $x$ and $y$ are irrational numbers, determine whether the following numbers are always rational, sometimes rational, or never rational. Explain how you know.
    a. $x + y$
    b. $x - y$
    c. $xy$
    d. $\frac{x}{y}$

# Lesson 17: Graphing the Logarithm Function

## Classwork

### Opening Exercise

Graph the points in the table for your assigned function $f(x) = \log(x)$, $g(x) = \log_2(x)$, or $h(x) = \log_5(x)$ for $0 < x \leq 16$. Then, sketch a smooth curve through those points and answer the questions that follow.

| 10-team $f(x) = \log(x)$ | |
|---|---|
| $x$ | $f(x)$ |
| 0.0625 | −1.20 |
| 0.125 | −0.90 |
| 0.25 | −0.60 |
| 0.5 | −0.30 |
| 1 | 0 |
| 2 | 0.30 |
| 4 | 0.60 |
| 8 | 0.90 |
| 16 | 1.20 |

| 2-team $g(x) = \log_2(x)$ | |
|---|---|
| $x$ | $g(x)$ |
| 0.0625 | −4 |
| 0.125 | −3 |
| 0.25 | −2 |
| 0.5 | −1 |
| 1 | 0 |
| 2 | 1 |
| 4 | 2 |
| 8 | 3 |
| 16 | 4 |

| 5-team $h(x) = \log_5(x)$ | |
|---|---|
| $x$ | $h(x)$ |
| 0.0625 | −1.72 |
| 0.125 | −1.29 |
| 0.25 | −0.86 |
| 0.5 | −0.43 |
| 1 | 0 |
| 2 | 0.43 |
| 4 | 0.86 |
| 8 | 1.29 |
| 16 | 1.72 |

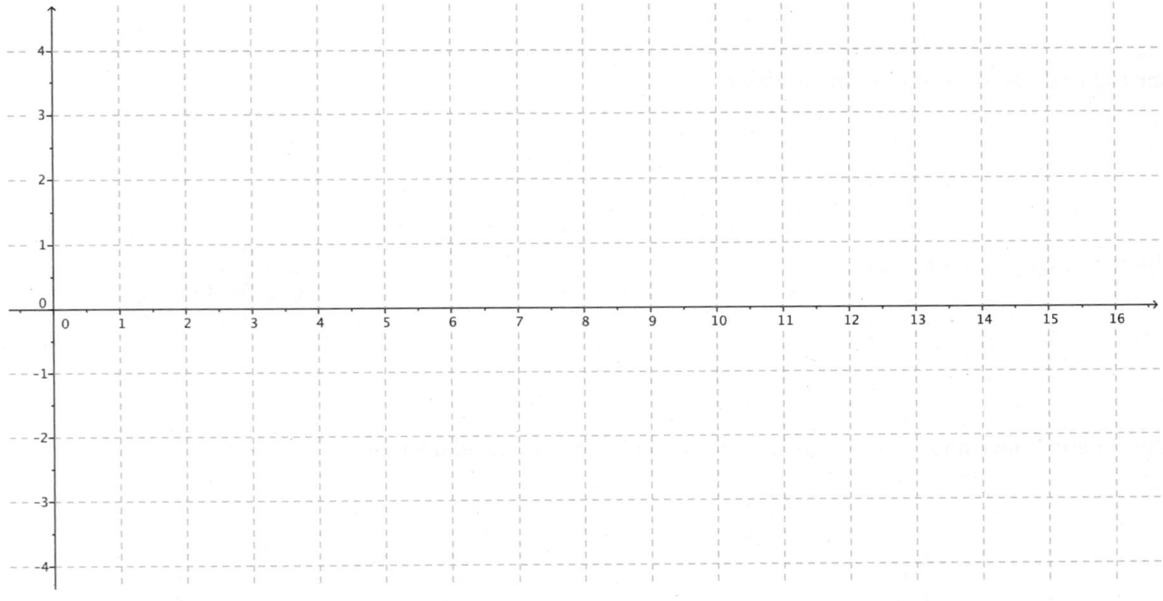

a. What does the graph indicate about the domain of your function?

b. Describe the $x$-intercepts of the graph.

c. Describe the $y$-intercepts of the graph.

d. Find the coordinates of the point on the graph with $y$-value $1$.

e. Describe the behavior of the function as $x \to 0$.

f. Describe the end behavior of the function as $x \to \infty$.

g. Describe the range of your function.

h. Does this function have any relative maxima or minima? Explain how you know.

## Exercises

1. Graph the points in the table for your assigned function $r(x) = \log_{\frac{1}{10}}(x)$, $s(x) = \log_{\frac{1}{2}}(x)$, or $t(x) = \log_{\frac{1}{5}}(x)$ for $0 < x \leq 16$. Then sketch a smooth curve through those points, and answer the questions that follow.

| 10-team<br>$r(x) = \log_{\frac{1}{10}}(x)$ | |
|---|---|
| $x$ | $r(x)$ |
| 0.0625 | 1.20 |
| 0.125 | 0.90 |
| 0.25 | 0.60 |
| 0.5 | 0.30 |
| 1 | 0 |
| 2 | −0.30 |
| 4 | −0.60 |
| 8 | −0.90 |
| 16 | −1.20 |

| 2-team<br>$s(x) = \log_{\frac{1}{2}}(x)$ | |
|---|---|
| $x$ | $s(x)$ |
| 0.0625 | 4 |
| 0.125 | 3 |
| 0.25 | 2 |
| 0.5 | 1 |
| 1 | 0 |
| 2 | −1 |
| 4 | −2 |
| 8 | −3 |
| 16 | −4 |

| $e$-team<br>$t(x) = \log_{\frac{1}{5}}(x)$ | |
|---|---|
| $x$ | $t(x)$ |
| 0.0625 | 1.72 |
| 0.125 | 1.29 |
| 0.25 | 0.86 |
| 0.5 | 0.43 |
| 1 | 0 |
| 2 | −0.43 |
| 4 | −0.86 |
| 8 | −1.29 |
| 16 | −1.72 |

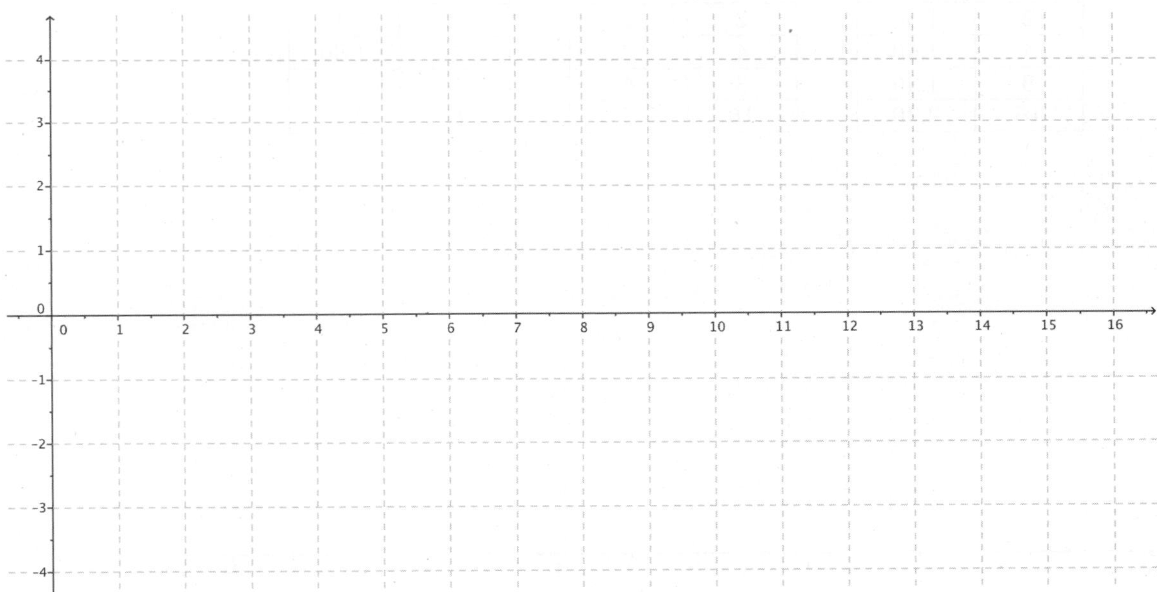


a. What is the relationship between your graph in the Opening Exercise and your graph from this exercise?

b. Why does this happen? Use the change of base formula to justify what you have observed in part (a).

2. In general, what is the relationship between the graph of a function $y = f(x)$ and the graph of $y = f(kx)$ for a constant $k$?

3. Graph the points in the table for your assigned function $u(x) = \log(10x)$, $v(x) = \log_2(2x)$, or $w(x) = \log_5(5x)$ for $0 < x \leq 16$. Then sketch a smooth curve through those points, and answer the questions that follow.

| 10-team | |
|---|---|
| $u(x) = \log(10x)$ | |
| $x$ | $u(x)$ |
| 0.0625 | −0.20 |
| 0.125 | 0.10 |
| 0.25 | 0.40 |
| 0.5 | 0.70 |
| 1 | 1 |
| 2 | 1.30 |
| 4 | 1.60 |
| 8 | 1.90 |
| 16 | 2.20 |

| 2-team | |
|---|---|
| $v(x) = \log_2(2x)$ | |
| $x$ | $v(x)$ |
| 0.0625 | −3 |
| 0.125 | −2 |
| 0.25 | −1 |
| 0.5 | 0 |
| 1 | 1 |
| 2 | 2 |
| 4 | 3 |
| 8 | 4 |
| 16 | 5 |

| 5-team | |
|---|---|
| $w(x) = \log_5(5x)$ | |
| $x$ | $w(x)$ |
| 0.0625 | −0.72 |
| 0.125 | −0.29 |
| 0.25 | 0.14 |
| 0.5 | 0.57 |
| 1 | 1 |
| 2 | 1.43 |
| 4 | 1.86 |
| 8 | 2.29 |
| 16 | 2.72 |

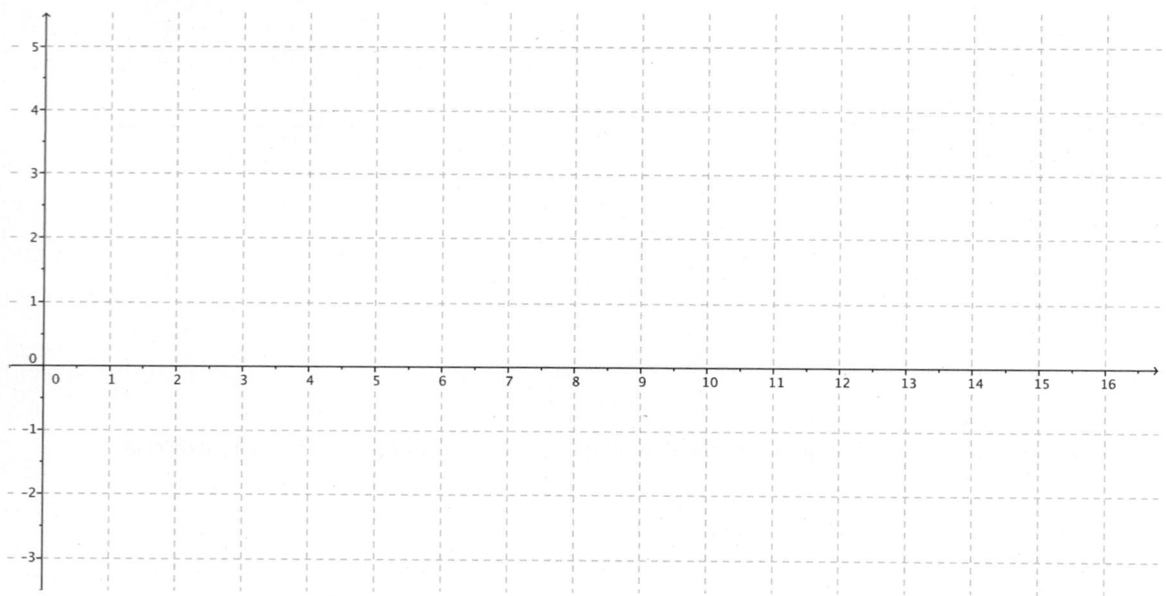

a. Describe a transformation that takes the graph of your team's function in this exercise to the graph of your team's function in the Opening Exercise.

b. Do your answers to Exercise 2 and part (a) agree? If not, use properties of logarithms to justify your observations in part (a).

**Lesson Summary**

The function $f(x) = \log_b(x)$ is defined for irrational and rational numbers. Its domain is all positive real numbers. Its range is all real numbers.

The function $f(x) = \log_b(x)$ goes to negative infinity as $x$ goes to zero. It goes to positive infinity as $x$ goes to positive infinity.

The larger the base $b$, the more slowly the function $f(x) = \log_b(x)$ increases.

By the change of base formula, $\log_{\frac{1}{b}}(x) = -\log_b(x)$.

## Problem Set

1. The function $Q(x) = \log_b(x)$ has function values in the table at right.
   a. Use the values in the table to sketch the graph of $y = Q(x)$.
   b. What is the value of $b$ in $Q(x) = \log_b(x)$? Explain how you know.
   c. Identify the key features in the graph of $y = Q(x)$.

| $x$ | $Q(x)$ |
|---|---|
| 0.1 | 1.66 |
| 0.3 | 0.87 |
| 0.5 | 0.50 |
| 1.00 | 0.00 |
| 2.00 | −0.50 |
| 4.00 | −1.00 |
| 6.00 | −1.29 |
| 10.00 | −1.66 |
| 12.00 | −1.79 |

2. Consider the logarithmic functions $f(x) = \log_b(x)$, $g(x) = \log_5(x)$, where $b$ is a positive real number, and $b \neq 1$. The graph of $f$ is given at right.
   a. Is $b > 5$, or is $b < 5$? Explain how you know.
   b. Compare the domain and range of functions $f$ and $g$.
   c. Compare the $x$-intercepts and $y$-intercepts of $f$ and $g$.
   d. Compare the end behavior of $f$ and $g$.

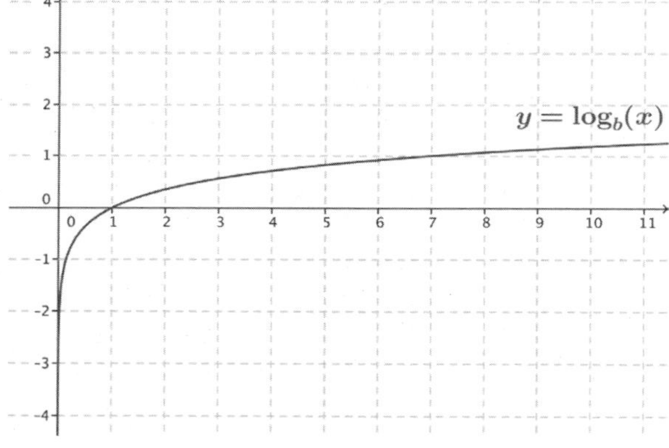

3. Consider the logarithmic functions $f(x) = \log_b(x)$, $g(x) = \log_{\frac{1}{2}}(x)$, where $b$ is a positive real number and $b \neq 1$. A table of approximate values of $f$ is given below.

| $x$ | $f(x)$ |
|---|---|
| $\frac{1}{4}$ | $0.86$ |
| $\frac{1}{2}$ | $0.43$ |
| $1$ | $0$ |
| $2$ | $-0.43$ |
| $4$ | $-0.86$ |

a. Is $b > \frac{1}{2}$, or is $b < \frac{1}{2}$? Explain how you know.

b. Compare the domain and range of functions $f$ and $g$.

c. Compare the $x$-intercepts and $y$-intercepts of $f$ and $g$.

d. Compare the end behavior of $f$ and $g$.

4. On the same set of axes, sketch the functions $f(x) = \log_2(x)$ and $g(x) = \log_2(x^3)$.

a. Describe a transformation that takes the graph of $f$ to the graph of $g$.

b. Use properties of logarithms to justify your observations in part (a).

5. On the same set of axes, sketch the functions $f(x) = \log_2(x)$ and $g(x) = \log_2\left(\frac{x}{4}\right)$.

a. Describe a transformation that takes the graph of $f$ to the graph of $g$.

b. Use properties of logarithms to justify your observations in part (a).

6. On the same set of axes, sketch the functions $f(x) = \log_{\frac{1}{2}}(x)$ and $g(x) = \log_2\left(\frac{1}{x}\right)$.

a. Describe a transformation that takes the graph of $f$ to the graph of $g$.

b. Use properties of logarithms to justify your observations in part (a).

7. The figure below shows graphs of the functions $f(x) = \log_3(x)$, $g(x) = \log_5(x)$, and $h(x) = \log_{11}(x)$.

a. Identify which graph corresponds to which function. Explain how you know.

b. Sketch the graph of $k(x) = \log_7(x)$ on the same axes.

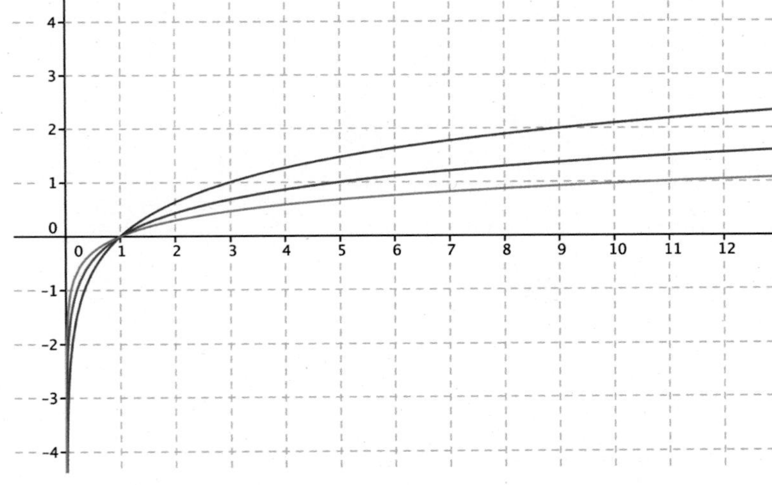

8. The figure below shows graphs of the functions $f(x) = \log_{\frac{1}{3}}(x)$, $g(x) = \log_{\frac{1}{5}}(x)$, and $h(x) = \log_{\frac{1}{11}}(x)$.

a. Identify which graph corresponds to which function. Explain how you know.

b. Sketch the graph of $k(x) = \log_{\frac{1}{7}}(x)$ on the same axes.

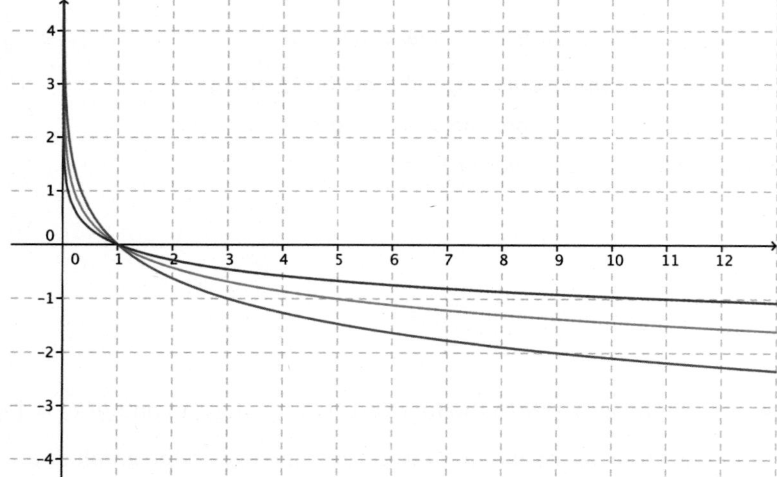

9. For each function $f$, find a formula for the function $h$ in terms of $x$. Part (a) has been done for you.

a. If $f(x) = x^2 + x$, find $h(x) = f(x + 1)$.

b. If $f(x) = \sqrt{x^2 + \frac{1}{4}}$, find $h(x) = f\left(\frac{1}{2}x\right)$.

c. If $f(x) = \log(x)$, find $h(x) = f\left(\sqrt[3]{10x}\right)$ when $x > 0$.

d. If $f(x) = 3^x$, find $h(x) = f(\log_3(x^2 + 3))$.

e. If $f(x) = x^3$, find $h(x) = f\left(\frac{1}{x^3}\right)$ when $x \neq 0$.

f. If $f(x) = x^3$, find $h(x) = f\left(\sqrt[3]{x}\right)$.

g. If $f(x) = \sin(x)$, find $h(x) = f\left(x + \frac{\pi}{2}\right)$.

h. If $f(x) = x^2 + 2x + 2$, find $h(x) = f(\cos(x))$.

10. For each of the functions $f$ and $g$ below, write an expression for (i) $f(g(x))$, (ii) $g(f(x))$, and (iii) $f(f(x))$ in terms of $x$. Part (a) has been done for you.

 a. $f(x) = x^2$, $g(x) = x + 1$

  i. $f(g(x)) = f(x+1)$
  $= (x+1)^2$

  ii. $g(f(x)) = g(x^2)$
  $= x^2 + 1$

  iii. $f(f(x)) = f(x^2)$
  $= (x^2)^2$
  $= x^4$

 b. $f(x) = \frac{1}{4}x - 8$, $g(x) = 4x + 1$

 c. $f(x) = \sqrt[3]{x+1}$, $g(x) = x^3 - 1$

 d. $f(x) = x^3$, $g(x) = \frac{1}{x}$

 e. $f(x) = |x|$, $g(x) = x^2$

**Extension:**

11. Consider the functions $f(x) = \log_2(x)$ and $(x) = \sqrt{x-1}$.

 a. Use a calculator or other graphing utility to produce graphs of $f(x) = \log_2(x)$ and $g(x) = \sqrt{x-1}$ for $x \leq 17$.

 b. Compare the graph of the function $f(x) = \log_2(x)$ with the graph of the function $g(x) = \sqrt{x-1}$. Describe the similarities and differences between the graphs.

 c. Is it always the case that $\log_2(x) > \sqrt{x-1}$ for $x > 2$?

12. Consider the functions $f(x) = \log_2(x)$ and $(x) = \sqrt[3]{x-1}$.

 a. Use a calculator or other graphing utility to produce graphs of $f(x) = \log_2(x)$ and $h(x) = \sqrt[3]{x-1}$ for $x \leq 28$.

 b. Compare the graph of the function $f(x) = \log_2(x)$ with the graph of the function $h(x) = \sqrt[3]{x-1}$. Describe the similarities and differences between the graphs.

 c. Is it always the case that $\log_2(x) > \sqrt[3]{x-1}$ for $x > 2$?

# Lesson 18: Graphs of Exponential Functions and Logarithmic Functions

## Classwork

### Opening Exercise

Complete the following table of values of the function $f(x) = 2^x$. We want to sketch the graph of $y = f(x)$ and then reflect that graph across the diagonal line with equation $y = x$.

| $x$ | $y = 2^x$ | Point $(x, y)$ on the graph of $y = 2^x$ |
|---|---|---|
| $-3$ | | |
| $-2$ | | |
| $-1$ | | |
| $0$ | | |
| $1$ | | |
| $2$ | | |
| $3$ | | |

On the set of axes below, plot the points from the table and sketch the graph of $y = 2^x$. Next, sketch the diagonal line with equation $y = x$, and then reflect the graph of $y = 2^x$ across the line.

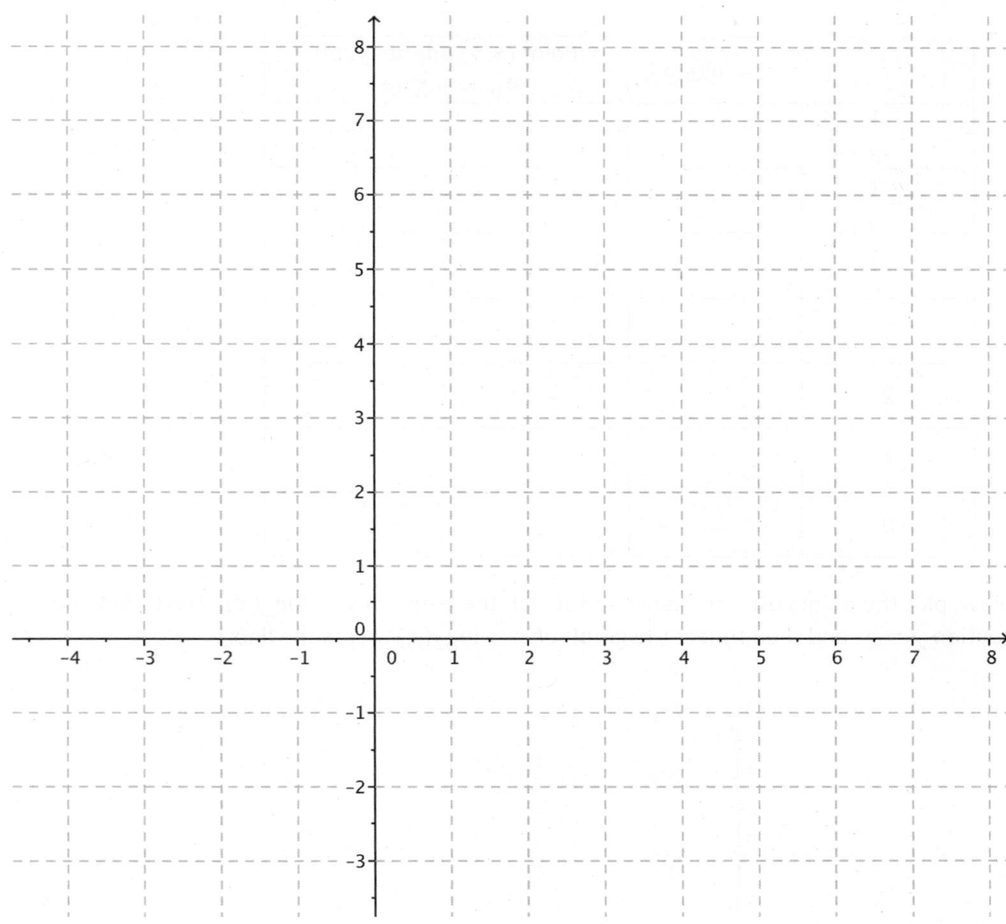

## Exercises

1. Complete the following table of values of the function $g(x) = \log_2(x)$. We want to sketch the graph of $y = g(x)$ and then reflect that graph across the diagonal line with equation $y = x$.

| $x$ | $y = \log_2(x)$ | Point $(x, y)$ on the graph of $y = \log_2(x)$ |
|---|---|---|
| $-\frac{1}{8}$ | | |
| $-\frac{1}{4}$ | | |
| $-\frac{1}{2}$ | | |
| 1 | | |
| 2 | | |
| 4 | | |
| 8 | | |

On the set of axes below, plot the points from the table and sketch the graph of $y = \log_2(x)$. Next, sketch the diagonal line with equation $y = x$, and then reflect the graph of $y = \log_2(x)$ across the line.

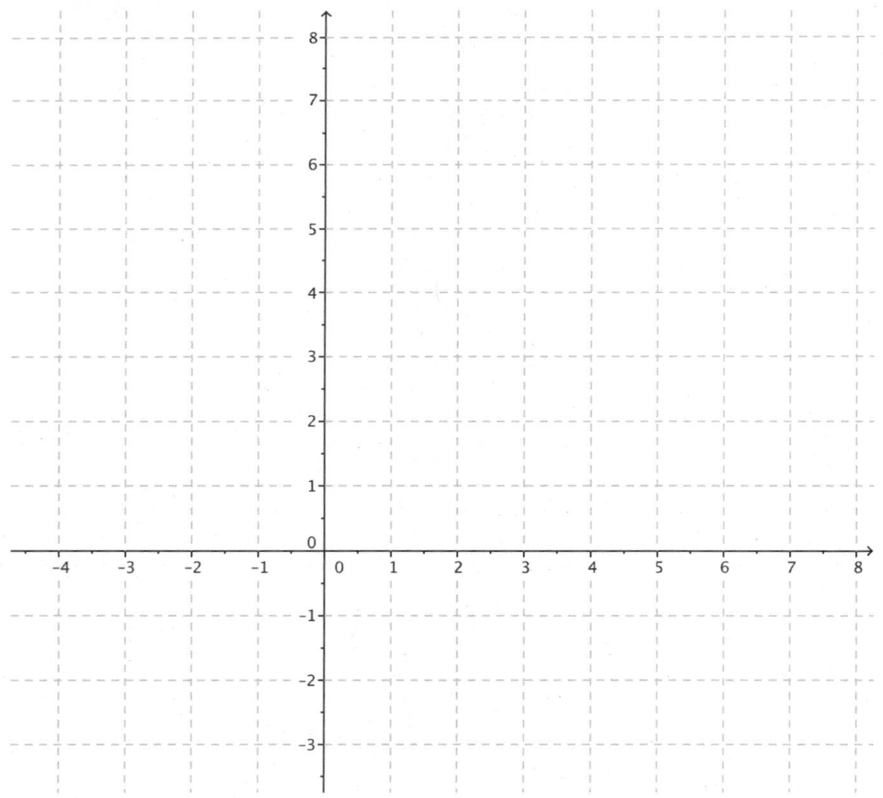

2. Working independently, predict the relation between the graphs of the functions $f(x) = 3^x$ and $g(x) = \log_3(x)$. Test your predictions by sketching the graphs of these two functions. Write your prediction in your notebook, provide justification for your prediction, and compare your prediction with that of your neighbor.

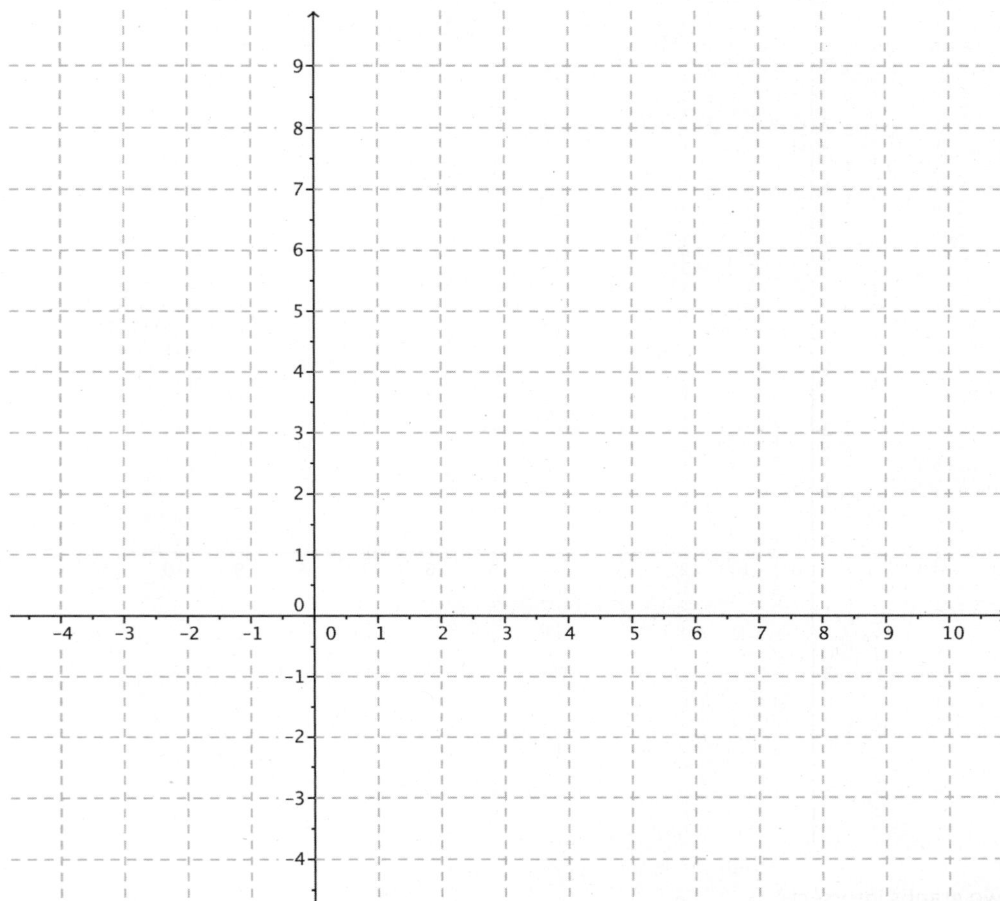

3. Now let's compare the graphs of the functions $f_2(x) = 2^x$ and $f_3(x) = 3^x$; sketch the graphs of the two exponential functions on the same set of axes; then, answer the questions below.

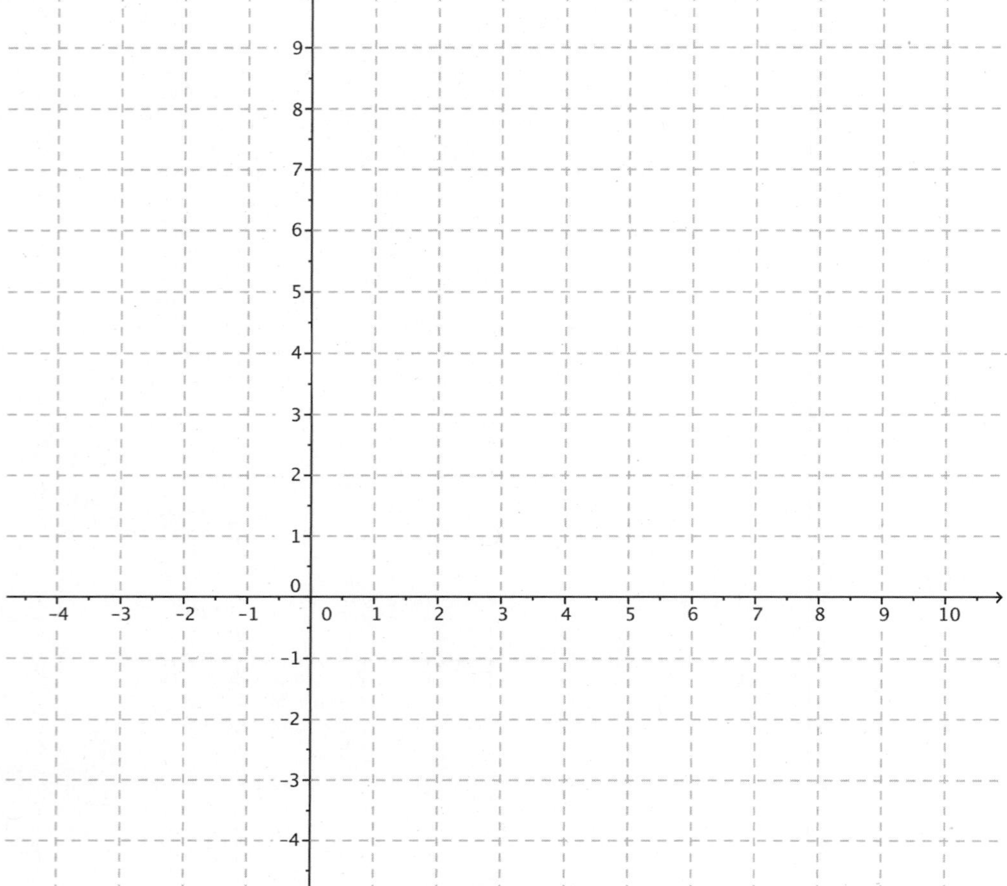

a. Where do the two graphs intersect?

b. For which values of $x$ is $2^x < 3^x$?

c. For which values of $x$ is $2^x > 3^x$?

d. What happens to the values of the functions $f_2$ and $f_3$ as $x \to \infty$?

e. What happens to the values of the functions $f_2$ and $f_3$ as $x \to -\infty$?

f. Does either graph ever intersect the $x$-axis? Explain how you know.

4. Add sketches of the two logarithmic functions $g_2(x) = \log_2(x)$ and $g_3(x) = \log_3(x)$ to the axes with the graphs of the exponential functions; then, answer the questions below.

   a. Where do the two logarithmic graphs intersect?

   b. For which values of $x$ is $\log_2(x) < \log_3(x)$?

   c. For which values of $x$ is $\log_2(x) > \log_3(x)$?

   d. What happens to the values of the functions $f_2$ and $f_3$ as $x \to \infty$?

   e. What happens to the values of the functions $f_2$ and $f_3$ as $x \to 0$?

   f. Does either graph ever intersect the $y$-axis? Explain how you know.

   g. Describe the similarities and differences in the behavior of $f_2(x)$ and $g_2(x)$ as $x \to \infty$.

## Problem Set

1. Sketch the graphs of the functions $f(x) = 5^x$ and $g(x) = \log_5(x)$.

2. Sketch the graphs of the functions $f(x) = \left(\frac{1}{2}\right)^x$ and $g(x) = \log_{\frac{1}{2}}(x)$.

3. Sketch the graphs of the functions $f_1(x) = \left(\frac{1}{2}\right)^x$ and $f_2(x) = \left(\frac{3}{4}\right)^x$ on the same sheet of graph paper and answer the following questions.
   a. Where do the two exponential graphs intersect?
   b. For which values of $x$ is $\left(\frac{1}{2}\right)^x < \left(\frac{3}{4}\right)^x$?
   c. For which values of $x$ is $\left(\frac{1}{2}\right)^x > \left(\frac{3}{4}\right)^x$?
   d. What happens to the values of the functions $f_1$ and $f_2$ as $x \to \infty$?
   e. What are the domains of the two functions $f_1$ and $f_2$?

4. Use the information from Problem 3 together with the relationship between graphs of exponential and logarithmic functions to sketch the graphs of the functions $g_1(x) = \log_{\frac{1}{2}}(x)$ and $g_2(x) = \log_{\frac{3}{4}}(x)$ on the same sheet of graph paper. Then, answer the following questions.
   a. Where do the two logarithmic graphs intersect?
   b. For which values of $x$ is $\log_{\frac{1}{2}}(x) < \log_{\frac{3}{4}}(x)$?
   c. For which values of $x$ is $\log_{\frac{1}{2}}(x) > \log_{\frac{3}{4}}(x)$?
   d. What happens to the values of the functions $g_1$ and $g_2$ as $x \to \infty$?
   e. What are the domains of the two functions $g_1$ and $g_2$?

5. For each function $f$, find a formula for the function $h$ in terms of $x$.
   a. If $f(x) = x^3$, find $h(x) = 128f\left(\frac{1}{4}x\right) + f(2x)$.
   b. If $f(x) = x^2 + 1$, find $h(x) = f(x + 2) - f(2)$.
   c. If $f(x) = x^3 + 2x^2 + 5x + 1$, find $h(x) = \frac{f(x) + f(-x)}{2}$.
   d. If $f(x) = x^3 + 2x^2 + 5x + 1$, find $h(x) = \frac{f(x) - f(-x)}{2}$.

6. In Problem 5, parts (c) and (d), list at least two aspects about the formulas you found as they relate to the function $f(x) = x^3 + 2x^2 + 5x + 1$.

7. For each of the functions $f$ and $g$ below, write an expression for (i) $f(g(x))$, (ii) $g(f(x))$, and (iii) $f(f(x))$ in terms of $x$.

   a. $f(x) = x^{\frac{2}{3}},\ g(x) = x^{12}$

   b. $f(x) = \frac{b}{x-a},\ g(x) = \frac{b}{x} + a$ for two numbers $a$ and $b$, when $x$ is not equal to 0 or $a$

   c. $f(x) = \frac{x+1}{x-1},\ g(x) = \frac{x+1}{x-1}$, when $x$ is not equal to 1 or $-1$

   d. $f(x) = 2^x,\ g(x) = \log_2(x)$

   e. $f(x) = \ln(x),\ g(x) = e^x$

   f. $f(x) = 2 \cdot 100^x,\ g(x) = \frac{1}{2}\log\left(\frac{1}{2}x\right)$

# Lesson 19: The Inverse Relationship Between Logarithmic and Exponential Functions

## Classwork

### Opening Exercise

a. Consider the mapping diagram of the function $f$ below. Fill in the blanks of the mapping diagram of $g$ to construct a function that "undoes" each output value of $f$ by returning the original input value of $f$. (The first one is done for you.)

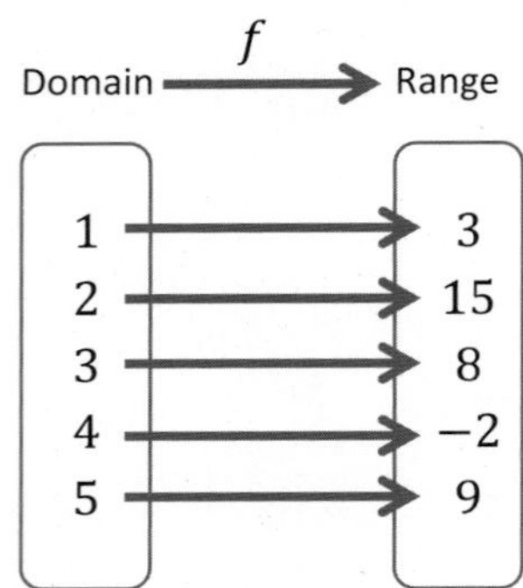


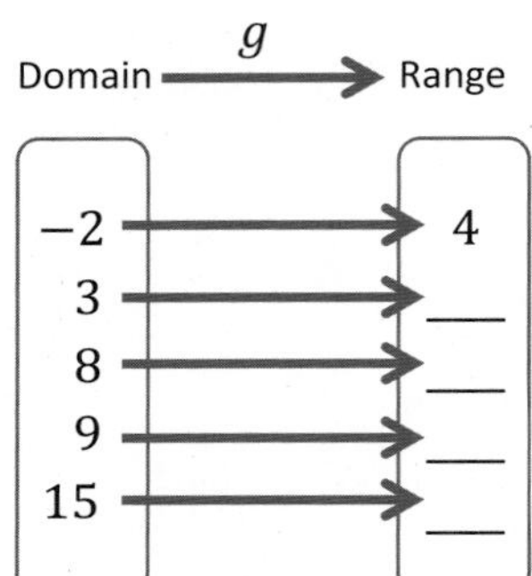


b. Write the set of input-output pairs for the functions $f$ and $g$ by filling in the blanks below. (The set $F$ for the function $f$ has been done for you.)
$F = \{(1,3), (2,15), (3,8), (4,-2), (5,9)\}$

$G = \{(-2,4),$ ________, ________, ________, ________$\}$

c. How can the points in the set $G$ be obtained from the points in $F$?

d. Peter studied the mapping diagrams of the functions $f$ and $g$ above and exclaimed, "I can get the mapping diagram for $g$ by simply taking the mapping diagram for $f$ and reversing all of the arrows!" Is he correct?

## Exercises

For each function $f$ in Exercises 1–5, find the formula for the corresponding inverse function $g$. Graph both functions on a calculator to check your work.

1. $f(x) = 1 - 4x$

2. $f(x) = x^3 - 3$

3. $f(x) = 3\log(x^2)$ for $x > 0$

4. $f(x) = 2^{x-3}$

5. $f(x) = \frac{x+1}{x-1}$ for $x \neq 1$

6. Cindy thinks that the inverse of $f(x) = x - 2$ is $g(x) = 2 - x$. To justify her answer, she calculates $f(2) = 0$ and then substitutes the output $0$ into $g$ to get $g(0) = 2$, which gives back the original input. Show that Cindy is incorrect by using other examples from the domain and range of $f$.

7. After finding the inverse for several functions, Henry claims that every function must have an inverse. Rihanna says that his statement is not true and came up with the following example: If $f(x) = |x|$ has an inverse, then because $f(3)$ and $f(-3)$ both have the same output $3$, the inverse function $g$ would have to map $3$ to both $3$ and $-3$ simultaneously, which violates the definition of a function. What is another example of a function without an inverse?

## Example

Consider the function $f(x) = 2^x + 1$, whose graph is shown at right.

a. What are the domain and range of $f$?

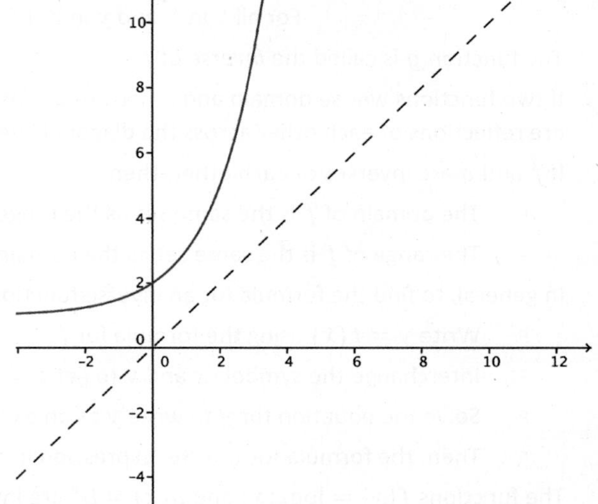


b. Sketch the graph of the inverse function $g$ on the graph. What type of function do you expect $g$ to be?

c. What are the domain and range of $g$? How does that relate to your answer in part (a)?

d. Find the formula for $g$.

### Lesson Summary

- **INVERTIBLE FUNCTION:** Let $f$ be a function whose domain is the set $X$ and whose image is the set $Y$. Then $f$ is *invertible* if there exists a function $g$ with domain $Y$ and image $X$ such that $f$ and $g$ satisfy the property:

  For all $x$ in $X$ and $y$ in $Y$, $f(x) = y$ if and only if $g(y) = x$.
- The function $g$ is called the *inverse* of $f$.
- If two functions whose domain and range are a subset of the real numbers are inverses, then their graphs are reflections of each other across the diagonal line given by $y = x$ in the Cartesian plane.
- If $f$ and $g$ are inverses of each other, then
  - The domain of $f$ is the same set as the range of $g$.
  - The range of $f$ is the same set as the domain of $g$.
- In general, to find the formula for an inverse function $g$ of a given function $f$:
  - Write $y = f(x)$ using the formula for $f$.
  - Interchange the symbols $x$ and $y$ to get $x = f(y)$.
  - Solve the equation for $y$ to write $y$ as an expression in $x$.
  - Then, the formula for $g$ is the expression in $x$ found in step (iii).
- The functions $f(x) = \log_b(x)$ and $g(x) = b^x$ are inverses of each other.

## Problem Set

1. For each function $h$ below, find two functions $f$ and $g$ such that $h(x) = f\big(g(x)\big)$. (There are many correct answers.)

   a. $h(x) = (3x + 7)^2$

   b. $h(x) = \sqrt[3]{x^2 - 8}$

   c. $h(x) = \frac{1}{2x - 3}$

   d. $h(x) = \frac{4}{(2x - 3)^3}$

   e. $h(x) = (x + 1)^2 + 2(x + 1)$

   f. $h(x) = (x + 4)^{\frac{4}{5}}$

   g. $h(x) = \sqrt[3]{\log(x^2 + 1)}$

   h. $h(x) = \sin(x^2 + 2)$

   i. $h(x) = \ln(\sin(x))$

2. Let $f$ be the function that assigns to each student in your class his or her biological mother.
   a. Use the definition of function to explain why $f$ is a function.
   b. In order for $f$ to have an inverse, what condition must be true about the students in your class?
   c. If we enlarged the domain to include all students in your school, would this larger domain function have an inverse?

3. The table below shows a partially filled-out set of input-output pairs for two functions $f$ and $h$ that have the same finite domain of $\{0, 5, 10, 15, 20, 25, 30, 35, 40\}$.

| $x$ | 0 | 5 | 10 | 15 | 20 | 25 | 30 | 35 | 40 |
|---|---|---|---|---|---|---|---|---|---|
| $f(x)$ | 0 | 0.3 | 1.4 | | 2.1 | | 2.7 | 6 | |
| $h(x)$ | 0 | 0.3 | 1.4 | | 2.1 | | 2.7 | 6 | |

   a. Complete the table so that $f$ is invertible but $h$ is definitely not invertible.
   b. Graph both functions and use their graphs to explain why $f$ is invertible and $h$ is not.

4. Find the inverse of each of the following functions. In each case, indicate the domain and range of both the original function and its inverse.
   a. $f(x) = \frac{3x - 7}{5}$
   b. $f(x) = \frac{5 + x}{6 - 2x}$
   c. $f(x) = e^{x-5}$
   d. $f(x) = 2^{5-8x}$
   e. $f(x) = 7\log(1 + 9x)$
   f. $f(x) = 8 + \ln(5 + \sqrt[3]{x})$
   g. $f(x) = \log\left(\frac{100}{3x+2}\right)$
   h. $f(x) = \ln(x) - \ln(x + 1)$
   i. $f(x) = \frac{2^x}{2^x+1}$

5. Unlike square roots that do not have any real principal square roots for negative numbers, principal cube roots do exist for negative numbers: $\sqrt[3]{-8}$ is the real number $-2$ since it satisfies $-2 \cdot -2 \cdot -2 = -8$. Use the identities $\sqrt[3]{x^3} = x$ and $\left(\sqrt[3]{x}\right)^3 = x$ for any real number $x$ to find the inverse of each of the functions below. In each case, indicate the domain and range of both the original function and its inverse.
   a. $f(x) = \sqrt[3]{2x}$ for any real number $x$.
   b. $f(x) = \sqrt[3]{2x - 3}$ for any real number $x$.
   c. $f(x) = (x - 1)^3 + 3$ for any real number $x$.

6. Suppose that the inverse of a function is the function itself. For example, the inverse of the function $f(x) = \frac{1}{x}$ (for $x \neq 0$) is just itself again, $g(x) = \frac{1}{x}$ (for $x \neq 0$). What symmetry must the graphs of all such functions have? (Hint: Study the graph of Exercise 5 in the lesson.)

7. When traveling abroad, you will find that daily temperatures in other countries are often reported in Celsius. The sentence, "It will be 25°C today in Paris," does not mean it will be freezing in Paris. It will often be necessary for you to convert temperatures reported in degrees Celsius to degrees Fahrenheit, the scale we use in the U.S. for reporting daily temperatures.

   Let $f$ be the function that inputs a temperature measure in degrees Celsius and outputs the corresponding temperature measure in degrees Fahrenheit.

   a. Assuming that $f$ is linear, we can use two points on the graph of $f$ to determine a formula for $f$. In degrees Celsius, the freezing point of water is $0$, and its boiling point is $100$. In degrees Fahrenheit, the freezing point of water is $32$, and its boiling point is $212$. Use this information to find a formula for the function $f$. (Hint: Plot the points and draw the graph of $f$ first, keeping careful track of the meaning of values on the $x$-axis and $y$-axis.)

   b. What temperature will Paris be in degrees Fahrenheit if it is reported that it will be 25°C?

   c. Find the inverse of the function $f$ and explain its meaning in terms of degree scales that its domain and range represent.

   d. The graphs of $f$ and its inverse are two lines that intersect in one point. What is that point? What is its significance in terms of degrees Celsius and degrees Fahrenheit?

**Extension:** Use the fact that, for $b > 1$, the functions $f(x) = b^x$ and $g(x) = \log_b(x)$ are increasing to solve the following problems. Recall that an increasing function $f$ has the property that if both $a$ and $b$ are in the domain of $f$ and $a < b$, then $f(a) < f(b)$.

8. For which values of $x$ is $2^x < \frac{1}{1{,}000{,}000}$?

9. For which values of $x$ is $\log_2(x) < -1{,}000{,}000$?

# Lesson 20: Transformations of the Graphs of Logarithmic and Exponential Functions

## Classwork

### Opening Exercise

a. Sketch the graphs of the three functions $f(x) = x^2$, $g(x) = (2x)^2 + 1$, and $h(x) = 4x^2 + 1$.

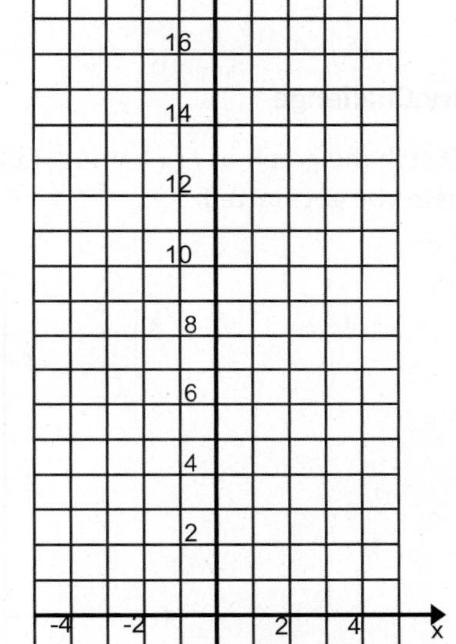


i. Describe the transformations that will take the graph of $f(x) = x^2$ to the graph of $g(x) = (2x)^2 + 1$.

ii. Describe the transformations that will take the graph of $f(x) = x^2$ to the graph of $h(x) = 4x^2 + 1$.

iii. Explain why $g$ and $h$ from parts (i) and (ii) are equivalent functions.

b. Describe the transformations that will take the graph of $f(x) = \sin(x)$ to the graph of $g(x) = \sin(2x) - 3$.

c. Describe the transformations that will take the graph of $f(x) = \sin(x)$ to the graph of $h(x) = 4\sin(x) - 3$.

d. Explain why $g$ and $h$ from parts (b)–(c) are *not* equivalent functions.

## Exploratory Challenge

a. Sketch the graph of $f(x) = \log_2(x)$ by identifying and plotting at least five key points. Use the table below to help you get started.

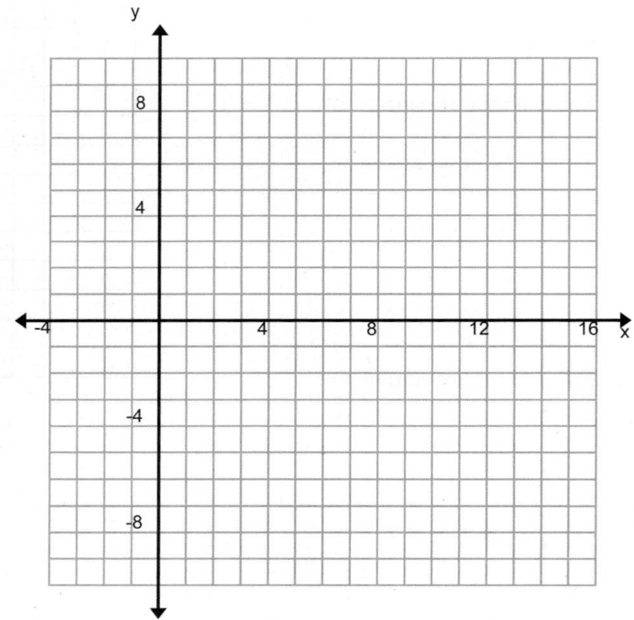


b. Describe the transformations that will take the graph of $f$ to the graph of $g(x) = \log_2(4x)$.

c. Describe the transformations that will take the graph of $f$ to the graph of $h(x) = 2 + \log_2(x)$.

d. Complete the table below for $f$, $g$, and $h$ and describe any patterns that you notice.

| $x$ | $f(x)$ | $g(x)$ | $h(x)$ |
|---|---|---|---|
| $\frac{1}{4}$ | | | |
| $\frac{1}{2}$ | | | |
| $1$ | | | |
| $2$ | | | |
| $4$ | | | |
| $8$ | | | |

e. Graph the three functions on the same coordinate axes and describe any patterns that you notice. Use a property of logarithms to show that $g$ and $h$ are equivalent.

f. Describe the graph of $g(x) = \log_2\left(\frac{x}{4}\right)$ as a vertical translation of the graph of $f(x) = \log_2(x)$. Justify your response.

g. Describe the graph of $h(x) = \log_2(x) + 3$ as a horizontal scaling of the graph of $f(x) = \log_2(x)$. Justify your response.

h. Do the functions $f(x) = \log_2(x) + \log_2(4)$ and $g(x) = \log_2(x + 4)$ have the same graphs? Justify your reasoning.

i. Use the properties of exponents to explain why the graphs of $f(x) = 4^x$ and $g(x) = 2^{2x}$ are identical.

j. Use the properties of exponents to predict what the graphs of $f(x) = 4 \cdot 2^x$ and $g(x) = 2^{x+2}$ will look like compared to one another. Describe the graphs of $f$ and $g$ as transformations of the graph of $f = 2^x$. Confirm your prediction by graphing $f$ and $g$ on the same coordinate axes.

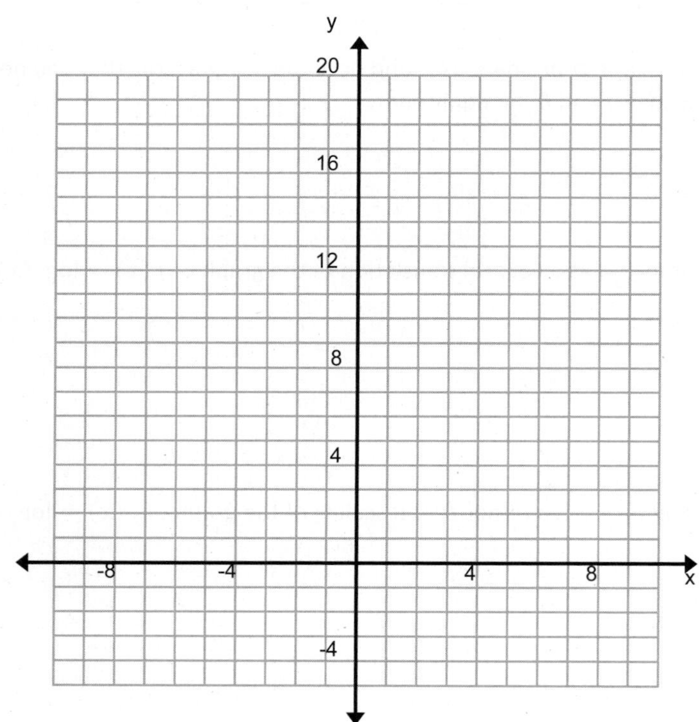

k. Graph $f(x) = 2^x$, $g(x) = 2^{-x}$, and $h(x) = \left(\frac{1}{2}\right)^x$ on the same coordinate axes. Describe the graphs of $g$ and $h$ as transformations of the graph of $f$. Use the properties of exponents to explain why $g$ and $h$ are equivalent.

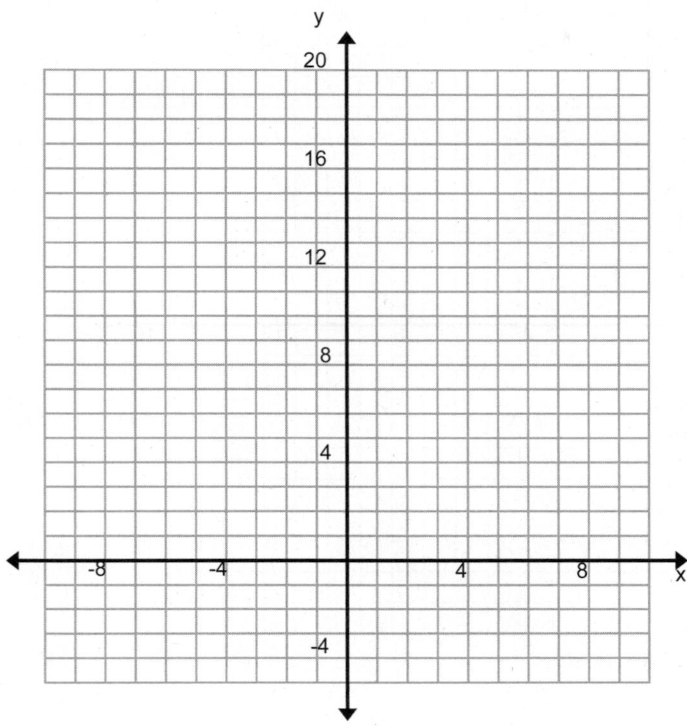


## Example 1: Graphing Transformations of the Logarithm Functions

The general form of a logarithm function is given by $f(x) = k + a\log_b(x - h)$, where $a$, $b$, $k$, and $h$ are real numbers such that $b$ is a positive number not equal to $1$, and $x - h > 0$.

a. Given $g(x) = 3 + 2\log(x - 2)$, describe the graph of $g$ as a transformation of the common logarithm function.

b. Graph the common logarithm function and $g$ on the same coordinate axes.

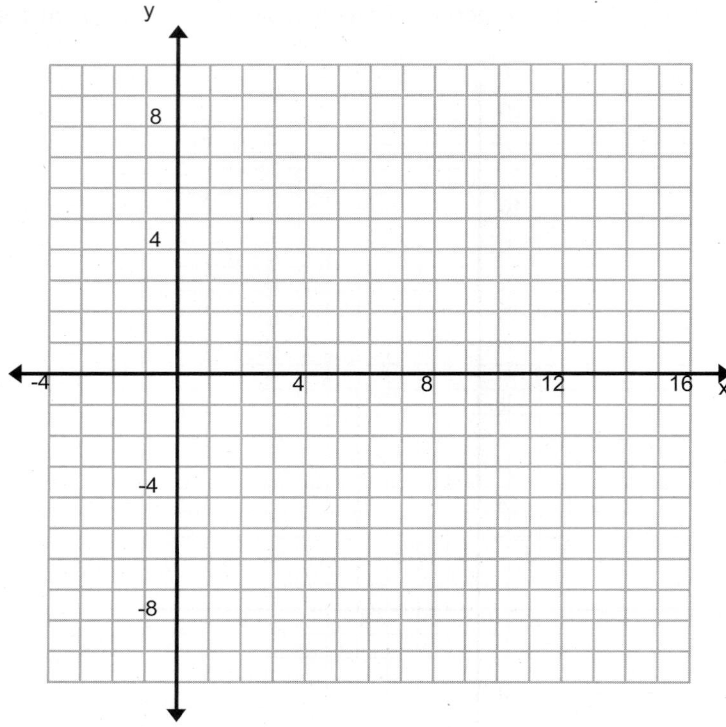


**Example 2: Graphing Transformations of Exponential Functions**

The general form of the exponential function is given by $f(x) = a \cdot b^x + k$, where $a$, $b$, and $k$ are real numbers such that $b$ is a positive number not equal to 1.

a. Use the properties of exponents to transform the function $g(x) = 3^{2x+1} - 2$ to the general form, and then graph it. What are the values of $a$, $b$, and $k$?

b. Describe the graph of $g$ as a transformation of the graph of $h(x) = 9^x$.

c. Describe the graph of $g$ as a transformation of the graph of $h(x) = 3^x$.

d. Graph $g$ using transformations.

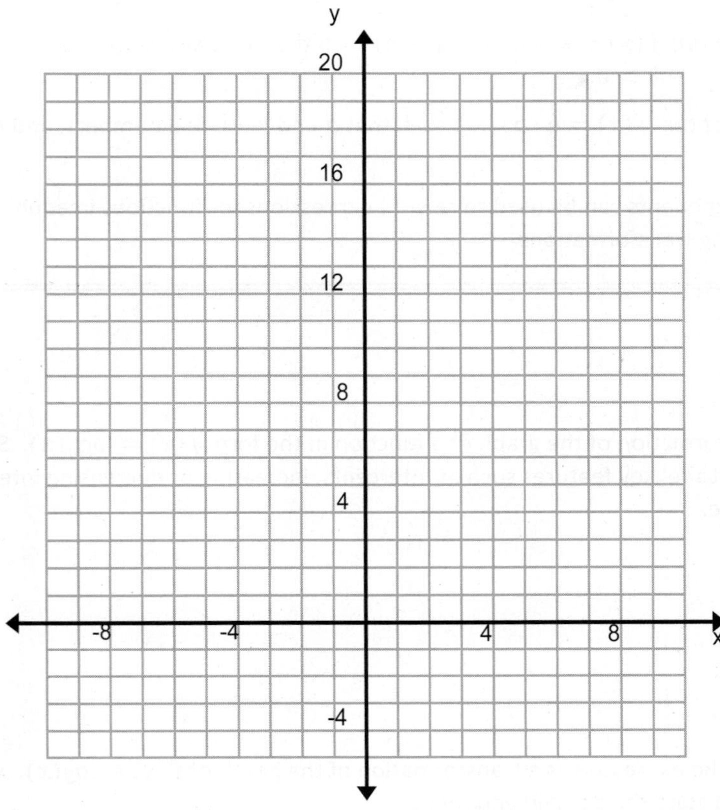


## Exercises 1–4

Graph each pair of functions by first graphing $f$ and then graphing $g$ by applying transformations of the graph of $f$. Describe the graph of $g$ as a transformation of the graph of $f$.

1. $f(x) = \log_3(x)$ and $g(x) = 2\log_3(x - 1)$

2. $f(x) = \log(x)$ and $g(x) = \log(100x)$

3. $f(x) = \log_5 x$ and $g(x) = -\log_5\big(5(x + 2)\big)$

4. $f(x) = 3^x$and $g(x) = -2 \cdot 3^{x-1}$

## Lesson Summary

**GENERAL FORM OF A LOGARITHMIC FUNCTION:** $f(x) = k + a\log_b(x - h)$ such that $a$, $h$, and $k$ are real numbers, $b$ is any positive number not equal to 1, and $x - h > 0$.

**GENERAL FORM OF AN EXPONENTIAL FUNCTION:** $f(x) = a \cdot b^x + k$ such that $a$ and $k$ are real numbers, and $b$ is any positive number not equal to 1.

The properties of logarithms and exponents can be used to rewrite expressions for functions in equivalent forms that can then be graphed by applying transformations.

## Problem Set

1. Describe each function as a transformation of the graph of a function in the form $f(x) = \log_b(x)$. Sketch the graph of $f$ and the graph of $g$ by hand. Label key features such as intercepts, increasing or decreasing intervals, and the equation of the vertical asymptote.

   a. $g(x) = \log_2(x - 3)$

   b. $g(x) = \log_2(16x)$

   c. $g(x) = \log_2\left(\frac{8}{x}\right)$

   d. $g(x) = \log_2((x - 3)^2)$

2. Each function graphed below can be expressed as a transformation of the graph of $f(x) = \log(x)$. Write an algebraic function for $g$ and $h$ and state the domain and range.

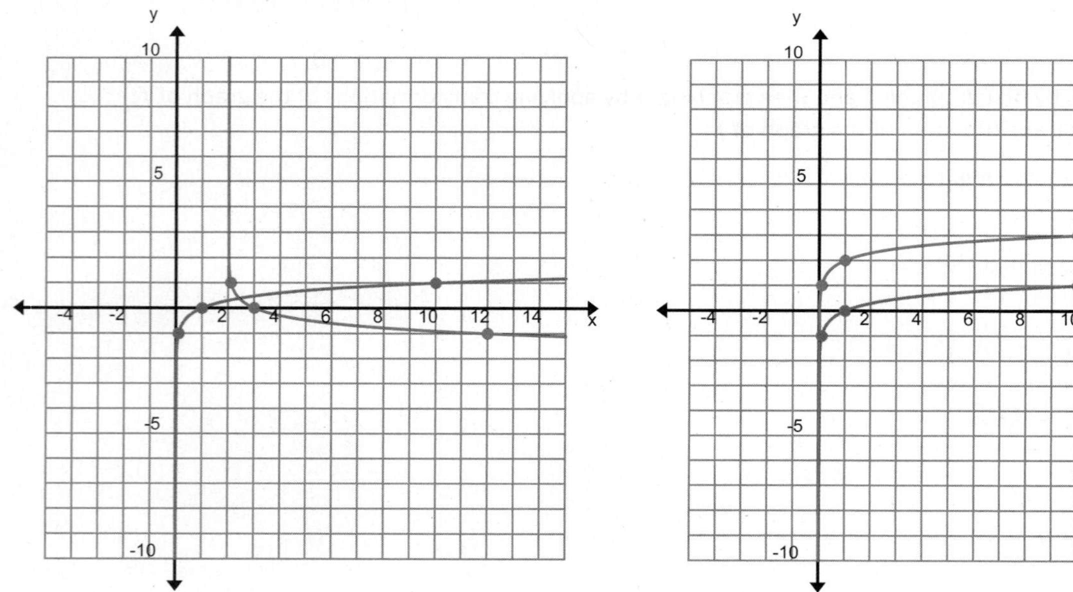


**Figure 1: Graphs of $f(x) = \log(x)$ and the function $g$**

**Figure 2: Graphs of $f(x) = \log(x)$ and the function $h$**

3. Describe each function as a transformation of the graph of a function in the form $f(x) = b^x$. Sketch the graph of $f$ and the graph of $g$ by hand. Label key features such as intercepts, increasing or decreasing intervals, and the horizontal asymptote. (Estimate when needed from the graph.)

   a. $g(x) = 2 \cdot 3^x - 1$
   b. $g(x) = 2^{2x} + 3$
   c. $g(x) = 3^{x-2}$
   d. $g(x) = -9^{\frac{x}{2}} + 1$

4. Using the function $f(x) = 2^x$, create a new function $g$ whose graph is a series of transformations of the graph of $f$ with the following characteristics:

   - The graph of $g$ is decreasing for all real numbers.
   - The equation for the horizontal asymptote is $y = 5$.
   - The $y$-intercept is $7$.

5. Using the function $f(x) = 2^x$, create a new function $g$ whose graph is a series of transformations of the graph of $f$ with the following characteristics:

   - The graph of $g$ is increasing for all real numbers.
   - The equation for the horizontal asymptote is $y = 5$.
   - The $y$-intercept is $4$.

6. Given the function $g(x) = \left(\frac{1}{4}\right)^{x-3}$:

   a. Write the function $g$ as an exponential function with base $4$. Describe the transformations that would take the graph of $f(x) = 4^x$ to the graph of $g$.
   b. Write the function $g$ as an exponential function with base $2$. Describe two different series of transformations that would take the graph of $f(x) = 2^x$ to the graph of $g$.

7. Explore the graphs of functions in the form $f(x) = \log(x^n)$ for $n > 1$. Explain how the graphs of these functions change as the values of $n$ increase. Use a property of logarithms to support your reasoning.

8. Use a graphical approach to solve each equation. If the equation has no solution, explain why.

   a. $\log(x) = \log(x - 2)$
   b. $\log(x) = \log(2x)$
   c. $\log(x) = \log\left(\frac{2}{x}\right)$
   d. Show algebraically that the exact solution to the equation in part (c) is $\sqrt{2}$.

9. Make a table of values for $f(x) = x^{\frac{1}{\log(x)}}$ for $x > 1$. Graph this function for $x > 1$. Use properties of logarithms to explain what you see in the graph and the table of values.

# Lesson 21: The Graph of the Natural Logarithm Function

## Classwork

### Exploratory Challenge

Your task is to compare graphs of base $b$ logarithm functions to the graph of the common logarithm function $f(x) = \log(x)$ and summarize your results with your group. Recall that the base of the common logarithm function is 10. A graph of $f$ is provided below.

a. Select at least one base value from this list: $\frac{1}{10}, \frac{1}{2}, 2, 5, 20, 100$. Write a function in the form $g(x) = \log_b(x)$ for your selected base value, $b$.

b. Graph the functions $f$ and $g$ in the same viewing window using a graphing calculator or other graphing application, and then add a sketch of the graph of $g$ to the graph of $f$ shown below.

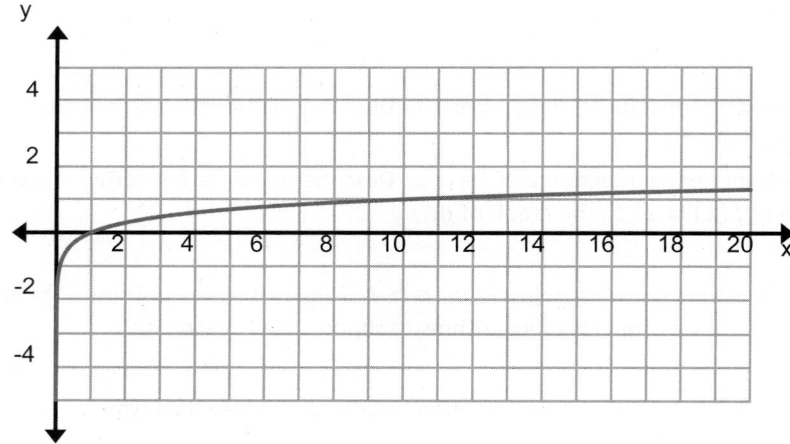


c. Describe how the graph of $g$ for the base you selected compares to the graph of $f(x) = \log(x)$.

d. Share your results with your group and record observations on the graphic organizer below. Prepare a group presentation that summarizes the group's findings.

| **How does the graph of $g(x) = \log_b(x)$ compare to the graph of $f(x) = \log(x)$ for various values of $b$?** | |
|---|---|
| $0 < b < 1$ | |
| $1 < b < 10$ | |
| $b > 10$ | |

### Exercise 1

Use the change of base property to rewrite each function as a common logarithm.

<u>Base $b$</u> <u>Base 10 (Common Logarithm)</u>

$g(x) = \log_{\frac{1}{4}}(x)$

$g(x) = \log_{\frac{1}{2}}(x)$

$g(x) = \log_2(x)$

$g(x) = \log_5(x)$

$g(x) = \log_{20}(x)$

$g(x) = \log_{100}(x)$

## Example 1: The Graph of the Natural Logarithm Function $f(x) = \ln(x)$

Graph the natural logarithm function below to demonstrate where it sits in relation to the base 2 and base 10 logarithm functions.

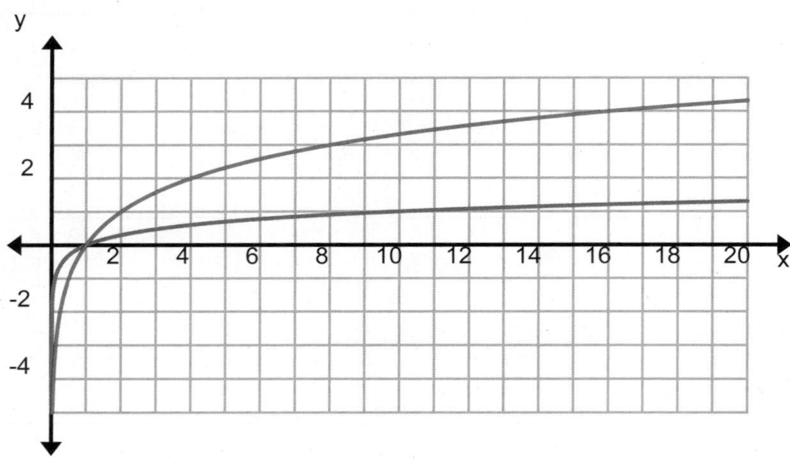


## Example 2

Graph each function by applying transformations of the graphs of the natural logarithm function.

a. $f(x) = 3\ln(x - 1)$

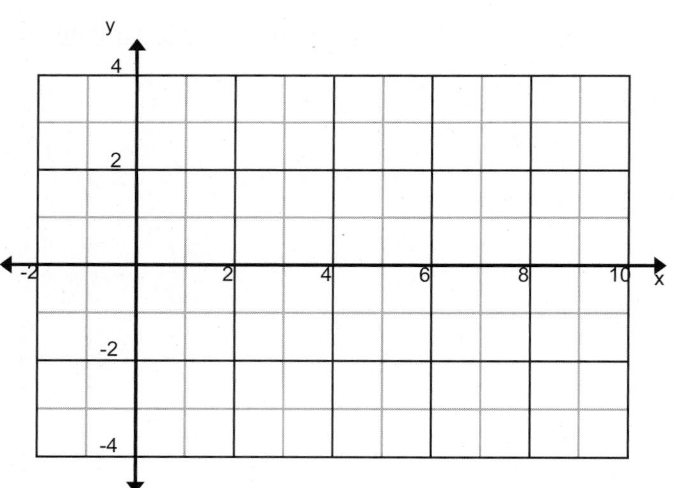

b. $g(x) = \log_6(x) - 2$

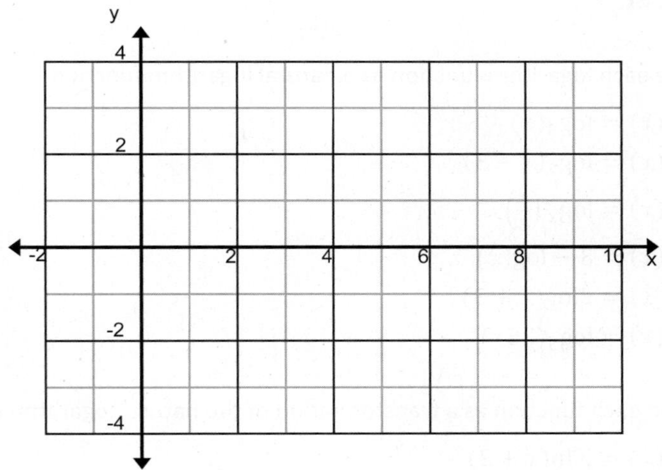

## Problem Set

1. Rewrite each logarithm function as a natural logarithm function.
   a. $f(x) = \log_5(x)$
   b. $f(x) = \log_2(x - 3)$
   c. $f(x) = \log_2\left(\frac{x}{3}\right)$
   d. $f(x) = 3 - \log(x)$
   e. $f(x) = 2\log(x + 3)$
   f. $f(x) = \log_5(25x)$

2. Describe each function as a transformation of the natural logarithm function $f(x) = \ln(x)$.
   a. $g(x) = 3\ln(x + 2)$
   b. $g(x) = -\ln(1 - x)$
   c. $g(x) = 2 + \ln(e^2 x)$
   d. $g(x) = \log_5(25x)$

3. Sketch the graphs of each function in Problem 2 and identify the key features including intercepts, decreasing or increasing intervals, and the vertical asymptote.

4. Solve the equation $e^{-x} = \ln(x)$ graphically.

5. Use a graphical approach to explain why the equation $\log(x) = \ln(x)$ has only one solution.

6. Juliet tried to solve this equation as shown below using the change of base property and concluded there is no solution because $\ln(10) \neq 1$. Construct an argument to support or refute her reasoning.

$$\log(x) = \ln(x)$$
$$\frac{\ln(x)}{\ln(10)} = \ln(x)$$
$$\left(\frac{\ln(x)}{\ln(10)}\right)\frac{1}{\ln(x)} = (\ln(x))\frac{1}{\ln(x)}$$
$$\frac{1}{\ln(10)} = 1$$

7. Consider the function $f$ given by $f(x) = \log_x(100)$ for $x > 0$ and $x \neq 1$.
   a. What are the values of $f(100)$, $f(10)$, and $f(\sqrt{10})$?
   b. Why is the value $1$ excluded from the domain of this function?
   c. Find a value $x$ so that $f(x) = 0.5$.
   d. Find a value $w$ so that $f(w) = -1$.
   e. Sketch a graph of $y = \log_x(100)$ for $x > 0$ and $x \neq 1$.

# Lesson 22: Choosing a Model

## Classwork

### Opening Exercise

a. You are working on a team analyzing the following data gathered by your colleagues:

$$(-1.1, 5), (0, 105), (1.5, 178), (4.3, 120).$$

Your coworker Alexandra says that the model you should use to fit the data is

$$k(t) = 100 \cdot \sin(1.5t) + 105.$$

Sketch Alexandra's model on the axes at left on the next page.

b. How does the graph of Alexandra's model $k(t) = 100 \cdot \sin(1.5t) + 105$ relate to the four points? Is her model a good fit to this data?

c. Another teammate Randall says that the model you should use to fit the data is

$$g(t) = -16t^2 + 72t + 105.$$

Sketch Randall's model on the axes at right on the next page.

d. How does the graph of Randall's model $g(t) = -16t^2 + 72t + 105$ relate to the four points? Is his model a good fit to the data?

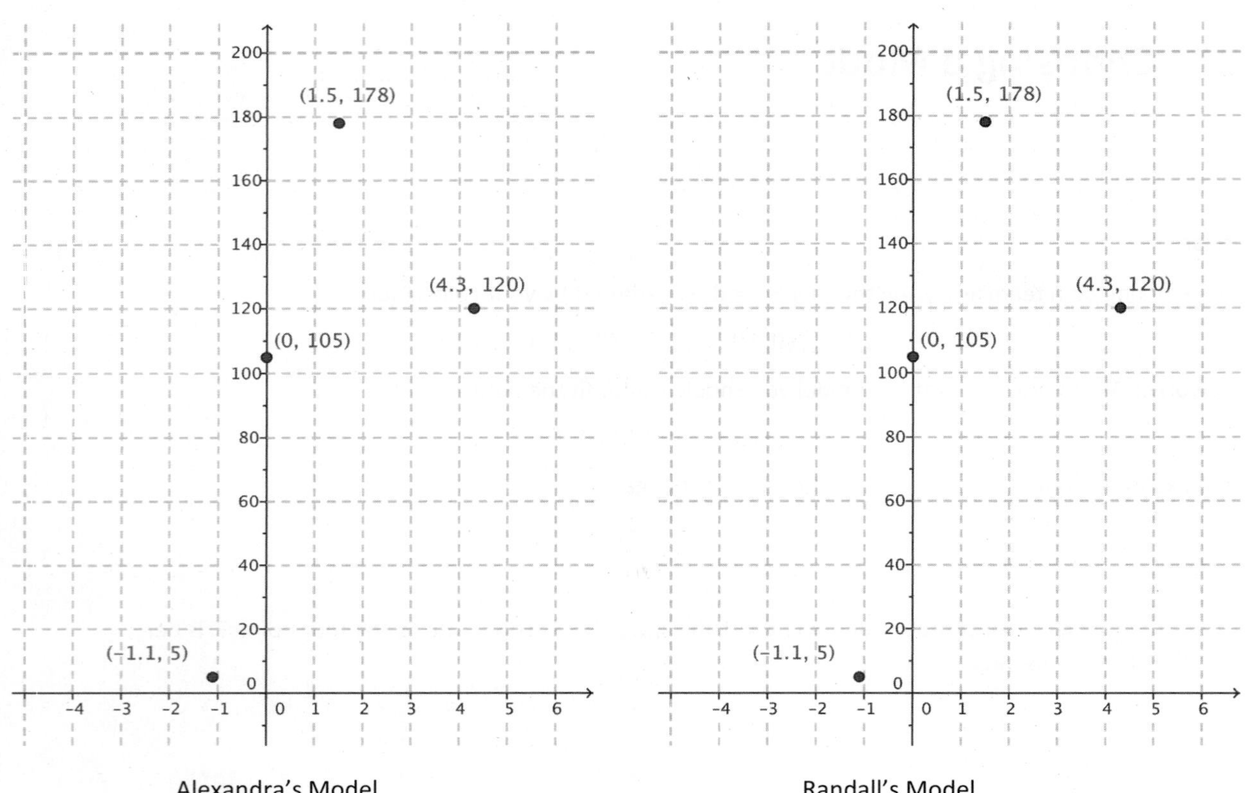


e. Suppose the four points represent positions of a projectile fired into the air. Which of the two models is more appropriate in that situation, and why?

f. In general, how do we know which model to choose?

## Exercises

1. The table below contains the number of daylight hours in Oslo, Norway, on the specified dates.

| Date | Hours and Minutes | Hours |
|---|---|---|
| August 1 | $16:56$ | $16.82$ |
| September 1 | $14:15$ | $14.25$ |
| October 1 | $11:33$ | $11.55$ |
| November 1 | $8:50$ | $8.90$ |

a. Plot the data on the grid provided. You will need to decide how to best represent it.

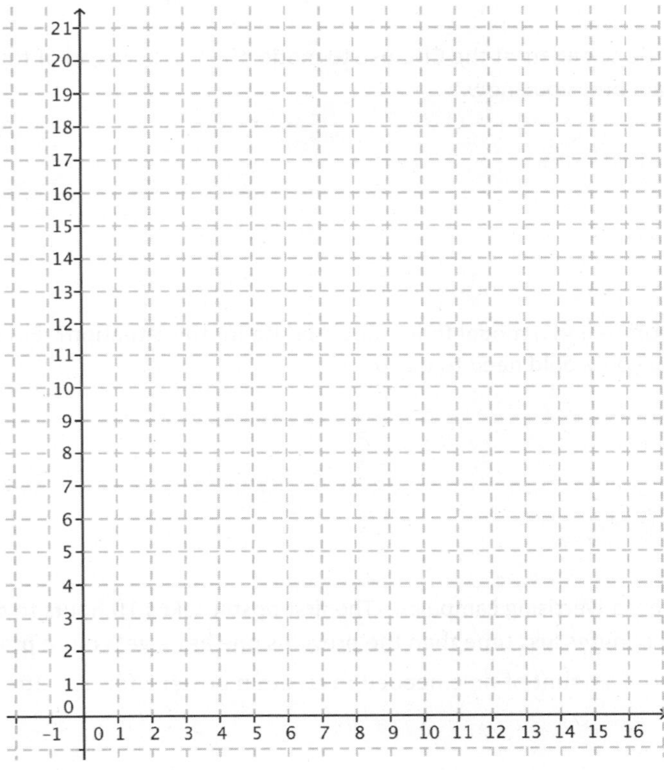

b. Looking at the data, what type of function appears to be the best fit?

c. Looking at the context in which the data was gathered, what type of function should be used to model the data?

d. Do you have enough information to find a model that is appropriate for this data? Either find a model or explain what other information you would need to do so.

2. The goal of the U.S. Centers for Disease Control and Prevention (CDC) is to protect public health and safety through the control and prevention of disease, injury, and disability. Suppose that $45$ people have been diagnosed with a new strain of the flu virus, and scientists estimate that each person with the virus will infect $5$ people every day with the flu.

   a. What type of function should the scientists at the CDC use to model the initial spread of this strain of flu to try to prevent an epidemic? Explain how you know.

   b. Do you have enough information to find a model that is appropriate for this situation? Either find a model or explain what other information you would need to do so.

3. An artist is designing posters for a new advertising campaign. The first poster takes $10$ hours to design, but each subsequent poster takes roughly $15$ minutes less time than the previous one as he gets more practice.

   a. What type of function models the amount of time needed to create $n$ posters, for $n \leq 20$? Explain how you know.

   b. Do you have enough information to find a model that is appropriate for this situation? Either find a model or explain what other information you would need to do so.

4. A homeowner notices that her heating bill is the lowest in the month of July and increases until it reaches its highest amount in the month of February. After February, the amount of the heating bill slowly drops back to the level it was in July, when it begins to increase again. The amount of the bill in February is roughly four times the amount of the bill in July.

   a. What type of function models the amount of the heating bill in a particular month? Explain how you know.

   b. Do you have enough information to find a model that is appropriate for this situation? Either find a model or explain what other information you would need to do so.

5. An online merchant sells used books for $5 each, and the sales tax rate is 6% of the cost of the books. Shipping charges are a flat rate of $4 plus an additional $1 per book.

   a. What type of function models the total cost, including the shipping costs, of a purchase of $x$ books? Explain how you know.

   b. Do you have enough information to find a model that is appropriate for this situation? Either find a model or explain what other information you would need to do so.

6. A stunt woman falls from a tall building in an action-packed movie scene. Her speed increases by 32 ft/s for every second that she is falling.

   a. What type of function models her distance from the ground at time $t$ seconds? Explain how you know.

   b. Do you have enough information to find a model that is appropriate for this situation? Either find a model or explain what other information you would need to do so.

## Lesson Summary

- If we expect from the context that each new term in the sequence of data is a constant added to the previous term, then we try a linear model.
- If we expect from the context that the second differences of the sequence are constant (meaning that the rate of change between terms either grows or shrinks linearly), then we try a quadratic model.
- If we expect from the context that each new term in the sequence of data is a constant multiple of the previous term, then we try an exponential model.
- If we expect from the context that the sequence of terms is periodic, then we try a sinusoidal model.

| Model | Equation of Function | Rate of Change |
|---|---|---|
| Linear | $f(t) = at + b$ for $a \neq 0$ | Constant |
| Quadratic | $g(t) = at^2 + bt + c$ for $a \neq 0$ | Changing linearly |
| Exponential | $h(t) = ab^{ct}$ for $0 < b < 1$ or $b > 1$ | A multiple of the current value |
| Sinusoidal | $k(t) = A\sin\big(w(t-h)\big) + k$ for $A, w \neq 0$ | Periodic |

## Problem Set

1. A new car depreciates at a rate of about $20\%$ per year, meaning that its resale value decreases by roughly $20\%$ each year. After hearing this, Brett said that if you buy a new car this year, then after $5$ years the car has a resale value of $\$0$. Is his reasoning correct? Explain how you know.

2. Alexei just moved to Seattle, and he keeps track of the average rainfall for a few months to see if the city deserves its reputation as the rainiest city in the United States.

| Month | Average rainfall |
|---|---|
| July | 0.93 in. |
| September | 1.61 in. |
| October | 3.24 in. |
| December | 6.06 in. |

What type of function should Alexei use to model the average rainfall in month $t$?

3. Sunny, who wears her hair long and straight, cuts her hair once per year on January 1, always to the same length. Her hair grows at a constant rate of $2$ cm per month. Is it appropriate to model the length of her hair with a sinusoidal function? Explain how you know.

4. On average, it takes 2 minutes for a customer to order and pay for a cup of coffee.
   a. What type of function models the amount of time you will wait in line as a function of how many people are in front of you? Explain how you know.
   b. Find a model that is appropriate for this situation.

5. An online ticket-selling service charges $\$50$ for each ticket to an upcoming concert. In addition, the buyer must pay $8\%$ sales tax and a convenience fee of $\$6.00$ for the purchase.
   a. What type of function models the total cost of the purchase of $n$ tickets in a single transaction?
   b. Find a model that is appropriate for this situation.

6. In a video game, the player must earn enough points to pass one level and progress to the next as shown in the table below.

| To pass this level ... | You need this many total points ... |
|---|---|
| 1 | 5,000 |
| 2 | 15,000 |
| 3 | 35,000 |
| 4 | 65,000 |

   That is, the increase in the required number of points increases by 10,000 points at each level.
   a. What type of function models the total number of points you need to pass to level $n$? Explain how you know.
   b. Find a model that is appropriate for this situation.

7. The southern white rhinoceros reproduces roughly once every three years, giving birth to one calf each time. Suppose that a nature preserve houses 100 white rhinoceroses, 50 of which are female. Assume that half of the calves born are female and that females can reproduce as soon as they are 1 year old.
   a. What type of function should be used to model the population of female white rhinoceroses in the preserve?
   b. Assuming that there is no death in the rhinoceros population, find a function to model the population of female white rhinoceroses in the preserve.
   c. Realistically, not all of the rhinoceroses will survive each year, so we will assume a $5\%$ death rate of all rhinoceroses. Now what type of function should be used to model the population of female white rhinoceroses in the preserve?
   d. Find a function to model the population of female white rhinoceroses in the preserve, taking into account the births of new calves and the $5\%$ death rate.

# Lesson 23: Bean Counting

## Classwork

### Mathematical Modeling Exercises

1. Working with a partner, you are going to gather some data, analyze it, and find a function to use to model it. Be prepared to justify your choice of function to the class.

   a. Gather your data: For each trial, roll the beans from the cup to the paper plate. Count the number of beans that land marked side up, and add that many beans to the cup. Record the data in the table below. Continue until you have either completed 10 trials or the number of beans at the start of the trial exceeds the number that you have.

| Trial Number, $t$ | Number of Beans at Start of Trial | Number of Beans That Landed Marked-Side Up |
|---|---|---|
| 1 | 1 | |
| 2 | | |
| 3 | | |
| 4 | | |
| 5 | | |
| 6 | | |
| 7 | | |
| 8 | | |
| 9 | | |
| 10 | | |

   b. Based on the context in which you gathered this data, which type of function would best model your data points?

c. Plot the data: Plot the trial number on the horizontal axis and the number of beans in the cup at the start of the trial on the vertical axis. Be sure to label the axes appropriately and to choose a reasonable scale for the axes.

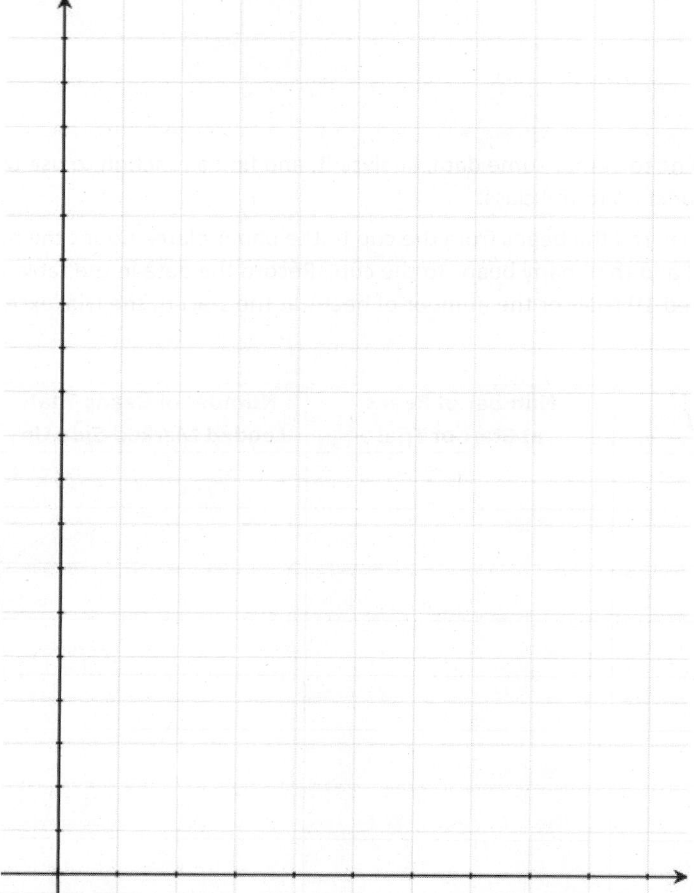

d. Analyze the data: Which type of function would best fit your data? Explain your reasoning.

e. Model the data: Enter the data into the calculator and use the appropriate type of regression to find an equation that fits this data. Round the constants to two decimal places.

2. This time, we are going to start with $50$ beans in your cup. Roll the beans onto the plate and remove any beans that land marked-side up. Repeat until you have no beans remaining.

   a. Gather your data: For each trial, roll the beans from the cup to the paper plate. Count the number of beans that land marked-side up, and remove that many beans from the plate. Record the data in the table below. Repeat until you have no beans remaining.

| Trial Number, $t$ | Number of Beans at Start of Trial | Number of Beans That Landed Marked-Side Up |
|---|---|---|
| $1$ | $50$ | |
| $2$ | | |
| $3$ | | |
| $4$ | | |
| $5$ | | |
| $6$ | | |
| $7$ | | |
| $8$ | | |
| $9$ | | |
| $10$ | | |
| | | |
| | | |
| | | |
| | | |

b. Plot the data: Plot the trial number on the horizontal axis and the number of beans in the cup at the start of the trial on the vertical axis. Be sure to label the axes appropriately and choose a reasonable scale for the axes.

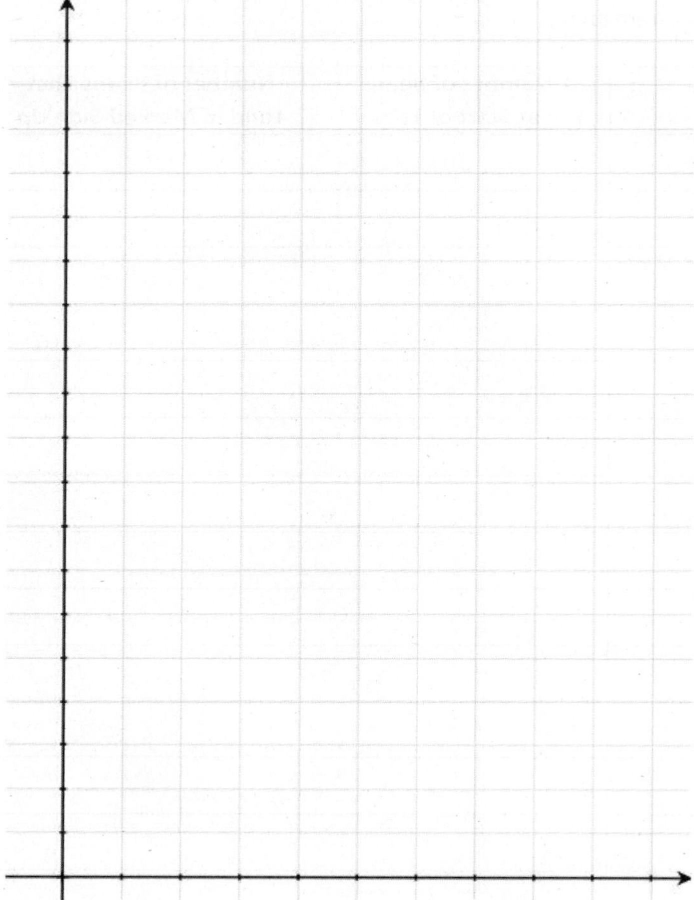

c. Analyze the data: Which type of function would best fit your data? Explain your reasoning.

d. Make a prediction: What do you expect the values of $a$ and $b$ to be for your function? Explain your reasoning.

e. Model the data: Enter the data into the calculator. Do not enter your final data point of 0 beans. Use the appropriate type of regression to find an equation that fits this data. Round the constants to two decimal places.

## Problem Set

1. For this exercise, we will consider three scenarios for which data has been collected and functions have been found to model the data, where $a$, $b$, $c$, $d$, $p$, $q$, $r$, $s$, $t$, and $u$ are positive real number constants.

   (i) The function $f(t) = a \cdot b^t$ models the original bean activity (Mathematical Modeling Exercise 1). Each bean is painted or marked on one side, and we start with one bean in the cup. A trial consists of throwing the beans in the cup and adding one more bean for each bean that lands marked side up.

   (ii) The function $g(t) = c \cdot d^t$ models a modified bean activity. Each bean is painted or marked on one side, and we start with one bean in the cup. A trial consists of throwing the beans in the cup and adding two more beans for each bean that lands marked side up.

   (iii) The function $h(t) = p \cdot q^t$ models the dice activity from the Exit Ticket. Start with one six-sided die in the cup. A trial consists of rolling the dice in the cup and adding one more die to the cup for each die that lands with a 6 showing.

   (iv) The function $j(t) = r \cdot s^t$ models a modified dice activity. Start with one six-sided die in the cup. A trial consists of rolling the dice in the cup and adding one more die to the cup for each die that lands with a 5 or a 6 showing.

   (v) The function $k(t) = u \cdot v^t$ models a modified dice activity. Start with one six-sided die in the cup. A trial consists of rolling the dice in the cup and adding one more die to the cup for each die that lands with an even number showing.

   a. What values do you expect for $a$, $c$, $p$, $r$, and $u$?

   b. What value do you expect for the base $b$ in the function $f(t) = a \cdot b^t$?

   c. What value do you expect for the base $d$ in the function $g(t) = c \cdot d^t$?

   d. What value do you expect for the base $q$ in the function $h(t) = p \cdot q^t$?

   e. What value do you expect for the base $s$ in the function $j(t) = r \cdot s^t$?

   f. What value do you expect for the base $v$ in the function $k(t) = u \cdot v^t$?

   g. The following graphs represent the four functions $f$, $g$, $h$, and $j$. Identify which graph represents which function.

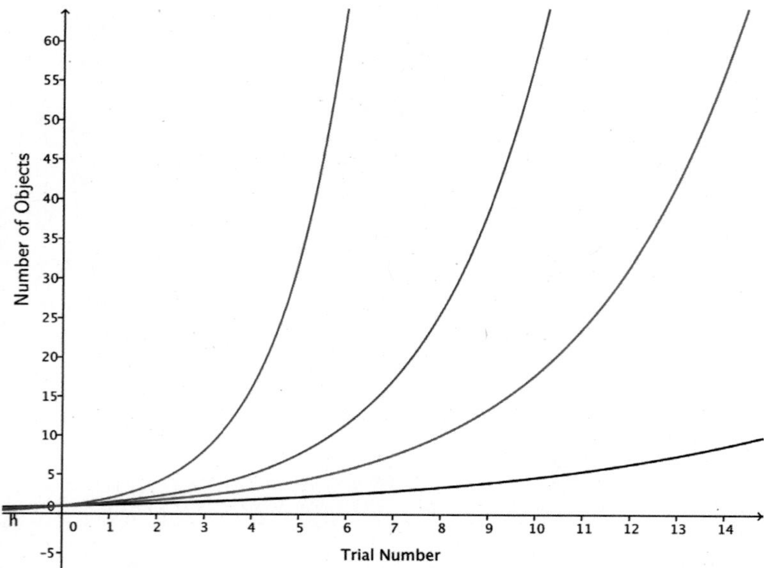

2. Teams 1, 2, and 3 gathered data as shown in the tables below, and each team modeled their data using an exponential function of the form $f(t) = a \cdot b^t$.

   a. Which team should have the highest value of $b$? Which team should have the lowest value of $b$? Explain how you know.

| Team 1 | | Team 2 | | Team 3 | |
|---|---|---|---|---|---|
| Trial Number, $t$ | Number of Beans | Trial Number, $t$ | Number of Beans | Trial Number, $t$ | Number of Beans |
| 0 | 1 | 0 | 1 | 0 | 1 |
| 1 | 1 | 1 | 1 | 1 | 2 |
| 2 | 2 | 2 | 1 | 2 | 3 |
| 3 | 2 | 3 | 2 | 3 | 5 |
| 4 | 4 | 4 | 2 | 4 | 8 |
| 5 | 6 | 5 | 3 | 5 | 14 |
| 6 | 8 | 6 | 5 | 6 | 26 |
| 7 | 14 | 7 | 7 | 7 | 46 |
| 8 | 22 | 8 | 12 | 8 | 76 |
| 9 | 41 | 9 | 18 | 9 | |
| 10 | 59 | 10 | 27 | 10 | |

   b. Use a graphing calculator to find the equation that best fits each set of data. Do the equations of the functions provide evidence that your answer in part (a) is correct?

3. Omar has devised an activity in which he starts with 15 dice in his cup. A trial consists of rolling the dice in the cup and adding one more die to the cup for each die that lands with a 1, 2, or 3 showing.

   a. Find a function $f(t) = a(b^t)$ that Omar would expect to model his data.

   b. Solve the equation $f(t) = 30$. What does the solution mean?

   c. Omar wants to know in advance how many trials it should take for his initial quantity of 15 dice to double. He uses properties of exponents and logarithms to rewrite the function from part (a) as an exponential function of the form $f(t) = a\left(2^{t \cdot \log_2(b)}\right)$.

   d. Has Omar correctly applied the properties of exponents and logarithms to obtain an equivalent expression for his original equation in part (a)? Explain how you know.

   e. Explain how the modified formula from part (c) allows Omar to easily find the expected amount of time, $t$, for the initial quantity of dice to double.

4. Brenna has devised an activity in which she starts with $10$ dice in her cup. A trial consists of rolling the dice in the cup and adding one more die to the cup for each die that lands with a $6$ showing.

   a. Find a function $f(t) = a(b^t)$ that you would expect to model her data.

   b. Solve the equation $f(t) = 30$. What does your solution mean?

   c. Brenna wants to know in advance how many trials it should take for her initial quantity of $10$ dice to triple. Use properties of exponents and logarithms to rewrite your function from part (a) as an exponential function of the form $f(t) = a(3^{ct})$.

   d. Explain how your formula from part (c) allows you to easily find the expected amount of time, $t$, for the initial quantity of dice to triple.

   e. Rewrite the formula for the function $f$ using a base $10$ exponential function.

   f. Use your formula from part (e) to find out how many trials it should take for the quantity of dice to grow to $100$ dice.

5. Suppose that one bacteria population can be modeled by the function $P_1(t) = 500(2^t)$ and a second bacteria population can be modeled by the function $P_2(t) = 500(2.83^t)$, where $t$ measures time in hours. Keep four digits of accuracy for decimal approximations of logarithmic values.

   a. What does the $500$ mean in each function?

   b. Which population should double first? Explain how you know.

   c. How many hours and minutes will it take until the first population doubles?

   d. Rewrite the formula for $P_2(t)$ in the form $P_2(t) = a(2^{ct})$, for some real numbers $a$ and $c$.

   e. Use your formula in part (d) to find the time, $t$, in hours and minutes until the second population doubles.

6. Copper has antibacterial properties, and it has been shown that direct contact with copper alloy C11000 at $20°$C kills $99.9\%$ of all methicillin-resistant *Staphylococcus aureus* (MRSA) bacteria in about $75$ minutes. Keep four digits of accuracy for decimal approximations of logarithmic values.

   a. A function that models a population of $1{,}000$ MRSA bacteria $t$ minutes after coming in contact with copper alloy C11000 is $P(t) = 1000(0.912)^t$. What does the base $0.912$ mean in this scenario?

   b. Rewrite the formula for $P$ as an exponential function with base $\frac{1}{2}$.

   c. Explain how your formula from part (b) allows you to easily find the time it takes for the population of MRSA to be reduced by half.

# Lesson 24: Solving Exponential Equations

## Classwork

### Opening Exercise

In Lesson 7, we modeled a population of bacteria that doubled every day by the function $P(t) = 2^t$, where $t$ was the time in days. We wanted to know the value of $t$ when there were $10$ bacteria. Since we didn't yet know about logarithms, we approximated the value of $t$ numerically and we found that $P(t) = 10$ at approximately $t \approx 3.32$ days.

Use your knowledge of logarithms to find an exact value for $t$ when $P(t) = 10$, and then use your calculator to approximate that value to four decimal places.

### Exercises

1. Fiona modeled her data from the bean-flipping experiment in Lesson 23 by the function $f(t) = 1.263(1.357)^t$, and Gregor modeled his data with the function $g(t) = 0.972(1.629)^t$.

   a. Without doing any calculating, determine which student, Fiona or Gregor, accumulated $100$ beans first. Explain how you know.

   b. Using Fiona's model …

      i. How many trials would be needed for her to accumulate $100$ beans?

ii. How many trials would be needed for her to accumulate $1{,}000$ beans?

c. Using Gregor's model …

i. How many trials would be needed for him to accumulate $100$ beans?

ii. How many trials would be needed for him to accumulate $1{,}000$ beans?

d. Was your prediction in part (a) correct? If not, what was the error in your reasoning?

2. Fiona wants to know when her model $f(t) = 1.263(1.357)^t$ predicts accumulations of 500, 5,000, and 50,000 beans, but she wants to find a way to figure it out without doing the same calculation three times.

   a. Let the positive number $c$ represent the number of beans that Fiona wants to have. Then solve the equation $1.263(1.357)^t = c$ for $t$.

   b. Your answer to part (a) can be written as a function $M$ of the number of beans $c$, where $c > 0$. Explain what this function represents.

   c. When does Fiona's model predict that she will accumulate ...

      i. 500 beans?

      ii. 5000 beans?

      iii. 50,000 beans?

3. Gregor states that the function $g$ that he found to model his bean-flipping data can be written in the form $g(t) = 0.972\left(10^{\log(1.629)t}\right)$. Since $\log(1.629) \approx 0.2119$, he is using $g(t) = 0.972(10^{0.2119t})$ as his new model.

   a. Is Gregor correct? Is $g(t) = 0.972\left(10^{\log(1.629)t}\right)$ an equivalent form of his original function? Use properties of exponents and logarithms to explain how you know.

   b. Gregor also wants to find a function that will help him to calculate the number of trials his function $g$ predicts it will take to accumulate 500, 5,000, and 50,000 beans. Let the positive number $c$ represent the number of beans that Gregor wants to have. Solve the equation $0.972(10^{0.2119t}) = c$ for $t$.

   c. Your answer to part (b) can be written as a function $N$ of the number of beans $c$, where $c > 0$. Explain what this function represents.

   d. When does Gregor's model predict that he will accumulate …

      i. 500 beans?

ii. 5,000 beans?

iii. 50,000 beans?

4. Helena and Karl each change the rules for the bean experiment. Helena started with four beans in her cup and added one bean for each that landed marked-side up for each trial. Karl started with one bean in his cup but added two beans for each that landed marked-side up for each trial.

   a. Helena modeled her data by the function $h(t) = 4.127(1.468^t)$. Explain why her values of $a = 4.127$ and $b = 1.468$ are reasonable.

   b. Karl modeled his data by the function $k(t) = 0.897(1.992^t)$. Explain why his values of $a = 0.897$ and $b = 1.992$ are reasonable.

c. At what value of $t$ do Karl and Helena have the same number of beans?

d. Use a graphing utility to graph $y = h(t)$ and $y = k(t)$ for $0 < t < 10$.

e. Explain the meaning of the intersection point of the two curves $y = h(t)$ and $y = k(t)$ in the context of this problem.

f. Which student reaches 20 beans first? Does the reasoning you used with whether Gregor or Fiona would get to 100 beans first hold true here? Why or why not?

For the following functions $f$ and $g$, solve the equation $f(x) = g(x)$. Express your solutions in terms of logarithms.

5. $f(x) = 10(3.7)^{x+1}$, $g(x) = 5(7.4)^{x}$

6. $f(x) = 135(5)^{3x+1}$, $g(x) = 75(3)^{4-3x}$

7. $f(x) = 100^{x^3+x^2-4x}$, $g(x) = 10^{2x^2-6x}$

8. $f(x) = 48\left(4^{x^2+3x}\right),\ g(x) = 3\left(8^{x^2+4x+4}\right)$

9. $f(x) = e^{\sin^2(x)},\ g(x) = e^{\cos^2(x)}$

10. $f(x) = (0.49)^{\cos(x)+\sin(x)},\ g(x) = (0.7)^{2\sin(x)}$

## Problem Set

1. Solve the following equations.
   a. $2 \cdot 5^{x+3} = 6250$
   b. $3 \cdot 6^{2x} = 648$
   c. $5 \cdot 2^{3x+5} = 10240$
   d. $4^{3x-1} = 32$
   e. $3 \cdot 2^{5x} = 216$
   f. $5 \cdot 11^{3x} = 120$
   g. $7 \cdot 9^{x} = 5405$
   h. $\sqrt{3} \cdot 3^{3x} = 9$
   i. $\log(400) \cdot 8^{5x} = \log(160000)$

2. Lucy came up with the model $f(t) = 0.701(1.382)^t$ for the first bean activity. When does her model predict that she would have 1,000 beans?

3. Jack came up with the model $g(t) = 1.033(1.707)^t$ for the first bean activity. When does his model predict that he would have 50,000 beans?

4. If instead of beans in the first bean activity you were using fair coins, when would you expect to have \$1,000,000?

5. Let $f(x) = 2 \cdot 3^x$ and $g(x) = 3 \cdot 2^x$.
   a. Which function is growing faster as $x$ increases? Why?
   b. When will $f(x) = g(x)$?

6. A population of *E. coli* bacteria can be modeled by the function $E(t) = 500(11.547)^t$, and a population of *Salmonella* bacteria can be modeled by the function $S(t) = 4000(3.668)^t$, where $t$ measures time in hours.
   a. Graph these two functions on the same set of axes. At which value of $t$ does it appear that the graphs intersect?
   b. Use properties of logarithms to find the time $t$ when these two populations are the same size. Give your answer to two decimal places.

7. Chain emails contain a message suggesting you will have bad luck if you do not forward the email to others. Suppose a student started a chain email by sending the message to 10 friends and asking those friends to each send the same email to 3 more friends exactly one day after receiving the message. Assuming that everyone that gets the email participates in the chain, we can model the number of people who will receive the email on the $n^{\text{th}}$ day by the formula $E(n) = 10(3^n)$, where $n = 0$ indicates the day the original email was sent.
   a. If we assume the population of the United States is 318 million people and everyone who receives the email sends it to 3 people who have not received it previously, how many days until there are as many emails being sent out as there are people in the United States?
   b. The population of Earth is approximately 7.1 billion people. On what day will 7.1 billion emails be sent out?

8. Solve the following exponential equations.
   a. $10^{(3x-5)} = 7^x$
   b. $3^{\frac{x}{5}} = 2^{4x-2}$
   c. $10^{x^2+5} = 100^{2x^2+x+2}$
   d. $4^{x^2-3x+4} = 2^{5x-4}$

9. Solve the following exponential equations.
   a. $(2^x)^x = 8^x$
   b. $(3^x)^x = 12$

10. Solve the following exponential equations.
   a. $10^{x+1} - 10^{x-1} = 1287$
   b. $2(4^x) + 4^{x+1} = 342$

11. Solve the following exponential equations.
   a. $(10^x)^2 - 3(10^x) + 2 = 0$ Hint: Let $u = 10^x$ and solve for $u$ before solving for $x$.
   b. $(2^x)^2 - 3(2^x) - 4 = 0$
   c. $3(e^x)^2 - 8(e^x) - 3 = 0$
   d. $4^x + 7(2^x) + 12 = 0$
   e. $(10^x)^2 - 2(10^x) - 1 = 0$

12. Solve the following systems of equations.
   a. $2^{x+2y} = 8$
      $4^{2x+y} = 1$
   b. $2^{2x+y-1} = 32$
      $4^{x-2y} = 2$
   c. $2^{3x} = 8^{2y+1}$
      $9^{2y} = 3^{3x-9}$

13. Because $f(x) = \log_b(x)$ is an increasing function, we know that if $p < q$, then $\log_b(p) < \log_b(q)$. Thus, if we take logarithms of both sides of an inequality, then the inequality is preserved. Use this property to solve the following inequalities.
   a. $4^x > \frac{5}{3}$
   b. $\left(\frac{2}{7}\right)^x > 9$
   c. $4^x > 8^{x-1}$
   d. $3^{x+2} > 5^{3-2x}$
   e. $\left(\frac{3}{4}\right)^x > \left(\frac{4}{3}\right)^{x+1}$

# Lesson 25: Geometric Sequences and Exponential Growth and Decay

## Classwork

### Opening Exercise

Suppose a ball is dropped from an initial height $h_0$ and that each time it rebounds, its new height is $60\%$ of its previous height.

a. What are the first four rebound heights $h_1$, $h_2$, $h_3$, and $h_4$ after being dropped from a height of $h_0 = 10$ ft.?

b. Suppose the initial height is $A$ ft. What are the first four rebound heights? Fill in the following table:

| Rebound | Height (ft.) |
|---|---|
| 1 | |
| 2 | |
| 3 | |
| 4 | |

c. How is each term in the sequence related to the one that came before it?

d. Suppose the initial height is $A$ ft. and that each rebound, rather than being $60\%$ of the previous height, is $r$ times the previous height, where $0 < r < 1$. What are the first four rebound heights? What is the $n^{\text{th}}$ rebound height?

e. What kind of sequence is the sequence of rebound heights?

f. Suppose that we define a function $f$ with domain all real numbers so that $f(1)$ is the first rebound height, $f(2)$ is the second rebound height, and continuing so that $f(k)$ is the $k^{\text{th}}$ rebound height for positive integers $k$. What type of function would you expect $f$ to be?

g. On the coordinate plane below, sketch the height of the bouncing ball when $A = 10$ and $r = 0.60$, assuming that the highest points occur at $x = 1, 2, 3, 4, \ldots$.

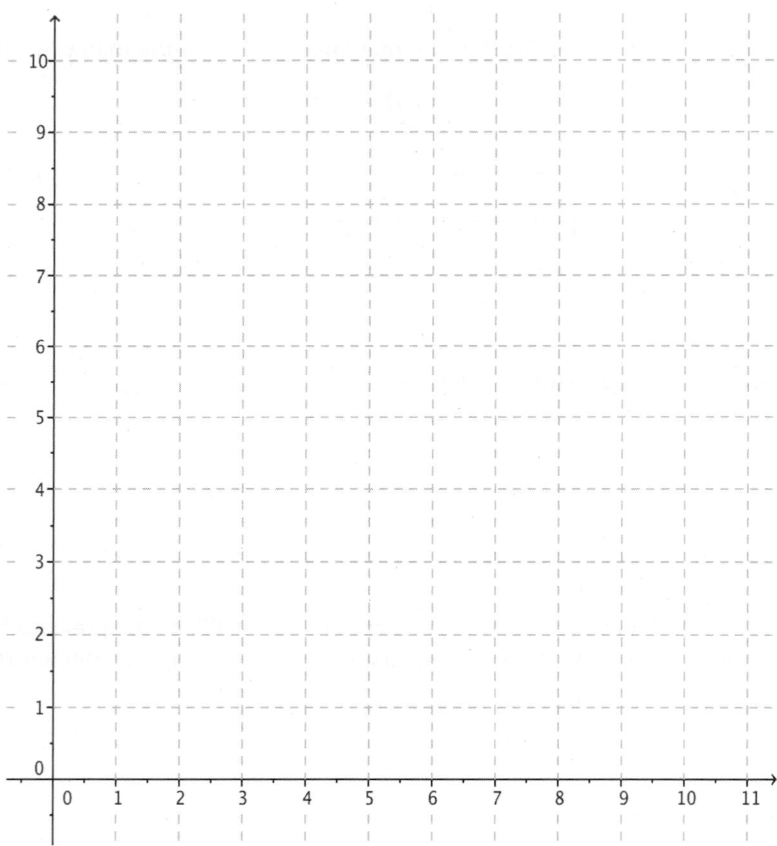

h. Does the exponential function $f(x) = 10(0.60)^x$ for real numbers $x$ model the height of the bouncing ball? Explain how you know.

i. What does the function $f(n) = 10(0.60)^n$ for integers $n \geq 0$ model?

## Exercises

1.

a. Jane works for a video game development company that pays her a starting salary of $\$100$ a day, and each day she works, she earns $\$100$ more than the day before. How much does she earn on day $5$?

b. If you were to graph the growth of her salary for the first $10$ days she worked, what would the graph look like?

c. What kind of sequence is the sequence of Jane's earnings each day?

2. A laboratory culture begins with $1{,}000$ bacteria at the beginning of the experiment, which we will denote by time $0$ hours. By time $2$ hours, there were $2{,}890$ bacteria.

   a. If the number of bacteria is increasing by a common factor each hour, how many bacteria were there at time $1$ hour? At time $3$ hours?

   b. Find the explicit formula for term $P_n$ of the sequence in this case.

   c. How would you find term $P_{n+1}$ if you know term $P_n$? Write a recursive formula for $P_{n+1}$ in terms of $P_n$.

   d. If $P_0$ is the initial population, the growth of the population $P_n$ at time $n$ hours can be modeled by the sequence $P_n = P(n)$, where $P$ is an exponential function with the following form:

   $$P(n) = P_0 2^{kn}, \text{ where } k > 0.$$

   Find the value of $k$ and write the function $P$ in this form. Approximate $k$ to four decimal places.

   e. Use the function in part (d) to determine the value of $t$ when the population of bacteria has doubled.

f. If $P_0$ is the initial population, the growth of the population $P$ at time $t$ can be expressed in the following form:

$$P(n) = P_0 e^{kn}, \text{ where } k > 0.$$

Find the value of $k$, and write the function $P$ in this form. Approximate $k$ to four decimal places.

g. Use the formula in part (d) to determine the value of $t$ when the population of bacteria has doubled.

3. The first term $a_0$ of a geometric sequence is $-5$, and the common ratio $r$ is $-2$.

a. What are the terms $a_0, a_1$, and $a_2$?

b. Find a recursive formula for this sequence.

c. Find an explicit formula for this sequence.

d. What is term $a_9$?

e. What is term $a_{10}$?

4. Term $a_4$ of a geometric sequence is $5.8564$, and term $a_5$ is $-6.44204$.
   a. What is the common ratio $r$?

   b. What is term $a_0$?

   c. Find a recursive formula for this sequence.

   d. Find an explicit formula for this sequence.

5. The recursive formula for a geometric sequence is $a_{n+1} = 3.92(a_n)$ with $a_0 = 4.05$. Find an explicit formula for this sequence.

6. The explicit formula for a geometric sequence is $a_n = 147(2.1)^{3n}$. Find a recursive formula for this sequence.

## Lesson Summary

**ARITHMETIC SEQUENCE:** A sequence is called *arithmetic* if there is a real number $d$ such that each term in the sequence is the sum of the previous term and $d$.

- *Explicit formula:* Term $a_n$ of an arithmetic sequence with first term $a_0$ and common difference $d$ is given by $a_n = a_0 + nd$, for $n \geq 0$.
- *Recursive formula:* Term $a_{n+1}$ of an arithmetic sequence with first term $a_0$ and common difference $d$ is given by $a_{n+1} = a_n + d$, for $n \geq 0$.

**GEOMETRIC SEQUENCE:** A sequence is called *geometric* if there is a real number $r$ such that each term in the sequence is a product of the previous term and $r$.

- *Explicit formula:* Term $a_n$ of a geometric sequence with first term $a_0$ and common ratio $r$ is given by $a_n = a_0 r^n$, for $n \geq 0$.
- *Recursive formula:* Term $a_{n+1}$ of a geometric sequence with first term $a_0$ and common ratio $r$ is given by $a_{n+1} = a_n r$.

## Problem Set

1. Convert the following recursive formulas for sequences to explicit formulas.

   a. $a_{n+1} = 4.2 + a_n$ with $a_0 = 12$

   b. $a_{n+1} = 4.2a_n$ with $a_0 = 12$

   c. $a_{n+1} = \sqrt{5}\, a_n$ with $a_0 = 2$

   d. $a_{n+1} = \sqrt{5} + a_n$ with $a_0 = 2$

   e. $a_{n+1} = \pi\, a_n$ with $a_0 = \pi$

2. Convert the following explicit formulas for sequences to recursive formulas.

   a. $a_n = \frac{1}{5}(3^n)$ for $n \geq 0$

   b. $a_n = 16 - 2n$ for $n \geq 0$

   c. $a_n = 16\left(\frac{1}{2}\right)^n$ for $n \geq 0$

   d. $a_n = 71 - \frac{6}{7}n$ for $n \geq 0$

   e. $a_n = 190(1.03)^n$ for $n \geq 0$

3. If a geometric sequence has $a_1 = 256$ and $a_8 = 512$, find the exact value of the common ratio $r$.

4. If a geometric sequence has $a_2 = 495$ and $a_6 = 311$, approximate the value of the common ratio $r$ to four decimal places.

5. Find the difference between the terms $a_{10}$ of an arithmetic sequence and a geometric sequence, both of which begin at term $a_0$ and have $a_2 = 4$ and $a_4 = 12$.

6. Given the geometric series defined by the following values of $a_0$ and $r$, find the value of $n$ so that $a_n$ has the specified value.

   a. $a_0 = 64, r = \frac{1}{2}, a_n = 2$

   b. $a_0 = 13, r = 3, a_n = 85293$

   c. $a_0 = 6.7, r = 1.9, a_n = 7804.8$

   d. $a_0 = 10958, r = 0.7, a_n = 25.5$

7. Jenny planted a sunflower seedling that started out 5 cm tall, and she finds that the average daily growth is $3.5$ cm.

   a. Find a recursive formula for the height of the sunflower plant on day $n$.

   b. Find an explicit formula for the height of the sunflower plant on day $n \geq 0$.

8. Kevin modeled the height of his son (in inches) at age $n$ years for $n = 2, 3, \dots, 8$ by the sequence $h_n = 34 + 3.2(n - 2)$. Interpret the meaning of the constants $34$ and $3.2$ in his model.

9. Astrid sells art prints through an online retailer. She charges a flat rate per order for an order processing fee, sales tax, and the same price for each print. The formula for the cost of buying $n$ prints is given by $P_n = 4.5 + 12.6\,n$.

   a. Interpret the number $4.5$ in the context of this problem.

   b. Interpret the number $12.6$ in the context of this problem.

   c. Find a recursive formula for the cost of buying $n$ prints.

10. A bouncy ball rebounds to $90\%$ of the height of the preceding bounce. Craig drops a bouncy ball from a height of $20$ ft.

    a. Write out the sequence of the heights $h_1$, $h_2$, $h_3$, and $h_4$ of the first four bounces, counting the initial height as $h_0 = 20$.

    b. Write a recursive formula for the rebound height of a bouncy ball dropped from an initial height of $20$ ft.

    c. Write an explicit formula for the rebound height of a bouncy ball dropped from an initial height of $20$ ft.

    d. How many bounces will it take until the rebound height is under $6$ ft.?

    e. Extension: Find a formula for the minimum number of bounces needed for the rebound height to be under $y$ ft., for a real number $0 < y < 20$.

11. Show that when a quantity $a_0 = A$ is increased by $x\%$, its new value is $a_1 = A\left(1 + \frac{x}{100}\right)$. If this quantity is again increased by $x\%$, what is its new value $a_2$? If the operation is performed $n$ times in succession, what is the final value of the quantity $a_n$?

12. When Eli and Daisy arrive at their cabin in the woods in the middle of winter, the internal temperature is $40°\text{F}$.
    a. Eli wants to turn up the thermostat by $2°\text{F}$ every $15$ minutes. Find an explicit formula for the sequence that represents the thermostat settings using Eli's plan.
    b. Daisy wants to turn up the thermostat by $4\%$ every $15$ minutes. Find an explicit formula for the sequence that represents the thermostat settings using Daisy's plan.
    c. Which plan will get the thermostat to $60°\text{F}$ most quickly?
    d. Which plan will get the thermostat to $72°\text{F}$ most quickly?

13. In nuclear fission, one neutron splits an atom causing the release of two other neutrons, each of which splits an atom and produces the release of two more neutrons, and so on.
    a. Write the first few terms of the sequence showing the numbers of atoms being split at each stage after a single atom splits. Use $a_0 = 1$.
    b. Find the explicit formula that represents your sequence in part (a).
    c. If the interval from one stage to the next is one-millionth of a second, write an expression for the number of atoms being split at the end of one second.
    d. If the number from part (c) were written out, how many digits would it have?

# Lesson 26: Percent Rate of Change

## Classwork

### Exercise

Answer the following questions.

The youth group from Example 1 is given the option of investing their money at $2.976\%$ interest per year, compounded monthly.

a. After two years, how much would be in each account with an initial deposit of $\$800$?

b. Compare the total amount from part (a) to how much they would have made using the interest rate of $3\%$ compounded yearly for two years. Which account would you recommend the youth group invest its money in? Why?

## Lesson Summary

- For application problems involving a percent rate of change represented by the unit rate $r$, we can write $F(t) = P(1+r)^t$, where $F$ is the future value (or ending amount), $P$ is the present amount, and $t$ is the number of time units. When the percent rate of change is negative, $r$ is negative and the quantity decreases with time.
- The nominal APR is the percent rate of change per compounding period times the number of compounding periods per year. If the nominal APR is given by the unit rate $r$ and is compounded $n$ times a year, then function $F(t) = P\left(1+\frac{r}{n}\right)^{nt}$ describes the future value at time $t$ of an account given that nominal APR and an initial value of $P$.
- For continuous compounding, we can write $F = Pe^{rt}$, where $e$ is Euler's number and $r$ is the unit rate associated to the percent rate of change.

## Problem Set

1. Write each recursive sequence in explicit form. Identify each sequence as arithmetic, geometric, or neither.
   a. $a_1 = 3, a_{n+1} = a_n + 5$
   b. $a_1 = -1, a_{n+1} = -2a_n$
   c. $a_1 = 30, a_{n+1} = a_n - 3$
   d. $a_1 = \sqrt{2}, a_{n+1} = \frac{a_n}{\sqrt{2}}$
   e. $a_1 = 1, a_{n+1} = \cos(\pi a_n)$

2. Write each sequence in recursive form. Assume the first term is when $n = 1$.
   a. $a_n = \frac{3}{2}n + 3$
   b. $a_n = 3\left(\frac{3}{2}\right)^n$
   c. $a_n = n^2$
   d. $a_n = \cos(2\pi n)$

3. Consider two bank accounts. Bank A gives simple interest on an initial investment in savings accounts at a rate of 3% per year. Bank B gives compound interest on savings accounts at a rate of 2.5% per year. Fill out the following table.

| Number of Years, $n$ | Bank A Balance, $a_n$ | Bank B Balance, $b_n$ |
|---|---|---|
| 0 | $1000.00 | $1000.00 |
| 1 | | |
| 2 | | |
| 3 | | |
| 4 | | |
| 5 | | |

   a. What type of sequence do the Bank A balances represent?
   b. Give both a recursive and an explicit formula for the Bank A balances.
   c. What type of sequence do the Bank B balances represent?
   d. Give both a recursive and an explicit formula for the Bank B balances.
   e. Which bank account balance is increasing faster in the first five years?
   f. If you were to recommend a bank account for a long-term investment, which would you recommend?
   g. At what point is the balance in Bank B larger than the balance in Bank A?

4. You decide to invest your money in a bank that uses continuous compounding at 5.5% interest per year. You have $500.
   a. Ja'mie decides to invest $1,000 in the same bank for one year. She predicts she will have double the amount in her account than you will have. Is this prediction correct? Explain.
   b. Jonas decides to invest $500 in the same bank as well, but for two years. He predicts that after two years he will have double the amount of cash that you will after one year. Is this prediction correct? Explain.

5. Use the properties of exponents to identify the percent rate of change of the functions below, and classify them as representing exponential growth or decay. (The first two problems are done for you.)
   a. $f(t) = (1.02)^t$
   b. $(t) = (1.01)^{12t}$
   c. $f(t) = (0.97)^t$
   d. $f(t) = 1000(1.2)^t$
   e. $f(t) = \frac{(1.07)^t}{1000}$
   f. $f(t) = 100 \cdot 3^t$
   g. $f(t) = 1.05 \cdot \left(\frac{1}{2}\right)^t$
   h. $f(t) = 80 \cdot \left(\frac{49}{64}\right)^{\frac{1}{2}t}$
   i. $f(t) = 1.02 \cdot (1.13)^{\pi t}$

6. The effective rate of an investment is the percent rate of change per year associated to the nominal APR. The effective rate is very useful in comparing accounts with different interest rates and compounding periods. In general, the effective rate can be found with the following formula: $r_E = \left(1 + \frac{r}{k}\right)^k - 1$. The effective rate presented here is the interest rate needed for annual compounding to be equal to compounding $n$ times per year.

   a. For investing, which account is better: an account earning a nominal APR of $7\%$ compounded monthly or an account earning a nominal APR of $6.875\%$ compounded daily? Why?

   b. The effective rate formula for an account compounded continuously is $r_E = e^r - 1$. Would an account earning $6.875\%$ interest compounded continuously be better than the accounts in part (a)?

7. Radioactive decay is the process in which radioactive elements decay into more stable elements. A half-life is the time it takes for half of an amount of an element to decay into a more stable element. For instance, the half-life for half of an amount of uranium-$235$ to transform into lead-$207$ is $704$ million years. Thus, after $704$ million years, only half of any sample of uranium-$235$ will remain, and the rest will have changed into lead-$207$. We will assume that radioactive decay is modeled by exponential decay with a constant decay rate.

   a. Suppose we have a sample of $A$ g of uranium-$235$. Write an exponential formula that gives the amount of uranium-$235$ remaining after $m$ half-lives.

   b. Does the formula that you wrote in part (a) work for any radioactive element? Why?

   c. Suppose we have a sample of $A$ g of uranium-$235$. What is the decay rate per million years? Write an exponential formula that gives the amount of uranium-$235$ remaining after $t$ million years.

   d. How would you calculate the number of years it takes to get to a specific percentage of the original amount of material? For example, how many years will it take for there to be $80\%$ of the original amount of uranium-$235$ remaining?

   e. How many millions of years would it take $2.35$ kg of uranium-$235$ to decay to $1$ kg of uranium?

8. Doug drank a cup of tea with $130$ mg of caffeine. Each hour, the caffeine in Doug's body diminishes by about $12\%$. (This rate varies between $6\%$ and $14\%$ depending on the person.)

   a. Write a formula to model the amount of caffeine remaining in Doug's system after each hour.

   b. About how long will it take for the level of caffeine in Doug's system to drop below $30$ mg?

   c. The time it takes for the body to metabolize half of a substance is called a half-life. To the nearest $5$ minutes, how long is the half-life for Doug to metabolize caffeine?

   d. Write a formula to model the amount of caffeine remaining in Doug's system after $m$ half-lives.

9. A study done from $1950$ through $2000$ estimated that the world population increased on average by $1.77\%$ each year. In $1950$, the world population was $2.519$ billion.

   a. Write a function $p$ for the world population $t$ years after $1950$.

   b. If this trend continued, when should the world population have reached 7 billion?

   c. The world population reached 7 billion October $31$, $2011$, according to the United Nations. Is the model reasonably accurate?

   d. According to the model, when will the world population be greater than $12$ billion people?

10. A particular mutual fund offers $4.5\%$ nominal APR compounded monthly. Trevor wishes to deposit $\$1{,}000$.
    a. What is the percent rate of change per month for this account?
    b. Write a formula for the amount Trevor will have in the account after $m$ months.
    c. *Doubling time* is the amount of time it takes for an investment to double. What is the doubling time of Trevor's investment?

11. When paying off loans, the monthly payment first goes to any interest owed before being applied to the remaining balance. Accountants and bankers use tables to help organize their work.
    a. Consider the situation that Fred is paying off a loan of $\$125{,}000$ with an interest rate of $6\%$ per year compounded monthly. Fred pays $\$749.44$ every month. Complete the following table:

| Payment | Interest Paid | Principal Paid | Remaining Principal |
|---|---|---|---|
| $749.44 | | | |
| $749.44 | | | |
| $749.44 | | | |

    b. Fred's loan is supposed to last for $30$ years. How much will Fred end up paying if he pays $\$749.44$ every month for $30$ years? How much of this is interest if his loan was originally for $\$125{,}000$?

# Lesson 27: Modeling with Exponential Functions

## Classwork

### Opening Exercise

The following table contains U.S. population data for the two most recent census years, $2000$ and $2010$.

| Census Year | U.S. Population (in millions) |
|---|---|
| $2000$ | $281.4$ |
| $2010$ | $308.7$ |

a. Steve thinks the data should be modeled by a linear function.

i. What is the average rate of change in population per year according to this data?

ii. Write a formula for a linear function, $L$, that will estimate the population $t$ years since the year $2000$.

b. Phillip thinks the data should be modeled by an exponential function.

i. What is the growth rate of the population per year according to this data?

ii. Write a formula for an exponential function, $E$, that will estimate the population $t$ years since the year $2000$.

c. Who has the correct model? How do you know?

## Mathematical Modeling Exercise/Exercises 1–14

In this challenge, you will continue to examine U.S. census data to select and refine a model for the population of the United States over time.

1. The following table contains additional U.S. census population data. Would it be more appropriate to model this data with a linear or an exponential function? Explain your reasoning.

| Census Year | U.S. Population (in millions of people) |
|---|---|
| 1900 | 76.2 |
| 1910 | 92.2 |
| 1920 | 106.0 |
| 1930 | 122.8 |
| 1940 | 132.2 |
| 1950 | 150.7 |
| 1960 | 179.3 |
| 1970 | 203.3 |
| 1980 | 226.5 |
| 1990 | 248.7 |
| 2000 | 281.4 |
| 2010 | 308.7 |

2. Use a calculator's regression capability to find a function, $f$, that models the U.S. Census Bureau data from 1900 to 2010.

3. Find the growth factor for each 10-year period and record it in the table below. What do you observe about these growth factors?

| Census Year | U.S. Population (in millions of people) | Growth Factor (10-year period) |
|---|---|---|
| 1900 | 76.2 | -- |
| 1910 | 92.2 | |
| 1920 | 106.0 | |
| 1930 | 122.8 | |
| 1940 | 132.2 | |
| 1950 | 150.7 | |
| 1960 | 179.3 | |
| 1970 | 203.3 | |
| 1980 | 226.5 | |
| 1990 | 248.7 | |
| 2000 | 281.4 | |
| 2010 | 308.7 | |

4. For which decade is the 10-year growth factor the lowest? What factors do you think caused that decrease?

5. Find an average 10-year growth factor for the population data in the table. What does that number represent? Use the average growth factor to find an exponential function, $g$, that can model this data.

6. You have now computed three potential models for the population of the United States over time: functions $E$, $f$, and $g$. Which one do you expect would be the most accurate model based on how they were created? Explain your reasoning.

7. Summarize the three formulas for exponential models that you have found so far: Write the formula, the initial populations, and the growth rates indicated by each function. What is different between the structures of these three functions?

8. Rewrite the functions $E$, $f$, and $g$ as needed in terms of an annual growth rate.

9. Transform the functions as needed so that the time $t = 0$ represents the same year in functions $E$, $f$, and $g$. Then compare the values of the initial populations and annual growth rates indicated by each function.

10. Which of the three functions is the best model to use for the U.S. census data from 1900 to 2010? Explain your reasoning.

11. The U.S. Census Bureau website http://www.census.gov/popclock displays the current estimate of both the United States and world populations.

 a. What is today's current estimated population of the U.S.?

 b. If time $t = 0$ represents the year $1900$, what is the value of $t$ for today's date? Give your answer to two decimal places.

 c. Which of the functions $E$, $f$, and $g$ gives the best estimate of today's population? Does that match what you expected? Justify your reasoning.

 d. With your group, discuss some possible reasons for the discrepancy between what you expected in Exercise $8$ and the results of part (c) above.

12. Use the model that most accurately predicted today's population in Exercise 9, part (c) to predict when the U.S. population will reach half a billion.

13. Based on your work so far, do you think this is an accurate prediction? Justify your reasoning.

14. Here is a graph of the U.S. population since the census began in $1790$. Which type of function would best model this data? Explain your reasoning.

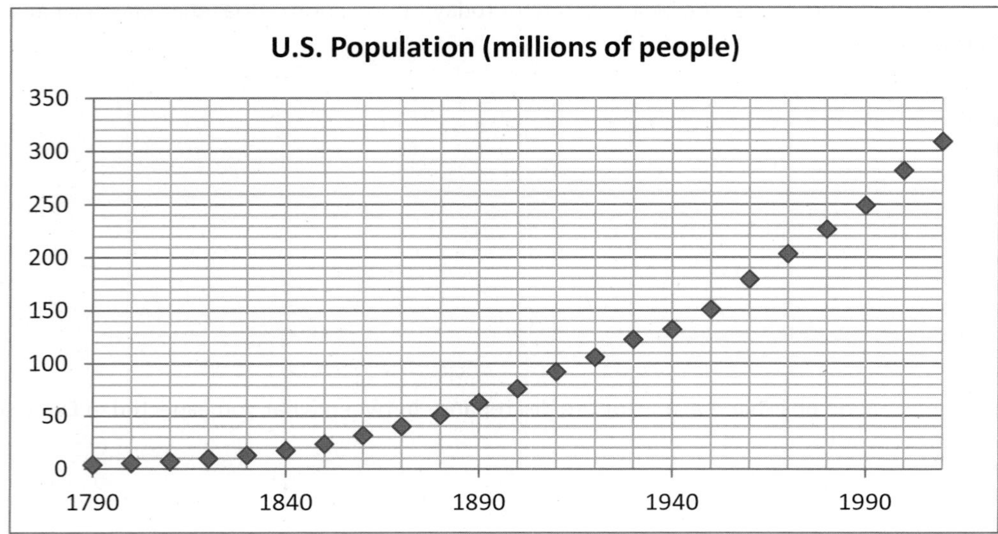


**Figure 1: Source U.S. Census Bureau**

15. The graph below shows the population of New York City during a time of rapid population growth.

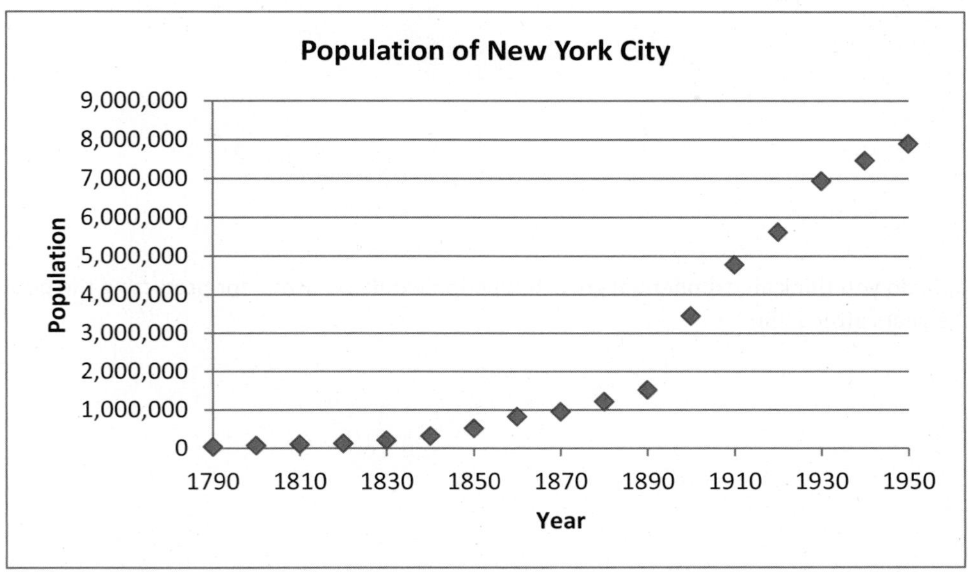


Finn averaged the 10-year growth rates and wrote the function $f(t) = 33131(1.44)^{\frac{t}{10}}$, where $t$ is the time in years since 1790.

Gwen used the regression features on a graphing calculator and got the function $g(t) = 48661(1.036)^t$, where $t$ is the time in years since 1790.

a. Rewrite each function to determine the annual growth rate for Finn's model and Gwen's model.

b. What is the predicted population in the year 1790 for each model?

c. Lenny calculated an exponential regression using his graphing calculator and got the same growth rate as Gwen, but his initial population was very close to 0. Explain what data Lenny may have used to find his function.

d. When does Gwen's function predict the population will reach $1{,}000{,}000$? How does this compare to the graph?

e. Based on the graph, do you think an exponential growth function would be useful for predicting the population of New York in the years after $1950$?

16. Suppose each function below represents the population of a different U.S. city since the year $1900$.

a. Complete the table below. Use the properties of exponents to rewrite expressions as needed to help support your answers.

| City Population Function ($t$ is years since $1900$) | Population in the Year $1900$ | Annual Growth/Decay Rate | Predicted in $2000$ | Between Which Years Did the Population Double? |
|---|---|---|---|---|
| $A(t) = 3000(1.1)^{\frac{t}{5}}$ | | | | |
| $B(t) = \dfrac{(1.5)^{2t}}{2.25}$ | | | | |
| $C(t) = 10000(1 - 0.01)^t$ | | | | |
| $D(t) = 900(1.02)^t$ | | | | |

b. Could the function $(t) = 6520(1.219)^{\frac{t}{10}}$, where $t$ is years since $2000$ also represent the population of one of these cities? Use the properties of exponents to support your answer.

c. Which cities are growing in size and which are decreasing according to these models?

d. Which of these functions might realistically represent city population growth over an extended period of time?

## Lesson Summary

To model data with an exponential function:

- Examine the data to see if there appears to be a constant growth or decay factor.
- Determine a growth factor and a point in time to correspond to $t = 0$.
- Create a function to model the situation $f(t) = a \cdot b^{ct}$, where $b$ is the growth factor every $\frac{1}{c}$ years and $a$ is the value of $f$ when $t = 0$.

Logarithms can be used to solve for $t$ when you know the value of $f(t)$ in an exponential function model.

## Problem Set

1. Does each pair of formulas described below represent the same sequence? Justify your reasoning.
   a. $a_{n+1} = \frac{2}{3}a_n$, $a_0 = -1$ and $b_n = -\left(\frac{2}{3}\right)^n$ for $n \geq 0$.
   b. $a_n = 2a_{n-1} + 3$, $a_0 = 3$ and $b_n = 2(n-1)^3 + 4(n-1) + 3$ for $n \geq 1$.
   c. $a_n = \frac{1}{3}(3)^n$ for $n \geq 0$ and $b_n = 3^{n-2}$ for $n \geq 0$.

2. Tina is saving her babysitting money. She has $\$500$ in the bank, and each month she deposits another $\$100$. Her account earns $2\%$ interest compounded monthly.
   a. Complete the table showing how much money she has in the bank for the first four months.

| Month | Amount |
|---|---|
| 1 | |
| 2 | |
| 3 | |
| 4 | |

   b. Write a recursive sequence for the amount of money she has in her account after $n$ months.

3. Assume each table represents values of an exponential function of the form $f(t) = a(b)^{ct}$, where $b$ is a positive real number and $a$ and $c$ are real numbers. Use the information in each table to write a formula for $f$ in terms of $t$ for parts (a)–(d).

a.

| $t$ | $f(t)$ |
|---|---|
| 0 | 10 |
| 4 | 50 |

b.

| $t$ | $f(t)$ |
|---|---|
| 0 | 1000 |
| 5 | 750 |

c.

| $t$ | $f(t)$ |
|---|---|
| 6 | 25 |
| 8 | 45 |

d.

| $t$ | $f(t)$ |
|---|---|
| 3 | 50 |
| 6 | 40 |

e. Rewrite the expressions for each function in parts (a)–(d) to determine the annual growth or decay rate.

f. For parts (a) and (c), determine when the value of the function is double its initial amount.

g. For parts (b) and (d), determine when the value of the function is half its initial amount.

4. When examining the data in Example 1, Juan noticed the population doubled every five years and wrote the formula $P(t) = 100(2)^{\frac{t}{5}}$. Use the properties of exponents to show that both functions grow at the same rate per year.

5. The growth of a tree seedling over a short period of time can be modeled by an exponential function. Suppose the tree starts out 3 ft. tall and its height increases by $15\%$ per year. When will the tree be 25 ft. tall?

6. Loggerhead turtles reproduce every 2–4 years, laying approximately 120 eggs in a clutch. Studying the local population, a biologist records the following data in the second and fourth years of her study:

| Year | Population |
|---|---|
| 2 | 50 |
| 4 | 1250 |

a. Find an exponential model that describes the loggerhead turtle population in year $t$.

b. According to your model, when will the population of loggerhead turtles be over 5,000? Give your answer in years and months.

7. The radioactive isotope seaborgium-266 has a half-life of 30 seconds, which means that if you have a sample of $A$ g of seaborgium-266, then after 30 seconds half of the sample has decayed (meaning it has turned into another element) and only $\frac{A}{2}$ g of seaborgium-266 remain. This decay happens continuously.

a. Define a sequence $a_0, a_1, a_2, \ldots$ so that $a_n$ represents the amount of a 100 g sample that remains after $n$ minutes.

b. Define a function $a(t)$ that describes the amount of seaborgium-266 that remains of a 100 g sample after $t$ minutes.

c. Does your sequence from part (a) and your function from part (b) model the same thing? Explain how you know.

d. How many minutes does it take for less than 1 g of seaborgium-266 to remain from the original 100 g sample? Give your answer to the nearest minute.

8. Compare the data for the amount of substance remaining for each element: strontium-90, magnesium-28, and bismuth.

Strontium-90 (grams) vs. time (hours)

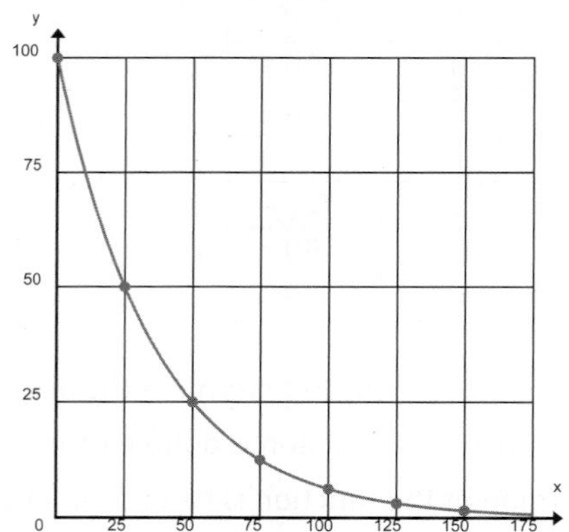


| Radioactive Decay of Magnesium-28 | |
|---|---|
| $R$ | $t$ hours |
| 1 | 0 |
| 0.5 | 21 |
| 0.25 | 42 |
| 0.125 | 63 |
| 0.0625 | 84 |

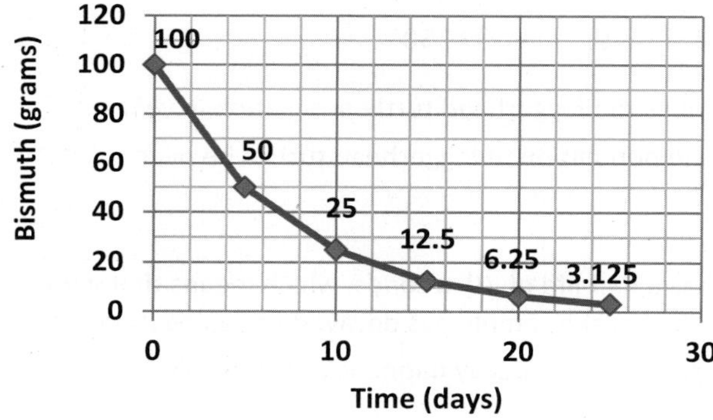


a. Which element decays most rapidly? How do you know?

b. Write an exponential function for each element that shows how much of a $100$ g sample will remain after $t$ days. Rewrite each expression to show precisely how their exponential decay rates compare to confirm your answer to part (a).

9. The growth of two different species of fish in a lake can be modeled by the functions shown below where $t$ is time in months since January $2000$. Assume these models will be valid for at least $5$ years.

Fish A: $f(t) = 5000(1.3)^t$

Fish B: $g(t) = 10{,}000(1.1)^t$

According to these models, explain why the fish population modeled by function $f$ will eventually catch up to the fish population modeled by function $g$. Determine precisely when this will occur.

10. When looking at U.S. minimum wage data, you can consider the nominal minimum wage, which is the amount paid in dollars for an hour of work in the given year. You can also consider the minimum wage adjusted for inflation. Below is a table showing the nominal minimum wage and a graph of the data when the minimum wage is adjusted for inflation. Do you think an exponential function would be an appropriate model for either situation? Explain your reasoning.

| Year | Nominal Minimum Wage |
|---|---|
| 1940 | \$0.30 |
| 1945 | \$0.40 |
| 1950 | \$0.75 |
| 1955 | \$0.75 |
| 1960 | \$1.00 |
| 1965 | \$1.25 |
| 1970 | \$1.60 |
| 1975 | \$2.10 |
| 1980 | \$3.10 |
| 1985 | \$3.35 |
| 1990 | \$3.80 |
| 1995 | \$4.25 |
| 2000 | \$5.15 |
| 2005 | \$5.15 |
| 2010 | \$7.25 |

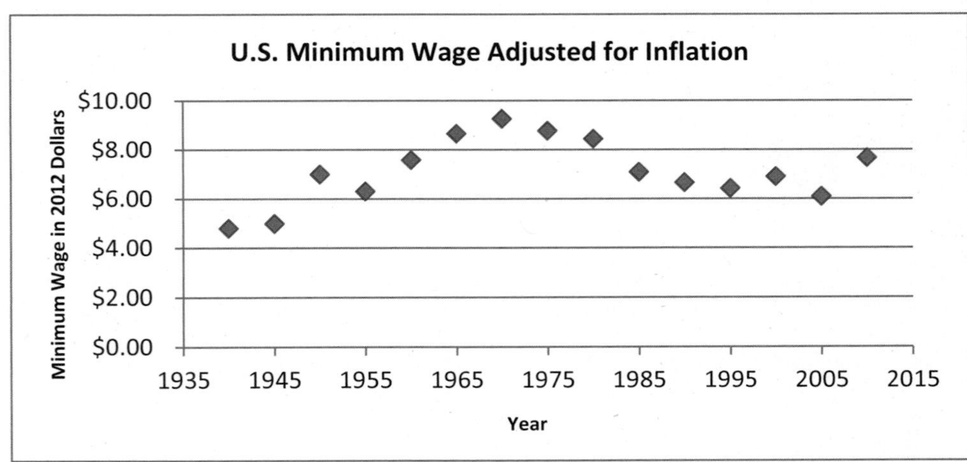

11. A dangerous bacterial compound forms in a closed environment but is immediately detected. An initial detection reading suggests the concentration of bacteria in the closed environment is one percent of the fatal exposure level. Two hours later, the concentration has increased to four percent of the fatal exposure level.

    a. Develop an exponential model that gives the percent of fatal exposure level in terms of the number of hours passed.

    b. Doctors and toxicology professionals estimate that exposure to two-thirds of the bacteria's fatal concentration level will begin to cause sickness. Provide a rough time limit (to the nearest 15 minutes) for the inhabitants of the infected environment to evacuate in order to avoid sickness.

    c. A prudent and more conservative approach is to evacuate the infected environment before bacteria concentration levels reach 45% of the fatal level. Provide a rough time limit (to the nearest 15 minutes) for evacuation in this circumstance.

    d. When will the infected environment reach 100% of the fatal level of bacteria concentration (to the nearest minute)?

12. Data for the number of users at two different social media companies is given below. Assuming an exponential growth rate, which company is adding users at a faster annual rate? Explain how you know.

| Social Media Company A | |
|---|---|
| Year | Number of Users (Millions) |
| 2010 | 54 |
| 2012 | 185 |

| Social Media Company B | |
|---|---|
| Year | Number of Users (Millions) |
| 2009 | 360 |
| 2012 | 1,056 |

# Lesson 28: Newton's Law of Cooling, Revisited

## Classwork

*Newton's law of cooling* is used to model the temperature of an object of some temperature placed in an environment of a different temperature. The temperature of the object $t$ hours after being placed in the new environment is modeled by the formula

$$T(t) = T_a + (T_0 - T_a) \cdot e^{-kt},$$

where:

$T(t)$ is the temperature of the object after a time of $t$ hours has elapsed,

$T_a$ is the ambient temperature (the temperature of the surroundings), assumed to be constant and not impacted by the cooling process,

$T_0$ is the initial temperature of the object, and

$k$ is the decay constant.

### Mathematical Modeling Exercise 1

A crime scene investigator is called to the scene of a crime where a dead body has been found. He arrives at the scene and measures the temperature of the dead body at $9{:}30$ p.m. to be $78.3°\text{F}$. He checks the thermostat and determines that the temperature of the room has been kept at $74°\text{F}$. At $10{:}30$ p.m., the investigator measures the temperature of the body again. It is now $76.8°\text{F}$. He assumes that the initial temperature of the body was $98.6°\text{F}$ (normal body temperature). Using this data, the crime scene investigator proceeds to calculate the time of death. According to the data he collected, what time did the person die?

a. Can we find the time of death using only the temperature measured at $9{:}30$ p.m.? Explain.

b. Set up a system of two equations using the data.

d. Find the value of the decay constant, $k$.

e. What was the time of death?

**Mathematical Modeling Exercise 2**

A pot of tea is heated to $90°C$. A cup of the tea is poured into a mug and taken outside where the temperature is $18°C$. After $2$ minutes, the temperature of the cup of tea is approximately $65°C$.

a. Determine the value of the decay constant, $k$.

b. Write a function for the temperature of the tea in the mug, $T$, in °C, as a function of time, $t$, in minutes.

c. Graph the function $T$.

d. Use the graph of $T$ to describe how the temperature decreases over time.

e. Use properties of exponents to rewrite the temperature function in the form $T(t) = 18 + 72(1 + r)^t$.

f. In Lesson 26, we saw that the value of $r$ represents the percent change of a quantity that is changing according to an exponential function of the form $f(t) = A(1 + r)^t$. Describe what $r$ represents in the context of the cooling tea.

g. As more time elapses, what temperature does the tea approach? Explain using both the context of the problem and the graph of the function $T$.

## Mathematical Modeling Exercise 3

Two thermometers are sitting in a room that is 22°C. When each thermometer reads 22°C, the thermometers are placed in two different ovens. Select data for the temperature $T$ of each thermometer (in °C) $t$ minutes after being placed in the oven is provided below.

Thermometer 1:

| $t$ (minutes) | 0 | 2 | 5 | 8 | 10 | 14 |
|---|---|---|---|---|---|---|
| $T$ (°C) | 22 | 75 | 132 | 173 | 175 | 176 |

Thermometer 2:

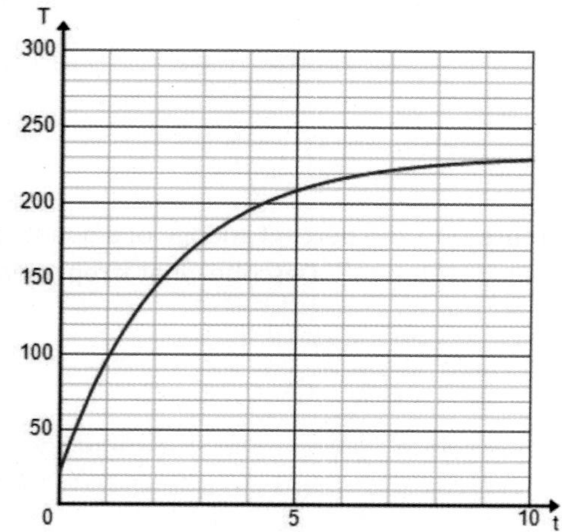


a. Do the table and graph given for each thermometer support the statement that Newton's law of cooling also applies when the surrounding temperature is warmer? Explain.

b. Which thermometer was placed in a hotter oven? Explain.

c. Using a generic decay constant, $k$, without finding its value, write an equation for each thermometer expressing the temperature as a function of time.

d. How do the equations differ when the surrounding temperature is warmer than the object rather than cooler as in previous examples?

e. How do the graphs differ when the surrounding temperature is warmer than the object rather than cooler as in previous examples?

## Problem Set

1. Experiments with a covered cup of coffee show that the temperature (in degrees Fahrenheit) of the coffee can be modeled by the following equation:

$$f(t) = 112e^{-0.08t} + 68,$$

where the time is measured in minutes after the coffee was poured into the cup.

   a. What is the temperature of the coffee at the beginning of the experiment?

   b. What is the temperature of the room?

   c. After how many minutes is the temperature of the coffee $140°\text{F}$? Give your answer to 3 decimal places.

   d. What is the temperature of the coffee after how many minutes have elapsed?

   e. What is the percent rate of change of the difference between the temperature of the room and the temperature of the coffee?

2. Suppose a frozen package of hamburger meat is removed from a freezer that is set at $0°\text{F}$ and placed in a refrigerator that is set at $38°\text{F}$. Six hours after being placed in the refrigerator, the temperature of the meat is $12°\text{F}$.

   a. Determine the decay constant, $k$.

   b. Write a function for the temperature of the meat, $T$ in Fahrenheit, as a function of time, $t$ in hours.

   c. Graph the function $T$.

   d. Describe the transformations required to graph the function $T$ beginning with the graph of the natural exponential function $f(t) = e^t$.

   e. How long will it take the meat to thaw (reach a temperature above $32°\text{F}$)? Give answer to three decimal places.

   f. What is the percent rate of change of the difference between the temperature of the refrigerator and the temperature of the meat?

3. The table below shows the temperature of biscuits that were removed from an oven at time $t = 0$.

| $t$ (min) | 0 | 10 | 20 | 30 | 40 | 50 | 60 |
|---|---|---|---|---|---|---|---|
| $T$ (°C) | 100 | 34.183 | 22.514 | 20.446 | 20.079 | 20.014 | 20.002 |

   a. What is the initial temperature of the biscuits?

   b. What does the ambient temperature (room temperature) appear to be?

   c. Use the temperature at $t = 10$ minutes to find the decay constant, $k$.

   d. Confirm the value of $k$ by using another data point from the table.

   e. Write a function for the temperature of the biscuits (in Celsius) as a function of time in minutes.

   f. Graph the function $T$.

4. Match each verbal description with its correct graph and write a possible equation expressing temperature as a function of time.

   a. A pot of liquid is heated to a boil and then placed on a counter to cool.

   b. A frozen dinner is placed in a preheated oven to cook.

   c. A can of room-temperature soda is placed in a refrigerator.

(i)

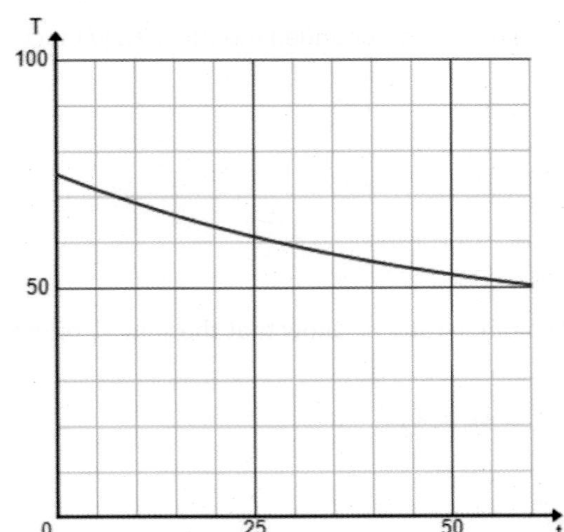


(ii)

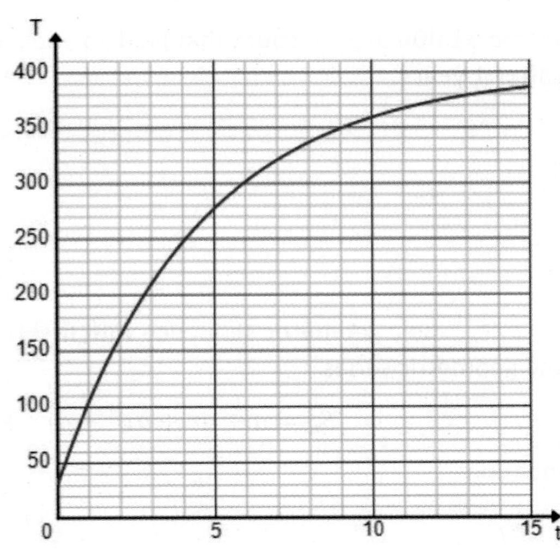


(iii)

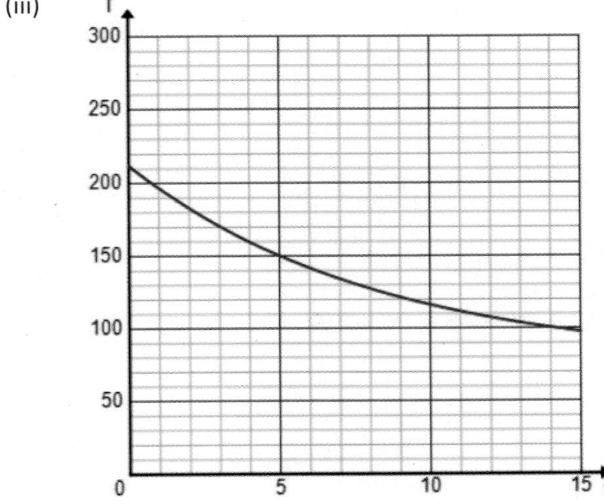

# Lesson 29: The Mathematics Behind a Structured Savings Plan

## Classwork

### Opening Exercise

Suppose you invested $1,000 in an account that paid an annual interest rate of 3% compounded monthly. How much would you have after 1 year?

### Example 1

Let $a, ar, ar^2, ar^3, ar^4, \ldots$ be a geometric sequence with first term $a$ and common ratio $r$. Show that the sum $S_n$ of the first $n$ terms of the geometric series

$$S_n = a + ar + ar^2 + ar^3 + \cdots + ar^{n-1} \qquad (r \neq 1)$$

is given by the equation

$$S_n = a\frac{1-r^n}{1-r}.$$

### Exercises 1–3

1. Find the sum of the geometric series $3 + 6 + 12 + 24 + 48 + 96 + 192$.

2. Find the sum of the geometric series $40 + 40(1.005) + 40(1.005)^2 + \cdots + 40(1.005)^{11}$.

3. Describe a situation that might lead to calculating the sum of the geometric series in Exercise 2.

### Example 2

A $\$100$ deposit is made at the end of every month for $12$ months in an account that earns interest at an annual interest rate of $3\%$ compounded monthly. How much will be in the account immediately after the last payment?

### Discussion

An *annuity* is a series of payments made at fixed intervals of time. Examples of annuities include structured savings plans, lease payments, loans, and monthly home mortgage payments. The term annuity sounds like it is only a yearly payment, but annuities are often monthly, quarterly, or semiannually. The *future amount of the annuity,* denoted $A_f$, is the sum of all the individual payments made plus all the interest generated from those payments over the specified period of time.

We can generalize the structured savings plan example above to get a generic formula for calculating the future value of an annuity $A_f$ in terms of the recurring payment $R$, interest rate $i$, and number of payment periods $n$. In the example above, we had a recurring payment of $R = 100$, an interest rate per time period of $i = 0.025$, and 12 payments, so $n = 12$. To make things simpler, we always assume that the payments and the time period in which interest is compounded are at the same time. That is, we do not consider plans where deposits are made halfway through the month with interest compounded at the end of the month.

In the example, the amount $A_f$ of the structured savings plan annuity was the sum of all payments plus the interest accrued for each payment:

$$A_f = R + R(1+i)^1 + R(1+i)^2 + \cdots + R(1+i)^{n-1}.$$

This, of course, is a geometric series with $n$ terms, $a = R$, and $r = 1 + i$, which after substituting into the formula for a geometric series and rearranging is

$$A_f = R \cdot \frac{(1+i)^n - 1}{i}.$$

## Exercises 4–5

4. Write the sum without using summation notation, and find the sum.

   a. $\sum_{k=0}^{5} k$

   b. $\sum_{j=5}^{7} j^2$

   c. $\sum_{i=2}^{4} \frac{1}{i}$

5. Write each sum using summation notation.

   a. $1^4 + 2^4 + 3^4 + 4^4 + 5^4 + 6^4 + 7^4 + 8^4 + 9^4$

   b. $1 + \cos(\pi) + \cos(2\pi) + \cos(3\pi) + \cos(4\pi) + \cos(5\pi)$

   c. $2 + 4 + 6 + \cdots + 1000$

## Lesson Summary

- **SERIES**: Let $a_1, a_2, a_3, a_4, \ldots$ be a sequence of numbers. A sum of the form

$$a_1 + a_2 + a_3 + \cdots + a_n$$

for some positive integer $n$ is called a *series* (or *finite series*) and is denoted $S_n$. The $a_i$'s are called the *terms* of the series. The number $S_n$ that the series adds to is called the *sum* of the series.

- **GEOMETRIC SERIES:** A *geometric series* is a series whose terms form a geometric sequence.
- **SUM OF A FINITE GEOMETRIC SERIES**: The sum $S_n$ of the first $n$ terms of the geometric series $S_n = a + ar + \cdots + ar^{n-1}$ (when $r \neq 1$) is given by

$$S_n = a\frac{1-r^n}{1-r}.$$

The sum of a finite geometric series can be written in summation notation as

$$\sum_{k=0}^{n-1} ar^k = a \cdot \frac{1-r^n}{1-r}.$$

- The generic formula for calculating the future value of an annuity $A_f$ in terms of the recurring payment $R$, interest rate $i$, and number of periods $n$ is given by

$$A_f = R \cdot \frac{(1+i)^n - 1}{i}.$$

## Problem Set

1. A car loan is one of the first secured loans most Americans obtain. Research used car prices and specifications in your area to find a reasonable used car that you would like to own (under $10,000). If possible, print out a picture of the car you selected.

   a. What is the year, make, and model of your vehicle?

   b. What is the selling price for your vehicle?

c. The following table gives the monthly cost per $1000 financed on a 5-year auto loan. Assume you can get a 5% annual interest rate. What is the monthly cost of financing the vehicle you selected? (A formula will be developed to find the monthly payment of a loan in Lesson 30.)

| Five-Year (60-month) Loan | |
|---|---|
| Interest Rate | Amount per $1,000 Financed |
| 1.0% | $17.09 |
| 1.5% | $17.31 |
| 2.0% | $17.53 |
| 2.5% | $17.75 |
| 3.0% | $17.97 |
| 3.5% | $18.19 |
| 4.0% | $18.41 |
| 4.5% | $18.64 |
| 5.0% | $18.87 |
| 5.5% | $19.10 |
| 6.0% | $19.33 |
| 6.5% | $19.56 |
| 7.0% | $19.80 |
| 7.5% | $20.04 |
| 8.0% | $20.28 |
| 8.5% | $20.52 |
| 9.0% | $20.76 |

d. What is the gas mileage for your vehicle?

e. If you drive 120 miles per week and gas is $4 per gallon, then how much will gas cost per month?

2. Write the sum without using summation notation, and find the sum.

a. $\sum_{k=1}^{8} k$

b. $\sum_{k=-8}^{8} k$

c. $\sum_{k=1}^{4} k^3$

d. $\sum_{m=0}^{6} 2m$

e. $\sum_{m=0}^{6} 2m + 1$

f. $\sum_{k=2}^{5} \frac{1}{k}$

g. $\sum_{j=0}^{3} (-4)^{j-2}$

h. $\sum_{m=1}^{4} 16\left(\frac{3}{2}\right)^m$

j. $\sum_{j=0}^{3} \frac{105}{2j+1}$

k. $\sum_{p=1}^{3} p \cdot 3^p$

l. $\sum_{j=1}^{6} 100$

m. $\sum_{k=0}^{4} \sin\left(\frac{k\pi}{2}\right)$

n. $\sum_{k=1}^{9} \log\left(\frac{k}{k+1}\right)$

(Hint: You do not need a calculator to find the sum.)

3. Write the sum without using sigma notation (you do not need to find the sum).

   a. $\sum_{k=0}^{4} \sqrt{k+3}$

   b. $\sum_{i=0}^{8} x^i$

   c. $\sum_{j=1}^{6} jx^{j-1}$

   d. $\sum_{k=0}^{9} (-1)^k x^k$

4. Write each sum using summation notation.

   a. $1 + 2 + 3 + 4 + \cdots + 1000$

   b. $2 + 4 + 6 + 8 + \cdots + 100$

   c. $1 + 3 + 5 + 7 + \cdots + 99$

   d. $\frac{1}{2} + \frac{2}{3} + \frac{3}{4} + \cdots + \frac{99}{100}$

   e. $1^2 + 2^2 + 3^2 + 4^2 + \cdots + 10{,}000^2$

   f. $1 + x + x^2 + x^3 + \cdots + x^{200}$

   g. $\frac{1}{1\cdot 2} + \frac{1}{2\cdot 3} + \frac{1}{3\cdot 4} + \cdots + \frac{1}{49\cdot 50}$

   h. $1\ln(1) + 2\ln(2) + 3\ln(3) + \cdots + 10\ln(10)$

5. Find the sum of the geometric series.

   a. $1 + 3 + 9 + \cdots + 2187$

   b. $1 + \frac{1}{2} + \frac{1}{4} + \frac{1}{8} + \cdots + \frac{1}{512}$

   c. $1 - \frac{1}{2} + \frac{1}{4} - \frac{1}{8} + \cdots - \frac{1}{512}$

   d. $0.8 + 0.64 + 0.512 + \cdots + 0.32768$

   e. $1 + \sqrt{3} + 3 + 3\sqrt{3} + \cdots + 243$

   f. $\sum_{k=0}^{5} 2^k$

   g. $\sum_{m=1}^{4} 5\left(\frac{3}{2}\right)^m$

   h. $1 - x + x^2 - x^3 + \cdots + x^{30}$ in terms of $x$

   i. $\sum_{m=0}^{11} 4^{\frac{m}{3}}$

   j. $\sum_{n=0}^{14} \left(\sqrt[5]{6}\right)^n$

   k. $\sum_{k=0}^{6} 2 \cdot \left(\sqrt{3}\right)^k$

6. Let $a_i$ represent the sequence of even natural numbers $\{2,4,6,8,\dots\}$, and evaluate the following expressions.

   a. $\displaystyle\sum_{i=1}^{5} a_i$

   b. $\displaystyle\sum_{i=1}^{4} a_{2i}$

   c. $\displaystyle\sum_{i=1}^{5} (a_i - 1)$

7. Let $a_i$ represent the sequence of integers giving the yardage gained per rush in a high school football game $\{3,-2,17,4,-8,19,2,3,3,4,0,1,-7\}$.

   a. Evaluate $\sum_{i=1}^{13} a_i$. What does this sum represent in the context of the situation?

   b. Evaluate $\frac{\sum_{i=1}^{13} a_i}{13}$. What does this expression represent in the context of the situation?

   c. In general, if $a_n$ describes any sequence of numbers, what does $\frac{\sum_{i=1}^{n} a_i}{n}$ represent?

8. Let $b_n$ represent the sequence given by the following recursive formula: $b_1 = 10$, $b_n = b_{n-1} \cdot 5$.

   a. Write the first 4 terms of this sequence.

   b. Expand the sum $\sum_{i=1}^{4} b_i$. Is it easier to add this series, or is it easier to use the formula for the sum of a finite geometric sequence? Explain your answer. Evaluate $\sum_{i=1}^{4} b_i$.

   c. Write an explicit form for $b_n$.

   d. Evaluate $\sum_{i=1}^{10} b_i$.

9. Consider the sequence given by $a_1 = 20$, $a_n = \frac{1}{2} \cdot a_{n-1}$.

   a. Evaluate $\sum_{i=1}^{10} a_i$, $\sum_{i=1}^{100} a_i$, and $\sum_{i=1}^{1000} a_i$.

   b. What value does it appear this series is approaching as $n$ continues to increase? Why might it seem like the series is bounded?

10. The sum of a geometric series with four terms is 60, and the common ratio is $r = \frac{1}{2}$. Find the first term.

11. The sum of the first four terms of a geometric series is 203, and the common ratio is 0.4. Find the first term.

12. The third term in a geometric series is $\frac{27}{2}$, and the sixth term is $\frac{729}{16}$. Find the common ratio.

13. The second term in a geometric series is 10, and the seventh term is 10240. Find the sum of the first six terms.

14. Find the interest earned and the future value of an annuity with monthly payments of \$200 for two years into an account that pays 6% interest per year compounded monthly.

15. Find the interest earned and the future value of an annuity with annual payments of $\$1{,}200$ for $15$ years into an account that pays $4\%$ interest per year.

16. Find the interest earned and the future value of an annuity with semiannual payments of $\$1{,}000$ for $20$ years into an account that pays $7\%$ interest per year compounded semiannually.

17. Find the interest earned and the future value of an annuity with weekly payments of $\$100$ for three years into an account that pays $5\%$ interest per year compounded weekly.

18. Find the interest earned and the future value of an annuity with quarterly payments of $\$500$ for $12$ years into an account that pays $3\%$ interest per year compounded quarterly.

19. How much money should be invested every month with $8\%$ interest per year compounded monthly in order to save up $\$10{,}000$ in $15$ months?

20. How much money should be invested every year with $4\%$ interest per year in order to save up $\$40{,}000$ in $18$ years?

21. Julian wants to save up to buy a car. He is told that a loan for a car will cost $\$274$ a month for five years, but Julian does not need a car presently. He decides to invest in a structured savings plan for the next three years. Every month Julian invests $\$274$ at an annual interest rate of $2\%$ compounded monthly.

    a. How much will Julian have at the end of three years?

    b. What are the benefits of investing in a structured savings plan instead of taking a loan out? What are the drawbacks?

22. An *arithmetic series* is a series whose terms form an arithmetic sequence. For example, $2 + 4 + 6 + \cdots + 100$ is an arithmetic series since $2,4,6,8, \dots, 100$ is an arithmetic sequence with constant difference $2$.

    The most famous arithmetic series is $1 + 2 + 3 + 4 + \cdots + n$ for some positive integer $n$. We studied this series in Algebra I and showed that its sum is $S_n = \frac{n(n+1)}{2}$. It can be shown that the general formula for the sum of an arithmetic series $a + (a + d) + (a + 2d) + \cdots + [a + (n - 1)d]$ is

    $$S_n = \frac{n}{2}[2a + (n - 1)d],$$

    where $a$ is the first term and $d$ is the constant difference.

    a. Use the general formula to show that the sum of $1 + 2 + 3 + \cdots + n$ is $S_n = \frac{n(n+1)}{2}$.

    b. Use the general formula to find the sum of $2 + 4 + 6 + 8 + 10 + \cdots + 100$.

23. The sum of the first five terms of an arithmetic series is $25$, and the first term is $2$. Find the constant difference.

24. The sum of the first nine terms of an arithmetic series is $135$, and the first term is $17$. Find the ninth term.

25. The sum of the first and $100^{\text{th}}$ terms of an arithmetic series is $101$. Find the sum of the first $100$ terms.

# Lesson 30: Buying a Car

## Classwork

### Opening Exercise

Write a sum to represent the future amount of a structured savings plan (i.e., annuity) if you deposit $\$250$ into an account each month for $5$ years that pays $3.6\%$ interest per year, compounded monthly. Find the future amount of your plan at the end of $3$ years.

### Example

Jack wanted to buy a $\$9{,}000$ 2-door sports coupe but could not pay the full price of the car all at once. He asked the car dealer if she could give him a loan where he paid a monthly payment. She told him she could give him a loan for the price of the car at an annual interest rate of $3.6\%$ compounded monthly for $60$ months ($5$ years).

The problems below exhibit how Jack's car dealer used the information above to figure out how much his monthly payment of $R$ dollars per month should be.

a. First, the car dealer imagined how much she would have in an account if she deposited $\$9{,}000$ into the account and left it there for $60$ months at an annual interest rate of $3.6\%$ compounded monthly. Use the compound interest formula $F = P(1+i)^n$ to calculate how much she would have in that account after $5$ years. This is the amount she would have in the account after $5$ years if Jack gave her $\$9{,}000$ for the car, and she immediately deposited it.

b. Next, she figured out how much would be in an account after $5$ years if she took each of Jack's payments of $R$ dollars and deposited it into a bank that earned $3.6\%$ per year (compounded monthly). Write a sum to represent the future amount of money that would be in the annuity after $5$ years in terms of $R$, and use the sum of a geometric series formula to rewrite that sum as an algebraic expression.

c. The car dealer then reasoned that, to be fair to her and Jack, the two final amounts in both accounts should be the same—that is, she should have the same amount in each account at the end of $60$ months either way. Write an equation in the variable $R$ that represents this equality.

d. She then solved her equation to get the amount $R$ that Jack would have to pay monthly. Solve the equation in part (c) to find out how much Jack needed to pay each month.

## Exercise

A college student wants to buy a car and can afford to pay $\$200$ per month. If she plans to take out a loan at $6\%$ interest per year with a recurring payment of $\$200$ per month for four years, what price car can she buy?

## Mathematical Modeling Exercise

In the Problem Set of Lesson 29, you researched the price of a car that you might like to own. In this exercise, you will determine how much a car payment would be for that price for different loan options.

If you did not find a suitable car, select a car and selling price from the list below:

| Car | Selling Price |
|---|---|
| 2005 Pickup Truck | $9000 |
| 2007 Two-Door Small Coupe | $7500 |
| 2003 Two-Door Luxury Coupe | $10,000 |
| 2006 Small SUV | $8000 |
| 2008 Four-Door Sedan | $8500 |

a. When you buy a car, you must pay sales tax and licensing and other fees. Assume that sales tax is 6% of the selling price and estimated license/title/fees will be 2% of the selling price. If you put a $1,000 down payment on your car, how much money will you need to borrow to pay for the car and taxes and other fees?

b. Using the loan amount you computed above, calculate the monthly payment for the different loan options shown below:

| | |
|---|---|
| Loan 1 | 36-month loan at 2% |
| Loan 2 | 48-month loan at 3% |
| Loan 3 | 60-month loan at 5% |

c. Which plan, if any, will keep your monthly payment under $175? Of the plans under $175 per month, why might you choose a plan with fewer months even though it costs more per month?

### Lesson Summary

The total cost of car ownership includes many different costs in addition to the selling price, such as sales tax, insurance, fees, maintenance, interest on loans, gasoline, etc.

The present value of an annuity formula can be used to calculate monthly loan payments given a total amount borrowed, the length of the loan, and the interest rate. The present value $A_p$ (i.e., loan amount) of an annuity consisting of $n$ recurring equal payments of size $R$ and interest rate $i$ per time period is

$$A_p = R \cdot \frac{1-(1+i)^{-n}}{i}.$$

Amortization tables and online loan calculators can also help you plan for buying a car.

The amount of your monthly payment depends on the interest rate, the down payment, and the length of the loan.

## Problem Set

1. Benji is 24 years old and plans to drive his new car about 200 miles per week. He has qualified for first-time buyer financing, which is a 60-month loan with 0% down at an interest rate of 4%. Use the information below to estimate the monthly cost of each vehicle.

   CAR A: 2010 Pickup Truck for \$12,000, 22 miles per gallon

   CAR B: 2006 Luxury Coupe for \$11,000, 25 miles per gallon

   Gasoline: \$4.00 per gallon    New vehicle fees: \$80    Sales Tax: 4.25%

   Maintenance Costs:

   (2010 model year or newer): 10% of purchase price annually

   (2009 model year or older): 20% of purchase price annually

   Insurance:

| Average Rate Ages 25–29 | \$100 per month |
|---|---|
| If you are male<br>If you are female | Add \$10 per month<br>Subtract \$10 per month |
| Type of Car<br>Pickup Truck<br>Small Two-Door Coupe or Four-Door Sedan<br>Luxury Two- or Four-Door Coupe | <br>Subtract \$10 per month<br>Subtract \$10 per month<br>Add \$15 per month |
| Ages 18–25 | Double the monthly cost |

   a. How much money will Benji have to borrow to purchase each car?
   b. What is the monthly payment for each car?
   c. What are the annual maintenance costs and insurance costs for each car?
   d. Which car should Benji purchase? Explain your choice.

2. Use the total initial cost of buying your car from the lesson to calculate the monthly payment for the following loan options.

| Option | Number of Months | Down Payment | Interest Rate | Monthly Payment |
|---|---|---|---|---|
| Option A | 48 months | \$0 | 2.5% | |
| Option B | 60 months | \$500 | 3.0% | |
| Option C | 60 months | \$0 | 4.0% | |
| Option D | 36 months | \$1000 | 0.9% | |

a. For each option, what is the total amount of money you will pay for your vehicle over the life of the loan?

b. Which option would you choose? Justify your reasoning.

3. Many lending institutions will allow you to pay additional money toward the principal of your loan every month. The table below shows the monthly payment for an \$8,000 loan using Option A above if you pay an additional \$25 per month.

| Month/ Year | Payment | Principal Paid | Interest Paid | Total Interest | Balance |
|---|---|---|---|---|---|
| Aug. 2014 | \$ 200.31 | \$ 183.65 | \$ 16.67 | \$ 16.67 | \$ 7,816.35 |
| Sept. 2014 | \$ 200.31 | \$ 184.03 | \$ 16.28 | \$ 32.95 | \$ 7,632.33 |
| Oct. 2014 | \$ 200.31 | \$ 184.41 | \$ 15.90 | \$ 48.85 | \$ 7,447.91 |
| Nov. 2014 | \$ 200.31 | \$ 184.80 | \$ 15.52 | \$ 64.37 | \$ 7,263.12 |
| Dec. 2014 | \$ 200.31 | \$ 185.18 | \$ 15.13 | \$ 79.50 | \$ 7,077.94 |
| Jan. 2015 | \$ 200.31 | \$ 185.57 | \$ 14.75 | \$ 94.25 | \$ 6,892.37 |
| Feb. 2015 | \$ 200.31 | \$ 185.95 | \$ 14.36 | \$ 108.60 | \$ 6,706.42 |
| Mar. 2015 | \$ 200.31 | \$ 186.34 | \$ 13.97 | \$ 122.58 | \$ 6,520.08 |
| April 2015 | \$ 200.31 | \$ 186.73 | \$ 13.58 | \$ 136.16 | \$ 6,333.35 |
| May 2015 | \$ 200.31 | \$ 187.12 | \$ 13.19 | \$ 149.35 | \$ 6,146.23 |
| June 2015 | \$ 200.31 | \$ 187.51 | \$ 12.80 | \$ 162.16 | \$ 5,958.72 |
| July 2015 | \$ 200.31 | \$ 187.90 | \$ 12.41 | \$ 174.57 | \$ 5,770.83 |
| Aug. 2015 | \$ 200.31 | \$ 188.29 | \$ 12.02 | \$ 186.60 | \$ 5,582.54 |
| Sept. 2015 | \$ 200.31 | \$ 188.68 | \$ 11.63 | \$ 198.23 | \$ 5,393.85 |
| Oct. 2015 | \$ 200.31 | \$ 189.08 | \$ 11.24 | \$ 209.46 | \$ 5,204.78 |
| Nov. 2015 | \$ 200.31 | \$ 189.47 | \$ 10.84 | \$ 220.31 | \$ 5,015.31 |
| Dec. 2015 | \$ 200.31 | \$ 189.86 | \$ 10.45 | \$ 230.75 | \$ 4,825.45 |

Note: The months from January 2016 to December 2016 are not shown.

| | | | | | |
|---|---|---|---|---|---|
| Jan. 2017 | $ 200.31 | $ 195.07 | $ 5.24 | $ 330.29 | $ 2,320.92 |
| Feb. 2017 | $ 200.31 | $ 195.48 | $ 4.84 | $ 335.12 | $ 2,125.44 |
| Mar. 2017 | $ 200.31 | $ 195.88 | $ 4.43 | $ 339.55 | $ 1,929.56 |
| April 2017 | $ 200.31 | $ 196.29 | $ 4.02 | $ 343.57 | $ 1,733.27 |
| May 2017 | $ 200.31 | $ 196.70 | $ 3.61 | $ 347.18 | $ 1,536.57 |
| June 2017 | $ 200.31 | $ 197.11 | $ 3.20 | $ 350.38 | $ 1,339.45 |
| July 2017 | $ 200.31 | $ 197.52 | $ 2.79 | $ 353.17 | $ 1,141.93 |
| Aug. 2017 | $ 200.31 | $ 197.93 | $ 2.38 | $ 355.55 | $ 944.00 |
| Sept. 2017 | $ 200.31 | $ 198.35 | $ 1.97 | $ 357.52 | $ 745.65 |
| Oct. 2017 | $ 200.31 | $ 198.76 | $ 1.55 | $ 359.07 | $ 546.90 |
| Nov. 2017 | $ 200.31 | $ 199.17 | $ 1.14 | $ 360.21 | $ 347.72 |
| Dec. 2017 | $ 200.31 | $ 199.59 | $ 0.72 | $ 360.94 | $ 148.13 |
| Jan. 2018 | $ 148.44 | $ 148.13 | $ 0.31 | $ 361.25 | $ 0.00 |

How much money would you save over the life of an $8,000 loan using Option A if you paid an extra $25 per month compared to the same loan without the extra payment toward the principal?

4. Suppose you can afford only $200 a month in car payments and your best loan option is a 60-month loan at 3%. How much money could you spend on a car? That is, calculate the present value of the loan with these conditions.

5. Would it make sense for you to pay an additional amount per month toward your car loan? Use an online loan calculator to support your reasoning.

6. What is the sum of each series?

   a. $900 + 900(1.01)^1 + 900(1.01)^2 + \cdots + 900(1.01)^{59}$

   b. $\sum_{n=0}^{47} 15{,}000\left(1 + \frac{0.04}{12}\right)^n$

7. Gerald wants to borrow $12,000 in order to buy an engagement ring. He wants to repay the loan by making monthly installments for two years. If the interest rate on this loan is $9\frac{1}{2}\%$ per year, compounded monthly, what is the amount of each payment?

8. Ivan plans to surprise his family with a new pool using his Christmas bonus of $4200 as a down payment. If the price of the pool is $9,500 and Ivan can finance it at an interest rate of $2\frac{7}{8}\%$ per year, compounded quarterly, how long is the loan for if he pays $285.45 per quarter?

9. Jenny wants to buy a car by making payments of $120 per month for three years. The dealer tells her that she will need to put a down payment of $3,000 on the car in order to get a loan with those terms at a 9% interest rate per year, compounded monthly. How much is the car that Jenny wants to buy?

10. Kelsey wants to refinish the floors in her house and estimates that it will cost $\$39{,}000$ to do so. She plans to finance the entire amount at $3\frac{1}{4}\%$ interest per year, compounded monthly for $10$ years. How much is her monthly payment?

11. Lawrence coaches little league baseball and needs to purchase all new equipment for his team. He has $\$489$ in donations, and the team's sponsor will take out a loan at $4\frac{1}{2}\%$ interest per year, compounded monthly for one year, paying up to $\$95$ per month. What is the most that Lawrence can purchase using the donations and loan?

# Lesson 31: Credit Cards

## Classwork

### Mathematical Modeling Exercise

You have charged $\$1,500$ for the down payment on your car to a credit card that charges $19.99\%$ annual interest, and you plan to pay a fixed amount toward this debt each month until it is paid off. We will denote the balance owed after the $n^{\text{th}}$ payment has been made as $b_n$.

a. What is the monthly interest rate, $i$? Approximate $i$ to $5$ decimal places.

b. You have been assigned to either the $50$-team, the $100$-team, or the $150$-team, where the number indicates the size of the monthly payment $R$ you will make toward your debt. What is your value of $R$?

c. Remember that you can make any size payment toward a credit card debt, as long as it is at least the minimum payment specified by the lender. Your lender calculates the minimum payment as the sum of $1\%$ of the outstanding balance and the total interest that has accrued over the month or $\$25$, whichever is greater. Under these stipulations, what is the minimum payment? Is your monthly payment $R$ at least as large as the minimum payment?

d. Complete the following table to show 6 months of payments.

| Month, $n$ | Interest Due | Payment, $R$ | Paid to Principal | Balance, $b_n$ |
|---|---|---|---|---|
| 0 | | | | 1500.00 |
| 1 | | | | |
| 2 | | | | |
| 3 | | | | |
| 4 | | | | |
| 5 | | | | |
| 6 | | | | |

e. Write a recursive formula for the balance $b_n$ in month $n$ in terms of the balance $b_{n-1}$.

f. Write an explicit formula for the balance $b_n$ in month $n$, leaving the expression $1 + i$ in symbolic form.

g. Rewrite your formula in part (f) using $r$ to represent the quantity $(1 + i)$.

h. What can you say about your formula in (g)? What term do we use to describe $r$ in this formula?

i. Write your formula from part (g) in summation notation using $\Sigma$.

j. Apply the appropriate formula from Lesson 29 to rewrite your formula from part (g).

k. Find the month when your balance is paid off.

l. Calculate the total amount paid over the life of the debt. How much was paid solely to interest?

## Problem Set

1. Suppose that you have a $2,000 balance on a credit card with a 29.99% annual interest rate, compounded monthly, and you can afford to pay $150 per month toward this debt.
   a. Find the amount of time it will take to pay off this debt. Give your answer in months and years.
   b. Calculate the total amount paid over the life of the debt.
   c. How much money was paid entirely to the interest on this debt?

2. Suppose that you have a $2,000 balance on a credit card with a 14.99% annual interest rate, and you can afford to pay $150 per month toward this debt.
   a. Find the amount of time it will take to pay off this debt. Give your answer in months and years.
   b. Calculate the total amount paid over the life of the debt.
   c. How much money was paid entirely to the interest on this debt?

3. Suppose that you have a $2,000 balance on a credit card with a 7.99% annual interest rate, and you can afford to pay $150 per month toward this debt.
   a. Find the amount of time it will take to pay off this debt. Give your answer in months and years.
   b. Calculate the total amount paid over the life of the debt.
   c. How much money was paid entirely to the interest on this debt?

4. Summarize the results of Problems 1, 2, and 3.

5. Brendan owes $1,500 on a credit card with an interest rate of 12%. He is making payments of $100 every month to pay this debt off. Maggie is also making regular payments to a debt owed on a credit card, and she created the following graph of her projected balance over the next 12 months.
   a. Who has the higher initial balance? Explain how you know.
   b. Who will pay their debt off first? Explain how you know.

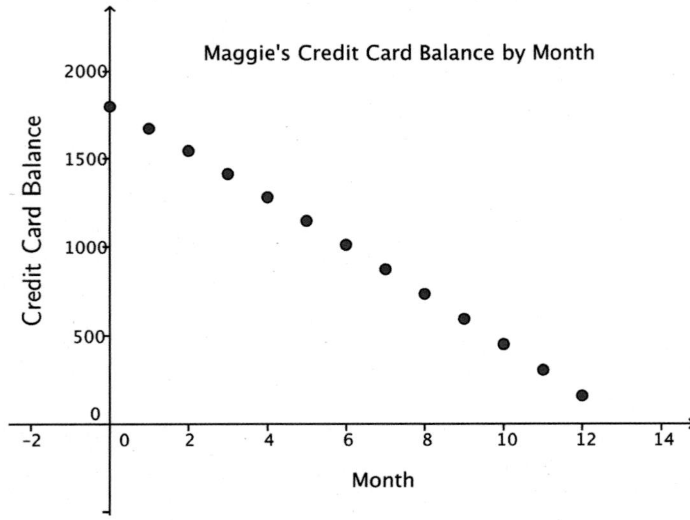

6. Alan and Emma are both making $200 monthly payments toward balances on credit cards. Alan has prepared a table to represent his projected balances, and Emma has prepared a graph.

| Alan's Credit Card Balance | | | |
|---|---|---|---|
| Month, $n$ | Interest | Payment | Balance, $b_n$ |
| 0 | —— | —— | 2000.00 |
| 1 | 41.65 | 200 | 1841.65 |
| 2 | 38.35 | 200 | 1680.00 |
| 3 | 34.99 | 200 | 1514.99 |
| 4 | 31.55 | 200 | 1346.54 |
| 5 | 28.04 | 200 | 1174.58 |
| 6 | 24.46 | 200 | 999.04 |
| 7 | 20.81 | 200 | 819.85 |
| 8 | 17.07 | 200 | 636.92 |
| 9 | 13.26 | 200 | 450.18 |
| 10 | 9.37 | 200 | 259.55 |
| 11 | 5.41 | 200 | 64.96 |

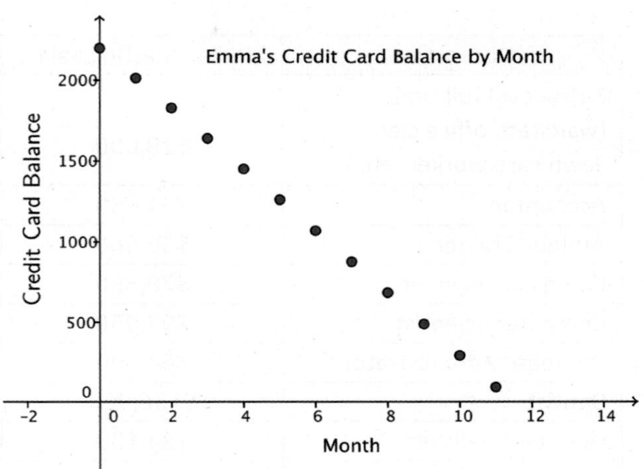


a. What is the annual interest rate on Alan's debt? Explain how you know.

b. Who has the higher initial balance? Explain how you know.

c. Who will pay their debt off first? Explain how you know.

d. What do your answers to parts (a), (b), and (c) tell you about the interest rate for Emma's debt?

7. Both Gary and Helena are paying regular monthly payments to a credit card balance. The balance on Gary's credit card debt can be modeled by the recursive formula $g_n = g_{n-1}(1.01666) - 200$ with $g_0 = 2500$, and the balance on Helena's credit card debt can be modeled by the explicit formula $h_n = 2000(1.01666)^n - 250\left(\frac{1.01666^n - 1}{0.01666}\right)$ for $n \geq 0$.

a. Who has the higher initial balance? Explain how you know.

b. Who has the higher monthly payment? Explain how you know.

c. Who will pay their debt off first? Explain how you know.

8. In the next lesson, we will apply the mathematics we have learned to the purchase of a house. In preparation for that task, you need to come to class prepared with an idea of the type of house you would like to buy.

a. Research the median housing price in the county where you live or where you wish to relocate.

b. Find the range of prices that are within 25% of the median price from part (a). That is, if the price from part (a) was $P$, then your range is $0.75P$ to $1.25P$.

c. Look at online real estate websites, and find a house located in your selected county that falls into the price range specified in part (b). You will be modeling the purchase of this house in Lesson 32, so bring a printout of the real estate listing to class with you.

9. Select a career that interests you from the following list of careers. If the career you are interested in is not on this list, check with your teacher to obtain permission to perform some independent research. Once it has been selected, you will use the career to answer questions in Lesson 32 and Lesson 33.

| Occupation | Median Starting Salary | Education Required |
|---|---|---|
| Entry-level full-time (waitstaff, office clerk, lawn care worker, etc.) | $\$18,000$ | High school diploma or GED |
| Accountant | $\$54,630$ | 4-year college degree |
| Athletic Trainer | $\$36,560$ | 4-year college degree |
| Chemical Engineer | $\$78,860$ | 4-year college degree |
| Computer Scientist | $\$93,950$ | 4-year college degree or more |
| Database Administrator | $\$64,600$ | 4-year college degree |
| Dentist | $\$136,960$ | Graduate degree |
| Desktop Publisher | $\$34,130$ | 4-year college degree |
| Electrical Engineer | $\$75,930$ | 4-year college degree |
| Graphic Designer | $\$39,900$ | 2- or 4-year college degree |
| HR Employment Specialist | $\$42,420$ | 4-year college degree |
| HR Compensation Manager | $\$66,530$ | 4-year college degree |
| Industrial Designer | $\$54,560$ | 4-year college degree or more |
| Industrial Engineer | $\$68,620$ | 4-year college degree |
| Landscape Architect | $\$55,140$ | 4-year college degree |
| Lawyer | $\$102,470$ | Law degree |
| Occupational Therapist | $\$60,470$ | Master's degree |
| Optometrist | $\$91,040$ | Master's degree |
| Physical Therapist | $\$66,200$ | Master's degree |
| Physician—Anesthesiology | $\$259,948$ | Medical degree |
| Physician—Family Practice | $\$137,119$ | Medical degree |
| Physician's Assistant | $\$74,980$ | 2 years college plus 2-year program |
| Radiology Technician | $\$47,170$ | 2-year degree |
| Registered Nurse | $\$57,280$ | 2- or 4-year college degree plus |
| Social Worker—Hospital | $\$48,420$ | Master's degree |
| Teacher—Special Education | $\$47,650$ | Master's degree |
| Veterinarian | $\$71,990$ | Veterinary degree |

# Lesson 32: Buying a House

## Classwork

### Mathematical Modeling Exercise

Now that you have studied the mathematics of structured savings plans, buying a car, and paying down a credit card debt, it's time to think about the mathematics behind the purchase of a house. In the Problem Set in Lesson 31, you selected a future career and a home to purchase. The question of the day is this: Can you buy the house you have chosen on the salary of the career you have chosen? You need to adhere to the following constraints:

- Mortgages are loans that are usually offered with $30$-, $20$-, or $15$-year repayment options. You will start with a $30$-year mortgage.
- The annual interest rate for your mortgage will be $5\%$.
- Your payment includes the payment of the loan for the house and payments into an account called an *escrow account*, which is used to pay for taxes and insurance on your home. We will approximate the annual payment to escrow as $1.2\%$ of the home's selling price.
- The bank will only approve a mortgage if the total monthly payment for your house, including the payment to the escrow account, does not exceed $30\%$ of your monthly salary.
- You have saved up enough money to put a $10\%$ down payment on this house.

1. Will the bank approve you for a $30$-year mortgage on the house that you have chosen?

2. Answer either (a) or (b) as appropriate.

   a. If your bank approved you for a 30-year mortgage, do you meet the criteria for a 20-year mortgage? If you could get a mortgage for any number of years that you want, what is the shortest term for which you would qualify?

   b. If your bank did not approve you for the 30-year mortgage, what is the maximum price of a house that fits your budget?

## Problem Set

1. Use the house you selected to purchase in the Problem Set from Lesson 31 for this problem.
   a. What was the selling price of this house?
   b. Calculate the total monthly payment, $R$, for a 15-year mortgage at $5\%$ annual interest, paying $10\%$ as a down payment and an annual escrow payment that is $1.2\%$ of the full price of the house.

2. In the summer of $2014$, the average listing price for homes for sale in the Hollywood Hills was $\$2{,}663{,}995$.
   a. Suppose you want to buy a home at that price with a 30-year mortgage at $5.25\%$ annual interest, paying $10\%$ as a down payment and with an annual escrow payment that is $1.2\%$ of the full price of the home. What is your total monthly payment on this house?
   b. How much is paid in interest over the life of the loan?

3. Suppose that you would like to buy a home priced at $\$200{,}000$. You will make a payment of $10\%$ of the purchase price and pay $1.2\%$ of the purchase price into an escrow account annually.
   a. Compute the total monthly payment and the total interest paid over the life of the loan for a 30-year mortgage at $4.8\%$ annual interest.
   b. Compute the total monthly payment and the total interest paid over the life of the loan for a 20-year mortgage at $4.8\%$ annual interest.
   c. Compute the total monthly payment and the total interest paid over the life of the loan for a 15-year mortgage at $4.8\%$ annual interest.

4. Suppose that you would like to buy a home priced at $\$180{,}000$. You will qualify for a 30-year mortgage at $4.5\%$ annual interest and pay $1.2\%$ of the purchase price into an escrow account annually.
   a. Calculate the total monthly payment and the total interest paid over the life of the loan if you make a $3\%$ down payment.
   b. Calculate the total monthly payment and the total interest paid over the life of the loan if you make a $10\%$ down payment.
   c. Calculate the total monthly payment and the total interest paid over the life of the loan if you make a $20\%$ down payment.
   d. Summarize the results of parts (a), (b), and (c) in the chart below.

| Percent down payment | Amount of down payment | Total interest paid |
|---|---|---|
| $3\%$ | | |
| $10\%$ | | |
| $20\%$ | | |

5. The following amortization table shows the amount of payments to principal and interest on a $100,000 mortgage at the beginning and the end of a 30-year loan. These payments do not include payments to the escrow account.

| Month/ Year | Payment | Principal Paid | Interest Paid | Total Interest | Balance |
|---|---|---|---|---|---|
| Sept. 2014 | $ 477.42 | $ 144.08 | $ 333.33 | $ 333.33 | $ 99,855.92 |
| Oct. 2014 | $ 477.42 | $ 144.56 | $ 332.85 | $ 666.19 | $ 99,711.36 |
| Nov. 2014 | $ 477.42 | $ 145.04 | $ 332.37 | $ 998.56 | $ 99,566.31 |
| Dec. 2014 | $ 477.42 | $ 145.53 | $ 331.89 | $ 1,330.45 | $ 99,420.78 |
| Jan. 2015 | $ 477.42 | $ 146.01 | $ 331.40 | $ 1,661.85 | $ 99,274.77 |
| | | | ⋮ | | |
| Mar. 2044 | $ 477.42 | $ 467.98 | $ 9.44 | $ 71,845.82 | $ 2,363.39 |
| April 2044 | $ 477.42 | $ 469.54 | $ 7.88 | $ 71,853.70 | $ 1,893.85 |
| May 2044 | $ 477.42 | $ 471.10 | $ 6.31 | $ 71,860.01 | $ 1,422.75 |
| June 2044 | $ 477.42 | $ 472.67 | $ 4.74 | $ 71,864.75 | $ 950.08 |
| July 2044 | $ 477.42 | $ 474.25 | $ 3.17 | $ 71,867.92 | $ 475.83 |
| Aug. 2044 | $ 477.42 | $ 475.83 | $ 1.59 | $ 71,869.51 | $ 0.00 |

a. What is the annual interest rate for this loan? Explain how you know.

b. Describe the changes in the amount of principal paid each month as the month $n$ gets closer to 360.

c. Describe the changes in the amount of interest paid each month as the month $n$ gets closer to 360.

6. Suppose you want to buy a $\$200{,}000$ home with a $30$-year mortgage at $4.5\%$ annual interest paying $10\%$ down with an annual escrow payment that is $1.2\%$ of the price of the home.

   a. Disregarding the payment to escrow, how much do you pay toward the loan on the house each month?

   b. What is the total monthly payment on this house?

   c. The graph below depicts the amount of your payment from part (b) that goes to the interest on the loan and the amount that goes to the principal on the loan. Explain how you can tell which graph is which.

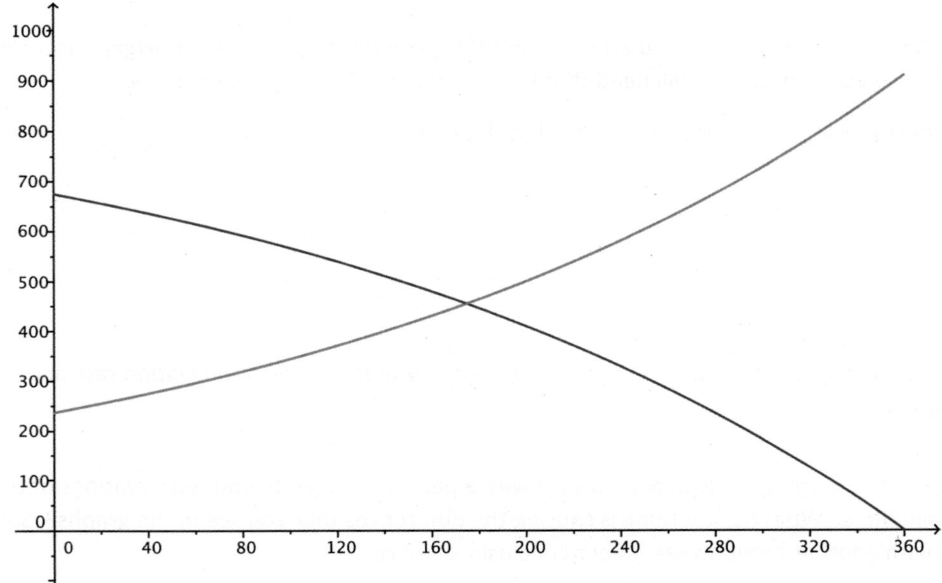


7. Student loans are very similar to both car loans and mortgages. The same techniques used for car loans and mortgages can be used for student loans. The difference between student loans and other types of loans is that usually students are not required to pay anything until $6$ months after they stop being full-time students.

   a. An unsubsidized student loan will accumulate interest while a student remains in school. Sal borrows $\$9{,}000$ his first term in school at an interest rate of $5.95\%$ per year compounded monthly and never makes a payment. How much will he owe $4\frac{1}{2}$ years later? How much of that amount is due to compounded interest?

   b. If Sal pays the interest on his student loan every month while he is in school, how much money has he paid?

   c. Explain why the answer to part (a) is different than the answer to part (b).

8. Consider the sequence $a_0 = 10000$, $a_n = a_{n-1} \cdot \frac{1}{10}$ for $n \geq 1$.

   a. Write the explicit form for the $n^{\text{th}}$ term of the sequence.

   b. Evaluate $\sum_{k=0}^{4} a_k$.

   c. Evaluate $\sum_{k=0}^{6} a_k$.

   d. Evaluate $\sum_{k=0}^{8} a_k$ using the sum of a geometric series formula.

   e. Evaluate $\sum_{k=0}^{10} a_k$ using the sum of a geometric series formula.

   f. Describe the value of $\sum_{k=0}^{n} a_k$ for any value of $n \geq 4$.

# Lesson 33: The Million Dollar Problem

## Classwork

### Opening Exercise

In Problem 1 of the Problem Set of Lesson 32, you calculated the monthly payment for a 15-year mortgage at a $5\%$ annual interest rate for the house you chose. You will need that monthly payment for these questions.

a. About how much do you expect your home to be worth in 15 years?

b. For $0 \leq x \leq 15$, plot the graph of the function $f(x) = P(1 + r)^x$ where $r$ is the appreciation rate and $P$ is the initial value of your home.

c. Compare the image of the graph you plotted in part (b) with a partner, and write your observations of the differences and similarities. What do you think is causing the differences that you see in the graphs? Share your observations with another group to see if your conclusions are correct.

Your friend Julia bought a home at the same time as you but chose to finance the loan over 30 years. Julia also was able to avoid a down payment and financed the entire value of her home. This allowed her to purchase a more expensive home, but 15 years later she still has not paid off the loan. Consider the following amortization table representing Julia's mortgage, and answer the following questions by comparing the table with your graph.

| Payment # | Beginning Balance | Payment on Interest | Payment on Principal |
|---|---|---|---|
| 1 | \$145000 | \$543.75 | \$190.94 |
| ⋮ | ⋮ | ⋮ | ⋮ |
| 178 | \$96,784.14 | \$362.94 | \$371.75 |
| 179 | \$96,412.38 | \$361.55 | \$373.15 |
| 180 | \$96,039.23 | \$360.15 | \$374.55 |

d. In Julia's neighborhood, her home has grown in value at around $2.95\%$ per year. Considering how much she still owes the bank, how much of her home does she own after 15 years (the equity in her home)? Express your answer in dollars to the nearest thousand and as a percent of the value of her home.

e. Reasoning from your graph in part (b) and the table above, if both you and Julia sell your homes in 15 years at the homes' appreciated values, who would have more equity?

f. How much more do you need to save over 15 years to have assets over \$1,000,000?

## Mathematical Modeling Exercises

Assume you can earn 7% interest annually, compounded monthly, in an investment account. Develop a savings plan so that you will have \$1 million in assets in 15 years (including the equity in your paid-off house).

1. Use your answer to Opening Exercise, part (g) as the future value of your savings plan.

   a. How much will you have to save every month to save up \$1 million in assets?

   b. Recall the monthly payment to pay off your home in 15 years (from Problem 1 of the Problem Set of Lesson 32). How much are the two together? What percentage of your monthly income is this for the profession you chose?

2. Write a report supported by the calculations you did above on how to save $1 million (or more) in your lifetime.

## Problem Set

1. Consider the following scenario: You would like to save up $\$50{,}000$ after $10$ years and plan to set up a structured savings plan to make monthly payments at $4.125\%$ interest annually, compounded monthly.
   a. What lump sum amount would you need to invest at this interest rate in order to have $\$50{,}000$ after $10$ years?
   b. Use an online amortization calculator to find the monthly payment necessary to take a loan for the amount in part (a) at this interest rate and for this time period.
   c. Use $A_f = R\left(\frac{(1+i)^n - 1}{i}\right)$ to solve for $R$.
   d. Compare your answers to part (b) and part (c). What do you notice? Why did this happen?

2. For structured savings plans, the future value of the savings plan as a function of the number of payments made at that point is an interesting function to examine. Consider a structured savings plan with a recurring payment of $\$450$ made monthly and an annual interest rate of $5.875\%$ compounded monthly.
   a. State the formula for the future value of this structured savings plan as a function of the number of payments made. Use $f$ for the function name.
   b. Graph the function you wrote in part (a) for $0 \leq x \leq 216$.
   c. State any trends that you notice for this function.
   d. What is the approximate value of the function $f$ for $x = 216$?
   e. What is the domain of $f$? Explain.
   f. If the domain of the function is restricted to natural numbers, is the function a geometric sequence? Why or why not?
   g. Recall that the $n^{\text{th}}$ partial sums of a geometric sequence can be represented with $S_n$. It is true that $f(x) = S_x$ for positive integers $x$, since it is a geometric sequence; that is, $S_x = \sum_{i=1}^{x} ar^i$. State the geometric sequence whose sums of the first $x$ terms are represented by this function.
   h. April has been following this structured savings plan for $18$ years. April says that taking out the money and starting over will not affect the total money earned because the interest rate does not change. Explain why April is incorrect in her reasoning.

3. Henry plans to have $\$195{,}000$ in property in $14$ years and would like to save up to $\$1$ million by depositing $\$3{,}068.95$ each month at $6\%$ interest per year, compounded monthly. Tina's structured savings plan over the same time span is described in the following table:
   a. Who has the higher interest rate? Who pays more every month?
   b. At the end of $14$ years, who has more money from their structured savings plan? Does this agree with what you expected? Why or why not?
   c. At the end of $40$ years, who has more money from their structured savings plan?

| Deposit # | Amount Saved |
|---|---|
| 30 | $110,574.77 |
| 31 | $114,466.39 |
| 32 | $118,371.79 |
| 33 | $122,291.02 |
| 34 | $126,224.14 |
| ⋮ | ⋮ |
| 167 | $795,266.92 |
| 168 | $801,583.49 |

4. Edgar and Paul are two brothers that both get an inheritance of $\$150{,}000$. Both plan to save up over $\$1{,}000{,}000$ in $25$ years. Edgar takes his inheritance and deposits the money into an investment account earning $8\%$ interest annually, compounded monthly, payable at the end of $25$ years. Paul spends his inheritance but uses a structured savings plan that is represented by the sequence $b_n = 1275 + b_{n-1} \cdot \left(1 + \frac{0.0775}{12}\right)$ with $b_0 = 1275$ in order to save up over $\$1{,}000{,}000$.

   a. Which of the two has more money at the end of $25$ years?

   b. What are the pros and cons of both brothers' plans? Which would you rather do? Why?

# Mathematics Curriculum

# Copy Ready Material

Name ____________________ Date ____________

# Lesson 1: Integer Exponents

## Exit Ticket

The following formulas for paper folding were discovered by Britney Gallivan in 2001 when she was a high school junior. The first formula determines the minimum width, $W$, of a square piece of paper of thickness $T$ needed to fold it in half $n$ times, alternating horizontal and vertical folds. The second formula determines the minimum length, $L$, of a long rectangular piece of paper of thickness $T$ needed to fold it in half $n$ times, always folding perpendicular to the long side.

$$W = \pi \cdot T \cdot 2^{\frac{3(n-1)}{2}} \qquad L = \frac{\pi T}{6}(2^n + 4)(2^n - 1)$$

1. Notebook paper is approximately $0.004$ in. thick. Using the formula for the width $W$, determine how wide a square piece of notebook paper would need to be to successfully fold it in half $13$ times, alternating horizontal and vertical folds.

2. Toilet paper is approximately $0.002$ in. thick. Using the formula for the length $L$, how long would a continuous sheet of toilet paper have to be to fold it in half $12$ times, folding perpendicular to the long edge each time?

3. Use the properties of exponents to rewrite each expression in the form $kx^n$. Then evaluate the expression for the given value of $x$.

   a. $2x^3 \cdot \frac{5}{4}x^{-1};\ x = 2$

   b. $\frac{9}{(2x)^{-3}};\ x = -\frac{1}{3}$

Name ______________________________ Date ______________

# Lesson 2: Base 10 and Scientific Notation

## Exit Ticket

1. A sheet of gold foil is $0.28$ millionths of a meter thick. Write the thickness of a gold foil sheet measured in centimeters using scientific notation.

2. Without performing the calculation, estimate which expression is larger. Explain how you know.

$$(4 \times 10^{10})(2 \times 10^{5}) \qquad \text{and} \qquad \frac{4 \times 10^{12}}{2 \times 10^{-4}}$$

Name ______________________________ Date ______________

# Lesson 3: Rational Exponents—What Are $2^{\frac{1}{2}}$ and $2^{\frac{1}{3}}$?

## Exit Ticket

1. Write the following exponential expressions as equivalent radical expressions.

   a. $2^{\frac{1}{2}}$

   b. $2^{\frac{3}{4}}$

   c. $3^{-\frac{2}{3}}$

2. Rewrite the following radical expressions as equivalent exponential expressions.

   a. $\sqrt{5}$

   b. $2\sqrt[4]{3}$

   c. $\frac{1}{\sqrt[3]{16}}$

3. Provide a written explanation for each question below.

   a. Is it true that $\left(4^{\frac{1}{2}}\right)^3 = (4^3)^{\frac{1}{2}}$? Explain how you know.

   b. Is it true that $\left(1000^{\frac{1}{3}}\right)^3 = (1000^3)^{\frac{1}{3}}$? Explain how you know.

   c. Suppose that $m$ and $n$ are positive integers and $b$ is a real number so that the principal $n^{\text{th}}$ root of $b$ exists. In general does $\left(b^{\frac{1}{n}}\right)^m = (b^m)^{\frac{1}{n}}$? Provide at least one example to support your claim.

Name ______________________________ Date ______________

# Lesson 4: Properties of Exponents and Radicals

## Exit Ticket

1. Find the exact value of $9^{\frac{11}{10}} \cdot 9^{\frac{2}{5}}$ without using a calculator.

2. Justify that $\sqrt[3]{8} \cdot \sqrt[3]{8} = \sqrt{16}$ using the properties of exponents in at least two different ways.

Name ______________________________ Date ______________

# Lesson 5: Irrational Exponents—What Are $2^{\sqrt{2}}$ and $2^{\pi}$?

## Exit Ticket

Use the process outlined in the lesson to approximate the number $2^{\sqrt{3}}$. Use the approximation $\sqrt{3} \approx 1.7320508$.

a. Find a sequence of five intervals that contain $\sqrt{3}$ whose endpoints get successively closer to $\sqrt{3}$.

b. Find a sequence of five intervals that contain $2^{\sqrt{3}}$ whose endpoints get successively closer to $2^{\sqrt{3}}$. Write your intervals in the form $2^r < 2^{\sqrt{3}} < 2^s$ for rational numbers $r$ and $s$.

c. Use your calculator to find approximations to four decimal places of the endpoints of the intervals in part (b).

d. Based on your work in part (c) what is your best estimate of the value of $2^{\sqrt{3}}$?

Name ______________________________ Date ______________

# Lesson 6: Euler's Number, $e$

## Exit Ticket

1. Suppose that water is entering a cylindrical water tank so that the initial height of the water is $3 \text{ cm}$ and the height of the water doubles every $30$ seconds. Write an equation of the height of the water at time $t$ seconds.

2. Explain how the number $e$ arose in our exploration of the average rate of change of the height of the water in the water tank.

Name ______________________________ Date ______________

# Lesson 7: Bacteria and Exponential Growth

## Exit Ticket

Loggerhead turtles reproduce every 2–4 years, laying approximately $120$ eggs in a clutch. Using this information, we can derive an approximate equation to model the turtle population. As is often the case in biological studies, we will count only the female turtles. If we start with a population of one female turtle in a protected area, and assume that all turtles survive, we can roughly approximate the population of female turtles by $T(t) = 5^t$. Use the methods of the Example to find the number of years, $Y$, it will take for this model to predict that there will be $300$ female turtles.

Name ______________________________ Date ______________

# Lesson 8: The "WhatPower" Function

## Exit Ticket

1. Explain why we need to specify $0 < b < 1$ and $b > 1$ as valid values for the base $b$ in the expression $\log_b(x)$.

2. Calculate the following logarithms.

   a. $\log_5(25)$

   b. $\log_{10}\left(\frac{1}{100}\right)$

   c. $\log_9(3)$

Name ______________________________ Date ______________

# Lesson 9: Logarithms—How Many Digits Do You Need?

## Exit Ticket

A brand new school district needs to generate ID numbers for its student body. The district anticipates a total enrollment of $75{,}000$ students within the next ten years. Will a five-digit ID number comprising the symbols $0, 1, \ldots, 9$ be enough? Explain your reasoning.

Name _______________________________________________ Date ____________________

# Lesson 10: Building Logarithmic Tables

## Exit Ticket

1. Use the log table below to approximate the following logarithms to four decimal places. Do not use a calculator.

| $x$ | $\log(x)$ |
|---|---|
| 1 | 0.0000 |
| 2 | 0.3010 |
| 3 | 0.4771 |
| 4 | 0.6021 |
| 5 | 0.6990 |

| $x$ | $\log(x)$ |
|---|---|
| 6 | 0.7782 |
| 7 | 0.8451 |
| 8 | 0.9031 |
| 9 | 0.9542 |
| 10 | 1.0000 |

a. $\log(500)$

b. $\log(0.0005)$

2. Suppose that $A$ is a number with $\log(A) = 1.352$.

a. What is the value of $\log(1000A)$?

b. Which of the following is true? Explain how you know.

i. $A < 0$

ii. $0 < A < 10$

iii. $10 < A < 100$

iv. $100 < A < 1000$

v. $A > 1000$

Name ______________________________ Date ______________

# Lesson 11: The Most Important Property of Logarithms

## Exit Ticket

1. Use the table below to approximate the following logarithms to 4 decimal places. Do not use a calculator.

   a. $\log(9)$

| $x$ | $\log(x)$ |
|---|---|
| 2 | 0.3010 |
| 3 | 0.4771 |
| 5 | 0.6990 |
| 7 | 0.8451 |

   b. $\log\left(\frac{1}{15}\right)$

   c. $\log(45{,}000)$

2. Suppose that $k$ is an integer, $a$ is a positive real number, and you know the value of $\log(a)$. Explain how to find the value of $\log(10^k \cdot a^2)$.

Name ______________________________ Date ______________

# Lesson 12: Properties of Logarithms

## Exit Ticket

1. State as many of the six properties of logarithms as you can.

2. Use the properties of logarithms to show that $\log\left(\frac{1}{x}\right) = -\log(x)$ for all $x > 0$.

3. Use the properties of logarithms to show that $\log\left(\frac{x}{y}\right) = \log(x) - \log(y)$ for $x > 0$ and $y > 0$.

Name ______________________________ Date ______________

# Lesson 13: Changing the Base

## Exit Ticket

1. Are there any properties that hold for base $10$ logarithms that would not be valid for the logarithm base $e$? Why? Are there any properties that hold for base $10$ logarithms that would not be valid for some positive base $b$, such that $b \neq 1$?

2. Write each logarithm as an equivalent expression involving only logarithms base $10$.

   a. $\log_3(25)$

   b. $\log_{100}(x^2)$

3. Rewrite each expression as an equivalent expression containing only one logarithm.

   a. $3\ln(p+q) - 2\ln(q) - 7\ln(p)$

   b. $\ln(xy) - \ln\left(\frac{x}{y}\right)$

Name ______________________________ Date ______________

# Lesson 14: Solving Logarithmic Equations

## Exit Ticket

Find all solutions to the following equations. Remember to check for extraneous solutions.

1. $5\log_2(3x+7)=0$

2. $\log(x-1)+\log(x-4)=1$

Name ______________________________ Date ____________________

# Lesson 15: Why Were Logarithms Developed?

## Exit Ticket

The surface area of Jupiter is $6.14 \times 10^{10}\ \mathrm{km}^2$, and the surface area of Earth is $5.10 \times 10^{8}\ \mathrm{km}^2$. Without using a calculator but using the table of logarithms, find how many times greater the surface area of Jupiter is than the surface area of Earth.

# Common Logarithm Table

| N | 0 | 1 | 2 | 3 | 4 | 5 | 6 | 7 | 8 | 9 |
|---|---|---|---|---|---|---|---|---|---|---|
| 1.0 | 0.0000 | 0.0043 | 0.0086 | 0.0128 | 0.0170 | 0.0212 | 0.0253 | 0.0294 | 0.0334 | 0.0374 |
| 1.1 | 0.0414 | 0.0453 | 0.0492 | 0.0531 | 0.0569 | 0.0607 | 0.0645 | 0.0682 | 0.0719 | 0.0755 |
| 1.2 | 0.0792 | 0.0828 | 0.0864 | 0.0899 | 0.0934 | 0.0969 | 0.1004 | 0.1038 | 0.1072 | 0.1106 |
| 1.3 | 0.1139 | 0.1173 | 0.1206 | 0.1239 | 0.1271 | 0.1303 | 0.1335 | 0.1367 | 0.1399 | 0.1430 |
| 1.4 | 0.1461 | 0.1492 | 0.1523 | 0.1553 | 0.1584 | 0.1614 | 0.1644 | 0.1673 | 0.1703 | 0.1732 |
| 1.5 | 0.1761 | 0.1790 | 0.1818 | 0.1847 | 0.1875 | 0.1903 | 0.1931 | 0.1959 | 0.1987 | 0.2014 |
| 1.6 | 0.2041 | 0.2068 | 0.2095 | 0.2122 | 0.2148 | 0.2175 | 0.2201 | 0.2227 | 0.2253 | 0.2279 |
| 1.7 | 0.2304 | 0.2330 | 0.2355 | 0.2380 | 0.2405 | 0.2430 | 0.2455 | 0.2480 | 0.2504 | 0.2529 |
| 1.8 | 0.2553 | 0.2577 | 0.2601 | 0.2625 | 0.2648 | 0.2672 | 0.2695 | 0.2718 | 0.2742 | 0.2765 |
| 1.9 | 0.2788 | 0.2810 | 0.2833 | 0.2856 | 0.2878 | 0.2900 | 0.2923 | 0.2945 | 0.2967 | 0.2989 |
| 2.0 | 0.3010 | 0.3032 | 0.3054 | 0.3075 | 0.3096 | 0.3118 | 0.3139 | 0.3160 | 0.3181 | 0.3201 |
| 2.1 | 0.3222 | 0.3243 | 0.3263 | 0.3284 | 0.3304 | 0.3324 | 0.3345 | 0.3365 | 0.3385 | 0.3404 |
| 2.2 | 0.3424 | 0.3444 | 0.3464 | 0.3483 | 0.3502 | 0.3522 | 0.3541 | 0.3560 | 0.3579 | 0.3598 |
| 2.3 | 0.3617 | 0.3636 | 0.3655 | 0.3674 | 0.3692 | 0.3711 | 0.3729 | 0.3747 | 0.3766 | 0.3784 |
| 2.4 | 0.3802 | 0.3820 | 0.3838 | 0.3856 | 0.3874 | 0.3892 | 0.3909 | 0.3927 | 0.3945 | 0.3962 |
| 2.5 | 0.3979 | 0.3997 | 0.4014 | 0.4031 | 0.4048 | 0.4065 | 0.4082 | 0.4099 | 0.4116 | 0.4133 |
| 2.6 | 0.4150 | 0.4166 | 0.4183 | 0.4200 | 0.4216 | 0.4232 | 0.4249 | 0.4265 | 0.4281 | 0.4298 |
| 2.7 | 0.4314 | 0.4330 | 0.4346 | 0.4362 | 0.4378 | 0.4393 | 0.4409 | 0.4425 | 0.4440 | 0.4456 |
| 2.8 | 0.4472 | 0.4487 | 0.4502 | 0.4518 | 0.4533 | 0.4548 | 0.4564 | 0.4579 | 0.4594 | 0.4609 |
| 2.9 | 0.4624 | 0.4639 | 0.4654 | 0.4669 | 0.4683 | 0.4698 | 0.4713 | 0.4728 | 0.4742 | 0.4757 |
| 3.0 | 0.4771 | 0.4786 | 0.4800 | 0.4814 | 0.4829 | 0.4843 | 0.4857 | 0.4871 | 0.4886 | 0.4900 |
| 3.1 | 0.4914 | 0.4928 | 0.4942 | 0.4955 | 0.4969 | 0.4983 | 0.4997 | 0.5011 | 0.5024 | 0.5038 |
| 3.2 | 0.5051 | 0.5065 | 0.5079 | 0.5092 | 0.5105 | 0.5119 | 0.5132 | 0.5145 | 0.5159 | 0.5172 |
| 3.3 | 0.5185 | 0.5198 | 0.5211 | 0.5224 | 0.5237 | 0.5250 | 0.5263 | 0.5276 | 0.5289 | 0.5302 |
| 3.4 | 0.5315 | 0.5328 | 0.5340 | 0.5353 | 0.5366 | 0.5378 | 0.5391 | 0.5403 | 0.5416 | 0.5428 |
| 3.5 | 0.5441 | 0.5453 | 0.5465 | 0.5478 | 0.5490 | 0.5502 | 0.5514 | 0.5527 | 0.5539 | 0.5551 |
| 3.6 | 0.5563 | 0.5575 | 0.5587 | 0.5599 | 0.5611 | 0.5623 | 0.5635 | 0.5647 | 0.5658 | 0.5670 |
| 3.7 | 0.5682 | 0.5694 | 0.5705 | 0.5717 | 0.5729 | 0.5740 | 0.5752 | 0.5763 | 0.5775 | 0.5786 |
| 3.8 | 0.5798 | 0.5809 | 0.5821 | 0.5832 | 0.5843 | 0.5855 | 0.5866 | 0.5877 | 0.5888 | 0.5899 |
| 3.9 | 0.5911 | 0.5922 | 0.5933 | 0.5944 | 0.5955 | 0.5966 | 0.5977 | 0.5988 | 0.5999 | 0.6010 |
| 4.0 | 0.6021 | 0.6031 | 0.6042 | 0.6053 | 0.6064 | 0.6075 | 0.6085 | 0.6096 | 0.6107 | 0.6117 |
| 4.1 | 0.6128 | 0.6138 | 0.6149 | 0.6160 | 0.6170 | 0.6180 | 0.6191 | 0.6201 | 0.6212 | 0.6222 |
| 4.2 | 0.6232 | 0.6243 | 0.6253 | 0.6263 | 0.6274 | 0.6284 | 0.6294 | 0.6304 | 0.6314 | 0.6325 |
| 4.3 | 0.6335 | 0.6345 | 0.6355 | 0.6365 | 0.6375 | 0.6385 | 0.6395 | 0.6405 | 0.6415 | 0.6425 |
| 4.4 | 0.6435 | 0.6444 | 0.6454 | 0.6464 | 0.6474 | 0.6484 | 0.6493 | 0.6503 | 0.6513 | 0.6522 |
| 4.5 | 0.6532 | 0.6542 | 0.6551 | 0.6561 | 0.6571 | 0.6580 | 0.6590 | 0.6599 | 0.6609 | 0.6618 |
| 4.6 | 0.6628 | 0.6637 | 0.6646 | 0.6656 | 0.6665 | 0.6675 | 0.6684 | 0.6693 | 0.6702 | 0.6712 |
| 4.7 | 0.6721 | 0.6730 | 0.6739 | 0.6749 | 0.6758 | 0.6767 | 0.6776 | 0.6785 | 0.6794 | 0.6803 |
| 4.8 | 0.6812 | 0.6821 | 0.6830 | 0.6839 | 0.6848 | 0.6857 | 0.6866 | 0.6875 | 0.6884 | 0.6893 |
| 4.9 | 0.6902 | 0.6911 | 0.6920 | 0.6928 | 0.6937 | 0.6946 | 0.6955 | 0.6964 | 0.6972 | 0.6981 |
| 5.0 | 0.6990 | 0.6998 | 0.7007 | 0.7016 | 0.7024 | 0.7033 | 0.7042 | 0.7050 | 0.7059 | 0.7067 |
| 5.1 | 0.7076 | 0.7084 | 0.7093 | 0.7101 | 0.7110 | 0.7118 | 0.7126 | 0.7135 | 0.7143 | 0.7152 |
| 5.2 | 0.7160 | 0.7168 | 0.7177 | 0.7185 | 0.7193 | 0.7202 | 0.7210 | 0.7218 | 0.7226 | 0.7235 |
| 5.3 | 0.7243 | 0.7251 | 0.7259 | 0.7267 | 0.7275 | 0.7284 | 0.7292 | 0.7300 | 0.7308 | 0.7316 |
| 5.4 | 0.7324 | 0.7332 | 0.7340 | 0.7348 | 0.7356 | 0.7364 | 0.7372 | 0.7380 | 0.7388 | 0.7396 |

| N | 0 | 1 | 2 | 3 | 4 | 5 | 6 | 7 | 8 | 9 |
|---|---|---|---|---|---|---|---|---|---|---|
| 5.5 | 0.7404 | 0.7412 | 0.7419 | 0.7427 | 0.7435 | 0.7443 | 0.7451 | 0.7459 | 0.7466 | 0.7474 |
| 5.6 | 0.7482 | 0.7490 | 0.7497 | 0.7505 | 0.7513 | 0.7520 | 0.7528 | 0.7536 | 0.7543 | 0.7551 |
| 5.7 | 0.7559 | 0.7566 | 0.7574 | 0.7582 | 0.7589 | 0.7597 | 0.7604 | 0.7612 | 0.7619 | 0.7627 |
| 5.8 | 0.7634 | 0.7642 | 0.7649 | 0.7657 | 0.7664 | 0.7672 | 0.7679 | 0.7686 | 0.7694 | 0.7701 |
| 5.9 | 0.7709 | 0.7716 | 0.7723 | 0.7731 | 0.7738 | 0.7745 | 0.7752 | 0.7760 | 0.7767 | 0.7774 |
| 6.0 | 0.7782 | 0.7789 | 0.7796 | 0.7803 | 0.7810 | 0.7818 | 0.7825 | 0.7832 | 0.7839 | 0.7846 |
| 6.1 | 0.7853 | 0.7860 | 0.7868 | 0.7875 | 0.7882 | 0.7889 | 0.7896 | 0.7903 | 0.7910 | 0.7917 |
| 6.2 | 0.7924 | 0.7931 | 0.7938 | 0.7945 | 0.7952 | 0.7959 | 0.7966 | 0.7973 | 0.7980 | 0.7987 |
| 6.3 | 0.7993 | 0.8000 | 0.8007 | 0.8014 | 0.8021 | 0.8028 | 0.8035 | 0.8041 | 0.8048 | 0.8055 |
| 6.4 | 0.8062 | 0.8069 | 0.8075 | 0.8082 | 0.8089 | 0.8096 | 0.8102 | 0.8109 | 0.8116 | 0.8122 |
| 6.5 | 0.8129 | 0.8136 | 0.8142 | 0.8149 | 0.8156 | 0.8162 | 0.8169 | 0.8176 | 0.8182 | 0.8189 |
| 6.6 | 0.8195 | 0.8202 | 0.8209 | 0.8215 | 0.8222 | 0.8228 | 0.8235 | 0.8241 | 0.8248 | 0.8254 |
| 6.7 | 0.8261 | 0.8267 | 0.8274 | 0.8280 | 0.8287 | 0.8293 | 0.8299 | 0.8306 | 0.8312 | 0.8319 |
| 6.8 | 0.8325 | 0.8331 | 0.8338 | 0.8344 | 0.8351 | 0.8357 | 0.8363 | 0.8370 | 0.8376 | 0.8382 |
| 6.9 | 0.8388 | 0.8395 | 0.8401 | 0.8407 | 0.8414 | 0.8420 | 0.8426 | 0.8432 | 0.8439 | 0.8445 |
| 7.0 | 0.8451 | 0.8457 | 0.8463 | 0.8470 | 0.8476 | 0.8482 | 0.8488 | 0.8494 | 0.8500 | 0.8506 |
| 7.1 | 0.8513 | 0.8519 | 0.8525 | 0.8531 | 0.8537 | 0.8543 | 0.8549 | 0.8555 | 0.8561 | 0.8567 |
| 7.2 | 0.8573 | 0.8579 | 0.8585 | 0.8591 | 0.8597 | 0.8603 | 0.8609 | 0.8615 | 0.8621 | 0.8627 |
| 7.3 | 0.8633 | 0.8639 | 0.8645 | 0.8651 | 0.8657 | 0.8663 | 0.8669 | 0.8675 | 0.8681 | 0.8686 |
| 7.4 | 0.8692 | 0.8698 | 0.8704 | 0.8710 | 0.8716 | 0.8722 | 0.8727 | 0.8733 | 0.8739 | 0.8745 |
| 7.5 | 0.8751 | 0.8756 | 0.8762 | 0.8768 | 0.8774 | 0.8779 | 0.8785 | 0.8791 | 0.8797 | 0.8802 |
| 7.6 | 0.8808 | 0.8814 | 0.8820 | 0.8825 | 0.8831 | 0.8837 | 0.8842 | 0.8848 | 0.8854 | 0.8859 |
| 7.7 | 0.8865 | 0.8871 | 0.8876 | 0.8882 | 0.8887 | 0.8893 | 0.8899 | 0.8904 | 0.8910 | 0.8915 |
| 7.8 | 0.8921 | 0.8927 | 0.8932 | 0.8938 | 0.8943 | 0.8949 | 0.8954 | 0.8960 | 0.8965 | 0.8971 |
| 7.9 | 0.8976 | 0.8982 | 0.8987 | 0.8993 | 0.8998 | 0.9004 | 0.9009 | 0.9015 | 0.9020 | 0.9025 |
| 8.0 | 0.9031 | 0.9036 | 0.9042 | 0.9047 | 0.9053 | 0.9058 | 0.9063 | 0.9069 | 0.9074 | 0.9079 |
| 8.1 | 0.9085 | 0.9090 | 0.9096 | 0.9101 | 0.9106 | 0.9112 | 0.9117 | 0.9122 | 0.9128 | 0.9133 |
| 8.2 | 0.9138 | 0.9143 | 0.9149 | 0.9154 | 0.9159 | 0.9165 | 0.9170 | 0.9175 | 0.9180 | 0.9186 |
| 8.3 | 0.9191 | 0.9196 | 0.9201 | 0.9206 | 0.9212 | 0.9217 | 0.9222 | 0.9227 | 0.9232 | 0.9238 |
| 8.4 | 0.9243 | 0.9248 | 0.9253 | 0.9258 | 0.9263 | 0.9269 | 0.9274 | 0.9279 | 0.9284 | 0.9289 |
| 8.5 | 0.9294 | 0.9299 | 0.9304 | 0.9309 | 0.9315 | 0.9320 | 0.9325 | 0.9330 | 0.9335 | 0.9340 |
| 8.6 | 0.9345 | 0.9350 | 0.9355 | 0.9360 | 0.9365 | 0.9370 | 0.9375 | 0.9380 | 0.9385 | 0.9390 |
| 8.7 | 0.9395 | 0.9400 | 0.9405 | 0.9410 | 0.9415 | 0.9420 | 0.9425 | 0.9430 | 0.9435 | 0.9440 |
| 8.8 | 0.9445 | 0.9450 | 0.9455 | 0.9460 | 0.9465 | 0.9469 | 0.9474 | 0.9479 | 0.9484 | 0.9489 |
| 8.9 | 0.9494 | 0.9499 | 0.9504 | 0.9509 | 0.9513 | 0.9518 | 0.9523 | 0.9528 | 0.9533 | 0.9538 |
| 9.0 | 0.9542 | 0.9547 | 0.9552 | 0.9557 | 0.9562 | 0.9566 | 0.9571 | 0.9576 | 0.9581 | 0.9586 |
| 9.1 | 0.9590 | 0.9595 | 0.9600 | 0.9605 | 0.9609 | 0.9614 | 0.9619 | 0.9624 | 0.9628 | 0.9633 |
| 9.2 | 0.9638 | 0.9643 | 0.9647 | 0.9652 | 0.9657 | 0.9661 | 0.9666 | 0.9671 | 0.9675 | 0.9680 |
| 9.3 | 0.9685 | 0.9689 | 0.9694 | 0.9699 | 0.9703 | 0.9708 | 0.9713 | 0.9717 | 0.9722 | 0.9727 |
| 9.4 | 0.9731 | 0.9736 | 0.9741 | 0.9745 | 0.9750 | 0.9754 | 0.9759 | 0.9763 | 0.9768 | 0.9773 |
| 9.5 | 0.9777 | 0.9782 | 0.9786 | 0.9791 | 0.9795 | 0.9800 | 0.9805 | 0.9809 | 0.9814 | 0.9818 |
| 9.6 | 0.9823 | 0.9827 | 0.9832 | 0.9836 | 0.9841 | 0.9845 | 0.9850 | 0.9854 | 0.9859 | 0.9863 |
| 9.7 | 0.9868 | 0.9872 | 0.9877 | 0.9881 | 0.9886 | 0.9890 | 0.9894 | 0.9899 | 0.9903 | 0.9908 |
| 9.8 | 0.9912 | 0.9917 | 0.9921 | 0.9926 | 0.9930 | 0.9934 | 0.9939 | 0.9943 | 0.9948 | 0.9952 |
| 9.9 | 0.9956 | 0.9961 | 0.9965 | 0.9969 | 0.9974 | 0.9978 | 0.9983 | 0.9987 | 0.9991 | 0.9996 |

Name ______________________________ Date ________________

1. Use properties of exponents to explain why it makes sense to define $16^{\frac{1}{4}}$ as $\sqrt[4]{16}$.

2. Use properties of exponents to rewrite each expression as either an integer or as a quotient of integers $\frac{p}{q}$ to show the expression is a rational number.

   a. $\sqrt[4]{2}\,\sqrt[4]{8}$

   b. $\dfrac{\sqrt[3]{54}}{\sqrt[3]{2}}$

   c. $16^{\frac{3}{2}} \cdot \left(\frac{1}{27}\right)^{\frac{2}{3}}$

3. Use properties of exponents to rewrite each expression with only positive, rational exponents. Then find the numerical value of each expression when $x = 9$, $y = 8$, and $z = 16$. In each case, the expression evaluates to a rational number.

   a. $\sqrt{\dfrac{xy^2}{(x^3z)^{\frac{1}{2}}}}$

   b. $\sqrt[11]{y^2z^4}$

   c. $x^{-\frac{3}{2}}y^{\frac{4}{3}}z^{-\frac{3}{4}}$

4. We can use finite approximations of the decimal expansion of $\pi = 3.141519 \ldots$ to find an approximate value of the number $3^{\pi}$.

   a. Fill in the missing exponents in the following sequence of inequalities that represents the recursive process of finding the value of $3^{\pi}$.

$$3^3 < 3^{\pi} < 3^4$$
$$3^{3.1} < 3^{\pi} < 3^{3.2}$$
$$3^{3.14} < 3^{\pi} < 3^{3.15}$$
$$3^{(\quad)} < 3^{\pi} < 3^{(\quad)}$$
$$3^{(\quad)} < 3^{\pi} < 3^{(\quad)}$$

   b. Explain how this recursive process leads to better and better approximations of the number $3^{\pi}$.

5. A scientist is studying the growth of a population of bacteria. At the beginning of her study, she has 800 bacteria. She notices that the population is quadrupling every hour.

   a. What quantities, including units, need to be identified to further investigate the growth of this bacteria population?

b. The scientist recorded the following information in her notebook, but she forgot to label each row. Label each row to show what quantities, including appropriate units, are represented by the numbers in the table, and then complete the table.

| | $0$ | $1$ | $2$ | $3$ | $4$ |
|---|---|---|---|---|---|
| | $8$ | $32$ | $128$ | | |

c. Write an explicit formula for the number of bacteria present after $t$ hours.

d. Another scientist studying the same population notices that the population is doubling every half an hour. Complete the table, and write an explicit formula for the number of bacteria present after $x$ half hours.

| Time, $t$ (hours) | $0$ | $\frac{1}{2}$ | $1$ | $\frac{3}{2}$ | $2$ | $\frac{5}{2}$ | $3$ |
|---|---|---|---|---|---|---|---|
| Time, $x$ (half-hours) | $0$ | $1$ | $2$ | $3$ | $4$ | $5$ | $6$ |
| Bacteria (hundreds) | $8$ | $16$ | $32$ | | | | |

e. Find the time, in hours, when there will be $5{,}120{,}000$ bacteria. Express your answer as a logarithmic expression.

f. A scientist calculated the average rate of change for the bacteria in the first three hours to be $168$. Which units should the scientist use when reporting this number?

6. Solve each equation. Express your answer as a logarithm, and then approximate the solution to the nearest thousandth.

   a. $3(10)^{-x} = \frac{1}{9}$

   b. $362\left(10^{\frac{t}{12}}\right) = 500$

   c. $(2)^{3x} = 9$

   d. $300e^{0.4t} = 900$

7. Because atoms and molecules are very small, they are counted in units of *moles*, where $1 \text{ mole} = 6.022 \times 10^{23}$. Concentration of molecules in a liquid is measured in units of moles per liter. The measure of the acidity of a liquid is called the $\text{pH}$ of the liquid and is given by the formula

$$\text{pH} = -\log(H),$$

where $H$ is the concentration of hydrogen ions in units of moles per liter.

a. Water has a $\text{pH}$ value of $7.0$. How many hydrogen ions are in one liter of water?

b. If a liquid has a $\text{pH}$ value larger than $7.0$, does one liter of that liquid contain more or less hydrogen ions than one liter of water? Explain.

c. Suppose that liquid A is more acidic than liquid B, and their $\text{pH}$ values differ by $1.2$. What is the ratio of the concentration of hydrogen ions in liquid A to the concentration of hydrogen ions in liquid B?

8. A social media site is experiencing rapid growth. The table below shows values of $V$, the total number of unique visitors in each month, $t$, for a 6-month period of time. The graph shows the average minutes per visit to the site, $M$, in each month, $t$, for the same 6-month period of time.

| $t$, Month | 1 | 2 | 3 | 4 | 5 | 6 |
|---|---|---|---|---|---|---|
| $V(t)$, Number of Unique Visitors | 418,000 | 608,000 | 1,031,000 | 1,270,000 | 2,023,000 | 3,295,000 |

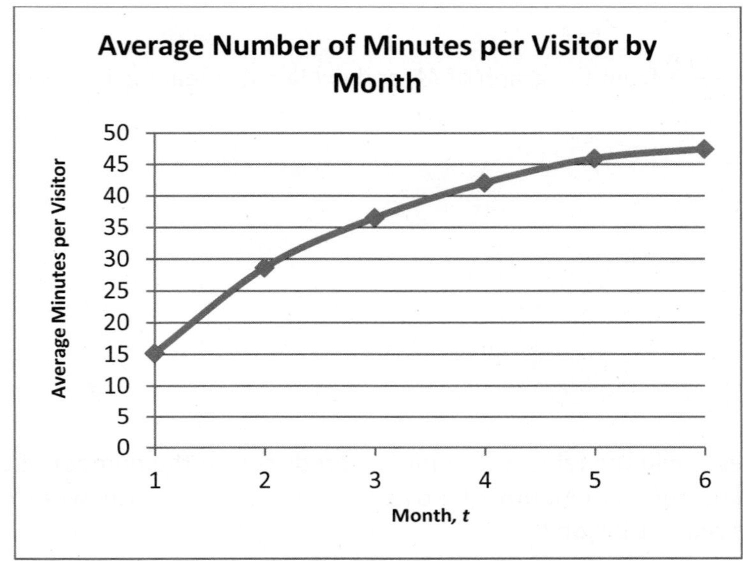


a. Between which two months did the site experience the most growth in total unique visitors? What is the average rate of change over this time interval?

b. Compute the value of $\frac{V(6)-V(1)}{6-1}$, and explain its meaning in this situation.

c. Between which two months did the average length of a visit change by the least amount? Estimate the average rate of change over this time interval.

d. Estimate the value of $\frac{M(3)-M(2)}{3-2}$ from the graph of $M$, and explain its meaning in this situation.

e. Based on the patterns they see in the table, the company predicts that the number of unique visitors will double each month after the sixth month. If growth continues at this pace, when will the number of unique visitors reach 1 billion?

Name ______________________________ Date ______________

# Lesson 16: Rational and Irrational Numbers

## Exit Ticket

The decimal expansion of $e$ and $\sqrt{5}$ are given below.

$$e \approx 2.71828182\ ...$$
$$\sqrt{5} \approx 2.23606797\ ...$$

a. Find an approximation of $\sqrt{5} + e$ to three decimal places. Do not use a calculator.

b. Explain how you can locate $\sqrt{5} + e$ on the number line. How is this different from locating $2.6 + 2.7$ on the number line?

Name ______________________________ Date ______________

# Lesson 17: Graphing the Logarithm Function

## Exit Ticket

Graph the function $f(x) = \log_3(x)$ without using a calculator, and identify its key features.

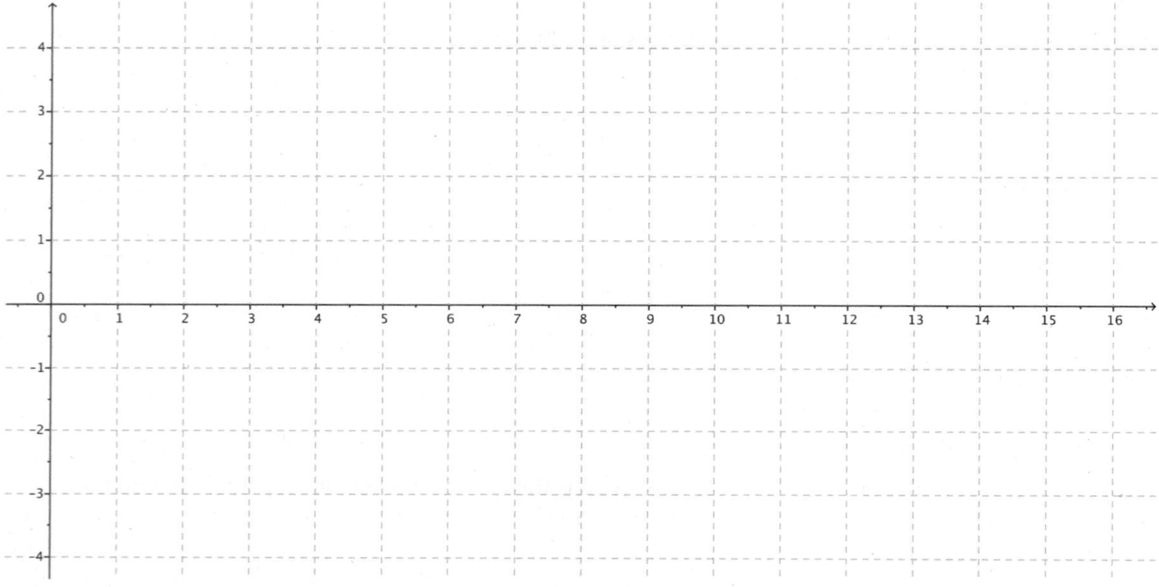

Name ________________________________ Date ________________

# Lesson 18: Graphs of Exponential Functions and Logarithmic Functions

## Exit Ticket

The graph of a logarithmic function $g(x) = \log_b(x)$ is shown below.

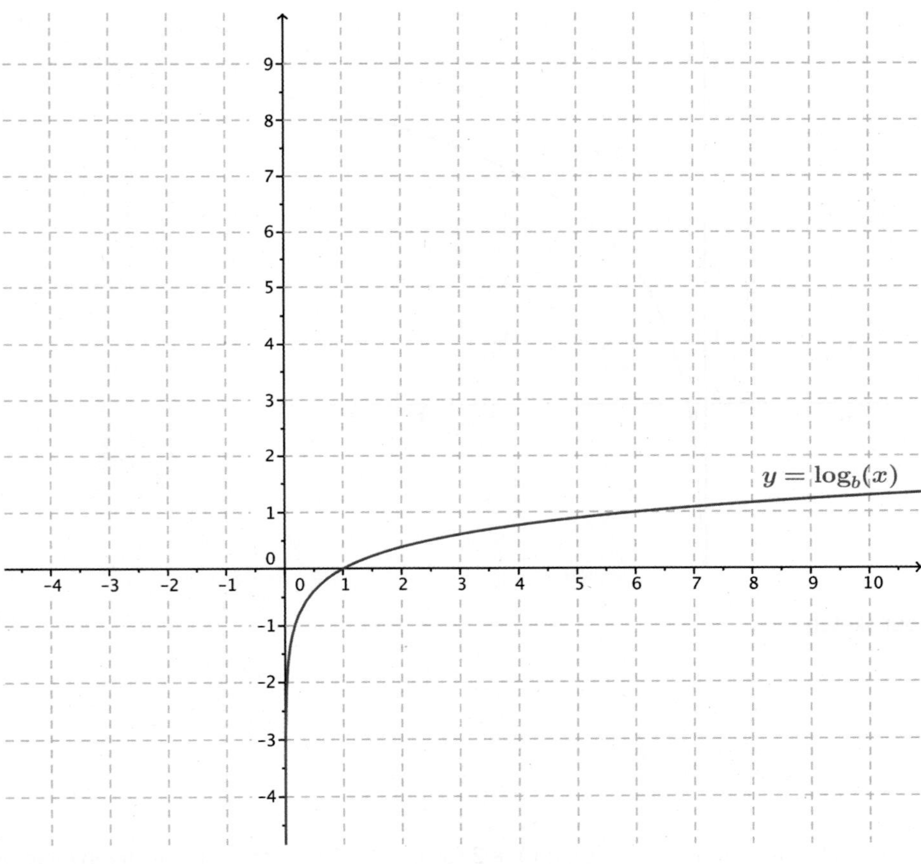


a. Explain how to find points on the graph of the function $f(x) = b^x$.

b. Sketch the graph of the function $f(x) = b^x$ on the same axes.

Name ______________________________ Date ______________

# Lesson 19: The Inverse Relationship Between Logarithmic and Exponential Functions

## Exit Ticket

1. The graph of a function $f$ is shown below. Sketch the graph of its inverse function $g$ on the same axes.

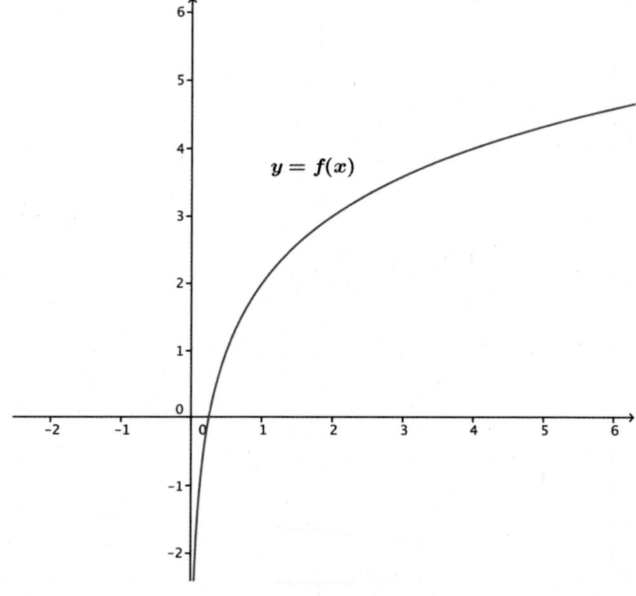


2. Explain how you made your sketch.

3. The function $f$ graphed above is the function $f(x) = \log_2(x) + 2$ for $x > 0$. Find a formula for the inverse of this function.

Name ____________________________________________ Date ______________

# Lesson 20: Transformations of the Graphs of Logarithmic and Exponential Functions

## Exit Ticket

1. Express $g(x) = -\log_4(2x)$ in the general form of a logarithmic function, $f(x) = k + a\log_b(x - h)$. Identify $a$, $b$, $h$, and $k$.

2. Use the structure of $g$ when written in general from to describe the graph of $g$ as a transformation of the graph of $h(x) = \log_4(x)$.

3. Graph $g$ anad $h$ on the same coordinate axes.

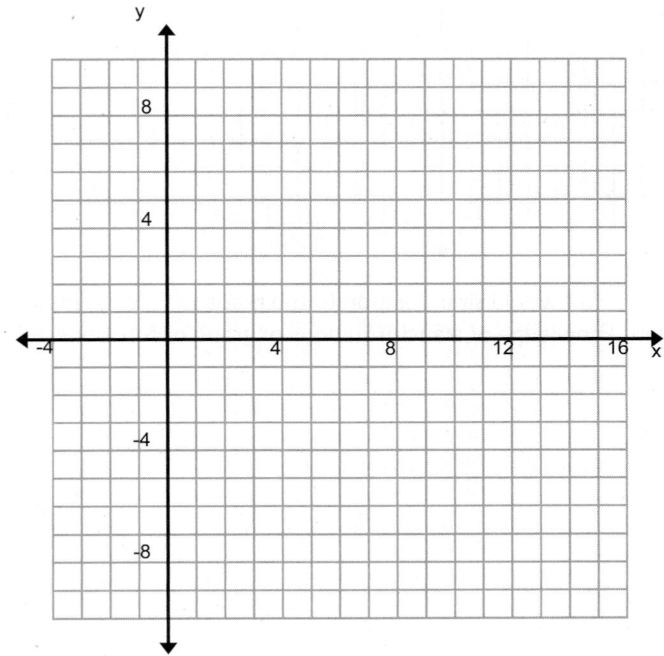

Name ______________________________ Date ______________

# Lesson 21: The Graph of the Natural Logarithm Function

## Exit Ticket

1. Describe the graph of $g(x) = 2 - \ln(x + 3)$ as a transformation of the graph of $f(x) = \ln(x)$.

2. Sketch the graphs of $f$ and $g$ by hand.

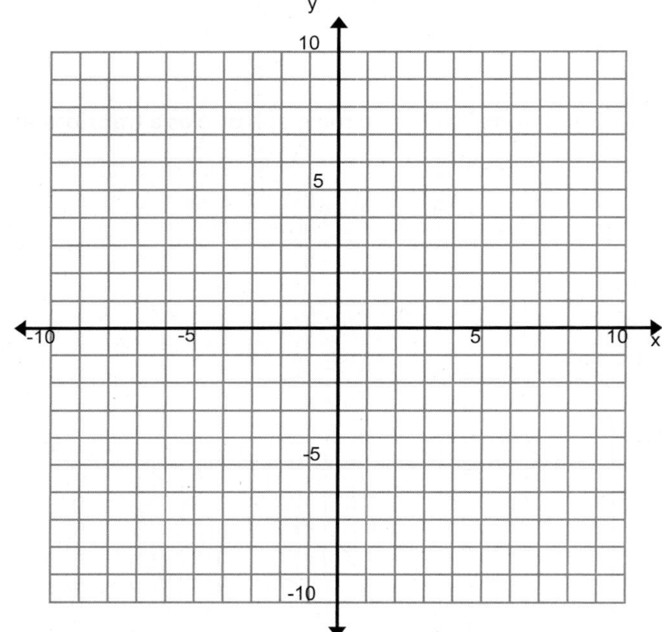


3. Explain where the graph of $g(x) = \log_3(2x)$ would sit in relation to the graph of $f(x) = \ln(x)$. Justify your answer using properties of logarithms and your knowledge of transformations of graph of functions.

Name ______________________________ Date ______________

# Lesson 22: Choosing a Model

## Exit Ticket

The amount of caffeine in a patient's bloodstream decreases by half every 3.5 hours. A latte contains 150 mg of caffeine, which is absorbed into the bloodstream almost immediately.

a. What type of function models the caffeine level in the patient's bloodstream at time $t$ hours after drinking the latte? Explain how you know.

b. Do you have enough information to find a model that is appropriate for this situation? Either find a model or explain what other information you would need to do so.

Name ____________________________________________ Date ____________________

# Lesson 23: Bean Counting

## Exit Ticket

Suppose that you were to repeat the bean activity, but in place of beans, you were to use six-sided dice. Starting with one die, each time a die is rolled with a 6 showing, you add a new die to your cup.

a. Would the number of dice in your cup grow more quickly or more slowly than the number of beans did? Explain how you know.

b. A sketch of one sample of data from the bean activity is shown below. On the same axes, draw a rough sketch of how you would expect the line of best fit from the dice activity to look.

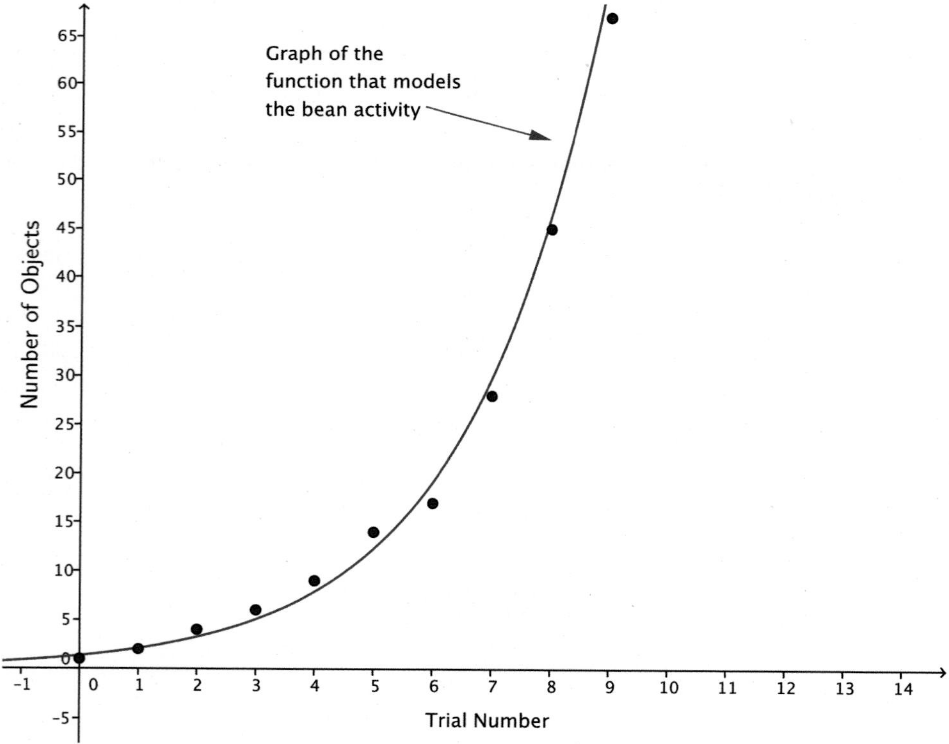

Name ______________________________ Date ______________

# Lesson 24: Solving Exponential Equations

## Exit Ticket

Consider the functions $f(x) = 2^{x+6}$ and $g(x) = 5^{2x}$.

a. Use properties of logarithms to solve the equation $f(x) = g(x)$. Give your answer as a logarithmic expression, and approximate it to two decimal places.

b. Verify your answer by graphing the functions $y = f(x)$ and $y = g(x)$ in the same window on a calculator, and sketch your graphs below. Explain how the graph validates your solution to part (a).

Name ______________________________ Date ______________

# Lesson 25: Geometric Sequences and Exponential Growth and Decay

## Exit Ticket

1. Every year, Mikhail receives a $3\%$ raise in his annual salary. His starting annual salary was $\$40{,}000$.
   a. Does a geometric or arithmetic sequence best model Mikhail's salary in year $n$? Explain how you know.

   b. Find a recursive formula for a sequence, $S_n$, which represents Mikhail's salary in year $n$.

2. Carmela's annual salary in year $n$ can be modeled by the recursive sequence $C_{n+1} = 1.05\, C_n$, where $C_0 = \$75{,}000$.
   a. What does the number $1.05$ represent in the context of this problem?

   b. What does the number $\$75{,}000$ represent in the context of this problem?

   c. Find an explicit formula for a sequence that represents Carmela's salary.

Name ______________________________ Date ______________

# Lesson 26: Percent Rate of Change

## Exit Ticket

April would like to invest $\$200$ in the bank for one year. Three banks all have a nominal APR of $1.5\%$, but compound the interest differently.

a. Bank A computes interest just once at the end of the year. What would April's balance be after one year with this bank?

b. Bank B compounds interest at the end of each six-month period. What would April's balance be after one year with this bank?

c. Bank C compounds interest continuously. What would April's balance be after one year with this bank?

d. Each bank decides to double the nominal APR it offers for one year. That is, they offer a nominal APR of $3\%$. Each bank advertises, "DOUBLE THE AMOUNT YOU EARN!" For which of the three banks, if any, is this advertised claim correct?

Name ______________________________ Date ____________

# Lesson 27: Modeling with Exponential Functions

## Exit Ticket

1. The table below gives the average annual cost (e.g., tuition, room, and board) for four-year public colleges and universities. Explain why a linear model might not be appropriate for this situation.

| Year | Average Annual Cost |
|---|---|
| 1981 | \$2,550 |
| 1991 | \$5,243 |
| 2001 | \$8,653 |
| 2011 | \$15,918 |

2. Algebraically determine an exponential function to model this situation.

3. Use the properties of exponents to rewrite the function from Problem 2 to determine an annual growth rate.

4. If this trend continues, when will the average annual cost of attendance exceed \$35,000?

Name ______________________________ Date ____________________

# Lesson 28: Newton's Law of Cooling, Revisited

## Exit Ticket

A pizza, heated to a temperature of $400°\text{F}$, is taken out of an oven and placed in a $75°\text{F}$ room at time $t = 0$ minutes. The temperature of the pizza is changing such that its decay constant, $k$, is $0.325$. At what time is the temperature of the pizza $150°\text{F}$ and, therefore, safe to eat? Give your answer in minutes.

Name ______________________________ Date ______________

# Lesson 29: The Mathematics Behind a Structured Savings Plan

## Exit Ticket

Martin attends a financial planning conference and creates a budget for himself, realizing that he can afford to put away $200 every month in savings and that he should be able to keep this up for two years. If Martin has the choice between an account earning an interest rate of 2.3% yearly versus an account earning an annual interest rate of 2.125% compounded monthly, which account will give Martin the largest return in two years?

Name ______________________________ Date ______________

# Lesson 30: Buying a Car

## Exit Ticket

Fran wants to purchase a new boat. She starts looking for a boat around $6,000. Fran creates a budget and thinks that she can afford $250 every month for 2 years. Her bank charges her 5% interest per year, compounded monthly.

1. What is the actual monthly payment for Fran's loan?

2. If Fran can only pay $250 per month, what is the most expensive boat she can buy without a down payment?

Name ____________________________________ Date ____________________

# Lesson 31: Credit Cards

## Exit Ticket

Suppose that you currently have one credit card with a balance of $\$10{,}000$ at an annual rate of $24.00\%$ interest. You have stopped adding any additional charges to this card and are determined to pay off the balance. You have worked out the formula $b_n = b_0 r^n - R(1 + r + r^2 + \cdots + r^{n-1})$, where $b_0$ is the initial balance, $b_n$ is the balance after you have made $n$ payments, $r = 1 + i$, where $i$ is the monthly interest rate, and $R$ is the amount you are planning to pay each month.

a. What is the monthly interest rate $i$? What is the growth rate, $r$?

b. Explain why we can rewrite the given formula as $b_n = b_0 r^n - R\left(\frac{1-r^n}{1-r}\right)$.

c. How long will it take you to pay off this debt if you can afford to pay a constant $\$250$ per month? Give your answer in years and months.

Name ______________________________ Date ______________

# Lesson 32: Buying a House

## Exit Ticket

1. Recall the present value of an annuity formula, where $A_p$ is the present value, $R$ is the monthly payment, $i$ is the monthly interest rate, and $n$ is the number of monthly payments:

$$A_p = R\left(\frac{1-(1+i)^{-n}}{i}\right).$$

Rewrite this formula to isolate $R$.

2. Suppose that you want to buy a house that costs $\$175{,}000$. You can make a $10\%$ down payment, and $1.2\%$ of the house's value is paid into the escrow account each month.
   a. Find the monthly payment for a $30$-year mortgage on this house.
   b. Find the monthly payment for a $15$-year mortgage on this house.

Name ______________________________ Date ______________

# Lesson 33: The Million Dollar Problem

## Exit Ticket

1. At age $25$, you begin planning for retirement at $65$. Knowing that you have $40$ years to save up for retirement and expecting an interest rate of $4\%$ per year compounded monthly throughout the $40$ years, how much do you need to deposit every month to save up $\$2$ million for retirement?

2. Currently, your savings for each month is capped at $\$400$. If you start investing all of this into a savings plan earning $1\%$ interest annually, compounded monthly, then how long will it take to save $\$160{,}000$? (Hint: Use logarithms.)

Name ______________________________ Date ____________________

1. For parts (a) to (c),
   - Sketch the graph of each pair of functions on the same coordinate axes showing end behavior and intercepts, and
   - Describe the graph of $g$ as a series of transformations of the graph of $f$.

   a. $f(x) = 2^x$, and $g(x) = 2^{-x} + 3$

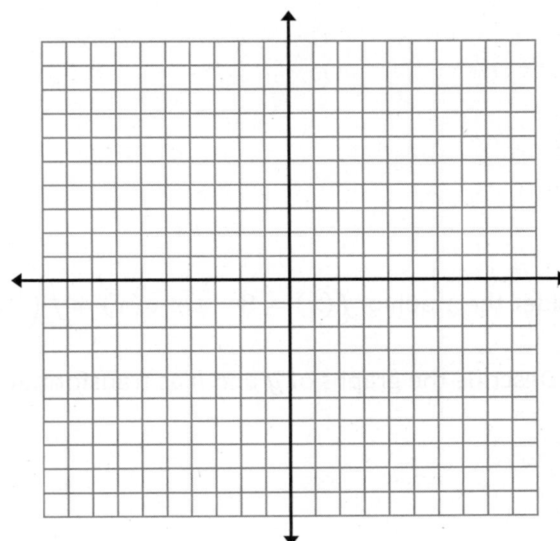

   b. $f(x) = 3^x$, and $g(x) = 9^{x-2}$

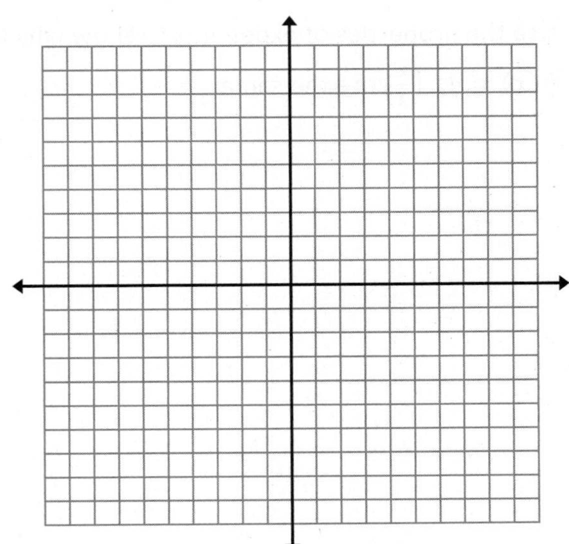

c. $f(x) = \log_2(x)$, and $g(x) = \log_2(x-1)^2$

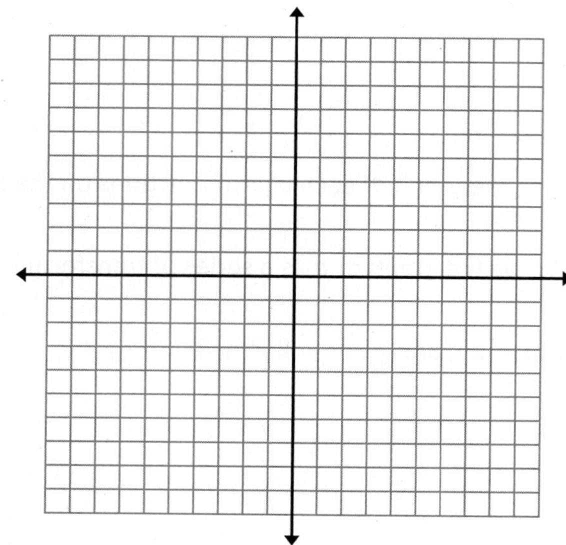

2. Consider the graph of $f(x) = 8^x$. Let $g(x) = f\left(\frac{1}{3}x + \frac{2}{3}\right)$ and $h(x) = 4f\left(\frac{x}{3}\right)$.

   a. Describe the graphs of $g$ and $h$ as transformations of the graph of $f$.

   b. Use the properties of exponents to show why the graphs of the functions $g(x) = f\left(\frac{1}{3}x + \frac{2}{3}\right)$ and $h(x) = 4f\left(\frac{x}{3}\right)$ are the same.

3. The graphs of the functions $f(x) = \ln(x)$ and $g(x) = \log_2(x)$ are shown to the right.

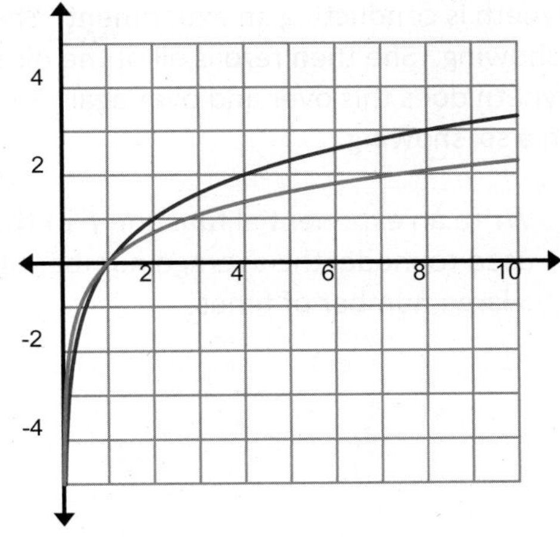


a. Which curve is the graph of $f$, and which curve is the graph of $g$? Explain.

b. Describe the graph of $g$ as a transformation of the graph of $f$.

c. By what factor has the graph of $f$ been scaled vertically to produce the graph of $g$? Explain how you know.

4. Gwyneth is conducting an experiment. She rolls $1{,}000$ dice simultaneously and removes any that have a six showing. She then rerolls all of the dice that remain and again removes any that have a six showing. Gwyneth does this over and over again—rerolling the remaining dice and then removing those that land with a six showing.

   a. Write an exponential function $f$ of the form $f(n) = a \cdot b^{cn}$ for any real number $n \geq 0$ that could be used to model the average number of dice she could expect on the $n^{\text{th}}$ roll if she ran her experiment a large number of times.

   b. Gwyneth computed $f(12) = 112.15\ldots$ using the function $f$. How should she interpret the number $112.15\ldots$ in the context of the experiment?

   c. Explain the meaning of the parameters in your function $f$ in terms of this experiment.

   d. Describe in words the key features of the graph of the function $f$ for $n \geq 0$. Be sure to describe where the function is increasing or decreasing, where it has maximums and minimums (if they exist), and the end behavior.

e. According to the model, on which roll does Gwyneth expect, on average, to find herself with only one die remaining? Write and solve an equation to support your answer to this question.

f. For all of the values in the domain of $f$, is there any value for which $f$ will predict an average number of 0 dice remaining? Explain why or why not. Be sure to use the domain of the function and the graph to support your reasoning.

Suppose the table below represents the results of one trial of Gwyneth's experiment.

| Roll | Number of Dice Left | Roll | Number of Dice Left | Roll | Number of Dice Left |
|---|---|---|---|---|---|
| 0 | 1000 | 10 | 157 | 20 | 26 |
| 1 | 840 | 11 | 139 | 21 | 22 |
| 2 | 692 | 12 | 115 | 22 | 15 |
| 3 | 581 | 13 | 90 | 23 | 13 |
| 4 | 475 | 14 | 78 | 24 | 10 |
| 5 | 400 | 15 | 63 | 25 | 6 |
| 6 | 341 | 16 | 55 | 26 | 2 |
| 7 | 282 | 17 | 43 | 27 | 1 |
| 8 | 232 | 18 | 40 | 28 | 0 |
| 9 | 190 | 19 | 33 | | |

g. Let $g$ be the function that is defined exactly by the data in the table, i.e., $g(0) = 1000$, $g(1) = 840$, $g(2) = 692$, and so forth, up to $g(28) = 0$. Describe in words how the graph of $g$ looks different from the graph of $f$. Be sure to use the domain of $g$ and the domain of $f$ to justify your description.

h. Gwyneth runs her experiment hundreds of times, and each time she generates a table like the one above. How are these tables similar to the function $f$? How are they different?

5. Find the inverse $g$ for each function $f$.

a. $f(x) = \frac{1}{2}x - 3$

b. $f(x) = \frac{x+3}{x-2}$

c. $f(x) = 2^{3x} + 1$

d. $f(x) = e^{x-3}$

e. $f(x) = \log(2x + 3)$

6. Dani has $\$1,000$ in an investment account that earns $3\%$ per year, compounded monthly.

   a. Write a recursive sequence for the amount of money in her account after $n$ months.

   b. Write an explicit formula for the amount of money in the account after $n$ months.

   c. Write an explicit formula for the amount of money in her account after $t$ years.

   d. Boris also has $\$1,000$, but in an account that earns $3\%$ per year, compounded yearly. Write an explicit formula for the amount of money in his account after $t$ years.

   e. Boris claims that the equivalent monthly interest rate for his account would be the same as Dani's. Use the expression you wrote in part (d) and the properties of exponents to show why Boris is incorrect.

7. Show that

$$\sum_{k=0}^{n} a \cdot r^k = a\left(\frac{1-r^n}{1-r}\right)$$

where $r \neq 1$.

8. Sami opens an account and deposits $\$100$ into it at the end of each month. The account earns $2\%$ per year compounded monthly. Let $S(n)$ denote the amount of money in her account at the end of $n$ months (just after she makes a deposit). For example, $S(1) = 100$ and $S(2) = 100\left(1 + \frac{0.02}{12}\right) + 100$.

   a. Write a geometric series for the amount of money in the account after 3, 4, and 5 months.

   b. Find a recursive description for $S(n)$.

c. Find an explicit function for $S(n)$, and use it to find $S(12)$.

d. When will Sami have at least $\$5{,}000$ in her account? Show work to support your answer.

9. Beatrice decides to deposit $\$100$ per month at the end of every month in a bank with an annual interest rate of $5.5\%$ compounded monthly.

   a. Write a geometric series to show how much she will accumulate in her account after one year.

b. Use the formula for the sum of a geometric series to calculate how much she will have in the bank after five years if she keeps on investing $\$100$ per month.

10. Nina has just taken out a car loan for $\$12,000$. She will pay an annual interest rate of $3\%$ through a series of monthly payments for $60$ months, which she pays at the end of each month. The amount of money she has left to pay on the loan at the end of the $n^{\text{th}}$ month can be modeled by the function $f(n) = 86248 - 74248(1.0025)^n$ for $0 \leq n \leq 60$.

    At the same time as her first payment (at the end of the first month), Nina placed $\$100$ into a separate investment account that earns $6\%$ per year compounded monthly. She placed $\$100$ into the account at the end of each month thereafter. The amount of money in her savings account at the end of the $n^{\text{th}}$ month can be modeled by the function $g(n) = 20000(1.005)^n - 20000$ for $n \geq 0$.

    a. Use the functions $f$ and $g$ to write an equation whose solution could be used to determine when Nina will have saved enough money to pay off the remaining balance on her car loan.

    b. Use a calculator or computer to graph $f$ and $g$ on the same coordinate plane. Sketch the graphs below, labeling intercepts and indicating end behavior on the sketch. Include the coordinates of any intersection points.

c. How would you interpret the end behavior of each function in the context of this situation?

d. What does the intersection point mean in the context of this situation? Explain how you know.

e. After how many months will Nina have enough money saved to pay off her car loan? Explain how you know.

11. Each function below models the growth of three different trees of different ages over a fixed time interval.

**Tree A:**

$f(t) = 15(1.69)^{\frac{t}{2}}$, where $t$ is time in years since the tree was 15 feet tall, $f(t)$ is the height of the tree in feet, and $0 \le t \le 4$.

**Tree B:**

| Years since the tree was 5 feet tall, $t$ | Height in feet after $t$ years, $g(t)$ |
|---|---|
| 0 | 5 |
| 1 | 6.3 |
| 2 | 7.6 |
| 3 | 8.9 |
| 4 | 10.2 |

**Tree C:** The graph of $h$ is shown where $t$ is years since the tree was 5 feet tall, and $h(t)$ is the height in feet after $t$ years.

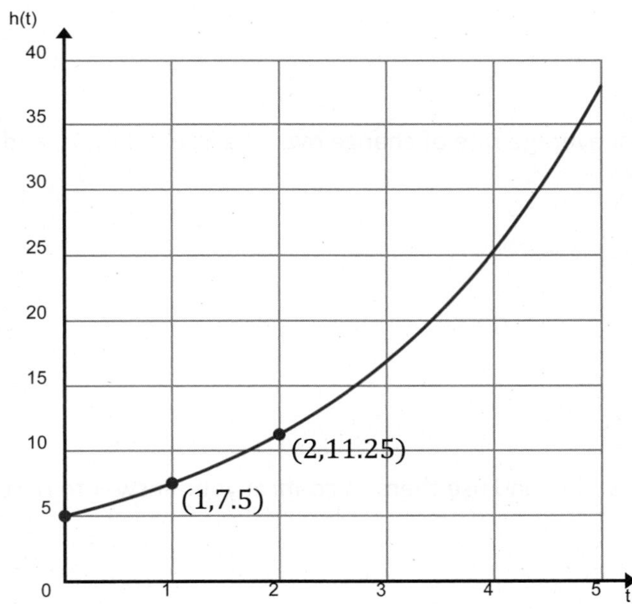

a. Classify each function $f$ and $g$ as representing a linear or nonlinear function. Justify your answers.

b. Use the properties of exponents to show that Tree A has a percent rate of change of $30\%$ per year.

c. Which tree, A or C, has the greatest percent rate of change? Justify your answer.

d. Which function has the greatest average rate of change over the interval $[0,4]$, and what does that mean in terms of tree heights?

e. Write formulas for functions $g$ and $h$, and use them to confirm your answer to part (c).

f. For the exponential models, if the average rate of change of one function over the interval $[0,4]$ is greater than the average rate of change of another function on the same interval, is the percent rate of change also greater? Why or why not?

12. Identify which functions are exponential. For the exponential functions, use the properties of exponents to identify the percent rate of change, and classify them as representing exponential growth or decay.

a. $f(x) = 3(1 - 0.4)^{-x}$

b. $g(x) = \frac{3}{4^x}$

c. $k(x) = 3x^{0.4}$

d. $h(x) = 3^{\frac{x}{4}}$

13. A patient in a hospital needs to maintain a certain amount of a medication in her bloodstream to fight an infection. Suppose the initial dosage is $10$ mg, and the patient is given an additional maintenance dosage of $4$ mg every hour. Assume that the amount of medication in the bloodstream is reduced by $25\%$ every hour.

   a. Write a function for the amount of the initial dosage that is in the bloodstream after $n$ hours.

   b. Complete the table below to track the amount of medication from the maintenance dosage in the patient's bloodstream for the first five hours.

| Hours since initial dose, $n$ | Amount of the medication in the bloodstream from the maintenance dosage at the beginning of each hour |
|---|---|
| $0$ | $0$ |
| $1$ | $4$ |
| $2$ | $4(1 + 0.75)$ |
| $3$ | |
| $4$ | |
| $5$ | |

   c. Write a function that models the total amount of medication in the bloodstream after $n$ hours.

d. Use a calculator to graph the function you wrote in part (c). According to the graph, will there ever be more than $16$ mg of the medication present in the patient's bloodstream after each dose is administered?

e. Rewrite this function as the difference of two functions (one a constant function and the other an exponential function), and use that difference to justify why the amount of medication in the patient's bloodstream will not exceed $16$ mg after each dose is administered.

**Staff**

Nashrah Ahmed, Coordinator of User Experience
Deirdre Bey, Formatter
Thomas Brasdefer, Formatter
Brenda Bryant, Program Operations Associate
Adrienne Burgess, Project Associate – Website Content
Alyson Burgess, Associate Director of Administration
Rose Calloway, Document Specialist
Adam Cardais, Copy Editor
Jamie Carruth, Formatter
Lauren Chapalee, Professional Learning Manager – English
Gregar Chapin, Website Project Coordinator
Chris Clary, Director of Branding and Marketing
Katelyn Colacino, Formatter
Julia Cooper, Formatter
Barbara Davidson, Deputy Director
Karen Elkins, Formatter
Jennifer George, Formatter
Erin Glover, Formatter
Laurie Gonsoulin, Formatter
Eric Halley, Formatter
Candice Hartley, Formatter
Thomas Haynes, Copy Editor
Robert Hunsicker, Program Operations Associate
Jennifer Hutchinson, Copy Editor
Anne Ireland, Print Edition Coordinator
Maggie Kay, Copy Editor
Liz LeBarron, Program Operations and Support Associate
Jeff LeBel, XML Developer
Tam Le, Document Production Manager
Natanya Levioff, Director of Program Operations and Support
Siena Mazero, Project Associate – Website Content
Stacie McClintock, XML Developer
Cindy Medici, Copy Editor
Elisabeth Mox, Executive Assistant to the President
Lynne Munson, President and Executive Director
Sarah Oyler, Document Specialist
Diego Quiroga, Accounts Specialist
Becky Robinson, Program Operations Associate
Amy Rome, Copy Editor
Rachel Rooney, Program Manager – English / History
Neela Roy, Print Edition Associate
Tim Shen, Customer Relations Associate
Kathleen Smith, Formatter
Leigh Sterten, Project Associate – Web Content
Wendy Taylor, Copy Editor
Megan Wall, Formatter
Marjani Warren, Account Manager
Sam Wertheim, Product Delivery Manager
Amy Wierzbicki, Assets and Permissions Manager
Sarah Woodard, Associate Director – English / History

***Eureka Math: A Story of Functions*** **Contributors**

Mimi Alkire, Algebra I Lead Writer / Editor
Michael Allwood, Curriculum Writer
Tiah Alphonso, Program Manager – Curriculum Production
Catriona Anderson, Program Manager – Implementation Support
Beau Bailey, Curriculum Writer
Scott Baldridge, Lead Mathematician and Lead Curriculum Writer
Christopher Bejar, Curriculum Writer
Andrew Bender, Curriculum Writer
Bonnie Bergstresser, Curriculum Writer and Math Auditor
Chris Black, Mathematician and Lead Writer, Algebra II
Gail Burrill, Curriculum Writer
Carlos Carrera, Curriculum Writer
Beth Chance, Statistician, Assessments – Statistics
Andrew Chen, Advising Mathematician
Melvin Damaolao, Curriculum Writer
Wendy DenBesten, Curriculum Writer
Jill Diniz, Program Director
Lori Fanning, Math Auditor
Joe Ferrantelli, Curriculum Writer
Ellen Fort, Curriculum Writer
Kathy Fritz, Curriculum Writer
Thomas Gaffey, Curriculum Writer
Sheri Goings, Curriculum Writer
Pam Goodner, Geometry & Precalculus Lead Writer / Editor
Stefanie Hassan, Curriculum Writer
Sherri Hernandez, Math Auditor
Patrick Hopfensperger, Curriculum Writer
Chih Ming Huang, Curriculum Writer
James Key, Curriculum Writer
Jeremy Kilpatrick, Mathematics Educator, Algebra II
Jenny Kim, Curriculum Writer
Brian Kotz, Curriculum Writer
Henry Kranendonk, Statistics Lead Writer / Editor
Yvonne Lai, Mathematician, Geometry
Connie Laughlin, Math Auditor
Athena Leonardo, Curriculum Writer
Jennifer Loftin, Program Manager – Professional Development
James Madden, Mathematician, Lead Writer, Geometry
Abby Mattern, Math Auditor
Nell McAnelly, Project Director
Ben McCarty, Mathematician, Lead Writer, Geometry
Robert Michelin, Curriculum Writer
Pia Mohsen, Geometry Lead Writer / Editor
Jerry Moreno, Statistician
Chris Murcko, Curriculum Writer
Selena Oswalt, Algebra I, Algebra II, & Precalculus Lead Writer / Editor
Roxy Peck, Mathematician, Lead Writer, Statistics
Noam Pillischer, Curriculum Writer
Terrie Poehl, Math Auditor
Rob Richardson, Curriculum Writer
Spencer Roby, Math Auditor
William Rorison, Curriculum Writer
Alex Sczesnak, Curriculum Writer
Michel Smith, Mathematician, Algebra II
Hester Sutton, Curriculum Writer
James Tanton, Advising Mathematician
Shannon Vinson, Statistics Lead Writer / Editor
Eric Weber, Mathematics Educator, Algebra II
David Wright, Mathematician, Geometry
Kristen Zimmerman, Document Production Manager